INTRODUCTION TO RADAR SYSTEMS

INTRODUCTION TO RADAR SYSTEMS

Second Edition

Merrill I. Skolnik

McGraw-Hill Book Company

New York St. Louis San Francisco Auckland Bogotá Düsseldorf
Johannesburg London Madrid Mexico Montreal New Delhi
Panama Paris São Paulo Singapore Sydney Tokyo Toronto

INTRODUCTION TO RADAR SYSTEMS

INTRODUCTION TO RADAR SYSTEMS

67890 DODO 898765

Library of Congress Cataloging in Publication Data

Skolnik, Merrill Ivan, date
    Introduction to radar systems.

    Includes bibliographical references and index.
    1. Radar. I. Title. II. Series.
TK6575.S477 1980   621.3848   79-15354
ISBN 0-07-057909-1

This book was set in Times Roman. The editor was Frank J. Cerra and
the production supervisor was Gayle Angelson.
R. R. Donnelley & Sons Company was printer and binder.

CONTENTS

PREFACE

Although the fundamentals of radar have changed little since the publication of the first edition, there has been continual development of new radar capabilities and continual improvements to the technology and practice of radar. This growth has necessitated extensive revisions and the introduction of topics not found in the original.

One of the major changes is in the treatment of MTI (moving target indication) radar (Chap. 4). Most of the basic MTI concepts that have been added were known at the time of the first edition, but they had not appeared in the open literature nor were they widely used in practice. Inclusion in the first edition would have been largely academic since the analog delay-line technology available at that time did not make it practical to build the sophisticated signal processors that were theoretically possible. However, subsequent advances in digital technology, originally developed for applications other than radar, have allowed the practical implementation of the multiple delay-line cancelers and multiple pulse-repetition-frequency MTI radars indicated by the basic MTI theory.

Automatic detection and tracking, or ADT (Secs. 5.10 and 10.7), is another important development whose basic theory was known for some time, but whose practical realization had to await advances in digital technology. The principle of ADT was demonstrated in the early 1950s, using vacuum-tube technology, as part of the United States Air Force's SAGE air-defense system developed by MIT Lincoln Laboratory. In this form ADT was physically large, expensive, and difficult to maintain. The commercial availability in the late 1960s of the solid-state minicomputer, however, permitted ADT to be relatively inexpensive, reliable, and of small size so that it can be used with almost any surveillance radar that requires it.

Another radar area that has seen much development is that of the electronically steered phased-array antenna. In the first edition, the radar antenna was the subject of a single chapter. In this edition, one chapter covers the conventional radar antenna (Chap. 7) and a separate chapter covers the phased-array antenna (Chap. 8). Devoting a single chapter to the array antenna is more a reflection of interest rather than recognition of extensive application.

The chapter on radar clutter (Chap. 13) has been reorganized to include methods for the detection of targets in the presence of clutter. Generally, the design techniques necessary for the detection of targets in a clutter background are considerably different from those necessary for detection in a noise background. Other subjects that are new or which have seen significant changes in the current edition include low-angle tracking, "on-axis" tracking, solid-state RF sources, the mirror-scan antenna, antenna stabilization, computer control of phased arrays, solid-state duplexers, CFAR, pulse compression, target classification, synthetic-aperture radar, over-the-horizon radar, air-surveillance radar, height-finder and 3D radar, and ECCM. The bistatic radar and millimeter-wave radar are also included even though their applications have been limited. Omitted from this second edition is the chapter on Radar Astronomy since interest in this subject has decreased with the availability of space probes that can explore the planets at close range. The basic material of the first edition that covers the radar equation, the detection of signals in noise, the extraction of information, and the propagation of radar waves has not changed significantly. The reader, however, will find only a few pages of the original edition that have not been modified in some manner.

One of the features of the first edition which has been continued is the inclusion of extensive references at the end of each chapter. These are provided to acknowledge the sources of material used in the preparation of the book, as well as to permit the interested reader to learn more about some particular subject. Some references that appeared in the first edition have been omitted since they have been replaced by more current references or appear in publications that are increasingly difficult to find. The references included in the first edition represented a large fraction of those available at the time. It would have been difficult to add to them extensively or to include many additional topics. This is not so with the second edition. The current literature is quite large; and, because of the limitations of space, only a much smaller proportion of what is available could be cited.

In addition to changes in radar technology, there have been changes also in style and nomenclature. For example, db has been changed to dB, and Mc is replaced by MHz. Also, the letter-band nomenclature widely employed by the radar engineer for designating the common radar frequency bands (such as L, S, and X) has been officially adopted as a standard by the IEEE.

The material in this book has been used as the basis for a graduate course in radar taught by the author at the Johns Hopkins University Evening College and, before that, at several other institutions. This course is different from those usually found in most graduate electrical engineering programs. Typical EE courses cover topics related to circuits, components, devices, and techniques that might make up an electrical or electronic system; but seldom is the student exposed to the system itself. It is the system application (whether radar, communications, navigation, control, information processing, or energy) that is the raison d'être for the electrical engineer. The course on which this book is based is a proven method for introducing the student to the subject of electronic systems. It integrates and applies the basic concepts found in the student's other courses and permits the inclusion of material important to the practice of electrical engineering not usually found in the traditional curriculum.

Instructors of engineering courses like to use texts that contain a variety of problems that can be assigned to students. Problems are not included in this book. Although the author assigns problems when using this book as a text, they are not considered a major learning technique. Instead, the comprehensive term paper, usually involving a radar design problem or a study in depth of some particular radar technology, has been found to be a better means for having the student reinforce what is covered in class and in the text. Even more important, it allows the student to research the literature and to be a bit more creative than is possible by simply solving standard problems.

A book of this type which covers a wide variety of topics cannot be written in isolation. It would not have been possible without the many contributions on radar that have appeared in the open literature and which have been used here as the basic source material. A large measure of gratitude must be expressed to those radar engineers who have taken the time and energy to ensure that the results of their work were made available by publication in recognized journals.

On a more personal note, neither edition of this book could have been written without the complete support and patience of my wife Judith and my entire family who allowed me the time necessary to undertake this work.

Merrill I. Skolnik

ONE

THE NATURE OF RADAR

1.1 INTRODUCTION

Radar is an electromagnetic system for the detection and location of objects. It operates by transmitting a particular type of waveform, a pulse-modulated sine wave for example, and detects the nature of the echo signal. Radar is used to extend the capability of one's senses for observing the environment, especially the sense of vision. The value of radar lies not in being a substitute for the eye, but in doing what the eye cannot do. Radar cannot resolve detail as well as the eye, nor is it capable of recognizing the " color " of objects to the degree of sophistication of which the eye is capable. However, radar can be designed to see through those conditions impervious to normal human vision, such as darkness, haze, fog, rain, and snow. In addition, radar has the advantage of being able to measure the distance or range to the object. This is probably its most important attribute.

An elementary form of radar consists of a transmitting antenna emitting electromagnetic radiation generated by an oscillator of some sort, a receiving antenna, and an energy-detecting device, or receiver. A portion of the transmitted signal is intercepted by a reflecting object (target) and is reradiated in all directions. It is the energy reradiated in the back direction that is of prime interest to the radar. The receiving antenna collects the returned energy and delivers it to a receiver, where it is processed to detect the presence of the target and to extract its location and relative velocity. The distance to the target is determined by measuring the time taken for the radar signal to travel to the target and back. The direction, or angular position, of the target may be determined from the direction of arrival of the reflected wavefront. The usual method of measuring the direction of arrival is with narrow antenna beams. If relative motion exists between target and radar, the shift in the carrier frequency of the reflected wave (doppler effect) is a measure of the target's relative (radial) velocity and may be used to distinguish moving targets from stationary objects. In radars which continuously track the movement of a target, a continuous indication of the rate of change of target position is also available.

1

The name *radar* reflects the emphasis placed by the early experimenters on a device to detect the presence of a target and measure its range. *Radar* is a contraction of the words *radio detection and ranging*. It was first developed as a detection device to warn of the approach of hostile aircraft and for directing antiaircraft weapons. Although a well-designed modern radar can usually extract more information from the target signal than merely range, the measurement of range is still one of radar's most important functions. There seem to be no other competitive techniques which can measure range as well or as rapidly as can a radar.

The most common radar waveform is a train of narrow, rectangular-shape pulses modulating a sinewave carrier. The distance, or range, to the target is determined by measuring the time T_R taken by the pulse to travel to the target and return. Since electromagnetic energy propagates at the speed of light $c = 3 \times 10^8$ m/s, the range R is

$$R = \frac{cT_R}{2} \tag{1.1}$$

The factor 2 appears in the denominator because of the two-way propagation of radar. With the range in kilometers or nautical miles, and T_R in microseconds, Eq. (1.1) becomes

$$R(\text{km}) = 0.15\,T_R(\mu s) \qquad \text{or} \qquad R(\text{nmi}) = 0.081\,T_R(\mu s)$$

Each microsecond of round-trip travel time corresponds to a distance of 0.081 nautical mile, 0.093 statute mile, 150 meters, 164 yards, or 492 feet.

Once the transmitted pulse is emitted by the radar, a sufficient length of time must elapse to allow any echo signals to return and be detected before the next pulse may be transmitted. Therefore the rate at which the pulses may be transmitted is determined by the longest range at which targets are expected. If the pulse repetition frequency is too high, echo signals from some targets might arrive after the transmission of the next pulse, and ambiguities in measuring

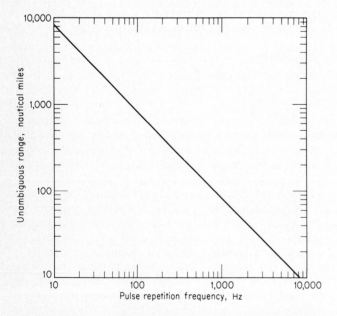

Figure 1.1 Plot of maximum unambiguous range as a function of the pulse repetition frequency.

range might result. Echoes that arrive after the transmission of the next pulse are called *second-time-around* (or multiple-time-around) echoes. Such an echo would appear to be at a much shorter range than the actual and could be misleading if it were not known to be a second-time-around echo. The range beyond which targets appear as second-time-around echoes is called the *maximum unambiguous range* and is

$$R_{\text{unamb}} = \frac{c}{2f_p} \tag{1.2}$$

where f_p = pulse repetition frequency, in Hz. A plot of the maximum unambiguous range as a function of pulse repetition frequency is shown in Fig. 1.1.

Although the typical radar transmits a simple pulse-modulated waveform, there are a number of other suitable modulations that might be used. The pulse carrier might be frequency- or phase-modulated to permit the echo signals to be compressed in time after reception. This achieves the benefits of high range-resolution without the need to resort to a short pulse. The technique of using a long, modulated pulse to obtain the resolution of a short pulse, but with the energy of a long pulse, is known as *pulse compression*. Continuous waveforms (CW) also can be used by taking advantage of the doppler frequency shift to separate the received echo from the transmitted signal and the echoes from stationary clutter. Unmodulated CW waveforms do not measure range, but a range measurement can be made by applying either frequency- or phase-modulation.

1.2 THE SIMPLE FORM OF THE RADAR EQUATION

The radar equation relates the range of a radar to the characteristics of the transmitter, receiver, antenna, target, and environment. It is useful not just as a means for determining the maximum distance from the radar to the target, but it can serve both as a tool for understanding radar operation and as a basis for radar design. In this section, the simple form of the radar equation is derived.

If the power of the radar transmitter is denoted by P_t, and if an isotropic antenna is used (one which radiates uniformly in all directions), the *power density* (watts per unit area) at a distance R from the radar is equal to the transmitter power divided by the surface area $4\pi R^2$ of an imaginary sphere of radius R, or

$$\text{Power density from isotropic antenna} = \frac{P_t}{4\pi R^2} \tag{1.3}$$

Radars employ directive antennas to channel, or direct, the radiated power P_t into some particular direction. The *gain G* of an antenna is a measure of the increased power radiated in the direction of the target as compared with the power that would have been radiated from an isotropic antenna. It may be defined as the ratio of the maximum radiation intensity from the subject antenna to the radiation intensity from a lossless, isotropic antenna with the same power input. (The radiation intensity is the power radiated per unit solid angle in a given direction.) The power density at the target from an antenna with a transmitting gain G is

$$\text{Power density from directive antenna} = \frac{P_t G}{4\pi R^2} \tag{1.4}$$

The target intercepts a portion of the incident power and reradiates it in various directions.

The measure of the amount of incident power intercepted by the target and reradiated back in the direction of the radar is denoted as the radar cross section σ, and is defined by the relation

$$\text{Power density of echo signal at radar} = \frac{P_t G}{4\pi R^2} \frac{\sigma}{4\pi R^2} \tag{1.5}$$

The radar cross section σ has units of area. It is a characteristic of the particular target and is a measure of its size *as seen by the radar*. The radar antenna captures a portion of the echo power. If the effective area of the receiving antenna is denoted A_e, the power P_r received by the radar is

$$P_r = \frac{P_t G}{4\pi R^2} \frac{\sigma}{4\pi R^2} A_e = \frac{P_t G A_e \sigma}{(4\pi)^2 R^4} \tag{1.6}$$

The maximum radar range R_{max} is the distance beyond which the target cannot be detected. It occurs when the received echo signal power P_r just equals the minimum detectable signal S_{min}. Therefore

$$R_{max} = \left[\frac{P_t G A_e \sigma}{(4\pi)^2 S_{min}} \right]^{1/4} \tag{1.7}$$

This is the fundamental form of the radar equation. Note that the important antenna parameters are the transmitting gain and the receiving effective area.

Antenna theory gives the relationship between the transmitting gain and the receiving effective area of an antenna as

$$G = \frac{4\pi A_e}{\lambda^2} \tag{1.8}$$

Since radars generally use the same antenna for both transmission and reception, Eq. (1.8) can be substituted into Eq. (1.7), first for A_e then for G, to give two other forms of the radar equation

$$R_{max} = \left[\frac{P_t G^2 \lambda^2 \sigma}{(4\pi)^3 S_{min}} \right]^{1/4} \tag{1.9}$$

$$R_{max} = \left[\frac{P_t A_e^2 \sigma}{4\pi \lambda^2 S_{min}} \right]^{1/4} \tag{1.10}$$

These three forms (Eqs. 1.7, 1.9, and 1.10) illustrate the need to be careful in the interpretation of the radar equation. For example, from Eq. (1.9) it might be thought that the range of a radar varies as $\lambda^{1/2}$, but Eq. (1.10) indicates a $\lambda^{-1/2}$ relationship, and Eq. (1.7) shows the range to be independent of λ. The correct relationship depends on whether it is assumed the gain is constant or the effective area is constant with wavelength. Furthermore, the introduction of other constraints, such as the requirement to scan a specified volume in a given time, can yield a different wavelength dependence.

These simplified versions of the radar equation do not adequately describe the performance of practical radar. Many important factors that affect range are not explicitly included. In practice, the observed maximum radar ranges are usually much smaller than what would be predicted by the above equations, sometimes by as much as a factor of two. There are many reasons for the failure of the simple radar equation to correlate with actual performance, as discussed in Chap. 2.

1.3 RADAR BLOCK DIAGRAM AND OPERATION

The operation of a typical pulse radar may be described with the aid of the block diagram shown in Fig. 1.2. The transmitter may be an oscillator, such as a magnetron, that is "pulsed" (turned on and off) by the modulator to generate a repetitive train of pulses. The magnetron has probably been the most widely used of the various microwave generators for radar. A typical radar for the detection of aircraft at ranges of 100 or 200 nmi might employ a peak power of the order of a megawatt, an average power of several kilowatts, a pulse width of several microseconds, and a pulse repetition frequency of several hundred pulses per second. The waveform generated by the transmitter travels via a transmission line to the antenna, where it is radiated into space. A single antenna is generally used for both transmitting and receiving. The receiver must be protected from damage caused by the high power of the transmitter. This is the function of the duplexer. The duplexer also serves to channel the returned echo signals to the receiver and not to the transmitter. The duplexer might consist of two gas-discharge devices, one known as a TR (transmit-receive) and the other an ATR (anti-transmit-receive). The TR protects the receiver during transmission and the ATR directs the echo signal to the receiver during reception. Solid-state ferrite circulators and receiver protectors with gas-plasma TR devices and/or diode limiters are also employed as duplexers.

The receiver is usually of the superheterodyne type. The first stage might be a low-noise RF amplifier, such as a parametric amplifier or a low-noise transistor. However, it is not always desirable to employ a low-noise first stage in radar. The receiver input can simply be the mixer stage, especially in military radars that must operate in a noisy environment. Although a receiver with a low-noise front-end will be more sensitive, the mixer input can have greater dynamic range, less susceptibility to overload, and less vulnerability to electronic interference.

The mixer and local oscillator (LO) convert the RF signal to an intermediate frequency (IF). A "typical" IF amplifier for an air-surveillance radar might have a center frequency of 30 or 60 MHz and a bandwidth of the order of one megahertz. The IF amplifier should be designed as a *matched filter*; i.e., its frequency-response function $H(f)$ should maximize the peak-signal–to–mean-noise-power ratio at the output. This occurs when the magnitude of the frequency-response function $|H(f)|$ is equal to the magnitude of the echo signal spectrum $|S(f)|$, and the phase spectrum of the matched filter is the negative of the phase spectrum of the echo signal (Sec. 10.2). In a radar whose signal waveform approximates a rectangular pulse, the conventional IF filter bandpass characteristic approximates a matched filter when the product of the IF bandwidth B and the pulse width τ is of the order of unity, that is, $B\tau \simeq 1$.

After maximizing the signal-to-noise ratio in the IF amplifier, the pulse modulation is extracted by the second detector and amplified by the video amplifier to a level where it can be

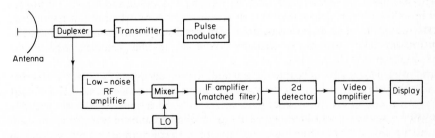

Figure 1.2 Block diagram of a pulse radar.

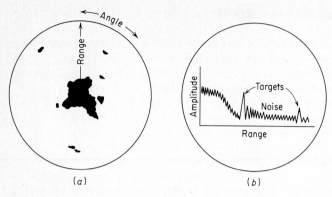

Figure 1.3 (*a*) PPI presentation displaying range vs. angle (intensity modulation); (*b*) A-scope presentation displaying amplitude vs. range (deflection modulation).

properly displayed, usually on a cathode-ray tube (CRT). Timing signals are also supplied to the indicator to provide the range zero. Angle information is obtained from the pointing direction of the antenna. The most common form of cathode-ray tube display is the plan position indicator, or *PPI* (Fig. 1.3*a*), which maps in polar coordinates the location of the target in azimuth and range. This is an intensity-modulated display in which the amplitude of the receiver output modulates the electron-beam intensity (z axis) as the electron beam is made to sweep outward from the center of the tube. The beam rotates in angle in response to the antenna position. A *B-scope* display is similar to the PPI except that it utilizes rectangular, rather than polar, coordinates to display range vs. angle. Both the B-scope and the PPI, being intensity modulated, have limited dynamic range. Another form of display is the *A-scope*, shown in Fig. 1.3*b*, which plots target amplitude (y axis) vs. range (x axis), for some fixed direction. This is a deflection-modulated display. It is more suited for tracking-radar application than for surveillance radar.

The block diagram of Fig. 1.2 is a simplified version that omits many details. It does not include several devices often found in radar, such as means for automatically compensating the receiver for changes in frequency (AFC) or gain (AGC), receiver circuits for reducing interference from other radars and from unwanted signals, rotary joints in the transmission lines to allow movement of the antenna, circuitry for discriminating between moving targets and unwanted stationary objects (MTI), and pulse compression for achieving the resolution benefits of a short pulse but with the energy of a long pulse. If the radar is used for tracking, some means are necessary for sensing the angular location of a moving target and allowing the antenna automatically to lock-on and to track the target. Monitoring devices are usually included to ensure that the transmitter is delivering the proper shape pulse at the proper power level and that the receiver sensitivity has not degraded. Provisions may also be incorporated in the radar for locating equipment failures so that faulty circuits can be easily found and replaced.

Instead of displaying the "raw-video" output of a surveillance radar directly on the CRT, it might first be processed by an automatic detection and tracking (ADT) device that quantizes the radar coverage into range-azimuth resolution cells, adds (or integrates) all the echo pulses received within each cell, establishes a threshold (on the basis of these integrated pulses) that permits only the strong outputs due to target echoes to pass while rejecting noise, establishes and maintains the tracks (trajectories) of each target, and displays the processed information to the operator. These operations of an ADT are usually implemented with digital computer technology.

A common form of radar antenna is a reflector with a parabolic shape, fed (illuminated) from a point source at its focus. The parabolic reflector focuses the energy into a narrow beam, just as does a searchlight or an automobile headlamp. The beam may be scanned in space by mechanical pointing of the antenna. Phased-array antennas have also been used for radar. In a phased array, the beam is scanned by electronically varying the phase of the currents across the aperture.

1.4 RADAR FREQUENCIES

Conventional radars generally have been operated at frequencies extending from about 220 MHz to 35 GHz, a spread of more than seven octaves. These are not necessarily the limits, since radars can be, and have been, operated at frequencies outside either end of this range. Skywave HF over-the-horizon (OTH) radar might be at frequencies as low as 4 or 5 MHz, and groundwave HF radars as low as 2 MHz. At the other end of the spectrum, millimeter radars have operated at 94 GHz. Laser radars operate at even higher frequencies.

The place of radar frequencies in the electromagnetic spectrum is shown in Fig. 1.4. Some of the nomenclature employed to designate the various frequency regions is also shown.

Early in the development of radar, a letter code such as S, X, L, etc., was employed to designate radar frequency bands. Although its original purpose was to guard military secrecy, the designations were maintained, probably out of habit as well as the need for some convenient short nomenclature. This usage has continued and is now an accepted practice of radar engineers. Table 1.1 lists the radar-frequency letter-band nomenclature adopted by the IEEE.[15] These are related to the specific bands assigned by the International Telecommunications Union for radar. For example, although the nominal frequency range for L band is 1000 to 2000 MHz, an L-band radar is thought of as being confined within the region from 1215 to 1400 MHz since that is the extent of the assigned band. Letter-band nomenclature is not a

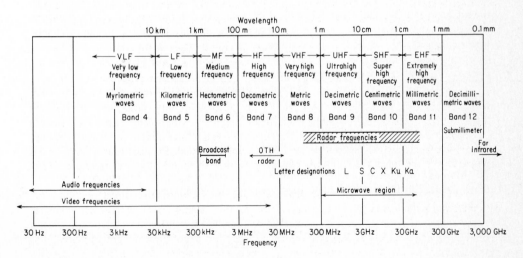

Figure 1.4 Radar frequencies and the electromagnetic spectrum.

Table 1.1 Standard radar-frequency letter-band nomenclature

Band designation	Nominal frequency range	Specific radiolocation (radar) bands based on ITU assignments for region 2
HF	3–30 MHz	
VHF	30–300 MHz	138–144 MHz
		216–225
UHF	300–1000 MHz	420–450 MHz
		890–942
L	1000–2000 MHz	1215–1400 MHz
S	2000–4000 MHz	2300–2500 MHz
		2700–3700
C	4000–8000 MHz	5250–5925 MHz
X	8000–12,000 MHz	8500–10,680 MHz
K_u	12.0–18 GHz	13.4–14.0 GHz
		15.7–17.7
K	18–27 GHz	24.05–24.25 GHz
K_a	27–40 GHz	33.4–36.0 GHz
mm	40–300 GHz	

substitute for the actual numerical frequency limits of radars. The specific numerical frequency limits should be used whenever appropriate, but the letter designations of Table 1.1 may be used whenever a short notation is desired.

1.5 RADAR DEVELOPMENT PRIOR TO WORLD WAR II

Although the development of radar as a full-fledged technology did not occur until World War II, the basic principle of radar detection is almost as old as the subject of electromagnetism itself. Heinrich Hertz, in 1886, experimentally tested the theories of Maxwell and demonstrated the similarity between radio and light waves. Hertz showed that radio waves could be reflected by metallic and dielectric bodies. It is interesting to note that although Hertz's experiments were performed with relatively short wavelength radiation (66 cm), later work in radio engineering was almost entirely at longer wavelengths. The shorter wavelengths were not actively used to any great extent until the late thirties.

In 1903 a German engineer by the name of Hülsmeyer experimented with the detection of radio waves reflected from ships. He obtained a patent in 1904 in several countries for an obstacle detector and ship navigational device.[2] His methods were demonstrated before the German Navy, but generated little interest. The state of technology at that time was not sufficiently adequate to obtain ranges of more than about a mile, and his detection technique was dismissed on the grounds that it was little better than a visual observer.

Marconi recognized the potentialities of short waves for radio detection and strongly urged their use in 1922 for this application. In a speech delivered before the Institute of Radio Engineers, he said:[3]

As was first shown by Hertz, electric waves can be completely reflected by conducting bodies. In some of my tests I have noticed the effects of reflection and detection of these waves by metallic objects miles away.

It seems to me that it should be possible to design apparatus by means of which a ship could radiate or project a divergent beam of these rays in any desired direction, which rays, if coming across a metallic object, such as another steamer or ship, would be reflected back to a receiver screened from the local transmitter on the sending ship, and thereby, immediately reveal the presence and bearing of the other ship in fog or thick weather.

Although Marconi predicted and successfully demonstrated radio communication between continents, he was apparently not successful in gaining support for some of his other ideas involving very short waves. One was the radar detection mentioned above; the other was the suggestion that very short waves are capable of propagation well beyond the optical line of sight—a phenomenon now known as tropospheric scatter. He also suggested that radio waves be used for the transfer of power from one point to the other without the use of wire or other transmission lines.

In the autumn of 1922 A. H. Taylor and L. C. Young of the Naval Research Laboratory detected a wooden ship using a CW wave-interference radar with separated receiver and transmitter. The wavelength was 5 m. A proposal was submitted for further work but was not accepted.

The first application of the pulse technique to the measurement of distance was in the basic scientific investigation by Breit and Tuve in 1925 for measuring the height of the ionosphere.[4,16] However, more than a decade was to elapse before the detection of aircraft by pulse radar was demonstrated.

The first experimental radar systems operated with CW and depended for detection upon the interference produced between the direct signal received from the transmitter and the doppler-frequency-shifted signal reflected by a moving target. This effect is the same as the rhythmic flickering, or flutter, observed in an ordinary television receiver, especially on weak stations, when an aircraft passes overhead. This type of radar originally was called *CW wave-interference radar*. Today, such a radar is called a *bistatic CW radar*. The first experimental detections of aircraft used this radar principle rather than a monostatic (single-site) pulse radar because CW equipment was readily available. Successful pulse radar had to await the development of suitable components, especially high-peak-power tubes, and a better understanding of pulse receivers.

The first detection of aircraft using the wave-interference effect was made in June, 1930, by L. A. Hyland of the Naval Research Laboratory.[1] It was made accidentally while he was working with a direction-finding apparatus located in an aircraft on the ground. The transmitter at a frequency of 33 MHz was located 2 miles away, and the beam crossed an air lane from a nearby airfield. When aircraft passed through the beam, Hyland noted an increase in the received signal. This stimulated a more deliberate investigation by the NRL personnel, but the work continued at a slow pace, lacking official encouragement and funds from the government, although it was fully supported by the NRL administration. By 1932 the equipment was demonstrated to detect aircraft at distances as great as 50 miles from the transmitter. The NRL work on aircraft detection with CW wave interference was kept classified until 1933, when several Bell Telephone Laboratories engineers reported the detection of aircraft during the course of other experiments.[5] The NRL work was disclosed in a patent filed and granted to Taylor, Young, and Hyland[6] on a "System for Detecting Objects by Radio." The type of radar described in this patent was a CW wave-interference radar. Early in 1934, a 60-MHz CW wave-interference radar was demonstrated by NRL.

The early CW wave-interference radars were useful only for detecting the *presence* of the target. The problem of extracting target-position information from such radars was a difficult one and could not be readily solved with the techniques existing at that time. A proposal was made by NRL in 1933 to employ a chain of transmitting and receiving stations along a line to be guarded, for the purpose of obtaining some knowledge of distance and velocity. This was never carried out, however. The limited ability of CW wave-interference radar to be anything more than a trip wire undoubtedly tempered what little official enthusiasm existed for radar.

It was recognized that the limitations to obtaining adequate position information could be overcome with pulse transmission. Strange as it may now seem, in the early days pulse radar encountered much skepticism. Nevertheless, an effort was started at NRL in the spring of 1934 to develop a pulse radar. The work received low priority and was carried out principally by R. M. Page, but he was not allowed to devote his full time to the effort.

The first attempt with pulse radar at NRL was at a frequency of 60 MHz. According to Guerlac,[1] the first tests of the 60-MHz pulse radar were carried out in late December, 1934, and early January, 1935. These tests were "hopelessly unsuccessful and a grievous disappointment." No pulse echoes were observed on the cathode-ray tube. The chief reason for this failure was attributed to the receiver's being designed for CW communications rather than for pulse reception. The shortcomings were corrected, and the first radar echoes obtained at NRL using pulses occurred on April 28, 1936, with a radar operating at a frequency of 28.3 MHz and a pulse width of 5 μs. The range was only $2\frac{1}{2}$ miles. By early June the range was 25 miles.

It was realized by the NRL experimenters that higher radar frequencies were desired, especially for shipboard application, where large antennas could not be tolerated. However, the necessary components did not exist. The success of the experiments at 28 MHz encouraged the NRL experimenters to develop a 200-MHz equipment. The first echoes at 200 MHz were received July 22, 1936, less than three months after the start of the project. This radar was also the first to employ a duplexing system with a common antenna for both transmitting and receiving. The range was only 10 to 12 miles. In the spring of 1937 it was installed and tested on the destroyer *Leary*. The range of the 200-MHz radar was limited by the transmitter. The development of higher-powered tubes by the Eitel-McCullough Corporation allowed an improved design of the 200-MHz radar known as XAF. This occurred in January, 1938. Although the power delivered to the antenna was only 6 kW, a range of 50 miles—the limit of the sweep—was obtained by February. The XAF was tested aboard the battleship *New York*, in maneuvers held during January and February of 1939, and met with considerable success. Ranges of 20 to 24 kiloyards were obtained on battleships and cruisers. By October, 1939, orders were placed for a manufactured version called the CXAM. Nineteen of these radars were installed on major ships of the fleet by 1941.

The United States Army Signal Corps also maintained an interest in radar during the early 1930s.[7] The beginning of serious Signal Corps work in pulse radar apparently resulted from a visit to NRL in January, 1936. By December of that year the Army tested its first pulse radar, obtaining a range of 7 miles. The first operational radar used for antiaircraft fire control was the SCR-268, available in 1938.[8] The SCR-268 was used in conjunction with searchlights for radar fire control. This was necessary because of its poor angular accuracy. However, its range accuracy was superior to that obtained with optical methods. The SCR-268 remained the standard fire-control equipment until January, 1944, when it was replaced by the SCR-584 microwave radar. The SCR-584 could control an antiaircraft battery without the necessity for searchlights or optical angle tracking.

In 1939 the Army developed the SCR-270, a long-range radar for early warning. The attack on Pearl Harbor in December, 1941, was detected by an SCR-270, one of six in Hawaii at the time.[1] (There were also 16 SCR-268s assigned to units in Honolulu.) But unfortunately, the true significance of the blips on the scope was not realized until after the bombs had fallen. A modified SCR-270 was also the first radar to detect echoes from the moon in 1946.

The early developments of pulse radar were primarily concerned with military applications. Although it was not recognized as being a radar at the time, the frequency-modulated aircraft radio altimeter was probably the first commercial application of the radar principle. The first equipments were operated in aircraft as early as 1936 and utilized the same principle of operation as the FM-CW radar described in Sec. 3.3. In the case of the radio altimeter, the target is the ground.

.In Britain the development of radar began later than in the United States.[8-11] But because they felt the nearness of war more acutely and were in a more vulnerable position with respect to air attack, the British expended a large amount of effort on radar development. By the time the United States entered the war, the British were well experienced in the military applications of radar. British interest in radar began in early 1935, when Sir Robert Watson-Watt was asked about the possibility of producing a death ray using radio waves. Watson-Watt concluded that this type of death ray required fantastically large amounts of power and could be regarded as not being practical at that time. Instead, he recommended that it would be more promising to investigate means for radio detection as opposed to radio destruction. (The only available means for locating aircraft prior to World War II were sound locators whose maximum detection range under favorable conditions was about 20 miles.) Watson-Watt was allowed to explore the possibilities of radio detection, and in February, 1935, he issued two memoranda outlining the conditions necessary for an effective radar system. In that same month the detection of an aircraft was carried out, using 6-MHz communication equipment, by observing the beats between the echo signal and the directly received signal (wave interference). The technique was similar to the first United States radar-detection experiments. The transmitter and receiver were separated by about 5.5 miles. When the aircraft receded from the receiver, it was possible to detect the beats to about an 8-mile range.

By June, 1935, the British had demonstrated the pulse technique to measure range of an aircraft target. This was almost a year sooner than the successful NRL experiments with pulse radar. By September, ranges greater than 40 miles were obtained on bomber aircraft. The frequency was 12 MHz. Also, in that month, the first radar measurement of the height of aircraft above ground was made by measuring the elevation angle of arrival of the reflected signal. In March, 1936, the range of detection had increased to 90 miles and the frequency was raised to 25 MHz.

A series of CH (Chain Home) radar stations at a frequency of 25 MHz were successfully demonstrated in April, 1937. Most of the stations were operating by September, 1938, and plotted the track of the aircraft which flew Neville Chamberlain, the British Prime Minister at that time, to Munich to confer with Hitler and Mussolini. In the same month, the CH radar stations began 24-hour duty, which continued until the end of the war.

The British realized quite early that ground-based search radars such as CH were not sufficiently accurate to guide fighter aircraft to a complete interception at night or in bad weather. Consequently, they developed, by 1939, an aircraft-interception radar (AI), mounted on an aircraft, for the detection and interception of hostile aircraft. The AI radar operated at a frequency of 200 MHz. During the development of the AI radar it was noted that radar could be used for the detection of ships from the air and also that the character of echoes from the ground was dependent on the nature of the terrain. The former phenomenon was quickly exploited for the detection and location of surface ships and submarines. The latter effect was not exploited initially, but was later used for airborne mapping radars.

Until the middle of 1940 the development of radar in Britain and the United States was carried out independently of one another. In September of that year a British technical mission visited the United States to exchange information concerning the radar developments in the two countries. The British realized the advantages to be gained from the better angular resolution possible at the microwave frequencies, especially for airborne and naval applications. They suggested that the United States undertake the development of a microwave AI radar and a microwave antiaircraft fire-control radar. The British technical mission demonstrated the cavity-magnetron power tube developed by Randell and Boot and furnished design information so that it could be duplicated by United States manufacturers. The Randell and Boot magnetron operated at a wavelength of 10 cm and produced a power output of about 1 kW, an improvement by a factor of 100 over anything previously achieved at centimeter wavelengths. The development of the magnetron was one of the most important contributions to the realization of microwave radar.

The success of microwave radar was by no means certain at the end of 1940. Therefore the United States Service Laboratories chose to concentrate on the development of radars at the lower frequencies, primarily the very high frequency (VHF) band, where techniques and components were more readily available. The exploration of the microwave region for radar application became the responsibility of the Radiation Laboratory, organized in November, 1940, under the administration of the Massachusetts Institute of Technology.

In addition to the developments carried out in the United States and Great Britain, radar was developed essentially independently in Germany, France, Russia, Italy, and Japan during the middle and late thirties.[12] The extent of these developments and their subsequent military deployment varied, however. All of these countries carried out experiments with CW wave interference, and even though the French and the Japanese deployed such radars operationally, they proved of limited value. Each country eventually progressed to pulse radar operation and the advantages pertaining thereto. Although the advantages of the higher frequencies were well recognized, except for the United States and Great Britain none of the others deployed radar at frequencies higher than about 600 MHz during the war.

The Germans deployed several different types of radars during World War II. Ground-based radars were available for air search and height finding so as to perform ground control of intercept (GCI). Coastal, shipboard, and airborne radar were also employed successfully in significant numbers. An excellent description of the electronic battle in World War II between the Germans and the Allies, with many lessons to offer, is the book "*Instruments of Darkness*" by Price.[13]

The French efforts in radar, although they got an early start, were not as energetically supported as in Britain or the United States, and were severely disrupted by the German occupation in 1940.[12] The development of radar in Italy also started early, but was slow. There were only relatively few Italian-produced radars operationally deployed by the time they left the war in September, 1943. The work in Japan was also slow but received impetus from disclosures by their German allies in 1940 and from the capture of United States pulse radars in the Philippines early in 1942. The development of radar in the Soviet Union was quite similar to the experience elsewhere. By the summer of 1941 they had deployed operationally a number of 80-MHz air-search radars for the defense of Moscow against the German invasion.[14] Their indigenous efforts were interrupted by the course of the war.

Thus, radar developed independently and simultaneously in several countries just prior to World War II. It is not possible to single out any one individual as the inventor; there were many fathers of radar. This was brought about not only by the spread of radio technology to many countries, but by the maturing of the airplane during this same time and the common recognition of its military threat and the need to defend against it.

1.6 APPLICATIONS OF RADAR

Radar has been employed on the ground, in the air, on the sea, and in space. Ground-based radar has been applied chiefly to the detection, location, and tracking of aircraft or space targets. Shipboard radar is used as a navigation aid and safety device to locate buoys, shore lines, and other ships, as well as for observing aircraft. Airborne radar may be used to detect other aircraft, ships, or land vehicles, or it may be used for mapping of land, storm avoidance, terrain avoidance, and navigation. In space, radar has assisted in the guidance of spacecraft and for the remote sensing of the land and sea.

The major user of radar, and contributor of the cost of almost all of its development, has been the military; although there have been increasingly important civil applications, chiefly for marine and air navigation. The major areas of radar application, in no particular order of importance, are briefly described below.

Air Traffic Control (ATC). Radars are employed throughout the world for the purpose of safely controlling air traffic en route and in the vicinity of airports. Aircraft and ground vehicular traffic at large airports are monitored by means of high-resolution radar. Radar has been used with GCA (ground-control approach) systems to guide aircraft to a safe landing in bad weather. In addition, the microwave landing system and the widely used ATC radar-beacon system are based in large part on radar technology.

Aircraft Navigation. The weather-avoidance radar used on aircraft to outline regions of precipitation to the pilot is a classical form of radar. Radar is also used for terrain avoidance and terrain following. Although they may not always be thought of as radars, the radio altimeter (either FM/CW or pulse) and the doppler navigator are also radars. Sometimes ground-mapping radars of moderately high resolution are used for aircraft navigation purposes.

Ship Safety. Radar is used for enhancing the safety of ship travel by warning of potential collision with other ships, and for detecting navigation buoys, especially in poor visibility. In terms of numbers, this is one of the larger applications of radar, but in terms of physical size and cost it is one of the smallest. It has also proven to be one of the most reliable radar systems. Automatic detection and tracking equipments (also called plot extractors) are commercially available for use with such radars for the purpose of collision avoidance. Shore-based radar of moderately high resolution is also used for the surveillance of harbors as an aid to navigation.

Space. Space vehicles have used radar for rendezvous and docking, and for landing on the moon. Some of the largest ground-based radars are for the detection and tracking of satellites. Satellite-borne radars have also been used for remote sensing as mentioned below.

Remote Sensing. All radars are remote sensors; however, as this term is used it implies the sensing of geophysical objects, or the "environment." For some time, radar has been used as a remote sensor of the weather. It was also used in the past to probe the moon and the planets (radar astronomy). The ionospheric sounder, an important adjunct for HF (short wave) communications, is a radar. Remote sensing with radar is also concerned with Earth resources, which includes the measurement and mapping of sea conditions, water resources, ice cover, agriculture, forestry conditions, geological formations, and environmental pollution. The platforms for such radars include satellites as well as aircraft.

Law Enforcement. In addition to the wide use of radar to measure the speed of automobile traffic by highway police, radar has also been employed as a means for the detection of intruders.

Military. Many of the civilian applications of radar are also employed by the military. The traditional role of radar for military application has been for surveillance, navigation, and for the control and guidance of weapons. It represents, by far, the largest use of radar.

REFERENCES

1. Guerlac, H. E.: "OSRD Long History," vol. V, Division 14, "Radar," available from Office of Technical Services, U.S. Department of Commerce.
2. British Patent 13,170, issued to Christian Hülsmeyer, Sept. 22, 1904, entitled "Hertzian-wave Projecting and Receiving Apparatus Adapted to Indicate or Give Warning of the Presence of a Metallic Body, Such as a Ship or a Train, in the Line of Projection of Such Waves."
3. Marconi, S. G.: Radio Telegraphy, *Proc. IRE*, vol. 10, no. 4, p. 237, 1922.
4. Breit, G., and M. A. Tuve: A Test of the Existence of the Conducting Layer, *Phys. Rev.*, vol. 28, pp. 554–575, September, 1926.
5. Englund, C. R., A. B. Crawford, and W. W. Mumford: Some results of a Study of Ultra-short-wave Transmission Phenomena, *Proc. IRE*, vol. 21, pp. 475–492, March, 1933.
6. U.S. Patent 1,981,884, "System for Detecting Objects by Radio," issued to A. H. Taylor, L. C. Young, and L. A. Hyland, Nov. 27, 1934.
7. Vieweger, A. L.: Radar in the Signal Corps, *IRE Trans.*, vol. MIL-4, pp. 555–561, October, 1960.
8. Origins of Radar: Background to the Awards of the Royal Commission, *Wireless World*, vol. 58, pp. 95–99, March, 1952.
9. Wilkins, A. F.: The Story of Radar, *Research (London)*, vol. 6, pp. 434–440, November, 1953.
10. Rowe, A. P.: "One Story of Radar," Cambridge University Press, New York, 1948. A very readable description of the history of radar development at TRE (Telecommunications Research Establishment, England) and how TRE went about its business from 1935 to the end of World War II.
11. Watson-Watt, Sir Robert: "Three Steps to Victory," Odhams Press, Ltd., London, 1957; "The Pulse of Radar," The Dial Press, Inc., New York, 1959.
12. Susskind, C.: "The Birth of the Golden Cockerel: The Development of Radar," in preparation.
13. Price, A.: "Instruments of Darkness," Macdonald and Janes, London, 1977.
14. Lobanov, M. M.: "Iz Proshlovo Radiolokatzii" (Out of the Past of Radar), Military Publisher of the Ministry of Defense, USSR, Moscow, 1969.
15. IEEE Standard Letter Designations for Radar-Frequency Bands, IEEE Std 521-1976, Nov. 30, 1976.
16. Villard, O. G., Jr.: The Ionospheric Sounder and Its Place in the History of Radio Science, *Radio Science*, vol. 11, pp. 847–860, November, 1976.

TWO

THE RADAR EQUATION

2.1 PREDICTION OF RANGE PERFORMANCE

The simple form of the radar equation derived in Sec. 1.2 expressed the maximum radar range R_{max} in terms of radar and target parameters:

$$R_{max} = \left[\frac{P_t G A_e \sigma}{(4\pi)^2 S_{min}} \right]^{1/4} \tag{2.1}$$

where P_t = transmitted power, watts
$\quad G$ = antenna gain
$\quad A_e$ = antenna effective aperture, m^2
$\quad \sigma$ = radar cross section, m^2
$\quad S_{min}$ = minimum detectable signal, watts

All the parameters are to some extent under the control of the radar designer, except for the target cross section σ. The radar equation states that if long ranges are desired, the transmitted power must be large, the radiated energy must be concentrated into a narrow beam (high transmitting antenna gain), the received echo energy must be collected with a large antenna aperture (also synonymous with high gain), and the receiver must be sensitive to weak signals.

In practice, however, the simple radar equation does not predict the range performance of actual radar equipments to a satisfactory degree of accuracy. The predicted values of radar range are usually optimistic. In some cases the actual range might be only half that predicted.[1] Part of this discrepancy is due to the failure of Eq. (2.1) to explicitly include the various losses that can occur throughout the system or the loss in performance usually experienced when electronic equipment is operated in the field rather than under laboratory-type conditions. Another important factor that must be considered in the radar equation is the statistical or unpredictable nature of several of the parameters. The minimum detectable signal S_{min} and the target cross section σ are both statistical in nature and must be expressed in statistical terms.

Other statistical factors which do not appear explicitly in Eq. (2.1) but which have an effect on the radar performance are the meteorological conditions along the propagation path and the performance of the radar operator, if one is employed. The statistical nature of these several parameters does not allow the maximum radar range to be described by a single number. Its specification must include a statement of the probability that the radar will detect a certain type of target at a particular range.

In this chapter, the simple radar equation will be extended to include most of the important factors that influence radar range performance. If all those factors affecting radar range were known, it would be possible, in principle, to make an accurate prediction of radar performance. But, as is true for most endeavors, the quality of the prediction is a function of the amount of effort employed in determining the quantitative effects of the various parameters. Unfortunately, the effort required to specify completely the effects of all radar parameters to the degree of accuracy required for range prediction is usually not economically justified. A compromise is always necessary between what one would like to have and what one can actually get with reasonable effort. This will be better appreciated as we proceed through the chapter and note the various factors that must be taken into account.

A complete and detailed discussion of all those factors that influence the prediction of radar range is beyond the scope of a single chapter. For this reason many subjects will appear to be treated only lightly. This is deliberate and is necessitated by brevity. More detailed information will be found in some of the subsequent chapters or in the references listed at the end of the chapter.

2.2 MINIMUM DETECTABLE SIGNAL

The ability of a radar receiver to detect a weak echo signal is limited by the noise energy that occupies the same portion of the frequency spectrum as does the signal energy. The weakest signal the receiver can detect is called the *minimum detectable signal*. The specification of the minimum detectable signal is sometimes difficult because of its statistical nature and because the criterion for deciding whether a target is present or not may not be too well defined.

Detection is based on establishing a threshold level at the output of the receiver. If the receiver output exceeds the threshold, a signal is assumed to be present. This is called *threshold detection*. Consider the output of a typical radar receiver as a function of time (Fig. 2.1). This might represent one sweep of the video output displayed on an A-scope. The envelope has a fluctuating appearance caused by the random nature of noise. If a large signal is present such as at A in Fig. 2.1, it is greater than the surrounding noise peaks and can be recognized on the basis of its amplitude. Thus, if the threshold level were set sufficiently high, the envelope would not generally exceed the threshold if noise alone were present, but would exceed it if a strong signal were present. If the signal were small, however, it would be more difficult to recognize its presence. The threshold level must be low if weak signals are to be detected, but it cannot be so low that noise peaks cross the threshold and give a false indication of the presence of targets.

The voltage envelope of Fig. 2.1 is assumed to be from a matched-filter receiver (Sec. 10.2). A matched filter is one designed to maximize the output peak signal to average noise (power) ratio. It has a frequency-response function which is proportional to the complex conjugate of the signal spectrum. (This is not the same as the concept of "impedance match" of circuit theory.) The ideal matched-filter receiver cannot always be exactly realized in practice, but it is possible to approach it with practical receiver circuits. A matched filter for a radar transmitting a rectangular-shaped pulse is usually characterized by a bandwidth B approximately the reciprocal of the pulse width τ, or $B\tau \approx 1$. The output of a matched-filter receiver is

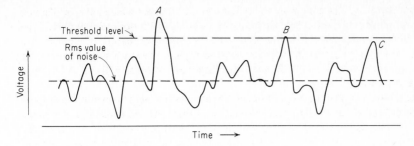

Figure 2.1 Typical envelope of the radar receiver output as a function of time. *A*, and *B*, and *C* represent signal plus noise. *A* and *B* would be valid detections, but *C* is a missed detection.

the cross correlation between the received waveform and a replica of the transmitted waveform. Hence it does not preserve the shape of the input waveform. (There is no reason to wish to preserve the shape of the received waveform so long as the output signal-to-noise ratio is maximized.)

Let us return to the receiver output as represented in Fig. 2.1. A threshold level is established, as shown by the dashed line. A target is said to be detected if the envelope crosses the threshold. If the signal is large such as at *A*, it is not difficult to decide that a target is present. But consider the two signals at *B* and *C*, representing target echoes of equal amplitude. The noise voltage accompanying the signal at *B* is large enough so that the combination of signal plus noise exceeds the threshold. At *C* the noise is not as large and the resultant signal plus noise does not cross the threshold. Thus the presence of noise will sometimes enhance the detection of weak signals but it may also cause the loss of a signal which would otherwise be detected.

Weak signals such as *C* would not be lost if the threshold level were lower. But too low a threshold increases the likelihood that noise alone will rise above the threshold and be taken for a real signal. Such an occurrence is called a *false alarm*. Therefore, if the threshold is set too low, false target indications are obtained, but if it is set too high, targets might be missed. The selection of the proper threshold level is a compromise that depends upon how important it is if a mistake is made either by (1) failing to recognize a signal that is present (probability of a miss) or by (2) falsely indicating the presence of a signal when none exists (probability of a false alarm).

When the target-decision process is made by an operator viewing a cathode-ray-tube display, it would seem that the criterion used by the operator for detection ought to be analogous to the setting of a threshold, either consciously or subconsciously. The chief difference between the electronic and the operator thresholds is that the former may be determined with some logic and can be expected to remain constant with time, while the latter's threshold might be difficult to predict and may not remain fixed. The individual's performance as part of the radar detection process depends upon the state of the operator's fatigue and motivation, as well as training.

The capability of the human operator as part of the radar detection process can be determined only by experiment. Needless to say, in experiments of this nature there are likely to be wide variations between different experimenters. Therefore, for the purposes of the present discussion, the operator will be considered the same as an electronic threshold detector, an assumption that is generally valid for an alert, trained operator.

The signal-to-noise ratio necessary to provide adequate detection is one of the important parameters that must be determined in order to compute the minimum detectable signal. Although the detection decision is usually based on measurements at the video output, it is easier to consider maximizing the signal-to-noise ratio at the output of the IF amplifier rather than in the video. The receiver may be considered linear up to the output of the IF. It is shown by Van Vleck and Middleton[3] that maximizing the signal-to-noise ratio at the output of the IF is equivalent to maximizing the video output. The advantage of considering the signal-to-noise ratio at the IF is that the assumption of linearity may be made. It is also assumed that the IF filter characteristic approximates the matched filter, so that the output signal-to-noise ratio is maximized.

2.3 RECEIVER NOISE

Since noise is the chief factor limiting receiver sensitivity, it is necessary to obtain some means of describing it quantitatively. Noise is unwanted electromagnetic energy which interferes with the ability of the receiver to detect the wanted signal. It may originate within the receiver itself, or it may enter via the receiving antenna along with the desired signal. If the radar were to operate in a perfectly noise-free environment so that no external sources of noise accompanied the desired signal, and if the receiver itself were so perfect that it did not generate any excess noise, there would still exist an unavoidable component of noise generated by the thermal motion of the conduction electrons in the ohmic portions of the receiver input stages. This is called *thermal noise*, or *Johnson noise*, and is directly proportional to the temperature of the ohmic portions of the circuit and the receiver bandwidth.[60] The available thermal-noise power generated by a receiver of bandwidth B_n (in hertz) at a temperature T (degrees Kelvin) is equal to

$$\text{Available thermal-noise power} = kTB_n \tag{2.2}$$

where k = Boltzmann's constant = 1.38×10^{-23} J/deg. If the temperature T is taken to be 290 K, which corresponds approximately to room temperature (62°F), the factor kT is 4×10^{-21} W/Hz of bandwidth. If the receiver circuitry were at some other temperature, the thermal-noise power would be correspondingly different.

A receiver with a reactance input such as a parametric amplifier need not have any significant ohmic loss. The limitation in this case is the thermal noise seen by the antenna and the ohmic losses in the transmission line.

For radar receivers of the superheterodyne type (the type of receiver used for most radar applications), the receiver bandwidth is approximately that of the intermediate-frequency stages. It should be cautioned that the bandwidth B_n of Eq. (2.2) is not the 3-dB, or half-power, bandwidth commonly employed by electronic engineers. It is an integrated bandwidth and is given by

$$B_n = \frac{\int_{-\infty}^{\infty} |H(f)|^2 \, df}{|H(f_0)|^2} \tag{2.3}$$

where $H(f)$ = frequency-response characteristic of IF amplifier (filter) and f_0 = frequency of maximum response (usually occurs at midband).

When $H(f)$ is normalized to unity at midband (maximum-response frequency), $H(f_0) = 1$. The bandwidth B_n is called the *noise bandwidth* and is the bandwidth of an equivalent rectangular filter whose noise-power output is the same as the filter with characteristic

$H(f)$. The 3-dB bandwidth is defined as the separation in hertz between the points on the frequency-response characteristic where the response is reduced to 0.707 (3 dB) from its maximum value. The 3-dB bandwidth is widely used, since it is easy to measure. The measurement of noise bandwidth, however, involves a complete knowledge of the response characteristic $H(f)$. The frequency-response characteristics of many practical radar receivers are such that the 3-dB and the noise bandwidths do not differ appreciably. Therefore the 3-dB bandwidth may be used in many cases as an approximation to the noise bandwidth.[2]

The noise power in practical receivers is often greater than can be accounted for by thermal noise alone. The additional noise components are due to mechanisms other than the thermal agitation of the conduction electrons. For purposes of the present discussion, however, the exact origin of the extra noise components is not important except to know that it exists. No matter whether the noise is generated by a thermal mechanism or by some other mechanism, the total noise at the output of the receiver may be considered to be equal to the thermal-noise power obtained from an "ideal" receiver multiplied by a factor called the *noise figure*. The noise figure F_n of a receiver is defined by the equation

$$F_n = \frac{N_o}{kT_0 B_n G_a} = \frac{\text{noise out of practical receiver}}{\text{noise out of ideal receiver at std temp } T_0} \qquad (2.4a)$$

where N_o = noise output from receiver, and G_a = available gain. The standard temperature T_0 is taken to be 290 K, according to the Institute of Electrical and Electronics Engineers definition. The noise N_o is measured over the linear portion of the receiver input-output characteristic, usually at the output of the IF amplifier before the nonlinear second detector. The receiver bandwidth B_n is that of the IF amplifier in most receivers. The available gain G_a is the ratio of the signal out S_o to the signal in S_i, and $kT_0 B_n$ is the input noise N_i in an ideal receiver. Equation (2.4a) may be rewritten as

$$F_n = \frac{S_i/N_i}{S_o/N_o} \qquad (2.4b)$$

The noise figure may be interpreted, therefore, as a measure of the degradation of signal-to-noise-ratio as the signal passes through the receiver.

Rearranging Eq. (2.4b), the input signal may be expressed as

$$S_i = \frac{kT_0 B_n F_n S_o}{N_o} \qquad (2.5)$$

If the minimum detectable signal S_{min} is that value of S_i corresponding to the minimum ratio of output (IF) signal-to-noise ratio $(S_o/N_o)_{min}$ necessary for detection, then

$$S_{min} = kT_0 B_n F_n \left(\frac{S_o}{N_o} \right)_{min} \qquad (2.6)$$

Substituting Eq. (2.6) into Eq. (2.1) results in the following form of the radar equation:

$$R_{max}^4 = \frac{P_t G A_e \sigma}{(4\pi)^2 kT_0 B_n F_n (S_o/N_o)_{min}} \qquad (2.7)$$

Before continuing the discussion of the factors involved in the radar equation, it is necessary to digress and review briefly some topics in probability theory in order to describe the signal-to-noise ratio in statistical terms.

2.4 PROBABILITY-DENSITY FUNCTIONS

The basic concepts of probability theory needed in solving noise problems may be found in any of several references.[4-8] In this section we shall briefly review probability and the probability-density function and cite some examples.

Noise is a random phenomenon. Predictions concerning the average performance of random phenomena are possible by observing and classifying occurrences, but one cannot predict exactly what will occur for any particular event. Phenomena of a random nature can be described with the aid of probability theory.

Probability is a measure of the likelihood of occurrence of an event. The scale of probability ranges from 0 to 1.† An event which is certain is assigned the probability 1. An impossible event is assigned the probability 0. The intermediate probabilities are assigned so that the more likely an event, the greater is its probability.

One of the more useful concepts of probability theory needed to analyze the detection of signals in noise is the *probability-density function*. Consider the variable x as representing a typical measured value of a random process such as a noise voltage or current. Imagine each x to define a point on a straight line corresponding to the distance from a fixed reference point. The distance of x from the reference point might represent the value of the noise current or the noise voltage. Divide the line into small equal segments of length Δx and count the number of times that x falls in each interval. The probability-density function $p(x)$ is then defined as

$$p(x) = \lim_{\substack{\Delta x \to 0 \\ N \to \infty}} \frac{(\text{number of values in range } \Delta x \text{ at } x)/\Delta x}{\text{total number of values} = N} \tag{2.8}$$

The probability that a particular measured value lies within the infinitesimal width dx centered at x is simply $p(x)\,dx$. The probability that the value of x lies within the finite range from x_1 to x_2 is found by integrating $p(x)$ over the range of interest, or

$$\text{Probability } (x_1 < x < x_2) = \int_{x_1}^{x_2} p(x)\,dx \tag{2.9}$$

By definition, the probability-density function is positive. Since every measurement must yield some value, the integral of the probability density over all values of x must be equal to unity; that is,

$$\int_{-\infty}^{\infty} p(x)\,dx = 1 \tag{2.10}$$

The average value of a variable function, $\phi(x)$, that is described by the probability-density function, $p(x)$, is

$$\langle \phi(x) \rangle_{\text{av}} = \int_{-\infty}^{\infty} \phi(x)p(x)\,dx \tag{2.11}$$

This follows from the definition of an average value and the probability-density function. The mean, or average, value of x is

$$\langle x \rangle_{\text{av}} = m_1 = \int_{-\infty}^{\infty} xp(x)\,dx \tag{2.12}$$

---

† Probabilities are sometimes expressed in percent (0 to 100) rather than 0 to 1.

and the mean square value is

$$\langle x^2 \rangle_{av} = m_2 = \int_{-\infty}^{\infty} x^2 p(x) \, dx \qquad (2.13)$$

The quantities m_1 and m_2 are sometimes called the first and second moments of the random variable x. If x represents an electric voltage or current, m_1 is the d-c component. It is the value read by a direct-current voltmeter or ammeter. The mean square value (m_2) of the current when multiplied by the resistance† gives the mean power. The mean square value of voltage times the conductance is also the mean power. The *variance* is defined as

$$\mu_2 = \sigma^2 = \langle (x - m_1)^2 \rangle_{av} = \int_{-\infty}^{\infty} (x - m_1)^2 p(x) \, dx = m_2 - m_1^2 = \langle x^2 \rangle_{av} - \langle x \rangle_{av}^2 \quad (2.14)$$

The variance is the mean square deviation of x about its mean and is sometimes called the *second central moment*. If the random variable is a noise current, the product of the variance and resistance gives the mean power of the a-c component. The square root of the variance σ is called the *standard deviation* and is the root-mean-square (rms) value of the a-c component.

We shall consider four examples of probability-density functions: the uniform, gaussian, Rayleigh, and exponential. The uniform probability-density (Fig. 2.2a) is defined as

$$p(x) = \begin{cases} k & \text{for } a < x < a + b \\ 0 & \text{for } x < a \text{ and } x > a + b \end{cases}$$

† In noise theory it is customary to take the resistance as 1 ohm or the conductance as 1 mho.

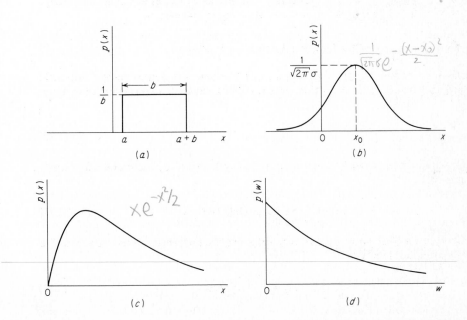

Figure 2.2 Examples of probability-density functions. (*a*) Uniform; (*b*) Gaussian; (*c*) Rayleigh (voltage); (*d*) Rayleigh (power) or exponential.

where k is a constant. A rectangular, or uniform, distribution describes the phase of a random sine wave relative to a particular origin of time; that is, the phase of the sine wave may be found, with equal probability, anywhere from 0 to 2π, with $k = 1/2\pi$. It also applies to the distribution of the round-off (quantizing) error in numerical computations and in analog-to-digital converters.

The constant k may be found by applying Eq. (2.10); that is,

$$\int_{-\infty}^{\infty} p(x)\,dx = \int_{a}^{a+b} k\,dx = 1 \qquad \text{or} \qquad k = \frac{1}{b}$$

The average value of x is

$$m_1 = \int_{a}^{a+b} \frac{1}{b} x\,dx = a + \frac{b}{2}$$

This result could have been determined by inspection. The second-moment, or mean square, value is

$$m_2 = \int_{a}^{a+b} \frac{x^2}{b}\,dx = a^2 + ab + \frac{b^2}{3}$$

and the variance is

$$\sigma^2 = m_2 - m_1^2 = \frac{b^2}{12}$$

$$\sigma = \text{standard deviation} = \frac{b}{2\sqrt{3}}$$

The gaussian, or normal, probability density (Fig. 2.2b) is one of the most important in noise theory, since many sources of noise, such as thermal noise or shot noise, may be represented by gaussian statistics. Also, a gaussian representation is often more convenient to manipulate mathematically. The gaussian density function has a bell-shaped appearance and is defined by

$$p(x) = \frac{1}{\sqrt{2\pi\sigma^2}} \exp \frac{-(x - x_0)^2}{2\sigma^2} \tag{2.15}$$

where exp [ ] is the exponential function, and the parameters have been adjusted to satisfy the normalizing condition of Eq. (2.10). It can be shown that

$$m_1 = \int_{-\infty}^{\infty} xp(x)\,dx = x_0 \qquad m_2 = \int_{-\infty}^{\infty} x^2 p(x)\,dx = x_0^2 + \sigma^2 \qquad \mu_2 = m_2 - m_1^2 = \sigma^2 \tag{2.16}$$

The probability density of the sum of a large number of independently distributed quantities approaches the gaussian probability-density function no matter what the individual distributions may be, provided that the contribution of any one quantity is not comparable with the resultant of all others. This is the *central limit theorem*. Another property of the gaussian distribution is that no matter how large a value x we may choose, there is always some finite probability of finding a greater value. If the noise at the input of the threshold detector were truly gaussian, then no matter how high the threshold were set, there would always be a chance that it would be exceeded by noise and appear as a false alarm. However, the probability diminishes rapidly with increasing x, and for all practical purposes the probability of obtaining an exceedingly high value of x is negligibly small.

The Rayleigh probability-density function is also of special interest to the radar systems engineer. It describes the envelope of the noise output from a narrowband filter (such as the IF filter in a superheterodyne receiver), the cross-section fluctuations of certain types of complex radar targets, and many kinds of clutter and weather echoes. The Rayleigh density function is

$$p(x) = \frac{2x}{\langle x^2 \rangle_{av}} \exp\left(-\frac{x^2}{\langle x^2 \rangle_{av}}\right) \qquad x \geq 0 \tag{2.17}$$

This is plotted in Fig. 2.2c. The parameter x might represent a voltage, and $\langle x^2 \rangle_{av}$ the mean, or average, value of the voltage squared. If x^2 is replaced by w, where w represents power instead of voltage (assuming the resistance is 1 ohm), Eq. (2.17) becomes

$$p(w) = \frac{1}{w_0} \exp\left(-\frac{w}{w_0}\right) \qquad w \geq 0 \tag{2.18}$$

where w_0 is the average power. This is the exponential probability-density function, but it is sometimes called the Rayleigh-power probability-density function. It is plotted in Fig. 2.2d. The standard deviation of the Rayleigh density of Eq. (2.17) is equal to $\sqrt{(4/\pi) - 1}$ times the mean value, and for the exponential density of Eq. (2.18) the standard deviation is equal to w_0. There are other probability-density functions of interest in radar, such as the Rice, log normal, and the chi square. These will be introduced as needed.

Another mathematical description of statistical phenomena is the *probability distribution function* $P(x)$, defined as the probability that the value x is less than some specified value

$$P(x) = \int_{-\infty}^{x} p(x)\, dx \qquad \text{or} \qquad p(x) = \frac{d}{dx} P(x) \tag{2.19}$$

In some cases, the distribution function may be easier to obtain from an experimental set of data than the density function. The density function may be found from the distribution function by differentiation.

2.5 SIGNAL-TO-NOISE RATIO

In this section the results of statistical noise theory will be applied to obtain the signal-to-noise ratio at the output of the IF amplifier necessary to achieve a specified probability of detection without exceeding a specified probability of false alarm. The output signal-to-noise ratio thus obtained may be substituted into Eq. (2.6) to find the minimum detectable signal, which, in turn, is used in the radar equation, as in Eq. (2.7).

Consider an IF amplifier with bandwidth B_{IF} followed by a second detector and a video amplifier with bandwidth B_v (Fig. 2.3). The second detector and video amplifier are assumed to form an envelope detector, that is, one which rejects the carrier frequency but passes the modulation envelope. To extract the modulation envelope, the video bandwidth must be wide enough to pass the low-frequency components generated by the second detector, but not so wide as to pass the high-frequency components at or near the intermediate frequency. The video bandwidth B_v must be greater than $B_{IF}/2$ in order to pass all the video modulation. Most radar receivers used in conjunction with an operator viewing a CRT display meet this condition and

Figure 2.3 Envelope detector.

may be considered envelope detectors. Either a square-law or a linear detector may be assumed since the effect on the detection probability by assuming one instead of the other is usually small.

The noise entering the IF filter (the terms filter and amplifier are used interchangeably) is assumed to be gaussian, with probability-density function given by

$$p(v) = \frac{1}{\sqrt{2\pi\psi_0}} \exp \frac{-v^2}{2\psi_0} \tag{2.20}$$

where $p(v)\,dv$ is the probability of finding the noise voltage v between the values of v and $v + dv$, ψ_0 is the variance, or mean-square value of the noise voltage, and the mean value of v is taken to be zero. If gaussian noise were passed through a narrowband IF filter—one whose bandwidth is small compared with the midfrequency—the probability density of the envelope of the noise voltage output is shown by Rice[9] to be

$$p(R) = \frac{R}{\psi_0} \exp \left(-\frac{R^2}{2\psi_0} \right) \tag{2.21}$$

where R is the amplitude of the envelope of the filter output. Equation (2.21) is a form of the Rayleigh probability-density function.

The probability that the envelope of the noise voltage will lie between the values of V_1 and V_2 is

$$\text{Probability } (V_1 < R < V_2) = \int_{V_1}^{V_2} \frac{R}{\psi_0} \exp \left(-\frac{R^2}{2\psi_0} \right) dR \tag{2.22}$$

The probability that the noise voltage envelope will exceed the voltage threshold V_T is

$$\text{Probability } (V_T < R < \infty) = \int_{V_T}^{\infty} \frac{R}{\psi_0} \exp \left(-\frac{R^2}{2\psi_0} \right) dR \tag{2.23}$$

$$= \exp \left(-\frac{V_T^2}{2\psi_0} \right) = P_{\text{fa}} \tag{2.24}$$

Whenever the voltage envelope exceeds the threshold, a target detection is considered to have occurred, by definition. Since the probability of a false alarm is the probability that noise will cross the threshold, Eq. (2.24) gives the probability of a false alarm, denoted P_{fa}.

The average time interval between crossings of the threshold by noise alone is defined as the *false-alarm time* T_{fa},

$$T_{\text{fa}} = \lim_{N \to \infty} \frac{1}{N} \sum_{k=1}^{N} T_k$$

where T_k is the time between crossings of the threshold V_T by the noise envelope, when the slope of the crossing is positive. The false-alarm probability may also be defined as the ratio of the duration of time the envelope is actually above the threshold to the total time it *could have been* above the threshold, or

$$P_{\text{fa}} = \frac{\sum\limits_{k=1}^{N} t_k}{\sum\limits_{k=1}^{N} T_k} = \frac{\langle t_k \rangle_{\text{av}}}{\langle T_k \rangle_{\text{av}}} = \frac{1}{T_{\text{fa}} B} \tag{2.25}$$

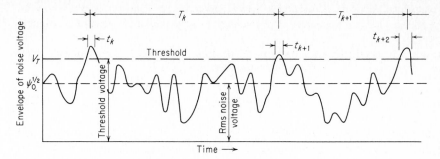

Figure 2.4 Envelope of receiver output illustrating false alarms due to noise.

where t_k and T_k are defined in Fig. 2.4. The average duration of a noise pulse is approximately the reciprocal of the bandwidth B, which in the case of the envelope detector is B_{IF}. Equating Eqs. (2.24) and (2.25), we get

$$T_{\text{fa}} = \frac{1}{B_{\text{IF}}} \exp \frac{V_T^2}{2\psi_0} \qquad (2.26)$$

A plot of Eq. (2.26) is shown in Fig. 2.5, with $V_T^2/2\psi_0$ as the abscissa. If, for example, the bandwidth of the IF amplifier were 1 MHz and the average false-alarm time that could be tolerated were 15 min, the probability of a false alarm is 1.11×10^{-9}. From Eq. (2.24) the threshold voltage necessary to achieve this false-alarm time is 6.45 times the rms value of the noise voltage.

The false-alarm probabilities of practical radars are quite small. The reason for this is that the false-alarm probability is the probability that a noise pulse will cross the threshold during an interval of time approximately equal to the reciprocal of the bandwidth. For a 1-MHz bandwidth, there are of the order of 10^6 noise pulses per second. Hence the false-alarm probability of any one pulse must be small ($< 10^{-6}$) if false-alarm times greater than 1 s are to be obtained.

The specification of a tolerable false-alarm time usually follows from the requirements desired by the customer and depends on the nature of the radar application. The exponential relationship between the false-alarm time T_{fa} and the threshold level V_T results in the false-alarm time being sensitive to variations or instabilities in the threshold level. For example, if the bandwidth were 1 MHz, a value of $10 \log (V_T^2/2\psi_0) = 12.95$ dB results in an average false-alarm time of 6 min, while a value of 14.72 dB results in a false-alarm time of 10,000 h. Thus a change in the threshold of only 1.77 dB changes the false-alarm time by five orders of magnitude. Such is the nature of gaussian noise. In practice, therefore, the threshold level would probably be adjusted slightly above that computed by Eq. (2.26), so that instabilities which lower the threshold slightly will not cause a flood of false alarms.

If the receiver were turned off (gated) for a fraction of time (as in a tracking radar with a servo-controlled range gate or a radar which turns off the receiver during the time of transmission), the false-alarm probability will be increased by the fraction of time the receiver is not operative assuming that the average false-alarm time remains the same. However, this is usually not important since small changes in the probability of false alarm result in even smaller changes in the threshold level because of the exponential relationship of Eq. (2.26).

Thus far, a receiver with only a noise input has been discussed. Next, consider a sine-wave signal of amplitude A to be present along with noise at the input to the IF filter. The frequency

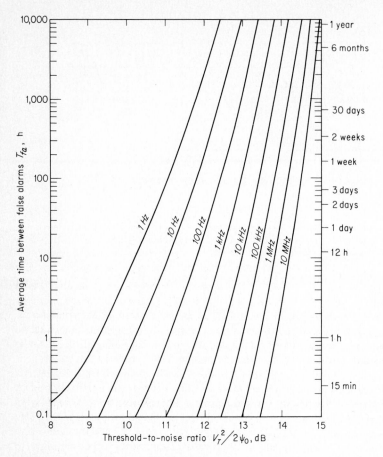

Figure 2.5 Average time between false alarms as a function of the threshold level V_T and the receiver bandwidth B; ψ_0 is the mean square noise voltage.

of the signal is the same as the IF midband frequency f_{IF}. The output of the envelope detector has a probability-density function given by[9]

$$p_s(R) = \frac{R}{\psi_0} \exp\left(-\frac{R^2 + A^2}{2\psi_0}\right) I_0\left(\frac{RA}{\psi_0}\right) \qquad (2.27)$$

where $I_0(Z)$ is the modified Bessel function of zero order and argument Z. For Z large, an asymptotic expansion for $I_0(Z)$ is

$$I_0(Z) \approx \frac{e^Z}{\sqrt{2\pi Z}}\left(1 + \frac{1}{8Z} + \cdots\right)$$

When the signal is absent, $A = 0$ and Eq. (2.27) reduces to Eq. (2.21), the probability-density function for noise alone. Equation (2.27) is sometimes called the Rice probability-density function.

The probability that the signal will be detected (which is the *probability of detection*) is the same as the probability that the envelope R will exceed the predetermined threshold V_T. The probability of detection P_d is therefore

$$P_d = \int_{V_T}^{\infty} p_s(R) \, dR = \int_{V_T}^{\infty} \frac{R}{\psi_0} \exp\left(-\frac{R^2 + A^2}{2\psi_0}\right) I_0\left(\frac{RA}{\psi_0}\right) dR \qquad (2.28)$$

This cannot be evaluated by simple means, and numerical techniques or a series approximation must be used. A series approximation valid when $RA/\psi_0 \gg 1$, $A \gg |R - A|$, and terms in A^{-3} and beyond can be neglected is[9]

$$P_d = \frac{1}{2}\left(1 - \text{erf} \frac{V_T - A}{\sqrt{2\psi_0}}\right) + \frac{\exp\left[-(V_T - A)^2/2\psi_0\right]}{2\sqrt{2\pi}(A/\sqrt{\psi_0})}$$

$$\times \left[1 - \frac{V_T - A}{4A} + \frac{1 + (V_T - A)^2/\psi_0}{8A^2/\psi_0} - \cdots\right] \qquad (2.29)$$

where the error function is defined as

$$\text{erf } Z = \frac{2}{\sqrt{\pi}} \int_0^Z e^{-u^2} \, du$$

A graphic illustration of the process of threshold detection is shown in Fig. 2.6. The probability density for noise alone [Eq. (2.21)] is plotted along with that for signal and noise [Eq. (2.27)] with $A/\psi_0^{1/2} = 3$. A threshold voltage $V_T/\psi_0^{1/2} = 2.5$ is shown. The crosshatched area to the right of $V_T/\psi_0^{1/2}$ under the curve for signal-plus-noise represents the probability of detection, while the double-crosshatched area under the curve for noise alone represents the probability of a false alarm. If $V_T/\psi_0^{1/2}$ is increased to reduce the probability of a false alarm, the probability of detection will be reduced also.

Equation (2.29) may be used to plot a family of curves relating the probability of detection to the threshold voltage and to the amplitude of the sine-wave signal. Although the receiver designer prefers to operate with voltages, it is more convenient for the radar system engineer to employ power relationships. Equation (2.29) may be converted to power by replacing the signal-to-rms-noise-voltage ratio with the following:

$$\frac{A}{\psi_0^{1/2}} = \frac{\text{signal amplitude}}{\text{rms noise voltage}} = \frac{\sqrt{2}(\text{rms signal voltage})}{\text{rms noise voltage}} = \left(2\frac{\text{signal power}}{\text{noise power}}\right)^{1/2} = \left(\frac{2S}{N}\right)^{1/2}$$

We shall also replace $V_T^2/2\psi_0$ by $\ln(1/P_{\text{fa}})$ [from Eq. (2.24)]. Using the above relationships, the probability of detection is plotted in Fig. 2.7 as a function of the signal-to-noise ratio with the probability of a false alarm as a parameter.

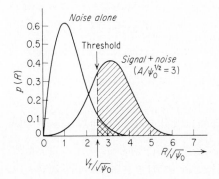

Figure 2.6 Probability-density function for noise alone and for signal-plus-noise, illustrating the process of threshold detection.

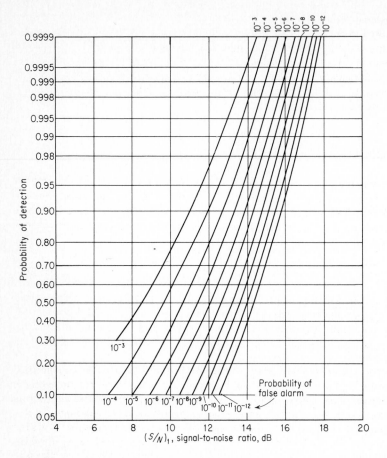

Figure 2.7 Probability of detection for a sine wave in noise as a function of the signal-to-noise (power) ratio and the probability of false alarm.

Both the false-alarm time and the detection probability are specified by the system requirements. The radar designer computes the probability of the false alarm and from Fig. 2.7 determines the signal-to-noise ratio. This is the signal-to-noise ratio that is used in the equation for minimum detectable signal [Eq. (2.6)]. The signal-to-noise ratios of Fig. 2.7 apply to a single radar pulse. For example, suppose that the desired false-alarm time was 15 min and the IF bandwidth was 1 MHz. This gives a false-alarm probability of 1.11×10^{-9}. Figure 2.7 indicates that a signal-to-noise ratio of 13.1 dB is required to yield a 0.50 probability of detection, 14.7 dB for 0.90, and 16.5 dB for 0.999.

There are several interesting facts illustrated by Fig. 2.7. At first glance, it might seem that the signal-to-noise ratio required for detection is higher than that dictated by intuition, even for a probability of detection of 0.50. One might be inclined to say that so long as the signal is greater than noise, detection should be accomplished. Such reasoning may not be correct when the false-alarm probability is properly taken into account. Another interesting effect to be noted from Fig. 2.7 is that a change of only 3.4 dB can mean the difference between reliable detection (0.999) and marginal detection (0.50). (When the target cross section fluctuates, the change in signal-to-noise ratio is much greater than this for a given change in detection probability, as discussed in Sec. 2.8.) Also, the signal-to-noise ratio required for detection is not a sensitive function of the false-alarm time. For example, a radar with a 1-MHz bandwidth requires a signal-to-noise ratio of 14.7 dB for a 0.90 probability of detection and a 15-min false-alarm time. If the false-alarm time were increased from 15 min to 24 h, the signal-to-noise ratio would be increased to 15.4 dB. If the false-alarm time were as high as 1 year, the required signal-to-noise ratio would be 16.2 dB.

2.6 INTEGRATION OF RADAR PULSES

The relationship between the signal-to-noise ratio, the probability of detection, and the probability of false alarm as given in Fig. 2.7 applies for a single pulse only. However, many pulses are usually returned from any particular target on each radar scan and can be used to improve detection. The number of pulses n_B returned from a point target as the radar antenna scans through its beamwidth is

$$n_B = \frac{\theta_B f_p}{\dot{\theta}_s} = \frac{\theta_B f_p}{6\omega_m} \tag{2.30}$$

where θ_B = antenna beamwidth, deg
f_p = pulse repetition frequency, Hz , PRF
$\dot{\theta}_s$ = antenna scanning rate, deg/s
ω_m = antenna scan rate, rpm

Typical parameters for a ground-based search radar might be pulse repetition frequency 300 Hz, 1.5° beamwidth, and antenna scan rate 5 rpm (30°/s). These parameters result in 15 hits from a point target on each scan. The process of summing all the radar echo pulses for the purpose of improving detection is called *integration*. Many techniques might be employed for accomplishing integration, as discussed in Sec. 10.7. All practical integration techniques employ some sort of storage device. Perhaps the most common radar integration method is the cathode-ray-tube display combined with the integrating properties of the eye and brain of the radar operator. The discussion in this section is concerned primarily with integration performed by electronic devices in which detection is made automatically on the basis of a threshold crossing.

Integration may be accomplished in the radar receiver either before the second detector (in the IF) or after the second detector (in the video). A definite distinction must be made between these two cases. Integration before the detector is called *predetection*, or *coherent*, integration, while integration after the detector is called *postdetection*, or *noncoherent*, integration. Predetection integration requires that the phase of the echo signal be preserved if full benefit is to be obtained from the summing process. On the other hand, phase information is destroyed by the second detector; hence postdetection integration is not concerned with preserving RF phase. For this convenience, postdetection integration is not as efficient as predetection integration.

If n pulses, all of the same signal-to-noise ratio, were integrated by an ideal predetection integrator, the resultant, or integrated, signal-to-noise (power) ratio would be exactly n times that of a single pulse. If the same n pulses were integrated by an ideal postdetection device, the resultant signal-to-noise ratio would be less than n times that of a single pulse. This loss in integration efficiency is caused by the nonlinear action of the second detector, which converts some of the signal energy to noise energy in the rectification process.

The comparison of predetection and postdetection integration may be briefly summarized by stating that although postdetection integration is not as efficient as predetection integration, it is easier to implement in most applications. Postdetection integration is therefore preferred, even though the integrated signal-to-noise ratio may not be as great. As mentioned in Sec. 10.6, an alert, trained operator viewing a properly designed cathode-ray tube display is a close approximation to the theoretical postdetection integrator.

The efficiency of postdetection integration relative to ideal predetection integration has been computed by Marcum[10] when all pulses are of equal amplitude. The integration efficiency may be defined as follows:

$$E_i(n) = \frac{(S/N)_1}{n(S/N)_n} \tag{2.31}$$

where n = number of pulses integrated
  $(S/N)_1$ = value of signal-to-noise ratio of a single pulse required to produce given probability of detection (for $n = 1$)
  $(S/N)_n$ = value of signal-to-noise ratio per pulse required to produce same probability of detection when n pulses are integrated

The improvement in the signal-to-noise ratio when n pulses are integrated postdetection is $nE_i(n)$ and is the *integration-improvement factor*. It may also be thought of as the effective number of pulses integrated by the postdetection integrator. The improvement with ideal predetection integration would be equal to n. Examples of the postdetection integration-improvement factor $I_i(n) = nE_i(n)$ are shown in Fig. 2.8a. These curves were derived from data given by Marcum. The integration loss is shown in Fig. 2.8b, where integration loss in decibels is defined as $L_i(n) = 10 \log [1/E_i(n)]$. The integration-improvement factor (or the integration loss) is not a sensitive function of either the probability of detection or the probability of false alarm.

The parameter n_f for the curves of Fig. 2.8 is the *false-alarm number*, as introduced by Marcum.[10] It is equal to the reciprocal of the false-alarm probability P_{fa} defined previously by Eqs. (2.24) and (2.25). Some authors, like Marcum, prefer to use the false-alarm number instead of the false-alarm probability. On the average, there will be one false decision out of n_f possible decisions within the false-alarm time T_{fa}. Thus the average number of possible decisions between false alarms is defined to be n_f. If τ is the pulse width, T_p the pulse repetition period, and $f_p = 1/T_p$ is the pulse repetition frequency, then the number of decisions n_f in time T_{fa} is equal to the number of range intervals per pulse period $\eta = T_p/\tau = 1/f_p\tau$ times the number of pulse periods per second f_p, times the false-alarm time T_{fa}. Therefore, the number of possible decisions is $n_f = T_{fa} f_p \eta = T_{fa}/\tau$. Since $\tau \simeq 1/B$, where B is the bandwidth, the false-alarm number is $n_f = T_{fa} B = 1/P_{fa}$.

Note that $P_{fa} = 1/T_{fa} B$ is the probability of false alarm assuming that independent decisions as to the presence or absence of a target are made B times a second. As the radar scans by a target it receives n pulses. If these n pulses are integrated before a target decision is made, then there are B/n possible decisions per second. The false-alarm probability is thus n times as great. This does not mean that there will be more false alarms, since it is the *rate* of detection-decisions that is reduced rather than the average time between alarms. This is another reason the average false-alarm time T_{fa} is a more significant parameter than the false-alarm probability. In this text, P_{fa} will be taken as the reciprocal of $T_{fa} B = n_f$, unless stated otherwise. Some authors[11] prefer to define a false-alarm number n_f' that takes account the number of pulses integrated, such that $n_f' = n_f/n$. Therefore, caution should be exercised when using different authors' computations for the signal-to-noise ratio as a function of probability of detection

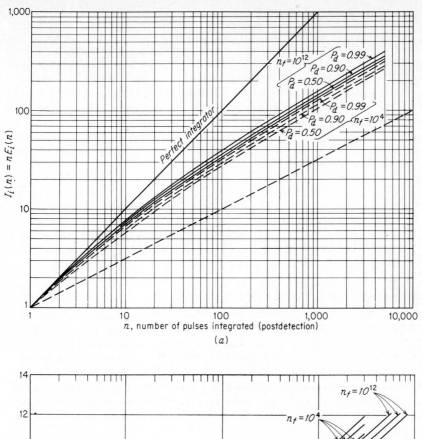

(a)

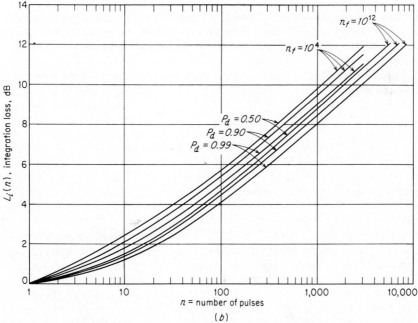

(b)

Figure 2.8 (a) Integration-improvement factor, square law detector, P_d = probability of detection, $n_f = T_{fa} B$ = false alarm number, T_{fa} = average time between false alarms, B = bandwidth; (b) integration loss as a function of n, the number of pulses integrated, P_d, and n_f. (*After Marcum,*[10] *courtesy IRE Trans.*)

non-fluctuating target.

and probability of false alarm, or false-alarm number, since there is no standardization of definitions. They all can give the correct values for use in the radar equation provided the assumptions used by each author are understood.

The original false-alarm time of Marcum[10] is different from that used in this text. He defined it as the time in which the probability is 1/2 that a false alarm will not occur. A comparison of the two definitions is given by Hollis.[12] Marcum's definition of false-alarm time is seldom used, although his definition of false-alarm number is often found.

The solid straight line plotted in Fig. 2.8a represents a perfect predetection integrator with $E_i(n) = 1$. It is hardly ever achieved in practice. When only a few pulses are integrated postdetection (large signal-to-noise ratio per pulse), Fig. 2.8a shows that the integration-improvement factor is not much different from a perfect predetection integrator. When a large number of pulses are integrated (small signal-to-noise ratio per pulse), the difference between postdetection and predetection integration is more pronounced.

The dashed straight line applies to an integration-improvement factor proportional to $n^{1/2}$. As discussed in Sec. 10.6, data obtained during World War II seemed to indicate that this described the performance of an operator viewing a cathode-ray tube display. More recent experiments, however, show that the operator-integration performance when viewing a properly designed PPI or B-scope display is better represented by the theoretical postdetection integrator as given by Fig. 2.8, rather than by the $n^{1/2}$ law.

The radar equation with n pulses integrated can be written

$$R^4_{max} = \frac{P_t G A_e \sigma}{(4\pi)^2 k T_0 B_n F_n (S/N)_n} \tag{2.32}$$

where the parameters are the same as that of Eq. (2.7) except that $(S/N)_n$ is the signal-to-noise ratio of one of the n equal pulses that are integrated to produce the required probability of detection for a specified probability of false alarm. To use this form of the radar equation it is necessary to have a set of curves like those of Fig. 2.7 for each value of n. Such curves are available,[11] but are not necessary since only Figs. 2.7 and 2.8 are needed. Substituting Eq. (2.31) into (2.32) gives

$$R^4_{max} = \frac{P_t G A_e \sigma n E_i(n)}{(4\pi)^2 k T_0 B_n F_n (S/N)_1} \tag{2.33}$$

The value of $(S/N)_1$ is found from Fig. 2.7 as before, and $n E_i(n)$ is found from Fig. 2.8a.

The post-detection integration loss described by Fig. 2.8 assumes a perfect integrator. Many practical integrators, however, have a "loss of memory" with time. That is, the amplitude of a signal stored in such an integrator decays, so that the stored pulses are not summed with equal weight as assumed above. Practical analog integrators such as the recirculating delay line (also called a feedback integrator), the low-pass filter, and the electronic storage tube apply what is equivalent to an exponential weighting to the integrated pulses; that is, if n pulses are integrated, the voltage out of the integrator is

$$V = \sum_{i=1}^{N} V_i \exp\left[-(i-1)\gamma\right] \tag{2.34}$$

where V_i is the voltage amplitude of the ith pulse and $\exp(-\gamma)$ is the attenuation per pulse. In a recirculating delay-line integrator, $e^{-\gamma}$ is the attenuation around the loop. In an RC low-pass filter $\gamma = T_p/RC$, where T_p is the pulse repetition period and RC is the filter time constant.

In order to find the signal-to-noise ratio for a given probability of detection and probability of false alarm, an analysis similar to that used to obtain Figs. 2.7 and 2.8 should be repeated for each value of γ and n.[13] This is not done here. Instead, for simplicity, an efficiency will be defined which is the ratio of the *average* signal-to-noise ratio for the exponential integrator to the *average* signal-to-noise ratio for the uniform integrator. For a dumped integrator, one which erases the contents of the integrator after n pulses and starts over, the efficiency is[14]

$$\rho = \frac{\tanh{(n\gamma/2)}}{n \tanh{(\gamma/2)}} \qquad (2.35a)$$

An example of an integrator that dumps is an electrostatic storage tube that is erased whenever it is read. The efficiency of an integrator that operates continuously without dumping is

$$\rho = \frac{[1 - \exp{(-n\gamma)}]^2}{n \tanh{(\gamma/2)}} \qquad (2.35b)$$

The maximum efficiency of a dumped integrator occurs for $\gamma = 0$, but for a continuous integrator the maximum efficiency occurs for $n\gamma = 1.257$.

2.7 RADAR CROSS SECTION OF TARGETS

The radar cross section of a target is the (fictional) area intercepting that amount of power which, when scattered equally in all directions, produces an echo at the radar equal to that from the target; or in other terms,

$$\sigma = \frac{\text{power reflected toward source/unit solid angle}}{\text{incident power density}/4\pi} = \lim_{R \to \infty} 4\pi R^2 \left| \frac{E_r}{E_i} \right|^2 \qquad (2.36)$$

where R = distance between radar and target
E_r = reflected field strength at radar
E_i = strength of incident field at target

This equation is equivalent to the radar range equation of Sec. 1.2. For most common types of radar targets such as aircraft, ships, and terrain, the radar cross section does not necessarily bear a simple relationship to the physical area, except that the larger the target size, the larger the cross section is likely to be.

Scattering and *diffraction* are variations of the same physical process.[15] When an object scatters an electromagnetic wave, the scattered field is defined as the difference between the total field in the presence of the object and the field that would exist if the object were absent (but with the sources unchanged). On the other hand, the diffracted field is the total field in the presence of the object. With radar backscatter, the two fields are the same, and one may talk about scattering and diffraction interchangeably.

In theory, the scattered field, and hence the radar cross section, can be determined by solving Maxwell's equations with the proper boundary conditions applied.[16] Unfortunately, the determination of the radar cross section with Maxwell's equations can be accomplished only for the most simple of shapes, and solutions valid over a large range of frequencies are not easy to obtain. The radar cross section of a simple sphere is shown in Fig. 2.9 as a function of its circumference measured in wavelengths ($2\pi a/\lambda$, where a is the radius of the sphere and λ is the wavelength).[17–19,34] The region where the size of the sphere is small compared with the wavelength ($2\pi a/\lambda \ll 1$) is called the *Rayleigh region*, after Lord Rayleigh who, in the early 1870s, first studied scattering by small particles. Lord Rayleigh was interested in the scattering of light by microscopic particles, rather than in radar. His work preceded the orginal electromagnetic echo experiments of Hertz by about fifteen years. The Rayleigh scattering region is of

Sphere

$$\sigma = \frac{2\Pi a}{\lambda^2}$$

Cone

Small σ

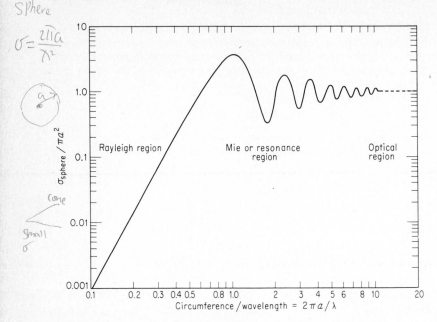

Figure 2.9 Radar cross section of the sphere. a = radius; λ = wavelength.

interest to the radar engineer because the cross sections of raindrops and other meteorological particles fall within this region at the usual radar frequencies. Since the cross section of objects within the Rayleigh region varies as λ^{-4}, rain and clouds are essentially invisible to radars which operate at relatively long wavelengths (low frequencies). The usual radar targets are much larger than raindrops or cloud particles, and lowering the radar frequency to the point where rain or cloud echoes are negligibly small will not seriously reduce the cross section of the larger desired targets. On the other hand, if it were desired to actually observe, rather than eliminate, raindrop echoes, as in a meteorological or weather-observing radar, the higher radar frequencies would be preferred.

At the other extreme from the Rayleigh region is the *optical region*, where the dimensions of the sphere are large compared with the wavelength ($2\pi a/\lambda \gg 1$). For large $2\pi a/\lambda$, the radar cross section approaches the optical cross section πa^2. In between the optical and the Rayleigh region is the *Mie*, or *resonance*, region. The cross section is oscillatory with frequency within this region. The maximum value is 5.6 dB greater than the optical value, while the value of the first null is 5.5 dB below the optical value. (The theoretical values of the maxima and minima may vary according to the method of calculation employed.) The behavior of the radar cross sections of other simple reflecting objects as a function of frequency is similar to that of the sphere.[15-25]

Since the sphere is a sphere no matter from what aspect it is viewed, its cross section will not be aspect-sensitive. The cross section of other objects, however, will depend upon the direction as viewed by the radar.

Figure 2.10 is a plot of the backscatter cross section of a long thin rod as a function of aspect.[26] The rod is 39λ long and $\lambda/4$ in diameter, and is made of silver. If the rod were of steel instead of silver, the first maximum would be about 5 dB below that shown. The radar cross section of the thin rod (and similar objects) is small when viewed end-on ($\theta = 0°$) since the physical area is small. However, at near end-on, waves couple onto the scatterer which travel

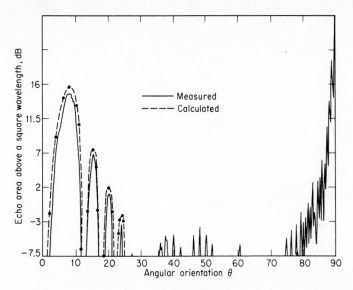

Figure 2.10 Backscatter cross section of a long thin rod. (*From Peters,*[26] *IRE Trans.*)

down the length of the object and reflect from the discontinuity at the far end. This gives rise to a traveling wave component that is not predicted by physical optics theory.[26,35]

An interesting radar scattering object is the cone-sphere, a cone whose base is capped with a sphere such that the first derivatives of the cone and sphere contours are equal at the join between the two. Figure 2.11 is a plot of the nose-on radar cross section. Figure 2.12 is a plot as a function of aspect. The cross section of the cone-sphere from the vicinity of the nose-on direction is quite low. Scattering from any object occurs from discontinuities. The discontinuities, and hence the backscattering, of the cone-sphere are from the tip and from the join between the cone and the sphere. There is also a backscattering contribution from a " creeping

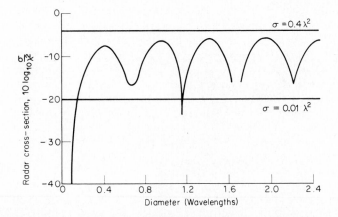

Figure 2.11 Radar cross section of a cone sphere with 15° half angle as a function of the diameter in wavelengths. (*After Blore,*[27] *IEEE Trans.*)

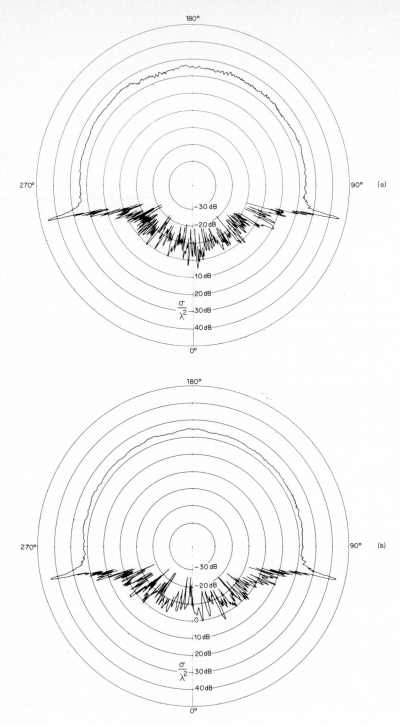

Figure 2.12 Measured radar cross section (σ/λ^2 given in dB) of a large cone-sphere with 12.5° half angle and radius of base = 10.4λ. (*a*) horizontal (perpendicular) polarization, (*b*) vertical (parallel) polarization. (*From Pannell et al.*[61])

wave" which travels around the base of the sphere. The nose-on radar cross section is small and decreases as the square of the wavelength. The cross section is small over a relatively large angular region. A large specular return is obtained when the cone-sphere is viewed at near perpendicular incidence to the cone surface, i.e., when $\theta = 90 - \alpha$, where $\alpha = $ cone half angle. From the rear half of the cone-sphere, the radar cross section is approximately that of the sphere.

The nose-on cross section of the cone-sphere varies, but its maximum value is approximately $0.4\lambda^2$ and its minimum is $0.01\lambda^2$ for a wide range of half-angles for frequencies above the Rayleigh region. The null spacing is also relatively insensitive to the cone half-angle. If a "typical" value of cross section is taken as $0.1\lambda^2$, the cross section at S band ($\lambda = 0.1$ m) is 10^{-3} m^2, and at X band ($\lambda = 3$ cm), the cross section is approximately 10^{-4} m^2. Thus, in theory, the cone-sphere can have very low backscatter energy. Suppose, for example, that the projected area of the cone-sphere were 1 m^2. The radar cross section of a sphere, with the same projected area, at S band is about 30 dB greater. A corner reflector at S band, also of the same projected area, has a radar cross section about 60 dB greater than the cone-sphere. Thus, objects with the same physical projected area can have considerably different radar cross sections.

In order to realize in practice the very low theoretical values of the radar cross section for a cone-sphere, the tip of the cone must be sharp and not rounded, the surface must be smooth (roughness small compared to a wavelength), the join between the cone and the sphere must have a continuous first derivative, and there must be no holes, windows, or protuberances on the surface. A comparison of the nose-on cross section of several cone-shaped objects is given in Fig. 2.13.

Shaping of the target, as with the cone-sphere, is a good method for reducing the radar cross section. Materials such as carbon-fiber composites, which are sometimes used in aerospace applications, can further reduce the radar cross section of targets as compared with that produced by highly reflecting metallic materials.[62]

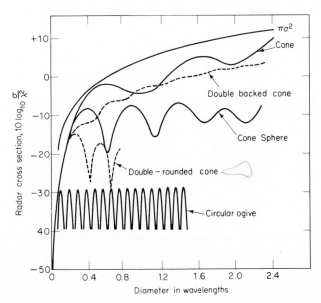

Figure 2.13 Radar cross section of a set of 40° cones, double-backed cones, cone-spheres, double-rounded cones, and circular ogives as a function of diameter in wavelengths. (*From Blore,*[27] *IEEE Trans.*)

Complex targets.[32,33] The radar cross section of complex targets such as ships, aircraft, cities, and terrain are complicated functions of the viewing aspect and the radar frequency. Target cross sections may be computed with the aid of digital computers, or they may be measured experimentally. The target cross section can be measured with full-scale targets, but it is more convenient to make cross-section measurements on scale models at the proper scaled frequency.[63]

A complex target may be considered as comprising a large number of independent objects that scatter energy in all directions. The energy scattered in the direction of the radar is of prime interest. The relative phases and amplitudes of the echo signals from the individual scattering objects as measured at the radar receiver determine the total cross section. The phases and amplitudes of the individual signals might add to give a large total cross section, or the relationships with one another might result in total cancellation. In general, the behavior is somewhere between total reinforcement and total cancellation. If the separation between the individual scattering objects is large compared with the wavelength—and this is usually true for most radar applications—the phases of the individual signals at the radar receiver will vary as the viewing aspect is changed and cause a scintillating echo.

Consider the scattering from a relatively " simple " complex target consisting of two equal, isotropic objects (such as spheres) separated a distance l (Fig. 2.14). By isotropic scattering is meant that the radar cross section of each object is independent of the viewing aspect. The separation l is assumed to be less than $c\tau/2$, where c is the velocity of propagation and τ is the pulse duration. With this assumption, both scatterers are illuminated simultaneously by the pulse packet. Another restriction placed on l is that it be small compared with the distance R from radar to target. Furthermore, $R_1 \approx R_2 \approx R$. The cross sections of the two targets are assumed equal and are designated σ_0. The composite cross section σ_r of the two scatterers is

$$\frac{\sigma_r}{\sigma_0} = 2\left[1 + \cos\left(\frac{4\pi l}{\lambda}\sin\theta\right)\right] \tag{2.37}$$

The ratio σ_r/σ_0 can be anything from a minimum of zero to a maximum of four times the cross section of an individual scatterer. Polar plots of σ_r/σ_0 for various values of l/λ are shown in Fig. 2.15. Although this is a rather simple example of a " complex " target, it is complicated enough to indicate the type of behavior to be expected with practical radar targets.

The radar cross sections of actual targets are far more complicated in structure than the simple two-scatterer target. Practical targets are composed of many individual scatterers, each with different scattering properties. Also, interactions may occur between the scatterers which affect the resultant cross section.

An example of the cross section as a function of aspect angle for a propeller-driven

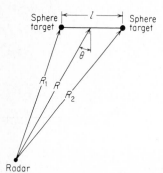

Figure 2.14 Geometry of the two-scatterer complex target.

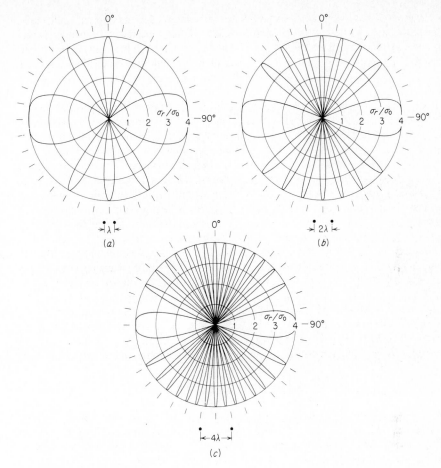

Figure 2.15 Polar plots of σ_r/σ_0 for the two-scatterer complex target [Eq. (2.37)]. (a) $l = \lambda$; (b) $l = 2\lambda$; (c) $l = 4\lambda$.

aircraft[28] is shown in Fig. 2.16. The aircraft is the B-26, a World War II medium-range two-engine bomber. The radar wavelength was 10 cm. These data were obtained experimentally by mounting the aircraft on a turntable in surroundings free from other reflecting objects and by observing with a nearby radar set. The propellers were running during the measurement and produced a modulation of the order of 1 to 2 kHz. The cross section can change by as much as 15 dB for a change in aspect of only $\frac{1}{3}°$. The maximum echo signal occurs in the vicinity of broadside, where the projected area of the aircraft is largest.

It is not usually convenient to obtain the radar cross section of aircraft by mounting the full-size aircraft on a rotating table. Measurements can be obtained with scale models on a pattern range.[29] An example of such model measurements is given by the dashed curves in Fig. 2.17. If care is taken in the construction of the model and in the pattern-range instrumentation, it is possible to achieve reasonably representative measurements.

The radar cross section of an aircraft can also be obtained by computation.[17] The target is broken up into a number of simple geometrical shapes, the contribution of each (taking

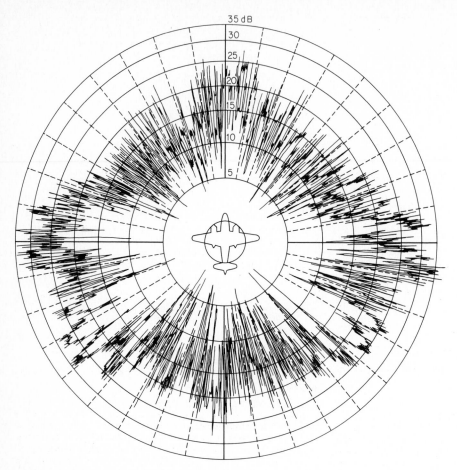

Figure 2.16 Experimental cross section of the B-26 two-engine bomber at 10-cm wavelength as a function of azimuth angle. (*From Ridenour*,[28] *courtesy McGraw-Hill Book Company, Inc.*)

account of aspect changes and shadowing of one component by another) is computed and the component cross sections are combined to yield the composite value. The "theoretical" values of Fig. 2.17 for B-47 were obtained by calculation.

The most realistic method for obtaining the radar cross section of aircraft is to measure the actual target in flight. There is no question about the authenticity of the target being measured. An example of such a facility is the dynamic radar cross-section range of the U.S. Naval Research Laboratory.[30] Radars at L, S, C and X bands illuminate the aircraft target in flight. The radar track data is used to establish the aspect angle of the target with respect to the radar. Pulse-to-pulse radar cross section is available, but for convenience in presenting the data the values plotted usually are an average of a large number of values taken within a 10 by 10° aspect angle interval. Examples of such data are given in Figs. 2.18 to 2.20. The radar cross section of the T-38 aircraft at head-on incidence is shown in Table 2.1. This data was also obtained from an aircraft in flight. (The T-38 is a twin-jet trainer with a 7.7 m wing span and a 14 m length.)

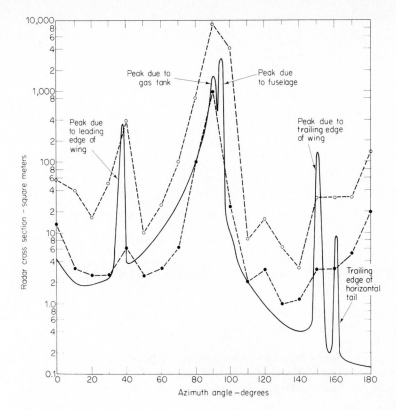

Figure 2.17 Comparison of the theoretical and model-measurement horizontal-polarization radar cross sections of the B-47 medium bomber jet aircraft with a wing span of 35 m and a length of 33 m. Solid curve is the average of the computed cross sections obtained by the University of Michigan Engineering Research Institute at a frequency of 980 MHz. Dashed curves are model measurements obtained by the Ohio State University Antenna Laboratory at a frequency of 600 MHz. Open circles are the maximum values averaged over 10° intervals; solid circles are median values. Radar is assumed to be in the same plane as the aircraft.[64]

It can be seen that the radar cross section of an aircraft is difficult to specify concisely. Slight changes in viewing aspect or frequency result in large fluctuations in cross section. Nevertheless, a single value of cross section is sometimes given for specific aircraft targets for use in computing the radar equation. There is no standard, agreed-upon method for specifying the single-valued cross section of an aircraft. The average value or the median might be taken. Sometimes it is a "minimum" value, perhaps the value exceeded 99 percent of the time or 95 percent of the time. It might also be the value which when substituted into the radar equation assures that the computed range agrees with the experimentally measured range.

Table 2.2 lists "example" values of cross section for various targets at microwave frequencies. Note that only a single value is given even though there can be a large variation. They should not be used for design purposes when actual data is available for the particular targets of interest.

A military propeller aircraft such as the AD-4B has a cross section of about 20 m² at *L* band, but a 100 m² cross section at VHF. The longer wavelengths at VHF result in greater

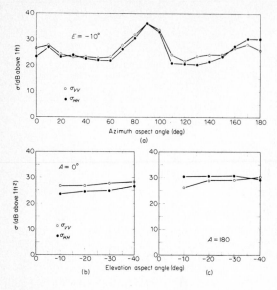

Figure 2.18 (*a*) Azimuth variation of radar cross section of a C-54 aircraft with constant elevation angle of − 10°. (The C-54 is the military version of the four-piston-engined DC-4 commercial aircraft with a wing span of 36 m and a length of 29 m.) Each point represents the average of medians obtained from samples within a 10 by 10° aspect cell. Frequency is 1300 MHz (*L* band) with linear polarization. *VV* = vertical polarization, *HH* = horizontal polarization. (*b*) Elevation-variation nose-on (azimuth = 0°) (*c*) Elevation-variation tail-on (azimuth = 180°). (*From Olin and Queen.*[30])

cross section than microwaves because the dimensions of the scattering objects are comparable to the wavelength and produce resonance effects.

An example of the measured radar cross section of a large ship (16,000 tons) is shown in Fig. 2.21. The aspect is at grazing incidence. When averages of the cross section are taken about the port and starboard bow and quarter aspects of a number of ships (omitting the peak at broadside), a simple empirical expression is obtained for the median (50th percentile) value of the cross section:

$$\sigma = 52f^{1/2}D^{3/2} \tag{2.38}$$

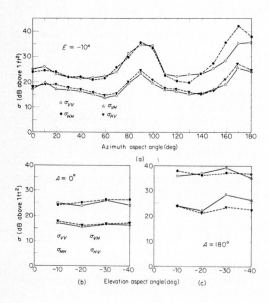

Figure 2.19 Same as Fig. 2.18 except frequency is 9225 MHz (*X* band). *VH* and *HV* represent cross-polarized components. (*From Olin and Queen.*[30])

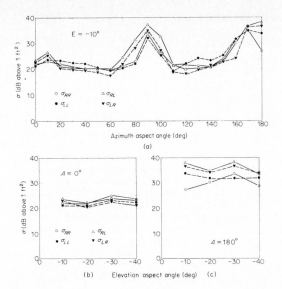

Figure 2.20 Same as Fig. 2.19, but for circular polarization. RR = right-hand polarization, LL = left-hand polarization; RL and LR are cross-polarized components. (*From Olin and Queen.*[30])

Table 2.1a Radar cross section (square meters) of the T-38[66]
Head-on aspect (± 1.0 degree)

	X_{LL}			X_{LR}			S_{VV}			L_{VV}		
Percentile	20	50	80	20	50	80	20	50	80	20	50	80
Landing gear up	0.33	0.83	1.7	0.21	0.53	1.2	1.6	3.1	4.7	1.5	1.8	2.2
Landing gear extended	0.53	1.6	3.5	0.24	0.80	2.1	1.1	2.3	4.4	0.99	2.6	4.5

Table 2.1b Median cross section (X_{LL}) for aspects near nose-on

Elevation angle	Azimuth angle (degrees)				Elevation angle	Azimuth angle (degrees)			
	0	2	5	7		0	2	5	7
0	0.83				16	0.72	0.45	0.90	1.2
6	0.68	0.61	0.99	0.82	18	0.43	0.44	1.3	0.84
8	1.2	0.94	1.7	2.2	20	0.43	0.52	0.63	0.64
10	1.4	1.6	3.1	1.4	22	0.45	0.66	1.1	1.1
12	0.70	1.0	1.6	2.1	24	0.65			
14	0.79	0.60	0.63	1.5					

X_{LL}: Transmit left circular polarization, receive left circular (X band). X_{LR}: Transmit left circular polarization, receive right circular (X band). S_{VV}: Transmit vertical polarization, receive vertical (S band). L_{VV}: Same for L band.

Table 2.2 Example radar cross sections at microwave frequencies

	Square meters
Conventional, unmanned winged missile	0.5
Small, single engine aircraft	1
Small fighter, or 4-passenger jet	2
Large fighter	6
Medium bomber or medium jet airliner	20
Large bomber or large jet airliner	40
Jumbo jet	100
Small open boat	0.02
Small pleasure boat	2
Cabin cruiser	10
Ship at zero grazing angle	See Eq. (2.38)
Ship at higher grazing angles	Displacement tonnage expressed in m^2
Pickup truck	200
Automobile	100
Bicycle	2
Man	1
Bird	0.01
Insect	10^{-5}

where σ = radar cross section in square meters, f = radar frequency in megahertz, and D is the ship's (full load) displacement in kilotons.[31] This expression was derived from measurements made at X, S, and L bands and for naval ships ranging from 2000 to 17,000 tons. Although it is probably valid outside this size and frequency range, it does not apply to elevation angles other than grazing incidence. At higher elevation angles, as might be viewed from aircraft, the cross sections of ships might be considerably less than at grazing incidence, perhaps by an order of magnitude. When no better information is available, a very rough order of magnitude estimate of the ship's cross section at other-than-grazing incidence can be had by taking the ship's displacement in tons to be equal to its cross section in square meters. The average cross section of small pleasure boats 20 to 30 ft in length might have a radar cross section in the vicinity of a few square meters at X band.[68] Boats from 40 to 50 ft in length might have a cross section of the order of 10 square meters.

The radar cross section of an automobile at X band is generally greater than that of an aircraft or a boat. From the front the cross section might vary from 10 to 200 m^2 at X band, with 100 m^2 being a typical value.[65] The cross section increases with increasing frequency (up to 60 GHz, the range of the measurements).

The measured radar cross section of a man has been reported[32] to be as follows:

Frequency, MHz	σ, m^2
410	0.033–2.33
1,120	0.098–0.997
2,890	0.140–1.05
4,800	0.368–1.88
9,375	0.495–1.22

The spread in cross-section values represents the variation with aspect and polarization.

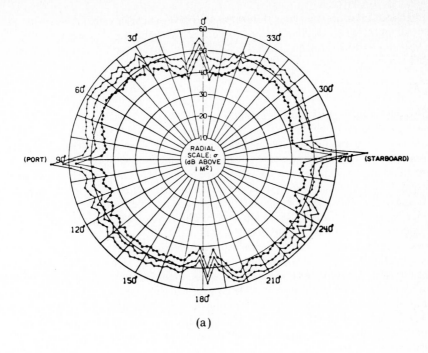

(a)

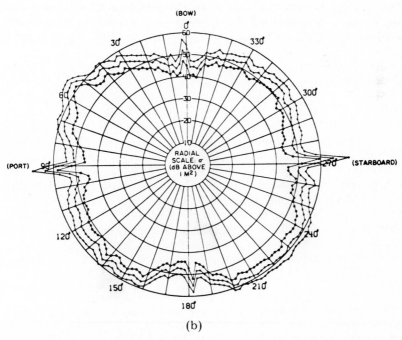

(b)

Figure 2.21 Azimuth variation of the radar cross-section of a large Naval Auxiliary Ship at (*a*) *S* band (2800 MHz) and (*b*) *X* band (9225 MHz), both with horizontal polarization.

The cross-section data presented in this section lead to the conclusion that it would not be appropriate simply to select a single value and expect it to have meaning in the computation of the radar equation without further qualification. Methods for dealing with the cross sections of complicated targets are discussed in the next section.

2.8 CROSS-SECTION FLUCTUATIONS

The discussion of the minimum signal-to-noise ratio in Sec. 2.6 assumed that the echo signal received from a particular target did not vary with time. In practice, however, the echo signal from a target in motion is almost never constant. Variations in the echo signal may be caused by meteorological conditions, the lobe structure of the antenna pattern, equipment instabilities, or variations in the target cross section. The cross sections of complex targets (the usual type of radar target) are quite sensitive to aspect.[30,36] Therefore, as the target aspect changes relative to the radar, variations in the echo signal will result.

One method of accounting for a fluctuating cross section in the radar equation is to select a lower bound, that is, a value of cross section that is exceeded some specified (large) fraction of time. The fraction of time that the actual cross section exceeds the selected value would be close to unity (0.95 or 0.99 being typical). For all practical purposes the value selected is a minimum and the target will always present a cross section greater than that selected. This procedure results in a conservative prediction of radar range and has the advantage of simplicity. The minimum cross section of typical aircraft or missile targets generally occurs at or near the head-on aspect.

However, to properly account for target cross-section fluctuations, the probability-density function and the correlation properties with time must be known for the particular target and type of trajectory. Curves of cross section as a function of aspect and a knowledge of the trajectory with respect to the radar are needed to obtain a true description of the dynamical variations of cross section. The probability-density function gives the probability of finding any particular value of target cross section between the values of σ and $\sigma + d\sigma$, while the autocorrelation function describes the degree of correlation of the cross section with time or number of pulses. The spectral density of the cross section (from which the autocorrelation function can be derived) is also sometimes of importance, especially in tracking radars. It is usually not practical to obtain the experimental data necessary to compute the probability-density function and the autocorrelation function from which the overall radar performance is determined. Most radar situations are of too complex a nature to warrant obtaining complete data. A more economical method to assess the effects of a fluctuating cross section is to postulate a reasonable model for the fluctuations and to analyze it mathematically. Swerling[37] has calculated the detection probabilities for four different fluctuation models of cross section. In two of the four cases, it is assumed that the fluctuations are completely correlated during a particular scan but are completely uncorrelated from scan to scan. In the other two cases, the fluctuations are assumed to be more rapid and uncorrelated pulse to pulse. The four fluctuation models are as follows:

Case 1. The echo pulses received from a target on any one scan are of constant amplitude throughout the entire scan but are independent (uncorrelated) from scan to scan. This assumption ignores the effect of the antenna beam shape on the echo amplitude. An echo fluctuation of this type will be referred to as scan-to-scan fluctuation. The probability-density function for the cross section σ is given by the density function

$$p(\sigma) = \frac{1}{\sigma_{av}} \exp\left(-\frac{\sigma}{\sigma_{av}}\right) \qquad \sigma \geq 0 \tag{2.39a}$$

where σ_{av} is the average cross section over all target fluctuations.

Case 2. The probability-density function for the target cross section is also given by Eq. (2.39a), but the fluctuations are more rapid than in case 1 and are taken to be independent from pulse to pulse instead of from scan to scan.

Case 3. In this case, the fluctuation is assumed to be independent from scan to scan as in case 1, but the probability-density function is given by

$$p(\sigma) = \frac{4\sigma}{\sigma_{av}^2} \exp\left(-\frac{2\sigma}{\sigma_{av}}\right) \tag{2.39b}$$

Case 4. The fluctuation is pulse to pulse according to Eq. (2.39b)

The probability-density function assumed in cases 1 and 2 applies to a complex target consisting of many independent scatterers of approximately equal echoing areas. Although, in theory, the number of independent scatterers must be essentially infinite, in practice the number may be as few as four or five. The probability-density function assumed in cases 3 and 4 is more indicative of targets that can be represented as one large reflector together with other small reflectors. In all the above cases, the value of cross section to be substituted in the radar equation is the average cross section σ_{av}. The signal-to-noise ratio needed to achieve a specified probability of detection without exceeding a specified false-alarm probability can be calculated for each model of target behavior. For purposes of comparison, the nonfluctuating cross section will be called *case 5.*

A comparison of these five models for a false-alarm number $n_f = 10^8$ is shown in Fig. 2.22 for $n = 10$ hits integrated. When the detection probability is large, all four cases in which the target cross section is not constant require greater signal-to-noise ratio than the constant cross section of case 5. For example, if the desired probability of detection were 0.95, a signal-to-noise ratio of 6.2 dB/pulse is necessary if the target cross section were constant (case 5), but if the target cross section fluctuated with a Rayleigh distribution and were scan-to-scan uncorrelated (case 1), the signal-to-noise ratio would have to be 16.8 dB/pulse. This increase in signal-to-noise corresponds to a reduction in range by a factor of 1.84. Therefore, if the characteristics of the target cross section are not properly taken into account, the actual performance of the radar might not measure up to the performance predicted as if the target cross section were constant. Figure 2.22 also indicates that for probabilities of detection greater than about 0.30, a greater signal-to-noise ratio is required when the fluctuations are uncorrelated scan to scan (cases 1 and 3) than when the fluctuations are uncorrelated pulse to pulse (cases 2 and 4). In fact, the larger the number of pulses integrated, the more likely it will be for the fluctuations to average out, and cases 2 and 4 will approach the nonfluctuating case.

Curves exist[11,37] for various values of hits per scan, n, that give the signal-to-noise ratio per pulse as a function of P_d and n_f. The signal-to-noise ratio per pulse can be used in the form of the radar equation as given by Eq. (2.32). It is not necessary, however, to employ such an elaborate set of data since for most engineering purposes the curves of Figs. 2.23 and 2.24 may be used as corrections to the probability of detection (as found in Fig. 2.7) and as the integration improvement factor (Fig. 2.8a) for substitution into the radar equation of Eq. (2.33).

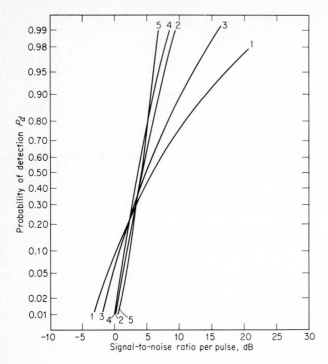

Figure 2.22 Comparison of detection probabilities for five different models of target fluctuation for $n = 10$ pulses integrated and false-alarm number $n_f = 10^8$. (*Adapted from Swerling.*[37])

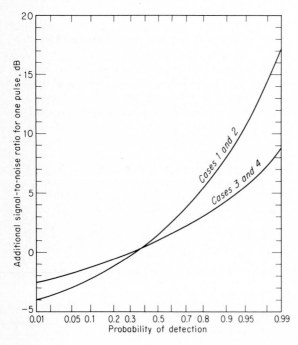

Figure 2.23 Additional signal-to-noise ratio required to achieve a particular probability of detection, when the target cross section fluctuates, as compared with a nonfluctuating target; single hit, $n = 1$. (To be used in conjunction with Fig. 2.7 to find $(S/N)_1$.)

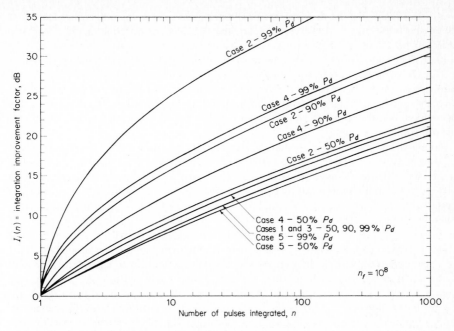

Figure 2.24 Integration-improvement factor as a function of the number of pulses integrated for the five types of target fluctuation considered.

The procedure for using the radar equation when the target is described by one of the Swerling models is as follows:

1. Find the signal-to-noise ratio from Fig. 2.7 corresponding to the desired value of detection probability P_d and false-alarm probability P_{fa}.
2. From Fig. 2.23 determine the correction factor for either cases 1 and 2 or cases 3 and 4 to be applied to the signal-to-noise ratio found from step 1 above. The resultant signal-to-noise ratio $(S/N)_1$ is that which would apply if detection were based upon a single pulse.
3. If n pulses are integrated, the integration-improvement factor $I_i(n) = nE_i(n)$ is found from Fig. 2.24. The parameters $(S/N)_1$ and $nE_i(n)$ are substituted into the radar equation (2.33) along with σ_{av}.

The integration-improvement factor in Fig. 2.24 is in some cases greater than n, or in other words, the integration efficiency factor $E_i(n) > 1$. One is not getting something for nothing, for in those cases in which the integration-improvement factor is greater than n, the signal-to-noise ratio required for $n = 1$ is larger than for a nonfluctuating target. The signal-to-noise per pulse will always be less than that of an ideal predetection integrator for reasonable values of P_d. It should also be noted that the data in Figs. 2.23 and 2.24 are essentially independent of the false-alarm number, at least over the range of 10^6 to 10^{10}.

The two probability-density functions of Eqs. (2.39a) and (2.39b) that describe the Swerling fluctuation models are special cases of the chi-square distribution of degree $2m$.[38] The probability density function is

$$p(\sigma) = \frac{m}{(m-1)! \, \sigma_{av}} \left(\frac{m\sigma}{\sigma_{av}}\right)^{m-1} \exp\left(-\frac{m\sigma}{\sigma_{av}}\right), \qquad \sigma > 0 \qquad (2.40)$$

It is also called the gamma distribution. In statistics texts, $2m$ is the number of degrees of freedom, and is an integer. However, when applied to target cross-section models, $2m$ is not required to be an integer. Instead, m can be any positive, real number. When $m = 1$, the chi-square distribution of Eq. (2.40) reduces to the exponential, or Rayleigh-power, distribution of Eq. (2.39a) that applies to Swerling cases 1 and 2. Cases 3 and 4, described by Eq. (2.39b), are equivalent to $m = 2$ in the chi-square distribution. The ratio of the standard deviation to the mean of the cross section is equal to $m^{-1/2}$ for the chi-square distribution. The larger the value of m, the more constrained will be the fluctuations. The limit of m equal to infinity corresponds to the nonfluctuating target.

The chi-square distribution is a mathematical model used to represent the statistics of the fluctuating radar cross section. These distributions might not always fit the observed data, but they are fair approximations in many cases and are used nevertheless for convenience. The chi-square distribution is described by two parameters: the average cross section σ_{av} and the number of degrees of freedom $2m$. Analysis[39] of measurements on actual aircraft flying straight, level courses shows that the cross-section fluctuations at a particular aspect are well fitted by the chi-square distribution with the parameter m ranging from 0.9 to approximately 2 and with σ_{av} varying approximately 15 dB from minimum to maximum. The parameters of the fitted distribution vary with aspect angle, type of aircraft, and frequency. The value of m is near unity for all aspects except at broadside; hence, the distribution is Rayleigh with a varying average value with the most variation at broadside aspect. It was also found that the average value has more effect on the calculation of the probability of detection than the value of m. Although the Rayleigh model might provide a good approximation to the radar cross sections of aircraft in many cases, it is not always applicable. Exceptions occur at broadside, as mentioned, and for smaller aircraft.[38] There are also examples where no chi-square distribution can be made to fit the experimental data.

The chi-square distribution has been used to approximate the statistics of other-than-aircraft targets. Weinstock[38,40] showed that this distribution can describe certain simple shapes, such as cylinders or cylinders with fins that are characteristic of some satellite objects. The parameter m varies between 0.3 and 2, depending on aspect. These have sometimes been called Weinstock cases.

The chi-square distribution with $m = 1$ (Swerling cases 1 and 2) is the Rayleigh, or exponential, distribution that results from a large number of independent scatterers, no one of which contributes more than a small fraction of the total backscatter energy. Although the chi-square distribution with other than $m = 1$ has been observed empirically to give a reasonable fit to the radar cross section distribution of many targets, there is no physical scattering mechanism on which it is based. It has been said that the chi-square distribution with $m = 2$ (Swerling cases 3 and 4) is indicative of scattering from one large dominant scatterer together with a collection of small independent scatterers. However, it is the Rice distribution that follows from such a model.[67] The Rice probability density function is

$$p(\sigma) = \frac{1+s}{\sigma_{av}} \exp\left[-s - \frac{\sigma}{\sigma_{av}}(1+s)\right] I_0\left(2\sqrt{\frac{\sigma}{\sigma_{av}}s(1+s)}\right), \qquad \sigma > 0 \qquad (2.41)$$

where s is the ratio of the radar cross section of the single dominant scatterer to the total cross section of the small scatterers, and $I_0(\ )$ is the modified Bessel function of zero order.[38] This is a more correct description of the single dominant scatterer model than the chi-square with $m = 2$. However it has been shown that the chi-square with $m = 2$ approximates the Rice when the dominant-scatterer power is equal to the total cross section of the other, small scatterers, and so long as the probability of detection is not large.[41]

THE RADAR EQUATION 51

The log-normal distribution has also been considered for representing target echo fluctuations. It can be expressed as

$$p(\sigma) = \frac{1}{\sqrt{2\pi}\, s_d \sigma} \exp\left\{-\frac{1}{2s_d^2}\left[\ln\left(\frac{\sigma}{\sigma_m}\right)\right]^2\right\}, \qquad \sigma > 0 \tag{2.42}$$

where s_d = standard deviation of $\ln\,(\sigma/\sigma_m)$, and σ_m = median of σ. The ratio of the mean to the median value of σ is $\exp\,(s_d^2/2)$. There is no theoretical model of target scattering that leads to the log-normal distribution, although it has been suggested that echoes from some satellite bodies, ships, cylinders, plates, and arrays can be approximated by a log-normal probability distribution.[42,43]

Figure 2.25 is a comparison of the several distribution models for a false alarm number of 10^6 when all pulses during a scan are perfectly correlated but with pulses in successive scans independent (scan-to-scan fluctuation).

The fluctuation models considered in this section assume either complete correlation between pulses in any particular scan but with scan-to-scan independence (slow fluctuations), or else complete independence from pulse to pulse (fast fluctuation). These represent two extreme cases. In some instances, there might be partial correlation of the pulses within a scan (moderate fluctuation). Schwartz[44] considered the effect of partial correlation for the case of two pulses per scan ($n = 2$). The results for partial correlation fall between the two extremes of completely uncorrelated and completely correlated, as might be expected. Thus to estimate

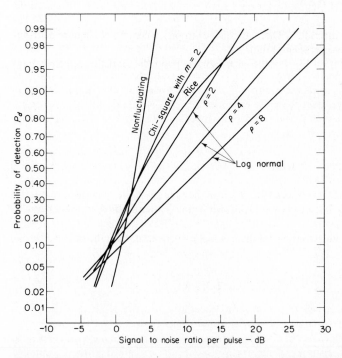

Figure 2.25 Comparison of detection probabilities for Rice, log normal, chi-square with $m = 2$ (Swerling case 3) and nonfluctuating target models with $n = 10$ hits and false-alarm number $n_f = 10^6$. Ratio of dominant-to-background equals unity ($s = 1$) for Rice distribution. Ratio of mean-to-median cross section for log-normal distribution = ρ.

performance for partially correlated pulses, interpolation between the results for the correlated and uncorrelated conditions can be used as an approximation. A more general treatment of partially correlated fluctuations has been given by Swerling.[45] His analysis applies to a large family of probability-density functions of the signal fluctuations and for very general correlation properties. Methods for the design of optimal receivers for the detection of moderately fluctuating signals have been considered.[46]

It is difficult to be precise about the statistical model to be applied to any particular target. Few, if any, real targets fit a mathematical model with any precision and in some cases it is not possible to approximate actual data with any mathematical model. Even if the statistical distribution of a target were known, it might be difficult to relate this to an actual radar measurement since a target generally travels on some well-defined trajectory rather than present a statistically independent cross section to the radar. Thus the various mathematical models cannot, in general, be expected to yield precise predictions of system performance.

It has been suggested[38,39] that if only one parameter is to be used to describe a complex target, it should be the median value of the cross section with Rayleigh statistics (Swerling cases 1 and 2). Quite often Swerling case 1 is specified for describing radar performance since it results in conservative values. The uncertainty regarding fluctuating target models makes the use of the constant (nonfluctuating) cross section in the radar equation an attractive alternative when a priori information about the target is minimal.

2.9 TRANSMITTER POWER

The power P_t in the radar equation (2.1) is called by the radar engineer the *peak* power. The peak pulse power as used in the radar equation is not the instantaneous peak power of a sine wave. It is defined as the power averaged over that carrier-frequency cycle which occurs at the maximum of the pulse of power. (Peak power is usually equal to one-half the maximum instantaneous power.) The *average* radar power P_{av} is also of interest in radar and is defined as the average transmitter power over the pulse-repetition period. If the transmitted waveform is a train of rectangular pulses of width τ and pulse-repetition period $T_p = 1/f_p$, the average power is related to the peak power by

$$P_{av} = \frac{P_t \tau}{T_p} = P_t \tau f_p \qquad (2.43)$$

The ratio P_{av}/P_t, τ/T_p, or τf_p is called the *duty cycle* of the radar. A pulse radar for detection of aircraft might have typically a duty cycle of 0.001, while a CW radar which transmits continuously has a duty cycle of unity.

Writing the radar equation in terms of the average power rather than the peak power, we get

$$R_{max}^4 = \frac{P_{av} G A_e \sigma n E_i(n)}{(4\pi)^2 k T_0 F_n(B_n \tau)(S/N)_1 f_p} \qquad (2.44a)$$

The bandwidth and the pulse width are grouped together since the product of the two is usually of the order of unity in most pulse-radar applications.

If the transmitted waveform is not a rectangular pulse, it is sometimes more convenient to express the radar equation in terms of the energy $E_\tau = P_{av}/f_p$ contained in the transmitted waveform:

$$R_{max}^4 = \frac{E_\tau G A_e \sigma n E_i(n)}{(4\pi)^2 k T_0 F_n(B_n \tau)(S/N)_1} \qquad (2.44b)$$

In this form, the range does not depend explicitly on either the wavelength or the pulse repetition frequency. The important parameters affecting range are the total transmitted energy nE_τ, the transmitting gain G, the effective receiving aperture A_e, and the receiver noise figure F_n.

2.10 PULSE REPETITION FREQUENCY AND RANGE AMBIGUITIES

The pulse repetition frequency (prf) is determined primarily by the maximum range at which targets are expected. If the prf is made too high, the likelihood of obtaining target echoes from the wrong pulse transmission is increased. Echo signals received after an interval exceeding the pulse-repetition period are called *multiple-time-around* echoes. They can result in erroneous or confusing range measurements. Consider the three targets labeled A, B, and C in Fig. 2.26a. Target A is located within the maximum unambiguous range R_{unamb} [Eq. (1.2)] of the radar, target B is at a distance greater than R_{unamb} but less than $2R_{unamb}$, while target C is greater than $2R_{unamb}$ but less than $3R_{unamb}$. The appearance of the three targets on an A-scope is sketched in Fig. 2.26b. The multiple-time-around echoes on the A-scope cannot be distinguished from proper target echoes actually within the maximum unambiguous range. Only the range measured for target A is correct; those for B and C are not.

One method of distinguishing multiple-time-around echoes from unambiguous echoes is to operate with a varying pulse repetition frequency. The echo signal from an unambiguous range target will appear at the same place on the A-scope on each sweep no matter whether the prf is modulated or not. However, echoes from multiple-time-around targets will be spread over a finite range as shown in Fig. 2.26c. The prf may be changed continuously within prescribed limits, or it may be changed discretely among several predetermined values. The number of separate pulse repetition frequencies will depend upon the degree of the multiple-time targets. Second-time targets need only two separate repetition frequencies in order to be resolved.

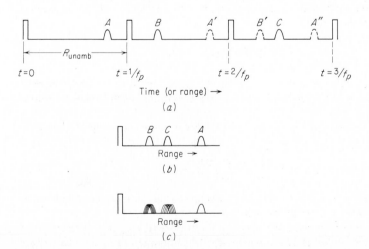

Figure 2.26 Multiple-time-around echoes that give rise to ambiguities in range. (a) Three targets A, B and C, where A is within R_{unamb}, and B and C are multiple-time-around targets; (b) the appearance of the three targets on the A-scope; (c) appearance of the three targets on the A-scope with a changing prf.

Instead of modulating the prf, other schemes that might be employed to "mark" successive pulses so as to identify multiple-time-around echoes include changing the pulse amplitude, pulse width, frequency, phase, or polarization of transmission from pulse to pulse. Generally, such schemes are not so successful in practice as one would like. One of the fundamental limitations is the foldover of nearby targets; that is, nearby strong ground targets (clutter) can be quite large and can mask weak multiple-time-around targets appearing at the same place on the display. Also, more time is required to process the data when resolving ambiguities.

Ambiguities may theoretically be resolved by observing the variation of the echo signal with time (range). This is not always a practical technique, however, since the echo-signal amplitude can fluctuate strongly for reasons other than a change in range. Instead, the range ambiguities in a multiple prf radar can be conveniently decoded and the true range found by the use of the Chinese remainder theorem[47] or other computational algorithms.[48]

2.11 ANTENNA PARAMETERS

Almost all radars use directive antennas for transmission and reception. On transmission, the directive antenna channels the radiated energy into a beam to enhance the energy concentrated in the direction of the target. The antenna gain G is a measure of the power radiated in a particular direction by a directive antenna to the power which would have been radiated in the same direction by an omnidirectional antenna with 100 percent efficiency. More precisely, the power gain of an antenna used for transmission is

$$G(\theta, \phi) = \frac{\text{power radiated per unit solid angle in azimuth } \theta \text{ and elevation } \phi}{\text{power accepted by antenna from its generator}/4\pi} \quad (2.45)$$

Note that the antenna gain is a function of direction. If it is greater than unity in some directions, it must be less than unity in other directions. This follows from the conservation of energy. When we speak of antenna gain in relation to the radar equation, we shall usually mean the *maximum gain* G, unless otherwise specified. One of the basic principles of antenna theory is that of *reciprocity*, which states that the properties of an antenna are the same no matter whether it is used for transmission or reception.

The antenna pattern is a plot of antenna gain as a function of the direction of radiation. (A typical antenna pattern plotted as a function of one angular coordinate is shown in Fig. 7.1.) Antenna beam shapes most commonly employed in radar are the pencil beam (Fig. 2.27a) and the fan beam (Fig. 2.27b). The pencil beam is axially symmetric, or nearly so. Beamwidths of typical pencil-beam antennas may be of the order of a few degrees or less. Pencil beams are commonly used where it is necessary to measure continuously the angular position of a target in both azimuth and elevation, as, for example, the target-tracking radar for the control of weapons or missile guidance. The pencil beam may be generated with a metallic reflector surface shaped in the form of a paraboloid of revolution with the electromagnetic energy fed from a point source placed at the focus.

Although a narrow beam can, if necessary, search a large sector or even a hemisphere, it is not always desirable to do so. Usually, operational requirements place a restriction on the maximum scan time (time for the beam to return to the same point in space) so that the radar cannot dwell too long at any one radar resolution cell. This is especially true if there is a large number of resolution cells to be searched. The number of resolution cells can be materially reduced if the narrow angular resolution cell of a pencil-beam radar is replaced by a beam in which one dimension is broad while the other dimension is narrow, that is, a fan-shaped pattern. One method of generating a fan beam is with a parabolic reflector shaped to yield the

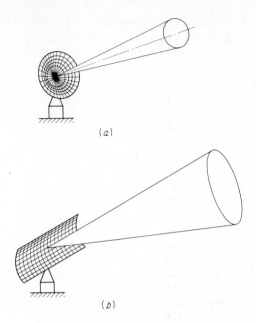

(a)

(b)

Figure 2.27 (a) Pencil-beam antenna pattern; (b) fan-beam-antenna pattern.

proper ratio between the azimuth and elevation beamwidths. Many long-range ground-based search radars use a fan-beam pattern narrow in azimuth and broad in elevation.

The rate at which a fan-beam antenna may be scanned is a compromise between the rate at which target-position information is desired (data rate) and the ability to detect weak targets (probability of detection). Unfortunately, the two are at odds with one another. The more slowly the radar antenna scans, the more pulses will be available for integration and the better the detection capability. On the other hand, a slow scan rate means a longer time between looks at the target. Scan rates of practical search radars vary from 1 to 60 rpm, 5 or 6 rpm being typical for the long-range surveillance of aircraft.

The coverage of a simple fan beam is usually inadequate for targets at high altitudes close to the radar. The simple fan-beam antenna radiates very little of its energy in this direction. However, it is possible to modify the antenna pattern to radiate more energy at higher angles. One technique for accomplishing this is to employ a fan beam with a shape proportional to the square of the cosecant of the elevation angle. In the cosecant-squared antenna (Sec. 7.7), the gain as a function of elevation angle is given by

$$G(\phi) = G(\phi_0) \frac{\csc^2 \phi}{\csc^2 \phi_0} \qquad \text{for } \phi_0 < \phi < \phi_m \qquad (2.46)$$

where $G(\phi)$ = gain at elevation angle ϕ, and ϕ_0 and ϕ_m are the angular limits between which the beam follows a $\csc^2$ shape. This applies to the airborne search radar observing ground targets as well as ground-based radars observing aircraft targets. (In the airborne case, the angle ϕ is the depression angle.) From $\phi = 0$ to $\phi = \phi_0$, the antenna pattern is similar to a normal antenna pattern, but from $\phi = \phi_0$ to $\phi = \phi_m$, the antenna gain varies as $\csc^2 \phi$. Ideally, the upper limit ϕ_m should be 90°, but it is always less than this with a single antenna because of practical difficulties. The cosecant-squared antenna may be generated by a distorted section of a parabola or by a true parabola with a properly designed set of multiple feed horns. The cosecant-squared pattern may also be generated with an array-type antenna.

The cosecant-squared antenna has the important property that the echo power P_r received from a target of constant cross section at constant altitude h is independent of the target's range R from the radar. Substituting the gain of the cosecant-squared antenna [Eq. (2.46)] into the simple radar equation gives

$$P_r = \frac{P_t G^2(\phi_0) \csc^4 \phi \lambda^2 \sigma}{(4\pi)^3 \csc^4 \phi_0 R^4} = K_1 \frac{\csc^4 \phi}{R^4} \tag{2.47}$$

where K_1 is a constant. The height h of the target is assumed constant, and since $\csc \phi = R/h$, the received power becomes

$$P_r = K_1/h^4 = K_2 \tag{2.48}$$

where K_2 is a constant. The echo signal is therefore independent of range for a constant-altitude target.

In practice, the power received from an antenna with a cosecant-squared pattern is not truly independent of range because of the simplifying assumptions made. The cross section σ varies with the viewing aspect, the earth is not flat, and the radiation pattern of any real antenna can be made to only approximate the desired cosecant-squared pattern. The gain of a typical cosecant-squared antenna used for ground-based search radar might be about 2 dB less than if a fan beam were generated by the same aperture.

The maximum gain of an antenna is related to its physical area A (aperture) by

$$G = \frac{4\pi A \rho}{\lambda^2} \tag{2.49}$$

where ρ = antenna efficiency and λ = wavelength of radiated energy. The antenna efficiency depends on the aperture illumination and the efficiency of the antenna feed. The product of ρA is the effective aperture A_e. A typical reflector antenna with a parabolic shape will produce a beamwidth approximately equal to

$$\theta° = \frac{65\lambda}{l} \tag{2.50}$$

where l is the dimension of the antenna in the plane of the angle θ, and λ and l are measured in the same units. The value of the constant, in this case taken to be 65, depends upon the distribution of energy (illumination) across the aperture.

2.12 SYSTEM LOSSES

At the beginning of this chapter it was mentioned that one of the important factors omitted from the simple radar equation was the losses that occur throughout the radar system. The losses reduce the signal-to-noise ratio at the receiver output. They may be of two kinds, depending upon whether or not they can be predicted with any degree of precision beforehand. The antenna beam-shape loss, collapsing loss, and losses in the microwave plumbing are examples of losses which can be calculated if the system configuration is known. These losses are very real and cannot be ignored in any serious prediction of radar performance. The loss due to the integration of many pulses (or integration efficiency) has already been mentioned in Sec. 2.6 and need not be discussed further. Losses not readily subject to calculation and which are less predictable include those due to field degradation and to operator fatigue or lack of operator motivation. Estimates of the latter type of loss must be made on the basis of prior experience and experimental observations. They may be subject to considerable variation and uncertainty. Although the loss associated with any one factor may be small, there are many possible loss mechanisms in a complete radar system, and their sum total can be significant.

In this section, loss (number greater than unity) and efficiency (number less than unity) are used interchangeably. One is simply the reciprocal of the other.

Plumbing loss. There is always some finite loss experienced in the transmission lines which connect the output of the transmitter to the antenna. The losses in decibels per 100 ft for radar transmission lines are shown in Fig. 2.28. At the lower radar frequencies the transmission line introduces little loss, unless its length is exceptionally long. At the higher radar frequencies, attenuation may not always be small and may have to be taken into account. In addition to the losses in the transmission line itself, an additional loss can occur at each connection or bend in the line and at the antenna rotary joint if used. Connector losses are usually small, but if the connection is poorly made, it can contribute significant attenuation. Since the same transmission line is generally used for both receiving and transmission, the loss to be inserted in the radar equation is twice the one-way loss.

The signal suffers attenuation as it passes through the duplexer. Generally, the greater the isolation required from the duplexer on transmission, the larger will be the insertion loss. By insertion loss is meant the loss introduced when the component, in this case the duplexer, is inserted into the transmission line. The precise value of the insertion loss depends to a large extent on the particular design. For a typical duplexer it might be of the order of 1 dB. A

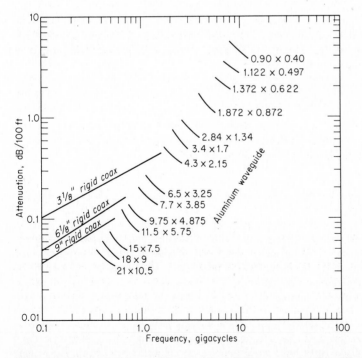

Figure 2.28 Theoretical (one-way) attenuation of RF transmission lines. Waveguide sizes are inches and are the inside dimensions. (*Data from Armed Services Index of R.F. Transmission Lines and Fittings, ASESA, 49-2B.*)

gas-tube duplexer also introduces loss when in the fired condition (arc loss); approximately 1 dB is typical.

In an S-band (3000 MHz) radar, for example, the plumbing losses might be as follows:

100 ft of RG-113/U A1 waveguide transmission line (two-way)	1.0 dB
Loss due to poor connections (estimate)	0.5 dB
Rotary-joint loss	0.4 dB
Duplexer loss	1.5 dB
Total plumbing loss	3.4 dB

Beam-shape loss. The antenna gain that appears in the radar equation was assumed to be a constant equal to the maximum value. But in reality the train of pulses returned from a target with a scanning radar is modulated in amplitude by the shape of the antenna beam. To properly take into account the pulse-train modulation caused by the beam shape, the computations of the probability of detection (as given in Secs. 2.5, 2.6, and 2.8) would have to be performed assuming a modulated train of pulses rather than constant-amplitude pulses. Some authors do indeed take account of the beam shape in this manner when computing the probability of detection. Therefore, when using published computations of probability of detection it should be noted whether the effect of the beam shape has been included. In this text, this approach is not used. Instead a *beam-shape loss* is added to the radar equation to account for the fact that the maximum gain is employed in the radar equation rather than a gain that changes pulse to pulse. This is a simpler, albeit less accurate, method. It is based on calculating the reduction in signal power and thus does not depend on the probability of detection. It applies for detection probabilities in the vicinity of 0.50, but it is used as an approximation with other values as a matter of convenience.

Let the one-way-power antenna pattern be approximated by the gaussian expression $\exp\left(-2.78\theta^2/\theta_B^2\right)$, where θ is the angle measured from the center of the beam and θ_B is the beamwidth measured between half-power points. If n_B is the number of pulses received within the half-power beamwidth θ_B, and n the total number of pulses integrated (n does not necessarily equal n_B), then the beam-shape loss (number greater than unity) relative to a radar that integrates all n pulses with an antenna gain corresponding to the maximum gain at the beam center is

$$\text{Beam-shape loss} = \frac{n}{1 + 2\sum_{k=1}^{(n-1)/2} \exp\left(-5.55k^2/n_B^2\right)} \tag{2.51}$$

For example, if we integrate 11 pulses, all lying uniformly between the 3-dB beamwidth, the loss is 1.96 dB.

The beam-shape loss considered above was for a beam shaped in one plane only. It applies to a fan beam, or to a pencil beam if the target passes through its center. If the target passes through any other point of the pencil beam, the maximum signal received will not correspond to the signal from the beam center. The beam-shape loss is reduced by the ratio of the square of the maximum antenna gain at which the pulses were transmitted divided by the square of the antenna gain at beam center. The ratio involves the square because of the two-way transit.

When there are a large number of pulses per beamwidth integrated, the scanning loss is generally taken to be 1.6 dB for a fan beam scanning in one coordinate and 3.2 dB when two-coordinate scanning is used.[49-52]

When the antenna scans rapidly enough that the gain on transmit is not the same as the gain on receive, an additional loss has to be computed, called the *scanning loss*. The technique for computing scanning loss is similar in principle to that for computing beam-shape loss. Scanning loss can be important for rapid-scan antennas or for very long range radars such as those designed to view extraterrestrial objects. A similar loss must be taken into account when covering a search volume with a step-scanning pencil beam, as with a phased array,[58] since not all regions of space are illuminated by the same value of antenna gain.

Limiting loss. Limiting in the radar receiver can lower the probability of detection. Although a well-designed and engineered receiver will not limit the received signal under normal circumstances, intensity modulated CRT displays such as the PPI or the B-scope have limited dynamic range and may limit. Some receivers, however, might employ limiting for some special purpose, as for pulse compression processing for example.

Limiting results in a loss of only a fraction of a decibel for a large number of pulses integrated, provided the limiting ratio (ratio of video limit level to rms noise level) is as large as 2 or 3.[10] Other analyses of bandpass limiters show that for small signal-to-noise ratio, the reduction in the signal-to-noise ratio of a sine-wave imbedded in narrowband gaussian noise is $\pi/4$ (about 1 dB).[53] However, by appropriately shaping the spectrum of the input noise, it has been suggested[54] that the degradation can be made negligibly small.

Collapsing loss. If the radar were to integrate additional noise samples along with the wanted signal-to-noise pulses, the added noise results in a degradation called the *collapsing loss*. It can occur in displays which collapse the range information, such as the C-scope which displays elevation vs. azimuth angle. The echo signal from a particular range interval must compete in a collapsed-range C-scope display, not only with the noise energy contained within that range interval, but with the noise energy from all other range intervals at the same elevation and azimuth. In some 3D radars (range, azimuth, and elevation) that display the outputs at all elevations on a single PPI (range, azimuth) display, the collapsing of the 3D radar information onto a 2D display results in a loss. A collapsing loss can occur when the output of a high-resolution radar is displayed on a device whose resolution is coarser than that inherent in the radar. A collapsing loss also results if the outputs of two (or more) radar receivers are combined and only one contains signal while the other contains noise.

The mathematical derivation of the collapsing loss, assuming a square-law detector, may be carried out as suggested by Marcum[10] who has shown that the integration of m noise pulses, along with n signal-plus-noise pulses with signal-to-noise ratio per pulse $(S/N)_n$, is equivalent to the integration of $m + n$ signal-to-noise pulses each with signal-to-noise ratio $n(S/N)_n/(m + n)$. The collapsing loss in this case is equal to the ratio of the integration loss L_i (Sec. 2.6) for $m + n$ pulses to the integration loss for n pulses, or

$$L_i(m,\ n) = \frac{L_i(m + n)}{L_i(n)} \tag{2.52}$$

For example, assume that 10 signal-plus-noise pulses are integrated along with 30 noise pulses and that $P_d = 0.90$ and $n_f = 10^8$. From Fig. 2.8b, $L_i(40)$ is 3.5 dB and $L_i(10)$ is 1.7 dB, so that the collapsing loss is 1.8 dB. It is also possible to account for the collapsing loss by substituting into the radar equation of Eq. (2.33) the parameter $E_i(m + n)$ for $E_i(n)$, since $E_i(n) = 1/L_i(n)$.

The above applies for a square-law detector. Trunk[55] has shown that the collapsing loss for a linear detector differs from that of the square-law detector, and it can be much greater. The comparison between the two is shown in Fig. 2.29 as a function of the collapsing ratio $(m + n)/n$. The difference between the two cases can be large. As the number of hits n increases, the difference becomes smaller.

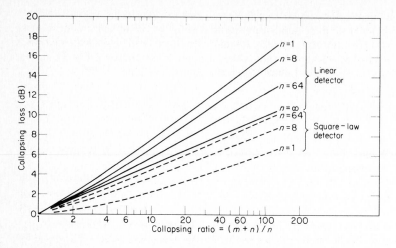

Figure 2.29 Collapsing loss versus collapsing ratio $(m + n)/n$, for a false alarm probability of 10^{-6} and a detection probability of 0.5. (*From Trunk,*[55] *courtesy Proc. IEEE.*)

Nonideal equipment. The transmitter power introduced into the radar equation was assumed to be the output power (either peak or average). However, transmitting tubes are not all uniform in quality, nor should it be expected that any individual tube will remain at the same level of performance throughout its useful life. Also the power is usually not uniform over the operating band of the device. Thus, for one reason or another, the transmitted power may be other than the design value. To allow for this, a loss factor may be introduced. This factor can vary with the application, but lacking a better number, a loss of about 2 dB might be used as an approximation.

Variations in the receiver noise figure over the operating band also are to be expected. Thus, if the best noise figure over the band is used in the radar equation, a loss factor has to be introduced to account for its poorer value elsewhere within the band.

If the receiver is not the exact matched filter for the transmitted waveform, a loss in signal-to-noise ratio will occur. Examples are given in Table 10.1. A typical value of loss for a nonmatched receiver might be about 1 dB. Because of the exponential relation between the false-alarm time and the threshold level [Eq. (2.26)], a slight change in the threshold can cause a significant change in the false alarm time. In practice, therefore, it may be necessary to set the threshold level slightly higher than calculated so as to insure a tolerable false alarm time in the event of circuit instabilities. This increase in the threshold is equivalent to a loss.

Operator loss. An alert, motivated, and well-trained operator should perform as well as described by theory. However, when distracted, tired, overloaded, or not properly trained, operator performance will decrease. There is little guidance available on how to account for the performance of an operator.

Based on both empirical and experimental results, one study[69] gives the operator-efficiency factor as

$$\rho_0 = 0.7(P_d)^2 \qquad (2.53)$$

where P_d is the single-scan probability of detection. This was said to apply to a good operator viewing a PPI under good conditions. Its degree of applicability, however, is not clear.

It is not unusual to find no account of the operator loss being taken in the radar equation. This is probably justified when operators are alert, motivated, and well trained. It is also justified when automatic (electronic) detections are made without the aid of an operator. When the operator does introduce loss into the system, it is not easy to select a proper value to account for it. The better action is to take steps to correct loss in operator performance rather than tolerate it by including it as a loss factor in the radar equation.

Field degradation. When a radar system is operated under laboratory conditions by engineering personnel and experienced technicians, the inclusion of the above losses into the radar equation should give a realistic description of the performance of the radar under normal conditions (ignoring anomalous propagation effects). However, when a radar is operated under field conditions, the performance usually deteriorates even more than can be accounted for by the above losses, especially when the equipment is operated and maintained by inexperienced or unmotivated personnel. It may even apply, to some extent, to equipment operated by professional engineers under adverse field conditions. Factors which contribute to field degradation are poor tuning, weak tubes, water in the transmission lines, incorrect mixer-crystal current, deterioration of receiver noise figure, poor TR tube recovery, loose cable connections, etc.

To minimize field degradation, radars should be designed with built-in automatic performance-monitoring equipment. Careful observation of performance-monitoring instruments and timely preventative maintenance can do much to keep radar performance up to design level. Radar characteristics that might be monitored include transmitter power, receiver noise figure, the spectrum and/or shape of the transmitted pulse, and the decay time of the TR tube.

A good estimate of the field degradation is difficult to obtain since it cannot be predicted and is dependent upon the particular radar design and the conditions under which it is operating. A degradation of 3 dB is sometimes assumed when no other information is available.

Other loss factors. A radar designed to discriminate between moving targets and stationary objects (MTI radar) may introduce additional loss over a radar without this facility. The MTI discrimination technique results in complete loss of sensitivity for certain values of target velocity relative to the radar. These are called *blind speeds*. The blind-speed problem and the loss resulting therefrom are discussed in more detail in Chap. 4.

In a radar with overlapping range gates, the gates may be wider than optimum for practical reasons. The additional noise introduced by the nonoptimum gate width will result in some degradation. The *straddling loss* accounts for the loss in signal-to-noise ratio for targets not at the center of the range gate or at the center of the filter in a multiple-filter-bank processor.

Another factor that has a profound effect on the radar range performance is the propagation medium discussed briefly in the next section and in Chap. 12.

There are many causes of loss and inefficiency in a radar. Not all have been included here. Although they may each be small, the sum total can result in a significant reduction in radar performance. It is important to understand the origins of these losses, not only for better predictions of radar range, but also for the purpose of keeping them to a minimum by careful radar design.

2.13 PROPAGATION EFFECTS

In analyzing radar performance it is convenient to assume that the radar and target are both located in free space. However, there are very few radar applications which approximate free-space conditions. In most cases of practical interest, the earth's surface and the medium in which radar waves propagate can have a significant effect on radar performance. In some instances the propagation factors might be important enough to overshadow all other factors that contribute to abnormal radar performance. The effects of non-free-space propagation on the radar are of three categories: (1) *attenuation* of the radar wave as it propagates through the earth's atmosphere, (2) *refraction* of the radar wave by the earth's atmosphere, and (3) *lobe structure* caused by interference between the direct wave from radar to target and the wave which arrives at the target via reflection from the ground.

In general, for most applications of radar at microwave frequencies, the attenuation in propagating through either the normal atmosphere or through precipitation is usually not sufficient to affect radar performance. However, the *reflection* of the radar signal from rain (clutter) is often a limiting factor in the performance of radar in adverse weather.

The decreasing density of the atmosphere with increasing altitude results in a bending, or refraction, of the radar waves in a manner analogous to the bending of light waves by an optical prism. This bending usually results in an increase in the radar line of sight. Normal atmospheric conditions can be accounted for in a relatively simple manner by considering the earth to have a larger radius than actual. A "typical" earth radius for refractive effects is four-thirds the actual radius. At times, atmospheric conditions might cause more than usual bending of the radar rays, with the result that the radar range will be considerably increased. This condition is called *superrefraction*, or *ducting*, and is a form of *anomalous propagation*. It is not necessarily a desirable condition since it cannot be relied upon. It can degrade the performance of MTI radar by extending the range at which ground clutter is seen.

The presence of the earth's surface not only restricts the line of sight, but it can cause major modification of the coverage within the line of sight by breaking up the antenna elevation pattern into many lobes. Energy propagates directly from the radar to the target, but there can also be energy that travels to the target via a path that includes a reflection from the ground. The direct and ground-reflected waves interfere at the target either destructively or constructively to produce nulls or reinforcements (lobes). The lobing that results causes non-uniform illumination of the coverage, and is an important factor that influences the capability of a radar system.

Most propagation effects that are of importance cannot be easily included into the radar equation. They must be properly taken into account, however, since they can have a major impact on performance. Further discussion of the effects of propagation on radar is given in Chap. 12.

2.14 OTHER CONSIDERATIONS

Prediction of radar range. In this chapter, some of the more important factors that enter into the radar equation for the prediction of range were briefly considered. The radar equation (2.1), with the modifications indicated in this chapter, becomes

$$R^4_{\max} = \frac{P_{av} G A \rho_a \sigma n E_i(n)}{(4\pi)^2 k T_0 F_n (B\tau) f_p (S/N)_1 L_s}$$

$$(2.54)$$

where R_{max} = maximum radar range, m

    G = antenna gain

    A = antenna aperture, m^2

    ρ_a = antenna efficiency

    n = number of hits integrated

    $E_i(n)$ = integration efficiency (less than unity)

    L_s = system losses (greater than unity) not included in other parameters

    σ = radar cross section of target, m^2

    F_n = noise figure

    k = Boltzmann's constant = 1.38×10^{-23} J/deg

    T_0 = standard temperature = 290 K

    B = receiver bandwidth, Hz

    τ = pulse width, s

    f_p = pulse repetition frequency, Hz

    $(S/N)_1$ = signal-to-noise ratio required at receiver output (based on single-hit detection)

This equation can also be written in terms of energy rather than power. The energy in the transmitted pulse is $E_\tau = P_{av}/f_p = P_t \tau$ and the signal-to-noise power ratio $(S/N)_1$ can be replaced by the signal-to-noise energy ratio $(E/N_0)_1$, where $E = S/\tau$ is the energy in the received signal, and $N_0 = N/B$ is the noise power per unit bandwidth, or the noise energy. Also note that $B\tau \simeq 1$, and $T_0 F_n = T_s$ is defined as the system noise temperature. Then the radar equation becomes

$$R_{max}^4 = \frac{E_\tau G A \rho_a \sigma n E_i(n)}{(4\pi)^2 k T_s (E/N_0)_1 L_s} \qquad (2.55)$$

Although Eq. (2.55) was derived for a rectangular pulse, it can be applied to other waveforms as well, provided matched-filter detection is employed. Most calculations for probability of detection with signal-to-noise ratio as the parameter apply equally well when the ratio of signal-energy–to–noise-power-per-hertz is used instead. The radar equation developed in this chapter for pulse radar can be readily modified to accommodate CW, FM-CW, pulse-doppler, MTI, or tracking radar.[47,56,57]

Radar performance figure. This is a figure of merit sometimes used to express the relative performance of radar. It is defined as the ratio of the pulse power of the radar transmitter to the minimum detectable signal power of the receiver. It is not often used.

Blip-scan ratio. This is the same as the single-scan probability of detection. It predates the widespread use of the term probability of detection and came about by the manner in which the performance of ground-based search-radars was checked. An aircraft would be flown on a radial course and on each scan of the antenna it would be recorded whether or not a target blip had been detected on the radar display. This was repeated many times until sufficient data was obtained to compute, as a function of range, the ratio of the average number of scans the target was seen at a particular range (blips) to the total number of times it could have been seen (scans). This is the blip-scan ratio and is the probability of detecting a target at a particular range, altitude, and aspect. The head-on and tail-on are the two easiest aspects to provide in field measurements. The experimentally found blip-scan ratio curves are subject to many limitations, but it represents one of the few methods for evaluating the performance of an actual radar equipment against real targets under somewhat controlled conditions.

Cumulative probability of detection. If the single scan probability of detection for a surveillance radar is P_d, the probability of detecting a target at least once during N scans is called the cumulative probability of detection, and may be written

$$P_c = 1 - (1 - P_d)^N \tag{2.56}$$

The variation of P_d with range might have to be taken into account when computing P_c. The variation of range based on the cumulative probability of detection can be as the third power rather than the more usual fourth power variation based on the single scan probability.[59]

The cumulative probability has sometimes been proposed as a measure of the detectability of a radar rather than the single-scan probability of detection, which is more conservative. In practice the use of the cumulative probability is not easy to apply. Furthermore, radar operators do not usually use such a criterion for reporting detections. They seldom report a detection the first time it is observed, as is implied in the definition of cumulative probability. Instead, the criterion for reporting a detection might be a threshold crossing on two successive scans, or threshold crossings on two out of three scans, three out of five, or so forth. In track-while-scan radars the measure of performance might be the probability of initiating a target track rather than a criterion based on detection alone.

Surveillance-radar range equation. The form of the radar equation described in this chapter applies to a radar that dwells on the target for n pulses. It is sometimes called the *searchlight* range equation. In a search or surveillance radar there is usually an additional constraint imposed that modifies the range equation significantly. This constraint is that the radar is required to search a specified volume of space within a specified time. Let Ω denote the angular region to be searched in the scan time t_s. (For example, Ω might represent a region 360° in azimuth and 30° in elevation.) The scan time is $t_s = t_0 \Omega/\Omega_0$, where t_0 is the time on target $= n/f_p$, and Ω_0 is the solid angular beamwidth which is approximately equal to the product of the azimuth beamwidth θ_a times the elevation beamwidth θ_e. (This assumes that θ_A/θ_e and θ_E/θ_e are integers, where θ_A is the total azimuth coverage and θ_E the total elevation coverage, such that $\Omega \simeq \theta_A \theta_E$.) The antenna gain can be written as $G = 4\pi/\Omega_0$. With the above substitutions into Eq. (2.54) the radar equation for a search radar becomes

$$R_{\max}^4 = \frac{P_{\text{av}} A_e \sigma E_i(n)}{4\pi k T_0 F_n (S/N)_1 L_s} \frac{t_s}{\Omega} \tag{2.57}$$

This indicates that the important parameters in a search radar are the average power and the antenna effective aperture. The frequency does not appear explicitly in the search-radar range equation. However, the lower frequencies are preferred for a search radar since large power and large aperture are easier to obtain at the lower frequencies, it is easier to build a good MTI capability, and there is little effect from adverse weather.

Different radar range equations can be derived for different applications, depending on the particular constraints imposed. The radar equation will also be considerably different if clutter echoes or external noise, rather than receiver noise, determine the background with which the radar signal must compete. Some of these other forms of the radar equation are given elsewhere in this text.

Accuracy of the radar equation. The predicted value of the range as found from the radar equation cannot be expected to be checked experimentally with any degree of accuracy. It is difficult to determine precisely all the important factors that must be included in the radar equation and it is difficult to establish a set of controlled, realistic experimental conditions in which to test the calculations. Thus it might not be worthwhile to try to obtain too great a precision in the individual parameters of the radar equation. Nevertheless, if a particular range is required of a radar, the systems engineer must provide it. The safest means to achieving a specified range performance is to design conservatively and add a safety factor. The inclusion of a safety factor in design is not always appreciated, especially in competitive procurements, but it is a standard procedure in many other engineering disciplines. In the few cases where this luxury was permitted, fine radars were obtained since they accomplished what was needed even under degraded conditions.

REFERENCES

1. Ridenour, L. N.: "Radar System Engineering," MIT Radiation Laboratory Series, vol. 1, p. 592, McGraw-Hill Book Company, New York, 1947.
2. Lawson, J. L., and G. E. Uhlenbeck (eds.): "Threshold Signals," MIT Radiation Laboratory Series, vol. 24, McGraw-Hill Book Company, New York, 1950, p. 177.
3. Van Vleck, J. H., and D. Middleton: A Theoretical Comparison of the Visual, Aural, and Meter Reception of Pulsed Signals in the Presence of Noise, *J. Appl. Phys.*, vol. 17, pp. 940–971, November, 1946.
4. Bennett, W. R.: Methods of Solving Noise Problems, *Proc. IRE*, vol. 44, pp. 609–638, May, 1956.
5. Davenport, W. B., and W. L. Root: "Introduction to Random Signals and Noise," McGraw-Hill Book Co., New York, 1958.
6. Bendat, J. S.: "Principles and Applications of Random Noise Theory," John Wiley & Sons, Inc., New York, 1958.
7. Helstrom, C. W.: "Statistical Theory of Signal Detection," Pergamon Press, New York, 1960.
8. Parzen, E.: "Modern Probability Theory and Its Applications," John Wiley & Sons, Inc., New York, 1960.
9. Rice, S. O.: Mathematical Analysis of Random Noise, *Bell System Tech. J.*, vol. 23, pp. 282–332, 1944, and vol. 24, pp. 46–156, 1945.
10. Marcum, J. I.: A Statistical Theory of Target Detection by Pulsed Radar, Mathematical Appendix, *IRE Trans.*, vol. IT-6, pp. 145–267, April, 1960.
11. Meyer, D. P., and H. A. Mayer: "Radar Target Detection," Academic Press, New York, 1973.
12. Hollis, R.: False Alarm Time in Pulse Radar, *Proc. IRE*, vol. 42, p. 1189, July, 1954.
13. Trunk, G. V.: Detection Results for Scanning Radars Employing Feedback Integration, *IEEE Trans.*, vol. AES-6, pp. 522–527, July, 1970.
14. Harrington, J. V., and T. F. Rogers: Signal-to-noise Improvement through Integration in a Storage Tube, *Proc. IRE*, vol. 38, pp. 1197–1203, October, 1950.
15. Mentzer, J. R.: "Scattering and Diffraction of Radio Waves," Pergamon Press, New York, 1955.
16. King, R. W. P., and T. T. Wu: "The Scattering and Diffraction of Waves," Harvard University Press, Cambridge, Mass., 1959.
17. Crispin, J. W., Jr., and K. M. Siegel: "Methods of Radar Cross-Section Analysis," Academic Press, New York, 1968.
18. Bowman, J. J., T. B. A. Senior, and P. L. E. Uslengh: "Electromagnetic and Acoustic Scattering by Simple Shapes," North-Holland Publishing Co., Amsterdam, and John Wiley & Sons, Inc., New York, 1969.
19. Ruck, G. T., D. E. Barrick, W. D. Stuart, and C. K. Krichbaum: "Radar Cross-Section Handbook," (2 vols.), Plenum Press, New York, 1970.
20. Siegel, K. M., F. V. Schultz, B. H. Gere, and F. B. Sleator: The Theoretical and Numerical Determination of the Radar Cross Section of a Prolate Spheroid, *IRE Trans.*, vol. AP-4, pp. 266–275, July, 1956.
21. Mathur, P. N., and E. A. Mueller: Radar Back-scattering Cross Sections for Nonspherical Targets, *IRE Trans.*, vol. AP-4, pp. 51–53, January, 1956.
22. Scharfman, H.: Scattering from Dielectric Coated Spheres in the Region of the First Resonance, *J. Appl. Phys.*, vol. 25, pp. 1352–1356, November, 1954.
23. Andreasen, M. G.: Back-scattering Cross Section of a Thin, Dielectric, Spherical Shell, *IRE Trans.*, vol. AP-5, pp. 267–270, July, 1957.
24. King, D. D.: The Measurement and Interpretation of Antenna Scattering, *Proc. IRE*, vol. 37, pp. 770–777, July, 1949.
25. Siegel, K. M.: Far Field Scattering from Bodies of Revolution, *Appl. Sci. Research*, sec. B, vol. 7, pp. 293–328, 1958.

26. Peters, L., Jr.: End-fire Echo Area of Long, Thin Bodies, *IRE Trans.*, vol. AP-6, pp. 133–139, January, 1958.
27. Blore, W. E.: The Radar Cross Section of Ogives, Double-Backed Cones, Double-Rounded Cones and Cone Spheres, *IEEE Trans.*, vol. AP-12, pp. 582–589, September, 1964.
28. Ridenour, L. N.: "Radar System Engineering," MIT Radiation Laboratory Series, vol. 1, fig. 3.8, McGraw-Hill Book Company, New York, 1947.
29. Blacksmith, P., Jr., R. E. Hiatt, and R. B. Mack: Introduction to Radar Cross-Section Measurements, *Proc. IEEE*, vol. 53, pp. 901–920, August, 1965.
30. Olin, I. D., and F. D. Queen: Dynamic Measurement of Radar Cross Sections, *Proc. IEEE*, vol. 53, pp. 954–961, August, 1965.
31. Skolnik, M. I.: An Empirical Formula for the Radar Cross Section of Ships at Grazing Incidence, *IEEE Trans.*, vol. AES-10, p. 292, March, 1974.
32. Schultz, F. V., R. C. Burgener, and S. King: Measurement of the Radar Cross Section of a Man, *Proc. IRE*, vol. 46, pp. 476–481, February, 1958.
33. Kell, R. E., and R. A. Ross: Radar Cross Section of Targets, chap. 27 of the "Radar Handbook," M. I. Skolnik (ed.), McGraw-Hill Book Co., New York, 1970.
34. Rheinstein, J.: Backscatter from Spheres: A Short Pulse View, *IEEE Trans.*, vol. AP-16, pp. 89–97, January, 1968.
35. Miller, E. K., G. J. Burke, B. J. Maxum, G. M. Pjerrou, and A. R. Neureuther: Radar Cross Section of a Long Wire, *IEEE Trans.*, vol. AP-17, pp. 381–384, May, 1969.
36. Dunn, J. H., and D. D. Howard: Target Noise, chap. 28 of "Radar Handbook," M. I. Skolnik (ed.), McGraw-Hill Book Co., New York, 1970.
37. Swerling, P.: Probability of Detection for Fluctuating Targets, *IRE Trans.*, vol. IT-6, pp. 269–308, April, 1960.
38. Nathanson, F. E.: "Radar Design Principles," McGraw-Hill Book Company, New York, 1969, chap. 5.
39. Wilson, J. D.: Probability of Detecting Aircraft Targets, *IEEE Trans.*, vol. AES-8, pp. 757–761, November, 1972.
40. Weinstock, W.: "Target Cross Section Models for Radar Systems Analysis," doctoral dissertation, University of Pennsylvania, Philadelphia, 1964.
41. Scholefield, P. H. R.: Statistical Aspects of Ideal Radar Targets, *Proc. IEEE*, vol. 55, pp. 587–589, April, 1967.
42. Pollon, G. E.: Statistical Parameters for Scattering from Randomly Oriented Arrays, Cylinders, and Plates, *IEEE Trans.*, vol. AP-18, pp. 68–75, January, 1970.
43. Heidbreder, G. R., and R. L. Mitchell: Detection Probabilities for Log-Normally Distributed Signals, *IEEE Trans.*, vol. AES-3, pp. 5–13, January, 1967.
44. Schwartz, M.: Effects of Signal Fluctuation on the Detection of Pulse Signals in Noise, *IRE Trans.*, vol. IT-2, pp. 66–71, June, 1956.
45. Swerling, P.: Detection of Fluctuating Pulsed Signals in the Presence of Noise, *IRE Trans.*, vol. IT-3, pp. 175–178, September, 1957.
46. Scholtz, R. A., J. J. Kappl, and N. E. Nahi: The Detection of Moderately Fluctuating Rayleigh Targets, *IEEE Trans.*, vol. AES-12, pp. 117–125, March, 1976.
47. Mooney, D. H., and W. A. Skillman: Pulse-doppler Radar, chap. 19 of "Radar Handbook," M. I. Skolnik (ed.), McGraw-Hill Book Co., New York, 1970.
48. Hovanessian, S. A.: An Algorithm for Calculation of Range in a Multiple PRF Radar, *IEEE Trans.*, vol. AES-12, pp. 287–290, March, 1976.
49. Blake, L. V.: The Effective Number of Pulses Per Beamwidth for a Scanning Radar, *Proc. IRE*, vol. 41, pp. 770–774, June, 1953.
50. Hall, W. M., and D. K. Barton: Antenna Pattern Loss Factor for Scanning Radars, *Proc. IEEE*, vol. 53, pp. 1257–1258, September, 1965.
51. Hall, W. M.: Antenna Beam-Shape Factor in Scanning Radars, *IEEE Trans.*, vol. AES-4, pp. 402–409, May, 1968.
52. Trunk, G. V.: Detection Results for Scanning Radars Employing Feedback Integration, *IEEE Trans.*, vol. AES-6, pp. 522–527, July, 1970.
53. Davenport, W. B., Jr.: Signal-to-noise Ratios in Band-pass Limiters, *J. Appl. Phys.*, vol. 24, pp. 720–727, June, 1953.
54. Manasse, R., R. Price, and R. M. Lerner: Loss of Signal Detectability in Band-pass Limiters, *IRE Trans.*, vol. IT-4, pp. 34–38, March, 1958.
55. Trunk, G. V.: Comparison of the Collapsing Losses in Linear and Square-law Detectors, *Proc. IEEE*, vol. 60, pp. 743–744, June, 1972.

56. Barton, D. K.: "The Radar Equation," Artech House, Dedham, Mass., 1974.
57. Blake, L. V.: Prediction of Radar Range, chap. 2 of "Radar Handbook," M. I. Skolnik (ed.), McGraw-Hill Book Co., New York, 1970.
58. Hahn, P. M., and S. D. Gross: Beam Shape Loss and Surveillance Optimization for Pencil Beam Arrays, *IEEE Trans.*, vol. AES-5, pp. 674–675, July, 1969.
59. Mallett, J. D., and L. E. Brennan: Cumulative Probability of Detection for Targets Approaching a Uniformly Scanning Search Radar, *Proc. IEEE*, vol. 51, 596–601, April, 1963, and vol. 52, pp. 708–709, June, 1964.
60. Johnson, J. B.: Thermal Agitation of Electricity in Conductors, *Phys. Rev.*, vol. 32, pp. 97–109, 1928.
61. Pannell, J. H., J. Rheinstein, and A. F. Smith: Radar Scattering from a Conducting Cone-Sphere, *MIT Lincoln Laboratory Tech. Rept.* no. 349, Mar. 2, 1964. (See also Chap. 6 of ref. 19.)
62. Elphick, B. L., M. J. Chappell, and R. Batty: The Measured and Calculated RCS of a Resistive Target Model, *IEE (London) Radar-77*, pp. 120–124, Oct. 25–28, 1977.
63. Sinclair, G.: Theory of Models of Electromagnetic Systems, *Proc. IRE*, vol. 36, pp. 1364–1370, November, 1948.
64. Schensted, C. E., J. W. Crispin, and K. M. Siegel: Studies in Radar Cross-Sections XV: Radar Cross-Sections of the B-47 and B-52 Aircraft, *Univ. of Michigan*, Radiation Laboratory Report 2260-1-T, August 1954 (unclassified), AD 46 741.
65. Chandler, R. A., and L. E. Wood: System Considerations for the Design of Radar Braking Sensors, *IEEE Trans.*, vol. VT-26, pp. 151–160, May, 1977.
66. Queen, F. D.: Radar Cross Sections of the T-38 Aircraft for the Head-On Aspects in *L, S,* and *X* Bands, *Naval Research Laboratory Report* 7951, Jan. 8, 1976.
67. Jao, J. K., and M. Elbaum: First-Order Statistics of a Non-Rayleigh Fading Signal and Its Detection, *Proc. IEEE*, vol. 66, pp. 781–789, July, 1978.
68. Williams, P. D. L., H. D. Cramp, and K. Curtis: Experimental Study of the Radar Cross-Section of Maritime Targets, *IEE (London) Journal on Electronic Circuits and Systems*, vol. 2, no. 4, pp. 121–136, July, 1978.
69. Varela, A. A.: The Operator Factor Concept, Its History and Present Status, *Symposium on Radar Detection Theory, ONR Symposium Report* ACR-10, Mar. 1–2, 1956, DDC No. 117533.

THREE

CW AND FREQUENCY-MODULATED RADAR

3.1 THE DOPPLER EFFECT

A radar detects the presence of objects and locates their position in space by transmitting electromagnetic energy and observing the returned echo. A pulse radar transmits a relatively short burst of electromagnetic energy, after which the receiver is turned on to listen for the echo. The echo not only indicates that a target is present, but the time that elapses between the transmission of the pulse and the receipt of the echo is a measure of the distance to the target. Separation of the echo signal and the transmitted signal is made on the basis of differences in time.

The radar transmitter may be operated continuously rather than pulsed if the strong transmitted signal can be separated from the weak echo. The received-echo-signal power is considerably smaller than the transmitter power; it might be as little as 10^{-18} that of the transmitted power—sometimes even less. Separate antennas for transmission and reception help segregate the weak echo from the strong leakage signal, but the isolation is usually not sufficient. A feasible technique for separating the received signal from the transmitted signal when there is relative motion between radar and target is based on recognizing the change in the echo-signal frequency caused by the doppler effect.

It is well known in the fields of optics and acoustics that if either the source of oscillation or the observer of the oscillation is in motion, an apparent shift in frequency will result. This is the *doppler effect* and is the basis of CW radar. If R is the distance from the radar to target, the total number of wavelengths λ contained in the two-way path between the radar and the target is $2R/\lambda$. The distance R and the wavelength λ are assumed to be measured in the same units. Since one wavelength corresponds to an angular excursion of 2π radians, the total angular excursion ϕ made by the electromagnetic wave during its transit to and from the target is $4\pi R/\lambda$ radians. If the target is in motion, R and the phase ϕ are continually changing. A change in ϕ with respect to time is equal to a frequency. This is the doppler angular frequency ω_d, given by

$$\omega_d = 2\pi f_d = \frac{d\phi}{dt} = \frac{4\pi}{\lambda}\frac{dR}{dt} = \frac{4\pi v_r}{\lambda} \tag{3.1}$$

where f_d = doppler frequency shift and v_r = relative (or radial) velocity of target with respect to radar. The doppler frequency shift is

$$f_d = \frac{2v_r}{\lambda} = \frac{2v_r f_0}{c} \tag{3.2a}$$

where f_0 = transmitted frequency and c = velocity of propagation $= 3 \times 10^8$ m/s. If f_d is in hertz, v_r in knots, and λ in meters,

$$f_d = \frac{1.03 v_r}{\lambda} \tag{3.2b}$$

A plot of this equation is shown in Fig. 3.1.

The relative velocity may be written $v_r = v \cos \theta$, where v is the target speed and θ is the angle made by the target trajectory and the line joining radar and target. When $\theta = 0$, the doppler frequency is maximum. The doppler is zero when the trajectory is perpendicular to the radar line of sight ($\theta = 90°$).

The type of radar which employs a continuous transmission, either modulated or unmodulated, has had wide application. Historically, the early radar experimenters worked almost exclusively with continuous rather than pulsed transmissions (Sec. 1.5). Two of the more

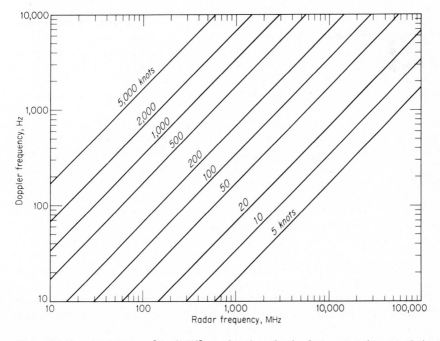

Figure 3.1 Doppler frequency [Eq. (3.2b)] as a function of radar frequency and target relative velocity.

important early applications of the CW radar principle were the proximity (VT) fuze and the FM-CW altimeter. The CW proximity fuze was first employed in artillery projectiles during World War II and greatly enhanced the effectiveness of both field and antiaircraft artillery. The first practical model of the FM-CW altimeter was developed by the Western Electric Company in 1938, although the principle of altitude determination using radio-wave reflections was known ten years earlier, in 1928.[1]

The CW radar is of interest not only because of its many applications, but its study also serves as a means for better understanding the nature and use of the doppler information contained in the echo signal, whether in a CW or a pulse radar (MTI) application. In addition to allowing the received signal to be separated from the transmitted signal, the CW radar provides a measurement of relative velocity which may be used to distinguish moving targets from stationary objects or clutter.

3.2 CW RADAR

Consider the simple CW radar as illustrated by the block diagram of Fig. 3.2a. The transmitter generates a continuous (unmodulated) oscillation of frequency f_0, which is radiated by the antenna. A portion of the radiated energy is intercepted by the target and is scattered, some of it in the direction of the radar, where it is collected by the receiving antenna. If the target is in motion with a velocity v_r relative to the radar, the received signal will be shifted in frequency from the transmitted frequency f_0 by an amount $\pm f_d$ as given by Eq. (3.2). The plus sign associated with the doppler frequency applies if the distance between target and radar is decreasing (closing target), that is, when the received signal frequency is greater than the transmitted signal frequency. The minus sign applies if the distance is increasing (receding target). The received echo signal at a frequency $f_0 \pm f_d$ enters the radar via the antenna and is heterodyned in the detector (mixer) with a portion of the transmitter signal f_0 to produce a doppler beat note of frequency f_d. The sign of f_d is lost in this process.

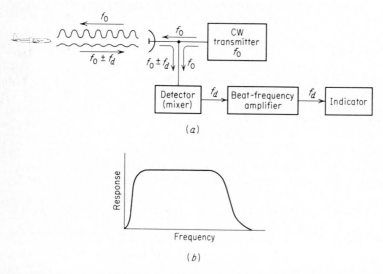

(a)

(b)

Figure 3.2 (a) Simple CW radar block diagram; (b) response characteristic of beat-frequency amplifier.

The purpose of the doppler amplifier is to eliminate echoes from stationary targets and to amplify the doppler echo signal to a level where it can operate an indicating device. It might have a frequency-response characteristic similar to that of Fig. 3.2b. The low-frequency cutoff must be high enough to reject the d-c component caused by stationary targets, but yet it must be low enough to pass the smallest doppler frequency expected. Sometimes both conditions cannot be met simultaneously and a compromise is necessary. The upper cutoff frequency is selected to pass the highest doppler frequency expected.

The indicator might be a pair of earphones or a frequency meter. If exact knowledge of the doppler frequency is not necessary, earphones are especially attractive provided the doppler frequencies lie within the audio-frequency response of the ear. Earphones are not only simple devices, but the ear acts as a selective bandpass filter with a passband of the order of 50 Hz centered about the signal frequency.[2] The narrow-bandpass characteristic of the ear results in an effective increase in the signal-to-noise ratio of the echo signal. With subsonic aircraft targets and transmitter frequencies in the middle range of the microwave frequency region, the doppler frequencies usually fall within the passband of the ear. If audio detection were desired for those combinations of target velocity and transmitter frequency which do not result in audible doppler frequencies, the doppler signal could be heterodyned to the audible range. The doppler frequency can also be detected and measured by conventional frequency meters, usually one that counts cycles. An example of the CW radar principle is the radio proximity (VT) fuze, used with great success during World War II for the fuzing of artillery projectiles. It may seem strange that the radio proximity fuze should be classified as a radar, but it fulfills the same basic function of a radar, which is the detection and location of reflecting objects by "radio" means.[3,4]

Isolation between transmitter and receiver. A single antenna serves the purpose of transmission and reception in the simple CW radar described above. In principle, a single antenna may be employed since the necessary isolation between the transmitted and the received signals is achieved via separation in frequency as a result of the doppler effect. In practice, it is not possible to eliminate completely the transmitter leakage. However, transmitter leakage is not always undesirable. A moderate amount of leakage entering the receiver along with the echo signal supplies the reference necessary for the detection of the doppler frequency shift. If a leakage signal of sufficient magnitude were not present, a sample of the transmitted signal would have to be deliberately introduced into the receiver to provide the necessary reference frequency.

There are two practical effects which limit the amount of transmitter leakage power which can be tolerated at the receiver. These are (1) the maximum amount of power the receiver input circuitry can withstand before it is physically damaged or its sensitivity reduced (burnout) and (2) the amount of transmitter noise due to hum, microphonics, stray pick-up, and instability which enters the receiver from the transmitter. The additional noise introduced by the transmitter reduces the receiver sensitivity. Except where the CW radar operates with relatively low transmitter power and insensitive receivers, additional isolation is usually required between the transmitter and the receiver if the sensitivity is not to be degraded either by burnout or by excessive noise.

The amount of isolation required depends on the transmitter power and the accompanying transmitter noise as well as the ruggedness and the sensitivity of the receiver. For example, if the safe value of power which might be applied to a receiver were 10 mW and if the transmitter power were 1 kW, the isolation between transmitter and receiver must be at least 50 dB.

The amount of isolation needed in a long-range CW radar is more often determined by the noise that accompanies the transmitter leakage signal rather than by any damage caused by high power.[5] For example, suppose the isolation between the transmitter and receiver were such that 10 mW of leakage signal appeared at the receiver. If the minimum detectable signal were 10^{-13} watt (100 dB below 1 mW), the transmitter noise must be at least 110 dB (preferably 120 or 130 dB) below the transmitted carrier.

The transmitter noise of concern in doppler radar includes those noise components that lie within the same frequency range as the doppler frequencies. The greater the desired radar range, the more stringent will be the need for reducing the noise modulation accompanying the transmitter signal. If complete elimination of the *direct* leakage signal at the receiver could be achieved, it might not entirely solve the isolation problem since echoes from nearby fixed targets (clutter) can also contain the noise components of the transmitted signal.[6,69]

It will be recalled (Sec. 1.3) that the receiver of a pulsed radar is isolated and protected from the damaging effects of the transmitted pulse by the duplexer, which short-circuits the receiver input during the transmission period. Turning off the receiver during transmission with a duplexer is not possible in a CW radar since the transmitter is operated continuously. Isolation between transmitter and receiver might be obtained with a single antenna by using a hybrid junction, circulator, turnstile junction, or with separate polarizations. Separate antennas for transmitting and receiving might also be used. The amount of isolation which can be readily achieved between the arms of practical hybrid junctions such as the magic-T, rat race, or short-slot coupler is of the order of 20 to 30 dB. In some instances, when extreme precision is exercised, an isolation of perhaps 60 dB or more might be achieved. One limitation of the hybrid junction is the 6-dB loss in overall performance which results from the inherent waste of half the transmitted power and half the received signal power. Both the loss in performance and the difficulty in obtaining large isolations have limited the application of the hybrid junction to short-range radars.

Ferrite isolation devices such as the circulator do not suffer the 6-dB loss inherent in the hybrid junction. Practical devices have isolation of the order of 20 to 50 dB. Turnstile junctions[7] achieve isolations as high as 40 to 60 dB.

The use of orthogonal polarizations for transmitting and receiving is limited to short-range radars because of the relatively small amount of isolation that can be obtained.[8]

An important factor which limits the use of isolation devices with a common antenna is the reflections produced in the transmission line by the antenna. The antenna can never be perfectly matched to free space, and there will always be some transmitted signal reflected back toward the receiver. The reflection coefficient from a mismatched antenna with a voltage-standing-wave ratio σ is $|\rho| = (\sigma - 1)/(\sigma + 1)$. Therefore, if an isolation of 20 dB is to be obtained, the VSWR must be less than 1.22. If 40 dB of isolation is required, the VSWR must be less than 1.02.

The largest isolations are obtained with two antennas—one for transmission, the other for reception—physically separated from one another. Isolations of the order of 80 dB or more are possible with high-gain antennas. The more directive the antenna beam and the greater the spacing between antennas, the greater will be the isolation. When the antenna designer is restricted by the nature of the application, large isolations may not be possible. For example, typical isolations between transmitting and receiving antennas on missiles might be about 50 dB at X band, 70 dB at K band and as low as 20 dB at L band.[9] Metallic baffles, as well as absorbing material, placed between the antennas can provide additional isolation.[10]

It has been reported[11] that the isolation between two X-band horn antennas of 22 dB gain can be increased from a normal value of 70 dB to about 120 dB by separating the two with a smooth surface covered by a sheet of radar-absorbing material and providing screening ridges at the edges of the horns. A common radome enclosing the two antennas should be avoided since it limits the amount of isolation that can be achieved.

The separate antennas of the AN/MPQ-46 CW tracker-illuminator of the Hawk missile system are shown in Fig. 3.3. The Cassegrain receiving antenna is on the right. Both antennas have a "tunnel" around the periphery for further shielding.

Additional isolation can be obtained by properly introducing a controlled sample of the transmitted signal directly into the receiver. The phase and amplitude of this "buck-off" signal are adjusted to cancel the portion of the transmitter signal that leaks into the receiver. An additional 10 dB of isolation might be obtained.[12] The phase and amplitude of the leakage signal, however, can vary as the antenna scans, which results in varying cancellation. Therefore, when additional isolation is necessary, as in the high-power CW tracker-illuminator, a dynamic canceler can be used that senses the proper phase and amplitude required of the nulling signal.[5,13] Dynamic cancelation of the leakage by this type of "feedthrough nulling" can exceed 30 dB.[31]

The transmitter signal is never a pure CW waveform. Minute variations in amplitude (AM) and phase (FM) can result in sideband components that fall within the doppler frequency band. These can generate false targets or mask the desired signals. Therefore both AM and FM modulations can result in undesired sidebands. AM sidebands are typically 120 dB below the carrier, as measured in a 1 kHz band, and are relatively constant across the

Figure 3.3 AN/MPQ-46 CW tracker-illuminator for the Hawk missile system showing separate transmitting and receiving antennas. (*Courtesy Raytheon Co.*).

usual doppler spectrum of interest.[31,69] The normal antenna isolation plus feedthrough nulling usually reduces the AM components below receiver noise in moderate power radars. FM sidebands are usually significantly greater than AM, but decrease with increasing offset from the carrier.[5] The character of FM noise in the leakage signal is also affected by stabilizing the output frequency of the CW transmitter and by active noise-degeneration using a microwave bridge circuit to extract the FM noise components. These are then fed back to the transmitter in such a manner as to reduce the original frequency deviation.[31] It has been said[69] that experience indicates that a satisfactory measurement of AM noise over the doppler frequency band provides assurance that both AM and FM noise generated by the tube are within required limits.

The transmitter noise that enters the radar receiver via backscatter from the clutter is sometimes called *transmitted clutter*.[69] It can appear at the same frequencies as the doppler shifts from moving targets and can mask desired targets or cause spurious responses. This extraneous noise is produced by ion oscillations in the tube (usually a klystron amplifier) rather than by the thermal noise, or noise figure. When the ion oscillations appear, they usually have a magnitude about 40 dB below the carrier, or else they are not measurable.[69] Thus there is no need to specify a measurement of this noise to a level better than 40 dB below the carrier to insure the required noise levels. Since ion oscillations may occur at some combination of tube parameters and not at others, the CW radar tube should be tested for noise-free operation over the expected range of beam voltage, heater voltage, RF drive level, and load VSWR. Noise-free operation also requires well-filtered beam power-supplies and a dc heater supply.

Since ion oscillations require a finite time to develop (tens of microseconds), a pulse-doppler radar with a pulse width of less than 10 μs should not experience this form of noise.

Intermediate-frequency receiver. The receiver of the simple CW radar of Fig. 3.2 is in some respects analogous to a superheterodyne receiver. Receivers of this type are called *homodyne* receivers, or superheterodyne receivers with zero IF.[14] The function of the local oscillator is replaced by the leakage signal from the transmitter. Such a receiver is simpler than one with a more conventional intermediate frequency since no IF amplifier or local oscillator is required. However, the simpler receiver is not as sensitive because of increased noise at the lower intermediate frequencies caused by flicker effect. Flicker-effect noise occurs in semiconductor devices such as diode detectors and cathodes of vacuum tubes. The noise power produced by the flicker effect varies as $1/f^{\alpha}$, where α is approximately unity. This is in contrast to shot noise or thermal noise, which is independent of frequency. Thus, at the lower range of frequencies (audio or video region), where the doppler frequencies usually are found, the detector of the CW receiver can introduce a considerable amount of flicker noise, resulting in reduced receiver sensitivity. For short-range, low-power applications this decrease in sensitivity might be tolerated since it can be compensated by a modest increase in antenna aperture and/or additional transmitter power. But for maximum efficiency with CW radar, the reduction in sensitivity caused by the simple doppler receiver with zero IF cannot be tolerated.

The effects of flicker noise are overcome in the normal superheterodyne receiver by using an intermediate frequency high enough to render the flicker noise small compared with the normal receiver noise. This results from the inverse frequency dependence of flicker noise. Figure 3.4 shows a block diagram of the CW radar whose receiver operates with a nonzero IF. Separate antennas are shown for transmission and reception. Instead of the usual local oscillator found in the conventional superheterodyne receiver, the local oscillator (or reference signal) is derived in this receiver from a portion of the transmitted signal mixed with a locally generated signal of frequency equal to that of the receiver IF. Since the output of the mixer

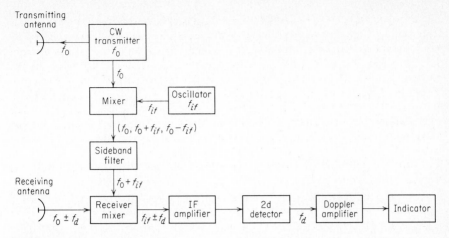

Figure 3.4 Block diagram of CW doppler radar with nonzero IF receiver, sometimes called *sideband superheterodyne.*

consists of two sidebands on either side of the carrier plus higher harmonics, a narrowband filter selects one of the sidebands as the reference signal. The improvement in receiver sensitivity with an intermediate-frequency superheterodyne might be as much as 30 dB over the simple receiver of Fig. 3.2.

Receiver bandwidth. One of the requirements of the doppler-frequency amplifier in the simple CW radar (Fig. 3.2) or the IF amplifier of the sideband superheterodyne (Fig. 3.4) is that it be wide enough to pass the expected range of doppler frequencies. In most cases of practical interest the expected range of doppler frequencies will be much wider than the frequency spectrum occupied by the signal energy. Consequently, the use of a wideband amplifier covering the expected doppler range will result in an increase in noise and a lowering of the receiver sensitivity. If the frequency of the doppler-shifted echo signal were known beforehand, a narrowband filter—one just wide enough to reduce the excess noise without eliminating a significant amount of signal energy—might be used. If the waveform of the echo signal were known, as well as its carrier frequency, the matched filter could be specified as outlined in Sec. 10.2.

Several factors tend to spread the CW signal energy over a finite frequency band. These must be known if an approximation to the bandwidth required for the narrowband doppler filter is to be obtained.

If the received waveform were a sine wave of infinite duration, its frequency spectrum would be a delta function (Fig. 3.5a) and the receiver bandwidth would be infinitesimal. But a sine wave of infinite duration and an infinitesimal bandwidth cannot occur in nature. The more normal situation is an echo signal which is a sine wave of finite rather than infinite duration. The frequency spectrum of a finite-duration sine wave has a shape of the form $[\sin \pi(f - f_0)\delta]/\pi(f - f_0)$, where f_0 and δ are the frequency and duration of the sine wave, respectively, and f is the frequency variable over which the spectrum is plotted (Fig. 3.5b). Practical receivers can only approximate this characteristic. (Note that this is the same as the spectrum of a pulse of sine wave, the only difference being the relative value of the duration δ.) In many instances, the echo is not a pure sine wave of finite duration but is perturbed by fluctuations in cross section, target accelerations, scanning fluctuations, etc., which tend to

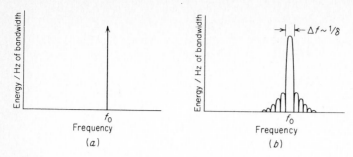

Figure 3.5 Frequency spectrum of CW oscillation of (*a*) infinite duration and (*b*) finite duration.

broaden the bandwidth still further. Some of these spectrum-broadening effects are considered below.

Assume a CW radar with an antenna beamwidth of θ_B deg scanning at the rate of $\dot\theta_s$ deg/s. The time on target (duration of the received signal) is $\delta = \theta_B/\dot\theta_s$ s. Thus the signal is of finite duration and the bandwidth of the receiver must be of the order of the reciprocal of the time on target $\dot\theta_s/\theta_B$. Although this is not an exact relation, it is a good enough approximation for purposes of the present discussion. If the antenna beamwidth were 2° and if the scanning rate were 36°/s (6 rpm), the spread in the spectrum of the received signal due to the finite time on target would be equal to 18 Hz, independent of the transmitted frequency.

In addition to the spread of the received signal spectrum caused by the finite time on target, the spectrum may be further widened if the target cross section fluctuates. The fluctuations widen the spectrum by modulating the echo signal. In a particular case, it has been reported[12] that the aircraft cross section can change by 15 dB for a change in target aspect of as little as $\frac{1}{3}°$. The echo signal from a propeller-driven aircraft can also contain modulation components at a frequency proportional to the propeller rotation.[15] The spectrum produced by propeller modulations is more like that produced by a sine-wave signal and its harmonics rather than a broad, white-noise spectrum. The frequency range of propeller modulation depends upon the shaft-rotation speed and the number of propeller blades. It is usually in the vicinity of 50 to 60 Hz for World War II aircraft engines. This could be a potential source of difficulty in a CW radar since it might mask the target's doppler signal or it might cause an erroneous measurement of doppler frequency. In some instances, propeller modulation can be of advantage. It might permit the detection of propeller-driven aircraft passing on a tangential trajectory, even though the doppler frequency shift is zero. The rotating blades of a helicopter and the compressor stages of a jet engine can also result in a modulation of the echo and a widening of the spectrum that can degrade the performance of CW doppler radar.

If the target's relative velocity is not constant, a further widening of the received signal spectrum can occur. If a_r is the acceleration of the target with respect to the radar, the signal will occupy a bandwidth

$$\Delta f_d = \left(\frac{2a_r}{\lambda}\right)^{1/2} \tag{3.3}$$

If, for example, a_r is twice the acceleration of gravity, the receiver bandwidth must be approximately 20 Hz when the radar's wavelength is 10 cm.

When the doppler-shifted echo signal is known to lie somewhere within a relatively wide band of frequencies, a bank of narrowband filters (Fig. 3.6) spaced throughout the frequency range permits a measurement of frequency and improves the signal-to-noise ratio. The

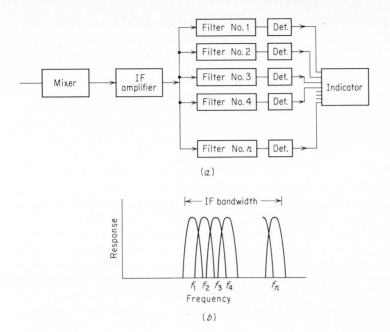

Figure 3.6 (*a*) Block diagram of IF doppler filter bank; (*b*) frequency-response characteristic of doppler filter bank.

bandwidth of each individual filter is wide enough to accept the signal energy, but not so wide as to introduce more noise than need be. The center frequencies of the filters are staggered to cover the entire range of doppler frequencies. If the filters are spaced with their half-power points overlapped, the maximum reduction in signal-to-noise ratio of a signal which lies midway between adjacent channels compared with the signal-to-noise ratio at midband is 3 dB. The more filters used to cover the band, the less will be the maximum loss experienced, but the greater the probability of false alarm.

A bank of narrowband filters may be used after the detector in the video of the simple CW radar of Fig. 3.2 instead of in the IF. The improvement in signal-to-noise ratio with a video filter bank is not as good as can be obtained with an IF filter bank, but the ability to measure the magnitude of doppler frequency is still preserved. Because of foldover, a frequency which lies to one side of the IF carrier appears, after detection, at the same video frequency as one which lies an equal amount on the other side of the IF. Therefore the sign of the doppler shift is lost with a video filter bank, and it cannot be directly determined whether the doppler frequency corresponds to an approaching or to a receding target. (The sign of the doppler may be determined in the video by other means, as described later.) One advantage of the foldover in the video is that only half the number of filters are required than in the IF filter bank. The equivalent of a bank of contiguous bandpass filters may also be obtained by converting the analog IF or video signal to a set of sampled, quantized signals which are processed with digital circuitry by means of the fast Fourier transform algorithm.[16]

A bank of overlapping doppler filters, whether in the IF or video, increases the complexity of the receiver. When the system requirements permit a time sharing of the doppler frequency range, the bank of doppler filters may be replaced by a single narrowband tunable filter which searches in frequency over the band of expected doppler frequencies until a signal is found.

After detecting and recognizing the signal, the filter may be programmed to continue its search in frequency for additional signals. One of the techniques for accomplishing this is similar to the tracking speed-gate mentioned in Sec. 5.8, the phase-locked filter,[17] or the phase-locked loop.

If, in any of the above techniques, moving targets are to be distinguished from stationary objects, the zero-doppler-frequency component must be removed. The zero-doppler-frequency component has, in practice, a finite bandwidth due to the finite time on target, clutter fluctuations, and equipment instabilities. The clutter-rejection band of the doppler filter must be wide enough to accommodate this spread. In the multiple-filter bank, removal of those filters in the vicinity of the RF or IF carrier removes the stationary-target signals.

Sign of the radial velocity. In some applications of CW radar it is of interest to know whether the target is approaching or receding. This might be determined with separate filters located on either side of the intermediate frequency. If the echo-signal frequency lies below the carrier, the target is receding; if the echo frequency is greater than the carrier, the target is approaching (Fig. 3.7).

Although the doppler-frequency spectrum "folds over" in the video because of the action of the detector, it is possible to determine its sign from a technique borrowed from single-sideband communications. If the transmitter signal is given by

$$E_t = E_0 \cos \omega_0 t \tag{3.4}$$

the echo signal from a moving target will be

$$E_r = k_1 E_0 \cos \left[(\omega_0 \pm \omega_d)t + \phi \right] \tag{3.5}$$

where E_0 = amplitude of transmitter signal
k_1 = a constant determined from the radar equation
ω_0 = angular frequency of transmitter, rad/s
ω_d = dopper angular frequency shift
ϕ = a constant phase shift, which depends upon range of initial detection

The sign of the doppler frequency, and therefore the direction of target motion, may be found by splitting the received signal into two channels as shown in Fig. 3.8. In channel A the signal is processed as in the simple CW radar of Fig. 3.2. The received signal and a portion of the transmitter heterodyne in the detector (mixer) to yield a difference signal

$$E_A = k_2 E_0 \cos \left(\pm \omega_d t + \phi \right) \tag{3.6}$$

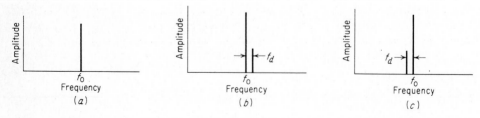

Figure 3.7 Spectra of received signals. (*a*) No doppler shift, no relative target motion; (*b*) approaching target; (*c*) receding target.

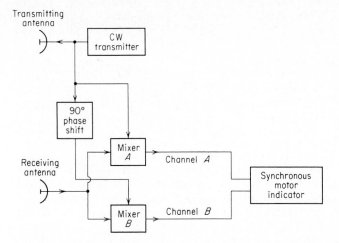

Figure 3.8 Measurement of doppler direction using synchronous, two-phase motor.

The other channel is similar, except for a 90° phase delay introduced in the reference signal. The output of the channel B mixer is

$$E_B = k_2 E_0 \cos \left(\pm \omega_d t + \phi + \frac{\pi}{2} \right) \tag{3.7}$$

If the target is approaching (positive doppler), the outputs from the two channels are

$$E_A(+) = k_2 E_0 \cos (\omega_d t + \phi) \qquad E_B(+) = k_2 E_0 \cos \left(\omega_d t + \phi + \frac{\pi}{2} \right) \tag{3.8a}$$

On the other hand, if the targets are receding (negative doppler),

$$E_A(-) = k_2 E_0 \cos (\omega_d t - \phi) \qquad E_B(-) = k_2 E_0 \cos \left(\omega_d t - \phi - \frac{\pi}{2} \right) \tag{3.8b}$$

The sign of ω_d and the direction of the target's motion may be determined according to whether the output of channel B leads or lags the output of channel A. One method of determining the relative phase relationship between the two channels is to apply the outputs to a synchronous two-phase motor.[18] The direction of motor rotation is an indication of the direction of the target motion.

Electronic methods may be used instead of a synchronous motor to sense the relative phase of the two channels. One application of this technique has been described for a rate-of-climb meter for vertical take-off aircraft to determine the velocity of the aircraft with respect to the ground during take-off and landing.[19] It has also been applied to the detection of moving targets in the presence of heavy foliage,[20] as discussed in Sec. 13.6.

The doppler frequency shift. The expression for the doppler frequency shift given previously by Eq. (3.2) is an approximation that is valid for most radar applications. The correct expression for the frequency f^* from a target moving with a relative velocity v, when the frequency f is transmitted is[21-23,70]

$$f^* = f \frac{(1 + v/c)}{(1 - v/c)} \tag{3.9}$$

where c is the velocity of propagation. When, as is usually the case, $v \ll c$, Eq. (3.9) reduces to the classical form of the doppler frequency shift. The phase shift associated with the return signal is $(4\pi f R_0 /c)/(1 - v/c)$, where R_0 is the range at time $t = 0$.

Applications of CW radar.[24,25] The chief use of the simple, unmodulated CW radar is for the measurement of the relative velocity of a moving target, as in the police speed monitor or in the previously mentioned rate-of-climb meter for vertical-take-off aircraft. In support of automobile traffic, CW radar has been suggested for the control of traffic lights, regulation of toll booths, vehicle counting, as a replacement for the "fifth-wheel" speedometer in vehicle testing, as a sensor in antilock braking systems, and for collision avoidance. For railways, CW radar can be used as a speedometer to replace the conventional axle-driven tachometer. In such an application it would be unaffected by errors caused by wheelslip on accelerating or wheelslide when braking. It has been used for the measurement of railroad-freight-car velocity during humping operations in marshalling yards, and as a detection device to give track maintenance personnel advance warning of approaching trains. CW radar is also employed for monitoring the docking speed of large ships. It has also seen application for intruder alarms and for the measurement of the velocity of missiles, ammunition, and baseballs.

The principal advantage of a CW doppler radar over other (nonradar) methods of measuring speed is that there need not be any physical contact with the object whose speed is being measured. In industry this has been applied to the measurement of turbine-blade vibration, the peripheral speed of grinding wheels, and the monitoring of vibrations in the cables of suspension bridges.

Most of the above applications can be satisfied with a simple, solid-state CW source with powers in the tens of milliwatts. High-power CW radars for the detection of aircraft and other targets have been developed and have been used in such systems as the Hawk missile systems (Fig. 3.3). However, the difficulty of eliminating the leakage of the transmitter signal into the receiver has limited the utility of unmodulated CW radar for many long-range applications. A notable exception is the Space Surveillance System (Spasur) for the detection of satellites.[26,27] The CW transmitter of Spasur at 216 MHz radiates a power of up to one megawatt from an antenna almost two miles long to produce a narrow, vertically looking fan beam. The receiver is separated from the transmitter by a distance of several hundred miles. Each receiver site consists of an interferometer antenna to obtain an angle measurement in the plane of the fan beam. There are three sets of transmitter-receiver stations to provide fence coverage of the southern United States.

The CW radar, when used for short or moderate ranges, is characterized by simpler equipment than a pulse radar. The amount of power that can be used with a CW radar is dependent on the isolation that can be achieved between the transmitter and receiver since the transmitter noise that finds its way into the receiver limits the receiver sensitivity. (The pulse radar has no similar limitation to its maximum range because the transmitter is not operative when the receiver is turned on.)

Perhaps one of the greatest shortcomings of the simple CW radar is its inability to obtain a measurement of range. This limitation can be overcome by modulating the CW carrier, as in the frequency-modulated radar described in the next section.

Some anti-air-warfare guided missile systems employ semiactive homing guidance in which a receiver in the missile receives energy from the target, the energy having been transmitted from an "illuminator" external to the missile. The illuminator, for example, might be at the launch platform. CW illumination has been used in many successful systems. An example is the Hawk tracking illuminator shown in Fig. 3.3. It is a tracking radar as well as an illuminator since it must be able to follow the target as it travels through space. The doppler

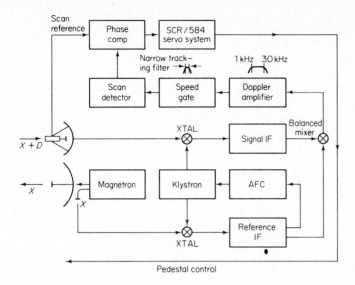

Figure 3.9 Block diagram of a CW tracking-illuminator.[31] (*Courtesy IEEE.*)

discrimination of a CW radar allows operation in the presence of clutter and has been well suited for low altitude missile defense systems. A block diagram of a CW tracking illuminator is shown in Fig. 3.9.[31] Note that following the wide-band doppler amplifier is a *speed gate*, which is a narrow-band tracking filter that acquires the target's doppler and tracks its changing doppler frequency shift.

3.3 FREQUENCY-MODULATED CW RADAR

The inability of the simple CW radar to measure range is related to the relatively narrow spectrum (bandwidth) of its transmitted waveform. Some sort of timing mark must be applied to a CW carrier if range is to be measured. The timing mark permits the time of transmission and the time of return to be recognized. The sharper or more distinct the mark, the more accurate the measurement of the transit time. But the more distinct the timing mark, the broader will be the transmitted spectrum. This follows from the properties of the Fourier transform. Therefore a finite spectrum must of necessity be transmitted if transit time or range is to be measured.

The spectrum of a CW transmission can be broadened by the application of modulation, either amplitude, frequency, or phase. An example of an amplitude modulation is the pulse radar. The narrower the pulse, the more accurate the measurement of range and the broader the transmitted spectrum. A widely used technique to broaden the spectrum of CW radar is to frequency-modulate the carrier. The timing mark is the changing frequency. The transit time is proportional to the difference in frequency between the echo signal and the transmitter signal. The greater the transmitter frequency deviation in a given time interval, the more accurate the measurement of the transit time and the greater will be the transmitted spectrum.

Range and doppler measurement. In the frequency-modulated CW radar (abbreviated FM-CW), the transmitter frequency is changed as a function of time in a known manner. Assume that the transmitter frequency increases linearly with time, as shown by the solid line in Fig. 3.10a. If there is a reflecting object at a distance R, an echo signal will return after a time $T = 2R/c$. The dashed line in the figure represents the echo signal. If the echo signal is heterodyned with a portion of the transmitter signal in a nonlinear element such as a diode, a beat note f_b will be produced. If there is no doppler frequency shift, the beat note (difference frequency) is a measure of the target's range and $f_b = f_r$, where f_r is the beat frequency due only to the target's range. If the rate of change of the carrier frequency is $\dot{f}_0$, the beat frequency is

$$f_r = \dot{f}_0 T = \frac{2R}{c} \dot{f}_0 \tag{3.10}$$

In any practical CW radar, the frequency cannot be continually changed in one direction only. Periodicity in the modulation is necessary, as in the triangular frequency-modulation waveform shown in Fig. 3.10b. The modulation need not necessarily be triangular; it can be sawtooth, sinusoidal, or some other shape. The resulting beat frequency as a function of time is shown in Fig. 3.10c for triangular modulation. The beat note is of constant frequency except at the turn-around region. If the frequency is modulated at a rate f_m over a range Δf, the beat frequency is

$$f_r = \frac{2R}{c} 2f_m \Delta f = \frac{4Rf_m \, \Delta f}{c} \tag{3.11}$$

Thus the measurement of the beat frequency determines the range R.

A block diagram illustrating the principle of the FM-CW radar is shown in Fig. 3.11. A portion of the transmitter signal acts as the reference signal required to produce the beat frequency. It is introduced directly into the receiver via a cable or other direct connection.

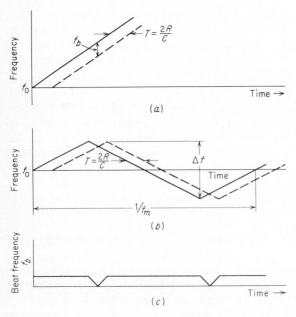

Figure 3.10 Frequency-time relationships in FM-CW radar. Solid curve represents transmitted signal; dashed curve represents echo. (*a*) Linear frequency modulation; (*b*) triangular frequency modulation; (*c*) beat note of (*b*).

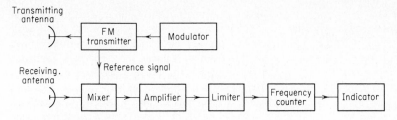

Figure 3.11 Block diagram of FM-CW radar.

Ideally, the isolation between transmitting and receiving antennas is made sufficiently large so as to reduce to a negligible level the transmitter leakage signal which arrives at the receiver via the coupling between antennas. The beat frequency is amplified and limited to remove any amplitude fluctuations. The frequency of the amplitude-limited beat note is measured with a cycle-counting frequency meter calibrated in distance.

In the above, the target was assumed to be stationary. If this assumption is not applicable, a doppler frequency shift will be superimposed on the FM range beat note and an erroneous range measurement results. The doppler frequency shift causes the frequency-time plot of the echo signal to be shifted up or down (Fig. 3.12a). On one portion of the frequency-modulation cycle, the beat frequency (Fig. 3.12b) is increased by the doppler shift, while on the other portion, it is decreased. If, for example, the target is approaching the radar, the beat frequency $f_b(\text{up})$ produced during the increasing, or up, portion of the FM cycle will be the difference between the beat frequency due to the range f_r and the doppler frequency shift f_d [Eq. (3.12a)]. Similarly, on the decreasing portion, the beat frequency $f_b(\text{down})$ is the sum of the two [Eq. (3.12b)].

$$f_b(\text{up}) = f_r - f_d \qquad (3.12a)$$

$$f_b(\text{down}) = f_r + f_d \qquad (3.12b)$$

The range frequency f_r may be extracted by measuring the average beat frequency; that is, $\frac{1}{2}[f_b(\text{up}) + f_b(\text{down})] = f_r$. If $f_b(\text{up})$ and $f_b(\text{down})$ are measured separately, for example, by switching a frequency counter every half modulation cycle, one-half the difference between the frequencies will yield the doppler frequency. This assumes $f_r > f_d$. If, on the other hand, $f_r < f_d$, such as might occur with a high-speed target at short range, the roles of the averaging and the difference-frequency measurements are reversed; the averaging meter will measure doppler velocity, and the difference meter, range. If it is not known that the roles of the meters are

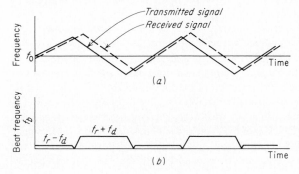

(a)

(b)

Figure 3.12 Frequency-time relationships in FM-CW radar when the received signal is shifted in frequency by the doppler effect (a) Transmitted (solid curve) and echo (dashed curve) frequencies; (b) beat frequency.

reversed because of a change in the inequality sign between f_r and f_d, an incorrect interpretation of the measurements may result.

When more than one target is present within the view of the radar, the mixer output will contain more than one difference frequency. If the system is linear, there will be a frequency component corresponding to each target. In principle, the range to each target may be determined by measuring the individual frequency components and applying Eq. (3.11) to each. To measure the individual frequencies, they must be separated from one another. This might be accomplished with a bank of narrowband filters, or alternatively, a single frequency corresponding to a single target may be singled out and continuously observed with a narrowband tunable filter. But if the motion of the targets were to produce a doppler frequency shift, or if the frequency-modulation waveform were nonlinear, or if the mixer were not operated in its linear region, the problem of resolving targets and measuring the range of each becomes more complicated.

If the FM-CW radar is used for single targets only, such as in the radio altimeter, it is not necessary to employ a linear modulation waveform. This is certainly advantageous since a sinusoidal or almost sinusoidal frequency modulation is easier to obtain with practical equipments than are linear modulations. The beat frequency obtained with sinusoidal modulation is not constant over the modulation cycle as it is with linear modulation. However, it may be shown that the *average* beat frequency measured over a modulation cycle, when substituted into Eq. (3.11) yields the correct value of target range. Any reasonable-shape modulation waveform can be used to measure the range, provided the average beat frequency is measured.[28,29] If the target is in motion and the beat signal contains a component due to the doppler frequency shift, the range frequency can be extracted, as before, if the average frequency is measured. To extract the doppler frequency, the modulation waveform must have equal upsweep and downsweep time intervals.

The FM-CW radar principle was known and used at about the same time as pulse radar, although the early development of these two radar techniques seemed to be relatively independent of each other. FM-CW was applied to the measurement of the height of the ionosphere in the 1920s[32] and as an aircraft altimeter in the 1930s.[33]

FM-CW altimeter. The FM-CW radar principle is used in the aircraft radio altimeter to measure height above the surface of the earth. The large backscatter cross section and the relatively short ranges required of altimeters permit low transmitter power and low antenna gain. Since the relative motion between the aircraft and ground is small, the effect of the doppler frequency shift may usually be neglected.

The band from 4.2 to 4.4 GHz is reserved for radio altimeters, although they have in the past operated at UHF. The transmitter power is relatively low and can be obtained from a CW magnetron, a backward-wave oscillator, or a reflex klystron, but these have been replaced by the solid state transmitter.

The altimeter can employ a simple homodyne[71] receiver, but for better sensitivity and stability the superheterodyne is to be prefered whenever its more complex construction can be tolerated. A block diagram of the FM-CW radar with a sideband superheterodyne receiver is shown in Fig. 3.13. A portion of the frequency-modulated transmitted signal is applied to a mixer along with the oscillator signal. The selection of the local-oscillator frequency is a bit different from that in the usual superheterodyne receiver. The local-oscillator frequency f_{IF} should be the same as the intermediate frequency used in the receiver, whereas in the conventional superheterodyne the LO frequency is of the same order of magnitude as the RF signal. The output of the mixer consists of the varying transmitter frequency $f_0(t)$ plus two sideband frequencies, one on either side of $f_0(t)$ and separated from $f_0(t)$ by the local-oscillator frequency

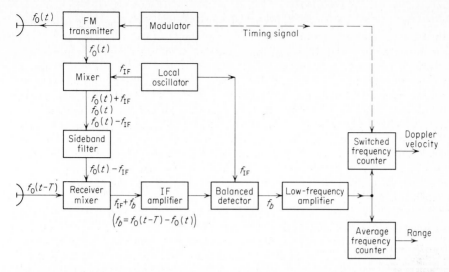

Figure 3.13 Block diagram of FM-CW radar using sideband superheterodyne receiver.

f_{IF}. The filter selects the lower sideband $f_0(t) - f_{IF}$ and rejects the carrier and the upper sideband. The sideband that is passed by the filter is modulated in the same fashion as the transmitted signal. The sideband filter must have sufficient bandwidth to pass the modulation, but not the carrier or other sideband. The filtered sideband serves the function of the local oscillator.

When an echo signal is present, the output of the receiver mixer is an IF signal of frequency $f_{IF} + f_b$, where f_b is composed of the range frequency f_r and the doppler velocity frequency f_d. The IF signal is amplified and applied to the balanced detector along with the local-oscillator signal f_{IF}. The output of the detector contains the beat frequency (range frequency and the doppler velocity frequency), which is amplified to a level where it can actuate the frequency-measuring circuits.

In Fig. 3.13, the output of the low-frequency amplifier is divided into two channels: one feeds an average-frequency counter to determine range, the other feeds a switched frequency counter to determine the doppler velocity (assuming $f_r > f_d$). Only the averaging frequency counter need be used in an altimeter application, since the rate of change of altitude is usually small.

A target at short range will generally result in a strong signal at low frequency, while one at long range will result in a weak signal at high frequency. Therefore the frequency characteristic of the low-frequency amplifier in the FM-CW radar may be shaped to provide attenuation at the low frequencies corresponding to short ranges and large echo signals. Less attentuation is applied to the higher frequencies, where the echo signals are weaker.

The echo signal from an isolated target varies inversely as the fourth power of the range, as is well known from the radar equation. With this as a criterion, the gain of the low-frequency amplifier should be made to increase at the rate of 12 dB/octave. The output of the amplifier would then be independent of the range, for constant target cross section. Amplifier response shaping is similar in function to sensitivity time control (STC) employed in conventional pulse radar. However, in the altimeter, the echo signal from an extended target such as the ground varies inversely as the square (rather than the fourth power) of the range, since the greater the range, the greater the echo area illuminated by the beam. For extended targets, therefore, the low-frequency amplifier gain should increase 6 dB/octave. A compromise between the isolated (12-dB slope) and extended (6-dB slope) target echoes might be a characteristic with a slope of 9 dB/octave. The constant output produced by shaping the doppler-amplifier frequency-response characteristic is not only helpful in lowering the dynamic range requirements of the frequency-measuring device, but the attenuation of the low frequencies effects a reduction of low-frequency interfering noise. Lowered gain at low altitudes also helps to reduce interference from unwanted reflections. The response at the upper end of the frequency characteristic is rapidly reduced for frequencies beyond that corresponding to maximum range. If there is a minimum target range, the response is also cut off at the low-frequency end, to further reduce the extraneous noise entering the receiver.

Another method of processing the range or height information from an altimeter so as to reduce the noise output from the receiver and improve the sensitivity uses a narrow-bandwidth low-frequency amplifier with a feedback loop to maintain the beat frequency constant.[30,34] When a fixed-frequency excursion (or deviation) is used, as in the usual altimeter, the beat frequency can vary over a considerable range of values. The low-frequency-amplifier bandwidth must be sufficiently wide to encompass the expected range of beat frequencies. Since the bandwidth is broader than need be to pass the signal energy, the signal-to-noise ratio is reduced and the receiver sensitivity degraded. Instead of maintaining the frequency excursion Δf constant and obtaining a varying beat frequency, Δf can be varied to maintain the beat frequency constant. The beat-frequency amplifier need only be wide enough to pass the received signal energy, thus reducing the amount of noise with which the signal must compete. The frequency excursion is maintained by a servomechanism to that value which permits the beat frequency to fall within the passband of the narrow filter. The value of the frequency excursion is then a measure of the altitude and may be substituted into Eq. (3.11).

When used in the FM altimeter, the technique of servo-controlling the frequency excursion is usually applied at all altitudes above a predetermined minimum. Since the frequency excursion Δf is inversely proportional to range, the radar is better operated at very low altitudes in the more normal manner with a fixed Δf, and hence a varying beat frequency.

Measurement errors. The absolute accuracy of radar altimeters is usually of more importance at low altitudes than at high altitudes. Errors of a few meters might not be of significance when cruising at altitudes of 10 km, but are important if the altimeter is part of a blind landing system.

The theoretical accuracy with which distance can be measured depends upon the bandwidth of the transmitted signal and the ratio of signal energy to noise energy. In addition, measurement accuracy might be limited by such practical restrictions as the accuracy of the frequency-measuring device, the residual path-length error caused by the circuits and transmission lines, errors caused by multiple reflections and transmitter leakage, and the frequency error due to the turn-around of the frequency modulation.

A common form of frequency-measuring device is the cycle counter, which measures the number of cycles or half cycles of the beat during the modulation period. The total cycle count is a discrete number since the counter is unable to measure fractions of a cycle. The discreteness of the frequency measurement gives rise to an error called the *fixed* error, or *step* error. It has also been called the *quantization* error, a more descriptive name. The average number of cycles N of the beat frequency f_b in one period of the modulation cycle f_m is $\bar{f}_b/f_m$, where the bar over f_b denotes time average. Equation (3.11) may be written as

$$R = \frac{cN}{4\,\Delta f} \tag{3.13}$$

where R = range (altitude), m
$\quad c$ = velocity of propagation, m/s
$\quad \Delta f$ = frequency excursion, Hz

Since the output of the frequency counter N is an integer, the range will be an integral multiple of $c/(4\ \Delta f)$ and will give rise to a quantization error equal to

$$\delta R = \frac{c}{4\ \Delta f} \qquad (3.14a)$$

or

$$\delta R\ (\text{m}) = \frac{75}{\Delta f\ (\text{MHz})} \qquad (3.14b)$$

Note that the fixed error is independent of the range and carrier frequency and is a function of the frequency excursion only. Large frequency excursions are necessary if the fixed error is to be small.

Since the fixed error is due to the discrete nature of the frequency counter, its effects can be reduced by wobbling the modulation frequency or the phase of the transmitter output. Wobbling the transmitter phase results in a wobbling of the phase of the beat signal so that an average reading of the cycle counter somewhere between N and $N + 1$ will be obtained on a normal meter movement. In one altimeter,[30] the modulation frequency was varied at a 10-Hz rate, causing the phase shift of the beat signal to vary cyclically with time. The indicating system was designed so that it did not respond to the 10-Hz modulation directly, but it caused the fixed error to be averaged. Normal fluctuations in aircraft altitude due to uneven terrain, waves on the water, or turbulent air can also average out the fixed error provided the time constant of the indicating device is large compared with the time between fluctuations. Over smooth terrain, such as an airport runway, the fixed error might not be averaged out. Note that even if the fixed error were not present, the accuracy with which the height can be measured will depend on the signal-to-noise ratio, as discussed in Sec. 11.3, and can be comparable to the fixed error as given by Eq. (3.14).

Other errors might be introduced in the CW radar if there are uncontrolled variations in the transmitter frequency, modulation frequency, or frequency excursion. Target motion can cause an error in range equal to $v_r T_0$, where v_r is the relative velocity and T_0 is the observation time. At short ranges the residual path error can also result in a significant error unless compensated for. The residual path error is the error caused by delays in the circuitry and transmission lines. Multipath signals also produce error. Figure 3.14 shows some of the unwanted signals that might occur in the FM altimeter.[36,37] The wanted signal is shown by the solid line, while the unwanted signals are shown by the broken arrows. The unwanted signals include:

1. The reflection of the transmitted signals at the antenna caused by impedance mismatch.
2. The standing-wave pattern on the cable feeding the reference signal to the receiver, due to poor mixer match.
3. The leakage signal entering the receiver via coupling between transmitter and receiver antennas. This can limit the ultimate receiver sensitivity, especially at high altitudes.
4. The interference due to power being reflected back to the transmitter, causing a change in the impedance seen by the transmitter. This is usually important only at low altitudes. It can be reduced by an attenuator introduced in the transmission line at low altitude or by a directional coupler or an isolator.
5. The double-bounce signal.

Reflections from the landing gear can also cause errors.

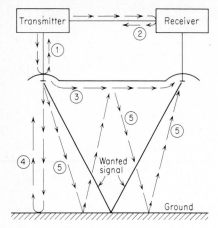

not a problem in practice

Figure 3.14 Unwanted signals in FM altimeter. (*From Capelli,[36] IRE Trans.*)

Transmitter leakage. The sensitivity of FM-CW radar is limited by the noise accompanying the transmitter signal which leaks into the receiver. Although advances have been made in reducing the AM and FM noise generated by high-power CW transmitters, the noise is usually of sufficient magnitude compared with the echo signal to require some means of minimizing the leakage that finds its way into the receiver. The techniques described previously for reducing leakage in the CW radar apply equally well to the FM-CW radar. Separate antennas and direct cancellation of the leakage signal are two techniques which give considerable isolation.

Filter bank can work on Bessel fcn

Sinusoidal modulation. The ability of the FM-CW radar to measure range provides an additional basis for obtaining isolation. Echoes from short-range targets—including the leakage signal—may be attenuated relative to the desired target echo from longer ranges by properly processing the difference-frequency signal obtained by heterodyning the transmitted and received signals.

If the CW carrier is frequency-modulated by a sine wave, the difference frequency obtained by heterodyning the returned signal with a portion of the transmitter signal may be expanded in a trigonometric series whose terms are the harmonics of the modulating frequency f_m.[9.28] Assume the form of the transmitted signal to be

$$\sin\left(2\pi f_0 t + \frac{\Delta f}{2f_m}\sin 2\pi f_m t\right) \tag{3.15}$$

where f_0 = carrier frequency
$\quad f_m$ = modulation frequency
$\quad \Delta f$ = frequency excursion (equal to twice the frequency derivation)

The difference frequency signal may be written

$$
\begin{aligned}
v_D = {} & J_0(D)\cos(2\pi f_d t - \phi_0) + 2J_1(D)\sin(2\pi f_d t - \phi_0)\cos(2\pi f_m t - \phi_m)\\
& - 2J_2(D)\cos(2\pi f_d t - \phi_0)\cos 2(2\pi f_m t - \phi_m)\\
& - 2J_3(D)\sin(2\pi f_d t - \phi_0)\cos 3(2\pi f_m t - \phi_m)\\
& + 2J_4(D)\cos(2\pi f_d t - \phi_0)\cos 4(2\pi f_m t - \phi_m) + 2J_5(D)\cdots
\end{aligned}
\tag{3.16}
$$

where J_0, J_1, J_2, etc = Bessel functions of first kind and order 0, 1, 2, etc., respectively

$D = (\Delta f/f_m) \sin 2\pi f_m R_0/c$

R_0 = distance to target at time $t = 0$ (distance that would have been measured if target were stationary)

c = velocity of propagation

$f_d = 2v_r f_0/c$ = doppler frequency shift

v_r = relative velocity of target with respect to radar

ϕ_0 = phase shift approximately equal to angular distance $4\pi f_0 R_0/c$

ϕ_m = phase shift approximately equal to $2\pi f_m R_0/c$

The difference-frequency signal of Eq. (3.16) consists of a doppler-frequency component of amplitude $J_0(D)$ and a series of cosine waves of frequency f_m, $2f_m$, $3f_m$, etc. Each of these harmonics of f_m is modulated by a doppler-frequency component with amplitude proportional to $J_n(D)$. The product of the doppler-frequency factor times the nth harmonic factor is equivalent to a suppressed-carrier double-sideband modulation (Fig. 3.15).

In principle, any of the J_n components of the difference-frequency signal can be extracted in the FM-CW radar. Consider first the d-c term $J_0(D) \cos (2\pi f_d t - \phi_0)$. This is a cosine wave at the doppler frequency with an amplitude proportional to $J_0(D)$. Figure 3.16 shows a plot of several of the Bessel functions. The argument D of the Bessel function is proportional to range. The J_0 amplitude applies maximum response to signals at zero range in a radar that extracts the d-c doppler-frequency component. This is the range at which the leakage signal and its noise components (including microphony and vibration) are found. At greater ranges, where the target is expected, the effect of the J_0 Bessel function is to reduce the echo-signal amplitude in comparison with the echo at zero range (in addition to the normal range attenuation). Therefore, if the J_0 term were used, it would enhance the leakage signal and reduce the target signal, a condition opposite to that desired.

An examination of the Bessel functions (Fig. 3.16) shows that if one of the modulation-frequency harmonics is extracted (such as the first, second, or third harmonic), the amplitude of the leakage signal at zero range may theoretically be made equal to zero. The higher the number of the harmonic, the higher will be the order of the Bessel function and the less will be the amount of microphonism-leakage feedthrough. This results from the property that $J_n(x)$ behaves as x^n for small x. Although higher-order Bessel functions may reduce the zero-range response, they may also reduce the response at the desired target range if the target happens to fall at or near a range corresponding to a zero of the Bessel function. When only a single target is involved, the frequency excursion Δf can be adjusted to obtain that value of D which places the maximum of the Bessel function at the target range.

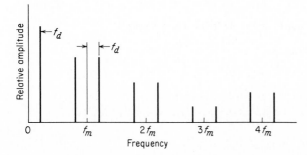

Figure 3.15 Spectrum of the difference-frequency signal obtained from an FM-CW radar sinusoidally modulated at a frequency f_m when the target motion produces a doppler frequency shift f_d. (*After Saunders,*[9] *IRE Trans.*)

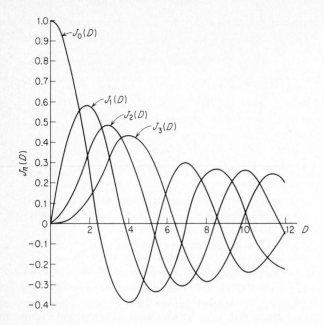

Figure 3.16 Plot of Bessel functions of order 0, 1, 2, and 3; $D = (\Delta f / f_m) \sin 2\pi f_m R_0/c$.

The technique of using higher-order Bessel functions has been applied to the type of doppler-navigation radar discussed in the next section. A block diagram of a CW radar using the third harmonic (J_3 term) is shown in Fig. 3.17. The transmitter is sinusoidally frequency-modulated at a frequency f_m to generate the waveform given by Eq. (3.15). The doppler-shifted echo is heterodyned with the transmitted signal to produce the beat-frequency signal of Eq. (3.16). One of the harmonics of f_m is selected (in this case the third) by a filter centered at the harmonic. The filter bandwidth is wide enough to pass both doppler-frequency sidebands. The filter output is mixed with the (third) harmonic of f_m. The doppler frequency is extracted by the low-pass filter.

Since the total energy contained in the beat-frequency signal is distributed among all the harmonics, extracting but one component wastes signal energy contained in the other harmonics and results in a loss of signal as compared with an ideal CW radar. However, the signal-to-noise ratio is generally superior in the FM radar designed to operate with the nth

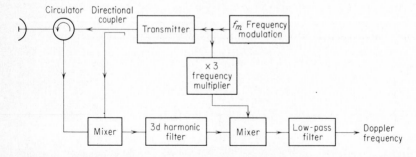

Figure 3.17 Sinusoidally modulated FM-CW radar extracting the third harmonic (J_3 Bessel component).

harmonic as compared with a practical CW radar because the transmitter leakage noise is suppressed by the nth-order Bessel function. The loss in signal energy when operating with the J_3 Bessel component is reported[9,35,54] to be from 4 to 12 dB. Although two separate transmitting and receiving antennas may be used, it is not necessary in many applications. A single antenna with a circulator is shown in the block diagram of Fig. 3.17. Leakage introduced by the circulator and by reflections from the antenna are at close range and thus are attenuated by the J_3 factor.

A plot of $J_3(D)$ as a function of distance is shown in Fig. 3.18. The curve is mirrored because of the periodicity of D. The nulls in the curve suggest that echoes from certain ranges can be suppressed if the modulation parameters are properly selected.

If the target is stationary (zero doppler frequency), the amplitudes of the modulation-frequency harmonics are proportional to either $J_n(D) \sin \phi_0$ or $J_n(D) \cos \phi_0$, where $\phi_0 = 4\pi f_0 R_0/c = 4\pi R_0/\lambda$. Therefore the amplitude depends on the range to the target in RF wavelengths. The sine or the cosine terms can take any value between $+1$ and -1, including zero, for a change in range corresponding to one RF wavelength. For this reason, the extraction of the higher-order modulation frequencies is not practical with a stationary target, such as in an altimeter.

In order to use the properties of the Bessel function to obtain isolation in an FM-CW altimeter, when the doppler frequency is essentially zero, the role of the doppler frequency shift may be artificially introduced by translating the reference frequency to some different value. This might be accomplished with a single-sideband generator (frequency translator) inserted between the directional coupler and the RF mixer of Fig. 3.17. The frequency translation in the reference signal path is equivalent to a doppler shift in the antenna path. The frequency excursion of the modulation waveform can be adjusted by a servomechanism to maintain the maximum of the Bessel function at the aircraft's altitude. The frequency translator is not needed in an airborne doppler navigator since the antenna beam is directed at a depression angle other than 90° and a doppler-shifted echo is produced by the motion of the aircraft.

Matched filter detection. The operation of the FM-CW radar described earlier in this section does not employ optimum signal processing, such as described in Chap. 10. The receiver is not designed as a *matched filter* for the particular transmitted waveform. Therefore, the sensitivity of the FM-CW receiver described here is degraded and the ability to operate with multiple targets is usually poor. For a radio altimeter whose target is the earth, these limitations usually present no problem. When used for the long-range detection of air targets, for example, the nonoptimum detection of the classical FM-CW radar will seriously hinder its performance when compared to a properly designed pulse radar. It is possible, however, to utilize matched-filter processing in the FM-CW radar in a manner similar to that employed with FM (chirp) pulse compression radar as described in Sec. 11.5, and thus overcome the lower sensitivity and multiple-target problems.[38,39] As the duty cycle of an FM pulse compression waveform increases it becomes more like the FM-CW waveform of unity duty cycle. Hence, the processing

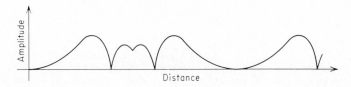

Figure 3.18 Plot of $J_3(D)$ as a function of distance. (*From Saunders,*[9] *IRE Trans.*)

techniques can be similar. Pseudorandom phase-coded waveforms[40] and random noise waveforms[41,42] may also be applied to CW transmission. When the modulation period is long, it may be desirable to utilize correlation detection instead of matched filter detection.

3.4 AIRBORNE DOPPLER NAVIGATION[43-56]

An important requirement of aircraft flight is for a self-contained navigation system capable of operating anywhere over the surface of the earth under any conditions of visibility or weather. It should provide the necessary data for piloting the aircraft from one position to another without the need of navigation information transmitted to the aircraft from a ground station One method of obtaining a self-contained aircraft navigation system is based on the CW doppler-radar principle. Doppler radar can provide the drift angle and true speed of the aircraft relative to the earth. The drift angle is the angle between the horizontal projection of the centerline of the aircraft (heading) and the horizontal component of the aircraft velocity vector (ground track). From the ground-speed and drift-angle measurements, the aircraft's present position can be computed by dead reckoning.

An aircraft with a doppler radar whose antenna beam is directed at an angle γ to the horizontal (Fig. 3.19a) will receive a doppler-shifted echo signal from the ground. The shift in

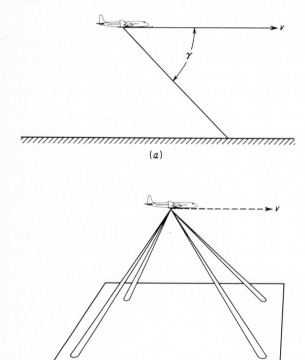

(a)

(b)

Figure 3.19 (a) Aircraft with single doppler navigation antenna beam at an angle γ to the horizontal; (b) aircraft employing four doppler-navigation beams to obtain vector velocity.

frequency is $f_d = 2(f_0/c)v \cos \gamma$, where f_0 is the carrier frequency, v is the aircraft velocity, and c is the velocity of propagation. Typically, the depression angle γ might be in the vicinity of 65 to 70°. A single antenna beam from a doppler radar measures one component of aircraft velocity relative to the direction of propagation. A minimum of three noncoplanar beams are needed to determine the vector velocity, that is, the speed and direction of travel.

Doppler-navigation radar measures the vector velocity relative to the frame of reference of the antenna assembly. To convert this vector velocity to a horizontal reference on the ground, the direction of the vertical must be determined by some auxiliary means. The heading of the aircraft, as might be obtained from a compass, must also be known for proper navigation. The vertical reference may be used either to stabilize the antenna beam system so as to align it with the horizontal, or alternatively, the antennas might be fixed relative to the aircraft and the ground velocity components calculated with a computer.

A practical form of doppler-navigation radar might have four beams oriented as in Fig. 3.19b. A doppler-navigation radar with forward and rearward beams is called a *Janus* system, after the Roman god who looked both forward and backward at the same time. Assume initially that the two forward and two backward beams are symmetrically disposed about the axis of the aircraft. If the aircraft's velocity is not in the same direction as the aircraft heading, the doppler frequency in the two forward beams will not be the same. This difference in frequency may be used to generate an error signal in a servomechanism which rotates the antennas until the doppler frequencies are equal, indicating that the axis of the antennas is aligned with the ground track of the aircraft. The angular displacement of the antenna from the aircraft heading is the drift angle, and the magnitude of the doppler is a measure of the speed along the ground track.

The use of the two rearward beams in conjunction with the two forward beams results in considerable improvement in accuracy. It eliminates the error introduced by vertical motion of the aircraft and reduces the error caused by pitching movements of the antenna.

Navigation may also be performed with only two antenna beams if some auxiliary means is used to obtain a third coordinate. Two beams give the two components of the aircraft velocity tangent to the surface of the earth. A third component, the vertical velocity, is needed and may be provided from some nondoppler source such as a barometric rate-of-climb meter. The primary advantage of the two-beam system is a reduction in equipment. However, the accuracy is not as good as with systems using three or four beams.

In principle, the CW radar would seem to be the ideal method of obtaining doppler-navigation information. However, in practice, leakage between transmitter and receiver can limit the sensitivity of the CW doppler-navigation radar, just as it does in any CW radar. One method of eliminating the ill effects of leakage is by pulsing the transmitter on and turning the receiver off for the duration of the transmitted pulse in a manner similar to the pulse-doppler radar described in Sec. 4.10. The pulse-doppler mode of operation has the further advantage in that each beam can operate with a single antenna for both transmitter and receiver, whereas a CW radar must usually employ two separate antennas in order to achieve the needed isolation. However, pulse systems suffer from loss of coverage and/or sensitivity because of " altitude holes." These are caused by the high prf commonly used with pulse-doppler radars when it is necessary to achieve unambiguous doppler measurements. The high prf, although it gives unambiguous doppler, usually results in ambiguous range. But more important, a pulse radar with ambiguous prf can result in lost targets. If the transmitter is pulsed just when the ground echo arrives back at the radar, it will not be detected. Thus, altitude holes exist at or near those altitudes where the echo time is an integral multiple of the pulse-repetition period. Techniques exist for reducing the undesired effects of altitude holes, but not without some inconvenience or possible loss in overall performance.[44]

The Janus system can be operated incoherently by using the same transmitter to feed a pair of beams simultaneously. Typically, one beam is directed ahead and to the right of the ground track, and the other aft and to the left. A forward-left and an aft-right are also fed by the transmitter as a second channel. The two channels may be operated simultaneously or timed-shared. By heterodyning in a mixing element the echo signal received in the fore and aft beams, the doppler frequency is extracted. The difference frequency resulting from the mixing operation is twice the doppler frequency. A stable transmitter frequency is not needed in this system as it is in the coherent system. Coherence is obtained on a relative basis in the process of comparing the signals received from the forward and backward direction. Changes in transmitter frequency affect the echo signals in the two directions equally and are therefore canceled when taking the difference frequency.

Another method of achieving the necessary isolation in a doppler-navigation radar is with a sinusoidal frequency-modulated CW system. By frequency-modulating the transmission, the leakage signal may be reduced relative to the signal from the ground by extracting a harmonic of the modulating frequency and taking advantage of one of the higher-order Bessel functions as described in the previous section. It has been claimed that a doppler navigation system based on this principle can provide 150 dB of isolation, the amount necessary to operate at altitudes of 50,000 ft.[49]

The frequency band from 13.25 to 13.4 GHz has been allocated for airborne doppler navigation radar, but such radars may also operate within the band from 8.75 to 8.85 MHz.

The doppler navigator over land can provide a measurement of ground speed with a standard deviation of from 0.13 to 0.5 percent.[55] The standard deviation of the drift angle can be less than 0.25°. Over water the accuracies are slightly worse. The backscatter energy over water, except for near vertical incidence, is generally less than over land, thus lowering the signal-to-noise ratio. Also there is an apparent increase in the angle of depression since the rapid variation of sea echo (σ^0) as a function of the depression angle γ favors returns from the lower half of the incident beam. It results in the center of the doppler spectrum being shifted, so that the direction as determined from the doppler spectrum will differ from that of the center of the antenna beam. This is sometimes called the *calibration-shift error*. This error can be corrected by lobe switching; that is, by periodically switching the antenna beam a small amount in the γ-direction at a low rate (perhaps 20 Hz).[55] Two displaced doppler-frequency spectra are produced at the two antenna positions. A narrow tracking filter tracks the crossover of the two spectra. The frequency of this crossover is the same for water as for land. Hence, there will be no bias error over water. A very narrow antenna beam will also reduce the over-water bias error without the need for lobe switching. Another source of error is the mass movement of water caused by tides, currents, and winds, which result in a doppler frequency shift in addition to that caused by the aircraft's motion.

In order to use the doppler radar for navigation purposes, a heading reference is required to refer the antenna direction indication to some navigational coordinate system, such as true north. A computer is also needed to integrate the velocity measurements so as to give the distance traveled.

The AN/APN-200 doppler navigation radar used in the Navy's S3A aircraft, and the similar AN/APN-213 in the Air Force's E3A (AWACS) operate at 13.3 GHz with 1 watt of CW power obtained from a solid-state IMPATT diode transmitter.[56] The waveform is unmodulated CW. The antenna is a fixed array with an integral radome producing four receive beams and four transmit beams with a minimum of 70 dB of isolation between transmit and receive. The one-way beamwidths are approximately 2.5° (fore-aft) by 5° (lateral). The four beams are displaced symmetrically at an outward angle of 25° from the vertical. The beams are time shared by a single channel which is switched through the four-beam sequence every

250 ms. A phase-monopulse technique is used to reduce the terrain bias effect.[68] The unit weighs 44 pounds including radome, and requires 165 VA of power. It is required to operate from 50 to 1000 knots at attitudes of $\pm 25°$ pitch and $\pm 45°$ roll of the aircraft, at an altitude of 40,000 ft over water. The rms accuracy is claimed to be within 0.13 percent ± 0.1 knots of the true ground velocity.

3.5 MULTIPLE-FREQUENCY CW RADAR[12,57-63]

Although it has been said in this chapter that CW radar does not measure range, it is possible under some circumstances to do so by measuring the phase of the echo signal relative to the phase of the transmitted signal. Consider a CW radar radiating a single-frequency sine wave of the form $\sin 2\pi f_0 t$. (The amplitude of the signal is taken to be unity since it does not influence the result.) The signal travels to the target at a range R and returns to the radar after a time $T = 2R/c$, where c is the velocity of propagation. The echo signal received at the radar is $\sin [2\pi f_0(t - T)]$. If the transmitted and received signals are compared in a phase detector, the output is proportional to the phase difference between the two and is $\Delta\phi = 2\pi f_0 T = 4\pi f_0 R/c$. The phase difference may therefore be used as a measure of the range, or

$$R = \frac{c\,\Delta\phi}{4\pi f_0} = \frac{\lambda}{4\pi}\,\Delta\phi \tag{3.17}$$

However, the measurement of the phase difference $\Delta\phi$ is unambiguous only if $\Delta\phi$ does not exceed 2π radians. Substituting $\Delta\phi = 2\pi$ into Eq. (3.17) gives the maximum unambiguous range as $\lambda/2$. At radar frequencies this unambiguous range is much too small to be of practical interest.

The region of unambiguous range may be extended considerably by utilizing two separate CW signals differing slightly in frequency. The unambiguous range in this case corresponds to a half wavelength at the difference frequency.

The transmitted waveform is assumed to consist of two continuous sine waves of frequency f_1 and f_2 separated by an amount Δf. For convenience, the amplitudes of all signals are set equal to unity. The voltage waveforms of the two components of the transmitted signal v_{1T} and v_{2T} may be written as

$$v_{1T} = \sin\left(2\pi f_1 t + \phi_1\right) \tag{3.18a}$$

$$v_{2T} = \sin\left(2\pi f_2 t + \phi_2\right) \tag{3.18b}$$

where ϕ_1 and ϕ_2 are arbitrary (constant) phase angles. The echo signal is shifted in frequency by the doppler effect. The form of the doppler-shifted signals at each of the two frequencies f_1 and f_2 may be written

$$v_{1R} = \sin\left[2\pi(f_1 \pm f_{d1})t - \frac{4\pi f_1 R_0}{c} + \phi_1\right] \tag{3.19a}$$

$$v_{2R} = \sin\left[2\pi(f_2 \pm f_{d2})t - \frac{4\pi f_2 R_0}{c} + \phi_2\right] \tag{3.19b}$$

where R_0 = range to target at a particular time $t = t_0$ (range that would be measured if target were not moving)

f_{d1} = doppler frequency shift associated with frequency f_1

f_{d2} = doppler frequency shift associated with frequency f_2

Since the two RF frequencies f_1 and f_2 are approximately the same (that is, $f_2 = f_1 + \Delta f$, where $\Delta f \ll f_1$) the doppler frequency shifts f_{d1} and f_{d2} are approximately equal to one another. Therefore we may write $f_{d1} = f_{d2} = f_d$.

The receiver separates the two components of the echo signal and heterodynes each received signal component with the corresponding transmitted waveform and extracts the two doppler-frequency components given below:

$$v_{1D} = \sin \left(\pm 2\pi f_d t - \frac{4\pi f_1 R_0}{c} \right) \tag{3.20a}$$

$$v_{2D} = \sin \left(\pm 2\pi f_d t - \frac{4\pi f_2 R_0}{c} \right) \tag{3.20b}$$

The phase difference between these two components is

$$\Delta\phi = \frac{4\pi(f_2 - f_1)R_0}{c} = \frac{4\pi \, \Delta f R_0}{c} \tag{3.21}$$

Hence

$$R_0 = \frac{c \, \Delta\phi}{4\pi \, \Delta f} \tag{3.22}$$

which is the same as that of Eq. (3.17), with Δf substituted in place of f_0.

The two-frequency CW technique for measuring range was described as using the doppler frequency shift. When the doppler frequency is zero, as with a stationary target, it is also possible, in principle, to extract the phase difference. If the target carries a beacon or some other form of echo-signal augmentor, the doppler frequency shift may be simulated by translating the echo frequency, as with a single-sideband modulator.

The two frequencies of the two-frequency radar were described as being transmitted simultaneously.[63] They may also be transmitted sequentially in some applications by rapidly switching a single RF source.

A large difference in frequency between the two transmitted signals improves the accuracy of the range measurement since large Δf means a proportionately large change in $\Delta\phi$ for a given range. However, there is a limit to the value of Δf, since $\Delta\phi$ cannot be greater than 2π radians if the range is to remain unambiguous. The maximum unambiguous range R_{unamb} is

$$R_{unamb} = \frac{c}{2 \, \Delta f} \tag{3.23}$$

Therefore Δf must be less than $c/2R_{unamb}$. Note that when Δf is replaced by the pulse repetition rate, Eq. (3.23) gives the maximum unambiguous range of a pulse radar.

A qualitative explanation of the operation of the two-frequency radar may be had by considering both carrier frequencies to be in phase at zero range. As they progress outward from the radar, the relative phase between the two increases because of their difference in frequency. This phase difference may be used as a measure of the elapsed time. When the two signals slip in phase by 1 cycle, the measurement of phase, and hence range, becomes ambiguous.

The two-frequency CW radar is essentially a single-target radar since only one phase difference can be measured at a time. If more than one target is present, the echo signal becomes complicated and the meaning of the phase measurement is doubtful. The theoretical accuracy with which range can be measured with the two-frequency CW radar can be found from the methods described in Sec. 11.3. It can be shown that the theoretical rms range error is

$$\delta R = \frac{c}{4\pi \, \Delta f \, (2E/N_0)^{1/2}} \tag{3.24}$$

where E = energy contained in received signal and N_0 = noise power per hertz of bandwidth. If this is compared with the rms range error theoretically possible with the linear FM pulse-compression waveform whose spectrum occupies the same bandwidth Δf, the error obtained with the two-frequency CW waveform is less by the factor 0.29.

Equation (3.24) indicates that the greater the separation Δf between the two frequencies, the less will be the rms error. However, the frequency difference must not be too large if unambiguous measurements are to be made. The selection of Δf represents a compromise between the requirements of accuracy and ambiguity. Both accurate and unambiguous range measurements can be made by transmitting three or more frequencies instead of just two. For example, if the three frequencies f_1, f_2, and f_3 are such that $f_3 - f_1 = k(f_2 - f_1)$, where k is a factor of the order of 10 or 20, the pair of frequencies f_3, f_1 gives an ambiguous but accurate range measurement while the pair of frequencies f_2, f_1 are chosen close to resolve the ambiguities in the f_3, f_1 measurement. Likewise, if further accuracy is required a fourth frequency can be transmitted and its ambiguities resolved by the less accurate but unambiguous measurement obtained from the three frequencies f_1, f_2, f_3. As more frequencies are added, the spectrum and target resolution approach that obtained with a pulse or an FM-CW waveform.

The measurement of range by measuring the phase difference between separated frequencies is analogous to the measurement of angle by measuring the phase difference between widely spaced antennas, as in an interferometer antenna. The interferometer antenna gives an accurate but ambiguous measurement of angle. The ambiguities may be resolved by additional antennas spaced closer together. The spacing between the individual antennas in the interferometer system corresponds to the separation between frequencies in the multiple-frequency distance-measuring technique. The minitrack system is an example of an interferometer in which angular ambiguities are resolved in a manner similar to that described.[64]

The multiple-frequency CW radar technique has been applied to the accurate measurement of distance in surveying and in missile guidance. The *Tellurometer* is the name given to a portable electronic surveying instrument which is based on this principle.[57-59] It consists of a master unit at one end of the line and a remote unit at the other end. The master unit transmits a carrier frequency of 3,000 MHz, with four single-sideband modulated frequencies separated from the carrier by 10.000, 9.990, 9.900, and 9.000 MHz. The 10-MHz difference frequency provides the basic accuracy measurement, while the difference frequencies of 1 MHz, 100 kHz, and 10 kHz permit the resolution of ambiguities.

The remote unit at the other end of the line receives the signals from the master unit and amplifies and retransmits them. The phases of the returned signals at the master unit are compared with the phases of the outgoing signals. Since the master and the remote units are stationary, there is no doppler frequency shift. The function of the doppler frequency is provided by modulating the retransmitted signals at the remote unit in such a manner that a 1-kHz beat frequency is obtained from the heterodyning process at the receiver of the master unit. The phase of the 1-kHz signals contains the same information as the phase of the multiple frequencies.

The MRB 201 Tellurometer is capable of measuring distances from 200 to 250 km, assuming reasonable line of sight conditions, with an accuracy of ± 0.5 m $\pm 3 \times 10^{-6} d$, where d is the distance being measured. The transmitter power is 200 mW and the antenna is a small paraboloid with crossed feeds to make the polarizations of the transmitted and received signals orthogonal. The MRA 5 operates within the 10 to 10.5 GHz band, from 100 m to 50 km range with an accuracy of ± 0.05 m $\pm$ 10^{-5} d, assuming a correction is made for meteorological conditions.

In addition to its use in surveying, the multiple CW frequency method of measuring range has been applied in range-instrumentation radar for the measurement of the distance to a transponder-equipped missile;[65] the distance to satellites;[66] in satellite navigation systems based on range measurement;[67] and for detecting the presence of an obstacle in the path of a moving automobile by measuring the distance, the doppler velocity, and the sign of the doppler (whether the target is approaching or receding).[63]

REFERENCES

1. Sandretto, P. C.: The Long Quest, *IRE Trans.*, vol. ANE-1, no. 2, p. 2, June, 1954.
2. Van Vleck, J. H., and D. Middleton: A Theoretical Comparison of the Visual, Aural, and Meter Reception of Pulsed Signals in the Presence of Noise, *J. Appl. Phys.*, vol. 17, pp. 940–971, November, 1946.
3. Selridge, H.: Proximity Fuzes for Artillery, *Electronics*, vol. 19, pp. 104–109, February, 1946.
4. Bonner, H. M.: The Radio Proximity Fuse, *Elec. Eng.*, vol. 66, pp. 888–893, September, 1947.
5. Saunders, W. K.: CW and FM Radar, chap. 16 of " Radar Handbook," M. I. Skolnik (ed.), McGraw-Hill Book Co., New York, 1970.
6. Barlow, E. J.: Doppler Radar, *Proc. IRE*, vol. 37, pp. 340–355, April, 1949.
7. Meyer, M. A., and A. B. Goldberg: Applications of the Turnstile Junction, *IRE Trans.*, vol. MTT-3, pp. 40–45, December, 1955.
8. Roberts, W. B.: Rotating Wave Radar, *Electronics*, vol. 19, pp. 130–133, July, 1946.
9. Sanders, W. K.: Post-war Developments in Continuous-wave and Frequency-modulated Radar, *IRE Trans.*, vol. ANE-8, pp. 7-19, March, 1961.
10. Saunders, W. K.: Control of Surface Currents by the Use of Channels, *IRE Trans.*, vol. AP-4, pp. 85–87, January, 1956.
11. Agar, W. O., and M. Morgan: Isolation of Separate Transmitter and Receiver Aerials for Continuous Wave Radars, *Marconi Review*, vol. 26, pp. 25–34, 1st qtr., 1963.
12. Ridenour, L. N.: " Radar System Engineering," MIT Radiation Laboratory Series, vol. 1, McGraw-Hill Book Co., New York, 1947.
13. O'Hara, F. J., and G. M. Moore: A High Performance CW Receiver Using Feedthru Nulling, *Microwave J.*, vol. 6, pp. 63–71, September, 1963.
14. Greene, J. C., and J. F. Lyons: Receivers with Zero Intermediate Frequency, *Proc. IRE*, vol. 47, pp. 335–336, February, 1959.
15. Kerr, D. E. (ed.): " Propagation of Short Radio Waves," MIT Radiation Laboratory Series, vol. 13, pp. 539–543, McGraw-Hill Book Company, New York, 1951.
16. Rabiner, L. R., and C. M. Rader (eds.): " Digital Signal Processing," IEEE Press, New York, 1972.
17. Gardner, F. M.: DOPLOC Uses Phase-locked Filter, *Electronic Ind.*, vol. 18, pp. 96–99, October, 1959.
18. Kalmus, H. P.: Direction Sensitive Doppler Device, *Proc. IRE*, vol. 43, pp. 698–700, June, 1955.
19. Logue, S. H.: Rate-of-climb Meter Uses Doppler Radar, *Electronics*, vol. 30, no. 6, pp. 150–152, June 1, 1957.
20. Kalmus, H. P.: Doppler Wave Recognition with High Clutter Rejection, *IEEE Trans.*, vol. AES-3, no. 6 Suppl., pp. 334–339, November, 1967.
21. Temes, C. L.: Relativistic Considerations of Doppler Shift, *IRE Trans.*, vol. ANE-6, p. 37, March, 1959.
22. Knop, C. M.: A Note on Doppler-Shifted Signals, *Proc. IEEE*, vol. 54, pp. 807–808, May, 1966.
23. Gill, T. P.: " The Doppler Effect," Academic Press, New York, 1965.
24. Merriman, R. H., and J. W. Rush: Microwave Doppler Sensors, *Microwave J.*, vol. 17, pp. 27–30, July, 1974.
25. Whetton, C. P.: Industrial and Scientific Applications of Doppler Radar, *Microwave J.*, vol. 18, pp. 39–42, November, 1975.
26. Easton, R. L., and J. J. Fleming: The Navy Space Surveillance System, *Proc. IRE*, vol. 48, pp. 663–669, April, 1960.

27. Klass, P. J.: Navy Improves Accuracy, Detection Range of Space Surveillance Chain, *Aviation Week*, Aug. 16, 1965.
28. Ismail, M. A. W.: A Study of the Double Modulated FM Radar, *Inst. Hockfrequenztech. an der E.T.H. Rept.* 21, Verlag Leemann, Zurich, 1955.
29. Luck, D. G. C.: "Frequency Modulated Radar," McGraw-Hill Book Company, New York, 1949.
30. Wimberley, F. T., and J. F. Lane, Jr.: The AN/APN-22 Radio Altimeter, *IRE Trans.*, vol. ANE-1, no. 2, pp. 8–14, June, 1954.
31. Banks, D. S.: Continuous Wave (CW) Radar, *EASCON '75 Record*, IEEE Publ. 75 CHO 998-5 EASCON, pp. 107-A to 107-G, 1975.
32. Appleton, E. V., and M. A. F. Barnett: On Some Direct Evidence for Downward Atmospheric Reflection of Electric Rays, *Proc. Roy. Soc.*, vol. A109, p. 621, 1925.
33. Espenshied, L., and R. C. Newhouse: A Terrain Clearance Indicator, *Bell Sys. Tech. J.*, vol. 18, pp. 222–234, 1939.
34. Block, A., K. E. Buecks, and A. G. Heaton: Improved Radio Altimeter, *Wireless World*, vol. 60, pp. 138–140, March, 1954.
35. Glegg, K. C. M.: A Low Noise CW Doppler Technique, *Natl. Conf. Proc. Aeronaut. Electronics* (Dayton, Ohio), pp. 133–144, 1958.
36. Capelli, M.: Radio Altimeter, *IRE Trans.*, vol. ANE-1, no. 2, pp. 3–7, June, 1954.
37. Sollenberger, T. E.: Multipath Phase Errors in CW-FM Tracking Systems, *IRE Trans.*, vol. AP-3, pp. 185–192, October, 1955.
38. Withers, M. J.: Matched Filter for Frequency-Modulated Continuous-Wave Radar Systems, *Proc. IEE (London)*, vol. 113, pp. 405–412, March, 1966.
39. Barrick, D. E.: FM/CW Radar Signals and Digital Processing, *National Oceanic and Atmospheric Administration*, Boulder, Colorado, NOAA Technical Report ERL 283-WPL 26, July, 1973.
40. Craig, S. E., W. Fishbein, and O. E. Rittenback: Continuous-Wave Radar with High Range Resolution and Unambiguous Velocity Determination, *IRE Trans.*, vol. MIL-6, pp. 153–161, April, 1962.
41. Horton, B. M.: Noise-Modulated Distance Measuring System, *Proc. IRE*, vol. 47, pp. 821-828, May, 1959.
42. Chadwick, R. B., and G. R. Cooper: Measurement of Distributed Targets with the Random Signal Radar, *IEEE Trans.*, vol. AES-8, pp. 743–750, November, 1972.
43. Berger, F. B.: The Nature of Doppler Velocity Measurement, *IRE Trans.*, vol. ANE-4, pp. 103–112, September, 1957.
44. Berger, F. B.: The Design of Airborne Doppler Velocity Measuring Systems, *IRE Trans.*, vol. ANE-4, pp. 157–175, December, 1957.
45. Fried, W. R.: Principles and Performance Analysis of Doppler Navigation Systems, *IRE Trans.*, vol. ANE-4, pp. 176–196, December, 1957.
46. Condie, M. A.: Basic Design Considerations: Automatic Navigator AN/APN-67, *IRE Trans.*, vol. ANE-4, pp. 197–201, December, 1957.
47. McMahon, F. A.: The AN/APN-81 Doppler Navigation System, *IRE Trans.*, vol. ANE-4, pp. 202–211, December, 1957.
48. McKay, M. W.: The AN/APN-96 Doppler Radar Set, *IRE Natl. Conv. Record*, vol. 6, pt. 5, pp. 71–79, 1958.
49. Brown, R. K., N. F. Moody, P. M. Thompson, R. J. Bibby, C. A. Franklin, J. H. Ganton, and J. Mitchell: A Lightweight and Self-contained Airborne Navigational System, *Proc. IRE*, vol. 47, pp. 778–807, May, 1959.
50. Glegg, K. C. M.: A Low Noise CW Doppler Technique, *Natl. Conf. Proc. Aeronaut, Electronics* (Dayton, Ohio), pp. 133–144, 1958.
51. Tollefson, R. D.: Application of Frequency Modulation Techniques to Doppler Radar Sensors, *Natl. Conf. Proc. Aeronaut. Electronics* (Dayton, Ohio), pp. 683–687, 1959.
52. Kramer, E.: A Historical Survey of the Application of the Doppler Principle for Radio Navigation, *IEEE Trans.*, vol. AES-8, pp. 258–263, May, 1972.
53. Fried, W. R.: New Developments in Radar and Radio Sensors for Aircraft Navigation, *IEEE Trans.*, vol. AES-10, pp. 25–33, January, 1974.
54. Gray, T.: Airborne Doppler Navigation Techniques, chap. 13 of "Radar Techniques for Detection, Tracking, and Navigation," W. T. Blackband (ed.), Gordon and Breach Science Publishers, New York, 1966.
55. Fried, W. R.: Doppler Navigation, chap. 6 of "Avionics Navigation Systems," M. Kayton and W. R. Fried (eds.), John Wiley and Sons, Inc., New York, 1969.
56. Andreone, V. M., and C. N. Bates: Doppler Radar Boast Design Innovations, *Microwaves*, vol. 13, pp. 72–82, October, 1974.

57. Wadley, T. L.: Electronic Principles of the Tellurometer, *Trans. South African Inst. Elec. Engrs.*, vol. 49, pp. 143–161, May, 1958; discussion, pp. 161–172.
58. Poling, A. C.: Tellurometer Manual, *U.S. Dept. Commerce Publ.* 62-1, 1959.
59. Robinson, T. A.: Application of Electronic Distance Measuring Equipment in Surveying, *IRE Trans.*, vol. MIL-4, pp. 263–267, April–July, 1960.
60. Varian, R. H., W. W. Hansen, and J. R. Woodyard: Object Detecting and Locating System, U.S. Patent 2,435,615, Feb. 10, 1948.
61. Skolnik, M. I.: An Analysis of Bistatic Radar, Appendix, *IRE Trans.*, vol. ANE-8, pp. 19–27, March, 1961.
62. Hastings, C. E.: Raydist: A Radio Navigation and Tracking System, *Tele-Tech*, vol. 6, pp. 30–33, 100–103, June, 1947.
63. Boyer, W. D.: A Diplex, Doppler Phase Comparison Radar, *IEEE Trans.*, vol. ANE-10, pp. 27–33, March, 1963.
64. Mengel, J. T.: Tracking the Earth Satellite and Data Transmission by Radio, *Proc. IRE*, vol. 44, pp. 755–760, June, 1956.
65. Mertens, L. E., and R. H. Tabeling: Tracking Instrumentation and Accuracy on the Eastern Test Range, *IEEE Trans.*, vol. SET-11, pp. 14–23, March, 1965.
66. McCaskill, T. B., J. A. Buisson, and D. W. Lynch: Principles and Techniques of Satellite Navigation Using the Timation II Satellite, *Naval Research Laboratory Rep.* 7252, June 17, 1971, Washington, D.C.
67. Easton, R. L.: The Role of Time/Frequency in Navy Navigation Satellites, *Proc. IEEE*, vol. 60, pp. 557–563, May, 1972.
68. Bates, C. N.: New Advances in Doppler Radar, *Countermeasures*, vol. 2, pp. 34–37, April, 1976.
69. Acker, A. E.: Eliminating Transmitted Clutter in Doppler Radar Systems, *Microwave J.*, vol. 18, pp. 47–50, November, 1975.
70. Gupta, P. D.: Exact Derivation of the Doppler Shift Formula for a Radar Echo Without Using Transformation Equations, *Am. J. Phys.*, vol. 45, pp. 674–675, July, 1977.
71. King, R. J.: " Microwave Homodyne Systems," Peter Peregrinus Ltd., Stevenage, Herts, England, 1978 (an Inst. Elect. Engs. publ.).
72. Blomfield, D. L. H.: Low-Noise Microwave Sources, *International Conference on Radar—Present and Future*, Oct. 23–25, 1973, *IEE (London) Publ. no. 105*, pp. 178–183.
73. Ashley, J. R., T. A. Barley, and G. J. Rast, Jr.: The Measurement of Noise in Microwave Transmitters, *IEEE Trans.*, vol. MTT-25, pp. 294–318, April, 1977.

FOUR

MTI AND PULSE DOPPLER RADAR

4.1 INTRODUCTION

The doppler frequency shift [Eq. (3.2)] produced by a moving target may be used in a pulse radar, just as in the CW radar discussed in Chap. 3, to determine the relative velocity of a target or to separate desired moving targets from undesired stationary objects (clutter). Although there are applications of pulse radar where a determination of the target's relative velocity is made from the doppler frequency shift, the use of doppler to separate small moving targets in the presense of large clutter has probably been of far greater interest. Such a pulse radar that utilizes the doppler frequency shift as a means for discriminating moving from fixed targets is called an *MTI* (moving target indication) or a *pulse doppler* radar. The two are based on the same physical principle, but in practice there are generally recognizable differences between them (Sec. 4.10). The MTI radar, for instance, usually operates with ambiguous doppler measurement (so-called *blind speeds*) but with unambiguous range measurement (no second-time-around echoes). The opposite is generally the case for a pulse doppler radar. Its pulse repetition frequency is usually high enough to operate with unambiguous doppler (no blind speeds) but at the expense of range ambiguities. The discussion in this chapter, for the most part, is based on the MTI radar, but much of what applies to MTI can be extended to pulse doppler radar as well.

MTI is a necessity in high-quality air-surveillance radars that operate in the presence of clutter. Its design is more challenging than that of a simple pulse radar or a simple CW radar. An MTI capability adds to a radar's cost and complexity and often system designers must accept compromises they might not wish to make. The basic MTI concepts were introduced during World War II, and most of the signal processing theory on which MTI (and pulse doppler) radar depends was formulated during the mid-1950s. However, the reduction of theory to practice was paced by the availability of the necessary signal-processing technology. It took almost twenty years for the full capabilities offered by MTI signal-processing theory to be converted into practical and economical radar equipment. The chief factor that made this possible was the introduction of reliable, small, and inexpensive digital processing hardware.

Description of operation. A simple CW radar such as was described in Sec. 3.2 is shown in Fig. 4.1*a*. It consists of a transmitter, receiver, indicator, and the necessary antennas. In principle, the CW radar may be converted into a pulse radar as shown in Fig. 4.1*b* by providing a power amplifier and a modulator to turn the amplifier on and off for the purpose of generating pulses. The chief difference between the pulse radar of Fig. 4.1*b* and the one described in Chap. 1 is that a small portion of the CW oscillator power that generates the transmitted pulses is diverted to the receiver to take the place of the local oscillator. However, this CW signal does more than function as a replacement for the local oscillator. It acts as the coherent reference needed to detect the doppler frequency shift. By *coherent* it is meant that the phase of the transmitted signal is preserved in the reference signal. The reference signal is the distinguishing feature of coherent MTI radar.

If the CW oscillator voltage is represented as $A_1 \sin 2\pi f_t t$, where A_1 is the amplitude and f_t the carrier frequency, the reference signal is

$$V_{\text{ref}} = A_2 \sin 2\pi f_t t \qquad (4.1)$$

and the doppler-shifted echo-signal voltage is

$$V_{\text{echo}} = A_3 \sin \left[2\pi(f_t \pm f_d)t - \frac{4\pi f_t R_0}{c} \right] \qquad (4.2)$$

where A_2 = amplitude of reference signal
$\quad A_3$ = amplitude of signal received from a target at a range R_0
$\quad f_d$ = doppler frequency shift
$\quad t$ = time
$\quad c$ = velocity of propagation

The reference signal and the target echo signal are heterodyned in the mixer stage of the receiver. Only the low-frequency (difference-frequency) component from the mixer is of interest and is a voltage given by

$$V_{\text{diff}} = A_4 \sin \left(2\pi f_d t - \frac{4\pi f_t R_0}{c} \right) \qquad (4.3)$$

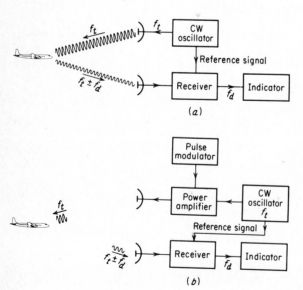

(a)

(b)

Figure 4.1 (*a*) Simple CW radar; (*b*) pulse radar using doppler information.

Note that Eqs. (4.1) to (4.3) represent sine-wave carriers upon which the pulse modulation is imposed. The difference frequency is equal to the doppler frequency f_d. For stationary targets the doppler frequency shift f_d will be zero; hence V_{diff} will not vary with time and may take on any constant value from $+A_4$ to $-A_4$, including zero. However, when the target is in motion relative to the radar, f_d has a value other than zero and the voltage corresponding to the difference frequency from the mixer [Eq. (4.3)] will be a function of time.

An example of the output from the mixer when the doppler frequency f_d is large compared with the reciprocal of the pulse width is shown in Fig. 4.2b. The doppler signal may be readily discerned from the information contained in a single pulse. If, on the other hand, f_d is small compared with the reciprocal of the pulse duration, the pulses will be modulated with an amplitude given by Eq. (4.3) (Fig. 4.2c) and many pulses will be needed to extract the doppler information. The case illustrated in Fig. 4.2c is more typical of aircraft-detection radar, while the waveform of Fig. 4.2b might be more applicable to a radar whose primary function is the detection of extraterrestrial targets such as ballistic missiles or satellites. Ambiguities in the measurement of doppler frequency can occur in the case of the discontinuous measurement of Fig. 4.2c, but not when the measurement is made on the basis of a single pulse. The video signals shown in Fig. 4.2 are called *bipolar*, since they contain both positive and negative amplitudes.

Moving targets may be distinguished from stationary targets by observing the video output on an A-scope (amplitude vs. range). A single sweep on an A-scope might appear as in Fig. 4.3a. This sweep shows several fixed targets and two moving targets indicated by the two arrows. On the basis of a single sweep, moving targets cannot be distinguished from fixed targets. (It may be possible to distinguish extended ground targets from point targets by the stretching of the echo pulse. However, this is not a reliable means of discriminating moving from fixed targets since some fixed targets can look like point targets, e.g., a water tower. Also, some moving targets such as aircraft flying in formation can look like extended targets.) Successive A-scope sweeps (pulse-repetition intervals) are shown in Fig. 4.3b to e. Echoes from fixed targets remain constant throughout, but echoes from moving targets vary in amplitude from sweep to sweep at a rate corresponding to the doppler frequency. The superposition of

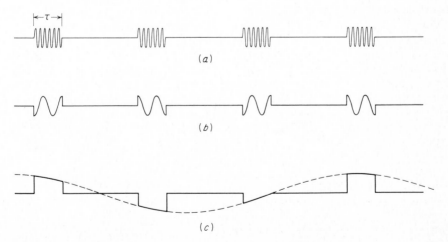

Figure 4.2 (*a*) RF echo pulse train; (*b*) video pulse train for doppler frequency $f_d > 1/\tau$; (*c*) video pulse train for doppler freuqncy $f_d < 1/\tau$.

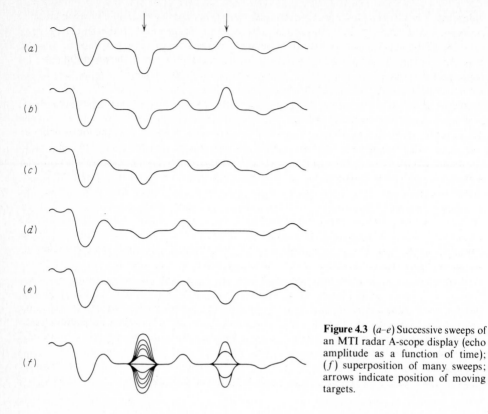

(a)

(b)

(c)

(d)

(e)

(f)

Figure 4.3 (*a–e*) Successive sweeps of an MTI radar A-scope display (echo amplitude as a function of time); (*f*) superposition of many sweeps; arrows indicate position of moving targets.

the successive A-scope sweeps is shown in Fig. 4.3*f*. The moving targets produce, with time, a "butterfly" effect on the A-scope.

Although the butterfly effect is suitable for recognizing moving targets on an A-scope, it is not appropriate for display on the PPI. One method commonly employed to extract doppler information in a form suitable for display on the PPI scope is with a delay-line canceler (Fig. 4.4). The delay-line canceler acts as a filter to eliminate the d-c component of fixed targets and to pass the a-c components of moving targets. The video portion of the receiver is divided into two channels. One is a normal video channel. In the other, the video signal experiences a time delay equal to one pulse-repetition period (equal to the reciprocal of the pulse-repetition frequency). The outputs from the two channels are subtracted from one another. The fixed targets with unchanging amplitudes from pulse to pulse are canceled on subtraction. However, the amplitudes of the moving-target echoes are not constant from pulse to pulse, and subtraction results in an uncanceled residue. The output of the subtraction circuit is bipolar

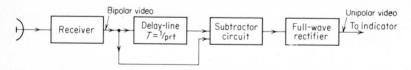

Figure 4.4 MTI receiver with delay-line canceler.

video, just as was the input. Before bipolar video can intensity-modulate a PPI display, it must be converted to unipotential voltages (unipolar video) by a full-wave rectifier.

The simple MTI radar shown in Fig. 4.1*b* is not necessarily the most typical. The block diagram of a more common MTI radar employing a power amplifier is shown in Fig. 4.5. The significant difference between this MTI configuration and that of Fig. 4.1*b* is the manner in which the reference signal is generated. In Fig. 4.5, the coherent reference is supplied by an oscillator called the *coho*, which stands for *coh*erent *o*scillator. The coho is a stable oscillator whose frequency is the same as the intermediate frequency used in the receiver. In addition to providing the reference signal, the output of the coho f_c is also mixed with the local-oscillator frequency f_l. The local oscillator must also be a stable oscillator and is called *stalo*, for *sta*ble *lo*cal *o*scillator. The RF echo signal is heterodyned with the stalo signal to produce the IF signal just as in the conventional superheterodyne receiver. The stalo, coho, and the mixer in which they are combined plus any low-level amplification are called the *receiver-exciter* because of the dual role they serve in both the receiver and the transmitter.

The characteristic feature of coherent MTI radar is that the transmitted signal must be coherent (in phase) with the reference signal in the receiver. This is accomplished in the radar system diagramed in Fig. 4.5 by generating the transmitted signal from the coho reference signal. The function of the stalo is to provide the necessary frequency translation from the IF to the transmitted (RF) frequency. Although the phase of the stalo influences the phase of the transmitted signal, any stalo phase shift is canceled on reception because the stalo that generates the transmitted signal also acts as the local oscillator in the receiver. The reference signal from the coho and the IF echo signal are both fed into a mixer called the *phase detector*. The phase detector differs from the normal amplitude detector since its output is proportional to the phase difference between the two input signals.

Any one of a number of transmitting-tube types might be used as the power amplifier. These include the triode, tetrode, klystron, traveling-wave tube, and the crossed-field amplifier.

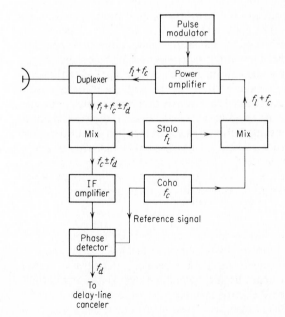

Figure 4.5 Block diagram of MTI radar with power-amplifier transmitter.

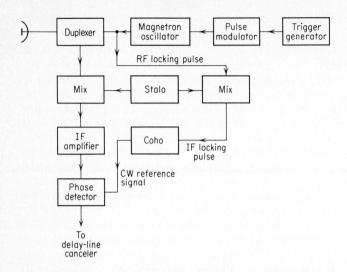

Figure 4.6 Block diagram of MTI radar with power-oscillator transmitter.

Each of these has its advantages and disadvantages, which are discussed in Chap. 6. A transmitter which consists of a stable low-power oscillator followed by a power amplifier is sometimes called MOPA, which stands for *master-oscillator power amplifier*.

Before the development of the klystron amplifier, the only high-power transmitter available at microwave frequencies for radar application was the magnetron oscillator. In an oscillator the phase of the RF bears no relationship from pulse to pulse. For this reason the reference signal cannot be generated by a continuously running oscillator. However, a coherent reference signal may be readily obtained with the power oscillator by readjusting the phase of the coho at the beginning of each sweep according to the phase of the transmitted pulse. The phase of the coho is locked to the phase of the transmitted pulse each time a pulse is generated.

A block diagram of an MTI radar (with a power oscillator) is shown in Fig. 4.6. A portion of the transmitted signal is mixed with the stalo output to produce an IF beat signal whose phase is directly related to the phase of the transmitter. This IF pulse is applied to the coho and causes the phase of the coho CW oscillation to "lock" in step with the phase of the IF reference pulse. The phase of the coho is then related to the phase of the transmitted pulse and may be used as the reference signal for echoes received from that particular transmitted pulse. Upon the next transmission another IF locking pulse is generated to relock the phase of the CW coho until the next locking pulse comes along. The type of MTI radar illustrated in Fig. 4.6 has had wide application.

4.2 DELAY-LINE CANCELERS

The simple MTI delay-line canceler shown in Fig. 4.4 is an example of a time-domain filter. The capability of this device depends on the quality of the medium used as the delay line. The delay line must introduce a time delay equal to the pulse repetition interval. For typical ground-based air-surveillance radars this might be several milliseconds. Delay times of this magnitude cannot be achieved with practical electromagnetic transmission lines. By converting the electromagnetic signal to an acoustic signal it is possible to utilize delay lines of a reasonable physical length since the velocity of propagation of acoustic waves is about 10^{-5} that of electromagnetic waves. After the necessary delay is introduced by the acoustic line, the signal is converted back to an electromagnetic signal for further processing. The early acoustic delay lines developed during World War II used liquid delay lines filled with either water or mercury.[1] Liquid delay lines were large and inconvenient to use. They were replaced in the mid-1950s by the solid fused-quartz delay line that used multiple internal reflections to obtain a compact device. These analog acoustic delay lines were, in turn supplanted in the early 1970s by storage devices based on digital computer technology. The use of digital delay lines requires that the output of the MTI receiver phase-detector be quantized into a sequence of digital words. The compactness and convenience of digital processing allows the implementation of more complex delay-line cancelers with filter characteristics not practical with analog methods.

One of the advantages of a time-domain delay-line canceler as compared to the more conventional frequency-domain filter is that a single network operates at all ranges and does not require a separate filter for each range resolution cell. Frequency-domain doppler filter-banks are of interest in some forms of MTI and pulse-doppler radar.

Filter characteristics of the delay-line canceler. The delay-line canceler acts as a filter which rejects the d-c component of clutter. Because of its periodic nature, the filter also rejects energy in the vicinity of the pulse repetition frequency and its harmonics.

The video signal [Eq. (4.3)] received from a particular target at a range R_0 is

$$V_1 = k \sin (2\pi f_d t - \phi_0) \qquad (4.4)$$

where $\phi_0 =$ phase shift and $k =$ amplitude of video signal. The signal from the previous transmission, which is delayed by a time $T =$ pulse repetition interval, is

$$V_2 = k \sin [2\pi f_d(t - T) - \phi_0] \qquad (4.5)$$

Everything else is assumed to remain essentially constant over the interval T so that k is the same for both pulses. The output from the subtractor is

$$V = V_1 - V_2 = 2k \sin \pi f_d T \cos \left[2\pi f_d\left(t - \frac{T}{2}\right) - \phi_0\right] \qquad (4.6)$$

It is assumed that the gain through the delay-line canceler is unity. The output from the canceler [Eq. (4.6)] consists of a cosine wave at the doppler frequency f_d with an amplitude $2k \sin \pi f_d T$. Thus the amplitude of the canceled video output is a function of the doppler frequency shift and the pulse-repetition interval, or prf. The magnitude of the relative frequency-response of the delay-line canceler [ratio of the amplitude of the output from the delay-line canceler, $2k \sin (\pi f_d T)$, to the amplitude of the normal radar video k] is shown in Fig. 4.7.

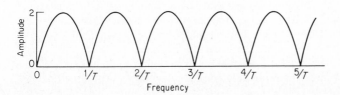

Figure 4.7 Frequency response of the single delay-line canceler; $T =$ delay time $= 1/f_p$.

Blind speeds. The response of the single-delay-line canceler will be zero whenever the argument $\pi f_d T$ in the amplitude factor of Eq. (4.6) is $0, \pi, 2\pi, \ldots$, etc., or when

$$f_d = \frac{n}{T} = nf_p \tag{4.7}$$

where $n = 0, 1, 2, \ldots$, and f_p = pulse repetition frequency. The delay-line canceler not only eliminates the d-c component caused by clutter ($n = 0$), but unfortunately it also rejects any moving target whose doppler frequency happens to be the same as the prf or a multiple thereof. Those relative target velocities which result in zero MTI response are called *blind speeds* and are given by

$$v_n = \frac{n\lambda}{2T} = \frac{n\lambda f_p}{2} \qquad n = 1, 2, 3, \ldots \tag{4.8}$$

where v_n is the nth blind speed. If λ is measured in meters, f_p in Hz, and the relative velocity in knots, the blind speeds are

$$v_n = \frac{n\lambda f_p}{1.02} \approx n\lambda f_p \tag{4.9}$$

The blind speeds are one of the limitations of pulse MTI radar which do not occur with CW radar. They are present in pulse radar because doppler is measured by discrete samples (pulses) at the prf rather than continuously. If the first blind speed is to be greater than the maximum radial velocity expected from the target, the product λf_p must be large. Thus the MTI radar must operate at long wavelengths (low frequencies) or with high pulse repetition frequencies, or both. Unfortunately, there are usually constraints other than blind speeds which determine the wavelength and the pulse repetition frequency. Therefore blind speeds might not be easy to avoid. Low radar frequencies have the disadvantage that antenna beamwidths, for a given-size antenna, are wider than at the higher frequencies and would not be satisfactory in applications where angular accuracy or angular resolution is important. The pulse repetition frequency cannot always be varied over wide limits since it is primarily determined by the unambiguous range requirement. In Fig. 4.8, the first blind speed v_1 is plotted as a function of the maximum unambiguous range ($R_{\text{unamb}} = cT/2$), with radar frequency as the parameter. If the first blind speed were 600 knots, the maximum unambiguous range would be 130 nautical miles at a frequency of 300 MHz (UHF), 13 nautical miles at 3000 MHz (S band), and 4 nautical miles at 10,000 MHz (X band). Since commercial jet aircraft have speeds of the order of 600 knots, and military aircraft even higher, blind speeds in the MTI radar can be a serious limitation.

In practice, long-range MTI radars that operate in the region of L or S band or higher and are primarily designed for the detection of aircraft must usually operate with ambiguous doppler and blind speeds if they are to operate with unambiguous range. The presence of blind speeds within the doppler-frequency band reduces the detection capabilities of the radar. Blind speeds can sometimes be traded for ambiguous range, so that in systems applications which require good MTI performance, the first blind speed might be placed outside the range of expected doppler frequencies if ambiguous range can be tolerated. (Pulse-doppler radars usually operate in this manner). As will be described later, the effect of blind speeds can be significantly reduced, without incurring range ambiguities, by operating with more than one pulse repetition frequency. This is called a *staggered-prf MTI*. Operating at more than one RF frequency can also reduce the effect of blind speeds.

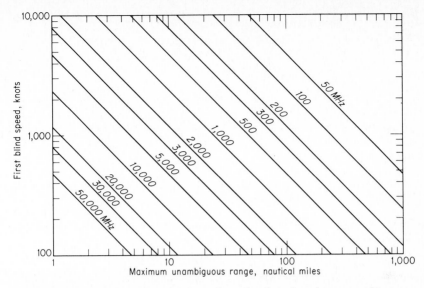

Figure 4.8 Plot of MTI radar first blind speed as a function of maximum unambiguous range.

Double cancellation. The frequency response of a single-delay-line canceler (Fig. 4.7) does not always have as broad a clutter-rejection null as might be desired in the vicinity of d-c. The clutter-rejection notches may be widened by passing the output of the delay-line canceler through a second delay-line canceler as shown in Fig. 4.9a. The output of the two single-delay-line cancelers in cascade is the square of that from a single canceler. Thus the frequency response is $4 \sin^2 \pi f_d T$. The configuration of Fig. 4.9a is called a double-delay-line canceler, or simply a *double canceler*. The relative response of the double canceler compared with that of a single-delay-line canceler is shown in Fig. 4.10. The finite width of the clutter spectrum is also shown in this figure so as to illustrate the additional cancellation of clutter offered by the double canceler.

The two-delay-line configuration of Fig. 4.9b has the same frequency-response characteristic as the double-delay-line canceler. The operation of the device is as follows. A signal $f(t)$ is inserted into the adder along with the signal from the preceding pulse period, with its amplitude weighted by the factor -2, plus the signal from two pulse periods previous. The output of the adder is therefore

$$f(t) - 2f(t + T) + f(t + 2T)$$

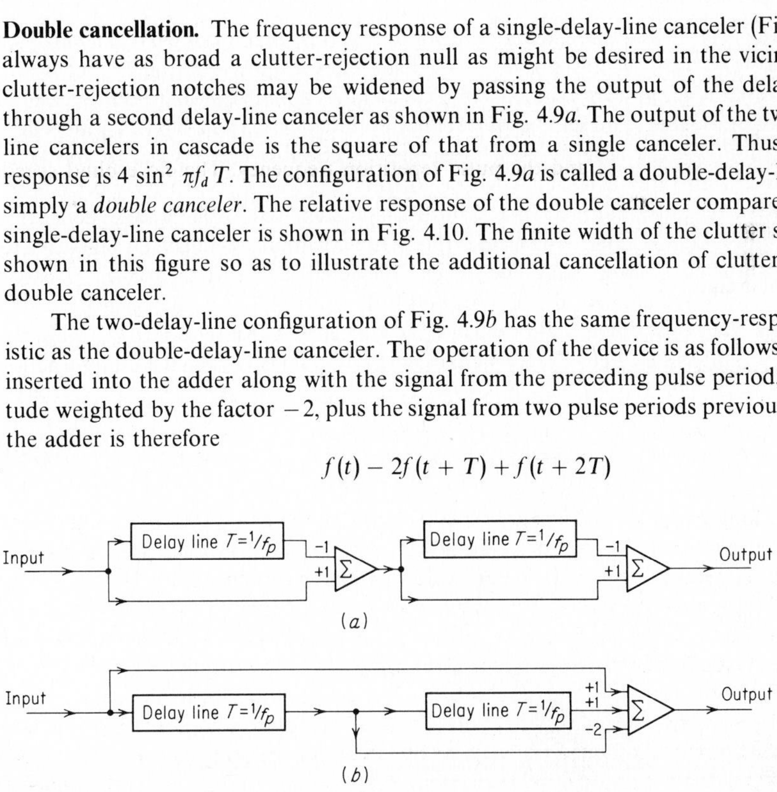

Figure 4.9 (a) Double-delay-line canceler; (b) three-pulse canceler.

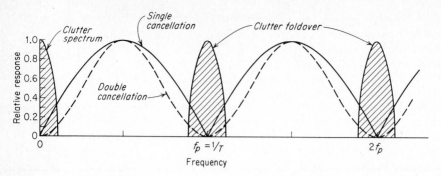

Figure 4.10 Relative frequency response of the single-delay-line canceler (solid curve) and the double-delay-line canceler (dashed curve). Shaded area represents clutter spectrum.

which is the same as the output from the double-delay-line canceler.

$$f(t) - f(t + T) - f(t + T) + f(t + 2T)$$

This configuration is commonly called the *three-pulse canceler*.

Transversal filters. The three-pulse canceler shown in Fig. 4.9b is an example of a *transversal filter*. Its general form with N pulses and $N - 1$ delay lines is shown in Fig. 4.11. It is also sometimes known as a *feedforward filter*, a *nonrecursive filter*, a *finite memory filter* or a *tapped delay-line filter*. The weights w_i for a three-pulse canceler utilizing two delay lines arranged as a transversal filter are 1, -2, 1. The frequency response function is proportional to $\sin^2 \pi f_d T$. A transversal filter with three delay lines whose weights are 1, -3, 3, -1 gives a $\sin^3 \pi f_d T$ response. This is a four-pulse canceler. Its response is equivalent to a triple canceler consisting of a cascade of three single-delay-line cancelers. (Note the potentially confusing nomenclature. A cascade configuration of *three* delay lines, each connected as a single canceler, is called a *triple* canceler, but when connected as a transversal filter it is called a *four-pulse* canceler.) The weights for a transversal filter with n delay lines that gives a response $\sin^n \pi f_d T$ are the coefficients of the expansion of $(1 - x)^n$, which are the binomial coefficients with alternating signs:

$$w_i = (-1)^{i-1} \frac{n!}{(n - i + 1)! \, (i - 1)!}, \qquad i = 1, 2, \ldots, n + 1 \qquad (4.10)$$

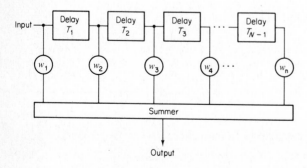

Figure 4.11 General form of a transversal (or nonrecursive) filter for MTI signal processing.

The transversal filter with alternating binomial weights is closely related to the filter which maximizes the average of the ratio $I_c = (S/C)_{out}/(S/C)_{in}$, where $(S/C)_{out}$ is the signal-to-clutter ratio at the output of the filter, and $(S/C)_{in}$ is the signal-to-clutter ratio at the input.[3,4] The average is taken over the range of doppler frequencies. This criterion was first formulated in a limited-distribution report by Emerson.[5] The ratio I_c was called in the early literature the *reference gain*, but it is now called the *improvement factor* for clutter. It is independent of the target velocity and depends only on the weights w_i, the autocorrelation function (or power spectrum) describing the clutter, and the number of pulses. For the two-pulse canceler (a single delay line), the optimum weights based on the above criterion are the same as the binomial weights, when the clutter spectrum is represented by a gaussian function.[6] The difference between a transversal filter with optimal weights and one with binomial weights for a three-pulse canceler (two delay lines) is less than 2 dB.[4,5] The difference is also small for higher-order cancelers. Thus the improvement obtained with optimal weights as compared with binomial weights is relatively small. This applies over a wide range of clutter spectral widths. Similarly, it is found that the use of a criterion which maximizes the clutter attenuation (ratio of input clutter power to the output clutter power) is also well approximated by a transversal filter with binomial weights of alternating sign when the clutter spectrum can be represented by a gaussian function whose spectral width is small compared to the pulse repetition frequency.[7] Thus the delay line cancelers with response $\sin^n \pi f_d T$ are "optimum" in the sense that they approximate the filters which maximize the average signal-to-clutter ratio or the average clutter attentuation. It also approximates the filter which maximizes the probability of detection for a target at the midband doppler frequency or its harmonics.[6]

In spite of the fact that such filters are "optimum" in several senses as mentioned above, they do not necessarily have characteristics that are always desirable for an MTI filter. The notches at dc, at the prf, and the harmonics of the prf are increasingly broad with increasing number of delay lines. Although added delay lines reduce the clutter, they also reduce the number of moving targets detected because of the narrowing of the passband. For example, if the -10 dB response of the filter characteristic is taken as the threshold for detection and if the targets are distributed uniformly across the doppler frequency band, 20 percent of all targets will be rejected by a two-pulse canceler (single delay-line canceler), 38 percent will be rejected by a three-pulse canceler (double canceler), and 48 percent by a four-pulse canceler (triple canceler). These filters might be "optimum" in that they satisfy the specified criterion, but the criterion might not be the best for satisfying MTI requirements. (Optimum is not a synonym for *best*; it means the best under the given set of assumptions.) Maximizing the signal-to-clutter ratio over all doppler frequencies, which leads to the binomial weights and $\sin^n \pi f_d T$ filters, is not necessarily a pertinent criterion for design of an MTI filter since this criterion is independent of the target signal characteristics.[3–5] It would seem that the MTI filter should be shaped to reject the clutter at d-c and around the prf and its harmonics, but have a flat response over the region where no clutter is expected. That is, it would be desirable to have the freedom to shape the filter response, just as with any conventional filter.

The transversal, or nonrecursive, filter of Fig. 4.11 can be designed to achieve filter responses suitable for MTI[7–10] but a relatively large number of delay lines are needed for filters with desirable characteristics. An $N-1$ delay-line canceler requires N pulses, which sets a restriction on the radar's pulse repetition frequency, beamwidth, and antenna rotation rate, or dwell time.

Figure 4.12[26] shows the amplitude response for (1) a classical three-pulse canceler with $\sin^2 \pi f_d T$ response, (2) a five-pulse "optimum" canceler designed to maximize the improvement factor[3] and (3) a 15-pulse canceler with a Chebyshev filter characteristic. (The amplitude is normalized by dividing the output of each tap by the square root of $\sum_{i=1}^{N} w_i^2$, where

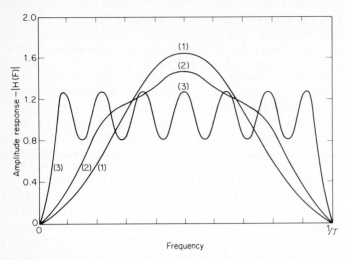

Figure 4.12 Amplitude responses for three MTI delay-line cancelers. (1) Classical three-pulse canceler, (2) five-pulse delay-line canceler with "optimum" weights, and (3) 15-pulse Chebyshev design. (*After Houts and Burlage.*[26])

w_i = weight at the ith tap.) A large number of delay lines are seen to be required of a nonrecursive canceler if highly-shaped filter responses are desired. It has been suggested,[26] however, that even with only a five-pulse canceler, a five-pulse Chebyshev design provides significantly wider bandwidth than the five-pulse "optimum" design. To achieve the wider band the Chebyshev design has a lower improvement factor (since it is not "optimum"), but in many cases the trade is worthwhile especially if the clutter spectrum is narrow. However, when only a few pulses are available for processing there is probably little that can be done to control the shape of the filter characteristic. Thus, there is not much to be gained in trying to shape the nonrecursive filter response for three- or four-pulse cancelers other than to use the classical $\sin^2$ or $\sin^3$ response of the "optimum" canceler.

The N-pulse nonrecursive delay-line canceler allows the designer N zeros for synthesizing the frequency response. The result is that many delay lines are required for highly-shaped filter responses. There are limits to the number of delay lines (and pulses) that can be employed. Therefore other approaches to MTI filter implementation are sometimes desired.

Shaping the frequency response. Nonrecursive filters employ only feedforward loops. If feedback loops are used, as well as feedforward loops, each delay line can provide one pole as well as one zero for increased design flexibility. The *canonical* configuration of a time-domain filter with feedback as well as feedforward loops is illustrated in Fig. 4.13. When feedback loops are

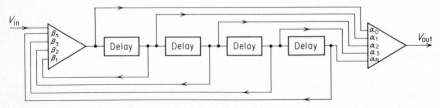

Figure 4.13 Canonical-configuration comb filter. (*After White and Ruvin,*[2] *IRE Natl. Conv. Record.*)

used the filter is called *recursive*. Using the Z-transform as the basis for design it is possible in principle to synthesize almost any frequency-response function.[2,11-13]

The canonical configuration is useful for conceptual purposes, but it may not always be desirable to design a filter in this manner. White and Ruvin[2] state that the canonical configuration may be broken into cascaded sections, no section having more than two delay elements. Thus no feedback or feedforward path need span more than two delay elements. This type of configuration is sometimes preferred to the canonical configuration.

The synthesis technique described by White and Ruvin may be applied with any known low-pass filter characteristic, whether it is a Butterworth, Chebyshev, or Bessel filter or one of the filters based on the elliptic-function transformation which has equal ripple in the rejection band as well as in the passband. An example of the use of these filter characteristics applied to the design of a delay-line periodic filter is given in either of White's papers.[2,12] Consider the frequency-response characteristic of a three-pole Chebyshev low-pass filter having 0.5 dB ripple in the passband (Fig. 4.14). The three different delay-line-filter frequency-response characteristics shown in Fig. 4.14b to d were derived from the low-pass filter characteristic of Fig. 4.14a. This type of filter characteristic may be obtained with a single delay line in cascade with a double delay line as shown in Fig. 4.15. The weighting factors shown on the feedback paths apply to the characteristic of Fig. 4.14c.

The additional degrees of freedom available in the design of recursive delay-line filters offer a steady-state response that is superior to that of comparable nonrecursive filters. However the feedback loops in the recursive filter result in a poor transient response. The presence

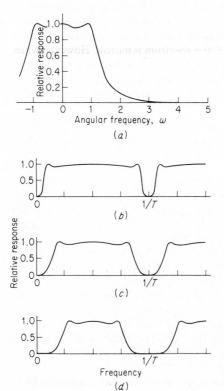

Figure 4.14 (a) Three-pole Chebyshev low-pass filter characteristic with 0.5 dB ripple in the passband; (b–d) delay-line filter characteristics derived from (a). (*After White.*[12])

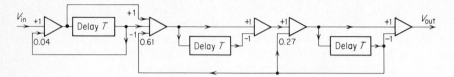

Figure 4.15 Form of the delay-line filter required to achieve the characteristic of Fig. 4.14c.

of large clutter returns can effectively appear as a large step input to the filter with the result that severe ringing is produced in the filter output. The ringing can mask the target signal until the transient response dies out. In the usual surveillance radar application, the number of pulses from any target is limited, hence the operation of an MTI filter is almost always in the transient state for most recursive filters.[16] It has been said that a frequency response characteristic with steep sides might allow 15 to 30 or more pulses to be generated at the filter output because of the feedback.[14] This might make the system unusable in situations in the presence of large discrete clutter, and where interference from nearby radars or intentional jamming is encountered. The poor transient response might also be undesirable in a radar with a step-scan antenna (as in a phased array), since these extra pulses might have to be gated out after the beam is moved to a new position. In the step-scan antenna it has been suggested that the undesirable transient effects can be mitigated by using the initial return from each new beam position, or an average of the first few returns, to apply initial conditions to the MTI filter for processing the remaining returns from that position.[15] The clutter returns can be approximated by a step-input equal in magnitude to the first return for that beam position so that the steady-state values that would normally appear in the filter memory elements after an infinitely long sequence of these inputs can be immediately calculated and loaded into the filter to suppress the transient response.

Although the undesirable transient effects of a recursive filter can be reduced to some extent, other approaches to MTI filtering are often desired. One alternative the designer has available is the use of multiple pulse-repetition frequencies for achieving the desired MTI filter characteristics, as described next.

4.3 MULTIPLE, OR STAGGERED, PULSE REPETITION FREQUENCIES[8,17–22,81,82]

The use of more than one pulse repetition frequency offers additional flexibility in the design of MTI doppler filters. It not only reduces the effect of the blind speeds [Eq. (4.8)], but it also allows a sharper low-frequency cutoff in the frequency response than might be obtained with a cascade of single-delay-line cancelers with $\sin^n \pi f_d T$ response.

The blind speeds of two independent radars operating at the same frequency will be different if their pulse repetition frequencies are different. Therefore, if one radar were " blind " to moving targets, it would be unlikely that the other radar would be " blind " also. Instead of using two separate radars, the same result can be obtained with one radar which time-shares its pulse repetition frequency between two or more different values (*multiple prf's*). The pulse repetition frequency might be switched every other scan or every time the antenna is scanned a half beamwidth, or the period might be alternated on every other pulse. When the switching is pulse to pulse, it is known as a *staggered* prf.

An example of the composite (average) response of an MTI radar operating with two separate pulse repetition frequencies on a time-shared basis is shown in Fig. 4.16. The pulse

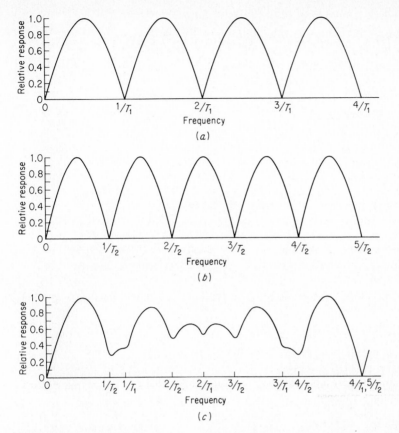

Figure 4.16 (a) Frequency-response of a single-delay-line canceler for $f_p = 1/T_1$; (b) same for $f_p = 1/T_2$; (c) composite response with $T_1/T_2 = \frac{4}{5}$.

repetition frequencies are in the ratio of 5 : 4. Note that the first blind speed of the composite response is increased several times over what it would be for a radar operating on only a single pulse repetition frequency. Zero response occurs only when the blind speeds of each prf coincide. In the example of Fig. 4.16, the blind speeds are coincident for $4/T_1 = 5/T_2$. Although the first blind speed may be extended by using more than one prf, regions of low sensitivity might appear within the composite passband.

The closer the ratio $T_1 : T_2$ approaches unity, the greater will be the value of the first blind speed. However, the first null in the vicinity of $f_d = 1/T_1$ becomes deeper. Thus the choice of T_1/T_2 is a compromise between the value of the first blind speed and the depth of the nulls within the filter pass band. The depth of the nulls can be reduced and the first blind speed increased by operating with more than two interpulse periods. Figure 4.17 shows the response of a five-pulse stagger (four periods) that might be used with a long-range air traffic control radar.[8] In this example the periods are in the ratio 25 : 30 : 27 : 31 and the first blind speed is 28.25 times that of a constant prf waveform with the same average period. If the periods of the staggered waveforms have the relationship $n_1/T_1 = n_2/T_2 = \cdots = n_N/T_N$, where $n_1, n_2, \ldots, n_N$ are integers, and if v_B is equal to the first blind speed of a nonstaggered waveform with a

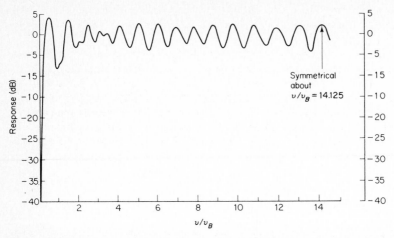

Figure 4.17 Frequency response of a five-pulse (four-period) stagger. (*From Shrader,*[8] *Courtesy McGraw-Hill Book Co.*)

constant period equal to the average period $T_{av} = (T_1 + T_2 + \cdots T_N)/N$, then the first blind speed v_1 is

$$\frac{v_1}{v_B} = \frac{n_1 + n_2 + \cdots + n_N}{N} \tag{4.11}$$

It is also possible to apply weighting to the received pulses of a staggered prf waveform. An example is shown in Fig. 4.18.[22] The dashed curve is the response of a five-pulse canceler with fixed prf and with weightings of $\frac{7}{8}$, 1, $-3\frac{3}{4}$, 1, $\frac{7}{8}$. The solid curve is for a staggered prf with

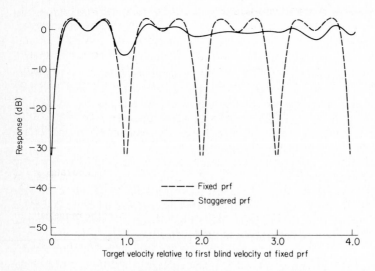

Figure 4.18 Response of a weighted five-pulse canceler. Dashed curve, constant prf; solid curve, staggered prf's. (*From Zverev,*[22] *Courtesy IEEE.*)

the same weightings, but with 4 interpulse periods of -15 percent, -5 percent, $+5$ percent, and $+15$ percent of the fixed period. The response at the first blind speed of the fixed period waveform is down only 6.6 dB.

A disadvantage of the staggered prf is its inability to cancel second-time-around clutter echoes. Such clutter does not appear at the same range from pulse to pulse and thus produces uncanceled residue. Second-time-around clutter echoes can be removed by use of a constant prf, providing there is pulse-to-pulse coherence as in the power amplifier form of MTI. The constant prf might be employed only over those angular sectors where second-time-around clutter is expected (as in the ARSR-3 of Sec. 14.3), or by changing the prf each time the antenna scans half-a-beamwidth (as in the MTD of Sec. 4.7), or by changing the prf every scan period (rotation of the antenna).

4.4 RANGE-GATED DOPPLER FILTERS

The delay-line canceler, which can be considered as a time-domain filter, has been widely used in MTI radar as the means for separating moving targets from stationary clutter. It is also possible to employ the more usual frequency-domain bandpass filters of conventional design in MTI radar to sort the doppler-frequency-shifted targets. The filter configuration must be more complex, however, than the single, narrow-bandpass filter. A narrowband filter with a passband designed to pass the doppler frequency components of moving targets will "ring" when excited by the usual short radar pulse. That is, its passband is much narrower than the reciprocal of the input pulse width so that the output will be of much greater duration than the input. The narrowband filter "smears" the input pulse since the impulse response is approximately the reciprocal of the filter bandwidth. This smearing destroys the range resolution. If more than one target is present they cannot be resolved. Even if only one target were present, the noise from the other range cells that do not contain the target will interfere with the desired target signal. The result is a reduction in sensitivity due to a collapsing loss (Sec. 2.12).

The loss of the range information and the collapsing loss may be eliminated by first quantizing the range (time) into small intervals. This process is called *range gating*. The width of the range gates depends upon the range accuracy desired and the complexity which can be tolerated, but they are usually of the order of the pulse width. Range resolution is established by gating. Once the radar return is quantized into range intervals, the output from each gate may be applied to a narrowband filter since the pulse shape need no longer be preserved for range resolution. A collapsing loss does not take place since noise from the other range intervals is excluded.

A block diagram of the video of an MTI radar with multiple range gates followed by clutter-rejection filters is shown in Fig. 4.19. The output of the phase detector is sampled sequentially by the range gates. Each range gate opens in sequence just long enough to sample the voltage of the video waveform corresponding to a different range interval in space. The range gate acts as a switch or a gate which opens and closes at the proper time. The range gates are activated once each pulse-repetition interval. The output for a stationary target is a series of pulses of constant amplitude. An echo from a moving target produces a series of pulses which vary in amplitude according to the doppler frequency. The output of the range gates is stretched in a circuit called the *boxcar generator*, or *sample-and-hold* circuit, whose purpose is to aid in the filtering and detection process by emphasizing the fundamental of the modulation frequency and eliminating harmonics of the pulse repetition frequency (Sec. 5.3). The clutter-rejection filter is a bandpass filter whose bandwidth depends upon the extent of the expected clutter spectrum.

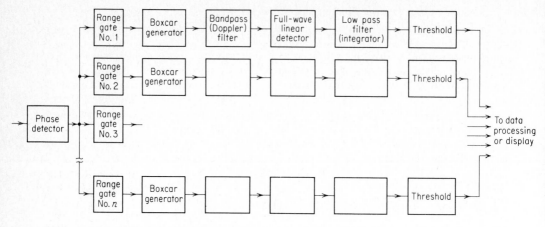

Figure 4.19 Block diagram of MTI radar using range gates and filters.

Following the doppler filter is a full-wave linear detector and an integrator (a low-pass filter). The purpose of the detector is to convert the bipolar video to unipolar video. The output of the integrator is applied to a threshold-detection circuit. Only those signals which cross the threshold are reported as targets. Following the threshold detector, the outputs from each of the range channels must be properly combined for display on the PPI or A-scope or for any other appropriate indicating or data-processing device. The CRT display from this type of MTI radar appears "cleaner" than the display from a normal MTI radar, not only because of better clutter rejection, but also because the threshold device eliminates many of the unwanted false alarms due to noise. The frequency-response characteristic of the range-gated MTI might appear as in Fig. 4.20. The shape of the rejection band is determined primarily by the shape of the bandpass filter of Fig. 4.19.

The bandpass filter can be designed with a variable low-frequency cutoff that can be selected to conform to the prevailing clutter conditions. The selection of the lower cutoff might be at the option of the operator or it can be done adaptively. A variable lower cutoff might be advantageous when the width of the clutter spectrum changes with time as when the radar receives unwanted echoes from birds. A relatively wide notch at zero frequency is needed to remove moving birds. If the notch were set wide enough to remove the birds, it might be wider than necessary for ordinary clutter and desired targets might be removed. Since the appearance of birds varies with the time of day and the season, it is important that the width of the notch be controlled according to the local conditions.

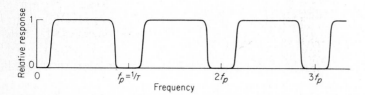

Figure 4.20 Frequency-response characteristic of an MTI using range gates and filters.

MTI radar using range gates and filters is usually more complex than an MTI with a single-delay-line canceler. The additional complexity is justified in those applications where good MTI performance and the flexibility of the range gates and filter MTI are desired. The better MTI performance results from the better match between the clutter filter characteristic and the clutter spectrum.

4.5 DIGITAL SIGNAL PROCESSING

The introduction of practical and economical digital processing to MTI radar allowed a significant increase in the options open to the signal processing designer. The convenience of digital processing means that multiple delay-line cancelers with tailored frequency-response characteristics can be readily achieved. A digital MTI processor does not, in principle, do any better than a well-designed analog canceler; but it is more dependable, it requires less adjustments and attention, and can do some tasks easier. Most of the advantages of a digital MTI processor are due to its use of digital delay lines, rather than analog delay lines which are characterized by variations due to temperature, critical gains, and poor on-line availability.

A simple block diagram of a digital MTI processor is shown in Fig. 4.21. From the output of the IF amplifier the signal is split into two channels. One is denoted I, for *in-phase channel*. The other is denoted Q, for *quadrature channel*, since a 90° phase change ($\pi/2$ radians) is introduced into the coho reference signal at the phase detector. This causes the outputs of the two detectors to be 90° out of phase. The purpose of the quadrature channel is to eliminate the effects of *blind phases*, as will be described later. It is desirable to eliminate blind phases in any MTI processor, but it is seldom done with analog delay-line cancelers because of the complexity of the added analog delay lines of the second channel. The convenience of digital processing allows the quadrature channel to be added without significant burden so that it is often included in digital processing systems. It is for this reason it is shown in this block diagram, but was not included in the previous discussion of MTI delay-line cancelers.

Following the phase detector the bipolar video signal is sampled at a rate sufficient to obtain one or more samples within each range resolution cell. These voltage samples are converted to a series of digital words by the analog-to-digital (A/D) converter.[23,24] The digital

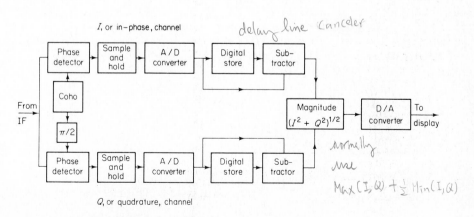

Figure 4.21 Block diagram of a simple digital MTI signal processor.

words are stored in a digital memory for one pulse repetition period and are then subtracted from the digital words of the next sweep. The digital outputs of the I and Q channels are combined by taking the square root of $I^2 + Q^2$. An alternative method of combining, which is adequate for most cases, is to take $|I| + |Q|$. The combined output is then converted to an analog signal by the digital-to-analog (D/A) converter. The unipolar video output is then ready to be displayed. The digital MTI processor depicted in Fig. 4.21 is that of a single-delay-line canceler. Digital processors are likely to employ more complex filtering schemes, but the simple canceler is shown here for convenience. Almost any type of digital storage device can be used. A shift register is the direct digital analogy of a delay line, but other digital computer memories can also be used effectively.

The A/D converter has been, in the past, one of the critical parts of the MTI signal processor. It must operate at a speed high enough to preserve the information content of the radar signal, and the number of bits into which it quantizes the signal must be sufficient for the precision required. The number of bits in the A/D converter determines the maximum *improvement factor* the MTI radar can achieve.[8,75,76] Generally the A/D converter is designed to cover the peak excursion of the phase detector output. A limiter may be necessary to ensure this. An N-bit converter divides the output of the phase detector into $2^N - 1$ discrete intervals. According to Shrader,[8] the quantization noise introduced by the discrete nature of the A/D converter causes, on the average, a limit to the improvement factor which is

$$I_{QN} = 20 \log \left[(2^N - 1)\sqrt{0.75} \right] \quad \text{(dB)} \qquad (4.12)$$

This is approximately equal to 6 dB per bit since each bit represents a factor of two in amplitude resolution.[22] When a fixed signal of maximum level is present, a possible error of one quantization interval is possible. A 9-bit A/D converter therefore has a maximum discrimination of 1 out of 511 levels; or approximately 54 dB. (Equation (4.12) on the other hand, predicts 52.9 dB for 9-bit quantization.)

In the above it was said that the addition of the Q channel removed the problem of reduced sensitivity due to *blind phases*. This is different from the blind speeds which occur when the pulse sampling appears at the same point in the doppler cycle at each sampling instant, as shown in Fig. 4.22a. Figure 4.22b shows the in-phase, or I, channel with the pulse train such

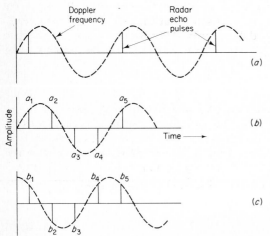

(a)

(b)

Time →

Amplitude

(c)

Figure 4.22 (a) Blind speed in an MTI radar. The target doppler frequency is equal to the prf. (b) Effect of blind phase in the I channel, and (c) in the Q channel.

that the signals are of the same amplitude and with a spacing such that when pulse a_1 is subtracted from pulse a_2, the result is zero. However, a residue is produced when pulse a_2 is subtracted from pulse a_3, but not when a_3 is subtracted from a_4, and so on. In the quadrature channel, the doppler-frequency signal is shifted 90° so that those pulse pairs that were lost in the I channel are recovered in the Q channel, and vice versa. The combination of the I and Q channels thus results in a uniform signal with no loss. The phase of the pulse train relative to that of the doppler signal in Fig. 4.22b and c is a special case to illustrate the effect. With other phase and frequency relationships, there is still a loss with a single channel MTI that can be recovered by the use of both the I and Q channels. An extreme case where the blind phase with only a single channel results in a complete loss of signal is when the doppler frequency is half the prf and the phase relationship between the two is such that the echo pulses lie on the zeros of the doppler-frequency sine wave. This is not the condition for a blind speed but nevertheless there is no signal. However, if the phase relationship is shifted 90°, as it is in the Q channel, then all the echo pulses occur at the peaks of the doppler-frequency sine wave. Thus, to ensure the signal will be obtained without loss, both I and Q channels are desired.

Digital signal processing has some significant advantages over analog delay lines, particularly those that use acoustic devices. As with most digital technology, it is possible to achieve greater stability, repeatability, and precision with digital processing than with analog delay-line cancelers. Thus the reliability is better. No special temperature control is required, and it can be packaged in convenient size. The dynamic range is greater since digital MTI processors do not experience the spurious responses which limit signals in acoustic delay lines to about 35 to 40 dB above minimum detectable signal level.[25] (A major restriction on dynamic range in a digital MTI is that imposed by the A/D converter.) In an analog delay-line canceler the delay time and the pulse repetition period must be made equal. This is simplified in a digital MTI since the timing of the sampling of the bipolar video can be controlled readily by the timing of the transmitted pulse. Thus, different pulse repetition periods can be used without the necessity of switching delay lines of various lengths in and out. The echo signals for each interpulse period can be stored in the digital memory with reference to the time of transmission. This allows more elaborate stagger periods. The flexibility of the digital processor also permits more freedom in the selection and application of amplitude weightings for shaping the filters. It has also allowed the ready incorporation of the quadrature channel for elimination of blind phases. In short, digital MTI has allowed the radar designer the freedom to take advantage of the full theoretical capabilities of doppler processing in practical radar systems.

The development of digital processing technology has not only made the delay-line canceler a more versatile tool for the MTI radar designer, but it has also allowed the application of the contiguous filter bank for added flexibility in MTI radar design. One of the major factors in this regard has been the introduction of digital devices for conveniently computing the Fourier transform.

Digital filter banks and the FFT. A transversal filter with N outputs (N pulses and $N - 1$ delay lines) can be made to form a bank of N contiguous filters covering the frequency range from 0 to f_p, where f_p = pulse repetition frequency. A filter bank covering the doppler frequency range is of advantage in some radar applications and offers another option in the design of MTI signal processors. Consider the transversal filter that was shown in Fig. 4.11 to have $N - 1$ delay lines each with a delay time $T = 1/f_p$. Let the weights applied to the outputs of the N taps be:

$$w_{ik} = e^{-j[2\pi(i-1)k/N]} \tag{4.13}$$

where $i = 1, 2, \ldots, N$ represents the ith tap, and k is an index from 0 to $N - 1$. Each value of k corresponds to a different set of N weights, and to a different doppler-filter response. The N filters generated by the index k constitute the filter bank. Note that the weights of Eq. (4.13) and the diagram of Fig. 4.11 are similar in form to the phased array antenna as described in Sec. 8.2.

The impulse response of the transversal filter of Fig. 4.11 with the weights given by Eq. (4.13) is

$$h_k(t) = \sum_{i=1}^{N} \delta[t - (i - 1)T]e^{-j2\pi(i-1)k/N} \tag{4.14}$$

The Fourier transform of the impulse response is the frequency response function

$$H_k(f) = e^{-j2\pi ft} \sum_{i=1}^{N} e^{j2\pi(i-1)[fT-k/N]} \tag{4.15}$$

The magnitude of the frequency response function is the amplitude passband characteristic of the filter. Therefore

$$|H_k(f)| = \left| \sum_{i=1}^{N} e^{j2\pi(i-1)[fT-k/N]} \right| = \left| \frac{\sin[\pi N(fT - k/N)]}{\sin[\pi(fT - k/N)]} \right| \tag{4.16}$$

By analogy to the discussion of the array antenna in Sec. 8.2, the peak response of the filter occurs when the denominator of Eq. (4.16) is zero, or when $\pi(fT - k/N) = 0, \pi, 2\pi, \ldots$. For $k = 0$, the peak response of the filter occurs at $f = 0, 1/T, 2/T, \ldots$, which defines a filter centered at dc, the prf, and its harmonics. This filter passes the clutter component at dc, hence it has no clutter rejection capability. (Its output is useful, however, in some MTI radars for providing a map of the clutter.) The first null of the filter response occurs when the numerator is zero, or when $f = 1/NT$. The bandwidth between the first nulls is $2/NT$ and the half-power bandwidth is approximately $0.9/NT$ (Fig. 4.23).

When $k = 1$, the peak response occurs at $f = 1/NT$ as well as $f = 1/T + 1/NT, 2/T + 1/NT$, etc. For $k = 2$, the peak response is at $f = 2/NT$, and so forth. Thus each value of the index k defines a separate filter response, as indicated in Fig. 4.23, with the total response covering the region from $f = 0$ to $f = 1/T = f_p$. Each filter has a bandwidth of $2/NT$ as measured between the first nulls. Because of the sampled nature of the signals, the remainder of the frequency band is also covered with similar response, but with ambiguity. A bank of filters, as in Fig. 4.23, is sometimes called a *coherent integration filter*.

To generate the N filters simultaneously, each of the taps of the transversal filter of Fig. 4.11 would have to be divided into N separate outputs with separate weights correspond-

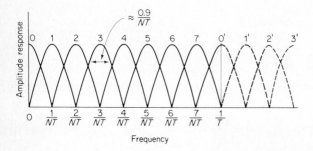

Figure 4.23 MTI doppler filter bank resulting from the processing of $N = 8$ pulses with the weights of Eq. (4.13), yielding the response of Eq. (4.16). Filter sidelobes not shown.

ing to the $k = 0$ to $N - 1$ weighting as given by Eq. (4.13). (This is analogous to generating N independent beams from an N-element array by use of the Blass multiple-beam array as in Fig. 8.26.) When generating the filter bank by digital processing it is not necessary literally to subdivide each of the N taps. The equivalent can be accomplished in the digital computations.

The generation by digital processing of N filters from the outputs of N taps of a transversal filter requires a total of $(N - 1)^2$ multiplications. The process is equivalent to the computation of a *discrete Fourier transform*. However, when N is some power of 2, that is, $N = 2, 4, 8, 16, \ldots$, an algorithm is available that requires approximately $(N/2) \log_2 N$ multiplications instead of $(N - 1)^2$, which results in a considerable saving in processing time. This is the *fast Fourier transform* (FFT), which has been widely described in the literature.[11,27–29]

Since each filter occupies approximately $1/N$th the bandwidth of a delay-line canceler, its signal-to-noise ratio will be greater than that of a delay-line canceler. The dividing of the frequency band into N independent parts by the N filters also allows a measure of the doppler frequency to be made. Furthermore if moving clutter, such as from birds or weather, appears at other-than-zero frequency, the threshold of each filter may be individually adjusted so as to adapt to the clutter contained within it. This selectivity allows clutter to be removed which would be passed by a delay-line canceler. The first sidelobe of the filters described by Eq. (4.16) has a value of -13.2 dB with respect to the peak response of the filter. If this is too high for proper clutter rejection, it may be reduced at the expense of wider bandwidth by applying amplitude weights w_i. Just as is done in antenna design, the filter sidelobe can be reduced by using Chebyshev, Taylor, $\sin^2 x/x^2$, or other weightings.[39] (In the digital signal processing literature, filter weighting is called *windowing*.[80] Two popular forms of windows are the Hamming and hanning.) It may not be convenient to display the outputs of all the doppler filters of the filter bank. One approach is to connect the output of the filters to a *greatest-of* circuit so that only a single output is obtained, that of the largest signal. (The filter at dc which contains clutter would not be included.)

The improvement factor for each of the 8 filters of an 8-pulse filter bank is shown in Fig. 4.24 as a function of the standard deviation of the clutter spectrum assuming the spectrum to be represented by the gaussian function described by Eq. (4.19). The average improvement for all filters is indicated by the dotted curve. For comparison, the improvement factor for an N-pulse canceler is shown in Fig. 4.25. Note that the improvement factor of a two-pulse canceler is almost as good as that of the 8-pulse doppler-filter bank. The three-pulse canceler is even better. (As mentioned previously, maximizing the average improvement factor might not be the only criterion used in judging the effectiveness of MTI doppler processors.)

If a two- or three-pulse canceler is cascaded with a doppler-filter bank, better clutter rejection is provided. Figure 4.26 shows the improvement factor for a three-pulse canceler and an eight-pulse filter bank in cascade, as a function of the clutter spectral width. The upper figure assumes uniform amplitude weighting (-13.2 dB first sidelobe) and the lower figure shows the effect of Chebyshev weighting designed to produce equal sidelobes with a peak value of -25 dB. It is found that doubling to 16 the number of pulses in the filter bank does not offer significant advantage over an eight-pulse filter.[30] More pulses do not necessarily mean more gain in signal-to-clutter ratio, because the filter widths and sidelobe levels change relative to the clutter spectrum as the number of pulses (and number of filters) is varied. If the sidelobes of the individual filters of the doppler filter bank can be made low enough, the inclusion of the delay-line canceler ahead of it might not be needed.

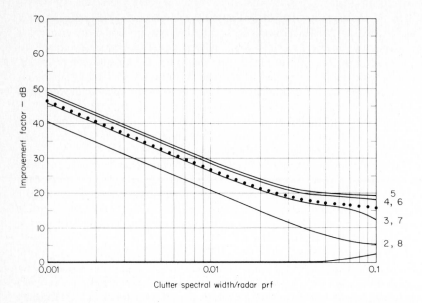

Figure 4.24 Improvement factor for each filter of an 8-pulse doppler filter bank with uniform weighting as a function of the clutter spectral width (standard deviation). The average improvement for all filters is indicated by the dotted curve. (*From Andrews.*[30])

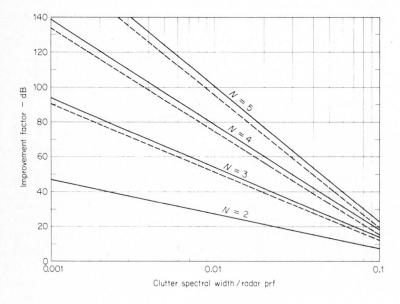

Figure 4.25 Improvement factor for an N-pulse delay-line canceler with optimum weights (solid curves) and binomial weights (dashed curves), as a function of the clutter spectral width. (*After Andrews.*[31,32])

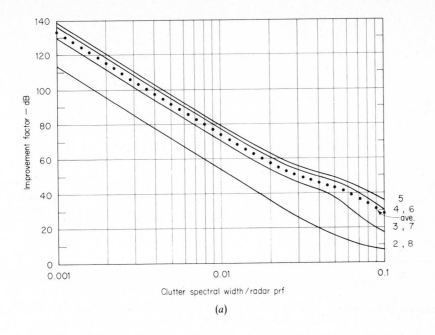

(a)

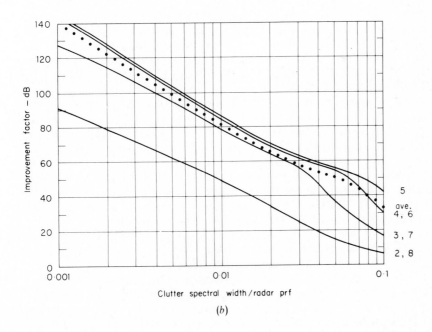

(b)

Figure 4.26 Improvement factor for a 3-pulse (double-canceler) MTI cascaded with an 8-pulse doppler filter bank, or integrator. (a) Uniform amplitude weights and (b) 25-dB Chebyshev weights. The average improvement for all filters is indicated by the dotted curve. (*From Andrews.*[30])

4.6 OTHER MTI DELAY LINES

There are delay lines other than digital devices that have been used in MTI signal processors. Originally, MTI radar used acoustic delay lines in which electromagnetic signals were converted into acoustic waves; the acoustic signals were delayed, and then converted back into electromagnetic signals. The process was lossy (50 to 70 dB might be typical), of limited dynamic range, and spurious responses were generated that could be confused for legitimate echoes. Since acoustic waves travel with a speed about 10^{-5} that of electromagnetic waves, an acoustic line can be of practical size whereas an electromagnetic delay is not. However, acoustic delay lines are larger and heavier than digital lines and must usually be kept in a temperature-controlled environment to prevent unwanted changes in delay time. They typically operate at frequencies from 5 to 60 MHz. The first MTI acoustic delay lines were liquid and used water or mercury. The development of the solid quartz delay line for MTI application in the 1950s offered greater convenience than the liquid lines.[33] The solid line is constructed of many facets so that a relatively long delay time can be obtained with a small volume by means of multiple internal reflections. In one popular design the quartz was cut as a 15-sided polygon in which the acoustic signal made 31 internal passes to achieve a total delay time of 1 ms.

Electrostatic storage tubes have also been used for MTI delay lines.[34,40] The signals are stored in the form of electrostatic charges on a mosaic or mesh similar to that of a TV camera tube. The first sweep reads the signal on the storage tube. The next sweep is written on the same space and generates the difference between it and the first sweep, as required for two-pulse MTI cancellation. Two storage tubes are required: one to write and store the new sweep, the other to subtract the new sweep from the old sweep. One advantage of a storage tube compared to an acoustic delay line is that it can be used with different pulse repetition frequencies. That is, a constant prf is not needed. The matching of the delay time and the pulse repetition period is not critical as in an acoustic line because the timing of the transmitter pulse starts the sweep of the storage tube.

The charge transfer device (CTD) is a sampled-data, analog shift-register that may be used in MTI processors as delay lines, transversal filters, and recursive filters.[35–38] There are two basic types of CTDs. One is the bucket-brigade device (BBD) which consists of capacitive storage elements separated by switches. Analog information is transferred through the BBD at a rate determined by the rate at which the switches are operated. The delay time can be varied by changing either the number of stages (a capacitive storage element and its associated switch) or by changing the switching frequency controlled by a digital clock. CCDs provide a similar function except that charge packets are transferred to adjacent potential wells by clocked voltage gradients. The sampling frequency must be such as to obtain one or two samples per range resolution cell. Since the timing is controlled by a digital clock, the delay can be made very stable, just as with digital processors. However, no A/D or D/A converters are required as in digital systems. It has been claimed that such devices are more economical in cost, consume less power, are of less size, and of greater reliability than digital processing.

The MTI cancellation may also be done at IF rather than in the video after the phase detector.[8] Although there may be some advantages in IF cancellation (such as no blind phases), it is more complicated than video cancellation. It is difficult for IF cancelers to compete with the flexibility and convenience of digital processing containing I and Q channels. FM delay-line cancelers have also been considered for MTI application. The signals are frequency modulated so as to avoid the problems associated with the amplitude stability of acoustic delay-line cancelers.[41]

4.7 EXAMPLE OF AN MTI RADAR PROCESSOR

The Moving Target Detector (MTD) is an MTI radar processor originally developed by the MIT Lincoln Laboratory for the FAA's Airport Surveillance Radars (ASR).[42-44] The ASR is a medium range (60 nmi) radar located at most major United States airports. It operates at S band (2.7–2.9 GHz) with a pulse width of less than 1 μs, a 1.4° azimuth beamwidth, an antenna rotation rate of from 12.5 to 15 rpm depending on the model, a prf from 700 to 1200 Hz (1030 Hz typical), and an average power of from 400 to 600 W. The MTD processor employs several techniques for the increased detection of moving targets in clutter. Its implementation is based on the application of digital technology. It utilizes a three-pulse canceler followed by an 8-pulse FFT doppler filter-bank with weighting in the frequency domain to reduce the filter sidelobes, alternate prf's to eliminate blind speeds, adaptive thresholds, and a clutter map that is used in detecting crossing targets with zero radial velocity. The measured MTI improvement factor of the MTD on an ASR radar was about 45 dB, which represented a 20 dB increase over the ASR's conventional three-pulse MTI processor with a limiting IF amplifier. In addition, the MTD achieves a narrower notch at zero velocity and at the blind speeds.

The processor is preceded by a large dynamic range receiver to avoid the reduction in improvement factor caused by limiting (Sec. 4.8). The output of the IF amplifier is fed to I and Q phase detectors. The analog signals from the phase detectors are converted into 10-bit digital words by A/D converters. The range coverage, which totaled 47.5 nmi in the original implementation, is divided into $\frac{1}{16}$ nmi intervals and the azimuth into $\frac{3}{4}$-degree intervals, for a total of 365,000 range-azimuth resolution cells. In each $\frac{3}{4}$-degree azimuth interval (about one-half the beamwidth) ten pulses are transmitted at a constant prf. On receive, this is called a coherent processing interval (CPI). These ten pulses are processed by the delay-line canceler and the doppler filter-bank, to form eight doppler filters. Thus, the radar output is divided into approximately 2,920,000 range-azimuth-doppler cells. Each of these cells has its own adaptive threshold. In the alternate $\frac{3}{4}$-degree azimuth intervals another 10 pulses are transmitted at a different prf to eliminate blind speeds and to unmask moving targets hidden by weather. Changing the prf every 10 pulses, or every half beamwidth, eliminates second-time-around clutter returns that would normally degrade an MTI with pulse-to-pulse variation of the prf.

A block diagram of the MTD processor is shown in Fig. 4.27. The input on the left is from the output of the I and Q A/D converters. The three-pulse canceler and the eight-pulse doppler filter-bank eliminate zero-velocity clutter and generate eight overlapping filters covering the doppler interval, as described in the previous section. The use of a three-pulse canceler ahead of the filter-bank eliminates stationary clutter and thereby reduces the dynamic range required

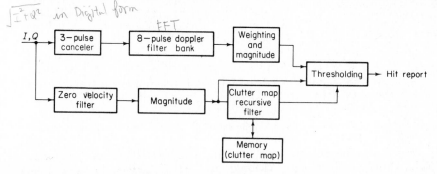

Figure 4.27 Simple block diagram of the Moving Target Detector (MTD) signal processor.

of the doppler filter-bank. The fast Fourier transform algorithm is used to implement the doppler filter-bank. Since the first two pulses of a three-pulse canceler are meaningless, only the last eight of the ten pulses output from the canceler are passed to the filter-bank. Following the filter-bank, weighting is applied in the frequency domain to reduce the filter sidelobes. Although unweighted filters with -13.2-dB sidelobes might be satisfactory for most ground clutter, the frequency spectrum of rain and other windblown clutter often requires lower sidelobe levels. The magnitude operation forms $(I^2 + Q^2)^{1/2}$. Separate thresholds are applied to each filter. The thresholds for the nonzero-velocity resolution cells are established by summing the detected outputs of the signals in the same velocity filter in 16 range cells, eight on either side of the cell of interest. Thus, each filter output is averaged over one mile in range to establish the statistical mean level of nonzero-velocity clutter (such as rain) or noise. The filter thresholds are determined by multiplying the mean levels by an appropriate constant to obtain the desired false-alarm probability. This application of an adaptive threshold to each doppler filter at each range cell provides a constant false-alarm rate (CFAR) and results in *subweather visibility* in that an aircraft with a radial velocity sufficiently different from the rain so as to fall into another filter can be seen even if the aircraft echo is substantially less than the weather echo.

A digital clutter map is generated which establishes the thresholds for the zero-velocity cells. The map is implemented with one word for each of the 365,000 range-azimuth cells. The original MTD stored the map on a magnetic disc memory. The purpose of the zero-velocity filter is to recover the clutter signal eliminated by the MTI delay-line canceler and to use this signal as a means for detecting targets on crossing trajectories with zero velocities that would normally be lost in the usual MTI. Only targets larger than the clutter would be so detected. In each cell of the ground clutter map is stored the average value of the output of the zero-velocity filter for the past eight scans (32 seconds). On each scan, one-eighth of the output of the zero-velocity filter is added to seven-eighths of the value stored in the map. Thus, the map is built up in a recursive manner. About 10 to 20 scans are required to establish steady-clutter values. As rain moves into the area, or as propagation conditions result in changing ground-clutter levels, the clutter map changes accordingly. The values stored in the map are multiplied by an appropriate constant to establish the threshold for zero-relative-velocity targets. This eliminates the usual MTI blind speed at zero radial velocity and permits the detection of crossing targets in clutter if the target cross section is sufficiently large.

Thus, in Fig. 4.28 the zero-velocity filter (No. 1) threshold is determined by the output of the clutter map. The thresholds for filters 3 through 7 are obtained from the mean level of the signals in the 16 range cells centered around the range cell of interest. Both a mean-level

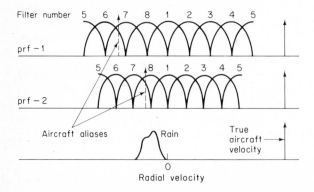

Figure number   5   6   7   8   1   2   3   4   5

prf − 1

5   6   7   8   1   2   3   4   5

prf − 2

Aircraft aliases   Rain   True aircraft velocity

0
Radial velocity

Figure 4.28 Detection of aircraft in rain using two prf's with a doppler filter bank, illustrating the effect of doppler foldover. (*From Muehe,*[43] *Courtesy IEEE.*)

threshold from the 16 range cells and a clutter threshold from the clutter map are calculated for filters 2 and 8 adjacent to the zero-velocity filter, and the larger of the two is used as the threshold.

The advantage of using two prf's to detect targets in rain is illustrated by Fig. 4.28. The prf's differ by about 20 percent. A typical rain spectrum with a nonzero average velocity is depicted on the bottom line. (Precipitation at S band might typically have a spectral width of about 25 to 30 knots centered anywhere from -60 to $+60$ knots, depending on the wind conditions and the antenna pointing.[42]) The narrow spectrum of the moving aircraft is at the right of the figure. Because of foldover it is shown as occupying filters 6 and 7 on prf-1, and filters 7 and 8 on prf-2. With prf-2, the aircraft velocity is shown competing with the rain clutter; but with prf-1 it appears in one filter (No. 6) without any rain clutter. Thus, by using two prf's which are alternated every 10 pulses (one-half beamwidth), aircraft targets will usually appear in at least one filter free from rain except for a small region (approximately 30 knots wide) when the target's radial velocity is exactly that of the rain.

Interference from other radars might appear as one large return among the ten returns normally processed. In the MTD, an interference eliminator compares the magnitude of each of the 10 pulses against the average magnitude of the ten. If any pulse is greater than five times the average, all information from that range-azimuth cell is discarded. A saturation detector detects whether any of the 10 pulses within a processing interval saturates the A/D converter and if it does, the entire 10 pulses are discarded.

The output of the MTD is a hit report which contains the azimuth, range, and amplitude of the target return as well as the filter number and prf. On a particular scan, a large aircraft might be reported from more than one doppler filter, from several coherent processing intervals, and from adjacent range gates. As many as 20 hit-reports might be generated by a single large target.[43] A post processor groups together all reports which appear to originate from the same target and interpolates to find the best azimuth, range, amplitude, and radial velocity. The target amplitude and doppler are used to eliminate small cross section and low-speed angel echoes before the target reports are delivered to the automatic tracking circuits. The tracking circuits also eliminate false hit-reports which do not form logical tracks. The output of the automatic tracker is what is displayed on the scope. Since the MTD processor eliminates a large amount of the clutter and has a low false-detection rate, its output can be reliably remoted via narrow bandwidth telephone circuits.

4.8 LIMITATIONS TO MTI PERFORMANCE

The improvement in signal-to-clutter ratio of an MTI is affected by factors other than the design of the doppler signal processor. Instabilities of the transmitter and receiver, physical motions of the clutter, the finite time on target (or scanning modulation), and limiting in the receiver can all detract from the performance of an MTI radar. Before discussing these effects, some definitions will be stated.

MTI improvement factor. The signal-to-clutter ratio at the output of the MTI system divided by the signal-to-clutter ratio at the input, averaged uniformly over all target radial velocities of interest.

Subclutter visibility. The ratio by which the target echo power may be weaker than the coincident clutter echo power and still be detected with specified detection and false-alarm probabilities. All target radial velocities are assumed equally likely. A subclutter visibility of, for example, 30 dB implies that a moving target can be detected in the presence of clutter even though the clutter echo power is 1000 times the target echo power. Two radars with the same subclutter visibility might not have the same ability to detect targets in clutter if the resolution cell of one is greater than the other and accepts a greater clutter signal power; that is, both radars might reduce the clutter power equally, but one starts with greater clutter power because its resolution cell is greater and "sees" more clutter targets.

Clutter visibility factor. The signal-to-clutter ratio, after cancellation or doppler filtering, that provides stated probabilities of detection and false alarm.

Clutter attenuation. The ratio of clutter power at the canceler input to the clutter residue at the output, normalized to the attenuation of a single pulse passing through the unprocessed channel of the canceler. (The *clutter residue* is the clutter power remaining at the output of an MTI system.)

Cancellation ratio. The ratio of canceler voltage amplification for the fixed-target echoes received with a fixed antenna, to the gain for a single pulse passing through the unprocessed channel of the canceler.

The improvement factor (I) is equal to the subclutter visibility (SCV) times the clutter visibility factor (V_{oc}). In decibels, $I(dB) = SCV(dB) + V_{oc}(dB)$. When the MTI is limited by noiselike system instabilities, the clutter visibility factor should be chosen as is the signal-to-noise ratio of Chap. 2. When the MTI performance is limited by the antenna scanning fluctuations, Shrader[8] suggests letting $V_{oc} = 6$ dB. (Although not stated by Shrader, it seems this value is for a single pulse.) The improvement factor is the preferred measure of MTI radar performance.

Still another term sometimes employed in MTI radar is the *interclutter visibility*. This describes the ability of an MTI radar to detect moving targets which occur in the relatively clear resolution cells between patches of strong clutter. Clutter echo power is not uniform, so if a radar has sufficient resolution it can see targets in the clear areas between clutter patches. The higher the radar resolution, the better the interclutter visibility. Radars with "moderate" resolution might require only enough improvement factor to deal with the median clutter power, which may be 20 dB less than the average clutter power.[45] According to Shrader a medium-resolution radar with a 2 μs pulse width and a 1.5° beamwidth, is of sufficient resolution to achieve a 20 dB advantage over low-resolution radars for the detection of targets in ground clutter.[50]

Equipment instabilities. Pulse-to-pulse changes in the amplitude, frequency, or phase of the transmitter signal, changes in the stalo or coho oscillators in the receiver, jitter in the timing of the pulse transmission, variations in the time delay through the delay lines, and changes in the pulse width can cause the apparent frequency spectrum from perfectly stationary clutter to broaden and thereby lower the improvement factor of an MTI radar. The stability of the equipment in an MTI radar must be considerably better than that of an ordinary radar. It can limit the performance of an MTI radar if sufficient care is not taken in design, construction, and maintenance.

Consider the effect of phase variations in an oscillator. If the echo from stationary clutter on the first pulse is represented by $A \cos \omega t$ and from the second pulse is $A \cos (\omega t + \Delta\phi)$, where $\Delta\phi$ is the change in oscillator phase between the two, then the difference between the two after subtraction is $A \cos \omega t - A \cos (\omega t + \Delta\phi) = 2A \sin (\Delta\phi/2) \sin (\omega t + \Delta\phi/2)$. For small phase errors, the amplitude of the resultant difference is $2A \sin \Delta\phi/2 \simeq A \Delta\phi$. Therefore the limitation on the improvement factor due to oscillator instability is

$$I = \frac{1}{(\Delta\phi)^2} \tag{4.17}$$

This would apply to the coho locking or to the phase change introduced by a power amplifier. A phase change pulse-to-pulse of 0.01 radians results in an improvement-factor limitation of 40 dB. The limits to the improvement factor imposed by pulse-to-pulse instability are listed below:[8,46,47]

Transmitter frequency	$(\pi \Delta f \tau)^{-2}$
Stalo or coho frequency	$(2\pi \Delta f T)^{-2}$
Transmitter phase shift	$(\Delta\phi)^{-2}$
Coho locking	$(\Delta\phi)^{-2}$
Pulse timing	$\tau^2/(\Delta t)^2 2B\tau$
Pulse width	$\tau^2/(\Delta\tau)^2 B\tau$
Pulse amplitude	$(A/\Delta A)^2$

where Δf = interpulse frequency change, τ = pulse width, T = transmission time to and from target, $\Delta\phi$ = interpulse phase change, Δt = time jitter, $B\tau$ = time-bandwidth product of pulse compression system (= unity for simple pulses,) $\Delta\tau$ = pulse-width jitter, A = pulse amplitude, ΔA = interpulse amplitude change. In a digital signal processor the improvement factor is also limited by the quantization noise introduced by the A/D converter, as was discussed in Sec. 4.5. The digital processor, however, does not experience degradation due to time jitter of the transmitted pulse since the system clock controlling the processor timing may be started from the detected rf envelope of the transmitted pulse.

Internal fluctuation of clutter.[48] Although clutter targets such as buildings, water towers, bare hills, or mountains produce echo signals that are constant in both phase and amplitude as a function of time, there are many types of clutter that cannot be considered as absolutely stationary. Echoes from trees, vegetation, sea, rain, and chaff fluctuate with time, and these fluctuations can limit the performance of MTI radar.

Because of its varied nature, it is difficult to describe precisely the clutter echo signal. However, for purposes of analysis, most fluctuating clutter targets may be represented by a model consisting of many independent scatterers located within the resolution cell of the radar. The echo at the radar receiver is the vector sum of the echo signals received from each of the individual scatters; that is, the relative phase as well as the amplitude from each scatterer influences the resultant composite signal. If the individual scatters remain fixed from pulse to pulse, the resultant echo signal will also remain fixed. But any motion of the scatterers relative to the radar will result in different phase relationships at the radar receiver. Hence the phase and amplitude of the new resultant echo signal will differ pulse to pulse.

Examples of the power spectra of typical clutter are shown in Fig. 4.29. These data apply at a frequency of 1000 MHz. The experimentally measured power spectra of clutter signals may be approximated by

$$W(f) = |g(f)|^2 = |g_0|^2 \exp\left[-a\left(\frac{f}{f_0}\right)^2\right] \tag{4.18}$$

where $W(f)$ = clutter-power spectrum as a function of frequency
  $g(f)$ = Fourier transform of input waveform (clutter echo)
  f_0 = radar carrier frequency
  a = a parameter dependent upon clutter

Values of the parameter a which correspond to the clutter spectra in Fig. 4.29 are given in the caption.

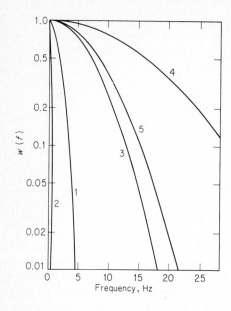

Figure 4.29 Power spectra of various clutter targets. (1) Heavily wooded hills, 20 mi/h wind blowing $(a = 2.3 \times 10^{17})$; (2) sparsely wooded hills, calm day $(a = 3.9 \times 10^{19})$; (3) sea echo, windy day $(a = 1.41 \times 10^{16})$; (4) rain clouds $(a = 2.8 \times 10^{15})$; (5) chaff $(a = 1 \times 10^{16})$. (*From Barlow,*[49] *Proc. IRE.*)

The clutter spectrum can also be expressed in terms of an rms clutter frequency spread σ_c in hertz or by the rms velocity spread σ_v in m/s.[46] Thus Eq. (4.18) can be written

$$W(f) = W_0 \exp\left(-\frac{f^2}{2\sigma_c^2}\right) = W_0 \exp\left(-\frac{f^2\lambda^2}{8\sigma_v^2}\right) \tag{4.19}$$

where $W_0 = |g_0|^2$, $\sigma_c = 2\sigma_v/\lambda$, λ = wavelength = c/f_0, and c = velocity of propagation. It can be seen that $a = c^2/8\sigma_v^2$. The rms velocity spread σ_v is usually the preferred method for describing the clutter fluctuation spectrum.

The improvement factor can be written as

$$I = \left(\frac{S_0/C_0}{S_i/C_i}\right)_{ave} = \left(\frac{S_0}{S_i}\right)_{ave} \times \frac{C_i}{C_0} = \left(\frac{S_0}{S_i}\right)_{ave} \times CA \tag{4.20}$$

where S_0/C_0 = output signal-to-clutter ratio, S_i/C_i = input signal-to-clutter ratio, and CA = clutter attentuation. The average is taken over all target doppler frequencies of interest. For a single-delay-line canceler, the clutter attentuation is

$$CA = \frac{\int_0^\infty W(f)\,df}{\int_0^\infty W(f)|H(f)|^2\,df} \tag{4.21}$$

where $H(f)$ is the frequency response function of the canceler. Since the frequency response function of a delay line of time delay T is $\exp(-j2\pi f T)$, $H(f)$ for the single-delay-line canceler is

$$H(f) = 1 - \exp(-j2\pi f T) = 2j \sin(\pi f T) \exp(-j\pi f T) \tag{4.22}$$

Substituting Eqs. (4.19) and (4.22) into Eq. (4.21) and assuming that $\sigma_c \ll 1/T$, the clutter attentuation is

$$CA = \frac{\int_0^\infty W_0 \exp(-f^2/2\sigma_c^2)\,df}{\int_0^\infty W_0 \exp(-f^2/2\sigma_c^2)4\sin^2 \pi f T\,df}$$

$$= \frac{0.5}{1 - \exp(-2\pi^2 T^2\sigma_c^2)} \tag{4.23}$$

If the exponent in the denominator of Eq. (4.23) is small, the exponential term can be replaced by the first two terms of a series expansion, or

$$CA = \frac{f_p^2}{4\pi^2\sigma_c^2} = \frac{f_p^2\lambda^2}{16\pi^2\sigma_v^2} = \frac{af_p^2}{2\pi^2f_0^2} \tag{4.24}$$

where f_p, the pulse repetition frequency, has been substituted for $1/T$. The average gain $(S_0/S_i)_{\text{ave}}$ of the single delay-line-canceler can be shown to be equal to 2. Therefore, the improvement factor is

$$I_{1C} = \frac{f_p^2}{2\pi^2\sigma_c^2} = \frac{f_p^2\lambda^2}{8\pi^2\sigma_v^2} = \frac{af_p^2}{\pi^2f_0^2} \tag{4.25}$$

Similarly, for a double canceler, whose average gain is 6, the improvement factor is

$$I_{2C} \simeq \frac{f_p^2}{8\pi^4\sigma_c^4} = \frac{f_p^4\lambda^4}{128\pi^4\sigma_v^4} = \frac{a^2f_p^4}{2\pi^4f_0^4} \tag{4.26}$$

A plot of Eq. (4.26) for the double canceler is shown in Fig. 4.30. The parameter describing the curves is $f_p\lambda$. Example prf's and frequencies are shown. Several "representative" examples of clutter are indicated, based on published data for σ_v, which for the most part dates back to World War II.[46,49] Although each type of clutter is shown at a particular value of σ_v, nature is more variable than this. Actual measurements cover a range of values. The spectral spread in velocity is with respect to the mean velocity, which for ground clutter is usually zero. Rain and

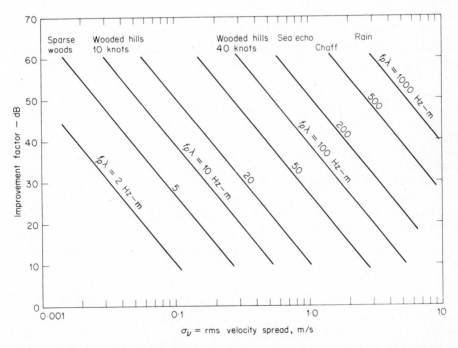

Figure 4.30 Plot of double-canceler clutter improvement factor [Eq. (4.26)] as a function of σ_v = rms velocity spread of the clutter. Parameter is the product of the pulse repetition frequency (f_p) and the radar wavelength (λ).

chaff, however, as well as sea echo, can have a nonzero mean velocity[47] which must be properly accounted for when designing MTI signal processors.

The frequency dependence of the clutter spectrum as given by Eqs. (4.25) and (4.26) cannot be extended over too great a frequency range since account is not taken of any variation in radar cross section of the individual scatterers as a function of frequency. The leaves and branches of trees, for example, might have considerably different reflecting properties at K_a band ($\lambda = 0.86$ cm), where the dimensions are comparable with the wavelength, from those at VHF ($\lambda = 1.35$ m), where the wavelength is long compared with the dimensions.

The general expression for improvement factor for an N-pulse canceler with $N_l = N - 1$ delay lines is[61]

$$I_{NC} = \frac{2^{N_l}}{N_l!} \left(\frac{f_p}{2\pi\sigma_c} \right)^{2N_l} \tag{4.27}$$

Antenna scanning modulation.[46,49-52] As the antenna scans by a target, it observes the target for a finite time equal to $t_0 = n_B/f_p = \theta_B/\theta_s$, where n_B = number of hits received, f_p = pulse repetition frequency, θ_B = antenna beamwidth and θ_s = antenna scanning rate. The received pulse train of finite duration t_0 has a frequency spectrum (which can be found by taking the Fourier transform of the waveform) whose width is proportional to $1/t_0$. Therefore, even if the clutter were perfectly stationary, there will still be a finite width to the clutter spectrum because of the finite time on target. If the clutter spectrum is too wide because the observation time is too short, it will affect the improvement factor. This limitation has sometimes been called *scanning fluctuations* or *scanning modulation*.

The computation of the limitation to the improvement factor can be found in a manner similar to that of the clutter fluctuations described previously. The clutter attentuation is first found using Eq. (4.21), except that the power spectrum $W_s(f)$ describing the spectrum produced by the finite time on target is used. The clutter attenuation is

$$CA = \frac{\int_0^\infty W_s(f)\, df}{\int_0^\infty W_s(f)|H(f)|^2\, df} \tag{4.28}$$

where $H(f)$ is the frequency response function of the MTI signal processor. If the antenna main-beam pattern is approximated by the gaussian shape, the spectrum will also be gaussian. Therefore, the results previously derived for a gaussian clutter spectrum can be readily applied. Equations 4.25 and 4.26 derived for the clutter fluctuation improvement factor apply for the antenna scanning fluctuations by proper interpretation of σ_c, the standard deviation, or the rms spread of the frequency spectrum about the mean.

The voltage waveform of the received signal is modulated by the square of the antenna electric-field-strength-pattern, which is equal to the (one-way) antenna power pattern $G(\theta)$, described by the gaussian function as

$$G(\theta) = G_0 \exp\left(-\frac{2.776\theta^2}{\theta_B^2} \right) \tag{4.29}$$

Since the antenna is scanning at a rate of θ_s deg/s the time waveform may be found from Eq. (4.29) by dividing both the numerator and denominator of the exponent by θ_s. Letting $\theta/\theta_s = t$, the time variable, and noting that $\theta_B/\theta_s = t_0$, the time on target, the modulation of the received signal due to the antenna pattern is

$$s_a(t) = k \exp\left(-\frac{2.776\, t^2}{t_0^2} \right) \tag{4.30}$$

where k = constant. The angular frequency spectrum of this time waveform is found by taking its Fourier transform, which is

$$S_a(f) = k \int_{-\infty}^{\infty} \exp\left(-\frac{2.776t^2}{t_0^2}\right) \exp\left(-j2\pi ft\right) dt$$

$$= k_1 \exp\left[-\frac{\pi^2 f^2 t_0^2}{2.776}\right] \tag{4.31}$$

where k_1 = constant. Since this is a gaussian function, the exponent is of the form $f^2/2\sigma_f^2$ where σ_f = standard deviation. Therefore

$$\sigma_f = \frac{1.178}{\pi t_0} \tag{4.32}$$

This applies to the voltage spectrum. Since the standard deviation of the power spectrum is less than that of the voltage spectrum by $\sqrt{2}$, the power spectrum due to antenna scanning can be described by a standard deviation

$$\sigma_s = \frac{1.178}{\sqrt{2}\pi t_0} = \frac{1}{3.77 t_0} \tag{4.33}$$

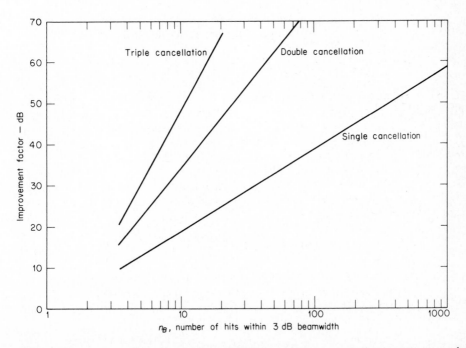

Figure 4.31 Limitation to improvement factor due to a scanning antenna. Antenna pattern assumed to be of gaussian shape.

This can be substituted for σ_c in Eqs. (4.25) and (4.26) to obtain the limitation to the improvement factor caused by antenna scanning. These are

$$I_{1s} = \frac{n_B^2}{1.388} \quad \text{(single canceler)} \tag{4.34}$$

$$I_{2s} = \frac{n_B^4}{3.853} \quad \text{(double canceler)} \tag{4.35}$$

These are plotted in Fig. 4.31.

A stepped-scan antenna that dwells at a particular region in space, rather than scan continuously, also is limited in MTI performance by the finite time on target t_0. The time waveform is constant so that it will have a different spectral shape and a different improvement factor than that produced by the gaussian beam assumed in the above.

Limiting in MTI radar.[8,53-55] A limiter is usually employed in the IF amplifier just before the MTI processor to prevent the residue from large clutter echoes from saturating the display. Ideally an MTI radar should reduce the clutter to a level comparable to receiver noise.

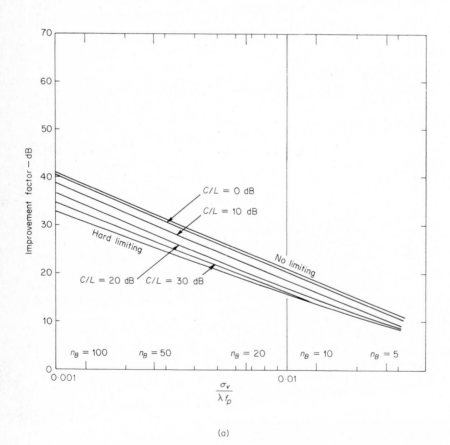

(a)

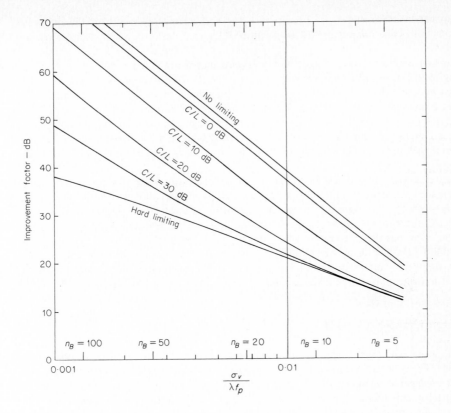

Figure 4.32 Effect of limit level on the improvement factor for (*a*) two-pulse delay-line canceler and (*b*) three-pulse delay-line canceler. C/L = ratio of rms clutter power to limit level. (*From Ward and Shrader,*[53] *Courtesy IEEE.*)

However, when the MTI improvement factor is not great enough to reduce the clutter sufficiently, the clutter residue will appear on the display and prevent the detection of aircraft targets whose cross sections are larger than the clutter residue. This condition may be prevented by setting the limit level L, relative to the noise N, equal to the MTI improvement factor I; or $L/N = I$. If the limit level relative to noise is set higher than the improvement factor, clutter residue obscures part of the display. If it is set too low there may be a "black hole" effect on the display. The limiter provides a constant false alarm rate (CFAR) and is essential to usable MTI performance.[50]

Unfortunately, nonlinear devices such as limiters have side-effects that can degrade performance. Limiters cause the spectrum of strong clutter to spread into the canceler passband, and result in the generation of additional residue that can significantly degrade MTI performance as compared with a perfect linear system.

An example of the effect of limiting is shown in Fig. 4.32, which plots the improvement factor for two-pulse and three-pulse cancelers with various levels of limiting.[53] The abscissa applies to a gaussian clutter spectrum that is generated either by clutter motion with standard deviation σ_v at a wavelength λ and a prf f_p, or by antenna scanning modulation with a gaussian-shaped beam and n_B pulses between the half-power beamwidth of the one-way antenna pattern. The parameter C/L is the ratio of the rms clutter power to the receiver-IF limit level.

The loss of improvement factor increases with increasing complexity of the canceler. Limiting in the three-pulse canceler will cause a 15 to 25 dB reduction in the performance predicted by linear theory.[50] A four-pulse canceler (not shown) with limiting is typically only 2 dB better than the three-pulse canceler in the presence of limiting clutter and offers little advantage. Thus the added complexity of higher-order cancelers is seldom justified in such situations. The linear analysis of MTI signal processors is therefore not adequate when limiting is employed and can lead to disappointing differences between theory and measurement of actual systems.

Limiters need not be used if the MTI is linear over the entire range of clutter signals and if the processor has sufficient improvement factor to reduce the largest clutter to the noise level. To accomplish this the signal processor must provide at least 60 dB improvement factor, which is a difficult task.[56] Not only must the signal processor be designed to reduce the clutter by this amount, but the receiver must be linear over this range, there must be at least eleven bits in the A/D converter of the digital processor, the equipment must be sufficiently stable, and the number of pulses processed (for reducing antenna scanning modulation) must be sufficient to achieve this large value of improvement factor.

4.9 NONCOHERENT MTI

The composite echo signal from a moving target and clutter fluctuates in both phase and amplitude. The coherent MTI and the pulse-doppler radar make use of the *phase* fluctuations in the echo signal to recognize the doppler component produced by a moving target. In these systems, amplitude fluctuations are removed by the phase detector. The operation of this type of radar, which may be called *coherent MTI*, depends upon a reference signal at the radar receiver that is coherent with the transmitter signal.

It is also possible to use the *amplitude* fluctuations to recognize the doppler component produced by a moving target. MTI radar which uses amplitude instead of phase fluctuations is called *noncoherent* (Fig. 4.33). It has also been called *externally coherent*, which is a more descriptive name. The noncoherent MTI radar does not require an internal coherent reference signal or a phase detector as does the coherent form of MTI. Amplitude limiting cannot be

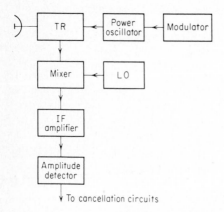

To cancellation circuits

Figure 4.33 Block diagram of a noncoherent MTI radar.

employed in the noncoherent MTI receiver, else the desired amplitude fluctuations would be lost. Therefore the IF amplifier must be linear, or if a large dynamic range is required, it can be logarithmic. A logarithmic gain characteristic not only provides protection from saturation, but it also tends to make the clutter fluctuations at its output more uniform with variations in the clutter input amplitude [Sec. (13.8)]. The detector following the IF amplifier is a conventional amplitude detector. The phase detector is not used since phase information is of no interest to the noncoherent radar. The local oscillator of the noncoherent radar does not have to be as frequency-stable as in the coherent MTI. The transmitter must be sufficiently stable over the pulse duration to prevent beats between overlapping ground clutter, but this is not as severe a requirement as in the case of coherent radar. The output of the amplitude detector is followed by an MTI processor such as a delay-line canceler. The doppler component contained in the amplitude fluctuations may also be detected by applying the output of the amplitude detector to an A-scope. Amplitude fluctuations due to doppler produce a butterfly modulation similar to that in Fig. 4.3, but in this case, they ride on top of the clutter echoes. Except for the inclusion of means to extract the doppler amplitude component, the noncoherent MTI block diagram is similar to that of a conventional pulse radar.

The advantage of the noncoherent MTI is its simplicity; hence it is attractive for those applications where space and weight are limited. Its chief limitation is that the target must be in the presence of relatively large clutter signals if moving-target detection is to take place. Clutter echoes may not always be present over the range at which detection is desired. The clutter serves the same function as does the reference signal in the coherent MTI. If clutter were not present, the desired targets would not be detected. It is possible, however, to provide a switch to disconnect the noncoherent MTI operation and revert to normal radar whenever sufficient clutter echoes are not present. If the radar is stationary, a map of the clutter might be stored in a digital memory and used to determine when to switch in or out the noncoherent MTI.

The improvement factor of a noncoherent MTI will not, in general, be as good as can be obtained with a coherent MTI that employs a reference oscillator (coho). The reference signal in the noncoherent case is the clutter itself, which will not be as stable as a reference oscillator because of the finite width of the clutter spectrum caused by its own internal motions. If a nonlinear IF amplifier is used, it will also limit the improvement factor that can be achieved.

4.10 PULSE DOPPLER RADAR[57]

A pulse radar that extracts the doppler frequency shift for the purpose of detecting moving targets in the presence of clutter is either an *MTI* radar or a *pulse doppler radar*. The distinction between them is based on the fact that in a sampled measurement system like a pulse radar, ambiguities can arise in both the doppler frequency (relative velocity) and the range (time delay) measurements. Range ambiguities are avoided with a *low* sampling rate (low pulse repetition frequency), and doppler frequency ambiguities are avoided with a *high* sampling rate. However, in most radar applications the sampling rate, or pulse repetition frequency, cannot be selected to avoid *both* types of measurement ambiguities. Therefore a compromise must be made and the nature of the compromise generally determines whether the radar is called an MTI or a pulse doppler. MTI usually refers to a radar in which the pulse repetition frequency is chosen low enough to avoid ambiguities in range (no multiple-time-around echoes), but with the consequence that the frequency measurement is ambiguous and results in blind speeds, Eq. (4.8). The pulse doppler radar, on the other hand, has a high pulse repetition frequency that avoids blind speeds, but it experiences ambiguities in range. It performs doppler filtering on a single spectral line of the pulse spectrum. (A radar which employs multiple pulse repetition frequencies to avoid blind speeds is usually classed as an MTI if its average prf would cause blind speeds. The justification for this definition is that the technology and design philosophy of a multiple prf radar are more like that of an MTI than a pulse doppler.)

The pulse doppler radar is more likely to use range-gated doppler filter-banks than delay-line cancelers. Also, a power amplifier such as a klystron is more likely to be used than a power oscillator like the magnetron. A pulse doppler radar operates at a higher duty cycle than does an MTI. Although it is difficult to generalize, the MTI radar seems to be the more widely used of the two, but pulse doppler is usually more capable of reducing clutter.

When the prf must be so high that the number of range ambiguities is too large to be easily resolved, the performance of the pulse-doppler radar approaches that of the CW doppler radar. The pulse-doppler radar, like the CW radar, may be limited in its ability to measure range under these conditions. Even so, the pulse-doppler radar has an advantage over the CW radar in that the detection performance is not limited by transmitter leakage or by signals reflected from nearby clutter or from the radome. The pulse-doppler radar avoids this difficulty since its receiver is turned off during transmission, whereas the CW radar receiver is always on. On the other hand, the detection capability of the pulse-doppler radar is reduced because of the blind spots in range resulting from the high prf.

One other method should be mentioned of achieving coherent MTI. If the number of cycles of the doppler frequency shift contained within the duration of a single pulse is sufficient, the returned echoes from moving targets may be separated from clutter by suitable RF or IF filters. This is possible if the doppler frequency shift is at least comparable with or greater than the spectral width of the transmitted signal. It is not usually applicable to aircraft targets, but it can sometimes be applied to radars designed to detect extraterrestrial targets such as satellites or astronomical bodies. In these cases, the transmitted pulse width is relatively wide and its spectrum is narrow. The high speed of extraterrestrial targets results in doppler shifts that are usually significantly greater than the spectral width of the transmitted signal.

4.11 MTI FROM A MOVING PLATFORM

When the radar itself is in motion, as when mounted on a ship or an aircraft, the detection of a moving target in the presence of clutter is more difficult than if the radar were stationary. The doppler frequency shift of the clutter is no longer at dc. It varies with the speed of the radar platform, the direction of the antenna in azimuth, and the elevation angle to the clutter. Thus the clutter rejection notch needed to cancel clutter cannot be fixed, but must vary. The design of an MTI is more difficult with an airborne radar than a shipborne radar because the higher speeds and the greater range of elevation angles result in a greater variation of the clutter spectrum.

In addition to shifting the center frequency of the clutter, its spectrum is also widened. An approximate measure of the spectrum width can be found by taking the differential of the doppler frequency $f_d = 2(v/\lambda) \cos \theta$, or

$$\Delta f_d = \frac{2v}{\lambda} \sin \theta \ \Delta\theta \tag{4.36}$$

where v = platform speed, λ = wavelength, and θ is the azimuth angle between the aircraft's velocity and the direction of the antenna beam. (The negative sign introduced by differentiation of $\cos \theta$ is ignored and the elevation angle is assumed to be zero.) If the beamwidth is taken as $\Delta\theta$, then Δf_d is a measure of the width of doppler frequency spectrum. When the antenna points in the direction of the platform velocity ($\theta = 0$), the doppler shift of the clutter is maximum, but the width of the doppler spectrum Δf_d is a minimum. On the other hand, when the antenna is directed perpendicular to the direction of the platform velocity ($\theta = 90°$), the clutter doppler center-frequency is zero, but the spread is maximum. This widening of the clutter spectrum can set a limit on the improvement factor.

Thus the effect of platform velocity can be considered as having two components. One is in the direction of antenna pointing and shifts the center frequency of the clutter doppler spectrum. The other is normal to the direction of antenna pointing and results in a widening of the clutter doppler spectrum. These two components are compensated by two different techniques.

An MTI radar on a moving platform is called AMTI. Although the "A" originally stood for *airborne*, the term is now often applied to an MTI radar on any moving platform. Most of the interest in AMTI, however, is for airborne radar.[58]

Compensation for clutter doppler shift. When the clutter doppler frequency is other than at dc, the null of the frequency response of the MTI processor must be shifted accordingly. The effect on the improvement factor when the center of the clutter doppler frequency is shifted by an amount f_e is shown in Fig. 4.34 for a three-pulse delay-line canceler.[61,69] There are two basic methods for providing the doppler frequency compensation. In one implementation the frequency of the coherent oscillator (coho) is changed to compensate for the shift in the clutter doppler frequency. This may be accomplished by mixing the output of the coho with a signal from a tunable oscillator, the frequency of which is made equal to the clutter doppler. The other implementation is to insert a phase shifter in one branch of the delay-line canceler and adjust its phase to shift the null of the frequency response. (A phase shift Ψ in one branch of the canceler corresponds to a frequency shift $2\pi f_d = \Psi/T_p$, where T_p = pulse repetition interval.)

The clutter-doppler-frequency compensation can, in some cases, be made open loop by using the a priori knowledge of the velocity of the platform carrying the radar and the direction of the antenna pointing. This is more practical with a shipborne radar rather than with an airborne radar. When the clutter-doppler-frequency compensation cannot be obtained

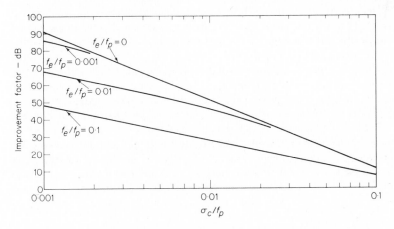

Figure 4.34 Effect of a nonzero clutter doppler frequency on the improvement factor of a three-pulse canceler. f_e = mean frequency of the clutter spectrum, σ_c = standard deviation of clutter spectrum, f_p = pulse repetition frequency. (*From Andrews.*[61])

in this fashion, the clutter frequency can be measured directly by sampling the received echo signal over an interval of range. The sampled range interval is selected so that clutter is likely to be the dominant signal. From this measurement of clutter doppler within the sampled range interval, compensation is made over the entire range of observation either by changing the reference signal from the coho or by adjusting the phase shifter inserted in one of the arms of the delay-line canceler. Generally, the average doppler frequency or phase shift is obtained by averaging the sampled range-interval over a number of pulse repetition periods. A single doppler measurement and subsequent compensation might not suffice over the entire range of the radar, especially if the radar is elevated as in an aircraft. The doppler shift from clutter will be range dependent with an elevated radar since the doppler frequency is a function of the elevation angle from the radar to the clutter cell. Thus more than one doppler measurement may be necessary to compensate for the variation of the clutter doppler with range.

An MTI radar that measures the average doppler frequency shift of clutter over a sampled range interval and uses this measurement to cause the clutter mean-doppler-frequency to coincide with the null of the MTI doppler-filter-frequency response over the remainder of the range of observation is called a *clutter-lock MTI*.

One version of a clutter-lock MTI is TACCAR, which stands for time-averaged-clutter coherent airborne radar.[8] Although the name was originally applied to a particular airborne MTI radar system developed by MIT Lincoln Laboratory, it continues to be used to refer to the clutter-lock technique that was the special feature of that system. The name is even retained when the technique is applied to a shipboard radar or when the radar is on land and used for the compensation of moving clutter. The chief feature of TACCAR is the use of a voltage-controlled oscillator arranged in a phase-lock loop. As with other clutter-lock methods, the correction for the clutter doppler is obtained from the averaged measurement of the clutter doppler frequency within a sampled range interval.

Other methods that have been proposed for compensating the clutter doppler shift seen by a moving radar include the "matrix MTI"[60] for implementation in digital MTI, and a "trial and error" technique[59,68] that provides simultaneously a number of possible doppler corrections and uses that one which produces the minimum residue over the sampled range interval.

Still another technique that is attractive when it can be applied, is to use a doppler processor with a rejection notch wide enough to reject the clutter doppler even when the radar platform is in motion. This is applicable only when the first blind speed is high (a low radar frequency and/or high prf) and when the platform speed is low, as it would be with a ship-mounted radar.

Generally, most clutter-lock MTI techniques do not adequately eliminate both stationary and moving clutter when they appear simultaneously within the same range resolution cell. A TACCAR, for example, might be designed to reject ground clutter close in, and weather or chaff at a different doppler at ranges beyond the ground clutter; but not to cancel two different clutter doppler frequencies simultaneously.[8] An exception is the adaptive MTI[62,65] which can adapt to any type of clutter. Using technology similar to that of the antenna sidelobe canceler, nulls are adaptively placed at those frequencies containing large clutter. A three-loop adaptive canceler, for example, can adaptively place three nulls at three different frequencies or it can place the three nulls so as to make a single wide notch, depending on the nature of the clutter spectrum.

Compensation for clutter doppler spread. The simple expression of Eq. (4.36) shows that the spread in the clutter spectrum is a function of the angle θ between the velocity vector of the moving platform and the antenna beam-pointing direction. It also depends on the wavelength.

Based on an analysis of antenna radiation patterns and experimental data, Staudaher[58] gives the standard deviation of the clutter spectrum due to platform motion as

$$\sigma_{\mathrm{pm}} \simeq 0.6 \frac{v_x}{a} \tag{4.37}$$

where v_x is the horizontal component of the velocity perpendicular to the antenna pointing-direction and a is the effective horizontal aperture width. The antenna beamwidth is assumed to be approximated by $\theta_B = \lambda/a$. [Equation (4.37) is not inconsistent with the simple derivation of Eq. (4.36).] If the mean doppler-frequency shift of the clutter echo is perfectly compensated, the limitations on the improvement factor due to clutter spread can be found by assuming a gaussian spectral shape and substituting the standard deviation of Eq. (4.37) into the expression of Eq. (4.27) to obtain

$$I_{\mathrm{pm}} = \frac{2^{N_l}}{N_l!} \left(\frac{1}{1.2\pi} \frac{a}{v_x T_p} \right)^{2N_l} \tag{4.38}$$

where N_l is the number of delay lines in the MTI processor. If the clutter spread σ_{pm} due to the platform motion combines with the clutter spread σ_c due to internal clutter motion such that the total standard deviation σ_T of the clutter spectrum is $\sigma_T^2 = \sigma_c^2 + \sigma_{\mathrm{pm}}^2$, the MTI improvement factor for the total clutter spectrum is

$$I_{\mathrm{cs}} = \frac{2^{N_l}}{N_l!} \left[\frac{f_p}{2\pi(\sigma_c^2 + \sigma_{\mathrm{pm}}^2)^{1/2}} \right]^{2N_l} \tag{4.39}$$

The solid curves of Fig. 4.35 plot this equation for a three-pulse delay-line canceler ($N_l = 2$).

If the widening of the spectrum is a result of the radar platform's velocity, its effects can be mitigated by making the radar antenna appear stationary. This might be accomplished with

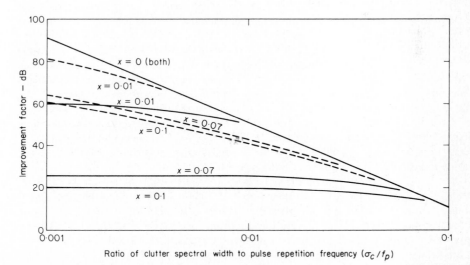

Figure 4.35 Solid curves show the improvement factor of a three-pulse canceler limited by platform motion [Eq. (4.39)]. Dashed curves show the effect of the DPCA compensation. x is the fraction of the antenna aperture that the antenna is displaced per interpulse period ($x = 0$ corresponds to no platform motion.) (*From Andrews.*[63])

two separate antennas with the distance between them equal to $T_p v_x = T_p v \sin \theta_a$, where $\theta_a =$ angle between velocity vector of the vehicle and the antenna beam-pointing direction, and $T_p =$ pulse repetition period. One pulse is transmitted on the forward antenna, and the other pulse is transmitted on the rear antenna so that the two pulses from the two different antennas are transmitted and received at the same point in space. The result is as if the radar antenna were stationary. The distance traveled between pulses is generally less than the antenna dimension so that the two antenna beams might be generated with two overlapping reflector antennas or with a phased array divided into two overlapping subarrays.[64] The effective separation between the antennas, $T_p v_x$, varies with the angle θ_a as well as the aircraft velocity v. With reflector antennas, it is not convenient to change the antenna physical separation to compensate for changes in θ_a or v. The pulse repetition period T_p might be varied to provide compensation, but this can introduce other complications into the radar design and the signal processing. With a phased array divided into two overlapping subarrays, a constant pulse repetition frequency can be used and the horizontal separation of the two overlapping subarrays can be controlled electronically to compensate for platform motion. However, it is possible to change the effective phase center of a reflector antenna by employing two feeds to produce two squinted overlapping beams, as in an amplitude-comparison monopulse radar (Sec. 5.4). The outputs of the two feeds are combined using a hybrid junction to produce a sum pattern Σ and a difference pattern Δ. By taking $\Sigma \pm jk\Delta$, the effective phase center can be shifted depending on the value of k. (The factor j multiplying the difference pattern signifies a 90° phase shift added to the difference signal relative to the sum signal.) The use of this technique in an AMTI radar to compensate for the effects of platform motion is called DPCA, which stands for *Displaced Phase Center Antenna*.

The sum and difference patterns can be obtained by connecting a hybrid junction to the outputs of the two antenna feeds as described in Sec. 5.4. The sum pattern is used on transmit and both the sum and the difference patterns are extracted on receive. The signal received on the difference pattern is weighted by the factor k, shifted in phase by 90° and is added to the sum-pattern signal in the delayed channel and subtracted from the sum-pattern signal in the undelayed channel. Because of the phase relationships between the lobes of the difference pattern and the sum pattern, the result is an apparent forward displacement of the pattern on the first transmission, and a displacement to the rear on the second transmission. When the gains of the sum and difference channels are properly adjusted, and when the distance between the phase centers of the two antenna beams is $2T_p v_x$ the combined sum and difference patterns on successive pulses illuminate the same region and the antenna appears stationary.[58] (The factor 2 appears in the distance between phase centers as a result of using both feeds for transmission. The phase center on transmit is half-way between the two feeds, and the phase center on receive alternates from one feed to the other.) As the antenna pointing-direction changes from the port to starboard side of the vehicle, the sign of the difference signal must be reversed to keep the displaced beams in the proper orientation.

The dashed curves of Fig. 4.35 show the improvement in MTI processing that is theoretically possible with DPCA and a three-pulse delay-line canceler. (Note that the DPCA corrects only one canceler of a multiple-stage MTI.[63]) The curve for $x = 0$ applies for no platform motion and represents the maximum improvement offered by an idea platform-motion cancellation method. It is seen that when the clutter spectral width is small, as for overland clutter, a significant improvement is offered by DPCA.

The limitation to the improvement factor due to antenna rotation, or scanning modulation, can be reduced by a method similar to DPCA.[51,58,66] DPCA applies the difference pattern in quadrature (90° phase shift) to the sum pattern while compensation for scanning modulation requires the difference pattern to be applied in phase with the sum pattern. Thus it is possible to combine the two techniques for compensating platform motion and scanning modulation.[58]

If the antenna sidelobes of an airborne MTI radar are not sufficiently low, the clutter that enters the receiver via the sidelobes can set a limit to the improvement factor equal to[58]

$$I_{sl} = \frac{K \int_{-\pi}^{\pi} G^2(\theta) \, d\theta}{\int_{sl} G^2(\theta) \, d\theta} \tag{4.40}$$

where $G(\theta)$ is the one-way power gain of the antenna in the plane of the ground surface. The lower integral is taken outside the main-beam region. This assumes the sidelobes are well distributed in azimuth. The constant K is the average gain of the delay line canceler ($K = 2$ for a two-pulse canceler and 6 for a three-pulse canceler.) The combined improvement factor for DPCA and the sidelobe limitation is

$$\frac{1}{I_{total}} = \frac{1}{I_{sl}} + \frac{1}{I_{DPCA}} \tag{4.41}$$

Adaptive array antennas may be employed to compensate for platform motion in an AMTI radar.[67] The full array is illuminated on transmission so that the transmit pattern is the same from pulse to pulse. On receive, the array is made adaptive by obtaining a separate output from each element. Each element output can be weighted separately and the outputs added together to form an aperture illumination function that adaptively permits motion of the antenna phase center so as to compensate for platform motion. Adaptive loops are also used in the delay-line canceler to control the doppler response of the canceler as well as the antenna angular response. In addition, compensation for scattering from near-field aircraft structure that distorts the antenna pattern and degrades AMTI performance can also be performed with this adaptive circuitry, as can the adaptive nulling of external interference sources.

For applications that cannot afford a fully adaptive array antenna, a design procedure can be formulated that applies an optimal correction to an arbitrary receive array antenna pattern based on the use of a least-mean-squares algorithm to minimize the total clutter residue of an AMTI radar averaged over all angles.[74]

Sidelobes and pulse-doppler radar. Since the pulse-doppler radar is capable of good MTI performance, it is also a good AMTI radar. However, if the antenna sidelobes are not low, the clutter that enters the radar via the sidelobes can limit the improvement factor, as mentioned previously [Eq. (4.40)]. The effect of the sidelobe clutter must often be considered in the design of the signal processor of an airborne pulse-doppler radar.

The spectrum of the signal received by an airborne pulse-doppler radar might appear as in Fig. 4.36. Only that portion of the spectrum in the vicinity of the carrier frequency f_0 is shown since the prf of a pulse-doppler radar is chosen to avoid overlap of target signals from adjacent spectral lines (no blind speeds). Thus the prf is at least twice the maximum target doppler-frequency. The leakage of the transmitter signal into the receiver produces the spike at a frequency f_0 and the spikes at $f_0 \pm n f_p$ where n is an integer and f_p is the pulse repetition frequency. Also in the vicinity of f_0 is the clutter energy from the sidelobes which illuminate the ground directly beneath the aircraft. The echo from the ground directly beneath the aircraft is called the *altitude return*. The altitude return is not shifted in frequency since the relative velocity between radar and ground is essentially zero. Clutter to either side of the perpendicular will have a relative-velocity component and hence some doppler frequency shift; consequently the clutter spectrum from the altitude return will be of finite width. The shape of the

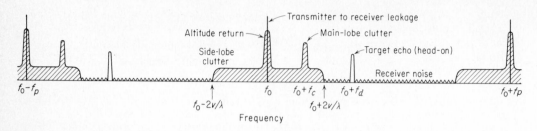

Figure 4.36 Portion of the received signal spectrum in the vicinity of the RF carrier frequency f_0, for a pulse-doppler AMTI radar. (*After Maquire,*[70] *Proc. Natl. Conf. on Aeronaut. Electronics.*)

altitude-return spectrum will depend upon the variation of the clutter cross section as a function of antenna depression angle. The cross section of the clutter directly beneath the aircraft for a depression angle of 90° can be quite large compared with that at small depression angles. The large cross section and the close range can result in considerable altitude return.

The clutter illuminated by the antenna sidelobes in directions other than directly beneath the aircraft may have any relative velocity from $+v$ to $-v$, depending on the angle made by the antenna beam and the aircraft vector velocity (v is the aircraft velocity). The clutter spectrum contributed by these sidelobes will extend $2v/\lambda$ Hz on either side of the transmitter frequency. The shape of the spectrum will depend upon the nature of the clutter illuminated and the shape of the antenna sidelobes. For purposes of illustration it is shown in Fig. 4.36 as a uniform spectrum.

The altitude return may be eliminated by turning the receiver off (gating) at that range corresponding to the altitude of the aircraft. Gating the altitude return has the disadvantage that targets at ranges corresponding to the aircraft altitude will also be eliminated from the receiver. Another method of suppressing the altitude return in the pulse radar is to eliminate the signal in the frequency domain, rather than in the time domain, by inserting a rejection filter at the frequency f_0. The same rejection filter will also suppress the transmitter-to-receiver leakage. The clutter energy from the main beam may also be suppressed by a rejection filter, but since the doppler frequency of this clutter component is not fixed, the rejection filter must be tunable and servo-controlled to track the main-beam clutter as it changes because of scanning or because of changes in aircraft velocity.

The position of the target echo in the frequency spectrum depends upon its velocity relative to that of the radar aircraft. If the target aircraft approaches the radar aircraft head on (from the forward sector), the doppler frequency shift of the target will be greater than the doppler shifts of the clutter echoes, as shown in Fig. 4.36. A filter can be used to exclude the clutter but pass the target echo. Similarly, if the targets are receding from one another along headings 180° apart, the target doppler frequency shift will again lie outside the clutter spectrum and may be readily separated from the clutter energy by filters. In other situations where the radar may be closing on the target from the tail or from the side, the relative velocities may be small and the target doppler will lie within the clutter doppler spectrum. In such situations the target echo must compete with the clutter energy for recognition. A large part of the clutter energy may be removed with a bank of fixed narrowband filters covering the expected range of doppler frequencies. The bandwidth of each individual filter must be wide enough to accept the energy contained in the target echo signal. The width of the filter will depend upon the time on target, equipment fluctuations, and other effects which broaden the echo-signal spectrum as discussed previously. Each of the doppler filters can have its own individually set threshold whose level is determined by the amount of noise or clutter within the filter. This can be done adaptively. A separate set of filters is required for each range gate. A consequence of the high prf of the pulse doppler radar is that there are fewer range gates so that fewer sets of filter-banks are needed. In some pulse-doppler radars, the duty factor might be as high as 0.3 to 0.5 so that only one set of filters is needed. Such a high-duty factor radar is sometimes called *interrupted CW*, or ICW.

Noncoherent AMTI. The noncoherent MTI principle that uses the clutter echo instead of a coho as the reference signal can also be applied to a radar on a moving platform. Although it is attractive for operation in aircraft where space and weight must be kept to a minimum, its MTI performance is limited, as is its ground-based counterpart, by the lack of spectral purity of the clutter when used as a reference signal, and by the spatial inhomogeneity (or patchiness) of the clutter.

Moving clutter and stationary radar. The discussion in this section has been concerned with the operation of an MTI radar when the clutter doppler-frequency was not at dc but had some finite value because of the motion of the vehicle carrying the radar. The clutter doppler frequency will also be other than at dc if the radar is stationary and the clutter has a component of velocity relative to the radar, as when the clutter is due to rain storms, chaff, or birds. Many of the AMTI techniques described above can be applied to the stationary MTI radar that must cope with moving clutter. Most conventional techniques fail, however, when both stationary and moving clutter are within the same range resolution cell. In such situations the MTI signal processor must have the capability of placing separate rejection notches at the doppler frequencies of the clutter.[71,72]

4.12 OTHER TYPES OF MTI

Two-frequency MTI.[78,79] The first blind speed of an MTI radar is inversely proportional to the carrier frequency, as described by Eq. (4.8). This can result in the appearance of many blind speeds in conventional MTI radar that operate at the higher microwave frequencies. One of the methods sometimes suggested for increasing the first blind speed is to transmit two carrier frequencies f_0 and $f_0 + \Delta f$ and extract the difference frequency Δf for MTI processing. The resulting blind speeds will be the same as if the radar transmitted the difference frequency rather than the carrier. For example, if $\Delta f = 0.1 f_0$, the first blind speed corresponding to the difference frequency is 10 times that of an MTI radar at the carrier frequency f_0. Thus, it would seem that the advantages of a VHF or UHF MTI might be obtained with radars operating at the higher microwave frequencies. A two-frequency MTI transmits a pair of pulses, either simultaneously or in close sequence, at two separate carrier frequencies. The two received signals are mixed in a nonlinear device and the difference frequency is extracted for normal MTI signal processing.

The advantage of the greater first blind speed obtained with the two-frequency MTI is accompanied by several disadvantages.[73] If the ratio of the two frequencies is $r < 1$, the standard deviation of the clutter doppler spectrum σ_c for a single-frequency MTI is increased to $\sigma_c (1 + r^2)^{1/2}$ in a two-frequency MTI. (This assumes that the clutter-velocity spectrum width σ_v is the same for both carriers, or $\sigma_1 = r\sigma_2$ where σ_1 and σ_2 are the clutter doppler-frequency spreads. This results in less improvement factor for a two-frequency MTI as compared with a single-frequency MTI. Note also that the clutter doppler spread of a radar which actually radiated a carrier frequency Δf would be less than that of a radar at carrier frequency f_0, by the amount $\Delta f/f_0$. The two-frequency MTI might have the blind speeds of a radar at the difference frequency but it has none of its other favorable clutter characteristics. Although the first blind speed is greater in a two-frequency MTI, there may be deeper nulls than one might desire in the doppler response characteristic, just as there would be in a staggered MTI with only two pulse repetition frequencies. The two-frequency MTI has the advantage of being less sensitive than a single-frequency MTI to a mean clutter-doppler-frequency other than dc, assuming the single-frequency MTI employs no compensation such as TACCAR. This also results, however, in the loss of detection of targets with low doppler-frequency shift that otherwise would have been detected with the single-frequency MTI.

In general, the two-frequency MTI does not offer any obvious net advantage over properly designed single-frequency MTI systems for most MTI radar applications.

Area MTI. This form of MTI does not use doppler information directly as do the other MTI techniques discussed in this chapter. The early area MTI systems stored a complete scan of radar video in a memory, such as a storage tube, and subtracted the stored video scan to scan. Instead of subtracting successive scans, the subtraction can be on a pulse-to-pulse basis with much less memory required, if a short pulse is used.[77] The pulse widths required for aircraft detection are of the order of nanoseconds. This technique relies on the fact that the echoes from moving targets change range from pulse to pulse and those from stationary and slowly moving targets do not. The short-pulse area MTI has no blind speeds and can be designed to have no range ambiguities. It is more attractive for application at the higher microwave frequencies where the available bandwidths are large and the normal MTI suffers from excessive blind speeds.

REFERENCES

1. Ridenour, L. N.: "Radar System Engineering," MIT Radiation Laboratory Series, vol. 1, chap. 16, McGraw-Hill Book Co., New York, 1947.
2. White, W. D., and A. E. Ruvin: Recent Advances in the Synthesis of Comb Filters, *IRE Natl. Conv. Record*, vol. 5, pt. 2, pp. 186–199, 1957.
3. Capon, J.: Optimum Weighting Functions for the Detection of Sampled Signals in Noise, *IEEE Trans.*, vol. IT-10, pp. 152–159, April, 1964.
4. Murakami, T., and R. S. Johnson: Clutter Suppression by Use of Weighted Pulse Trains, *RCA Rev.*, vol. 32, pp. 402–428, September, 1971.
5. Emerson, R. C.: Some Pulsed Doppler MTI and AMTI Techniques, *RAND Corp. Report no. R-274*, March, 1954.
6. Kretschmer, F. F.: MTI Weightings, *IEEE Trans.*, vol. AES-10, pp. 153–155, January, 1974.
7. Benning, C., and D. Hunt: Coefficients for Feed-Forward MTI Radar Filters, *Proc. IEEE*, vol. 57, pp. 1788–1789, October, 1969.
8. Shrader, W. W.: MTI Radar, chap. 17 of "Radar Handbook," M. I. Skolnik (ed.), McGraw-Hill Book Co., New York, 1970.
9. Shreve, J. S.: Digital Signal Processing, chap. 35 of "Radar Handbook," M. I. Skolnik (ed.), McGraw-Hill Book Co., New York, 1970.
10. Hill, J. J.: Design of Nonrecursive Digital Moving-Target-Indicator Radar Filters, *Electronics Letter*, vol. 8, pp. 359–360, July 13, 1972.
11. Rabiner, L. R., and C. M. Radar: "Digital Signal Processing," IEEE Press, New York, 1972.
12. White, W. D.: Synthesis of Comb Filters, *Proc. Natl. Conf. on Aeronaut. Electronics*, 1958, pp. 279–285.
13. Urkowitz, H.: Analysis and Synthesis of Delay Line Periodic Filters, *IRE Trans.*, vol. CT-4, pp. 41–53, June, 1957.
14. Ellis, J. G.: Digital MTI, A New Tool for the Radar User, *Marconi Rev.*, vol. 36, pp. 237–248, 4th qtr., 1973.
15. Fletcher, R. H., Jr., and D. W. Burlage: An Initialization Technique for Improved MTI Performance in Phased Array Radars, *Proc. IEEE*, vol. 60, pp. 1551–1552, December, 1972.
16. Hsiao, J. K.: Comb Filter Design, *Naval Research Laboratory, Washington, D.C. Memorandum Report 2433*, May, 1972.

17. Brennan, L. E., and I. S. Reed: Optimum Processing of Unequally Spaced Radar Pulse Trains for Clutter Rejection, *IEEE Trans.*, vol. AES-4, pp. 474–477, May, 1968.
18. Jacomini, O. J.: Weighting Factor and Transmission Time Optimization in Video MTI Radar, *IEEE Trans.*, vol. AES-8, pp. 517–527, July, 1972.
19. Prinsen, P. J. A.: Elimination of Blind Velocities of MTI Radar by Modulating the Interpulse Period, *IEEE Trans.*, vol. AES-9, pp. 714–724, September, 1973.
20. Hsiao, J. K., and F. F. Kretschmer, Jr.: Design of a Staggered-prf Moving Target Indication Filter, *The Radio and Electronic Engineer*, vol. 43, pp. 689–693, November, 1973.
21. Owen, P. L.: The Effect of Random PRF Staggering on MTI Performance, *IEEE 1975 International Radar Conference*, pp. 73–78, IEEE Publ. 75 CHO 938–1 AES.
22. Zverev, A. I.: Digital MTI Radar Filters, *IEEE Trans*, vol. AU-16, pp. 422–432, September, 1968.
23. Taylor, J. W., Jr. and J. Mattern: Receivers, chap. 5 of "Radar Handbook," M. I. Skolnik (ed.), McGraw-Hill Book Co., New York, 1970, sec. 5.12.
24. Sheingold, D. H., and R. A. Ferrero: Understanding A/D and D/A Converters, *IEEE Spectrum*, vol. 9, pp. 47–56, September, 1972.
25. Taylor, J. W., Jr.: Digital MTI Radar System, U.S. Patent No. 3,797,017, Mar. 12, 1974.
26. Houts, R. C., and D. W. Burlage: Design Procedure for Improving the Usable Bandwidth of an MTI Radar Signal Processor, *IEEE Conference Record 1976 International Conference on Acoustics, Speech, and Signal Processing*, Apr. 12–14, 1976, Phila. PA, IEEE Cat. no. 76 CH1067–8 ASSP. See also by same authors: Maximizing the Usable Bandwidth of MTI Signal Processors, *IEEE Trans.*, vol. AES-13, pp. 48–55, January, 1977.
27. Brigham, E. O., and R. W. Morrow: The Fast Fourier Transform, *IEEE Spectrum*, vol. 4, pp. 63–70, December, 1967.
28. Rabiner, L. R., et al.: Terminology in Digital Signal Processing, *IEEE Trans.*, vol. AU-20, pp. 322–337, December, 1972.
29. Cooley, J. W., and J. W. Tukey: An Algorithm for the Machine Computation of Complex Fourier Series, *Math. Comp.*, vol. 19, pp. 297–301, April, 1965.
30. Andrews, G. A., Jr.: Performance of Cascaded MTI and Coherent Integration Filters in a Clutter Environment, *NRL Report 7533, Naval Research Laboratory*, Washington, D.C., Mar. 27, 1973.
31. Andrews, G. A., Jr.: Optimization of Radar Doppler Filters to Maximize Moving Target Detection, *NAECON '74 Record*, pp. 279–283.
32. Andrews, G. A., Jr.: Optimal Radar Doppler Processors, *NRL Report 7727, Naval Research Laboratory*, Washington, D.C., May 29, 1974.
33. Arenberg, D. L.: Ultrasonic Solid Delay Lines, *J. Acous. Soc. Am.*, vol. 20, pp. 1–28, January, 1948.
34. Anonymous: Moving Target Radar Using Storage Tubes, *Electronics*, vol. 34, no. 33, pp. 30–31, Aug. 18, 1961.
35. Lobenstein, H., and D. N. Ludington: A Charge Transfer Device MTI Implementation, *IEEE 1975 International Radar Conference Record*, pp. 107–110, IEEE Publication 75 CHO 938-1 AES.
36. Roberts, J. B. G., R. Eames, and K. A. Roche: Moving-Target-Indicator Recursive Radar Filter using Bucket-Brigade Circuits, *Electronics Letters*, vol. 9, no. 4, pp. 89–90, Feb. 22, 1973.
37. Bounden, J. E., and M. J. Tomlinson: C.C.D. Chebyshev Filter for Radar MTI Applications, *Electronics Letters*, vol. 10, no. 7, pp. 89–90, Apr. 4th, 1974.
38. Barbe, D. F., and W. D. Baker: Signal Processing Devices Using the Charge-Coupled Concept, *Microelectronics*, vol. 7, no. 2, pp. 36–45, 1975.
39. Hsiao, J. K.: FFT Doppler Filter Performance Computations, *NRL Memorandum Report 2744, Naval Research Laboratory*, Washington, D.C., March, 1974.
40. Jensen, A. S.: The Radechon. A Barrier Grid Storage Tube, *RCA Rev.*, vol. 16, pp. 197–213, June, 1955.
41. McKee, D. A.: An FM MTI Cancellation System, *MIT Lincoln Lab. Tech. Rept.* 171, Jan. 8, 1958.
42. Meuehe, C. E.: Digital Signal Processor for Air Traffic Control Radars, *IEEE NEREM 74 Record, Part 4: Radar Systems and Components*, pp. 73–82, Oct. 28–31, 1974, IEEE Catalog no. 74 CHO 934-0 NEREM.
43. Muehe, C. E.: Advances in Radar Signal Processing, *IEEE Electro '76*, Boston, Mass, May 11–14, 1976.
44. Drury, W. H.: Improved MTI Radar Signal Processor, Report no. FAA-RD-74-185, *MIT Lincoln Laboratory*, Apr. 3, 1975. Available from National Technical Information Service, Catalog no. AD-A010 478.
45. Barton, D. K., and W. W. Shrader: Interclutter Visibility in MTI Systems, *IEEE EASCON '69 Convention Record*, pp. 294–297, Oct. 27–29, 1969, IEEE Publication 69 C 31–AES.
46. Barton, D. K.: "Radar System Analysis," Prentice-Hall, Inc., Englewood Cliffs, N.J., 1964.

47. Nathanson, F. E.: "Radar Design Principles," McGraw-Hill Book Co., New York, 1969.
48. Goldstein, H.: The Effect of Clutter Fluctuations on MTI, *MIT Radiation Lab. Rept.* 700, Dec. 27, 1945.
49. Barlow, E. J.: Doppler Radar, *Proc. IRE*, vol. 37, pp. 340–355, April, 1949.
50. Shrader, W. W.: Moving Target Indication Radar, *IEEE 74 Record NEREM*, Pt 4, pp. 18–26, Oct. 28–31, 1974, IEEE Catalog no. 74 CHO 934-0 NEREM.
51. Grisetti, R. S., M. M. Santa, and G. M. Kirkpatrick: Effect of Internal Fluctuations and Scanning on Clutter Attenuation in MTI Radar, *IRE Trans.*, vol. ANE-2, pp. 37–41, March, 1955.
52. Steinberg, B. D.: Target Clutter, and Noise Spectra, pt VI, chap. 1 of "Modern Radar," R. S. Berkowitz (ed.), John Wiley and Sons, New York, 1965.
53. Ward, H. R., and W. W. Shrader: MTI Performance Caused by Limiting, *EASCON '68 Record*, Supplement to *IEEE Trans*, vol. AES-4, pp. 168–174, November, 1968.
54. Grasso, G.: Improvement Factor of a Nonlinear MTI in Point Clutter, *IEEE Trans.*, vol. AES-4, pp. 640–644, July, 1968.
55. Grasso, G., and P. F. Guarguaglini: Clutter Residues of a Coherent MTI Radar Receiver, *IEEE Trans.*, vol. AES-5, pp. 195–204, March, 1969.
56. Shrader, W. W.: Radar Technology Applied to Air Traffic Control, *IEEE Trans.*, vol. COM-21, pp. 591–605, May, 1973.
57. Mooney, D. H., and W. A. Skillman: Pulse Doppler Radar, chap. 19 of "Radar Handbook," M. I. Skolnik (ed.), McGraw-Hill Book Co., New York, 1970.
58. Staudaher, F. M.: Airborne MTI, chap. 18 of "Radar Handbook," M. I. Skolnik (ed.), McGraw-Hill Book Co., New York, 1970.
59. Voles, R.: New Approach to MTI Clutter Locking, *Proc. IEE*, vol. 120, pp. 1383–1390, November, 1972.
60. See sec. 14.5 of ref. 47.
61. Andrews, G. A.: Airborne Radar Motion Compensation Techniques, Evaluation of TACCAR, *NRL Report 7407, Naval Research Laboratory*, Washington, D.C., Apr. 12, 1972.
62. Spafford, L. J.: Optimum Radar Signal Processing in Clutter, *IEEE Trans.*, vol. IT-14, pp. 734–743, September, 1968.
63. Andrews, G. A.: Airborne Radar Motion Compensation Techniques, Evaluation of DPCA, *NRL Report 7426, Naval Research Laboratory*, Washington, D.C., July 20, 1972.
64. Zeger, A. E., and L. R. Burgess: An Adaptive AMTI Radar Antenna Array, *Proc. IEEE 1974 Natl. Aerospace and Electronics Conf.*, May 13–15, 1974, Dayton, Ohio, pp. 126–133.
65. Johnson, M. A., and D. C. Stoner: ECCM from the Radar Designer's Viewpoint, paper 30–5, *IEEE ELECTRO-76*, Boston, Mass., May 11–14, 1976. Reprinted in *Microwave J.*, vol. 21, pp. 59–63, March, 1978.
66. Anderson, D. B.: A Microwave Technique to Reduce Platform Motion and Scanning Noise in Airborne Moving-Target Radar, *IRE WESCON Conv. Record*, vol. 2, pt. 1, pp. 202–211, 1958.
67. Brennan, L. E., J. D. Mallett, and I. S. Reed: Adaptive Arrays in Airborne MTI Radar, *IEEE Trans.*, vol. AP-24, pp. 607–615, September, 1976.
68. Voles, R.: New Techniques for MTI Clutter-Locking, *Radar—Present and Future*, IEE (London) Conf. Publ. no. 105, pp. 274–279, 1973.
69. Kroszczynski, J.: Efficiency of Attenuation of moving Clutter, *The Radio and Electronic Engr.*, vol. 34, no. 3, pp. 157–159, September, 1967.
70. Maguire, W. W.: Application of Pulsed Doppler Radar to Airborne Radar Systems, *Proc. Natl. Conf. on Aeronaut. Electronics* (Dayton, Ohio), pp. 291–295, 1958.
71. Hsiao, J. K.: MTI Optimization in a Multiple-Clutter Environment, *NRL Report 7860, Naval Research Laboratory*, Washington, D.C., Mar. 20, 1975.
72. Hsiao, J. K.: A Digital Mean-Clutter-Doppler Compensation System, *NRL Memorandum Report 2772, Naval Research Laboratory*, Washington, D.C., April, 1974.
73. Hsiao, J. K.: Analysis of a Dual Frequency Moving Target Indication System, *The Radio and Electronic Engineer*, vol. 45, pp. 351–356, July, 1975.
74. Andrews, G. A.: Radar Antenna Pattern Design for Platform Motion Compensation, *IEEE Trans.*, vol. AP-26, pp. 566–571, July, 1978.
75. Hadjifotiou, A.: Round-off Error Analysis in Digital MTI Processors for Radar, *The Radio and Electronic Engineer*, vol. 47, pp. 59–65, January/February, 1977.
76. Tong, P. S.: Quantization Requirements in Digital MTI, *IEEE Trans.*, vol. AES-13, pp. 512–521, September, 1977.
77. Cantrell, B. H.: A Short-Pulse Area MTI, *NRL Report 8162, Naval Research Laboratory*, Washington, D.C., Sept. 22, 1977.

78. Kroszczynski, J.: The Two-Frequency MTI System, *The Radio and Electronic Engineer*, vol. 39, pp. 172–176, March, 1970.
79. Kroszczynsky, J.: Efficiency of the Two-Frequency MTI System, *The Radio and Electronic Engineer*, vol. 41, pp. 77–80, February, 1971.
80. Rabiner, L. R., and B. Gold: "Theory and Application of Digital Signal Processing," Prentice-Hall, Inc., Englewood Cliffs, N.J., 1975.
81. Thomas, H. W., N. P. Lutte, and M. W. Jelffs: Design of MTI Filters with Staggered PRF: A Pole-Zero Approach, *Proc. IEE*, vol. 121, pp. 1460–1466, December, 1974.
82. Ewell, G. W., and A. M. Bush: Constrained Improvement MTI Radar Processors, *IEEE Trans.*, vol. AES-11, pp. 768–780, September, 1975.

CHAPTER
FIVE

TRACKING RADAR

5.1 TRACKING WITH RADAR

A tracking-radar system measures the coordinates of a target and provides data which may be used to determine the target path and to predict its future position. All or only part of the available radar data—range, elevation angle, azimuth angle, and doppler frequency shift—may be used in predicting future position; that is, a radar might track in range, in angle, in doppler, or with any combination. Almost any radar can be considered a tracking radar provided its output information is processed properly. But, in general, it is the method by which *angle tracking* is accomplished that distinguishes what is normally considered a tracking radar from any other radar. It is also necessary to distinguish between a *continuous tracking radar* and a *track-while-scan* (TWS) *radar*. The former supplies continuous tracking data on a particular target, while the track-while-scan supplies sampled data on one or more targets. In general, the continuous tracking radar and the TWS radar employ different types of equipment.

The antenna beam in the continuous tracking radar is positioned in angle by a servomechanism actuated by an error signal. The various methods for generating the error signal may be classified as *sequential lobing, conical scan,* and *simultaneous lobing* or *monopulse*. The range and doppler frequency shift can also be continuously tracked, if desired, by a servo-control loop actuated by an error signal generated in the radar receiver. The information available from a tracking radar may be presented on a cathode-ray-tube (CRT) display for action by an operator, or may be supplied to an automatic computer which determines the target path and calculates its probable future course.

The tracking radar must first find its target before it can track. Some radars operate in a search, or acquisition, mode in order to find the target before switching to a tracking mode. Although it is possible to use a single radar for both the search and the tracking functions, such a procedure usually results in certain operational limitations. Obviously, when the radar is used in its tracking mode, it has no knowledge of other potential targets. Also, if the antenna pattern is a narrow pencil beam and if the search volume is large, a relatively long time might be required to find the target. Therefore many radar tracking systems employ a separate search radar to provide the information necessary to position the tracker on the target. A search radar, when used for this purpose, it called an *acquisition radar*. The acquisition radar *designates* targets to the tracking radar by providing the coordinates where the targets are to be found. The tracking radar *acquires* a target by performing a limited search in the area of the designated target coordinates.

The scanning fan-beam search radar can also provide tracking information to determine the path of the target and predict its future position. Each time the radar beam scans past the target, its coordinates are obtained. If the change in target coordinates from scan to scan is not too large, it is possible to reconstruct the track of the target from the sampled data. This may be accomplished by providing the PPI-scope operator with a grease pencil to mark the target pips on the face of the scope. A line joining those pips that correspond to the same target provides the target track. When the traffic is so dense that operators cannot maintain pace with the information available from the radar, the target trajectory data may be processed automatically in a digital computer. The availability of small, inexpensive minicomputers has made it practical to obtain target tracks, not just target detections, from a surveillance radar. Such processing is usually called ADT (automatic detection and track). When the outputs from more than one radar are automatically combined to provide target tracks, the processing is called ADIT (automatic detection and integrated track) or IADT (integrated ADT).

A surveillance radar that provides target tracks is sometimes called a track-while-scan radar. This terminology is also applied to radars that scan a limited angular sector to provide tracking information at a high data rate on one or more targets within its field of view. Landing radars used for GCA (ground control of approach) and some missile control radars are of this type.

When the term *tracking radar* is used in this book, it generally refers to the continuous tracker, unless otherwise specified.

5.2 SEQUENTIAL LOBING

The antenna pattern commonly employed with tracking radars is the symmetrical pencil beam in which the elevation and azimuth beamwidths are approximately equal. However, a simple pencil-beam antenna is not suitable for tracking radars unless means are provided for determining the magnitude and direction of the target's angular position with respect to some reference direction, usually the axis of the antenna. The difference between the target position and the reference direction is the *angular error*. The tracking radar attempts to position the antenna to make the angular error zero. When the angular error is zero, the target is located along the reference direction.

One method of obtaining the direction and the magnitude of the angular error in one coordinate is by alternately switching the antenna beam between two positions (Fig. 5.1). This is called *lobe switching*, *sequential switching*, or *sequential lobing*. Figure 5.1a is a polar representation of the antenna beam (minus the sidelobes) in the two switched positions. A plot in rectangular coordinates is shown in Fig. 5.1b, and the error signal obtained from a target not on the switching axis (reference direction) is shown in Fig. 5.1c. The difference in amplitude between the voltages obtained in the two switched positions is a measure of the angular displacement of the target from the switching axis. The sign of the difference determines the

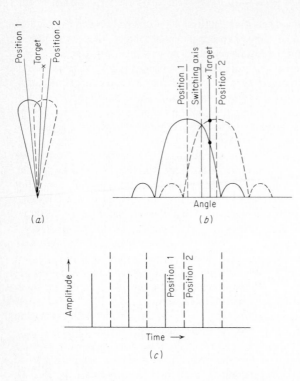

(a)

(b)

(c)

Figure 5.1 Lobe-switching antenna patterns and error signal (one dimension). (a) Polar representation of switched antenna patterns; (b) rectangular representation; (c) error signal.

direction the antenna must be moved in order to align the switching axis with the direction of the target. When the voltages in the two switched positions are equal, the target is on axis and its position may be determined from the axis direction.

Two additional switching positions are needed to obtain the angular error in the orthogonal coordinate. Thus a two-dimensional sequentially lobing radar might consist of a cluster of four feed horns illuminating a single antenna, arranged so that the right-left, up-down sectors are covered by successive antenna positions. Both transmission and reception are accomplished at each position. A cluster of five feeds might also be employed, with the central feed used for transmission while the outer four feeds are used for receiving. High-power RF switches are not needed since only the receiving beams, and not the transmitting beam, are stepped in this five-feed arrangement.

One of the limitations of a simple unswitched nonscanning pencil-beam antenna is that the angle accuracy can be no better than the size of the antenna beamwidth. An important feature of sequential lobing (as well as the other tracking techniques to be discussed) is that the target-position accuracy can be far better than that given by the antenna beamwidth. The accuracy depends on how well equality of the signals in the switched positions can be determined. The fundamental limitation to accuracy is system noise caused either by mechanical or electrical fluctuations.

Sequential lobing, or lobe switching, was one of the first tracking-radar techniques to be employed. Early applications were in airborne-interception radar, where it provided directional information for homing on a target, and in ground-based antiaircraft fire-control radars. It is not used as often in modern tracking-radar applications as some of the other techniques to be described.

5.3 CONICAL SCAN

A logical extension of the sequential lobing technique described in the previous section is to rotate continuously an offset antenna beam rather than discontinuously step the beam be-tween four discrete positions. This is known as *conical scanning* (Fig. 5.2). The angle between the axis of rotation (which is usually, but not always, the axis of the antenna reflector) and the axis of the antenna beam is called the squint angle. Consider a target at position *A*. The echo signal will be modulated at a frequency equal to the rotation frequency of the beam. The amplitude of the echo-signal modulation will depend upon the shape of the antenna pattern, the squint angle, and the angle between the target line of sight and the rotation axis. The phase of the modulation depends on the angle between the target and the rotation axis. The conical-scan modulation is extracted from the echo signal and applied to a servo-control system which continually positions the antenna on the target. [Note that two servos are required because the tracking problem is two-dimensional. Both the rectangular (az-el) and polar tracking coordinates may be used.] When the antenna is on target, as in *B* of Fig. 5.2, the line of sight to the target and the rotation axis coincide, and the conical-scan modulation is zero.

A block diagram of the angle-tracking portion of a typical conical-scan tracking radar is shown in Fig. 5.3. The antenna is mounted so that it can be positioned in both azimuth and elevation by separate motors, which might be either electric- or hydraulic-driven. The antenna beam is offset by tilting either the feed or the reflector with respect to one another.

One of the simplest conical-scan antennas is a parabola with an offset rear feed rotated about the axis of the reflector. If the feed maintains the plane of polarization fixed as it rotates, it is called a *nutating* feed. A *rotating* feed causes the polarization to rotate. The latter type of feed requires a rotary joint. The nutating feed requires a flexible joint. If the antenna is small, it may be easier to rotate the dish, which is offset, rather than the feed, thus avoiding the problem of a rotary or flexible RF joint in the feed. A typical conical-scan rotation speed might be 30 r/s. The same motor that provides the conical-scan rotation of the antenna beam also drives a two-phase reference generator with two outputs 90° apart in phase. These two outputs serve as a reference to extract the elevation and azimuth errors. The received echo signal is fed to the receiver from the antenna via two rotary joints (not shown in the block diagram). One rotary joint permits motion in azimuth; the other, in elevation.

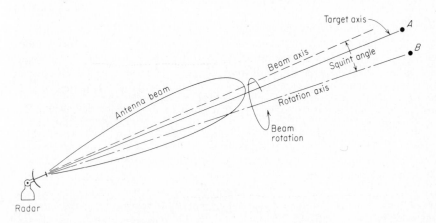

Figure 5.2 Conical-scan tracking.

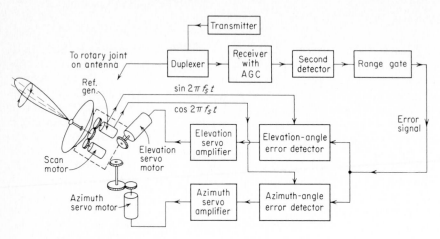

Figure 5.3 Block diagram of conical-scan tracking radar.

The receiver is a conventional superheterodyne except for features peculiar to the conical-scan tracking radar. One feature not found in other radar receivers is a means of extracting the conical-scan modulation, or error signal. This is accomplished after the second detector in the video portion of the receiver. The error signal is compared with the elevation and azimuth reference signals in the angle-error detectors, which are phase-sensitive detectors.[3-7] A phase-sensitive detector is a nonlinear device in which the input signal (in this case the angle-error signal) is mixed with the reference signal. The input and reference signals are of the same frequency. The output d-c voltage reverses polarity as the phase of the input signal changes through 180°. The magnitude of the d-c output from the angle-error detector is proportional to the error, and the sign (polarity) is an indication of the direction of the error. The angle-error-detector outputs are amplified and drive the antenna elevation and azimuth servo motors.

The angular position of the target may be determined from the elevation and azimuth of the antenna axis. The position can be read out by means of standard angle transducers such as synchros, potentiometers, or analog–to–digital-data converters.

Boxcar generator. When extracting the modulation imposed on a repetitive train of narrow pulses, it is usually convenient to stretch the pulses before low-pass filtering. This is called *boxcaring*, or *sample and hold*. Here the device is called the *boxcar generator.*[8] The boxcar generator was also mentioned in the discussion of the MTI receiver using range-gated filters (Sec. 4.4). In essence, it clamps or stretches the video pulses of Fig. 5.4a in time so as to cover the entire pulse-repetition period (Fig. 5.4b). This is possible only in a range-gated receiver. (Tracking radars are normally operated with range gates.) The boxcar generator consists of an electric circuit that clamps the potential of a storage element, such as a capacitor, to the video-pulse amplitude each time the pulse is received. The capacitor maintains the potential of the pulse during the entire repetition period and is altered only when a new video pulse appears whose amplitude differs from the previous one. The boxcar generator eliminates the pulse repetition frequency and reduces its harmonics. It also has the practical advantage that the magnitude of the conical-scan modulation is amplified because pulse stretching puts more of the available energy at the modulation frequency. The pulse repetition frequency must be sufficiently large compared with the conical-scan frequency for proper boxcar filtering. If

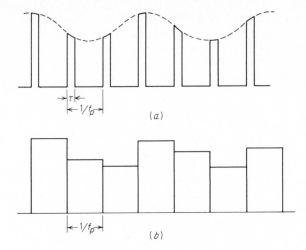

(a)

(b)

Figure 5.4 (a) Pulse train with conical-scan modulation; (b) same pulse train after passing through boxcar generator.

not, it may be necessary to provide additional filtering to attenuate undesired cross-modulation frequency components.

Automatic gain control.[9-11] The echo-signal amplitude at the tracking-radar receiver will not be constant but will vary with time. The three major causes of variation in amplitude are (1) the inverse-fourth-power relationship between the echo signal and range, (2) the conical-scan modulation (angle-error signal), and (3) amplitude fluctuations in the target cross section. The function of the automatic gain control (AGC) is to maintain the d-c level of the receiver output constant and to smooth or eliminate as much of the noiselike amplitude fluctuations as possible without disturbing the extraction of the desired error signal at the conical-scan frequency.

One of the purposes of AGC in any receiver is to prevent saturation by large signals. The scanning modulation and the error signal would be lost if the receiver were to saturate. In the conical-scan tracking radar an AGC that maintains the d-c level constant results in an error signal that is a true indication of the angular pointing error. The d-c level of the receiver must be maintained constant if the angular error is to be linearly related to the angle-error signal voltage.

An example of the AGC portion of a tracking-radar receiver is shown in Fig. 5.5. A portion of the video-amplifier output is passed through a low-pass or smoothing filter and fed back to control the gain of the IF amplifier. The larger the video output, the larger will be the

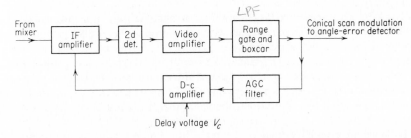

Figure 5.5 Block diagram of the AGC portion of a tracking-radar receiver.

feedback signal and the greater will be the gain reduction. The filter in the AGC loop should pass all frequencies from direct current to just below the conical-scan-modulation frequency. The loop gain of the AGC filter measured at the conical-scan frequency should be low so that the error signal will not be affected by AGC action. (If the AGC responds to the conical-scan frequency, the error signal might be lost.) The phase shift of this filter must be small if its phase characteristic is not to influence the error signal. A phase change of the error signal is equivalent to a rotation of the reference axes and introduces cross coupling, or "cross talk," between the elevation and azimuth angle-tracking loops. Cross talk affects the stability of the tracking and might result in an unwanted nutating motion of the antenna. In conventional tracking-radar applications the phase change introduced by the feedback-loop filter should be less than 10°, and in some applications it should be as little as 2°.[10] For this reason, a filter with a sharp attenuation characteristic in the vicinity of the conical-scan frequency might not be desirable because of the relatively large amount of phase shift which it would introduce.

The output of the feedback loop will be zero unless the feedback voltage exceeds a prespecified minimum value V_c. In the block diagram the feedback voltage and the voltage V_c are compared in the d-c amplifier. If the feedback voltage exceeds V_c, the AGC is operative, while if it is less, there is no AGC action. The voltage V_c is called the *delay voltage*. The terminology may be a bit misleading since the delay is not in time but in amplitude. The purpose of the delay voltage is to provide a reference for the constant output signal and permit receiver gain for weak signals. If the delay voltage were zero, any output which might appear from the receiver would be due to the failure of the AGC circuit to regulate completely.

In many applications of AGC the delay voltage is actually zero. This is called *undelayed AGC*. In such cases the AGC can still perform satisfactorily since the loop gain is usually low for small signals. Thus the AGC will not regulate weak signals. The effect is similar to having a delay voltage, but the performance will not be as good.

The required dynamic range of the AGC will depend upon the variation in range over which targets are tracked and the variations expected in the target cross section. If the range variation were 10 to 1, the contribution to the dynamic range would be 40 dB. The target cross section might also contribute another 40-dB variation. Another 10 dB ought to be allowed to account for variations in the other parameters of the radar equation. Hence the dynamic range of operation required of the receiver AGC might be of the order of 90 dB, or perhaps more.

It is found[10] in practice that the maximum gain variation which can be obtained with a single IF stage is of the order of 40 dB. Therefore two to three stages of the IF amplifier must be gain-controlled to accommodate the total dynamic range. The middle stages are usually the ones controlled since the first stage gain should remain high so as not to influence the noise figure of the mixer stage. It is also best not to control the last IF stage since the maximum undistorted output of an amplifying stage is reduced when its gain is reduced by the application of a control voltage.

An alternative AGC filter design would maintain the AGC loop gain up to frequencies much higher than the conical-scan frequency. The scan modulation would be effectively suppressed in the output of the receiver, and the output would be used to measure range in the normal manner. In this case, the error signal can be recovered from the AGC voltage since it varies at the conical-scan frequency. The AGC voltage will also contain any amplitude fluctuations that appear with the echo signal. The error signal may be recovered from the AGC voltage with a narrow bandpass filter centered at the scan-modulation frequency.

Squint angle. The angle-error-signal voltage is shown in Fig. 5.6 as a function of θ_T, the angle between the axis of rotation and the direction to the target.[12] The squint angle θ_q is the

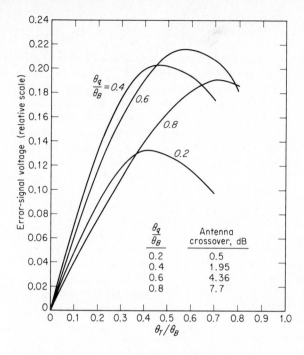

$\frac{\theta_q}{\theta_B}$	Antenna crossover, dB
0.2	0.5
0.4	1.95
0.6	4.36
0.8	7.7

Figure 5.6 Plot of the relative angle-error signal from the conical-scan radar as a function of target angle (θ_T/θ_B) and squint angle (θ_q/θ_B). θ_B = half-power beamwidth.

angle between the antenna-beam axis and the axis of rotation; and θ_B is the half-power beamwidth. The antenna beam shape is approximated by a gaussian function in the calculations leading to Fig. 5.6. The greater the slope of the error signal, the more accurate will be the tracking of the target. The maximum slope occurs for a value θ_q/θ_B slightly greater than 0.4. This corresponds to a point on the antenna pattern (the antenna crossover) about 2 dB down from the peak. It is the optimum crossover for maximizing the accuracy of angle tracking. The accuracy of range tracking, however, is affected by the loss in signal but not by the slope at the crossover point. Therefore, as a compromise between the requirements for accurate range and angle tracking, a crossover nearer the peak of the beam is usually selected rather than that indicated from Fig. 5.6. It has been suggested that the compromise value of θ_q/θ_B be about 0.28, corresponding to a point on the antenna pattern about 1.0 dB below the peak.[1,2]

Other considerations. In both the sequential-lobing and conical-scan techniques, the measurement of the angle error in two orthogonal coordinates (azimuth and elevation) requires that a minimum of three pulses be processed. In practice, however, the minimum number of pulses in sequential lobing is usually four—one per quadrant. Although a conical scan radar can also be operated with only four pulses per revolution, it is more usual to have ten or more per revolution. This allows the modulation due to the angle error to be more that of a continuous sine wave. Thus the prf is usually at least an order of magnitude greater than the conical-scan frequency. The scan frequency also must be at least an order of magnitude greater than the tracking bandwidth.

A conical-scan-on-receive-only (COSRO) tracking radar radiates a nonscanning transmit beam, but receives with a conical scanning beam to extract the angle error. The analogous operation with sequential lobing is called lobe-on-receive-only (LORO).

5.4 MONOPULSE TRACKING RADAR[1,2,13-18]

The conical-scan and sequential-lobing tracking radars require a minimum number of pulses in order to extract the angle-error signal. In the time interval during which a measurement is made with either sequential lobing or conical scan, the train of echo pulses must contain no amplitude-modulation components other than the modulation produced by scanning. If the echo pulse-train did contain additional modulation components, caused, for example, by a fluctuating target cross section, the tracking accuracy might be degraded, especially if the frequency components of the fluctuations were at or near the conical-scan frequency or the sequential-lobing rate. The effect of the fluctuating echo can be sufficiently serious in some applications to severely limit the accuracy of those tracking radars which require many pulses to be processed in extracting the error signal.

Pulse-to-pulse amplitude fluctuations of the echo signal have no effect on tracking accuracy if the angular measurement is made on the basis of *one* pulse rather than many. There are several methods by which angle-error information might be obtained with only a single pulse. More than one antenna beam is used simultaneously in these methods, in contrast to the conical-scan or lobe-switching tracker, which utilizes one antenna beam on a time-shared basis. The angle of arrival of the echo signal may be determined in a single-pulse system by measuring the relative phase or the relative amplitude of the echo pulse received in each beam. The names *simultaneous lobing* and *monopulse* are used to describe those tracking techniques which derive angle-error information on the basis of a single pulse.

An example of a simultaneous-lobing technique is *amplitude-comparison monopulse*, or more simply, *monopulse*. In this technique the RF signals received from two offset antenna beams are combined so that both the sum and the difference signals are obtained simultaneously. The sum and difference signals are multiplied in a phase-sensitive detector to obtain both the magnitude and the direction of the error signal. All the information necessary to determine the angular error is obtained on the basis of a single pulse; hence the name *monopulse* is quite appropriate.

Amplitude-comparison monopulse. The amplitude-comparison monopulse employs two overlapping antenna patterns (Fig. 5.7a) to obtain the angular error in one coordinate. The two overlapping antenna beams may be generated with a single reflector or with a lens antenna illuminated by two adjacent feeds. (A cluster of four feeds may be used if both elevation- and azimuth-error signals are wanted.) The sum of the two antenna patterns of Fig. 5.7a is shown in Fig. 5.7b, and the difference in Fig. 5.7c. The sum patterns is used for transmission, while both the sum pattern and the difference pattern are used on reception. The signal received with the difference pattern provides the magnitude of the angle error. The sum signal provides the range measurement and is also used as a reference to extract the sign of the error signal. Signals received from the sum and the difference patterns are amplified separately and combined in a phase-sensitive detector to produce the error-signal characteristic shown in Fig. 5.7d.

A block diagram of the amplitude-comparison-monopulse tracking radar for a single angular coordinate is shown in Fig. 5.8. The two adjacent antenna feeds are connected to the two arms of a hybrid junction such as a " magic T," a " rat race," or a short-slot coupler. The sum and difference signals appear at the two other arms of the hybrid. On reception, the outputs of the sum arm and the difference arm are each heterodyned to an intermediate frequency and amplified as in any superheterodyne receiver. The transmitter is connected to the sum arm. Range information is also extracted from the sum channel. A duplexer is included in the sum arm for the protection of the receiver. The output of the phase-sensitive detector is

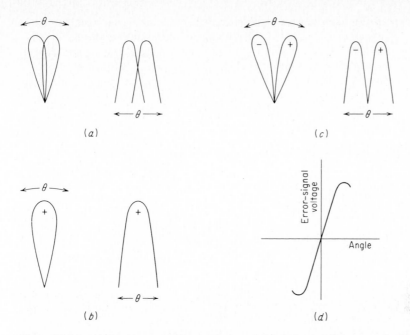

Figure 5.7 Monopulse antenna patterns and error signal. Left-hand diagrams in (*a–c*) are in polar coordinates; right-hand diagrams are in rectangular coordinates. (*a*) Overlapping antenna patterns; (*b*) sum pattern; (*c*) difference pattern; (*d*) product (error) signal.

an error signal whose magnitude is proportional to the angular error and whose sign is proportional to the direction.

The output of the monopulse radar is used to perform automatic tracking. The angular-error signal actuates a servo-control system to position the antenna, and the range output from the sum channel feeds into an automatic-range-tracking unit.

The sign of the difference signal (and the direction of the angular error) is determined by comparing the phase of the difference signal with the phase of the sum signal. If the sum signal

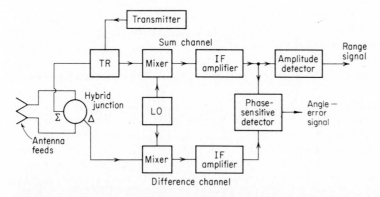

Figure 5.8 Block diagram of amplitude-comparison monopulse radar (one angular coordinate).

in the IF portion of the receiver were $A_s \cos \omega_{IF} t$, the difference signal would be either $A_d \cos \omega_{IF} t$ or $-A_d \cos \omega_{IF} t$ $(A_s > 0, A_d > 0)$, depending on which side of center is the target. Since $-A_d \cos \omega_{IF} t = A_d \cos \omega_{IF}(t + \pi)$, the sign of the difference signal may be measured by determining whether the difference signal is in phase with the sum or 180° out of phase.

Although a phase comparison is a part of the amplitude-comparison-monopulse radar, the angular-error signal is basically derived by comparing the echo *amplitudes* from simultaneous offset beams. The phase relationship between the signals in the offset beams is not used. The purpose of the phase-sensitive detector is to conveniently furnish the *sign* of the error signal.

A block diagram of a monopulse radar with provision for extracting error signals in both elevation and azimuth is shown in Fig. 5.9. The cluster of four feeds generates four partially overlapping antenna beams. The feeds might be used with a parabolic reflector, Cassegrain antenna, or a lens. All four feeds generate the sum pattern. The difference pattern in one plane is formed by taking the sum of two adjacent feeds and subtracting this from the sum of the other two adjacent feeds. The difference pattern in the orthogonal plane is obtained by adding the differences of the orthogonal adjacent pairs. A total of four hybrid junctions generate the sum channel, the azimuth difference channel, and the elevation difference channel. Three separate mixers and IF amplifiers are shown, one for each channel. All three mixers operate from a single local oscillator in order to maintain the phase relationships between the three channels. Two phase-sensitive detectors extract the angle-error information, one for azimuth, the other for elevation. Range information is extracted from the output of the sum channel after amplitude detection.

Since a phase comparison is made between the output of the sum channel and each of the difference channels, it is important that the phase shifts introduced by each of the channels be almost identical. According to Page,[13] the phase difference between channels must be maintained to within 25° or better for reasonably proper performance. The gains of the channels also must not differ by more than specified amounts.

An alternative approach to using three identical amplifers in the monopulse receiver is to use but one IF channel which amplifies the sum signal and the two difference signals on a time-shared basis.[17,19] The sum signal is passed through the single IF amplifier followed by the two difference signals delayed in time by a suitable amount. Most of the gain and gain

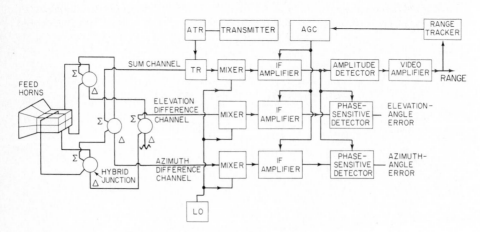

Figure 5.9 Block diagram of two-coordinate (azimuth and elevation) amplitude-comparison monopulse tracking radar.

control take place in the single IF amplifier. Any variations affect all three signals simultaneously. After amplification, compensating delays are introduced to unscramble the time sequence and bring the sum signal and the two difference signals in time coincidence. Phase detection occurs as in the conventional monopulse. Another single-channel system SCAMP converts the sum and the two difference signals to different IF frequencies and amplifies them simultaneously in a single, wide-band amplifier.[20] The output is hard-limited to provide the effect of an instantaneous AGC. The three signals after limiting are separated by narrowband filters and then converted to the same IF frequency for further processing. The hard-limiting, however, causes cross-coupling between the azimuth and the elevation error-signal channels and can result in significant error.[21] Two-channel monopulse receivers have also been used by combining the sum and the two difference signals in a manner such that they can be again resolved into three components after amplification.[1,22]

The purpose in using one- or two-channel monopulse receivers is to ease the problem associated with maintaining identical phase and amplitude balance among the three channels of the conventional receiver. These techniques provide some advantage in this regard but they can result in undesired coupling between the azimuth and elevation channels and a loss in signal-to-noise ratio.

The monopulse antenna must generate a sum pattern with high efficiency (maximum boresight gain), and a difference pattern with a large value of slope at the crossover of the offset beams. Furthermore, the sidelobes of both the sum and the difference patterns must be low. The antenna must be capable of the desired bandwidth, and the patterns must have the desired polarization characteristics. It is not surprising that the achievement of all these properties cannot always be fully satisfied simultaneously. Antenna design is an important part of the successful realization of a good monopulse radar.

The uniform aperture illumination maximizes the aperture efficiency and, hence, the directive gain. The aperture illumination that maximizes the slope of the difference pattern is the linear illumination function with odd symmetry about the center of the aperture.[23] Not only do both of these aperture illuminations produce patterns with relatively high sidelobes, but they cannot be readily generated by the simple four-horn feed. Independent control of the sum and the difference patterns is required in order to obtain simultaneously the optimum patterns.

The approximately "ideal" feed-illuminations for a monopulse radar is shown in Fig. 5.10. This has been approximated in some precision tracking radars by a five-horn feed consisting of one horn generating the sum pattern surrounded by four horns generating the difference patterns.[1] Other approximations to the ideal designs include a twelve-horn feed[24] and a feed consisting of four stacked horns in one plane with each horn generating three waveguide modes in the other plane.[25] Higher-order waveguide modes are used to obtain the desired sum and difference patterns from a single horn without the necessity of complex

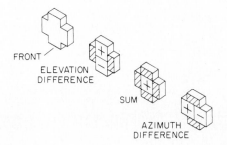

FRONT

ELEVATION
DIFFERENCE

SUM

AZIMUTH
DIFFERENCE

Figure 5.10 Approximately "ideal" feed-aperture illumination for monopulse sum and difference channels.[1]

microwave combining circuitry.[1,26] These are called *multimode feeds*. The use of higher-order waveguide modes in monopulse feeds to generate the required patterns results in feeds that are of high efficiency, compact, simple, low loss, light weight, low aperture blockage, and excellent boresight stability independent of frequency.

The greater the signal-to-noise ratio and the steeper the slope of the error signal in the vicinity of zero angular error, the more accurate is the measurement of angle. The error-signal slope as a function of the squint angle or beam crossover is shown in Fig. 5.11. The maximum slope occurs at a beam crossover of about 1.1 dB.

Automatic gain control (AGC) is required in order to maintain a stable closed-loop servo system for angle tracking.[1,2] The AGC in a monopulse radar is accomplished by employing a voltage proportional to the sum-channel IF output to control the gain of all three receiver channels. The AGC results in a constant angle sensitivity independent of target size and range. With AGC the output of the angle-error detector is proportional to the difference signal normalized (divided) by the sum signal. The output of the sum channel is constant.

Hybrid tracking system. Conopulse[86] and "scan with compensation"[87] are the names given to a hybrid tracking system that is a combination of monopulse and conical scan. Two squinted beams, similar to those of a single angle-coordinate amplitude-comparison monopulse, are scanned (rotated or nutated) in space around the boresight axis. (In the conical scan tracker, only one squinted beam is scanned.) In conopulse, the sum and difference signals received in the two squinted beams are extracted in a manner similar to that of the conventional monopulse system. Two-coordinate angle information is obtained by scanning the beams about the boresight axis similar to the extraction of azimuth and elevation angle information in the conical-scan tracker. Amplitude fluctuations of the target echo signal do not in principle degrade the tracking accuracy as in the conical-scan tracker because a difference signal is extracted from the simultaneous outputs of the two squinted beams. However, the need to scan the two beams means that a single pulse angle-estimate is not obtained as in a true monopulse.

An advantage claimed for this hybrid tracking technique is that, like monopulse, target amplitude fluctuations do not affect the tracking accuracy. It is also claimed that the simplicity of conical scan is retained. Both claims can be debated. Although only two receivers are required instead of the three used in a monopulse tracker, the mechanical rotation of the two

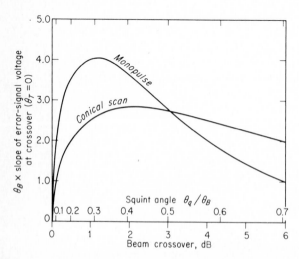

Figure 5.11 Slope of the angular-error signal at crossover for a monopulse and conical-scan tracking radar. θ_B = half-power beamwidth, θ_q = squint angle.

squinted beams is not an easy task. Nutation, which preserves the plane of polarization, is even more difficult. Rotation of the beams causes the plane of polarization to rotate which can cause undesirable target amplitude fluctuations at the scan rate. Although such fluctuations are theoretically removed by the monopulse type of processing, in practice the removal is not complete and angular error can result.[86] With modern solid-state technology, the need for a third receiver in the conventional monopulse tracker might not be as difficult to realize in practice as the complexity of properly scanning a pair of squinted beams. Generally it can be said that the amplitude-comparison monopulse tracking system is usually preferred over this hybrid technique.

Phase-comparison monopulse. The tracking techniques discussed thus far in this chapter were based on a comparison of the *amplitudes* of echo signals received from two or more antenna positions. The sequential-lobing and conical-scan techniques used a single, time-shared antenna beam, while the monopulse technique used two or more simultaneous beams. The difference in amplitudes in the several antenna positions was proportional to the angular error. The angle of arrival (in one coordinate) may also be determined by comparing the phase difference between the signals from two separate antennas. Unlike the antennas of amplitude-comparison trackers, those used in phase-comparison systems are not offset from the axis. The individual boresight axes of the antennas are parallel, causing the (far-field) radiation to illuminate the same volume in space. The amplitudes of the target echo signals are essentially the same from each antenna beam, but the phases are different.

The measurement of angle of arrival by comparison of the phase relationships in the signals from the separated antennas of a radio interferometer has been widely used by the radio astronomers for precise measurements of the positions of radio stars. The interferometer as used by the radio astronomer is a passive instrument, the source of energy being radiated by the target itself. A tracking radar which operates with phase information is similar to an active interferometer and might be called an *interferometer* radar. It has also been called *simultaneous-phase-comparison* radar, or *phase-comparison monopulse*. The latter term is the one which will be used here.

In Fig. 5.12 two antennas are shown separated by a distance *d*. The distance to the target is *R* and is assumed large compared with the antenna separation *d*. The line of sight to the

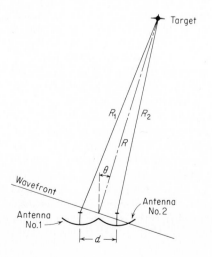

Figure 5.12 Wavefront phase relationships in phase-comparison monopulse radar.

target makes an angle θ to the perpendicular bisector of the line joining the two antennas. The distance from antenna 1 to the target is

$$R_1 = R + \frac{d}{2} \sin \theta$$

and the distance from antenna 2 to the target is

$$R_2 = R - \frac{d}{2} \sin \theta$$

The phase difference between the echo signals in the two antennas is approximately

$$\Delta\phi = \frac{2\pi}{\lambda} d \sin \theta \qquad (5.1)$$

For small angles where $\sin \theta \approx \theta$, the phase difference is a linear function of the angular error and may be used to position the antenna via a servo-control loop.

In the early versions of the phase-comparison monopulse radar, the angular error was determined by measuring the phase difference between the outputs of receivers connected to each antenna. The output from one of the antennas was used for transmission and for providing the range information. With such an arrangement it was difficult to obtain the desired aperture illuminations and to maintain a stable boresight. A more satisfactory method of operation is to form the sum and difference patterns in the RF and to process the signals as in a conventional amplitude-comparison monopulse radar.

In one embodiment of the phase-comparison principle as applied to missile guidance the phase difference between the signals in two fixed antennas is measured with a servo-controlled phase shifter located in one of the arms.[27] The servo loop adjusts the phase shifter until the difference in phase between the two channels is a null. The amount of phase shift which has to be introduced to make a null signal is a measure of the angular error.

The phase- and amplitude-comparison principles can be combined in a single radar to produce two-dimensional angle tracking with only two, rather than four, antenna beams.[28] The angle information in one plane (the azimuth) is obtained by two separate antennas placed side by side as in a phase-comparison monopulse. One of the beams is tilted slightly upward, while the other is tilted slightly downward, to achieve the squint needed for amplitude-comparison monopulse in elevation. Therefore the horizontal projection of the antenna patterns is that of a phase-comparison system, while the vertical projection is that of an amplitude-comparison system.

Both the amplitude-comparison-monopulse and the phase-comparison-monopulse trackers employ two antenna beams (for one coordinate tracking). The measurements made by the two systems are not the same; consequently, the characteristics of the antenna beams will also be different. In the amplitude-comparison monopulse the two beams are offset, that is, point in slightly different directions. This type of pattern may be generated by using one reflector dish with two feed horns side by side (four feed horns for two coordinate data). Since the feeds may be placed side by side, they could be as close as one-half wavelength. With such close spacing the phase difference between the signals received in the two feeds is negligibly small. Any difference in the amplitudes between the two antenna outputs in the amplitude-comparison system is a result of differences in amplitude and not phase. The phase-comparison monopulse, on the other hand, measures phase differences only and is not concerned with amplitude difference. Therefore the antenna beams are not offset, but are directed to illuminate a common volume in space. Separate antennas are needed since it is difficult to illuminate a single reflector with more than one feed and produce independent antenna patterns which illuminate the same volume in space.

Although tracking radars based upon the phase-comparison monopulse principle have been built and operated, this technique has not been as widely used as some of the other angle-tracking methods. The sum signal has higher sidelobes because the separation between the phase centers of the separate antennas is large. (These high sidelobes are the result of *grating lobes*, similar to those produced in phased arrays.) The problem of high sidelobes can be reduced by overlapping the antenna apertures. With reflector antennas, this results in a loss of angle sensitivity and antenna gain.

5.5 TARGET-REFLECTION CHARACTERISTICS AND ANGULAR ACCURACY[30,41,95]

The angular accuracy of tracking radar will be influenced by such factors as the mechanical properties of the radar antenna and pedestal, the method by which the angular position of the antenna is measured, the quality of the servo system, the stability of the electronic circuits, the noise level of the receiver, the antenna beamwidth, atmospheric fluctuations, and the reflection characteristics of the target. These factors can degrade the tracking accuracy by causing the antenna beam to fluctuate in a random manner about the true target path. These noiselike fluctuations are sometimes called *tracking noise*, or *jitter*.

A simple radar target such as a smooth sphere will not cause degradation of the angular-tracking accuracy. The radar cross section of a sphere is independent of the aspect at which it is viewed; consequently, its echo will not fluctuate with time. The same is true, in general, of a radar beacon if its antenna pattern is omnidirectional. However, most radar targets are of a more complex nature than the sphere. The amplitude of the echo signal from a complex target may vary over wide limits as the aspect changes with respect to the radar. In addition, the effective center of radar reflection may also change. Both of these effects—amplitude fluctuations and wandering of the radar center of reflection—as well as the limitation imposed by receiver noise can limit the tracking accuracy. These effects are discussed below.

Amplitude fluctuations. A complex target such as an aircraft or a ship may be considered as a number of independent scattering elements. The echo signal can be represented as the vector addition of the contributions from the individual scatterers. If the target aspect changes with respect to the radar—as might occur because of motion of the target, or turbulence in the case of aircraft targets—the relative phase and amplitude relationships of the contributions from the individual scatterers also change. Consequently, the vector sum, and therefore the amplitude, change with changing target aspect.

Amplitude fluctuations of the echo signal are important in the design of the lobe-switching radar and the conical-scan radar but are of little consequence to the monopulse tracker. Both the conical-scan tracker and the lobe-switching tracker require a finite time to obtain a measurement of the angle error. This time corresponds in the conical-scan tracker to at least one revolution of the antenna beam. With lobe switching, the minimum time is that necessary to obtain echoes at the four successive angular positions. In either case four pulse-repetition periods are required to make a measurement; in practice, many more than four are often used. If the target cross section were to vary during this observation time, the change might be erroneously interpreted as an angular-error signal. The monopulse radar, on the other hand, determines the angular error on the basis of a single pulse. Its accuracy will therefore not be affected by changes in amplitude with time.

To reduce the effect of amplitude noise on tracking, the conical-scan frequency should be chosen to correspond to a low value of amplitude noise. If considerable amplitude fluctuation noise were to appear at the conical-scan or lobe-switching frequencies, it could not be readily eliminated with filters or AGC. A typical scan frequency might be of the order of 30 Hz. Higher frequencies might also be used since target amplitude noise generally decreases with increasing frequency. However, this may not always be true. Propeller-driven aircraft produce modulation components at the blade frequency and harmonics thereof and can cause a substantial increase in the spectral energy density at certain frequencies. It has been found experimentally that the tracking accuracy of radars operating with pulse repetition frequencies from 1000 to 4000 Hz and a lobing or scan rate one-quarter of the prf are not limited by echo amplitude fluctuations.[29]

The percentage modulation of the echo signal due to cross-section amplitude fluctuations is independent of range if AGC is used. Consequently, the angular error as a result of amplitude fluctuations will also be independent of range.

Angle fluctuations.[29,30] Changes in the target aspect with respect to the radar can cause the apparent center of radar reflections to wander from one point to another. (The apparent center of radar reflection is the direction of the antenna when the error signal is zero.) In general, the apparent center of reflection might not correspond to the target center. In fact, it need not be confined to the physical extent of the target and may be off the target a significant fraction of the time. The random wandering of the apparent radar reflecting center gives rise to noisy or jittered angle tracking. This form of tracking noise is called *angle noise, angle scintillations, angle fluctuations,* or *target glint.* The angular fluctuations produced by small targets at long range may be of little consequence in most instances. However, at short range or with relatively large targets (as might be seen by a radar seeker on a homing missile), angular fluctuations may be the chief factor limiting tracking accuracy. Angle fluctuations affect all tracking radars whether conical-scan, sequential-lobing, or monopulse.

Consider a rather simplified model of a complex radar target consisting of two independent isotropic scatterers separated by an angular distance θ_D, as measured from the radar. Although such a target may be fictitious and used for reasons of mathematical simplicity, it might approximate a target such as a small fighter aircraft with wing-tip tanks or two aircraft targets flying in formation and located within the same radar resolution cell. It is also a close approximation to the low-angle tracking problem in which the radar sees the target plus its image reflected from the surface. The qualitative effects of target glint may be assessed from this model. The relative amplitude of the two scatterers is assumed to be a, and the relative phase difference is α. Differences in phase might be due to differences in range or to reflecting properties. The ratio a is defined as a number less than unity. The angular error $\Delta\theta$ as measured from the larger of the two targets is[31]

$$\frac{\Delta\theta}{\theta_D} = \frac{a^2 + a \cos \alpha}{1 + a^2 + 2a \cos \alpha} \tag{5.2}$$

This is plotted in Fig. 5.13. The position of the larger of the two scatterers corresponds to $\Delta\theta/\theta_D = 0$, while the smaller-scatterer position is at $\Delta\theta/\theta_D = +1$. Positive values of $\Delta\theta$ correspond to an apparent radar center which lies between the two scatterers; negative values lie outside the target. When the echo signals from both scatterers are in phase ($\alpha = 0$), the error reduces to $a/(a + 1)$, which corresponds to the so-called "center of gravity" of the two scatterers (not to be confused with the mechanical center of gravity).

Angle fluctuations are due to random changes in the relative distance from radar to the scatterers, that is, varying values of α. These changes may result from turbulence in the aircraft

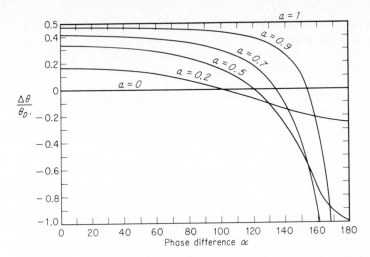

Figure 5.13 Plot of Eq. (5.2). Apparent radar center $\Delta\theta$ of two isotropic scatterers of relative amplitude a and relative phase shift α, separated by an angular extent θ_D.

flight path or from the changing aspect caused by target motion. In essence, angle fluctuations are a distortion of the phase front of the echo signal reflected from a complex target and may be visualized as the apparent tilt of this phase front as it arrives at the tracking system.

Equation (5.2) indicates that the tracking error $\Delta\theta$ due to glint for the two-scatterer target is directly proportional to the angular extent of the target θ_D. This is probably a reasonable approximation to the behavior of real targets, provided the angular extent of the target is not too large compared with the antenna beamwidth. Since θ_D varies inversely with distance for a fixed target size, the tracking error due to glint also varies inversely with distance.

A slightly more complex model than the two-scatterer target considered above is one consisting of many individual scatterers, each of the same cross section, arranged uniformly along a line of length L perpendicular to the line of sight from the radar. The resultant cross section from such a target is assumed to behave according to the Rayleigh probability distribution. The probability of the apparent radar center lying outside the angular region of L/R radians (in one tracking plane) is 0.134, where R is the distance to the target.[32] Thus 13.4 percent of the time the radar will not be directed to a point on the target. Similar results for a two-dimensional model consisting of equal-cross-section scatterers uniformly spaced over a circular area indicate that the probability that the apparent radar center lies outside this target is 0.20.

Angle fluctuations in a tracking radar are reduced by increasing the time constant of the AGC system (reducing the bandwidth).[29,33,34] However, this reduction in angle fluctuation is accompanied by a new component of noise caused by the amplitude fluctuations associated with the echo signal; that is, narrowing the AGC bandwidth generates additional noise in the vicinity of zero frequency, and poorer tracking results. Amplitude noise modulates the tracking-error signals and produces a new noise component, proportional to true tracking errors, that is enhanced with a slow AGC. Under practical tracking conditions it seems that a wide-bandwidth (short-time constant) AGC should be used to minimize the overall tracking noise. However, the servo bandwidth should be kept to a minimum consistent with tactical requirements in order to minimize the noise.

Receiver and servo noise. Another limitation on tracking accuracy is the receiver noise power. The accuracy of the angle measurement is inversely proportional to the square root of the signal-to-noise power ratio.[2] Since the signal-to-noise ratio is proportional to $1/R^4$ (from the radar equation), the angular error due to receiver noise is proportional to the square of the target distance.

Servo noise is the hunting action of the tracking servomechanism which results from backlash and compliance in the gears, shafts, and structures of the mount. The magnitude of servo noise is essentially independent of the target echo and will therefore be independent of range.

Summary of errors. The contributions of the various factors affecting the tracking error are summarized in Fig. 5.14. Angle-fluctuation noise varies inversely with range; receiver noise varies as the square of the range; and amplitude fluctuations and servo noise are independent of range. This is a qualitative plot showing the gross effects of each of the factors. Two different resultant curves are shown. Curve A is the sum of all effects and is representative of conical-scan and sequential-lobing tracking radars. Curve B does not include the amplitude fluctuations and is therefore representative of monopulse radars. In Fig. 5.14 the amplitude fluctuations are assumed to be larger than servo noise. If not, the improvement of monopulse tracking over conical scan will be negligible. In general, the tracking accuracy deteriorates at both short and long target ranges, with the best tracking occurring at some intermediate range.

Frequency agility and glint reduction.[35–40,85] The angular error due to glint, which affects all tracking radars, results from the radar receiving the vector sum of the echoes contributed by the individual scattering centers of a complex target, and processing it as if it were the return from a single scattering center. If the frequency is changed, the relative phases of the individual

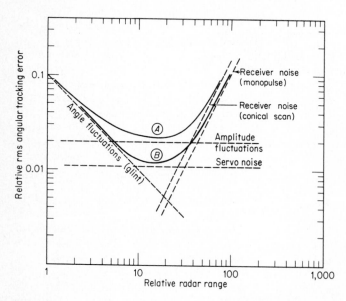

Figure 5.14 Relative contributions to angle tracking error due to amplitude fluctuations, angle fluctuations, receiver noise, and servo noise as a function of range. (A) Composite error for a conical-scan or sequential-lobing radar; (B) composite error for monopulse.

scatterers will change and a new resultant is obtained as well as a new angular measurement. Measurements are independent if the frequency is changed by an amount[36]

$$\Delta f_c = \frac{c}{2D} \tag{5.3}$$

where c = velocity of propagation and D = target depth. The glint error can be reduced by averaging the independent measurements obtained with frequency agility. (The depth D as seen by the radar might be less than the geometrical measurement of target depth if the extremities of the target result in small backscatter.)

The improvement I in the tracking accuracy when the frequency is changed pulse-to-pulse is approximately[37]

$$I^2 = \frac{1}{2\Delta f_c/B_{fa} + 2B_g/f_p} \approx \frac{DB_{fa}}{c} \tag{5.4}$$

where B_{fa} = the frequency agility bandwidth, D = target depth, c = velocity of propagation, B_g = glint bandwidth, and f_p = pulse repetition frequency. (The approximation holds for large prf's and for the usual glint bandwidths which are of the order of a few hertz to several tens of hertz.)[36] For example, with a target depth D of 7 m and a frequency-agile bandwidth of 300 MHz, the glint error is reduced by a factor of 2.6. According to the above, the improvement in tracking accuracy is proportional to the square root of the frequency agility bandwidth, or $I \sim \sqrt{B_{fa}}$.

A different glint model, based on the assumption that the angular motion of a complex target can be described by a gaussian random yaw motion of zero mean, yields the result that the reduction in angle error due to frequency agility asymptotically approaches a value of 3.1 with increasing agility bandwidth.[38] The model also gives the variance of the inherent glint for a frequency-agile radar as

$$\text{var} = 0.142\gamma_0^2 \tag{5.5}$$

where γ_0 is the lateral radius of gyration of the collection of scatterers comprising the target. The value of γ_0 for a "typical" twin-jet aircraft in level flight at near head-on or near tail-on aspect is said[38] to be equal to half the separation of the jet engines. For a ship at broadside, γ_0 is approximately 0.15 times the ship length.

When angle errors due to glint are large, the received signals are small; that is, the received signal amplitude and the glint error are negatively correlated. Thus, by transmitting a number of frequencies and using the angle error corresponding to that frequency with the largest signal, it is possible to eliminate the large angle errors associated with glint.[39,40] Those returns of low amplitude and, hence, of high error, are excluded in this technique. Instead of selecting only the largest signal for processing, the indicated position of the target at each frequency can be weighted according to the amplitude of the return.[40] Only a small number of frequencies is needed to reduce substantially the glint error. The reduction in rms tracking error by processing only that signal (frequency) with the largest amplitude is approximately[40]

$$\sigma_{\text{rms}} \approx \frac{\sigma_g}{N} \qquad 1 \leq N \leq 4 \tag{5.6}$$

where σ_g is the single-frequency glint error and N is the number of frequencies. The tracking error will not decrease significantly for more than four pulses. Each of the frequencies must be separated by at least Δf_c, as given by Eq. (5.3).

Frequency agility, as described here for the reduction of glint, applies to the monopulse tracking radar. It also reduces the glint in a conical scan or a sequential-lobing radar, but the changing frequency can result in amplitude fluctuations which can affect the angle tracking accuracy if the spectrum of the fluctuations at the conical scan or the lobing frequencies is increased. Thus, frequency agility might cause an increase in the angle error due to amplitude fluctuations in these systems while decreasing the error due to glint. The overall effect of frequency agility in conical scan or sequential lobing systems is therefore more complicated to analyze than monopulse systems which are unaffected by amplitude fluctuations.

It has also been suggested[54] that polarization agility can reduce the glint error. Since the individual echoes from the various scattering centers that make up a complex target are likely to be sensitive to the polarization of the incident radar signal, a change in the polarization can possibly result in an independent measure of the apparent target direction. By observing the target with a variable polarization producing independent measurements, the angle error due to glint is averaged and the effect of the large glint errors is reduced. Experimental measurements with an X-band, conical-scan, pulse doppler radar tracking an M-48 tank at approximately 500 m range reduced the angular tracking error by about one-half. In these tests, the best results were obtained when the plane of polarization was switched in small increments (5.6 to 22.5°) at a rate greater than 500 steps per second, which is more than an order of magnitude greater than the 40 Hz conical-scan rate.

Low-angle tracking.[41-53,90-94] A radar that tracks a target at a low elevation angle, near the surface of the earth, can receive two echo signals from the target, Fig. 5.15. One signal is reflected directly from the target, and the other arrives via the earth's surface. (This is similar to the description of surface reflections and its effect on the elevation coverage, as in Sec. 12.2.) The direct and the surface-reflected signals combine at the radar to yield an angle measurement that differs from the true measurement that would have been made with a single target in the absence of surface reflections. The result is an error in the measurement of elevation. The surface-reflected signal may be thought of as originating from the image of the target mirrored by the earth's surface. Thus, the effect on tracking is similar to the two-target model used to describe glint, as discussed previously. The surface-reflected signal is sometimes called a *multipath signal.*

An example of the elevation angle error at low angles is shown in Fig. 5.16 for a target at constant height.[43] At close range the target elevation angle is large and the antenna beam does not illuminate the surface; hence the tracking is smooth. At intermediate range, where the elevation angle is from 0.8 to as much as six beamwidths, the surface-reflected signal enters the radar by means of the antenna near-in sidelobes. The surface-reflected signal is small so that the antenna makes small oscillations about some mean position. At greater ranges (elevation angles less than about 0.8 beamwidth), where the antenna main beam illuminates the surface, the interference between the direct and the reflected signals can result in large errors in elevation angle. The angular excursions can be up (into the air) or down (into the ground). The peak errors are severe and can be many times the angular separation between the target and its

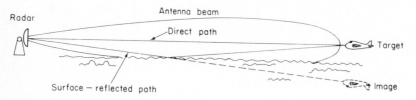

Figure 5.15 Low-angle tracking illustrating the surface-reflected signal path and the target image.

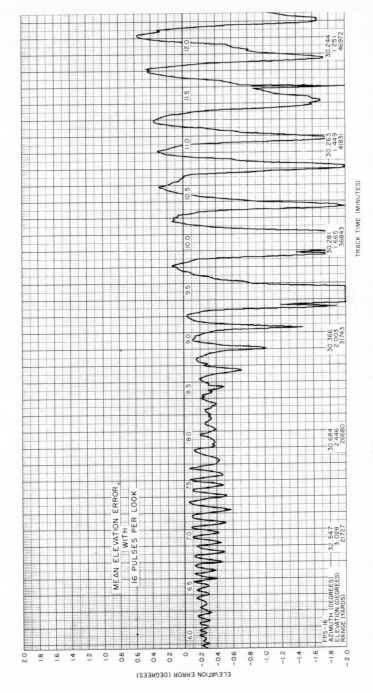

Figure 5.16 Example of the measured elevation tracking error using a phased-array radar with 2.7° beamwidth. Aircraft target flew out in range at a nearly constant altitude. The numbers along the zero error line indicate the track time in minutes. (*From Linde.*[46])

image. The tracking has been described as "wild" and can be great enough to cause the radar to break track. The effect is most pronounced over a smooth water surface where the reflected signal is strong. The effect can be so great that it can become impossible to track targets at low elevation angles with a conventional tracking radar. In addition to causing errors in the elevation-angle tracking, it is also possible for surface-reflected multipath to introduce errors in the azimuth-angle tracking; either by "cross-talk" in the radar between the azimuth and elevation error channels, or by the target-image plane departing from the vertical as when over sloping land or when the radar is on a rolling and pitching ship.

The surest method for avoiding tracking error due to multipath reflections via the surface of the earth is to use an antenna with such a narrow beamwidth that it doesn't illuminate the surface. This requires a large antenna and/or a high frequency.[44] Although such a solution eliminates the problem, there may be compelling reasons in some applications that mitigate against the large antenna needed for a narrow beamwidth or against operation at high frequency.

Prior knowledge of target behavior sometimes can be used to avoid the serious effects of low-angle multipath without overly complicating the radar. Since targets of interest will not go below the surface of the earth, and are limited in their ability to accelerate upward and downward, radar data indicative of unreasonable behavior can be recognized and rejected. In some situations, the target might be flying fast enough and the inertia of the antenna may be great enough to dampen the angle-error excursions caused by the multipath.[47] Another solution takes advantage of the fact that large errors are limited to a region of low elevation angles predictable from the antenna pattern and the terrain. It is thus possible to determine when the target is in the low-angle region by sensing large elevation-angle errors and locking the antenna in elevation at some small positive angle while continuing closed-loop azimuth tracking.[43] This is sometimes called *off-axis tracking*,[45] or *off-boresight tracking*.[47] The elevation angle at which the antenna is fixed depends on the terrain and the antenna pattern. Typically it might be about 0.7 to 0.8 beamwidth. With the beam fixed at a positive elevation angle, the elevation-angle error may then be determined open-loop from the error-signal voltage or the elevation measurement may simply be assumed to be halfway between the horizon and the antenna boresight. In the extreme, the peak-to-peak tracking error would not exceed 0.7 to 0.8 beamwidth and typically the rms error would be about 0.3 beamwidth. These relatively simple methods, combined with heavy smoothing of the error signal, can allow meaningful, but not necessarily accurate, tracking at low angles.

The surface-reflected signal travels a longer path than the direct signal so that it may be possible in some cases to separate the two in time (range). Tracking on the direct signal avoids the angle errors introduced by the multipath. The range-resolution required to separate the direct from the ground-reflected signal is

$$\Delta R = \frac{2h_a h_t}{R} \tag{5.7}$$

where h_a = radar antenna height, h_t = target height, and R = range to the target. For a radar height of 30 m, a target height of 100 m and a range of 10 km, the range-resolution must be 0.6 m, corresponding to a pulse width of 4 ns. This is a much shorter pulse than is commonly employed in radar. Although the required range-resolutions for a ground-based radar are achievable in principle, it is usually not applicable in practice.

The use of frequency diversity, as described previously for reducing glint, can also reduce the multipath tracking error. As seen in Fig. 5.16, the angle errors due to multipath at low angle are cyclical. This is a result of the direct and surface-reflected signals reinforcing and canceling each other as the relative phase between the two paths varies. A change in frequency also changes the phase relationship between the two signals. Thus the angle errors can be averaged by operating the radar over a wide frequency band or by sweeping the RF frequency and deducing the angle on the basis of the corresponding behavior of the error signal.[46] It turns out, however, that the bandwidth required to extract the target elevation angle is essentially the same required of a short pulse for separating the direct and reflected signals. Thus, the bandwidths needed to eliminate the multipath error are usually quite large for most applications.

The doppler frequency shift of the direct signal differs from that of the surface-reflected signal, but the difference is too small in most cases to be of use in reducing the errors due to multipath. Radar fences, properly located, can mask the surface-reflected signal from the near-in elevation sidelobes, but they are of limited utility when the main beam illuminates the top edge of the fence and creates diffracted energy.[53] Vertical polarization, often used in trackers, reduces the surface-reflected signal in the vicinity of the Brewster angle, but has no special advantage at low angles (less than $1.5°$ over water and $3°$ over land).[45] For a similar reason, circular polarization has no inherent advantage in improving multipath below elevation angles corresponding to the Brewster angle.

The basic reason for poor tracking at low angle results from the fact that the conventional tracking radar with a two-horn feed in elevation (or its equivalent for a conical-scan tracker) provides unambiguous information for only one target. At low elevation angles two "targets" are present, the real one and its image. Thus, the aperture must be provided with more degrees of freedom than are available from a simple two-element feed. One approach is to utilize multiple feeds in the vertical plane.[48–50] A minimum of three feeds (or elements) is necessary to resolve two targets, but antennas with from four to nine elements in the vertical dimension have been considered. Another approach that depends on additional antenna feeds, or degrees of freedom, employs a monopulse tracker with a difference pattern containing a second null which is independently steerable. This null is maintained in the direction of the image, as computed by Snell's law for the measured target range. The use of the classical maximum likelihood solution for the optimum double-null difference pattern results in relatively large sidelobes at the horizon and at angles below the image. By slightly sacrificing the accuracy of angle tracking (due to less slope sensitivity), lower sidelobes can be achieved. Such a pattern which sacrifices slope sensitivity for lower sidelobes has been called a *tempered double-null difference pattern*.[48]

The tempered double-null is more suited to mechanically steered trackers than to phased-array trackers since it is basically a null balance technique that does not achieve full accuracy until it goes through a settling process. A technique that can operate on the basis of a single pulse, and therefore is more applicable to the phased array, has been called the *fixed-beam* approach; or more descriptively, the *asymmetric monopulse*.[48] Two asymmetrical patterns are generated and squinted above the horizon to minimize response to the surface-reflected signal. The two patterns are constrained so that their ratio has an even-order symmetry about the horizon. Knowing the shape of the antenna patterns, the measurement of the ratio of signals in the two beams allows the target elevation angle to be deduced. A similar technique[41,51] utilizes sum (Σ) and difference (Δ) patterns whose ratio Δ/Σ is symmetrical. Just as in off-axis tracking, the antenna boresight is locked at some elevation angle. This results in the lower of the two peaks of the symmetrical Δ/Σ ratio being placed in the direction of the bisector of the target and image. Under this condition, the values of Δ/Σ in the direction of the target and its image are equal. The estimate of target elevation is unaffected by the phase and amplitude of the image. (The angle of the target/image bisector is approximately given by the ratio of the radar antenna-height–to–target-range.)

The normal monopulse radar receiver uses only the in-phase (or the out-of-phase) component of the difference signal. When there is a multipath signal present along with the direct signal, a quadrature component of the difference signal exists. The in-phase and the quadrature components of the error signal define a complex angle error-signal. In the complex plane, with the in-phase and quadrature components as the axes, the locus of the complex angle with target elevation as the parameter is a spiral path. By measuring the complex angle, the target elevation can be inferred.[52] In using the complex-angle technique, the radar antenna is fixed at some angle above the horizon and the open-loop measurement of complex angle is compared with a predicted set of values for the particular radar installation, antenna elevation-pointing angle, and terrain properties. A given in-phase and quadrature measurement does not give a unique value of elevation angle since the plot of the complex angle shows multiple, overlapping turns of a spiral with increasing target altitude. The ambiguities can be resolved with frequency diversity or by continuous tracking over an interval long enough to recognize the ambiguous spirals. In one simulation of the technique it was found that tracking over a smooth surface can be improved by at least a factor of two, but over rough surfaces the improvement was marginal.[61]

Thus there exist a number of possible techniques for reducing the angle errors found when tracking a target at low angle. Some of these require considerable modification to the conventional tracking radar but others require only modification in the processing of the angle-error signals. It has been said that these various techniques can avoid the large errors encountered by conventional trackers when the elevation angle is between 0.25 and 1.0 beamwidth. The rms accuracies are between 0.05 and 0.1 beamwidth in these regions.[45] The choice of technique to permit tracking of targets at low altitude with reduced error depends on the degree of complexity that can be tolerated for the amount of improvement obtained.

Electro-optical devices such as TV, IR, or lasers can be used in conjunction with radar to provide tracking at low target altitude when the radar errors are unacceptable. These devices, however, are of limited range and are not all-weather.

5.6 TRACKING IN RANGE[3,55-58]

In most tracking-radar applications the target is continuously tracked in range as well as in angle. Range tracking might be accomplished by an operator who watches an A-scope or J-scope presentation and manually positions a handwheel in order to maintain a marker over the desired target pip. The setting of the handwheel is a measure of the target range and may be converted to a voltage that is supplied to a data processor.

As target speeds increase, it is increasingly difficult for an operator to perform at the necessary levels of efficiency over a sustained period of time, and automatic tracking becomes a necessity. Indeed, there are many tracking applications where an operator has no place, as in a homing missile or in a small space vehicle.

The technique for automatically tracking in range is based on the split range gate. Two range gates are generated as shown in Fig. 5.17. One is the *early gate*, and the other is the *late gate*. The echo pulse is shown in Fig. 5.17a, the relative position of the gates at a particular instant in Fig. 5.17b, and the error signal in Fig. 5.17c. The portion of the signal energy contained in the early gate is less than that in the late gate. If the outputs of the two gates are subtracted, an error signal (Fig. 5.17c) will result which may be used to reposition the center of the gates.[55] The magnitude of the error signal is a measure of the difference between the center of the pulse and the center of the gates. The sign of the error signal determines the direction in

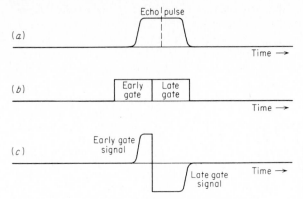

Figure 5.17 Split-range-gate tracking. (a) Echo pulse; (b) early-late range gates; (c) difference signal between early and late range gates.

which the gates must be repositioned by a feedback-control system. When the error signal is zero, the range gates are centered on the pulse.

The range gating necessary to perform automatic tracking offers several advantages as by-products. It isolates one target, excluding targets at other ranges. This permits the boxcar generator to be employed. Also, range gating improves the signal-to-noise ratio since it eliminates the noise from the other range intervals. Hence the width of the gate should be sufficiently narrow to minimize extraneous noise. On the other hand, it must not be so narrow that an appreciable fraction of the signal energy is excluded. A reasonable compromise is to make the gate width of the order of the pulse width.

A target of finite length can cause noise in range-tracking circuits in an analogous manner to angle-fluctuation noise (glint) in the angle-tracking circuits. Range-tracking noise depends on the length of the target and its shape. It has been reported[29] that the rms value of the range noise is approximately 0.8 of the target length when tracking is accomplished with a video split-range-gate error detector.

5.7 ACQUISITION

A tracking radar must first find and acquire its target before it can operate as a tracker. Therefore it is usually necessary for the radar to scan an angular sector in which the presence of the target is suspected. Most tracking radars employ a narrow pencil-beam antenna. Searching a volume in space for an aircraft target with a narrow pencil beam would be somewhat analogous to searching for a fly in a darkened auditorium with a flashlight. It must be done with some care if the entire volume is to be covered uniformly and efficiently. Examples of the common types of scanning patterns employed with pencil-beam antennas are illustrated in Fig. 5.18.

In the *helical* scan, the antenna is continuously rotated in azimuth while it is simultaneously raised or lowered in elevation. It traces a helix in space. Helical scanning was employed for the search mode of the SCR-584 fire-control radar, developed during World War II for the aiming of antiaircraft-gun batteries.[59] The SCR-584 antenna was rotated at the rate of 6 rpm and covered a 20° elevation angle in 1 min. The *Palmer* scan derives its name from the familiar penmanship exercises of grammar school days. It consists of a rapid circular scan (conical scan) about the axis of the antenna, combined with a linear movement of the axis of rotation. When the axis of rotation is held stationary, the Palmer scan reduces to the *conical*

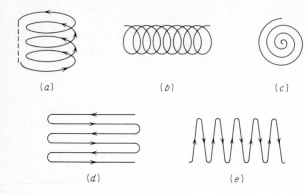

(a) (b) (c)

(d) (e)

Figure 5.18 Examples of acquisition search patterns. (a) Trace of helical scanning beam; (b) Palmer scan; (c) spiral scan; (d) raster, or TV, scan; (e) nodding scan. The raster scan is sometimes called an n-bar scan, where n is the number of horizontal rows.

scan. Because of this property, the Palmer scan is sometimes used with conical-scan tracking radars which must operate with a search as well as a track mode since the same mechanisms used to produce conical scanning can also be used for Palmer scanning.[60] Some conical-scan tracking radars increase the squint angle during search in order to reduce the time required to scan a given volume. The conical scan of the SCR-584 was operated during the search mode and was actually a Palmer scan in a helix. In general, conical scan is performed during the search mode of most tracking radars.

The Palmer scan is suited to a search area which is larger in one dimension than another. The *spiral* scan covers an angular search volume with circular symmetry. Both the spiral scan and the Palmer scan suffer from the disadvantage that all parts of the scan volume do not receive the same energy unless the scanning speed is varied during the scan cycle. As a consequence, the number of hits returned from a target when searching with a constant scanning rate depends upon the position of the target within the search area.

The *raster*, or *TV*, scan, unlike the Palmer or the spiral scan, paints the search area in a uniform manner. The raster scan is a simple and convenient means for searching a limited sector, rectangular in shape. Similar to the raster scan is the *nodding* scan produced by oscillating the antenna beam rapidly in elevation and slowly in azimuth. Although it may be employed to cover a limited sector—as does the raster scan—nodding scan may also be used to obtain hemispherical coverage, that is, elevation angle extending to 90° and the azimuth scan angle to 360°.

The helical scan and the nodding scan can both be used to obtain hemispheric coverage with a pencil beam. The nodding scan is also used with height-finding radars. The Palmer, spiral, and raster scans are employed in fire-control tracking radars to assist in the acquisition of the target when the search sector is of limited extent.

5.8 OTHER TOPICS

Servo system. The automatic tracking of the target coordinates in angle, range, and doppler frequency is usually accomplished with a so-called *Type II servo system*.[1,2,3,67,68] It is also referred to as a *zero velocity error* system since in theory no steady-state error exists for a constant velocity input. A steady-state error exists, however, for a step-acceleration input. Thus, the accelerations the system must handle need to be specified in order to select a suitable Type II system. The *tracking bandwidth* of the servo system is defined as the frequency where its open-loop filter transfer function is of unity gain. It represents the transition from closed loop to open loop operation. One of the functions of the servo system is to reduce the fluctuations of the input signal by filtering or smoothing. Therefore, the tracking bandwidth should be narrow to reduce the effects of noise or jitter, reject unwanted spectral components such as the conical-scan frequency or engine modulation, and to provide a smoothed output of the measurement. On the other hand, a wide tracking bandwidth is required to accurately follow, with minimum lag, rapid changes in the target trajectory or in the vehicle carrying the radar antenna. That is, a wide bandwidth is required for following changes in the target trajectory and a narrow bandwidth for sensitivity. A compromise must generally be made between these conflicting requirements. A target at long range has low angular rates and a low signal-to-noise ratio. A narrow tracking bandwidth is indicated in such a case to increase sensitivity and yet follow the target with minimum lag. At short range, however, the angular rates are likely to be large so that a wide tracking bandwidth is needed in order to follow the target properly. The loss in sensitivity due to the greater bandwidth is offset by the greater target signal at the shorter ranges. The bandwidth should be no wider than necessary in order to keep the angle errors due to target scintillation, or glint, from becoming excessive. The tracking bandwidth in some systems might be made variable or even adaptive to conform automatically to the target conditions.

Another restriction on the tracking bandwidth is that it should be small compared to the lowest natural resonant frequency of the antenna and servo system including the structure foundation, in order to prevent the system from oscillating at the resonant frequency. The shaded region in Fig. 5.19 describes the measured bounds of the lowest servo resonant frequency as a function of antenna size achieved with actual tracking radars.

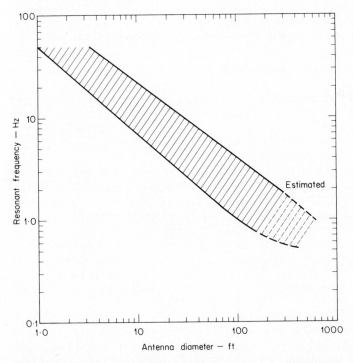

Figure 5.19 Lowest servo resonant-frequency as a function of antenna diameter for hemispherical scanning paraboloid reflector antennas. (*Based on measurements compiled by D. D. Pidhayny of the Aerospace Corporation.*)

Precision " on-axis " tracking.[62-65] Some of the most precise tracking radars are those associated with the instrumentation used at missile-testing ranges.[62] One such class of precision tracking radar has been called *on-axis tracking*.[63] The output of a conventional servo system lags its input. The result of the lag is a tracking error. The on-axis tracker accounts for this lag, as well as for other factors that can contribute to tracking error, so as to keep the target being tracked in the center of the beam or on the null axis of the difference pattern. On-axis tracking, as compared with trackers with a target lag, improves the accuracy by reducing the coupling between the azimuth and elevation angle-tracking channels, by minimizing the generation of cross polarization and by reducing the effects of system nonlinearities.

The processes that constitute on-axis tracking include (1) the use of adaptive tracking whose output updates a *stored prediction* of the target trajectory rather than control the antenna servo directly, (2) the removal by prior calibration of static and dynamic system biases and errors, and (3) the use of appropriate coordinate systems for filtering (smoothing) the target data.

The radar's angle-error signals are smoothed and compared to a predicted measurement based on a target-trajectory model updated by the results of previous measurements. (Prior knowledge of the characteristics of the trajectory can be incorporated in the model, as, for example, when the trajectory is known to be ballistic.) If the difference between the prediction and the measurement is zero, no adjustment is made and the antenna mount is pointed according to the stored prediction. If they do not agree, the target trajectory prediction is changed until they do. Thus, the pointing of the antenna is made open-loop based on the stored target-trajectory prediction updated by the radar measurements. The servo loop that points the antenna is made relatively wideband (high data rate) to permit a fast tracking response against targets with high angular acceleration. The process of adjusting the predicted position based on the measured position is performed with a narrow bandwidth. This error-signal bandwidth is adaptive and can be made very narrow to obtain good signal-to-noise ratio, yet the system will continue to point open-loop based on the stored target-trajectory prediction and the wide tracking-bandwidth of the antenna-pointing servos. For convenience the range, azimuth, and elevation (r, θ, ϕ) coordinates of the radar output are converted to rectilinear (x, y, z) target coordinates to perform the data smoothing and comparison with prediction. In radar coordinates, the track of a target on a straight-line trajectory is curvilinear and can generate apparent accelerations. This does not happen when tracking in rectilinear coordinates. The rectilinear coordinates of the updated target prediction are converted back to radar coordinates to drive the antenna.

Systematic errors are determined by prior measurement and are used to adjust the encoded antenna position to provide the correct target position. Systematic errors include (1) error in the zero reference of the encoders that indicate the orientation of the radar axes, (2) misalignment of the elevation axis with respect to the azimuth axis (nonorthogonality), (3) droop or flexing of the antenna and mount caused by gravity, (4) misalignment of the radar with respect to the elevation axis (skew), (5) noncoincidence of the azimuth plane of the mount to the local reference plane (mislevel), (6) dynamic lag in the servo system, (7) finite transit time that results in the target being at a different position by the time the echo is received by the radar, and (8) bending and additional time delay of the propagation path due to atmospheric refraction.

A boresight telescope mounted on the radar antenna permits calibration of the mechanical axis of the antenna with respect to a star field. This calibration accounts for bias in azimuth and elevation, mislevel, skew, droop, and nonorthogonality. Tracking a visible satellite with the radar permits the position of the RF axis relative to the mechanical (optical) axis to be determined. The difference between the position measured by the optics and that measured by the radar is obtained after correction is made for the difference in atmospheric refraction for optical and RF propagation. This type of dynamic calibration requires the radar to be large enough to track satellites.

There is nothing unique about any of the individual processes that enter into on-axis tracking. They can each be applied individually, if desired, to any tracking radar to improve the accuracy of track.

High-range-resolution monopulse.[66] It has been noted previously in this chapter that the presence of multiple scatterers within the range-resolution cell of the radar results in scintillation, or glint, which can introduce a significant error. The use of a radar with high range-

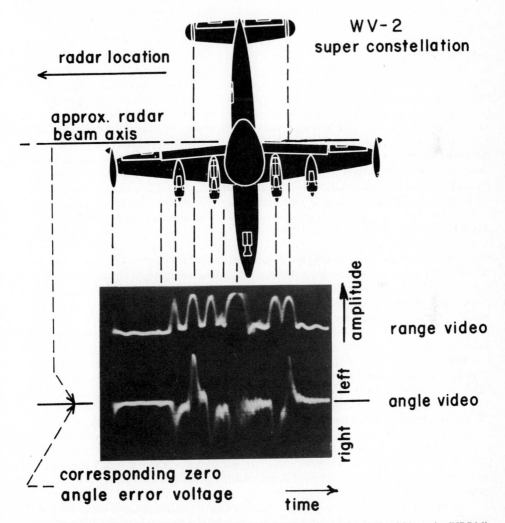

Figure 5.20 Range and angle video obtained with the NRL High Range Resolution Monopulse (HRRM) radar. The target is a Super Constellation aircraft in flight. Note that the angle video indicates to what side of the radar beam axis are located the individual scatterers which are resolved in range. The radar operated at X band with a 7-ft-diameter antenna and a pulse width of 3 nanoseconds.[66]

resolution isolates the individual scattering centers and therefore eliminates the glint problem. With aircraft targets an effective pulse width of several nanoseconds (less than a meter range-resolution) can resolve the individual scatterers. Once the scatterers are resolved in range, an angle measurement can be obtained on each one by use of monopulse. This provides a three-dimensional "image" of the target since the azimuth and elevation of each scatterer is given along with the range. Figure 5.20 is an example of the type of radar image that might be obtained with an aircraft. This technique not only improved significantly the tracking accuracy as compared with a longer pulse conventional tracker, but its unique measurement properties provide additional capabilities. For example, it can be used as a means of target classification or recognition, as a precision vector miss-distance indicator for uninstrumented targets, as a counter (ECCM) to repeater jammers, as an aid in low-angle tracking, and as a means for discriminating unwanted clutter or chaff from the desired target.

Tracking in doppler. Tracking radars that are based on CW, pulse-doppler, or MTI principles can also track the doppler frequency shift generated by a moving target. This may be accomplished with a frequency discriminator and a tunable oscillator to maintain the received signal in the center of a narrow-band filter. It is also possible to use a phase-lock loop, or phase-sensitive discriminator whose output drives a voltage-controlled oscillator to hold its phase in step with the input signal.[2] Tracking the doppler frequency shift with the equivalent of a narrow-band filter just wide enough to encompass the frequency spectrum of the received signal allows an improvement in the signal-to-noise ratio as compared with wideband processing. It can also provide resolution of the desired moving target from stationary clutter. The doppler tracking filter is sometimes called a *speed gate*.

5.9 COMPARISON OF TRACKERS[1]

Of the four continuous-tracking-radar techniques that have been discussed (sequential lobing, conical scan, amplitude-comparison monopulse, and phase-comparison monopulse), conical scan and amplitude-comparison monopulse have seen more application than the other two. The phase-comparison monopulse has not been too popular because of the relative awkwardness of its antenna (four separate antennas mounted to point their individual beams in the same direction), and because the sidelobe levels might be higher than desired. Although sequential lobing is similar to conical scan, the latter is preferred in most applications, since it suffers less loss and the antenna and feed systems are usually less complex. In this section, only the conical-scan radar and the amplitude-comparison monopulse will be compared. (The latter will be referred to simply as monopulse.)

When the target is being tracked, the signal-to-noise ratio available from the monopulse radar is greater than that of a conical-scan radar, all other things being equal, since the monopulse radar views the target at the peak of its sum pattern while the conical-scan radar views the target at an angle off the peak of the antenna beam. The difference in signal-to-noise ratio might be from 2 to 4 dB. For the same size aperture, the beamwidth of a conical-scan radar will be slightly greater than that of the monopulse because its feed is offset from the focus.

The tracking accuracy of a monopulse radar is superior to that of the conical-scan radar because of the absence of target amplitude-fluctuations and because of its greater signal-to-noise ratio. It is the preferred technique for precision tracking. However, both monopulse and conical-scan radars are degraded equally by the wandering of the apparent position of the target (glint).

The monopulse radar is the more complex of the two. Three separate receivers are necessary to derive the error signal in two orthogonal angular coordinates. Only one receiver is needed in the conical-scan radar. (There are certain monopulse implementations that can use either one or two receivers, but at some sacrifice in performance.) Since the monopulse radar compares the amplitudes of signals received in three separate channels, it is important that the gain and phase shift through these channels be identical. The RF circuitry that generates the sum and difference signals in a monopulse radar has been steadily improved, and can be realized without excessive physical bulk. A popular form of antenna for monopulse is the Cassegrain.

With the monopulse tracker it is possible to obtain a measure of the angular error in two coordinates on the basis of a single pulse. A minimum of four pulses are usually necessary with the conical-scan radar. However, a continuous-tracking radar seldom makes a measurement on a single pulse. (Phased-array radars and some surveillance radars, however, might use the monopulse principle to extract an angle measurement on the basis of a single pulse.) In practice, the two radars utilize essentially the same number of pulses to obtain an error signal if the servo tracking bandwidths and pulse repetition frequencies are the same. The monopulse radar first makes its angle measurement and then integrates a number of pulses to obtain the required signal-to-noise ratio and to smooth the error. The conical-scan radar, on the other hand, integrates a number of pulses first and then extracts the angle measurement.

Because a monopulse radar is not degraded by amplitude fluctuations, it is less susceptible to hostile electronic countermeasures than is conical scan.

In brief, the monopulse radar is the better tracking technique; but in many applications where the ultimate in performance is not needed, the conical-scan radar is used because it is less costly and less complex.

5.10 TRACKING WITH SURVEILLANCE RADAR

The track of a target can be determined with a surveillance radar from the coordinates of the target as measured from scan to scan. The quality of such a track will depend on the time between observations, the location accuracy of each observation, and the number of extraneous targets that might be present in the vicinity of the tracked target. A surveillance radar that develops tracks on targets it has detected is sometimes called a *track-while-scan* (TWS) radar.

One method of obtaining tracks with a surveillance radar is to have an operator manually mark with grease pencil on the face of the cathode-ray tube the location of the target on each scan. The simplicity of such a procedure is offset by the poor accuracy of the track. The accuracy of track can be improved by using a computer to determine the trajectory from inputs supplied by an operator. A human operator, however, cannot update target tracks at a rate greater than about once per two seconds.[69] Thus, a single operator cannot handle more than about six target tracks when the radar has a twelve-second scan rate (5 rpm antenna rotation rate). Furthermore an operator's effectiveness in detecting new targets decreases rapidly after about a half hour of operation. The radar operator's traffic handling limitation and the effects of fatigue can be mitigated by automating the target detection and tracking process with data processing called *automatic detection and tracking* (ADT). The availability of digital data processing technology has made ADT economically feasible. An ADT system performs the functions of target detection, track initiation, track association, track update, track smoothing (filtering) and track termination.

The *automatic detector* part of the ADT quantizes the range into intervals equal to the range resolution. At each range interval the detector integrates n pulses, where n is the number of pulses expected to be returned from a target as the antenna scans past. The integrated pulses are compared with a threshold to indicate the presence or absence of a target. An example is the commonly used *moving window detector* which examines continuously the last n samples within each quantized range interval and announces the presence of a target if m out of n of these samples cross a preset threshold. (This and other automatic detectors are described in Sec. 10.7 and in Ref. 70.) By locating the center of the n pulses, an estimate of the target's angular direction can be obtained. This is called *beam splitting.*

If there is but one target present within the radar's coverage, then detections on two scans are all that is needed to establish a target track and to estimate its velocity. However, there are usually other targets as well as clutter echoes present, so that three or more detections are needed to reliably establish a track without the generation of false or spurious tracks. Although a computer can be programmed to recognize and reject false tracks, too many false tracks can overload the computer and result in poor information. It is for this same reason of avoiding computer overload that the radar used with ADT should be designed to exclude unwanted signals, as from clutter and interference. A good ADT system therefore requires a radar with a good MTI and a good CFAR (constant false alarm rate) receiver. A clutter map, generated by the radar, is sometimes used to reduce the load on the tracking computer by blanking clutter areas and removing detections associated with large point clutter sources not rejected by the MTI. Slowly moving echoes that are not of interest can also be removed by the clutter map. The availability of some distinctive target characteristic, such as its altitude, might also prove of help when performing track association.[71] Thus, the quality of the ADT will depend significantly on the ability of the radar to reject unwanted signals.

When a new detection is received, an attempt is made to associate it with existing tracks.[96] This is aided by establishing for each track a small search region, or gate, within which a new detection is predicted based on the estimate of the target speed and direction. It is desired to make the gate as small as possible so as to avoid having more than one echo fall within it when the traffic density is high or when two tracks are close to one another. However, a large gate area is required if the tracker is to follow target turns or maneuvers. More than one size gate might therefore be used to overcome this dilemma. The size of the small gate would be determined by the accuracy of the track. When a target does not appear in the small gate, a larger gate would be used whose search area is determined by the maximum accelera-tion expected of the target during turns.

On the basis of the past detections the track-while-scan radar must make a smoothed estimate of a target's present position and velocity, as well as a predicted position and velocity. One method for computing this information is the so-called α-β tracker (also called the g-h tracker[84]), which computes the present smoothed target position $\bar{x}_n$ and velocity $\bar{\dot{x}}$ by the following equations[72]

$$\text{Smoothed position: } \bar{x}_n = x_{pn} + \alpha(x_n - x_{pn}) \qquad (5.8)$$

$$\text{Smoothed velocity: } \bar{\dot{x}}_n = \bar{\dot{x}}_{n-1} + \frac{\beta}{T_s}(x_n - x_{pn}) \qquad (5.9)$$

where x_{pn} = predicted position of the target at the nth scan, x_n = measured position at the nth scan, α = position smoothing parameter, β = velocity smoothing parameter, and T_s = time between observations. The predicted position at the $n + 1$st scan is $\bar{x}_n + \bar{\dot{x}}_n T_s$. (When accelera-tion is important a third equation can be added to describe an α-β-γ tracker, where γ = acceleration smoothing parameter.)[73] For $\alpha = \beta = 0$, the tracker uses no current informa-tion, only the smoothed data of prior observations. If $\alpha = \beta = 1$, no smoothing is included at all. The classical α-β filter is designed to minimize the mean square error in the smoothed (filtered) position and velocity, assuming small velocity changes between observations, or data samples. Benedict[72] suggests that to minimize the output noise variance at steady state and the transient response to a maneuvering target as modeled by a ramp function, the α-β coefficients are related by $\beta = \alpha^2/(2 - \alpha)$. The particular choice of α within the range of zero to one depends upon the system application, in particular the tracking bandwidth. A compromise usually must be made between good smoothing of the random measurement errors (requiring narrow bandwidth) and rapid response to maneuvering targets (requiring wide bandwidth). Another criterion for selecting the α-β coefficients is based on the best linear track fitted to the radar data in a least squares sense. This gives the values of α and β as[74]

$$\alpha = \frac{2(2n - 1)}{n(n + 1)} \qquad \beta = \frac{6}{n(n + 1)}$$

where n is the number of the scan or target observation ($n > 2$).

The standard α-β tracker does not handle the maneuvering target. However, an adaptive α-β tracker is one which varies the two smoothing parameters to achieve a variable bandwidth so as to follow maneuvers. The value of α can be set by observing the measurement error $x_n - x_{pn}$. At the start of tracking the bandwidth is made wide and then it narrows down if the target moves in a straight-line trajectory. As the target maneuvers or turns, the bandwidth is widened to keep the tracking error small.

The Kalman filter[78] is similar to the classical α-β tracker except that it inherently provides for the dynamical or maneuvering target. In the Kalman filter a model for the measurement error has to be assumed, as well as a model of the target trajectory and the disturbance or uncertainty of the trajectory.[81] Such disturbances in the trajectory might be due to neglect of higher-order derivatives in the model of the dynamics, random motions due to atmospheric turbulence, and deliberate target maneuvers. The Kalman filter can, in principle, utilize a wide variety of models for measurement noise and trajectory disturbance; however, it is often assumed that these are described by white noise with zero mean.[75] A maneuvering target does not always fit such an ideal model, since it is quite likely to produce correlated observations. The proper inclusion of realistic dynamical models increases the complexity of the calculations. Also, it is difficult to describe a priori the precise nature of the trajectory disturbances. Some form of adaptation to maneuvers is required.[76] The Kalman filter is sophisticated and accurate, but is more costly to implement than the several other methods commonly used for the smoothing and prediction of tracking data.[77] Its chief advantage over the classical α-β tracker is its inherent ability to take account of maneuver statistics. If, however, the Kalman filter were restricted to modeling the target trajectory as a straight line and if the measurement noise and the trajectory disturbance noise were modeled as white, gaussian noise with zero mean, the Kalman filter equations reduce to the α-β filter equations with the parameters α and β computed sequentially by the Kalman filter procedure.

The classical α-β tracking filter is relatively easy to implement. To handle the maneuvering target, some means may be included to detect maneuvers and change the values of α and β accordingly. In some radar systems, the data rate might also be increased during target maneuvers. As the means for choosing α and β become more sophisticated, the optimal α-β tracker becomes equivalent to a Kalman filter even for a target trajectory model with error. In this sense, the optimal α-β tracking filter is one in which the values of α and β require knowledge of the statistics of the measurement errors and the prediction errors, and in which α and β are determined in a recursive manner in that they depend on previous estimates of the mean square error in the smoothed position and velocity.[79]

(The above discussion has been in terms of a sampled-data system tracking targets detected by a surveillance radar. The concept of the α-β tracker or the Kalman filter also can be applied to a continuous, single-target tracking radar when the error signal is processed digitally rather than analog. Indeed, the equations describing the α-β tracker are equivalent to the type II servo system widely used to model the continuous tracker.)

If, for some reason, the track-while-scan radar does not receive target information on a particular scan, the smoothing and prediction operation can be continued by properly accounting for the missed data.[80] However, when data to update a track is missing for a sufficient number of consecutive scans, the track is terminated. Although the criterion for terminating a track depends on the application, one example suggests that when three target reports are used to establish a track, five consecutive misses is a suitable criterion for termination.[82]

One of the corollary advantages of ADT is that it effects a bandwidth reduction in the output of a radar so as to allow the radar data to be transmitted to another location via narrowband phone lines rather than wideband microwave links. This makes it more convenient to operate the radar at a remote site, and permits the outputs from many radars to be communicated economically to a central control point.

It should be noted that the adaptive thresholding of the automatic detector can cause a worsening of the range-resolution. By analogy to the angular resolution possible in the angle coordinate[88] it would seem a priori that two targets might be resolved in range if their separation is about 0.8 of the pulse width. However, it has been shown[89] that with automatic detection the probability of resolving targets in range does not rise above 0.9 until they are separated by 2.5 pulse widths. To achieve this resolution a log-video receiver should be used and the threshold should be proportional to the smaller of the two means calculated from a number of reference cells on either side of the test cell. It also assumes that the shape of the return pulse is not known. If it is, it should be possible with the proper processing to resolve targets within a pulse width.

When more than one radar, covering approximately the same volume in space, are located within the vicinity of each other, it is sometimes desirable to combine their outputs to form a single track file rather than form separate tracks.[83,97-99] Such an automatic detection and *integrated* tracking system (ADIT) has the advantage of a greater data rate than any single radar operating independently. The development of a single track file by use of the total available data from all radars reduces the likelihood of a loss of target detections as might be caused by antenna lobing, fading, interference, and clutter since integrated processing permits the favorable weighting of the better data and lesser weighting of the poorer data.

REFERENCES

1. Dunn, J. H., D. D. Howard, and K. B. Pendleton: Tracking Radar, chap. 21 of "Radar Handbook," M. I. Skolnik (ed.), McGraw-Hill Book Co., Inc., New York, 1970.
2. Barton. D. K. "Radar Systems Analysis," Prentice-Hall, Inc., New Jersey, 1964.
3. James, H. M., N. B. Nichols, and R. S. Phillips: "Theory of Servomechanisms," MIT Radiation Laboratory Series, vol. 25, McGraw-Hill Book Company, New York, 1947.
4. Schafer, C. R.: Phase-selective Detectors, *Electronics*, vol. 27, no. 2, pp. 188–190, February, 1954.
5. Greenwood, I. A., Jr., J. V. Holdam, Jr., and D. Macrae, Jr. (eds.): "Electronic Instruments," MIT Radiation Laboratory Series, vol. 21, pp. 383–386, McGraw-Hill Book Company, New York, 1948.
6. Palma-Vittorelli, M. M., M. U. Palma, and D. Palumbo: The Behavior of Phase-sensitive Detectors, *Nuovo cimento*, vol. 6, pp. 1211–1220, Nov. 1. 1957.
7. Krishnan, S.: Diode Phase Detectors, *Electronic and Radio Engr.*, vol. 36, pp. 45–50, February, 1959.
8. Lawson, J. L., and G. E. Uhlenbeck (eds.): "Threshold Signals," MIT Radiation Laboratory Series, vol. 24, McGraw-Hill Book Company, New York, 1950.
9. Oliver, B. M.: Automatic Volume Control as a Feedback Problem, *Proc. IRE*, vol. 36, pp. 466–473, April, 1948.

10. Field, J. C. G.: The Design of Automatic-gain-control Systems for Auto-tracking Radar Receivers, *Proc. IEE*, pt. C, vol. 105, pp. 93–108, March, 1958.
11. Locke, A. S.: "Guidance," pp. 402–408, D. Van Nostrand Company, Inc., Princeton, N.J., 1955.
12. Damonte, J. B., and D. J. Stoddard: An Analysis of Conical Scan Antennas for Tracking, *IRE Natl. Conv. Record*, vol. 4, pt. 1, pp. 39–47, 1956.
13. Page, R. M.: Monopulse Radar, *IRE Natl. Conv. Record*, vol. 3, pt. 8, pp. 132–134, 1955.
14. Dunn, J. H., and D. D. Howard: Precision Tracking with Monopulse Radar, *Electronics*, vol. 33, no. 17, pp. 51–56, Apr. 22, 1960.
15. Barton, D. K., and S. M. Sherman: Pulse Radar for Trajectory Instrumentation, paper presented at Sixth National Flight Test instrumentation Symposium, Instrument Society of America, San Diego, Calif., May 3, 1960.
16. Cohen, W., and C. M. Steinmetz: Amplitude and Phase-sensing Monopulse System Parameters, pts I and II, *Microwave J.*, vol. 2, pp. 27–33, October, 1959, and pp. 33–38, November, 1959; also discussion by F. J. Gardiner, vol. 3, pp. 18, 20, January, 1960.
17. Rhodes, D. R.: "Introduction to Monopulse," McGraw-Hill Book Company, New York, 1959.
18. Barton, D. K.: "Radars, vol. 1, Monopulse Radar," Artech House, Inc., Dedham, Mass., 1974 (a collection of 26 reprints from the literature of monopulse radar).
19. Downey, E. J., R. H. Hardin, and J. Munishian: A Time Duplexed Monopulse Receiver, *Conf. Proc. on Military Electronics Conv.* (IRE), 1958, p. 405.
20. Rubin, W. L., and S. K. Kamen: SCAMP—A New Ratio Computing Technique with Application to Monopulse, *Microwave J.*, vol. 7, pp. 83–90, December, 1964.
21. Abel, J. E., S. F. George, and O. D. Sledge: The Possibility of Cross Modulation in the SCAMP Signal Processor, *Proc. IEEE*, vol. 53, pp. 317–318, March, 1965
22. Noblit, R. S.: Reliability without Redundancy from a Radar Monopulse Receiver, *Microwaves*, vol. 6, pp. 56–60, December, 1967.
23. Kirkpatrick, G. M.: Final Engineering Report on Angular Accuracy Improvement, *General Electric Co. Report on Contract* D.A. 36-039-sc-194, Aug. 1., 1952. (Reprinted in ref. 18.)
24. Ricardi, L. J., and L. Niro: Design of a Twelve-horn Monopulse Feed, *IRE Conv. Rec.*, 1, pp. 49–56, 1961. (Reprinted in ref. 18.)
25. Hannan, P. W., and P. A. Loth: A Monopulse Antenna Having Independent Optimization of the Sum and Difference Modes, *IRE Conv. Rec.*, 1, pp. 56–60, 1961. (Reprinted in ref. 18.)
26. Howard, D. D.: Single Aperture Monopulse Radar Multi-Mode Antenna Feed and Homing Device, *Proc. 1964 International Conv. Military Electronics*, Sept. 14–16, pp. 259–263.
27. Sommer, H. W.: An Improved Simultaneous Phase Comparison Guidance Radar, *IRE Trans.*, vol. ANE-3, pp. 67–70, June, 1956.
28. Hausz, W., and R. A. Zachary: Phase-Amplitude Monopulse System, *IRE Trans.*, vol. MIL-6, pp. 140–146, April 1972. (Reprinted in ref. 18.)
29. Dunn, J. H., D. D. Howard, and A. M. King: Phenomena of Scintillation Noise in Radar Tracking Systems. *Proc. IRE*, vol. 47, pp. 855–863, May, 1959.
30. Dunn, J. H., and D. D. Howard: Target Noise, chap. 28 of "Radar Handbook," M. I. Skolnik (ed.), McGraw-Hill Book Co., New York, 1970.
31. Locke, A. S.: "Guidance," D. Van Nostrand Company, Inc., Princeton, N.J., 1955, pp. 440–442.
32. Delano, R. H.: A Theory of Target Glint or Angular Scintillation in Radar Tracking, *Proc. IRE*, vol. 41, pp. 1778–1784, December, 1953.
33. Delano, R. H., and I. Pfeffer: The Effect of AGC on Radar Tracking Noise, *Proc. IRE*, vol. 44, pp. 801–810, June, 1956.
34. Dunn, J. H., and D. D. Howard: The Effects of Automatic Gain Control Performance on the Tracking Accuracy of Monopulse Radar Systems, *Proc. IRE*, vol. 47, pp. 430–435, March, 1959.
35. Gustafson, B. G., and B. O. Ås: System Properties of Jumping Frequency Radar, *Philips Telecom. Rev.*, vol. 25, no. 1, pp. 70–65, July, 1964.
36. Lind, G.: Reduction of Radar Tracking Errors with Frequency Agility, *IEEE Trans.*, vol. AES-4, pp. 410–416, May, 1968.
37. Lind, G.: A Simple Approximate Formula for Glint Improvement with Frequency Agility, *IEEE Trans.*, vol. AES-8, pp. 854–855, November, 1972.
38. Nicholls, L. A.: Reduction of Radar Glint for Complex Targets by Use of Frequency Agility, *IEEE Trans.*, vol. AES-11, pp. 647–649, July, 1975. (See also comment on p. 325 of ref. 85.)
39. Sims, R. J., and E. R. Graf: The Reduction of Radar Glint by Diversity Techniques, *IEEE Trans.*, vol. AP-19, pp. 462–468, July, 1971.
40. Loomis, J. M., and E. R. Graf: Frequency-Agility Processing to Reduce Radar Glint Pointing Error, *IEEE Trans.*, vol. AES-10, pp. 811–820, November, 1974.

41. Barton, D. K., and H. R. Ward: "Handbook of Radar Measurement," Prentice-Hall, Inc., Englewood Cliffs, N.J., 1969.
42. Barton, D. K.: "Radars, vol. 4, Radar Resolution and Multipath Effects," Artech House, Inc., Dedham, Mass, 1975. (collection of 38 reprints.)
43. Howard, D. D., J. Nessmith, and S. M. Sherman: Monopulse Tracking Errors Due to Multipath, *IEEE EASCON '71 Record*, pp. 175–182.
44. Kittredge, F., E. Ornstein, and M. C. Licitra, Millimeter Radar for Low-Angle Tracking, *IEEE EASCON '74 Record*, pp. 72–75, IEEE Publication 74 CHO 883-1 AES.
45. Barton, D. K.: Low-Angle Radar Tracking, *Proc. IEEE*, vol. 62, pp. 687–704, June, 1974.
46. Linde, G. J.: Improved Low-Elevation-Angle Tracking with Use of Frequency Agility, *Naval Research Laboratory Rept.* 7378, Washington, D.C., Mar. 17, 1972.
47. Dax, P. R.: Keep Track of that Low-Flying Attack, *Microwaves*, vol. 15, pp. 36–42, 47–53, April, 1976.
48. White, W. D.: Low-Angle Radar Tracking in the Presence of Multipath, *IEEE Trans.*, vol. AES-10, pp. 835–852, November, 1974.
49. Peebles, P. Z., Jr.: Multipath Angle Error Reduction Using Multiple-Target Methods, *IEEE Trans.*, vol. AES-7, pp. 1123–1130, November, 1971.
50. Howard, J. E.: A Low Angle Tracking System for Fire Control Radars, *IEEE 1975 International Radar Conference*, pp. 412–417.
51. Dax, P. R.: Accurate Tracking of Low Elevation Targets Over the Sea with a Monopulse Radar, *IEEE Conf. Publ. no. 105*, "Radar—Present and Future," Oct. 23–25, 1973, pp. 160–165.
52. Sherman, S. M.: Complex Indicated Angles Applied to Unresolved Radar Targets and Multipath, *IEEE Trans.* vol. AES-7, pp. 160–170, January, 1971.
53. Albersheim, W. J.: Elevation Tracking Through Clutter Fences, *Supplement to IEEE Trans.*, vol. AES-3, pp. 366–373, Nov. 1967.
54. Hatcher, J. L., and C. Cash: Polarization Agility for Radar Glint Reduction, *IEEE Region 3 Convention*, Huntsville, Alabama, Nov. 19–21, 1969.
55. Locke, A. S.: "Guidance," pp. 408–413, D. Van Nostrand Company, Inc., Princeton, N.J., 1955.
56. Cross, D. C., and J. E. Evans: Target-Generated Range Errors, *Naval Research Laboratory Memorandum Report* 2719, January, 1974.
57. Cross, D. C.: Low Jitter High Performance Electronic Range Tracker, *IEEE 1975 International Radar Conference*, pp. 408–411, IEEE Publication 75 CHO 938-1 AES.
58. Galati, G.: Range-tracking in Tracking Radars: Analysis and Evaluation of the Accuracy, *Rivista Technica Selenia* (Rome, Italy), vol. 2, no. 2, pp. 29–41, 1975.
59. The SCR-584 Radar, *Electronics*, pt. 1, vol. 18, pp. 104–109, November, 1945; pt. 2, vol. 18, pp. 104–109, December, 1945; pt. 3, vol. 19, pp. 110–117, February, 1946.
60. Cady, W. M., M. B. Karelitz, and L. A. Turner (eds.): "Radar Scanners and Radomes," MIT Radiation Laboratory Series, vol. 26, p. 66, McGraw-Hill Book Company, New York, 1948.
61. Symonds, M. D., and J. M. Smith: Multi-Frequency Complex-Angle Tracking of Low-Level Targets, *Radar—Present and Future*, pp. 166–171, Oct. 23–25, 1973, IEE Conference Publication No. 105.
62. Nessmith, J. T.: Range Instrumentation Radars, *IEEE ELECTRO '76*, Boston, Mass., May 11–14, 1976.
63. Schelonka, E. P.: Adaptive Control Techniques for On-Axis Radars, *IEEE 1975 International Radar Conference*, Arlington, Va., April 21–23, 1975, pp. 396–401, IEEE Publication 75 CHO 938-1 AES.
64. Clark, B. L., and J. A. Gaston: On-Axis Pointing and the Maneuvering Target, *NAECON '75 Record*, pp. 163–170.
65. Pollock, E. J.: In Loop Integration Control (ILIC), *Technical Memorandum 118*, Feb. 1, 1976, Air Force Special Weapons Center, Kirtland Air Force Base, New Mexico.
66. Howard, D. D.: High Range-Resolution Monopulse Tracking Radar and Applications, *EASCON '74 Record*, pp. 86–91, IEEE Publication 74 CHO 883-1 AES.
67. Provejsil, D. J., R. S. Raven, and P. Waterman: "Airborne Radar," D. Van Nostrand Co., Inc., Princeton, N.J., 1961.
68. Hoare, E., and G. Hine: Servomechanisms, chap. 11 of "Mechanical Engineering in Radar and Communications," C. J. Richards (ed.), Van Nostrand Reinhold Co., New York, 1969.
69. Plowman, J. C.: Automatic Radar Data Extraction by Storage Tube and Delay Line Techniques, *J. Brit. IRE*, vol. 27, pp. 317–328, October, 1963.
70. Caspers, J. W.: Automatic Detection Theory, chap. 15 of "Radar Handbook," M. I. Skolnik (ed.), McGraw-Hill Book Co., Inc., New York, 1970.
71. Oakley, B. W.: Tracking in an Air Traffic Control Environment, chap. 30 of "Radar Techniques for Detection, Tracking and Navigation," W. T. Blackband (ed.), Gordon and Breach, New York, 1966.
72. Benedict, T. R., and G. W. Bordner: Synthesis of an Optimal Set of Radar Track-While-Scan Smoothing Equations, *IRE Trans.*, vol. AC-7, pp. 27–32, July, 1962.

73. Simpson, H. R.: Performance Measures and Optimization Condition for a Third-Order Sampled-Data Tracker, *IEEE Trans.*, vol. AC-8, pp. 182–183, April, 1963.
74. Quigley, A. L. C.: Tracking and Associated Problems, *International Conf. on Radar—Present and Future*, Oct. 23–25, 1973, London, pp. 352–359, IEEE Conference Publication no. 105.
75. Hampton, R. L. T., and J. R. Cooke: Unsupervised Tracking of Maneuvering Vehicles, *IEEE Trans.*, vol. AES-9, pp. 197–207, March, 1973.
76. Thorp, J. S.: Optimal Tracking of Maneuvering Targets, *IEEE Trans.*, vol. AES-9, pp. 512–519, July, 1973.
77. Singer, R. A., and K. W. Behnke: Real-Time Tracking Filter Evaluation and Selection for Tactical Applications, *IEEE Trans.*, vol. AES-7, pp. 100–110, Jan. 1971.
78. Kalman, R. E.: A New Approach to Linear Filtering and Prediction Problems, *Trans. ASME, J. Basic Engrg.*, vol., 82, pp. 34–45, March, 1960.
79. Schooler, C. C.: Optimal α-β Filters for Systems with Modeling Inaccuracies, *IEEE Trans.*, vol. AES-11, pp. 1300–1306, November, 1975.
80. Kanyuck, A. J.: Transient Response of Tracking Filters with Randomly Interrupted Data, *IEEE Trans.*, vol. AES-6, pp. 313–323, May, 1970.
81. Morgan, D. R.: A Target Trajectory Noise Model for Kalman Trackers, *IEEE Trans.*, vol. AES-12, pp. 405–408, May, 1976.
82. Leth-Espensen, L.: Evaluation of Track-While-Scan Computer Logics, chap. 29 of "Radar Techniques for Detection, Tracking, and Navigation," W. T. Blackband (ed.), Gordon and Breach, New York, 1966.
83. Cantrell, B. H., G. V. Trunk, J. D. Wilson, and J. J. Alter: Automatic Detection and Integrated Tracking, *IEEE 1975 International Radar Conference*, pp. 391–395, Arlington, Va., Apr. 21–23, 1975, IEEE Publication 75 CHO 938-1 AES.
84. Bhagavan, B. K., and R. J. Polge: Performance of the *g-h* Filter for Tracking Maneuvering Targets, *IEEE Trans.*, vol. AES-10, pp. 864–866, November, 1974.
85. Barton, D. K.: "Radars, vol. 6. Frequency Agility and Diversity," Artech House, Inc., Dedham, Mass., 1977.
86. Sakamoto, H., and P. Z. Peebles, Jr.: Conopulse Radar, *IEEE Trans.* vol. AES-14, pp. 199–208, January, 1978.
87. Bakut, P. A., and I. S. Bol'shakov: "Questions of the Statistical Theory of Radar, vol. II," chaps 10 and 11, Sovetskoye Radio, Moscow, 1963. Translation available from NTIS, AD 645775, June 28, 1966.
88. Skolnik, M. I.: Comment on the Angular Resolution of Radar, *Proc. IEEE*, vol. 63, pp. 1354–1355, September, 1975.
89. Trunk, G. V.: Range Resolution of Targets Using Automatic Detectors, *Naval Research Laboratory Rept.* 8178, Nov. 28, 1977, Washington, D.C. Also, *IEEE Trans.*, vol. AES-14, pp. 750–755, September, 1978.
90. Howard, D. D., S. M. Sherman, D. N. Thomson, and J. J. Campbell: Experimental Results of the Complex Indicated Angle Technique for Multipath Correction, *IEEE Trans.*, vol. AES-10, pp. 779–787, November, 1974.
91. White, W. D.: Double Null Technique for Low Angle Tracking, *Microwave J.*, vol. 19, pp. 35–38, 60, December, 1976.
92. Barton, D. K.: Low Angle Tracking, *Microwave J.*, vol. 19, pp. 19–24, December, 1976.
93. Howard, D. D.: Investigation and Application of Radar Techniques for Low-Altitude Target Tracking, *IEE International Radar Conference*, RADAR-77, London, Eng., Oct. 25–28, 1977, IEE Conference Publication No. 105, pp. 313-317, IEEE (New York) Publication No. 77 CH 1271-6 AES.
94. Mrstik, A. V., and P. G. Smith: Multipath Limitations on Low-Angle Radar Tracking, *IEEE Trans.*, vol. AES-14, pp. 85–102, January, 1978.
95. Howard, D. D.: Predicting Target-Caused Errors in Radar Measurements, *Electronic Warfare Defense Electronics*, vol. 10, pp. 88–95, February, 1978. See also same title and author in *IEEE EASCON '76 Record*, pp. 30-A to 30-H.
96. Trunk, G. V., and J. D. Wilson: Track Initiation in a Dense Detection Environment, *Naval Research Laboratory* (Washington, D.C.) Rept 8238, July 28, 1978.
97. Flad, E. H.: Tracking of Formation Flying Aircraft, *International Conference RADAR-77*, Oct. 25–28, 1977, pp. 160–163, IEE (London) Conference Publication No. 155.
98. Kossiakoff, A., and J. R. Austin: Automated Radar Data Processing System, U.S. Patent 4,005,415, Jan. 25, 1977.
99. Casner, P. G., and R. J. Prengaman: Integration and Automation of Multiple Co-located Radars, *International Conference RADAR-77*, Oct. 25–28, 1977, pp. 145–149, IEE (London) Conference Publication no. 155.

CHAPTER

SIX

RADAR TRANSMITTERS

6.1 INTRODUCTION

The radar system designer has a choice of several different transmitter types, each with its own distinctive characteristics. The first successful radars developed prior to World War II employed the conventional grid-controlled (triode or tetrode) vacuum tube adapted for operation at VHF, a relatively high frequency at that time. The magnetron oscillator, which triggered the development of microwave radar in World War II, has been one of the most widely used of radar transmitters, especially for mobile systems. The klystron amplifier, introduced to radar in the 1950s, offered the system designer higher power at microwaves than available from the magnetron. Being an amplifier, the klystron permitted the use of more sophisticated waveforms than the conventional rectangular pulse train. The klystron was followed by the traveling-wave tube, a close cousin with similar properties to the klystron except for its wider bandwidth. The 1960s saw the availability of the crossed-field amplifier, a tube related to the magnetron. There are several variations of crossed-field amplifiers, each with its special properties; but they are all characterized by wide bandwidth, modest gain, and a compactness more like that of the magnetron than the klystron or the traveling-wave tube. Solid-state devices such as the transistor and the bulk-effect and avalanche diodes can also be employed as radar transmitters. They have some interesting properties as compared to the microwave vacuum tube, but the individual devices are inherently of low power.

There is no one universal transmitter best suited for all radar applications. Each power-generating device has its own particular advantages and limitations that require the radar system designer to examine carefully all the available choices when configuring a new radar design.

The transmitter must be of adequate power to obtain the desired radar range, but it must also satisfy other requirements imposed by the system application. The special demands of MTI (moving target indication), pulse doppler, CW radar, phased-array radar, EMC

190

(electromagnetic compatibility), and ECCM (electronic counter-countermeasures) all influence the type of transmitter selected and its method of operation. The choice of transmitter also depends on whether the radar operates from fixed land sites, mobile land vehicles, ships, aircraft, or spacecraft. Other considerations include the size and weight, high-voltage and X-ray protection, modulation requirements, and the method of cooling. A transmitter is a major part of a radar system; hence, its size, cost, reliability, and maintainability can significantly affect the size, cost, reliability, and maintainability of the radar system of which it is a part. Not only does the transmitter represent a significant fraction of the initial cost of a radar system, but it can often take a large share of the operating costs because of the prime power and the needs of maintenance. The classical radar range equation (Chap. 2) shows that the transmitter power depends on the fourth power of the radar range. To double the range of a radar, the power has to be increased 16-fold. Buying radar range with transmitter power alone can therefore be costly.

Thus there are many diverse requirements and system constraints that enter into the selection and design of a transmitter. This chapter briefly reviews the various types of transmitter tubes and their characteristics. More complete descriptions will be found in the *Radar Handbook*.[1]

For the most part, this chapter discusses the tubes used in radar transmitters and not the transmitters themselves. A transmitter is far more than the tube alone. It includes the exciter and driver amplifiers if a power amplifier, the power supply for generating the necessary voltages and currents needed by the tube, the modulator, cooling for the tube, heat exchanger for the cooling system if liquid, protection devices (crowbar) for arc discharges, safety interlocks, monitoring devices, isolators, and X-ray shielding.

The efficiency quoted for most tubes is the *RF conversion efficiency*, defined as the RF power output available from the tube to the d-c power input of the electron stream. This is the efficiency of interest to the tube designer. The system engineer, however, is more concerned with the overall transmitter efficiency, which is the ratio of the RF power available from the transmitter to the total power needed to operate the transmitter. If, for example, the RF efficiency of a microwave tube were 40 to 50 percent, the transmitter efficiency might be 20 to 25 percent. (The actual number, of course, varies considerably with tube type and application.)

There are two basic radar-transmitter configurations. One is the self-excited oscillator, exemplified by the magnetron. The other is the power amplifier, which utilizes a low power, stable oscillator whose output is raised to the required power level by one or more amplifier stages. The klystron, traveling-wave tube, and the crossed-field amplifier are examples of microwave power-amplifier tubes. The choice between the power oscillator and the power amplifier is governed mainly by the particular radar application. Transmitters that employ the magnetron power-oscillator are usually smaller in physical size than transmitters that employ the power amplifier. The various amplifier transmitters, however, are generally capable of higher power than the magnetron oscillator. Amplifiers are of greater inherent stability, which is of importance for MTI and other doppler radars, and they can generate more conveniently than can power oscillators the modulated waveforms needed for pulse-compression radar. Power oscillators, therefore, are likely to be found in applications where small size and portability are important and when the stability and high power of the amplifier transmitter are not required.

The magnetron power oscillator has probably seen more application in radar than any other tube. It is the only power oscillator widely used in radar. The classical magnetron is of low cost, convenient size and weight, and high efficiency, and has an operating voltage low enough not to generate dangerous X-rays. The coaxial magnetron improves on the classical magnetron by providing greater reliability, longer life, and better stability.

The klystron amplifier provides the radar system designer with high power, high gain, good efficiency, and stability for MTI and pulse-compression applications. It is probably the preferred tube for most high-power radar applications if its high operating voltage and large size can be tolerated. The traveling-wave tube is similar to the klystron. It differs from the klystron in having wider bandwidth, but at the expense of less gain. The crossed-field amplifier is of the same general family as the magnetron and shares some of its properties, especially small size and weight, high efficiency, and an operating voltage more convenient than that of the klystron and the traveling-wave tube. Like the traveling-wave tube, it enjoys a wide bandwidth; but it is of relatively low gain and therefore requires more than one stage in the amplifier chain. The magnetron and the crossed-field amplifier are devices which utilize the properties of electron streams in crossed electric and magnetic fields. The klystron and the traveling wave tube are of a different family known as linear beam tubes.

Radars at VHF and UHF have often employed grid-controlled triode and tetrode tubes. These have usually been, in the past, competitive in cost to other power generation means at these frequencies. Solid-state devices, such as transistors and microwave diode generators, have also been considered for radar operation. Their properties differ significantly from microwave tubes and require major changes in system design philosophy when they are used.

6.2 THE MAGNETRON OSCILLATOR

More than any other single device, the high-power magnetron oscillator invented in 1939[2] made possible the successful development of microwave radar during World War II. The magnetron is a crossed-field device in that the electric field is perpendicular to a static magnetic field. Although the name *magnetron* has been applied in the past to several different electron devices,[3] it was the application of cavity resonators to the magnetron structure that permitted a workable microwave oscillator of high power and high efficiency.

Conventional magnetron. The basic structure of the classical form of the magnetron is shown in Fig. 6.1.[4] The anode (1) is a large block of copper into which are cut holes (2) and slots (3). The holes and slots function as the resonant circuits and serve a purpose similar to that of the lumped-constant LC resonant circuits used at lower frequencies. The holes correspond, roughly, to the inductance L, and the slots correspond to the capacity C. In the desired mode of operation (the so-called π mode) the individual C's and L's are in parallel, and the frequency of the magnetron is approximately that of an individual resonator. The cathode (4) is a fat cylinder of oxide-coated material. The cathode must be rugged to withstand the heating and disintegration caused by the back-bombardment of electrons. Back-bombardment increases the cathode temperature during operation and causes secondary electrons to be emitted. For this reason the heater power may be reduced or even turned off once the oscillations have started. The relatively fat cathode, required for theoretical reasons, can dissipate more heat than can a thin cathode.

In the interaction space (5) the electrons interact with the d-c electric field and the magnetic field in such a manner that the electrons give up their energy to the RF field. The magnetic field, which is perpendicular to the plane of the figure, passes through the interaction space parallel to the cathode and perpendicular to the d-c electric field. The crossed electric and magnetic fields cause the electrons to be completely bunched almost as soon as they are emitted from the cathode. After becoming bunched, the electrons move along in a traveling-wave field. This traveling-wave field moves at almost the same speed as the electrons, causing RF power to be delivered to the wave. The RF power is extracted by placing a coupling

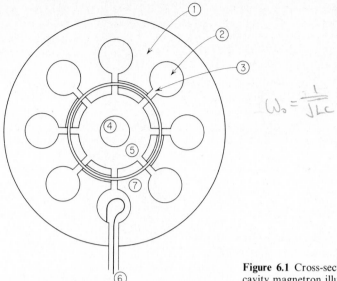

$$\omega_0 = \frac{1}{\sqrt{LC}}$$

Figure 6.1 Cross-sectional sketch of the classical cavity magnetron illustrating component parts.

loop (6) in one of the cavities or by coupling one cavity directly to a waveguide. Not shown in Fig. 6.1 are end-shield disks located at each end of the cathode for the purpose of confining the electrons to the interaction space.

The *straps* (7) are metal rings connected to alternate segments of the anode block. They improve the stability and efficiency of the tube. The preferred mode of magnetron operation corresponds to an RF field configuration in which the RF phase alternates 180° between adjacent cavities. This is called the π *mode*. The presence of N cavities in the magnetron results in $N/2$ possible modes of operation. Each of these $N/2$ modes corresponds to a different RF field configuration made up of a standing wave of charge. All the modes except the π mode are degenerate; that is, they can oscillate at two different frequencies corresponding to a rotation of the standing-wave pattern, where the positions of the nodes and antinodes are interchanged. Thus there are $N-1$ possible frequencies in which the magnetron can oscillate. The presence of more than one possible mode of operation means that the magnetron can oscillate in any one of these frequencies and can do so in an unpredictable manner. This is the essence of the stability problem. The magnetron must be designed with but one mode dominant. The π mode is usually preferred since it can be more readily separated from the others. The straps provide stability since they connect all those segments of the anode which have the same potential in the π mode and thus permit the tube more readily to operate in this preferred mode.

Instead of the hole and slot resonators of Fig. 6.1, vanes may be used, as in Fig. 6.2*a*. Slots have also been employed. When large and small slots are alternated, as in the rising-sun magnetron structure of Fig. 6.2*b*, stable oscillation can occur in the π mode without the need for straps. Since there are no straps, the rising-sun geometry is more suitable for the shorter wavelengths than are conventional resonators.

Coaxial magnetron. A significant improvement in power, efficiency, stability, and life over the conventional magnetron is obtained when the straps are removed and the π mode is controlled by coupling alternate resonators to a cavity surrounding the anode. This is known as a *coaxial magnetron* since the stabilizing cavity surrounds the conventional resonators, as sketched in

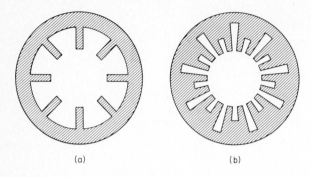

Figure 6.2 Magnetron resonators. (a) Vane type; (b) rising sun, with alternate slot lengths.

Fig. 6.3. The output power is coupled from the stabilizing coaxial cavity. The cavity operates in the TE_{011} mode with the electric field lines closed on themselves and concentric with the circular cavity. The RF current at every point on the circumference of the cavity has the same phase, so that the alternate slots which couple to the stabilizing cavity are of the same phase as required for π mode operation. The elimination of the straps allows the resonators to be designed for optimum efficiency rather than as a compromise between efficiency and mode control.

The power handling capability of a magnetron depends on its size. Increasing the size of a magnetron, however, requires increasing the number of resonators. The greater the number of resonators the more difficult the problem of mode separation. Since the mode separation in a coaxial magnetron is controlled in the TE_{011} stabilizing cavity rather than in the resonator area, the coaxial magnetron can operate stably with a large number of cavities. This allows a larger anode and cathode structure than with the conventional magnetron, and therefore a coaxial magnetron can operate at higher power levels. The larger structures permit more conservative design, with the result that coaxial magnetrons exhibit longer life and better

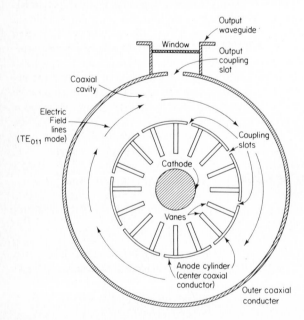

Figure 6.3 Cross-sectional sketch of the coaxial cavity magnetron.

Figure 6.4 Photograph of the SFD-341 mechanically tuned *C*-band coaxial magnetron for shipboard and ground-based radars. This tube delivers a peak power of 250 kW with a 0.001 duty cycle over the frequency range from 5.45 to 5.825 GHz. The efficiency is 40 to 45 percent. (*Courtesy Varian Associates, Inc., Beverly, MA.*)

reliability than conventional magnetrons. It has been said that the operating life of coaxial tubes can be between 5000 and 10,000 hours, a five- to twenty-fold improvement compared to conventional magnetrons.[44] Since most of the RF energy is stored in the TE_{011} cavity rather than in the resonator region, reliable broadband tuning of the magnetron may be accomplished by a noncontacting plunger in the cavity. Both the *pushing figure* (change in frequency with a change in anode current) and the *pulling figure* (change in frequency with a change in phase of the load) are much less in the coaxial magnetron than in the conventional configuration.

The external appearance of the coaxial magnetron, Fig. 6.4, is similar to that of the conventional magnetron, and the electron and RF operations that take place in the interaction space are the same for both types of magnetrons. The differences between the two are in the more effective mode control of the coaxial cavity as compared to that offered by conventional strapping.

In the *inverted coaxial magnetron* the cathode surrounds the anode. The stabilizing TE_{011} cavity is in the center of the magnetron with a vane-type resonator system arranged on the outside. The cathode is built as a ring surrounding the anode. Power is coupled from the end of the central stabilizing cavity by a circular waveguide. The geometry of the inverted coaxial magnetron makes it suitable for operation at the higher frequencies. Figure 6.5 is an example.

Performance chart and Rieke diagram. Four parameters determine the operation of the magnetron. These are (1) the magnetic field, (2) the anode current, (3) load conductance, and

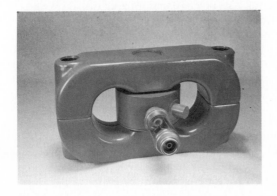

Figure 6.5 Photograph of the SFD-319 K_a-band fixed frequency inverted coaxial magnetron. This tube delivers a peak power of 100 kW with a 0.005 duty cycle at a frequency between 34,512 and 35.208 GHz. The efficiency is about 25 percent. (*Courtesy Varian Associates, Inc., Beverly, MA.*)

(4) load susceptance. The first two parameters are related to the input side of the tube, while the last two are related to the output side. In most magnetrons the magnetic field is fixed by the tube designer and may not be a variable the radar designer has under his control. The observed quantities are usually the output power, the wavelength, and the anode voltage. The problem of presenting the variation of the three quantities—power, wavelength, voltage— as a function of the four parameters mentioned above is greatly simplified since the input and output parameters operate nearly independently of each other. Thus it is possible to study the effect of the magnetic field and the anode current at some value of load susceptance and conductance chosen for convenience. The results will not be greatly dependent upon the particular values of susceptance and conductance chosen. Similarly, the variation of the observed quantities can be studied as a function of the load presented to the magnetron, with the input parameters—magnetic field and current—likewise chosen for convenience. The plot of the observed magnetron quantities as a function of the input circuit parameters, for some fixed load, is called the *performance chart*. The plot of the observed quantities as a function of the load conductance and susceptance, for a fixed magnetic field and anode current, is called a *Rieke diagram*, or a *load diagram*.

An example of the coaxial magnetron performance characteristics is shown in Fig. 6.6a. The power output, anode voltage, and efficiency are plotted as a function of the magnetron input power for a fixed frequency and with the magnetron waveguide-load matched. The peak voltage is seen to vary only slightly with a change in input power, but the power output varies almost linearly. Figure 6.6b plots the power output and voltage as the tube is tuned through its frequency range, when the current is held constant and the waveguide load is matched. The variation in magnetron efficiency is similar to that of the variation with power.

The change in the oscillator frequency produced by a change in the anode current for a fixed load is called the *pushing figure*. A plot of frequency vs. current, as in Fig. 6.6c, is called the *pushing characteristic* and the slope of the curve is the pushing figure. The lower the value of the pushing figure, the better the frequency stability. The coaxial magnetron has a lower pushing figure than a conventional magnetron because of the stabilizing effect (high Q) of its relatively large coaxial TE_{011} cavity. Pushing effects are more serious with longer pulses since their spectra are narrow. A given change in frequency with a narrow-spectrum pulse will be noticed more than with a wide spectrum pulse.

The effect of the load on the magnetron characteristics is shown by the Rieke diagram, whose coordinates are the load conductance and susceptance (or resistance and reactance). Plotted on the Rieke diagram are contours of constant power and constant frequency. Thus the Rieke diagram gives the power output and the frequency of oscillation for any specified load condition. Although a cartesian set of load coordinates could be used, it is usually more convenient to plot the power and frequency on a set of load coordinates known as the Smith chart. The Smith chart is a form of circle diagram widely used as an aid in transmission-line calculations. A point on the Smith chart may be expressed in conductance-susceptance coordinates or by a set of polar coordinates in which the voltage-standing-wave ratio (VSWR) is plotted as the radius, and the phase of the VSWR is plotted as the angular coordinate. The latter is the more usual of the two possible coordinate systems since it is easier for the microwave engineer to measure the VSWR and the position of the voltage-standing-wave minimum (or phase) than it is to measure the conductance and susceptance directly. The radial coordinate can also be specified by the reflection coefficient Γ of the load since the VSWR ρ and reflection coefficient are related by the equation $|\Gamma| = (\rho - 1)/(\rho + 1)$. The center of the Smith chart (Rieke diagram) corresponds to unity VSWR, or zero reflection coefficient. The circumference of the chart corresponds to infinite VSWR, or unity reflection coefficient. Thus the region of low standing-wave ratio is toward the center of the chart.

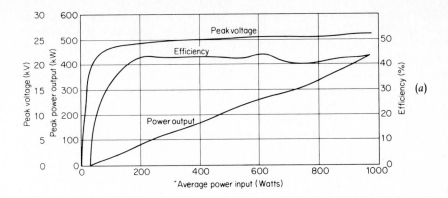

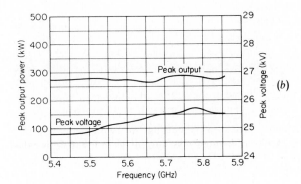

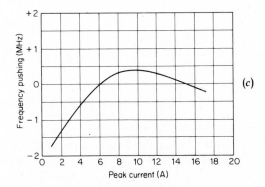

Figure 6.6 Performance characteristics of the coaxial magnetron. (*a*) Variation of power output, efficiency, and peak voltage of the SFD-341 as a function of the input average power for a fixed frequency (5.65 GHz); pulse width = 2.15 μs, duty cycle = 0.0009. (*b*) Variation of peak power output and peak voltage of the SFD-341 with frequency for a fixed current (23.9 A); pulse width = 1.8 μs, and duty cycle = 0.0009. (*c*) Variation of frequency with current for the SFD-377A *X*-band coaxial magnetron at a frequency of 9.373 GHz, with 0.001 duty cycle, and 1.0 μs pulse width (*Courtesy Varian Associates, Inc., Beverly, MA.*)

The standing-wave pattern along a transmission line repeats itself every half wavelength; therefore, 360° in the diagram is taken as a half wavelength. The reference axis in the Rieke diagram usually corresponds to the output terminals of the magnetron or the output flange of the waveguide. The angle in a clockwise direction from this reference axis is proportional to the distance (in wavelengths) of the standing-wave-pattern minimum from the reference point. An advantage of the Smith chart for plotting the effects of the load on the magnetron parameters is that the shapes of the curves are practically independent of the position of the reference point used for measuring the phase of the VSWR.

An example of a Rieke diagram for a coaxial magnetron is shown in Fig. 6.7. It is obtained by varying the magnitude and phase of the VSWR of the RF load, with the frequency and the peak current held constant. The region of highest power on the Rieke diagram is called the *sink* and represents the greatest coupling to the magnetron and the highest efficiency. Operation in the region of the sink, however, is not always desirable since it has poor frequency stability. Poor pulse shape and mode changes might result. The low-power region, where the magnetron is lightly loaded, is called the *antisink* or *opposite-sink* region. The build-up of oscillations in a lightly loaded magnetron is more ideal; however, the magnetron in this region may perform poorly by showing signs of instability which take the form of arcing and an increase in the number of missing pulses. Poor performance is a result of the higher RF voltages when operating in the antisink region, making RF discharges more likely.

In a radar with a rotating antenna, the phase and/or magnitude of the VSWR might vary because the antenna will experience a different load impedance depending on the environment it views. The Rieke diagram shows that a change in the VSWR which moves the operating point of the magnetron into either the sink or antisink regions can cause the magnetron to operate poorly. Ferrite isolators are sometimes used to avoid subjecting the magnetron to high VSWR. A 10-dB isolator, for example, lowers a VSWR of 1.5 to a value of 1.14. With a sufficiently low VSWR, the magnetron operation will not occur in either the sink or the antisink region, no matter what the phase angle of the load.

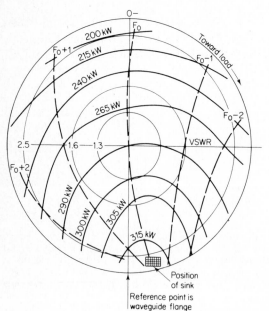

Figure 6.7 Rieke, or load, diagram for the SFD-341 coaxial magnetron. Duty cycle = 0.0009, peak current = 24 A, pulse width = 2.25 μs, frequency = 5.65 GHz. (*Courtesy Varian Associates, Inc., Beverly, MA.*)

A measure of the effect of the load on the magnetron frequency is the *pulling figure*, defined as the difference between the maximum and minimum frequencies when the phase angle of the load varies through 360° and the magnitude of the VSWR is fixed at 1.5 (or a reflection coefficient of 0.20). The pulling figure is readily obtained from an inspection of the Rieke diagram. For the magnetron whose Rieke diagram is shown in Fig. 6.7, the pulling figure is approximately 4 MHz. The pulling figures of coaxial magnetrons are lower by a factor of 3 to 5 than those of conventional magnetrons.

Tuning. The frequency of a conventional magnetron can be changed by mechanically inserting a tuning element, such as a rod, into the holes of the hole-and-slot resonators to change the inductance of the resonant circuit. A tuner that consists of a series of rods inserted into each cavity resonator so as to alter the inductance is called a *crown-of-thorns tuner*, or a *sprocket tuner*. The amount of mechanical motion of the tuning element need not be large (perhaps a fraction of an inch at L band) to tune the frequency over a 5 to 10 percent frequency range. A frequency change in a conventional magnetron can also be obtained with a change in capacity. One example is the *cookie cutter*, which consists of a metal ring inserted between the two rings of a double-ring-strapped magnetron, thereby increasing the strap capacitance. Because of the mechanical and voltage-breakdown problems associated with this tuner, it is more suited for use at the longer wavelengths. Either the cookie cutter or the crown-of-thorns mechanisms can achieve a 10 percent frequency change. The two can be used in combination to cover a larger tuning range than is possible with either one alone.

A limited tuning range, of the order of 1 percent, can be obtained by a screw inserted in the side of one of the resonator holes. This type of adjustment is useful when the normal scatter of frequencies expected of untuned magnetrons requires the frequency to be fixed to a specified value.

In frequency-agile radar systems, the magnetron frequency might be changed pulse-to-pulse in such a manner that the entire tuning range is covered. Such radars might be employed for ECCM, improving the detection of targets with fluctuating cross section and reducing the effects of glint (Sec. 5.5). With a high tuning rate, the pulse-to-pulse frequency can be made to appear pseudorandom, especially if the pulse repetition frequency or the tuning rate is varied rapidly. Many of the advantages of frequency agility can be obtained if the radar frequency shifts pulse-to-pulse the minimum amount required to decorrelate successive echoes. The minimum frequency shift is equal to the reciprocal of the pulse width. Slow tuning rates can be used, but the pulse-to-pulse frequency might not appear random.

One of the first techniques for achieving frequency-agile magnetrons was known as *spin-tuning*, or *rotary tuning*. In this device a rotating slotted disk is suspended above the anode resonators. Rotation of this disk alternately provides inductive or capacitive loading of the resonators to raise and lower the frequency.[1] The rotating disk is mounted on bearings inside the vacuum and coupled to a rotating mechanism outside the vacuum. The Amperex X-band DX-285 spin-tuned magnetron covers a 500-MHz band in an approximately sinusoidal manner at rates up to 1000 times per second, equivalent to frequency tuning rates of the order of 1 MHz per microsecond.

A coaxial magnetron may be tuned by mechanically positioning in the coaxial cavity a noncontacting washer-shaped metal ring, or tuning piston, as illustrated in Fig. 6.8. The tuning piston can be positioned mechanically from outside the vacuum by means of a vacuum bellows. This tuning mechanism may be adapted to provide narrowband frequency agility at a rapid tuning rate for frequency-agile radar. The RF frequency of the Varian SFD-354A X-band coaxial magnetron can be varied sinusoidally over a 60-MHz range at a rate of 70 times per second by this method.

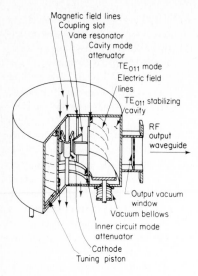

Magnetic field lines
Coupling slot
Vane resonator
Cavity mode
attenuator
TE$_{011}$ mode
Electric field
lines
TE$_{011}$ stabilizing
cavity
RF
output
waveguide
Output vacuum
window
Vacuum bellows
Inner circuit mode
attenuator
Cathode
Tuning piston

Figure 6.8 Schematic view of a coaxial magnetron showing the tuning piston mechanically actuated by a vacuum bellows.

The rapid tuning over a narrowband for purposes of providing frequency agility is sometimes called *dither tuning*.[39] In addition to being capable of rapid tuning over a narrowband, these tubes also can be tuned to a frequency over a broadband in the normal manner using a geared drive. The tuning mechanism may be controlled by a servo motor so as to select electrically any specific frequency within the operating band either manually or on an automatic, programmed basis. With servo-motor control, the tube can be tuned from one frequency to another in under 0.1 second.

Dither-tuning of a coaxial magnetron may also be obtained by a mechanical tuning element called a *ring tuner*.[5,6] This consists of a narrow ring installed in an annular groove cut into the outer wall of the cavity. The ring projects slightly into the cavity. The ring is split, with the split diametrically opposite the output cavity. The ring is also cut to accommodate the output coupling slot. Near the output, on both sides of the output coupling, the ring is firmly attached to the cavity wall, but is unattached otherwise. By deforming the ring inward from mechanical motion applied to the free-hanging ends of the ring, the tuning of the cavity is changed. The Raytheon QKH 1763 X-band coaxial magnetron tunes in this manner over a 100-MHz range at rates up to 200 Hz. Electronic tuning is also possible.[49]

6.3 KLYSTRON AMPLIFIER

The klystron amplifier is an example of a *linear beam tube*, or *O-type tube*. The characteristic feature of a linear beam tube is that the electrons emitted from the cathode are formed into a long cylindrical beam which receives the full potential energy of the electric field before the beam enters the RF interaction region. Transit-time effects, which limit the operation of conventional grid-controlled tubes at the higher frequencies, are used to good advantage in the klystron. As the electron beam of the klystron passes the input resonant cavity, the velocity of the electrons is modulated by the input signal. This velocity modulation of the beam electrons is then converted to density modulation. A resonant cavity at the output extracts the RF power from the density-modulated beam and delivers the power to a useful load.

The klystron has proven to be quite important for radar application. It is capable of high average and peak power, high gain, good efficiency, stable operation, low interpulse noise, and it can operate with the modulated waveforms required of sophisticated pulse-compression systems.

Description. A sketch of the principal parts of the klystron is shown in Fig. 6.9. At the left-hand portion of the figure is the cathode, which emits a stream of electrons that is focused into a narrow cylindrical beam by the *electron gun*. The electron gun consists of the cathode, modulating anode or control grid, and the anode. The electron emission density from the cathode is generally less than required for the electron beam, so a large-area cathode surface is used and the emitted electrons are caused to converge to a narrow beam of high electron density. The modulating anode, or other beam control electrode, is often included as part of the electron-gun structure to provide a means for pulsing the electron beam on and off. The RF cavities, which correspond to the LC resonant circuits of lower-frequency amplifiers, are at anode potential. Electrons are not intentionally collected by the anode as in other tubes; instead, the electrons are removed by the *collector* electrode (shown on the right-hand side of the diagram) after the beam has given up its RF energy to the output cavity.

The input signal is applied across the *interaction gap* of the first cavity. Low-power tubes might contain a grid structure at the gap to provide coupling to the beam. In high-power tubes, however, the gap does not usually contain a grid because a grid cannot accommodate high power. (The absence of a grid does not seriously impair the coupling between the gap and the beam.) Those electrons which arrive at the gap when the input signal voltage is at a maximum (peak of the sine wave) experience a voltage greater than those electrons which arrive at the gap when the input signal is at a minimum (trough of the sine wave). The process whereby a time variation in velocity is impressed upon the beam of electrons is called *velocity modulation*.

In the *drift space*, those electrons which were speeded up during the peak of one cycle catch up with those slowed down during the previous cycle. The result is that the electrons of the velocity-modulated beam become "bunched," or density modulated, after traveling through the drift space. If the interaction gap of the output cavity is placed at the point of maximum bunching, power can be extracted from the density-modulated beam. Most high-power klystrons for radar application have one or more cavities between the input and output

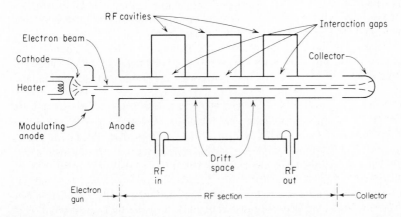

Figure 6.9 Diagrammatic representation of the principal parts of a three-cavity klystron.

cavities to provide additional bunching, and hence, higher gain. The gain of a klystron can be typically 15 to 20 dB per stage when synchronously tuned, so that a four-cavity (three-stage) klystron can provide over 50 dB of gain.

After the bunched electron beam delivers its RF power to the output cavity, the electrons are removed by the collector electrode which is at, or slightly below, the potential of the interaction structure. From 50 to 80 percent of the d-c input power might be converted to heat in typical linear-beam tubes, and most of this heat appears at the collector. Therefore the majority of the cooling required in the klystron is at the collector. Power is extracted from the output cavity and delivered to the load by a coupling loop (as shown in Fig. 6.9) for low-power tubes, or by waveguide in high-power tubes. A waveguide ceramic window is necessary to maintain the vacuum in the tube and yet couple power out efficiently. Waveguide arcing or thermal stresses are common causes of window failures. High-power tubes sometimes employ an arc detector looking directly at the output window that allows either the drive power to be removed or the beam to be shut off within a few tens of microseconds after the presence of an arc is detected.

In order to counteract the mutual repulsion of the electrons which constitute the electron beam, an axial magnetic field (not shown in Fig. 6.9) is generally employed. The magnetic field focuses, or confines, the electrons to a relatively long, thin beam, and prevents the beam from dispersing. The beam focusing can be provided by a uniform magnetic field generated by a long solenoid which has iron shielding around the outside diameter. Cooling might have to be provided for the electromagnets. The weight of the magnetic focusing system is a major portion of the total weight of a klystron. This weight can restrict the utility of klystrons for airborne and portable applications.

Klystrons can sometimes be focused with lightweight permanent magnets.[1,8] Permanent magnets require no power input or cooling, and the various protective circuits needed with solenoids are eliminated. A significant reduction in weight can be obtained in some tubes by replacing the solenoid with a periodic-permanent-magnetic (PPM) focusing system which consists of a series of magnetic lenses. PPM focusing is not suited to large average-power tubes. At X band, the maximum average power is probably under a kilowatt.[8] In some klystrons the electron beam may be confined by electrostatic fields designed into the tube structure so that external magnets are not required.[40]

In a high-power klystron, from 2 to 5 percent of the beam power might normally be intercepted by the interaction structure, or body of the tube. If the beam were not properly confined in a high-power klystron, the stray electrons that impinge upon the metal structure of the tube would cause it to overheat and possibly be destroyed. Since loss of the focusing magnetic field could cause the tube to fail, protective circuitry is normally employed to remove the beam voltage in the event of improper focusing or the complete loss of focusing. The collector of most high-power klystrons is insulated from the body (RF interaction circuit) of the tube so as to allow separate metering and overload protection for the body current and the collector current.

Pulse modulation. The klystron amplifier may be pulsed by turning on and off the beam accelerating voltage, similar to plate modulation of a triode or magnetron. The modulator in this case must be capable of handling the full power of the beam. When the tube is modulated by pulsing the RF input signal, the beam current must be turned on and off; otherwise beam power will be dissipated to no useful purpose in the collector in the interval between RF pulses, and the efficiency of the tube will be low. A common method for pulsing the beam of a klystron is with an electrode in the electron gun that controls the klystron-beam current. This is the *modulating anode*. The advantage of the modulating anode is that it requires little control power to modulate the beam. The power necessary is that required to charge and discharge the capacitance of the klystron gun and its associated circuitry, and this is independent of the pulse length. The cutoff characteristics of the modulating anode permit only a few electrons to escape from the electron gun during the interpulse period when the beam is turned off. This is important in radar application since the receiver sensitivity will be degraded if sufficient electrons are present during the interpulse period to cause the stray electron-current noise to exceed receiver noise.

The high-power klystron may also be pulsed with a grid in the electron gun so designed that the grid does not intercept the electrons.[7,8] Such nonintercepting gridded guns can switch the beam with low-power modulators. This technique actually uses two closely spaced and aligned grids. One is near the cathode and is at cathode potential. (In the so-called Unigrid,[42] this grid is physically placed on the cathode but is properly passivated to prevent emission.) The second grid is the control grid for the beam current. It is at a positive potential and is located in the shadow of the first grid. For this reason the first grid is also known as the *shadow grid*. The purpose of the shadow grid is to suppress electron emission from those portions of the cathode which otherwise would be intercepted by the second, or control, grid. The cathode surface under each opening of the first grid is dimpled. Each dimple is aligned with the openings of the grid so that beamlets are formed within each grid opening. The beam interception on the control grid of practical nonintercepting gridded guns might be less than a few hundredths of one percent of beam current.[8] The shadow grid can also be used in the traveling-wave tube.

Bandwidth. The frequency of a klystron is determined by the resonant cavities. When all the cavities are tuned to the same frequency, the gain of the tube is high, but the bandwidth is narrow, perhaps a fraction of one percent. This is known as *synchronous tuning*. Although maximum gain is obtained with all cavities tuned to the same frequency, klystrons are often operated with the next to the last cavity (the penultimate cavity) tuned outside the passband on the high-frequency side. The gain is reduced by about 10 dB in so doing, but the improved electron bunching results in greater efficiency and in 15 to 25 percent more output power.[7] Broadbanding of a multicavity klystron may be accomplished in a manner somewhat analogous to the methods used for broadbanding multistage IF amplifiers, that is, by tuning the individual cavities to different frequencies. This is known as *stagger tuning*. Stagger tuning of a klystron is not strictly analogous to stagger tuning a conventional IF amplifier because interactions among cavities can cause the tuning of one cavity to affect the tuning of the others. The S band VA-87, a four-cavity klystron amplifier, has a synchronously tuned half-power bandwidth of 20 MHz and a gain of 61 dB. When tuned for maximum power the bandwidth is increased to 27 MHz and the gain reduced to 57.6 dB.[9] By stagger tuning the various cavities the half-power bandwidth can be increased to 77 MHz (about 2.8 percent bandwidth), but with a concurrent decrease in gain to 44 dB. In practice, stagger tuning enables the bandwidth of the multicavity klystron amplifier to be increased from a synchronously tuned bandwidth of $\frac{1}{4}$ to $\frac{1}{2}$ percent to values of more than 5 percent. Multicavity klystrons can be designed with bandwidths as large as 10 to 12 percent or greater.[7,10,41,50]

Tuning. Although conventional klystrons are of narrow bandwidth, they may be tuned over a wide frequency range.[7] A simple tuning mechanism is a flexible wall in the resonant cavity. The tuning range is about 2 or 3 percent, and the tuner life is limited. Tuning ranges of 10 to 20 percent are possible with a movable capacitive element (paddle) in the cavity. Tuner life is no problem, but the tuner increases the cavity capacitance and thus decreases the cavity impedance so that there is reduced bandwidth at the low-frequency end of the tuning range. Tuning ranges of 10 to 15 percent can be obtained without a compromise in impedance by using a sliding-contact movable cavity wall, but at the expense of increased mechanical complexity.

To simplify the tuning of a klystron it is desirable to have a " gang tuner " by which all the cavities are controlled by a single knob when changing frequency. Gang tuning is complicated, however, since the resonant cavities do not generally have the same tuning rates. The *channel tuning mechanism* avoids the problem of the frequency tracking of the resonant cavities by pretuning the cavities (generally at the factory); and the tuning information is stored mechanically within the tuner mechanism. Thus when a particular frequency channel is desired, the tuner mechanism provides the correct tuner position for each cavity to achieve the desired klystron frequency response. The klystron differs from other tubes in that the higher the peak power, the greater can be its bandwidth.[48,50]

Other properties. The RF conversion efficiency of klystron amplifiers as used for radar might range from 35 to 50 percent. The less the bandwidth of the klystron, the greater can be its efficiency.[43] However, by use of harmonic bunching of the electron beam, high-power CW klystrons of wide bandwidth have demonstrated efficiencies as high as 70 to 75 percent.[8]

The advantage of the klystron over other microwave tubes in producing high power is due to its geometry. The regions of beam formation, RF interaction, and beam collection are separate and independent in the klystron. Each region can be designed to best perform its own particular function independently of the others. For example, the cathode is outside the RF field and need not be restricted to sizes small compared with a wavelength. Large cathode area and large interelectrode spacings may be used to keep the emission current densities and voltage gradients to reasonable values. The only function of the collector electrode in the klystron is to dissipate heat. It can be of a shape and size most suited for satisfying the average or peak power requirements without regard for conducting RF currents, since none are present.

The design flexibility available with the klystron is not present in other tube types considered in this chapter, except for the traveling-wave tube. In most other tubes the functions of electron emission, RF interaction, and collection of electrons usually occur in the same region. The design of such tubes must therefore be a compromise between good RF performance and good heat dissipation. Unfortunately, these requirements cannot always be satisfied simultaneously. Good RF performance usually requires the tube electrodes to be small compared with a wavelength, while good heat dissipation requires large structures.

The high-power capability of the klystron, like anything else, is not unlimited. One of the major factors which has restricted the power available from klystrons has been the problem of obtaining RF windows capable of coupling the output power from the vacuum envelope to the load. Other factors limiting large powers are the difficulty of operating with high voltages, of dissipating heat in the collector, and of obtaining sufficient cathode emission current.

Examples. Klystrons have seen wide application in radar. Several examples of radar pulse klystrons will be briefly mentioned. The VA-87E, shown in Fig. 6.10, is a 6-cavity, S-band klystron tunable over the range from 2.7 to 2.9 GHz. It was designed to meet the requirements for the ASR-8 Airport Surveillance Radar. It has a peak power of 0.5 to 2.0 MW and an average power of 0.5 to 3.5 kW. The gain is nominally 50 dB and its efficiency is greater than 45 percent. Its 1-dB bandwidth is 39 MHz, but it is inherently capable of greater values. A peak

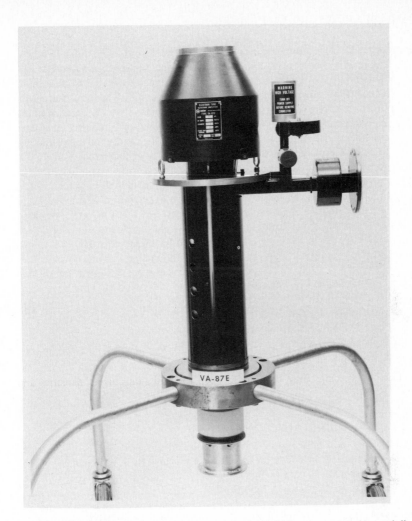

Figure 6.10 Photograph of the VA-87E 6-cavity S-band klystron mounted on a dolly. (*Courtesy Varian Associates, Inc., Palo Alto, CA.*)

beam voltage of 65 kV is required and its peak beam current is 34 amperes. The pulse duration can be from 0.5 to 6.0 μs.

The VA-812C is a wideband UHF klystron with a 12 percent bandwidth. It is capable of 8 MW of peak power and 30 kW of average power, with a pulse width of 6 μs. Its efficiency is 40 percent and gain is 30 dB. A peak beam voltage of 145 kV is required. The VA-812C formed the basis for the design of the VA-842, a tube used in the Ballistic Missile Early Warning System (BMEWS), with a demonstrated life in excess of 50,000 hours. The VA-812E, which was also derived from the same family as the VA-812C, has a peak power of 20 MW with an instantaneous 1-dB bandwidth of 25 MHz and a 40 dB gain. Its average power rating is 300 kW at a duty of 0.015 and a 40 μs pulse width.

6.4 TRAVELING-WAVE-TUBE AMPLIFIER

The traveling wave tube (TWT) is another example of a linear-beam, or O-type, tube. It differs from the klystron amplifier by the continuous interaction of the electron beam and the RF field over the entire length of the propagating structure of the traveling-wave tube rather than the interaction occurring at the gaps of a relatively few resonant cavities. The chief characteristic of the TWT of interest to the radar system engineer is its relatively wide bandwidth. A wide bandwidth is necessary in applications where good range-resolution is required or where it is desired to avoid deliberate jamming or mutual interference with nearby radars. Although low power TWTs are capable of octave bandwidths, bandwidths of the order of 10 to 20 percent are more typical at the power levels required for long-range radar applications. The gain, efficiency, and power levels of TWTs are like those of the klystron; but, in general, their values are usually slightly less than can be obtained with a klystron of comparable design.

A diagrammatic representation of a traveling-wave tube is shown in Fig. 6.11. The electron optics is similar to the klystron. Both employ the principle of velocity modulation to density-modulate the electron beam current. Electrons emitted by the cathode of the traveling-wave tube are focused into a beam and pass through the RF interaction circuit known as the slow-wave structure, or periodic delay line. An axial magnetic field is provided to maintain the electron-beam focus, just as in the klystron. A shadow grid to pulse-modulate the beam can also be included. After delivering their d-c energy to the RF field, the electrons are removed by the collector electrode. The RF signal to be amplified enters via the input coupler and propagates along the slow-wave structure. A helix is depicted as the slow-wave structure in Fig. 6.11, but TWTs for radar usually use a structure better suited for high power. The velocity of propagation of electromagnetic energy is slowed down by the periodic structure so that it is nearly equal to the velocity of the electron beam. It is for this reason that helix and similar microwave circuits are called slow-wave structures or delay lines. The synchronism between the electromagnetic wave propagating along the slow-wave structure and the d-c electron beam results in a cumulative interaction which transfers energy from the d-c electron beam to the RF wave, causing the RF wave to be amplified.

The simple helix was used as the slow-wave structure in the early TWTs and is still preferred in traveling-wave tubes at power levels up to a few kilowatts. It is capable of wider bandwidth than other slow-wave structures, but its power limitations do not make it suitable for most high-power radar applications. A modification of the helix known as the ring-bar circuit has been used in TWTs to achieve higher power and efficiencies between 35 and 50 percent.[11,12] The Raytheon QKW-1671A, which utilizes a ring-bar circuit, has a peak power of 160 kW, a duty cycle of 0.036, pulse width of 70 μs, gain of 45 dB, and a 200 MHz bandwidth at L band. This tube is suitable for air-search radar. Similar TWTs have been used in phased-array radar. The Air Force Cobra Dane phased-array radar, for example, uses 96

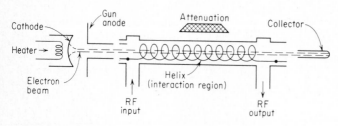

Figure 6.11 Diagrammatic representation of the traveling-wave tube.

QKW-1723 TWTs, each with a peak power of 175 kW and an average power of 10.5 kW operating over the frequency band of from 1175 to 1375 MHz. Since this radar is designed to monitor ballistic-missile reentry vehicles, its pulse width is as great as 2000 μs.

The ring-loop slow-wave circuit which consists of equally spaced rings and connecting bars, is also related to the helix and the ring-bar. It is claimed[14] to be preferred for tubes in the power range from 1 to 20 kW, as for lightweight airborne radar or as drivers for high-power tubes. The ring-loop circuit is not bothered by the backward-wave oscillations of the ordinary helix or the "rabbit ear" oscillations which can appear in coupled-cavity circuits.

The helix has been operated at high average power by passing cooling fluid through a helix constructed of copper tubing.[43] The bandwidth of this type of fluid-cooled helix TWT can be almost an octave, and it is capable of several tens of kilowatts average power at L band with a duty cycle suitable for radar applications.

A popular form of slow-wave structure for high-power TWTs is the *coupled-cavity* circuit.[7,11] It is not derived from the helix as are the ring-bar or ring-loop circuits. The individual unit cells of the coupled-cavity circuit resemble the ordinary klystron resonant cavities. There is no direct coupling between the cavities of a klystron; but in the traveling-wave tube, coupling is provided by a long slot in the wall of each cavity. The coupled-cavity circuit is quite compatible with the use of lightweight PPM focusing, a desired feature in some airborne applications.

Although the TWT and the klystron are similar in many respects, one of the major differences between the two is that feedback along the slow-wave structure is possible in the TWT, but the back coupling of RF energy in the klystron is negligible. If sufficient energy were fed back to the input, the TWT would produce undesired oscillations. Feedback energy might arise in the TWT from the reflection of a portion of the forward wave at the output coupler. The feedback must be eliminated if the traveling-wave amplifier is to function satisfactorily. Energy traveling in the backward direction may be reduced to an insignificant level in most tubes by the insertion of attenuation in the slow-wave structure. The attenuation may be distributed, or it may be lumped; but it is usually found within the middle third of the tube. Loss introduced to attenuate the backward wave also reduces the power of the forward wave, which results in a loss of efficiency. This loss in the forward wave can be avoided by the use of discontinuities called *severs*, which are short internal terminations designed to dissipate the reverse-directed power without seriously affecting the forward power.[13] The number of severs depends on the gain of the tube; one sever is used for each 15 to 20 dB of gain. In addition to reflection-type oscillations, backward-wave oscillations can occur. These frequently occur outside the passband so that they can be reduced by loss that is frequency selective.[13]

In principle, the traveling-wave tube should be capable of as large a power output as the klystron. The cathode, RF interaction region, and the collector are all separate and each can be designed to perform their required functions independently of the others. In practice, however, it is found that there are limitations to high power. The necessity for attenuation or severs in the structure, as mentioned above, tends to make the traveling-wave tube less efficient than the klystron. The slow-wave structure can also provide a limit to TWT capability. It seems that those slow-wave structures best suited for broad bandwidth (like the helix) have poor power capability and poor heat dissipation. A sacrifice in bandwidth must be made if high-power is required of a TWT. If the bandwidth is too small, however, there is little advantage to be gained with a traveling-wave tube as compared with multicavity klystrons.

A klystron can be operated over a fairly wide range of beam voltage without a large change in the gain characteristics. However, high-power traveling-wave tubes tend to oscillate at reduced beam voltage. Therefore the tolerance on the TWT beam voltage must be tighter with increasing bandwidth.[7] The protection requirements for traveling-wave tubes are similar to those of the klystron, but they are generally more difficult than for the klystron. In some traveling-wave tubes with coupled-cavity circuits, oscillations appear for an instant during the turn-on and turn-off portions of the pulse.[1] They are called *rabbit-ear* oscillations because of their characteristic appearance when the RF envelope of the pulse waveform is displayed visually on a CRT. These can be undesirable in some military applications since they might provide a distinctive feature for recognizing a particular radar.

An example of a traveling-wave tube designed for radar applications is the S-band Varian VA-125A. It is a broadband, liquid-cooled TWT which uses a clover leaf coupled-cavity slow-wave structure. It is capable of 3 MW of peak power over a 300 MHz bandwidth, with a 0.002 duty cycle, a 2-μs pulse width, and a gain of 33 dB. This tube is similar in many respects to the VA-87 klystron amplifier. It was originally designed to be used interchangeably with the VA-87 klystron, except that the VA-125 TWT has a broader bandwidth and requires a larger input power because of its lower gain.

In addition to being used as the power tube for high-power radar systems, the traveling-wave tube has also been employed, at lower power levels, as the driver for high-power tubes (such as the crossed-field amplifier), and in phased array radars which use many tubes to achieve the desired high-power levels.

6.5 HYBRID LINEAR-BEAM AMPLIFIER

By combining the advantages of the klystron and the traveling-wave tube into a single device it is possible to obtain a high-power amplifier with a bandwidth, efficiency, and gain flatness better than can be obtained with either the usual klystron or TWT.[7] One such device is the Varian Twystron (a trade name), which is a hybrid consisting of a multicavity klystron input section coupled to an extended interaction traveling-wave output section. The limitation to the bandwidth of a klystron is generally the output cavity. It cannot be made broadband without a decrease in efficiency. The slow-wave circuits of traveling-wave tubes have a broader bandwidth than klystron resonant cavities; and when used for the output of a klystron, as in the Twystron, a broad bandwidth can be achieved with the peak and average power capabilities of a klystron.

The C-band Varian VA-146 Twystron family of tubes provides peak power outputs from 200 kW to 9 MW with bandwidths greater than 10 percent. The VA-146N produces a 2.5 MW peak power with a 20 μs pulse at a 0.004 duty cycle over a $1\frac{1}{2}$ dB bandwidth of 500 MHz at C band. The gain is 31 dB and the efficiency at midband is 40 percent.

6.6 CROSSED-FIELD AMPLIFIERS

The crossed-field amplifier (CFA), like the magnetron oscillator, is characterized by magnetic and electric fields that are perpendicular to each other. Such tubes are of high efficiency (40 to 60 percent), relatively low voltage (compared with a linear-beam tube), and of light weight and small size so as to make them of interest for mobile applications. CFAs are capable of broad bandwidth (10 to 20 percent) with high peak power, but the gain is usually modest. They have good phase stability, and a number of CFAs can be operated in parallel for greater power. CFAs can be used as a power booster following a magnetron oscillator, as the high-power stage in master-oscillator power-amplifier (MOPA) transmitters with other CFAs or a TWT as the driver stage, or as the individual transmitters of a high-power phased-array radar.

The crossed-field amplifier is based on the same principle of electronic interaction as the magnetron. Thus, its characteristics resemble those of the magnetron, and many CFAs are even similar in physical appearance to a magnetron. The CFA also resembles the traveling-wave tube because the electronic interaction in both is with a traveling-wave.

There are several different types of crossed-field amplifiers. They all employ a slow-wave circuit, cathode, and input and output ports. A schematic of a CFA is as shown in Fig. 6.12. It resembles the magnetron oscillator except that there are two external couplings (an input and output). Electrons originate from the cylindrical cathode which is coaxial to the RF slow-wave circuit that acts as the anode. The cathode in a crossed-field tube is also known as the *sole*. (The term *sole*, which means *ground plate*, comes from the French, who did much of the early work on crossed-field devices.) In some crossed-field devices the sole does not emit electrons, and a separate emitting cathode is used. The two words *sole* and *cathode* are sometimes used interchangeably. The slow-wave structure is designed so that an RF signal propagates at a velocity near that of the electron beam. This permits an exchange of energy from the electron beam to the RF field to produce amplification. A d-c electric field is applied between the anode (slow-wave structure) and the cathode. A magnetic field is perpendicular to the plane of the paper. The electrons emitted from the cathode, under the action of the crossed electric and magnetic fields, form into rotating electron (space-charge) bunches, or spokes. These bunches drift along the slow-wave circuit in phase with the RF signal and transfer energy to the RF wave to provide amplification. Instead of the collector electrode found in the klystron and the TWT, the spent electrons terminate in the crossed-field amplifier on the slow-wave anode structure.

The CFA depicted in Fig. 6.12 is called *reentrant*, in that the electrons that are not collected after energy is extracted at the output port are permitted to reenter the RF interaction area at the input. This improves the efficiency of the tube. The reentering electrons, however, might contain modulation which will be amplified in the next pass through the circuit. To circumvent this the tube of Fig. 6.12 has a drift space to demodulate the electron stream so as to remove this RF feedback. Some crossed-field tubes, such as the Amplitron and the forward-wave magnetron, do not have the drift space and employ the feedback provided by the reentering electrons. If the drift space in Fig. 6.12 is replaced by a collector electrode which terminates the electron stream, the CFA is called *nonreentrant*.

The traveling-wave interaction of the electron beam and the RF signal may be with either a forward traveling-wave (as in the TWT) or with a backward traveling-wave. The type of interaction is determined by the slow-wave circuit employed. A forward-wave interaction takes place when the phase velocity and the group velocity of the propagating signal along the slow-wave circuit are in the same direction. (The group velocity is the velocity with which energy is propagated along the slow-wave circuit, and the phase velocity is the velocity of the RF signal on the slow-wave circuit as it appears to the electrons. To achieve RF

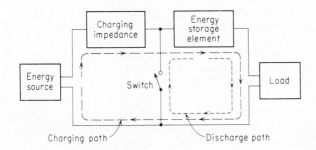

Figure 6.12 Simple representation of a crossed-field amplifier. (*From Clampitt*,[6] *Electronic Progress, Raythein.*)

amplification, the phase velocity must be near the velocity of the electron stream.) A *backward-wave* interaction takes place when the phase velocity is in the direction opposite that of the group velocity. The forward-wave CFA can operate over a broad range of frequency with a constant anode voltage, with only a small variation in the output power. On the other hand, the power output of a backward-wave CFA, with a constant anode voltage, varies with frequency. For a typical backward-wave CFA, the power output can vary 100 percent for a 10 percent change in frequency.[15] It is possible, however, with conventional modulator techniques, to operate the backward-wave CFA over a wide band with little change in output power. The type of modulators normally used with cathode-pulsed CFAs can readily compensate for the power variation with frequency of a backward-wave CFA and hold the variation of output power within acceptable levels. This is a result of the nature of the dispersion characteristics of such tubes. For example, with a particular X-band Varian backward-wave CFA,[16] a constant-voltage modulator produces a variation in output power from 505 kW at 9.0 GHz to 240 kW at 9.5 GHz, a change of 3.2 dB. A constant-current modulator with the same tube results in a power variation of only 0.3 dB over the same frequency range. Since most practical modulators are neither constant-current nor constant-voltage devices, actual performance is somewhere in between. A modulator with an internal impedance of 150 ohms will result in both current and voltage variations as a function of frequency when used with the above CFA, but the power output varies only 1 dB. Thus it does not matter significantly to the modulator designer whether the CFA is of the backward-wave or the forward-wave type as long as cathode pulsing is used with either a line-type modulator or a constant-current hard-tube modulator.

Electron emission in high-power crossed-field amplifiers can take place by cold-cathode emission without a thermally heated cathode. Some of the electrons emitted from the cathode are not collected by the anode but return to the cathode by the action of the RF field and the crossed electric and magnetic fields. When these electrons strike the cathode they produce secondary electrons that sustain the electron emission process. Cold-cathode emission requires the presence of *both* the RF drive signal applied to the tube as well as the d-c voltage between cathode and anode. The buildup of the current by means of this secondary emission process is very rapid (within nanoseconds). There is little pulse-to-pulse time jitter in the starting process. Although the name *cold-cathode emission* is used to describe the action by which electrons are produced in the CFA, the cathode might not actually be "cold." The bombardment of the cathode surface by returning electrons can raise the cathode temperature high enough to require liquid cooling to prevent its destruction.

The CFA can be pulse-modulated by turning on and off the high voltage between the anode (at ground potential) and the cathode (at a large negative potential). This is called *cathode pulsing*. The RF drive is usually applied to the CFA before the high voltage is applied since an RF signal is needed to start the emission process in a secondary-emission cathode. If high voltage is applied without RF drive, the amplifier appears as an open circuit to the modulator, causing the modulator voltage to double. The higher voltage could lead to arcing.

The d-c high-voltage can be applied before the RF signal if the design is such that voltage breakdown or arcing does not occur. This leads to the possibility of modulating the tube without the need for a high-power modulator as required with cathode pulsing. Such operation is possible with the forward-wave CFA using a cold secondary-emission cathode. The d-c operating voltage is applied continuously between cathode and anode. The tube remains inactive until the application of the RF input pulse starts the emission process, causing amplification to take place. At the end of the RF drive-pulse the electrons remaining in the tube must be cleared from the interaction area to avoid feedback which generates oscillations or noise. In reentrant CFAs, the electron stream can be collected after the removal of the RF drive-pulse by mounting an electrode in the cathode, but insulated from it, in the region of the drift space between the RF input and output ports. This is known as a *cutoff*, or *control*, *electrode*. A positive potential is applied to this cutoff electrode at the termination of the pulse to collect, or *quench*, the remaining electron current. The positive potential need be applied for only a short time; hence the energy requirements are low. This method of modulation in which the d-c anode-cathode voltage is applied continuously and the tube is turned on by the start of the RF drive-pulse and turned off at the end of the pulse by the aid of a cut-off electrode to remove the electrons, has been called *d-c operation*. It is applicable to forward-wave CFAs, but it is not usually used with backward-wave CFAs since the variation of output power with frequency at a constant d-c voltage that is characteristic of backward-wave tubes would limit the bandwidth to but a few percent.

It is also possible to turn the CFA on and off with the RF drive-pulse, without the need for a positive pulse applied to the cutoff electrode at the end of the drive pulse. The cutoff electrode can be designed to operate with an appropriate constant positive-bias that allows sufficient reentrance of the electrons when the RF drive is on but be unable to maintain the electron stream with the RF drive off.[17] This is called *RF keying*, or *self-keying*. In principle, it is the simplest way to modulate the CFA.

Although RF keying and d-c operation are simpler pulse-modulating methods than cathode pulsing, they have not been as widely used in the past as has the cathode pulser. Usually there have been factors other than modulator size that determine the method of modulation best used in practice.[15]

Unlike tubes that employ thermonic cathode emission for their supply of electrons, a tube with a cold secondary-electron cathode does not experience a droop, or decay, of the pulse amplitude with time. In addition to the insensitivity of electron emission with pulse duration, there is also an insensitivity to duty cycle if the cathode is cooled sufficiently so that the secondary emission properties of the cathode do not change with temperature.[17] Thus high and low duty-cycles can be employed interchangeably if the cathode temperature is maintained cool enough, usually below about 400 or 500°C.

The noise and spurious signals generated by a CFA can be quite low when the tube is locked in and controlled by an RF drive signal.[17] Measurements of noise made with the voltages applied but with the tube inactive because of no RF drive-signal, indicate the noise to approach what would be expected from its thermal level. Intrapulse noise is generally much higher. Spurious signals can also arise during the flat part of the pulse. The greater the RF drive the less will be the in-band noise. Thus the lower the gain the less will be the noise. Pedestals can appear at the leading and trailing edges of the pulse, somewhat like the "rabbit ears" found in traveling-wave tubes. They can be caused by the feedthrough of the RF drive-pulse when it is wider than the d-c modulator pulse, or they can be due to spurious oscillations, called *band-edge oscillations*, that occur just outside the normal tube-bandwidth.[16,17]

The insertion loss of a CFA is low and might be less than 0.5 dB. The RF drive will thus appear at the output of the tube with little attenuation. In a low-gain amplifier the input power which appears at the output can be a sizable fraction of the total. The conversion efficiency of a CFA is defined as

$$\text{Efficiency} = \frac{\text{RF power output} - \text{RF drive power}}{\text{d-c power input}} \tag{6.1}$$

This is a conservative definition since the RF drive power is not lost but appears as part of the output.

The low insertion loss of a CFA can be of advantage in systems that require more than one radiating power level. By omitting the application of d-c voltage to the final stage, the lower level RF drive-power can be fed through the final stage with little attenuation. This allows two power levels, depending on whether the final CFA stage has d-c voltage applied or not. The low insertion loss also means that an RF signal traveling in the reverse direction from output to input suffers little attenuation. A high-power circulator or other isolation device may be required between the CFA and its driver to prevent the power reflected from the output of the CFA building up into oscillations or interfering with the driver stage.

The gain of conventional pulsed crossed-field amplifiers is typically between 10 and 17 dB. By designing the cold cathode as a slow-wave circuit and introducing the RF drive at the cathode emitting surface itself, it has been possible to achieve about 30 dB of gain in a high-power pulse CFA with power, bandwidth, and efficiency commensurate with conventional designs.[43] The RF output is taken from the anode slow-wave circuit. This type of device has been called a *cathode-driven CFA*[6] or a *high-gain CFA*.[19]

The type of crossed-field amplifier principally considered in this section can be described as a distributed emission, or emitting sole, amplifier with a reentrant, circular format utilizing a forward-wave interaction without feedback. (The electron stream is debunched to remove the modulation before reentering the interaction space.) There are other types of CFAs, however. The Amplitron, one of the first successful CFAs, is similar to the above except it employs a backward-wave interaction with feedback. The circular reentrant CFA can be designed with or without feedback and with either forward- or backward-wave interaction. The nonreentrant tube can have either a linear or a circular format, with choice of forward- or backward-wave interaction. CFAs can also be built with an injected electron beam and a nonemitting sole, but these have not found much application in radar.

The crossed-field amplifier is capable of peak powers comparable to those of linear-beam tubes, and average powers only slightly less than those achieved with linear-beam tubes.[16] Efficiencies greater than 50 percent are common. The bandwidths are similar to those obtained with the traveling-wave tube. Forward-wave CFAs generally have bandwidths from 10 to 25 percent; backward-wave CFAs, less than 10 percent. The gains are low or modest, but the potential exists for higher gains, especially if trade-offs with other properties are permitted. Reasonably long life has been demonstrated with CFAs. Operation for greater than 10,000 hours is not unusual in some tubes.[17] The good phase stability and short electrical length of CFAs make it possible to operate tubes in parallel to achieve greater power than available from a single tube, as well as provide redundancy in the event of a single tube failure. The CFA behaves as a saturated amplifier rather than as a linear amplifier. The characteristic of a saturated amplifier is that, above a certain level, the RF output is independent of the RF input. A saturated amplifier is compatible with frequency and phase-modulated pulse compression waveforms.

When used in an amplifier chain, the CFA is generally found in only the one or two highest-power stages. It is often preceded by a medium-power traveling-wave tube. This combination takes advantage of the best qualities of both tube types. The TWT provides high gain, and the CFA allows high power to be obtained with high efficiency, good phase stability, and low voltage.[20]

The characteristics of a CFA are illustrated by the Varian SFD-257, a forward-wave tube used in the final high-power amplifier stage of the transmitter chain in the AN/MPS-36 C-band range-instrumentation tracking radar.[18] The SFD-257 operates over the frequency range 5.4 to 5.7 GHz with a peak power of 1 MW, 0.001 duty cycle, and an efficiency of over 50 percent. The tube is d-c operated in that the RF pulse turns the tube on and a control electrode turns it off. In this radar application the pulse widths are 0.25, 0.5, or 1.0 μs, but the tube can deliver a pulse as wide as 5 μs. A coded group, or burst, of five 0.25 μs pulses is also utilized. It requires 50 kW of drive power and operates with 30 kV anode voltage and 70 A of peak anode current. Liquid cooling is employed for both the anode and cathode. Approximately one gallon of water per minute with a pressure drop of 9 psi is required. The tube is housed in a magnetically shielded package that stands approximately 2 ft high with a diameter of 1 ft and weighs 210 lb. A Vac-Ion vacuum pump is included within the magnetically shielded package to monitor the vacuum and to provide continuous pumping action during operation. In the AN/MPS-36 transmitter chain the SFD-257 is preceded by an SFD-244 cathode-pulsed CFA with 11 dB gain, which is driven by a 904T1 cathode-pulsed TWT with 49 dB gain. The drive power to the TWT is 50 mW.

6.7 GRID-CONTROLLED TUBES

The early radars developed during the 1930s used conventional grid-controlled vacuum tubes since there existed no other source of large RF power. This limited the development of the early radars to the VHF and the lower UHF bands. These tubes were triodes or tetrodes designed to minimize the transit-time effects and other problems of operating at VHF and UHF.[21] The potential applied to the control grid of the tube acts as a gate, or valve, to control the number of electrons traveling from the cathode to the anode, or plate. The variation of potential applied to the grid is imparted to the current traveling to the anode. The process by which the electron stream is modulated in a grid-controlled tube to produce amplification is called *density modulation.*

When the discovery of the cavity magnetron led to the successful development of microwave radar early in World War II, interest in lower frequency radars waned. However, improvements in high-power, grid-controlled tubes continued to be made due to the needs of UHF-TV, tropospheric-scatter communications, and particle accelerators as well as radar. At VHF and UHF, grid-controlled power tubes have been widely used in high-power radar systems. They are capable of operation at frequencies as high as 1000 to 2000 MHz, but they do not seem to be competitive to the linear-beam and crossed-field microwave tubes at radar frequencies much above 450 MHz. At low-power levels, gridded tubes must compete with the solid-state transistor.

The grid-controlled tube is characterized as being capable of high power, broadband, low or moderate gain, good efficiency, and inherent long life. Unlike other microwave tubes, the grid-controlled tube can operate, if desired, with a linear rather than a saturated gain characteristic. In addition to being used in high-power amplifier chains, the grid-controlled tube has also seen service in large phased-array radars that operate in the lower radar-frequency bands.

An example of a grid-controlled tube that could be suitable for high-power radar application at UHF is the Coaxitron.[22] It has wider bandwidth and better performance than conventional UHF grid-control tubes because it integrates the complete RF input and output circuits and the electrical interaction system within the vacuum envelope. The RCA A15193, for example, operates from 406 to 450 MHz with 1.5 MW peak power, a duty cycle of 0.0039, 47 percent plate efficiency, and 13 dB gain. It requires a plate voltage of 17.5 kV, plate current of 183 A, 1.57 V filament voltage and 890 A filament current. The tube is 76 cm long, 42 cm in diameter and weighs 63.5 kg.

Low-power grid-control tubes have been used in phased array radars at UHF and have proven to be an economical and reliable source of power. In the AN/FPS-85 satellite-surveillance radar built by Bendix Corporation for the U.S. Air Force, the transmit array contains 5184 identical modules using the Eimac 4CPX250K ceramic tetrode.[23] Each module has a peak power of 10 kW and is 7 by 9 by 31 inches with a weight of nearly 50 lb. The array operates with a 10 MHz bandwidth centered at 442 MHz.

6.8 MODULATORS

The function of the modulator is to turn the transmitting tube on and off to generate the desired waveform. When the transmitted waveform is a pulse, the modulator is sometimes called a *pulser*. Each RF power tube has its own peculiar characteristics which determine the particular type of modulator to be used. The magnetron modulator, for instance, must be designed to handle the full pulse power. On the other hand, the full power of the klystron and the traveling-wave tube can be switched by a modulator handling only a small fraction of the total beam power, if the tubes are designed with a modulating anode or a shadow grid. The crossed-field amplifier (CFA) is often cathode-pulsed, requiring a full-power modulator. Some CFAs are d-c operated, which means they can be turned on by the start of the RF pulse and turned off by a short, low-energy pulse applied to a cutoff electrode. Some CFAs can be turned on and off by the start and stop of the RF pulse, thus requiring no modulator at all. Triode and tetrode grid-controlled tubes may be modulated by applying a low-power pulse to the grid. Plate modulation is also used when the radar application cannot tolerate the interpulse noise that results from those few electrons that escape the cutoff action of the grid.

The basic elements of one type of radar modulator are shown in Fig. 6.13. Energy from an external source is accumulated in the energy-storage element at a slow rate during the interpulse period. The charging impedence limits the rate at which energy can be delivered to the storage element. At the proper time, the switch is closed and the stored energy is quickly discharged through the load, or RF tube, to form the pulse. During the discharge part of the cycle, the charging impedance prevents energy from the storage element from being dissipated in the source.

Line-type modulator. A delay line, or pulse-forming network (PFN), is sometimes used as the storage element since it can produce a rectangular pulse and can be operated by a gas-tube switch. This combination of delay-line storage element and gas-tube switch is called a *line-type modulator*. It has seen wide application in radar because of its simplicity, compact size, and its ability to tolerate abnormal load conditions such as caused by magnetron sparking.[1,24,25] A diagram of a line-type pulse modulator is shown in Fig. 6.14. The charging impedance is shown as an inductance. The pulse-forming network is usually a lumped-constant delay line. It might consist of an air-core inductance with taps along its length to which are attached capacitance to ground. A transformer is used to match the impedance of the delay line to that of the load. A perfect match is not always possible because of the nonlinear impedance characteristic of microwave tubes.

The switch shown in Fig. 6.14 is a hydrogen thyratron, but it can also be a mercury ignitron, spark gap, silicon-controlled rectifier (SCR), or a saturable reactor. A gas tube such

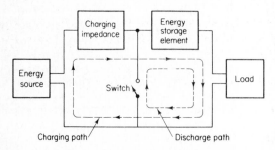

Figure 6.13 Basic elements of one type of radar pulse modulator.

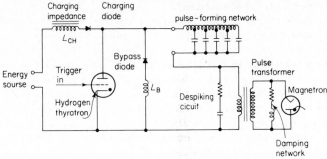

Figure 6.14 Diagram of a line-type modulator.

as a thyratron or ignitron is capable of handling high power and presents a low impedance when conducting. However, a gas tube cannot be turned off once it has been turned on unless the plate current is reduced to a small value. The switch initiates the start of the modulator pulse by discharging the pulse-forming network, and the shape and duration of the pulse are determined by the passive circuit elements of the pulse-forming network. Since the trailing edge of the pulse depends on how the pulse-forming network discharges into the nonlinear load, the trailing edge is usually not sharp and it may be difficult to achieve the desired pulse shape.

The charging inductance L_{ch} and the capacitance C of the pulse-forming network form a resonant circuit, whose frequency of oscillation approaches $f_0 = (2\pi)^{-1}(L_{ch}C)^{-1/2}$. (The inductance of the pulse-forming network and the load are assumed small.) With a d-c energy source the pulse repetition frequency f_p will be twice the resonant frequency if the thyratron is switched at the peak of maximum voltage. This method of operation, ignoring the effect of the charging diode, is called *d-c resonant charging*. A disadvantage of d-c resonant charging is that the pulse repetition frequency is fixed once the values of the charging inductance and the pulse-forming-network delay-line capacitance are fixed. However, the charging, or hold-off, diode inserted in series with the charging inductance permits the modulator to be operated at any pulse repetition frequency less than that determined by the resonant frequency f_0. The function of the diode is to hold the maximum voltage and keep the delay line from discharging until the thyratron is triggered.[26] Although the series diode is a convenient method for varying the prf, it is more difficult to change the pulse width since high-voltage switches in the pulse-forming network are required.

The bypass diode and the inductance L_B connected in parallel with the thyratron serve to dissipate any charge remaining in the capacitance due to tube mismatch. If this charge were allowed to remain, the peak voltage on the network would increase with each cycle and build up to a high value with the possibility of exceeding the permissible operating voltage of the thyratron. The mismatch of the pulse-forming network to the nonlinear impedance of the tube might also cause a spike to appear at the leading edge of the pulse. The *despiking circuit* helps minimize this effect. The *damping network* reduces the trailing edge of the pulse and prevents post-pulse oscillations which could introduce noise or false targets.

Hard-tube modulator.[1] The hard-tube modulator is essentially a high-power video pulse amplifier. It derives its name from the fact that the switching is accomplished with "hard-vacuum" tubes rather than gas tubes. Semiconductor devices such as the SCR (silicon-controlled rectifiers) can also be used in this application.[1] Therefore, the name *active-switch modulator* is sometimes used to reflect the fact that the function of a hard-tube modulator can be obtained without vacuum tubes. Active-switch pulse modulators can be *cathode pulsers* that control the full power of the RF tube, *mod-anode pulsers* that are required to switch at the full beam voltage of the RF tube but with little current, or *grid pulsers* that operate at a far smaller voltage than that of the RF beam.

The chief functional difference between a hard-tube modulator and a line-type modulator is that the switching device in the hard-tube modulator controls both the beginning and the end of the pulse. In the line-type modulator, the switch controls only the beginning of the pulse. The energy-storage element is a capacitor. To prevent droop in the pulse shape due to the exponential nature of a capacitor discharge, only a small fraction of the stored energy is extracted for the pulse delivered to the tube. In high-power transmitters with long pulses the capacitor must be very large. It is usually a collection of capacitors known as a *capacitor bank*.

The hard-tube modulator permits more flexibility and precision than the line-type modulator. It is readily capable of operating at various pulse widths and various pulse repetition frequencies, and it can generate closely spaced pulses. The hard-tube modulator, however, is generally of greater complexity and weight than a line-type modulator.

Tube protection.[1,27] Power tubes can develop internal flash arcs with little warning even though they are of good design. When a flash arc occurs in an unprotected tube, the capacitor-bank discharges large currents through the arc and the tube can be damaged. One method for protecting the tube is to direct the arc-discharge currents with a device called an *electronic crowbar*. It places a virtual short circuit across the capacitor bank to transfer the stored energy by means of a switch which is not damaged by the momentary short-circuit conditions. The name is derived from the analogous action of placing a heavy conductor, like a crowbar, directly across the capacitor bank. Hydrogen thyratrons, ignitrons, and spark-gaps have been used as switches. The sudden surge of current due to a fault in a protected power tube is sensed and the crowbar switching is actuated within a few microseconds. The current surge also causes the circuit breaker to open and deenergize the primary source of power. Crowbars are usually required for high-power, hard-tube modulators because of the large amounts of stored energy. They are also used with d-c operated crossed-field amplifiers and mod-anode pulsed linear-beam tubes which are connected directly across a capacitor bank. The line-type modulator does not usually require a crowbar since it stores less energy than the hard-tube modulator and it is designed to discharge safely all the stored energy each time it is triggered.

6.9 SOLID-STATE TRANSMITTERS

There have been two general classes of solid-state devices considered as potential sources of microwave power for radar applications. One is the transistor amplifier and the other is the single-port microwave diode that can operate as either an oscillator or as a negative-resistance amplifier. The silicon bipolar transistor has, in the past, been of interest at the lower microwave frequencies (*L* band or below), and the diodes have been of interest at the higher microwave frequencies. Gallium arsenide field-effect transistors (GaAs FET) have also been considered at the higher microwave frequencies. Both the transistor and the diode microwave generators are characterized by low power, as compared with the power capabilities of the microwave tubes discussed previously in this chapter. The low power, as well as other characteristics, make the application of solid-state devices to radar systems quite different from high-power microwave tubes. The almost total replacement of receiver-type vacuum tubes by solid-state devices in electronic systems has offered encouragement for replacing the power vacuum tube with an all solid-state transmitter to obtain the advantages offered by that technology. Although there have been significant advances in microwave solid-state devices and although they possess properties that differ from other microwave sources, the degree of application of these devices to radar systems has been limited.

Microwave transistor.[28-35,46] At L band the CW power that can be obtained from a single microwave transistor might be several tens of watts. Unlike vacuum tubes, the peak power that can be achieved with narrow pulse widths is only about twice the CW power.[28] This results in the microwave transistor being operated with relatively long pulse-widths and high duty cycles. For air-surveillance radar application, pulse widths might be many tens of microseconds or more. Duty cycles of the order of 0.1 are not unusual, which is significantly greater than the duty cycles typical of microwave tubes. The high duty cycles present special constraints on the radar system designer so that solid-state transmitters are not interchangeable with tube transmitters. A different system design philosophy usually must be employed with solid state.

To increase the power output, transistors may be operated in parallel. From 2 to 8 devices are usually combined into a single power *module*. Too large a number cannot be profitably paralleled because of losses that occur on combining. The gain of a transistor is only of the order of 10 dB, so that several stages of amplification are necessary to achieve a reasonable total gain. An L-band module using eight transistors (four paralleled in the final stage, two paralleled in the penultimate stage, and two series driver stages) might have a peak power greater than 200 W, 0.1 duty cycle, 200 MHz bandwidth, a gain of 30 dB, and an efficiency better than 30 percent.[31] (When examining the claims of power from solid-state devices, it should be kept in mind that the greater the junction temperature the less the life of the device. Since long life is a featured characteristic of such power sources, they should be operated conservatively.)

The power output of a microwave transistor theoretically decreases inversely as the square of the frequency, or 6 dB per octave.[29] For this reason the silicon bipolar transistor is unattractive for radar application at S band or above, especially when appreciable power is desired. Varactor frequency multipliers, however, generally have attenuation less than their frequency-multiplication ratio so that a transistor at some lower frequency followed by a varactor multiplier has sometimes been employed to obtain power at the higher microwave frequencies. In one experimental design,[30] X-band power was obtained by an S-band transistor followed by a four-times multiplier to obtain 1 watt of peak power at X band with a 0.05 duty cycle.

Bulk-effect and avalanche diodes. The Gunn and LSA *bulk-effect diodes* and the Trapatt and Impatt *avalanche diodes* can be operated as oscillators or single-port negative-resistance amplifiers.[29,45] They have been considered for use at the higher microwave frequencies where the transistor amplifier has reduced capability, but they can also operate at L band or below. Unfortunately, the peak and average powers of such devices are low, as is the efficiency and the gain. An S-band Trapatt amplifier, for example, might have a peak power of 150 W, 1 μs pulse width, 100 Hz pulse repetition frequency, 20 percent efficiency, 9 dB gain, and a 1 dB bandwidth of 6.25 percent.[36,37] LSA diodes may be capable of somewhat greater peak power, but their average power and efficiency are low. The performance of these diode microwave sources worsens with increasing frequency.

Electron-bombarded semiconductor device.[47] The EBS, or electron-bombarded semiconductor, is a hybrid device that combines semiconductor and vacuum-tube technology. It contains a heated cathode that generates an electron beam which strikes a semiconductor diode at high energy (typically 10 to 15 keV). On striking the diode, which is reverse biased well below the avalanche threshold, each impacting electron gives rise to thousands of additional carrier pairs to provide a current amplification of 2000 or more. A control grid is inserted between the cathode and the diode to density-modulate the electron beam. (The basic geometry is similar to that of the classic triode vacuum tube, but with a semiconductor diode as the plate.) The EBS can also be deflection-modulated by varying the position of the beam relative to two or more separated semiconductor targets whose outputs are combined. The EBS is claimed to be broadband and to have high gain (25 to 35 dB), high efficiency (50 percent), and long life. At 1 GHz, an EBS amplifier might be capable of a peak power of 2 to 3 kW, and an average power of about 200 W. The EBS has decreasing capability at the higher frequencies. A limitation of the EBS, not found with other semiconductor devices, is that it requires a power supply of 10 to 15 kV as well as a cathode heater supply.

Methods for employing solid-state transmitters. There are at least three methods for employing solid-state devices as radar transmitters: (1) direct replacement of a vacuum tube, (2) multiple modules in an electronically steered phased-array radar with the power combined in space, and (3) multiple modules in a mechanically scanned array with the power combined in space.

A significant restriction in attempting to utilize solid-state devices as a direct replacement for the conventional microwave-tube transmitter is the relatively low power available from a single solid-state device or even a single power module. The solid-state device or module finds application in those radars where high power is not needed, such as aircraft altimeters, CW police speedmeters, and short-range intrusion detectors. Higher power can be obtained by combining the outputs of a large number of individual devices. Unfortunately, for many radar applications hundreds or thousands of devices would have to be employed in order to achieve the requisite power levels. The resultant loss with the combining of a large number of devices in some microwave circuits can sometimes negate the advantage achieved by combining. Thus as a direct replacement for conventional microwave tubes, solid-state devices have been limited to low or moderate power applications.

A phased-array transmitting antenna combines in space the power from each of many transmitting sources. Solid-state sources at each radiating element of a large phased-array antenna can produce the total power required for many radar applications. The radiated beam is steered by electronic phase shifters at each element, usually on the input side of the individual power amplifiers. The transmitter, receiver, duplexer, and phase shifters for each element of the array antenna can be incorporated into a single integrated package, or module. Although there has been much development work in solid-state phased arrays with each element fed by its own integrated module, this approach usually results in a costly and complex system. This has tended, in the past, to weigh against the widespread use of such phased-array radars when the number of elements is large.

The combining of the power from a large number of individual solid-state devices is attractive when using an array antenna since the power is " combined in space," rather than by a lossy microwave network. A fixed phased array requires electronic beam steering that complicates the radar. If the flexibility of an electronically steered array is not needed, the advantages of an array antenna for combining the radiated power from many individual sources can be had by mechanically scanning the entire antenna. A separate solid-state source can be used at each element of the antenna, or a number of sources can be combined to feed each row (or each column) as in the AN/TPS-59, Fig. 6.15.[38] The mechanically rotating array antenna with solid-state transmitters has some special problems of its own that must be overcome. One such problem is the need to convey large power to the solid-state devices on the rotating antenna. Another is that the weight of the transmitter is now added to that of the

Figure 6.15 AN/TPS-59 *L*-band radar. The 30 ft by 15 ft antenna consists of 54 rows of 24 elements each. Each of the 54 rows contains its own transceiver with twenty-two 50-W modules along with phase shifter, low-noise amplifier, output filter, circulator, and 28-V power supply. The antenna rotates mechanically in azimuth and scans electronically in elevation. (*Courtesy General Electric Co.*)

antenna. Even if the solid-state transmitter were lighter than a conventional tube transmitter, the weight is found at a bad place, on the antenna itself. This is especially important for shipboard application where weight high on the mast must be minimized.

System considerations. The potential advantages claimed for solid-state sources in radar may be summarized as (1) long, failure-free life, (2) low transmitter voltage, which eliminates the risk of X-rays and electric shock, (3) amplitude control of the transmitted waveform by selectively switching modules or individual devices on or off, (4) wide bandwidth, (5) low projected volume-production costs, and (6) air cooling. There are some problems, however, in the use of solid-state devices for radar systems other than cost.

As mentioned, the solid-state transmitter has some significant differences as compared to the conventional tube transmitter. The basic power-generating unit comes in a relatively small size; hence, many units have to be combined in some manner to achieve the power levels required for radar. The higher the frequency, the less the power available from an individual solid-state device and the more difficult will be the combining problem because of the larger number of devices required. At the lower microwave frequencies, the transistor is one of the better available solid-state sources. Since the transistor cannot operate efficiently with the low duty-cycles characteristic of conventional radars, radars using transistors must operate with long pulses or high pulse repetition frequency. Neither is desirable except in special cases (as in pulse-doppler radar or CW radar). Long pulses result in a long minimum radar-range since the receiver cannot be turned on until the transmitted pulse is turned off. To see targets at ranges closer than that determined by the long pulse, a short pulse of lower energy and at a different frequency can be used in addition to the long pulse. This adds to the system complexity, however. Long pulses can be a disadvantage in military radars since they reduce the effectiveness of pulse-to-pulse frequency agility. A jammer can determine the radar frequency at the beginning of the long pulse and quickly tune a jammer to the correct frequency within the duration of the pulse. To achieve improved range-resolution with the long pulse, pulse compression is needed. This further complicates the radar. (Pulse compression does not relieve the minimum-range problem of long pulses.)

Although the solid-state transmitter does not require high voltage, it does require large current, perhaps several thousands of amperes in some applications. Wide bandwidth is one of the more favorable properties of such devices; but the overall transmitter efficiency, especially at the higher microwave frequencies, is not always as great as might be desired. Solid-state devices also are less "forgiving" than microwave tubes of transients due to lightning and nearby radars. The advantage of long life claimed for solid-state transmitters seems matched by the long life obtained with linear beam tube transmitters.

The transistor amplifier, and the bulk-effect and avalanche diodes have important applications in radar; but in the form in which they have been known, it is not likely that they will cause the high-power microwave tubes to disappear in the way their lower frequency counterparts have displaced the receiver vacuum tube.

REFERENCES

1. Weil, Thomas A.: Transmitters, Chap. 7 of the " Radar Handbook," M. I. Skolnik (ed.), McGraw-Hill Book Co., New York, 1970.
2. Boot, H. A. H., and J. T. Randall: The Cavity Magnetron, *J. Inst. Elect. Eng.*, vol. 93, pt. IIIA, pp. 928–938, 1946.
3. Wathen, R. L.: Genesis of a Generator: The Early History of the Magnetron, *J. Franklin Inst.*, vol. 255, pp. 271–288, April, 1953.
4. Collins, G. B. (ed.): "Microwave Magnetrons," MIT Radiation Laboratory Series, vol. 6, McGraw-Hill Book Co., New York, 1948.
5. Smith, W. A.: Ring-Tuned Agile Magnetron Improves Radar Performance, *Microwave System News*, vol. 3, pp. 97–102, February, 1974.
6. Clampitt, L. L.: Microwave Radar Tubes at Raytheon, *Electronic Progress*, vol. 17, no. 2, pp. 6–13, Summer, 1975.
7. Staprans, A., E. W. McCune, and J. A. Ruetz: High-Power Linear-Beam Tubes, *Proc. IEEE*, vol. 61, pp. 299–330, March, 1973.
8. Lien, E. L.: Advances in Klystron Amplifiers, *Microwave J.*, vol. 16, pp. 33–36, 39, December, 1973.
9. Dodds, W. J., T. Moreno, and W. J. McBride, Jr.: Methods of Increasing Bandwidth of High Power Microwave Amplifiers, *IRE WESCON Conv. Record*, vol. 1, pt. 3, pp. 101–110, 1957.

10. Beaver, W., G. Caryotakis, A. Staprans, and R. Symons: Wide Band High Power Klystrons, *IRE WESCON Conv. Record*, vol. 3, pt. 3, pp. 103–111, 1959.
11. Mendel, J. T.: Helix and Coupled-Cavity Traveling-Wave Tubes, *Proc. IEEE*, vol. 61, pp. 280–298, March, 1973.
12. Pittack, U., and R. A. Handy: Progress in Helix Traveling-wave Tubes for Radar Systems, *Microwave System News*, vol. 5, pp. 69–75, February/March, 1975.
13. Nalos, E. J.: Present State of Art in High Power Traveling-Wave Tubes, pt. I, *Microwave J.*, vol. 2, pp. 31–38, December, 1959; pt. II *Microwave J.*, vol. 3, pp. 46–52, January, 1960.
14. Phillips, R. M.: High Power Ring-Loop Traveling-Wave Tubes for Advanced Radar, *Microwave System News*, vol. 5, pp. 47–49, February/March, 1975.
15. Weil, T. A.: Comparison of CFA's for Pulsed-Radar Transmitters, *Microwave J.*, vol. 16, pp. 51–54, 72, June, 1973.
16. Anonymous: Introduction to Pulsed Crossed-Field Amplifiers, brochure published by Varian, Beverly, Mass., no date.
17. Skowron, J. F.: The Continuous-Cathode (Emitting-Sole) Crossed-Field Amplifier, *Proc. IEEE*, vol. 61, pp. 330–356, March, 1973.
18. Smith, W., and A. Wilczek: CFA Tube Enables New-Generation Coherent Radar, *Microwave J.*, vol. 16, pp. 39–42, 74, April, 1973.
19. Hill, R. T.: Shipboard Radar Systems, Research and Development, *EASCON-77 RECORD*, pp. 4–1A to 4–1 I, Sept. 26–28, 1977. Arlington, Va, IEEE Publication 77CH 1255-9 EASCON.
20. Barker, G. G.: CFA or TWT? Which Valve Will Best Meet Your Transmitter Output Stage Requirements; *IEE Conf. Publ. No. 105*, "Radar—Present and Future," *London*, Oct. 23–25, 1973, pp. 189–194.
21. Spangenberg, K. R.: "Vacuum Tubes," McGraw-Hill Book Company, New York, 1948.
22. Yingst, T. E., D. R. Carter, J. A. Eshleman, and J. M. Pawlikowski: High-Power Gridded Tubes—1972, *Proc. IEEE*, vol. 61, pp. 357–381, March, 1973.
23. Reed, J. E.: The AN/FPS-85 Radar System, *Proc. IEEE*, vol. 57, pp. 324–335, March, 1969.
24. Glasoe, G. N., and J. V. Lebacqz: "Pulse Generators," MIT Radiation Laboratory Series, vol. 5, McGraw-Hill Book Company, New York, 1948.
25. Weil, T. A.: Survey of Raytheon Pulsed Radar Transmitters, *Electronic Progress*, vol. 17, no. 2, pp. 14–24, Summer, 1975.
26. Reintjes, J. F., and G. T. Coate: "Principles of Radar," Chap. 3, McGraw-Hill Book Company, New York, 1952.
27. Parker, W. N., and M. V. Hoover: Gas Tubes Protect High-Power Transmitters, *Electronics*, vol. 29, pp. 144–147, January, 1956.
28. Hoft, D. J., and L. R. Lavallee: Solid-State Transmitters, *Electronic Progress*, vol. 17, no. 2, pp. 25–33, Summer, 1975.
29. Hyltin, T. M.: Solid-state Radar, Chap. 30 of "Radar Handbook," M. I. Skolnik (ed.), McGraw-Hill Book Company, New York, 1970.
30. Collins, J. R., and T. E. Harwell: RASSR Array Comes of Age, *Microwaves*, vol. 11, pp. 36–42, August, 1972.
31. Anonymous: Phased-Array Power Amplifier, *Microwave J.*, vol. 17, p. 28, February, 1974.
32. Harper, A. Y.: General Considerations in the Selection of an All Solid State or Power Tube Approach for a Phased Array Radar Application, *IEE Conf. Publ. no. 105*, "Radar—Present and Future," Oct. 23–25, 1973, pp. 112–117.
33. Gelnovatch, V. G.: A Review of Microwave Transistors for Radar Application, *NEREM Rec.*, 1974, pt. 4, pp. 116–125.
34. Ronconi, R.: *L*-band Solid-state Power Amplifier, *Revista Tecnica Selenia*, vol. 2, no. 3, pp. 21–32, 1975.
35. Davis, M. E., J. K. Smith, and C. E. Grove: *L*-band T/R Module for Airborne Phased Array, *Microwave J.*, vol. 20, pp. 54–61, February, 1977.
36. Kawamoto, H., H. J. Prager, E. L. Allen, Jr., and V. A. Mikenas: Advances in High Power, *S*-Band Trapatt-Diode Amplifier Design, *Microwave J.*, vol. 17, pp. 41–44, February, 1974.
37. Cohen, E. D.: Trapatts and Impatts: Current Status and Future Impact on Military Systems, *IEEE EASCON'75*, pp. 130-A to 130-G.
38. Lain, C. M., and E. J. Gersten: AN/TPS-59 System, *IEEE 1975 International Radar Conf.*, April 21–23, 1975, pp. 527–532, IEEE Publication 75 CHO 938-1 AES.
39. Gerard, W. A.: Frequency-Agile Coaxial Magnetrons, *Microwave J.*, vol. 16, pp. 29–33, March, 1973.
40. Day, W. R., Jr., and T. H. Luchsinger: New Developments in Electrostatically Focused Klystrons, *Microwave J.*, vol. 13, pp. 59–63, April, 1970.

41. Metivier, R. L.: Broadband Klystrons for Multimegawatt Radars, *Microwave J.*, vol. 14, pp. 29–32, April, 1971.
42. Acker, A.: Selecting Tubes for Airborne Radars, *Microwave System News*, vol. 7, pp. 55–60, May, 1977.
43. Kaisel, S. F.: Microwave Tube Technology Review, *Microwave J.*, vol. 20, pp. 23–42, July, 1977.
44. Butler, N.: The Microwave Tubes Reliability Problem, *Microwave J.*, vol. 16, pp. 41–42, March, 1973.
45. Cohen, E. D.: Trapatts and Impatts—State of the Art and Applications, *Microwave J.*, vol. 20, pp. 22–28, February, 1977.
46. DiLorenzo, J. V.: GaAs FET Development—Low Noise and High Power, *Microwave J.*, vol. 21, pp. 39–44, February, 1978.
47. Bates, D. J., R. I. Knight, and S. Spinella: Electron-Bombarded Semiconductor Devices, "Advances in Electronics and Electron Physics, vol. 44," Academic Press, Inc., New York, 1977.
48. Staprans, A.: High-Power Microwave Tubes, IEEE International Electron Devices Meeting Technical Digest 1976, Washington, D.C., pp. 245–248, IEEE Pub. no. 76 CH 1151-OED.
49. Pickering, A. H.: Electronic Tuning of Magnetrons, *Microwave J.*, vol. 22, pp. 73–78, July, 1979.
50. Staprans, A.: Linear Beam Tubes, chap. 22 of "Radar Technology," E. Brookner (ed.), Artech House, Inc., Dedham, Mass., 1977.

SEVEN

RADAR ANTENNAS

7.1 ANTENNA PARAMETERS[1,2]

The purpose of the radar antenna is to act as a transducer between free-space propagation and guided-wave (transmission-line) propagation. The function of the antenna during transmission is to concentrate the radiated energy into a shaped beam which points in the desired direction in space. On reception the antenna collects the energy contained in the echo signal and delivers it to the receiver. Thus the radar antenna is called upon to fulfill reciprocal but related roles. In the radar equation derived in Chap. 1 [Eq. (1.7)] these two roles were expressed by the transmitting gain and the effective receiving aperture. The two parameters are proportional to one another. An antenna with a large effective receiving aperture implies a large transmitting gain.

The large apertures required for long-range detection result in narrow beamwidths, one of the prime characteristics of radar. Narrow beamwidths are important if accurate angular measurements are to be made or if targets close to one another are to be resolved. The advantage of microwave frequencies for radar application is that with apertures of relatively small physical size, but large in terms of wavelengths, narrow beamwidths can be obtained conveniently.

Radar antennas are characterized by directive beams which are scanned, usually rapidly. The parabolic reflector, well known in optics, has been extensively employed in radar. The vast majority of radar antennas use the parabolic reflector in one form or another. Microwave lenses have also found some radar application, as have mechanically rotated array antennas. The electronically scanned phased array, described in Chap. 8, is an antenna with unique properties that has been of particular interest for radar application.

In this chapter, the radar antenna will be considered either as a transmitting or a receiving device, depending on which is more convenient for the particular discussion. Results obtained for one may be readily applied to the other because of the reciprocity theorem of antenna theory.[1]

Directive gain. A measure of the ability of an antenna to concentrate energy in a particular direction is called the *gain*. Two different, but related definitions of antenna gain are the *directive gain* and the *power gain*. The former is sometimes called the *directivity*, while the latter is often simply called the *gain*. Both definitions are of interest to the radar systems engineer. The directive gain is descriptive of the antenna pattern, but the power gain is more appropriate for use in the radar equation.

The directive gain of a transmitting antenna may be defined as

$$G_D = \frac{\text{maximum radiation intensity}}{\text{average radiation intensity}} \tag{7.1}$$

where the radiation intensity is the power per unit solid angle radiated in the direction (θ, ϕ) and is denoted $P(\theta, \phi)$. A plot of the radiation as a function of the angular coordinates is called a *radiation-intensity pattern*. The power density, or power per unit area, plotted as a function of angle is called a *power pattern*. The power pattern and the radiation-intensity pattern are identical when plotted on a relative basis, that is, when the maximum is normalized to a value of unity. When plotted on a relative basis both are called the *antenna radiation pattern*.

An example of an antenna radiation pattern for a paraboloid antenna is shown plotted in Fig. 7.1.[3] The main lobe is at zero degrees. The first irregularity in this particular radiation pattern is the vestigial lobe, or "shoulder," on the side of the main beam. The vestigial lobe does not always appear in antenna radiation patterns. It can result from an error in the aperture illumination and is generally undesired. In most antennas the first sidelobe appears instead. The first sidelobe is smeared into a vestigial lobe as in Fig. 7.1 if the phase distribution across the aperture is not constant. Following the first sidelobe are a series of minor lobes which decrease in intensity with increasing angular distance from the main lobe. In the vicinity of broadside (in this example 100 to 115°), spillover radiation from the feed causes the sidelobe level to rise. This is due to energy radiated from the feed which is not intercepted by the reflector. The radiation pattern also has a pronounced lobe in the backward direction (180°) due to diffraction effects of the reflector and to direct leakage through the mesh reflector surface.

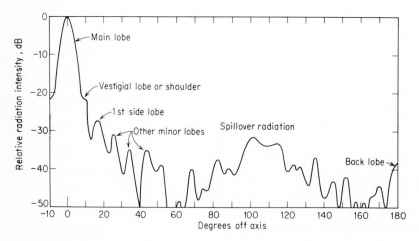

Figure 7.1 Radiation pattern for a particular paraboloid reflector antenna illustrating the main beam and the sidelobe radiation. (*After Cutler et al.,*[3] *Proc. IRE.*)

The radiation pattern shown in Fig. 7.1 is plotted as a function of one angular coordinate, but the actual pattern is a plot of the radiation intensity $P(\theta, \phi)$ as a function of the two angles θ and ϕ. The two angle coordinates commonly employed with ground-based antennas are azimuth and elevation, but any other convenient set of angles can be used.

A complete three-dimensional plot of the radiation pattern is not always necessary. For example, an antenna with a symmetrical pencil-beam pattern can be represented by a plot in one angular coordinate. The radiation-intensity pattern for rectangular apertures can often be written as the product of the radiation-intensity patterns in the two coordinate planes; for instance,

$$P(\theta, \phi) = P(\theta, 0)P(0, \phi)$$

The complete radiation pattern can be specified from the two single-coordinate radiation patterns in the θ plane and the ϕ plane.

Since the average radiation intensity over a solid angle of 4π radians is equal to the total power radiated divided by 4π, the directive gain as defined by Eq. (7.1) can be written as

$$G_D = \frac{4\pi(\text{maximum power radiated/unit solid angle})}{\text{total power radiated by the antenna}} \tag{7.2}$$

This equation indicates the procedure whereby the directive gain may be found from the radiation pattern. The maximum power per unit solid angle is obtained simply by inspection, and the total power radiated is found by integrating the volume contained under the radiation pattern. Equation (7.2) can be written as

$$G_D = \frac{4\pi P(\theta, \phi)_{\text{max}}}{\iint P(\theta, \phi)\, d\theta\, d\phi} = \frac{4\pi}{B} \tag{7.3}$$

where B is defined as the beam area:

$$B = \frac{\iint P(\theta, \phi)\, d\theta\, d\phi}{P(\theta, \phi)_{\text{max}}} \tag{7.4}$$

The beam area is the solid angle through which all the radiated power would pass if the power per unit solid angle were equal to $P(\theta, \phi)_{\text{max}}$ over the beam area. It defines, in effect, an equivalent antenna pattern. If θ_B and ϕ_B are the beamwidths in the two orthogonal planes, the beam area B is approximately equal to $\theta_B \phi_B$. Substituting into Eq. (7.3) gives

$$G_D \approx \frac{4\pi}{\theta_B \phi_B} \tag{7.5a}$$

Another expression for the gain sometimes used is

$$G_D \approx \frac{\pi^2}{\theta_B \phi_B} \tag{7.5b}$$

This was derived assuming a gaussian beamshape and with θ_B, ϕ_B defined as the half-power beamwidths.[137]

Power gain. The definition of directive gain is based primarily on the shape of the radiation pattern. It does not take account of dissipative losses. The power gain, which will be denoted by G, includes the antenna dissipative losses, but does not involve system losses arising from mismatch of impedance or of polarization. It can be defined similarly to the definition of directive gain in Eq. (7.2), except that the denominator is the net power *accepted* by the antenna from the connected transmitter, or

$$G = \frac{4\pi(\text{maximum power radiated/unit solid angle})}{\text{net power accepted by the antenna}} \qquad (7.6a)$$

An equivalent definition is

$$G = \frac{\text{maximum radiation intensity from subject antenna}}{\text{radiation intensity from (lossless) isotropic source with same power input}} \qquad (7.6b)$$

The power gain should be used in the radar equation since it includes the losses introduced by the antenna. The directive gain, which is always greater than the power gain, is of importance for coverage, accuracy, or resolution considerations and is more closely related to the antenna beamwidth. The difference between the two antenna gains is usually small. The power gain and the directive gain may be related by the radiation efficiency factor ρ_r as follows:

$$G = \rho_r G_D \qquad (7.7)$$

The radiation efficiency is also the ratio of the total power radiated by the antenna to the net power accepted by the antenna at its terminals. The difference between the total power radiated and the net power accepted is the power dissipated within the antenna. The radiation efficiency is an inherent property of an antenna and is not dependent on such factors as impedance or polarization match.

The relationship between the gain and the beamwidth of an antenna depends on the distribution of current across the aperture. For a "typical" reflector antenna the following expression is sometimes used:

$$G \simeq \frac{20,000}{\theta_B \phi_B} \qquad (7.8)$$

where θ_B and ϕ_B are the half-power beamwidths, in degrees, measured in the two principal planes. This is a rough rule of thumb that can be used when no other information is available, but it should not be a substitute for more exact expressions that acount for the actual aperture illumination.

The definitions of the directive and the power gains have been in terms of the maximum radiation intensity. Thus, the gains so defined describe the maximum concentration of radiated energy. It is also common to speak of the gain as a function of angle. Quite often the ordinate of a radiation pattern is given as the gain normalized to unity and called *relative gain*. Unfortunately the term *gain* is used to denote both the peak gain and the gain as a function of angle. Confusion as to which meaning is correct can usually be resolved from the context.

The definitions of power gain and directive gain were described above in terms of a transmitting antenna. One of the fundamental theorems of antenna theory concerns reciprocity. It states that under certain conditions (usually satisfied in radar practice) the transmitting and receiving patterns of an antenna are the same.[1] Thus the gain definitions apply equally well whether the antenna is used for transmission or for reception. The only practical distinction which must be made between transmitting and receiving antennas is that the transmitting antenna must be capable of withstanding greater power.

Effective aperture. Another useful antenna parameter related to the gain is the effective receiving aperture, or effective area. It may be regarded as a measure of the effective area presented by the antenna to the incident wave. The gain G and the effective area A_e of a *lossless* antenna are related by

$$G = \frac{4\pi A_e}{\lambda^2} = \frac{4\pi \rho_a A}{\lambda^2} \tag{7.9}$$

$$A_e = \rho_a A$$

where λ = wavelength
  A = physical area of antenna
  ρ_a = antenna aperture efficiency

Elliptical $\vec{E_2}\cos\omega t - \vec{E_1}\sin\omega t$

Polarization. The direction of polarization is defined as the direction of the electric field vector. Most radar antennas are linearly polarized; that is, the direction of the electric field vector is either vertical or horizontal. The polarization may also be elliptical or circular. Elliptical polarization may be considered as the combination of two linearly polarized waves of the same frequency, traveling in the same direction, which are perpendicular to each other in space. The relative amplitudes of the two waves and the phase relationship between them can assume any values. If the amplitudes of the two waves are equal, and if they are 90° out of (time) phase, the polarization is circular. Circular polarization and linear polarization are special cases of elliptical polarization. The degree of elliptical polarization is often described by the *axial ratio*, which is the ratio of the major axis to the minor axis of the polarization ellipse.

Linear polarization is most often used in conventional radar antennas since it is the easiest to achieve. The choice between horizontal and vertical linear polarization is often left to the discretion of the antenna designer, although the radar systems engineer might sometimes want to specify one or the other, depending upon the importance of ground reflections. For example, horizontal polarization might be employed with long range air-search radars operating at VHF or UHF so as to obtain longer range because of the reinforcement of the direct radiation by the ground-reflected radiation, Sec. 12.2. Circular polarization is often desirable in radars which must "see" through weather disturbances.

Sidelobe radiation. An example of sidelobe radiation from a typical antenna was shown in Fig. 7.1. Low sidelobes are generally desired for radar applications. If too large a portion of the radiated energy were contained in the sidelobes, there would be a reduction in the main-beam energy, with a consequent lowering of the maximum gain.

No general rule can be given for specifying the optimum sidelobe level. This depends upon the application and how difficult it is for the antenna designer to achieve low sidelobes. If the sidelobes are too high, strong echo signals can enter the receiver and appear as false targets. A high sidelobe level makes jamming of the radar easier. Also, the radar is more subject to interference from nearby friendly transmitters.

The first sidelobe nearest the main beam is generally the highest. A typical parabolic reflector antenna fed from a waveguide horn might have a first sidelobe 23 to 28 dB below the main beam. Lower first sidelobes require a highly tapered aperture illumination, one with the illumination at the edge of the aperture considerably less than that at the center. It is not easy to obtain with a reflector antenna the precisely tapered aperture illuminations necessary for low sidelobe radiation. Because of its many radiating elements, an array antenna is better suited for achieving the low-sidelobe aperture illumination than is a reflector. First sidelobes of from 40 to 50 dB below the main beam may be possible with the proper aperture illuminations and the proper care in implementation. A high-gain antenna is usually necessary to achieve such low sidelobes. (When referred to the radiation from an isotropic antenna, the peak sidelobes from a low-sidelobe antenna might be approximately 10 to 15 dB below the isotropic level.) The achievement of extremely low sidelobes requires the antenna to be well constructed so as to maintain the necessary mechanical and electrical tolerances. It takes only a small deviation of the antenna surface to have an increase of the peak sidelobe of a very low sidelobe antenna. Furthermore, there must be no obstructions in the vicinity of the antenna that can divert energy to the sidelobe regions and appear as high sidelobes. A low sidelobe antenna might have to be 20 to 30 percent larger than a conventional antenna to achieve the same beamwidth.

Aperture efficiency. The aperture efficiency is the ratio of the actual antenna directivity to the maximum possible directivity. Maximum directivity is achieved with a uniform aperture illumination.[1] Although it might seem that the higher the aperture efficiency the better, aperture efficiency is seldom a suitable measure of the quality of a radar antenna. Other factors are usually more important. For instance, the high sidelobes that accompany a uniform illumination are seldom desired, and the aperture efficiency is usually willingly sacrificed for lower sidelobes. When a shaped beam is desired in a surveillance radar, such as a cosecant-squared pattern, again it is more important to achieve the overall pattern required rather than simply maximize the directivity at the peak of the beam.

Aperture efficiency is a measure of the radiation intensity only at the center of the beam. In a search radar, however, the radiation intensity throughout the entire beam is of interest, not just that at the beam center. A number of hits are received as the antenna scans by each target. The detection decision is based on the energy from all the hits received and not just on the energy received when the center of the beam illuminates the target. Thus it is not the maximum directivity which is important, but the total directivity integrated over the number of hits processed by the radar. A better criterion for selecting the aperture illumination might be one which maximizes the radiated energy within a specified angular region.[5] Such illuminations are more typical of radar antenna practice than the uniformly illuminated aperture.

In a monopulse tracking antenna, uniform illumination might be desirable to maximize directivity when the target is being tracked by the center of the beam. However, a monopulse antenna with a uniform illumination for the sum-pattern will have a poor difference-pattern, even if the high sidelobes can be tolerated. A compromise must be made, and something other than a uniform illumination is generally selected.

Thus the parameter of aperture efficiency, which sometimes is held sacred by the antenna designer and to some who write antenna specifications, is often of secondary importance to the systems engineer wishing to optimize total radar performance. It can usually be traded for some more important characteristic.

7.2 ANTENNA RADIATION PATTERN AND APERTURE DISTRIBUTION

The electric-field intensity $E(\phi)$ produced by the radiation emitted from the antenna is a function of the amplitude and the phase of the current distribution across the aperture.[1,4,5] $E(\phi)$ may be found by adding vectorially the contribution from the various current elements constituting the aperture. The mathematical summation of all the contributions from the current elements contained within the aperture gives the field intensity in terms of an integral. This integral cannot be readily evaluated in the general case. However, approximations to the solution may be had by dividing the area about the antenna aperture into three regions as determined by the mathematical approximations that must be made. The demarcations among these three regions are not sharp and blend one into the other.

The region in the immediate neighborhood of the aperture is the *near field*. It extends several antenna diameters from the aperture and, for this reason, is usually of little importance to the radar engineer.

The near field is followed by the *Fresnel region*. In the Fresnel region, rays from the radiating aperture to the observation point (or target) are not parallel and the antenna radiation pattern is not constant with distance. Little application is made of the Fresnel region in radar. The *near field* and the *Fresnel region* have sometimes been called by antenna engineers the *reactive near-field region* and the *radiating near-field region*, respectively.[6]

The farthest region from the aperture is the *Fraunhofer*, or far-field, region. In the Fraunhofer region, the radiating source and the observation point are at a sufficiently large distance from each other so that the rays originating from the aperture may be considered parallel to one another at the target (observation point). Radar antennas operate in the Fraunhofer region.

The " boundary " R_F between Fresnel and Fraunhofer regions is usually taken to be either $R_F = D^2/\lambda$ or the distance $R_F = 2D^2/\lambda$, where D is the size of the aperture and λ is the wavelength, D and λ being measured in the same units. At a distance given by D^2/λ, the gain of a uniformly illuminated antenna is 0.94 that of the Fraunhofer gain at infinity. At a distance of $2D^2/\lambda$, the gain is 0.99 that at infinity.

The plot of the electric field intensity $|E(\theta, \phi)|$ is called the *field-intensity pattern* of the antenna. The plot of the square of the field intensity $|E(\theta, \phi)|^2$ is the *power radiation pattern* $P(\theta, \phi)$, defined in the previous section.

In the Fraunhofer region, the integral for electric field intensity in terms of current distribution across the aperture is given by a Fourier transform. Consider the rectangular aperture and coordinate system shown in Fig. 7.2. The width of the aperture in the z dimension is a, and the angle in the yz plane as measured from the y axis is ϕ. The far-field electric field intensity, assuming $a \gg \lambda$, is

$$F(\omega) = \int_{-\infty}^{\infty} f(t) e^{-j\omega t} dt \qquad E(\phi) = \int_{-a/2}^{a/2} A(z) \exp\left(j2\pi \frac{z}{\lambda} \sin \phi \right) dz \qquad (7.10)$$

where $A(z)$ = current at distance z, assumed to be flowing in x direction. $A(z)$, the *aperture distribution, or illumination*, may be written as a complex quantity, including both the amplitude and phase, or

$$A(z) = |A(z)| \exp j\Psi(z) \qquad (7.11)$$

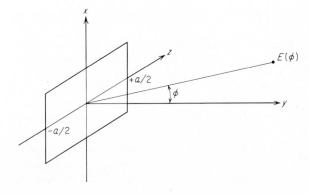

Figure 7.2 Rectangular aperture and coordinate system for illustrating the relationship between the aperture distribution and the far-field electric-field-intensity pattern.

where $|A(z)|$ = amplitude distribution and $\Psi(z)$ = phase distribution. [Equation (7.10) applies to a one-dimensional line source lying along the z axis. For the two-dimensional aperture of Fig. 7.2, $A(z)$ is the integral of $A(x, z)$ over the variable x.]

Equation (7.10) represents the summation, or integration, of the individual contributions from the current distribution across the aperture according to Huygens' principle. At an angle ϕ, the contribution from a particular point on the aperture will be advanced or retarded in phase by $2\pi(z/\lambda) \sin \phi$ radians. Each of these contributions is weighted by the factor $A(z)$. The field intensity is the integral of these individual contributions across the face of the aperture.

The aperture distribution has been defined in terms of the current i_x. It may also be defined in terms of the magnetic field component H_z for polarization in the x direction, or in terms of the electric field component E_z for polarization in the z direction, provided these field components are confined to the aperture.[7]

The expression for the electric field intensity [Eq. (7.10)] is mathematically similar to the inverse Fourier transform. Therefore the theory of Fourier transforms can be applied to the calculation of the radiation or field-intensity patterns if the aperture distribution is known. The Fourier transform of a function $f(t)$ is defined as

$$F(f) = \int_{-\infty}^{\infty} f(t) \exp(-j2\pi ft)\, dt \tag{7.12}$$

and the inverse Fourier transform is

$$f(t) = \int_{-\infty}^{\infty} F(f) \exp(j2\pi ft)\, df \tag{7.13}$$

The limits of Eq. (7.10) can be extended over the infinite interval from $-\infty$ to $+\infty$ since the aperture distribution is zero beyond $z = \pm a/2$.

The Fourier transform permits the aperture distribution $A(z)$ to be found for a given field-intensity pattern $E(\phi)$, since

$$A(z) = \frac{1}{\lambda} \int_{-\infty}^{\infty} E(\phi) \exp\left(-j2\pi \frac{z}{\lambda} \sin \phi\right) d(\sin \phi) \tag{7.14}$$

This may be used as a basis for synthesizing an antenna pattern, that is, finding the aperture distribution $A(z)$ which yields a desired antenna pattern $E(\phi)$.

In the remainder of this section, the antenna radiation pattern will be examined for various aperture distributions using Eq. (7.10). It will be assumed that the phase distribution across the aperture is constant and only the effects of the amplitude distribution need be considered.

The inverse Fourier transform gives the electric field intensity when the phase and amplitude of the distribution across the aperture are known. The aperture is defined as the projection of the antenna on a plane perpendicular to the direction of propagation. It does not matter whether the distribution is produced by a reflector antenna, a lens, or an array.

One-dimensional aperture distribution. Perhaps the simplest aperture distribution to conceive is the uniform, or rectangular, distribution. The uniform distribution is constant over the aperture extending from $-a/2$ to $+a/2$ and zero outside. For present purposes it will be assumed that the aperture extends in one dimension only. This might represent the distribution across a line source or the distribution in one plane of a rectangular aperture. If the constant value of the aperture distribution is equal to A_0 and if the phase distribution across the aperture is constant, the antenna pattern as computed from Eq. (7.10) is

$$E(\phi) = A_0 \int_{-a/2}^{a/2} \exp j\left(2\pi \frac{z}{\lambda} \sin \phi\right) dz$$

$$= \frac{A_0 \sin\left[\pi(a/\lambda)\sin\phi\right]}{(\pi/\lambda)\sin\phi} = A_0 a \frac{\left[\sin\pi(a/\lambda)\sin\phi\right]}{\pi(a/\lambda)\sin\phi} \qquad (7.15)$$

Normalizing to make $E(0) = 1$ results in $A_0 = 1/a$; therefore

$$E(\phi) = \frac{\sin\left[\pi(a/\lambda)\sin\phi\right]}{\pi(a/\lambda)\sin\phi} \qquad (7.16)$$

This pattern, which is of the form $(\sin x)/x$, is shown by the solid curve in Fig. 7.3. The intensity of the first sidelobe is 13.2 dB below that of the peak. The angular distance between the nulls adjacent to the peak is $2\lambda/a$ rad, and the beamwidth as measured between the half-power points is $0.88\lambda/a$ rad, or $51\lambda/a$ deg. The voltage pattern of Eq. (7.16) is positive over the entire main lobe, but changes sign in passing through the first zero, returning to a positive value in passing through the second zero, and so on. The odd-numbered sidelobes are therefore out of phase with the main lobe, and the even-numbered ones are in phase. Also shown in Fig. 7.3 is the radiation pattern for the cosine aperture distribution.

$$A(z) = \cos \frac{\pi z}{a} \qquad |z| < \frac{a}{2}$$

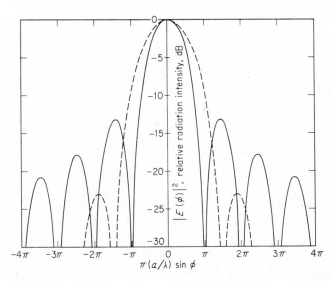

Figure 7.3 The solid curve is the antenna radiation pattern produced by a uniform aperture distribution; the dashed curve represents the antenna radiation pattern of an aperture distribution proportional to the cosine function.

The normalized radiation pattern is

$$E(\phi) = \frac{\pi}{4} \left[\frac{\sin (\psi + \pi/2)}{\psi + \pi/2} + \frac{\sin (\psi - \pi/2)}{\psi - \pi/2} \right] \tag{7.17}$$

where $\psi = \pi(a/\lambda) \sin \phi$. In Fig. 7.3 the gains of both patterns are normalized. However, the maximum gain of the pattern resulting from the cosine distribution is 0.9 dB less than the gain of a uniform distribution.

Table 7.1 lists some of the properties of the radiation patterns produced by various aperture distributions.[1] The aperture distributions are those which can be readily expressed in analytic form and for which the solution of the inverse Fourier transform of Eq. (7.10) can be conveniently carried out. Although these may not be the distributions employed with practical radar antennas, they serve to illustrate how the aperture distribution affects the antenna pattern. More complicated distributions which cannot be readily found from available tables of Fourier transforms or which cannot be expressed in analytical form may be determined by numerical computation methods or machine computation. (The Taylor distribution, mentioned in Sec. 7.6, is a more popular model for antenna design than the analytical models of Table 7.1.)

An examination of the information presented in this table reveals that the gain of the uniform distribution is greater than the gain of any other distribution. It is shown by Silver[1] that the uniform distribution is indeed the most efficient aperture distribution, that is, the one which maximizes the antenna gain. Therefore the relative-gain column may be considered as the efficiency of a particular aperture distribution as compared with the uniform, or most efficient, aperture distribution. The relative gain is also called the *aperture efficiency* [Eq. (7.9)]. The aperture efficiency times the physical area of the aperture is the *effective aperture*.

Table 7.1. Radiation-pattern characteristics produced by various aperture distributions

λ = wavelength; a = aperture width

Type of distribution, $\|z\| < 1$	Relative gain	Half-power beamwidth, deg	Intensity of first sidelobe, dB below maximum intensity
Uniform; $A(z) = 1$	1	$51\lambda/a$	13.2
Cosine; $A(z) = \cos^n (\pi z/2)$:			
$n = 0$	1	$51\lambda/a$	13.2
$n = 1$	0.810	$69\lambda/a$	23
$n = 2$	0.667	$83\lambda/a$	32
$n = 3$	0.575	$95\lambda/a$	40
$n = 4$	0.515	$111\lambda/a$	48
Parabolic; $A(z) = 1 - (1 - \Delta)z^2$:			
$\Delta = 1.0$	1	$51\lambda/a$	13.2
$\Delta = 0.8$	0.994	$53\lambda/a$	15.8
$\Delta = 0.5$	0.970	$56\lambda/a$	17.1
$\Delta = 0$	0.833	$66\lambda/a$	20.6
Triangular; $A(z) = 1 - \|z\|$	0.75	$73\lambda/a$	26.4
Circular; $A(z) = \sqrt{1 - z^2}$	0.865	$58.5\lambda/a$	17.6
Cosine-squared plus pedestal;			
$0.33 + 0.66 \cos^2(\pi z/2)$	0.88	$63\lambda/a$	25.7
$0.08 + 0.92 \cos^2(\pi z/2)$, Hamming	0.74	$76.5\lambda/a$	42.8

Uniform illumination. highest gain, narrowest beamwidth

Another property of the radiation patterns illustrated by Table 7.1 is that the antennas with the lowest sidelobes (adjacent to the main beam) are those with aperture distributions in which the amplitude tapers to a small value at the edges. The greater the amplitude taper, the lower the sidelobe level but the less the relative gain and the broader the beamwidth. Thus low sidelobes and good aperture efficiency run counter to one another.

A word of caution should be given concerning the ability to achieve in practice low sidelobe levels with extremely tapered illuminations. It was assumed in the computation of these radiation patterns that the distribution of the phase across the aperture was constant. In a practical antenna this will not necessarily be true since there will always be some unavoidable phase variations caused by the inability to fabricate the antenna as desired. Any practical device is never perfect; it will always be constructed with some error, albeit small. The phase variations due to the unavoidable errors can cause the sidelobe level to be raised and the gain to be lowered. There is a practical limit beyond which it becomes increasingly difficult to achieve low sidelobes even if a considerable amplitude taper is used.

Circular aperture.[5,8] The examples of aperture distribution presented previously in this section applied to distributions in one dimension. We shall consider here the antenna pattern produced by a two-dimensional distribution across a circular aperture. The polar coordinates (r, θ) are used to describe the aperture distribution $A(r, \theta)$, where r is the radial distance from the center of the circular aperture, and θ is the angle measured in the plane of the aperture with respect to a reference. Huygens' principle may be applied in the far field by dividing the plane wave across the circular aperture into a great many spherical wavelets, all of the same phase but of different amplitude. To find the field intensity at a point a distance R from the antenna, the amplitudes of all the waves are added at the point, taking account of the proper phase relationships due to the difference in path lengths. The field intensity at a distance R is thus proportional to

$$E(R) = \int_0^{2\pi} d\theta \int_0^{r_0} A(r, \theta) \exp\left(j\frac{2\pi R}{\lambda}\right) r \, dr \tag{7.18}$$

where r_0 is the radius of the aperture. For a circular aperture with uniform distribution, the field intensity is proportional to

$$E(\phi) = \int_0^{2\pi} d\theta \int_0^{r_0} \exp\left(j2\pi \frac{r}{\lambda} \sin \phi \cos \theta\right) r \, dr = \pi r_0^2 2J_1(\xi)/\xi \tag{7.19}$$

where $\xi = 2\pi(r_0/\lambda) \sin \phi$ and $J_1(\xi) =$ first-order Bessel function. A plot of the normalized radiation pattern is shown in Fig. 7.4. The first sidelobe is 17.5 dB below the main lobe, and the beamwidth is $58.5\lambda/D$.

The effect of tapering the amplitude distribution of a circular aperture is similar to tapering the distribution of a linear aperture. The sidelobes may be reduced, but at the expense of broader beamwidth and less antenna gain. One aperture distribution which has been considered in the past[1] is $[1 - (r/r_0)^2]^p$, where $p = 0, 1, 2, \ldots$. The radiation pattern is of the form $J_{p+1}(\xi)/\xi^{p+1}$. When $p = 0$, the distribution is uniform and the radiation pattern reduces to that given above. For $p = 1$, the gain is reduced 75 percent, the half-power beamwidth is $72.6\lambda/D$, and the first sidelobe is 24.6 dB below the maximum. The sidelobe level is 30.6 dB down for $p = 2$, but the gain relative to a uniform distribution is 56 percent. Additional properties of this distribution can be found in Ref. 1, table 6.2, and in Ref. 5, table 1.

Aperture blocking.[9-12] An obstacle in front of an antenna can alter the aperture illumination and the radiated pattern. This is called aperture blocking or shadowing. The chief example is

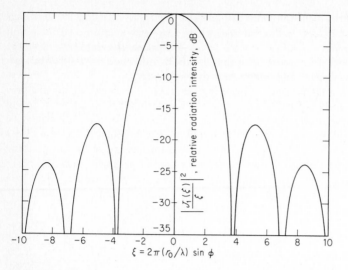

Figure 7.4 Radiation pattern for a uniformly illuminated circular aperture.

the blocking caused by the feed and its supports in reflector-type antennas. Aperture blocking degrades the performance of an antenna by lowering the gain, raising the sidelobes, and filling in the nulls. The effect of aperture blocking can be approximated by subtracting the antenna pattern produced by the obstacle from the antenna pattern of the undisturbed aperture. This procedure is possible because of the linearity of the Fourier-transform that relates the aperture illumination and the radiated pattern. An example[9] of the effect of aperture blocking caused by the feed in a paraboloid-reflector antenna is shown in Fig. 7.5. Buildings in the vicinity of ground-based radar, and masts and other obstructions in the vicinity of shipboard antennas can also degrade the radiation pattern because of aperture blocking.[131]

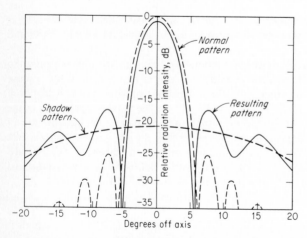

Figure 7.5 Effect of aperture blocking caused by the feed in a parabolic-reflector antenna. (*From C. Cutler,*[9] *Proc. IRE.*)

Broadband signals. The Fourier-integral-transform relationship between the radiation pattern $E(\phi)$ and the aperture distribution $A(z)$ as expressed in Eqs. (7.11) and (7.14) applies only when the signal is a CW sine wave. If the signal were a pulse or some other radar waveform with a spectrum of noninfinitesimal width, the simple Fourier integral which applies to a CW sine wave would not give the correct radiation pattern nor would it predict the transient behavior. In most cases of practical interest the spectral width of the signal is relatively small, with the consequence that the pattern is not affected appreciably and the Fourier-integral relationships are satisfactory approximations. However, when the reciprocal of the signal bandwidth is comparable with the time taken by a radar wave to transverse the antenna aperture, bandwidth effects can be important and signal distortion may result.

7.3 PARABOLIC-REFLECTOR ANTENNAS

One of the most widely used microwave antennas is the parabolic reflector (Fig. 7.6). The parabola is illuminated by a source of energy called the feed, placed at the focus of the parabola and directed toward the reflector surface. The parabola is well suited for microwave antennas because (1) any ray from the focus is reflected in a direction parallel to the axis of the parabola and (2) the distance traveled by any ray from the focus to the parabola and by reflection to a plane perpendicular to the parabola axis is independent of its path. Therefore a point source of energy located at the focus is converted into a plane wavefront of uniform phase.

The basic parabolic contour has been used in a variety of configurations. Rotating the parabolic curve shown in Fig. 7.6 about its axis produces a parabola of revolution called a circular parabola, or a paraboloid. When properly illuminated by a point source at the focus, the paraboloid generates a nearly symmetrical pencil-beam antenna pattern. Its chief application has been for tracking-radar antennas.

An asymmetrical beam shape can be obtained by using only a part of the paraboloid. This type of antenna, an example of which is shown in Fig. 7.25, is widely used when fan beams are desired.

Another means of producing either a symmetrical or an asymmetrical antenna pattern is with the parabolic cylinder.[1,7,13] The parabolic cylinder is generated by moving the parabolic contour parallel to itself. A line source such as a linear array, rather than a point source, must be used to feed the parabolic cylinder. The beamwidth in the plane containing the linear feed is

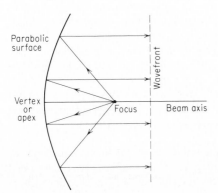

Figure 7.6 Parabolic-reflector antenna.

determined by the illumination of the line source, while the beamwidth in the perpendicular plane is determined by the illumination across the parabolic profile. The reflector is made longer than the linear feed to avoid spillover and diffraction effects. One of the advantages of the parabolic cylinder is that it can readily generate an asymmetrical fan beam with a much larger aspect ratio (length to width) than can a section of a paraboloid. It is not practical to use a paraboloidal reflector with a single horn feed for aspect ratios greater than about 8 : 1, although it is practical to use the parabolic cylinder for aspect ratios of this magnitude or larger. Another advantage of the parabolic cylinder antenna is that the line feed allows better control of the aperture illumination than does a single point source feeding a paraboloid. The patterns in the two orthogonal planes can be controlled separately, which is of importance for generating shaped beams. Also there is usually less depolarization in a parabolic cylinder than in a paraboloid fed from a point source. Since a directive feed is used with a parabolic cylinder, leakage through a mesh reflector will cause a higher backlobe than would a point-source feed. Therefore, solid reflector-surfaces are generally employed. When the feed must be pressurized in order to sustain high power it is often easier to do so with a paraboloid fed by a single point-source than with the larger line feed of a parabolic cylinder.

Still another variation of the parabola is the parabolic torus shown in Fig. 7.16 and discussed in Sec. 7.4. It is generated by moving the parabolic contour over an arc of a circle whose center is on the axis of the parabola. It is useful where a scan angle less than 120° is required and where it is not convenient to scan the reflector itself. Scanning is accomplished in the parabolic torus by moving the feed.

There are other variations of parabolic reflectors such as cheeses, pillboxes, and hoghorns, descriptions of which may be found in the literature.[1,7]

Feeds for paraboloids.[1,9,130] The ideal feed for a paraboloid consists of a point source of illumination with a pattern of proper shape to achieve the desired aperture distribution. It is important in a paraboloid that the phase of the radiation emitted by the feed be independent of the angle. The radiation pattern produced by the feed is called the *primary pattern*; the radiation pattern of the aperture when illuminated by the feed is called the *secondary pattern*.

A simple half-wave dipole or a dipole with a parasitic reflector can be used as the feed for a paraboloid. A dipole is of limited utility, however, because it is difficult to achieve the desired aperture illuminations, it has poor polarization properties in that some of the energy incident on the reflector is converted to the orthogonal polarization, and it is limited in power. The open-ended waveguide as the feed for a paraboloid directs the energy better than a dipole, and the phase characteristic is usually good if radiating in the proper mode. A circular paraboloid might be fed by a circular, open-ended waveguide operating in the TE_{11} mode. A rectangular guide operating in the TE_{10} mode does not give a circularly symmetric radiation pattern since the dimensions in the E and H planes, as well as the current distributions in these two planes, are different. As this is generally true of most waveguide feeds, a perfectly symmetrical antenna pattern is difficult to achieve in practice. The rectangular guide may be used, however, for feeding an asymmetrical section of a paraboloid that generates a fan beam wider in the H plane than in the E plane.

When more directivity is required than can be obtained with a simple open-ended waveguide, some form of waveguide horn may be used. The waveguide horn is probably the most popular method of feeding a paraboloid for radar application.

Optimum feed illumination angle. If the radiation pattern of the feed is known, the illumination of the aperture can be determined and the resulting secondary beam pattern can be found by evaluating a Fourier integral or performing a numerical calculation. The radiation pattern

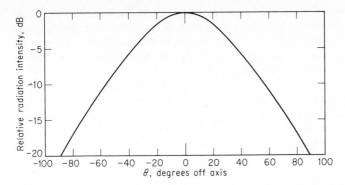

Figure 7.7 Radiation pattern of 0.84λ-diameter circular-waveguide aperture. (*From C. Cutler,*[9] *Proc. IRE.*)

of a 0.84λ-diameter circular waveguide is shown in Fig. 7.7. If one wished to obtain relatively uniform illumination across a paraboloid aperture with a feed of this type, only a small angular portion of the pattern should be used. An antenna with a large ratio of focal distance to antenna diameter would be necessary to achieve a relatively uniform illumination across the aperture. Also, a significant portion of the energy radiated by the feed would not intercept the paraboloid and would be lost. The lost "spillover" energy results in a lowering of the overall efficiency and defeats the purpose of the uniform illumination (maximum aperture efficiency). On the other hand, if the angle subtended by the paraboloid at the focus is large, more of the radiation from the feed will be intercepted by the reflector. The less the spillover, the higher the efficiency. However, the illumination is more tapered, causing a reduction in the aperture efficiency. Therefore, there will be some angle at which these two counteracting effects result in maximum efficiency. This is illustrated in Fig. 7.8 for the circular-waveguide feed whose pattern is shown in Fig. 7.7. The maximum of the curve is relatively broad, so that the optimum angle subtended by the antenna at the focus is not critical. The greatest efficiency is obtained with a reflector in which the radiation from the feed in the direction of the edges is between 8 and 12 dB below that at the center. As a rough rule of thumb, the intensity of the energy

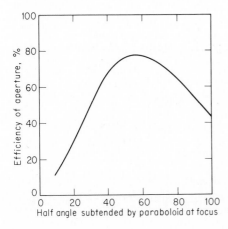

Figure 7.8 Efficiency of a paraboloid as a function of the half angle subtended by the paraboloid at the focus. (*From C. Cutler,*[9] *Proc. IRE.*)

radiated *toward* the edge of the reflector should usually be about one-tenth the maximum intensity. The aperture distribution at the edges will be even less than one-tenth the maximum because of the longer path from the feed to the edge of the reflector than from the feed to the center of the dish. When the primary feed pattern is 10 dB down at the edges, the first minor lobe in the secondary pattern is in the vicinity of 22 to 25 dB.

Calculations of the antenna efficiency based on the aperture distribution set up by the primary pattern as well as the spillover indicate theoretical efficiencies of about 80 percent for paraboloidal antennas when compared with an ideal, uniformly illuminated aperture. In practice, phase variations across the aperture, poor polarization characteristics, and antenna mismatch reduce the efficiency to the order of 55 to 65 percent for ordinary paraboloidal-reflector antennas.

Feed support. The resonant half-wave dipole and the waveguide horn can be arranged to feed the paraboloid as shown in Fig. 7.9a and b. These two arrangements are examples of rear feeds. The waveguide rear feed shown in Fig. 7.9b produces an asymmetrical pattern since the transmission line is not in the center of the dish. A rear feed not shown is the Cutler feed,[9] a dual-aperture rear feed in which the waveguide is in the center of the dish and the energy is made to bend 180° at the end of the waveguide by a properly designed reflecting plate. The rear feed has the advantage of compactness and utilizes a minimum length of transmission line. The antenna may also be fed in the manner shown in Fig. 7.9c. This is an example of a front feed. It is well suited for supporting horn feeds, but it obstructs the aperture.

Two basic limitations to any of the feed configurations mentioned above are aperture blocking and impedance mismatch in the feed. The feed, transmission line, and supporting structure intercept a portion of the radiated energy and alter the effective antenna pattern. Some of the energy reflected by the paraboloid enters the feed and acts as any other wave traveling in the reverse direction in the transmission line. Standing waves are produced along the line, causing an impedance mismatch and a degradation of the transmitter performance. The mismatch can be corrected by an impedance-matching device, but this remedy is effective only over a relatively narrow frequency band. Another technique for reducing the effect of the reflected radiation intercepted by the feed is to raise a portion of the reflecting surface at the center (apex) of the paraboloid. The raised surface is made of such a size and distance from the original reflector contour as to produce at the focus a reflected signal equal in amplitude but opposite in phase to the signal reflected from the remainder of the reflector. The two reflected signals cancel at the feed, so that there is no mismatch. The raised portion of the

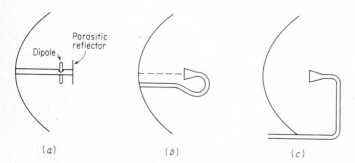

Figure 7.9 Examples of the placement of the feeds in parabolic reflectors. (a) Rear feed using half-wave dipole; (b) rear feed using horn; (c) front feed using horn.

reflector is called an *apex-matching plate*. Although the apex-matching plate has a broader bandwidth than matching devices inside the transmission line, it causes a slight reduction in the gain and increases the minor-lobe level of the radiation pattern.

Offset feed.[1] Both the aperture blocking and the mismatch at the feed are eliminated with the *offset-feed* parabolic antenna shown in Fig. 7.10. The center of the feed is placed at the focus of the parabola, but the horn is tipped with respect to the parabola's axis. The major portion of the lower half of the parabola is removed, leaving that portion shown by the solid curve in Fig. 7.10. For all practical purposes the feed is out of the path of the reflected energy, so that there is no pattern deterioration due to aperture blocking nor is there any significant amount of energy intercepted by the feed to produce an impedance mismatch.

It should be noted that the antenna aperture of an offset parabola (or any parabolic reflector) is the area projected on a plane perpendicular to its axis and is not the surface area.

The offset parabola eliminates two of the major limitations of rear or front feeds. However, it introduces problems of its own. Cross-polarization lobes are produced by the offset geometry, which may seriously deteriorate the radar system performance.[1,17] Also, it is usually more difficult to properly support and to scan an offset-feed antenna than a circular paraboloid with rear feed.

f/D ratio. An important design parameter for reflector antennas is the ratio of the focal length f to the antenna diameter D, or f/D ratio. The selection of the proper f/D ratio is based on both mechanical and electrical considerations. A small f/D ratio requires a deep-dish reflector, while a large f/D ratio requires a shallow reflector. The shallow reflector is easier to support and move mechanically since its center of gravity is closer to the vertex, but the feed must be supported farther from the reflector. The farther from the reflector the feed is placed, the narrower must be the primary-pattern beamwidth and the larger must be the feed. On the other hand, it is difficult to obtain a feed with uniform phase over the wide angle necessary to properly illuminate a reflector with small f/D. Most parabolic-reflector antennas seem to have f/D ratios ranging from 0.3 to 0.5.

Reflector surfaces. The reflecting surface may be made of a solid sheet material, but it is often preferable to use a wire screen, metal grating, perforated metal, or expanded metal mesh. The expanded metal mesh made from aluminum is a popular form. A nonsolid surface such as a mesh offers low wind resistance, light weight, low cost, ease of fabrication and assembly, and the ability to conform to variously shaped reflector surfaces.[14] However, a nonsolid surface

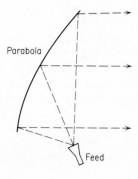

Figure 7.10 Parabolic reflector with offset feed.

may permit energy to leak through, with the result that the backlobe will increase and the antenna gain decrease.

Solid sheet surfaces may be of either aluminum or steel.[15,16] Steel is a popular choice, particularly where weight is not a controlling factor. It is cheap and strong but is relatively difficult to form. Stainless steel or plastic coated galvanized steel are used when the surface must be resistant to corrosion. Aluminum is light in weight but is more expensive than steel. Aluminum honeycomb with aluminum skin in a sandwich construction has been employed where highly accurate surfaces are required. Reflector surfaces may also be formed from fiberglass and asbestos resinated laminates with the reflecting surface made of embedded mesh or metal spray. Plastic structures have the advantage of being light, rigid, and capable of being made with highly accurate surfaces. Their thermal properties, however, require care in design and construction.

The presence of ice on the reflector surface is an important consideration for both the electrical and the mechanical design of the antenna. Ice adds to the weight of the antenna and makes it more difficult to rotate. In addition, if the ice were to close the holes of a mesh antenna so that a solid rather than an open surface is presented to the wind, bigger motors would be needed to operate the antenna. The structure also would have to be stronger.

The effect of ice on the electrical characteristics of a mesh reflecting surface is twofold.[18] On the one hand, ice which fills a part of the space between the mesh conductors may be considered a dielectric. A dielectric around the wires is equivalent to a shortening of the wavelength incident on the mesh. The spacing between wires appears wider, electrically, causing the transmission coefficient of the surface to increase. On the other hand, the total reflecting surface is increased by the presence of ice, reducing the transmission through the mesh. The relative importance of these two effects determines whether there is a net increase or a net decrease in transmission. In unfavorable cases, even strongly reflecting meshes can lose their reflecting properties almost completely.

Cassegrain antenna.[19-23] This is an adaptation to the microwave region of an optical technique invented in the seventeenth century by William Cassegrain, a contemporary of Isaac Newton. The Cassegrain principle is widely used in telescope design to obtain high magnification with a physically short telescope and allow a convenient rear location for the observer. Its application to microwave reflector antennas permits a reduction in the axial dimension of the antenna, just as in optics. But more importantly it eliminates the need for long transmission lines and allows greater flexibility in what can be placed at the focus of the antenna.

The principle of the Cassegrain antenna is shown in Fig. 7.11. It is a two-reflector system with the larger (primary) reflector having a parabolic contour and a (secondary) subreflector with a hyperbolic contour. One of the two foci of the hyperbola is the real focal point of the system. The feed is located at this point, which can be at the vertex of the parabola or in front of it. The other focus of the hyperbola is a virtual focal point and is located at the focus of the parabolic surface. Parallel rays coming from a target are reflected by the parabola as a convergent beam and are re-reflected by the hyperbolic subreflector, converging at the position of the feed. There exists a family of hyperbolic surfaces which can serve as the subreflector. The larger the subreflector, the nearer will it be to the main reflector and the shorter will be the axial dimension of the antenna assembly. However, a large subreflector results in large aperture blocking, which may be undesirable. A small subreflector reduces aperture blocking, but it has to be supported at a greater distance from the main reflector.

The geometry of the Cassegrain antenna is especially attractive for monopulse tracking radar since the RF plumbing can be placed behind the reflector to avoid blocking of the aperture. Also, the long runs of transmission line out to the feed at the focus of a conventional parabolid are avoided.

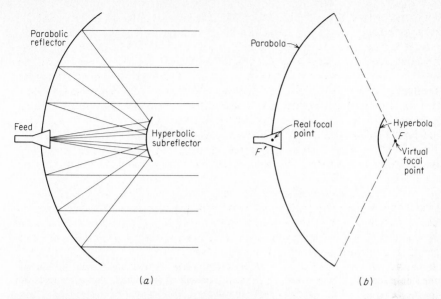

Figure 7.11 (*a*) Cassegrain antenna showing the hyperbolic subreflector and the feed at the vertex of the main parabolic reflector; (*b*) geometry of the Cassegrain antenna.

The Cassegrain antenna can, if desired, be designed to have a lower antenna noise temperature than the conventional paraboloid. This is due to the elimination of the long transmission lines from the feed to the receiver and by the fact that the spillover-sidelobes from the feed see the cold sky rather than the warm earth. Low antenna noise temperature is important in radio astronomy or space communications, but generally is of little interest in radar since extremely low noise receivers are not always desirable.

Still another advantage of the Cassegrain antenna is the flexibility with which different transmitters and receivers can be substituted because of the availability of the antenna feed system at the back of the reflector rather than out front at the normal focus. The Haystack microwave research system,[24] for example, is a fully steerable Cassegrain antenna 120 ft in diameter, enclosed by a 150-ft diameter metal space-frame randome. It was designed to operate at frequencies from 2 to 10 GHz for radar, radio astronomy, and space communications. To facilitate multifunction use, the RF components are installed in a number of plug-in modules, 8 by 8 by 12 feet in size, which mount directly behind the Cassegrain reflector.

The presence of the subreflector in front of the main reflector in the Cassegrain configuration causes aperture blocking. Part of the energy is removed, resulting in a reduction of the main-beam gain and an increase in the sidelobes. If the main reflector is circular and assumed to have a completely tapered parabolic illumination, a small circular obstacle in the center of the aperture will reduce the (power) gain by approximately $[1 - 2(D_b/D)^2]^2$, where D_b is the diameter of the obstacle (hyperbolic subreflector) and D is the diameter of the main aperture.[19] The relative (voltage) level of the first sidelobe is increased by $(2D_b/D)^2$. For example if $D_b/D = 0.122$, the gain would be lowered by about 0.3 dB and a -20 dB sidelobe would be increased to about -18 dB.

Aperture blocking may be reduced by decreasing the size of the subreflector. By making the feed more directive or by moving it closer to the subreflector, the size of the subreflector may be reduced without incurring a spillover loss. However, the feed cannot be made too large (too directive) since it partially shadows the energy reflected from the main parabolic reflector.

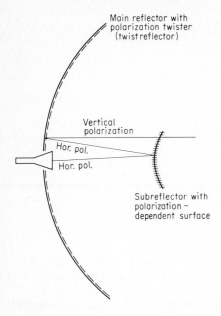

Figure 7.12 Polarization-twisting Cassegrain antenna. Aperture blocking by the subreflector is reduced with this design.

Minimum total aperture blocking occurs when the shadows produced by the subreflector and the feed are of equal area.[19] Aperture blocking is less serious with narrow beamwidths and small f/D ratios.

If operation with a single polarization is permissible, the technique diagramed in Fig. 7.12 can reduce aperture blocking. The subreflector consists of a horizontal grating of wires, called a *transreflector*, which passes vertically polarized waves with negligible attenuation but reflects the horizontally polarized wave radiated by the feed. The horizontally polarized wave reflected by the subreflector is rotated 90° by the *twist reflector* at the surface of the main reflector. The twist reflector consists of a grating of wires oriented 45° to the incident polarization and placed one-quarter wavelength from the reflector's surface.[132] (Modifications to this simple design can provide a 90° twist of the incident polarization over a broad frequency band and a wide range of incident angles.[25,133]) The wave reflected from the main reflector is vertically polarized and passes through the subreflector with negligible degradation. The subreflector is transparent to vertically polarized waves and does not block the aperture. Some aperture blocking by the feed does occur, but this can be made small. In one airborne monopulse radar antenna[23] using the polarization twisting technique, good performance was obtained over a 12 percent bandwidth. The subreflector was supported by a transparent (dielectric) cone, with resistive-card absorbers embedded in the support cone and oriented so as to reduce the cross-polarized wide-angle radiation by more than 20 dB.

Mechanical beam-scanning with planar twist-reflector. The antenna configuration shown in Fig. 7.13 may be employed to rapidly scan a beam over a relatively wide angle by mechanical motion of the planar mirror. This configuration has been called a *mirror scan antenna* (by the Naval Research Laboratory), *polarization twist Cassegrain* and *flat plate Cassegrain* (by Westinghouse Electric), and a *parabolic reflector with planar auxiliary mirror* (by Russian authors[138]). The parabolic reflector is made up of parallel wires spaced less than a half-

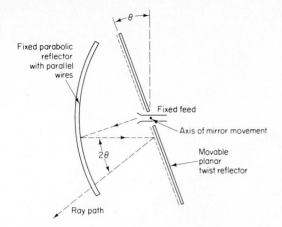

Figure 7.13 Geometry of the polarization-twist *mirror-scan antenna*, using a polarization-sensitive parabolic reflector and a planar polarization-rotating twist reflector. Scanning of the beam is accomplished by mechanical motion of the planar twist-reflector.

wavelength apart which are usually supported by low-loss dielectric material. The construction of the parabolic reflector with thin parallel wires makes it polarization sensitive. That is, it will completely reflect one sense of linear polarization and be transparent to the orthogonal sense of polarization. The sense of the linear polarization of the energy radiated by the feed is made the same as the orientation of the wires of the parabolic reflector. The feed in the center of the figure illuminates the parabolic reflector and the reflected energy is incident on a planar mirror constructed as a *twist reflector*. As mentioned previously a twist reflector reflects the incident energy with a 90° rotation of the plane of polarization. The reflected energy therefore will have a polarization perpendicular to the wires of the parabolic reflector and will pass through with negligible attenuation. The twist reflector can be made relatively broadband. An attractive feature of this antenna configuration is that the beam can be readily scanned over a wide angle by mechanical motion of the low inertia planar reflector. Another advantage is that a deflection of the planar mirror by an angle θ results in the beam scanning through an angle 2θ.

A similar principle is used in the configuration shown in Fig. 7.14 to obtain a 360° rotating-beam antenna without the need for RF rotary joints. In this cutaway sketch, four polarization sensitive parabolic reflectors surround a rotating twist-reflector mirror. For convenience, the entire structure is shown enclosed within a radome. Fixed feeds illuminate each reflector. A single radar transmitter and receiver is switched among the four sets of feeds on a time-shared basis every 90° rotation of the mirror. As in the antenna of Fig. 7.13, the parabolas consist of closely spaced parallel wires that reflect the polarization radiated by the feed, but pass the orthogonal polarization. The planar mirror rotates the plane of polarization 90° on reflection. The beam will illuminate the target twice for every revolution of the planar mirror. However, the scanning is not continuous in angle, and the time between observations alternates between two values. With a single-sided planar mirror, the two time intervals between observations correspond to the time it takes for the planar mirror to rotate $\frac{1}{8}$th and $\frac{3}{8}$th of a revolution. If the mirror is double-sided, the times between observations correspond to $\frac{3}{8}$th and $\frac{5}{8}$th of a revolution.

Three reflectors can be used instead of four to produce a uniform target observation rate, if the mirror is double-sided. Two target observations per rotation of the mirror are obtained, but the scanning is not continuous. That is, if the region from 0 to 120° is scanned first, the next region to be scanned is from 240 to 360°, followed by the region from 120 to 240°.

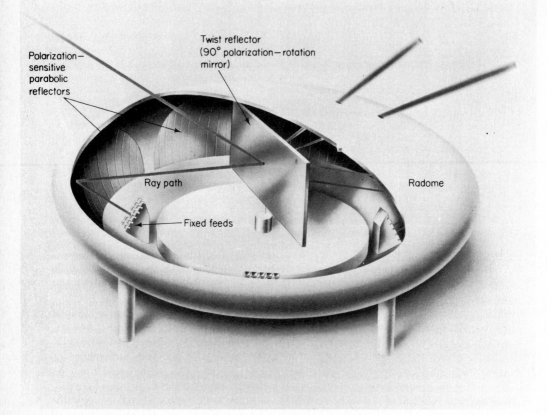

Polarization-sensitive parabolic reflectors

Twist reflector (90° polarization-rotation mirror)

Ray path

Fixed feeds

Radome

Figure 7.14 Configuration of the mirror-scan antenna for obtaining 360° scanning without RF rotating joints. (*From B. Lewis.*[139])

7.4 SCANNING-FEED REFLECTOR ANTENNAS

Large antennas are sometimes difficult to scan mechanically with as much flexibility as one might like. Some technique for scanning the beam of a large antenna must often be used other than the brute-force technique of mechanically positioning the entire structure. Phased array antennas and lens antennas offer the possibility of scanning the beam without the necessity for moving large mechanical masses. The present section considers the possibility of scanning the beam over a limited angle with a fixed reflector and a movable feed. It is much easier to mechanically position the feed than it is to position the entire antenna structure. In addition, large fixed reflectors are usually cheaper and easier to manufacture than antennas which must be moved about.

The beam produced by a simple paraboloid reflector can be scanned over a limited angle by positioning the feed.[1,26-29] However, the beam cannot be scanned too far without encountering serious deterioration of the antenna radiation pattern because of increasing coma and astigmatism. The gain of a paraboloid with $f/D = 0.25$ (f = focal distance, D = antenna diameter) is reduced to 80 percent of its maximum value when the beam is scanned ± 3 beamwidths off axis. A paraboloid with $f/D = 0.50$ can be scanned ± 6.5 beamwidths off axis before the gain is reduced to 80 percent of maximum (Ref. 1, p. 488). The antenna impedance also changes with a change in feed position. Hence scanning a simple paraboloid antenna by scanning the feed is possible, but is generally limited in angle because of the deterioration in the antenna pattern after scanning but a few beamwidths off axis.

Spherical reflectors. If the paraboloid reflector is replaced by a spherical-reflector surface, it is possible to achieve a wide scanning angle because of the symmetry of the sphere. However, a simple spherical reflector does not produce an equiphase radiation pattern (plane wave), and the pattern is generally poor. The term *spherical aberration* is used to describe the fact that the phase front of the wave radiated by a spherical reflector is not plane as it is with a wave radiated by an ideal parabolic reflector. There are at least three techniques which might be used to minimize the effect of spherical aberration. One is to employ a reflector of sufficiently large radius so that the portion of the sphere is a reasonable approximation to a paraboloid.[30-32] The second approach is to compensate for the spherical aberration with special feeds or correcting lenses.[33,34] These techniques yield only slightly larger scan angles than the single paraboloid reflector with movable feed.

A third technique to approximate the spherical surface and minimize the effects of spherical aberration is to step a parabolic reflector as shown in Fig. 7.15.[7,35,36] The focal length is reduced in half-wavelength steps, making a family of confocal paraboloids. It is possible to scan the stepped reflector to slightly wider angles than a simple paraboloid, but not as wide as with some other scanning techniques. Disadvantages of this reflector are the scattered radiation from the stepped portions and the narrow bandwidth.

If only a portion of the spherical reflector is illuminated at any one time, much wider scan angles are possible than if the entire aperture were illuminated. Li[32] has described experiments using a 10-ft-diameter spherical reflector at a frequency of 11.2 GHz. The focal length was 29.5 in. If the phase error from the sphere is to differ from that of a paraboloid by no more than $\lambda/16$, the maximum permissible diameter of the illuminated surface should be 3.56 ft. The beamwidth required of the primary feed pattern is determined by the illuminated portion of the aperture. Li used a square-aperture horn with diagonal polarization in order to obtain the required primary beamwidth and low primary-pattern sidelobes (better than 25 dB). The

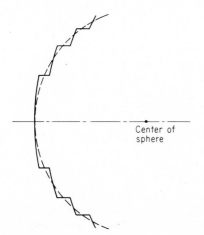

Center of
sphere

Figure 7.15 Stepped parabolic reflector.

resulting secondary beamwidth from the sphere was about 1.8° (39.4 dB gain) with a relative sidelobe level of 20 dB. A total useful scan angle of 140° was demonstrated. This type of antenna is similar in many respects to the torus antenna described below.

Parabolic Torus. Wide scan angles in one dimension can be obtained with a parabolic-torus configuration,[37-41] the principle of which is shown in Fig. 7.16. The parabolic torus is generated by rotating a section of a parabolic arc about an axis parallel to the latus rectum of the parabola. The cross section in one plane (the vertical plane in Fig. 7.16) is parabolic, while the cross section in the orthogonal plane is circular. The beam angle may be scanned by moving the feed along a circle whose radius is approximately half the radius of the torus circle. The radius of the torus is made large enough so that the portion of the circular cross section illuminated by the feed will not differ appreciably from the surface of a true parabola. Because of the circular symmetry of the reflector surface in the horizontal plane, the beam can be readily scanned in the plane without any deterioration in the pattern.

The wave reflected from the surface of the parabolic torus is not perfectly plane, but it can be made to approach a plane wave by proper choice of the ratio of focal length f to the radius of the torus R. The optimum ratio of f/R lies between 0.43 and 0.45.[39]

Good radiation patterns are possible in the principal planes with sidelobes only slightly worse than those of a conventional paraboloid. The larger the ratio of f/D, the better the radiation pattern. (The diameter D in the parabolic torus is the diameter of the illuminated area rather than the diameter of the torus itself.) The highest sidelobes produced by the parabolic torus do not lie within the principal planes. The inherent phase errors of the parabolic-torus surface due to its deviation from a true parabola can cause sidelobes on the order of 15 dB in intermediate planes.[38] These sidelobes usually lie in the 45° plane and are called *eyes*, because of their characteristic appearance on a contour plot of the radiation pattern.

In principle the parabolic torus can be scanned 180°, but because of beam spillover near the end of the scan and self-blocking by the opposite edge of the reflector, the maximum scan angle is usually limited to the vicinity of 120°.

Only a portion of the parabolic-torus is illuminated by the feed at any particular time. This may appear to result in low aperture utilization or poor efficiency since the total physical area is not related in a simple manner to the gain as it is in a fully illuminated antenna. However, the cost of the fixed reflector of the parabolic torus is relatively cheap compared with antennas which must be mechanically scanned. Nonutilization of the entire aperture is probably not too important a consideration when overall cost and feasibility are taken into account.

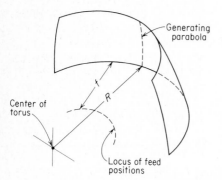

Figure 7.16 Principle of the parabolic-torus antenna.

The advantage of the parabolic torus is that it provides an economical method for rapidly scanning the beam of a physically large antenna aperture over a relatively wide scan angle with no deterioration of the pattern over this angle of scan. Its disadvantages are its relatively large physical size when compared with other means for scanning and the large sidelobes obtained in intermediate planes.

Organ-pipe scanner. Scanning the beam in the parabolic torus is accomplished by moving a single feed or by switching the transmitter between many fixed feeds. A single moving feed may be rotated about the center of the torus on an arm of length approximately one-half the radius of the torus. For example, a 120° torus antenna might be scanned by continuously rotating three feeds spaced 120° apart on the spokes of a wheel so that one feed is always illuminating the reflector. Although this may be practical in small-size antennas, it becomes a difficult mechanical problem if the radius of the rotating arm is large.

Scanning may also be accomplished by arranging a series of feeds on the locus of the focal points of the torus and switching the transmitter power from one feed to the next with an organ-pipe scanner.[128,129] The principle of the organ-pipe scanner is shown in Fig. 7.17. The transmission lines from the feeds are arranged to terminate on the periphery of a circle. A feed horn is rotated within this circle, transferring power from the transmitter to each feed or group of feeds in turn. The rotary horn may be flared to illuminate more than one elementary feed of the row of feeds. All the transmission lines in the organ-pipe scanner must be of equal length.

The radiation pattern from a torus with a well-designed organ-pipe scanner changes but little until the beam reaches one end of the scanning aperture. At this point the energy appears at both ends of the aperture and two beams are found in the secondary pattern. The antenna

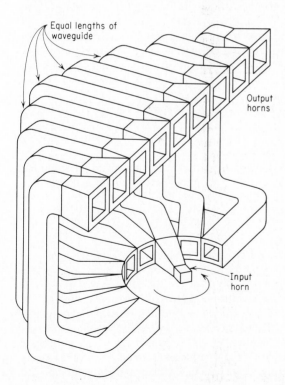

Figure 7.17 Principle of the organ-pipe scanner.

cannot be used during this period of ambiguity, called the *dead time*. In one model of the organ-pipe scanner, 36 elements were fed, three at a time.[128] The dead time for this model is equivalent to rotation past two of the 36 elements; consequently it was inoperative about 6 percent of the time.

In Fig. 7.17 the feeds are shown on a straight line, but in the parabolic torus they would lie on the arc of a circle.

The many feed horns plus all the transmission lines of the organ-pipe scanner result in a relatively large structure with significant aperture blocking. Aperture blocking can be minimized by designing the parabolic portion of the torus as an offset parabola, just as in the case of a paraboloid.

Other feed-motion scanners. There exist a number of antennas in which the beam is rapidly scanned over a limited sector by the mechanical motion of the feed. The motion of the feed is generally circular. The electromagnetic path within the antenna structure is designed to convert the circular motion of the feed to a linear motion of the antenna beam. In some designs the beam sweeps across a plane and then returns by the same route. In other designs the scan is saw-tooth with the beam sweeping across the plane of scan and then stepping back almost instantaneously to its starting point. The saw-tooth scan is generally preferred in radar applications. Such antennas would be used where a small angular sector needs to be covered with a rapid scan rate. These antennas can cover a sector of from 10 to 20 degrees or more, at a rate of 10 to 20 scans per second or greater. This is a rapid enough rate to permit essentially continuous observation of the target. A radar which uses such a sector-scanning antenna to track the path of airborne targets is sometimes called a *track-while-scan* radar.

The details of feed-motion scanners may be found in the literature.[1,42,43] Many of these, such as the Robinson, Foster,[44,134] Lewis, and Schwarzchild, were developed during World War II. Various types of lenses, pillboxes, and trough waveguides have also been used for mechanically scanning a beam over an angular sector. Another mechanical scanner is a coaxial line with radiating slots cut in its side.[1] The inner conductor is corrugated and eccentric with respect to the outer conductor. As the inner conductor rotates the wavelength in the line varies, causing the beam to scan.

7.5 LENS ANTENNAS

The most common type of radar antenna is the parabolic reflector in one of its various forms. The microwave paraboloid reflector is analogous to an automobile headlight or to a searchlight mirror. The analogy of an optical lens is also found in radar. Three types of microwave lenses applicable to radar are (1) dielectric lenses, (2) metal-plate lenses, and (3) lenses with nonuniform index of refraction.

Dielectric lenses. The homogeneous, solid, dielectric-lens antenna of Fig. 7.18a is similar to the conventional optical lens. A point at the focus of the lens produces a plane wave on the opposite side of the lens. Focusing action is a result of the difference in the velocity of propagation inside the dielectric as compared with the velocity of propagation in air. The index of refraction η of a dielectric is defined as the speed of light in free space to the speed of light in the dielectric medium. It is equal to the square root of the dielectric constant. Materials such as polyethylene, polystyrene, Plexiglas, and Teflon are suitable for small microwave lenses. They have low loss and may be easily shaped to the desired contour. Since the velocity of propagation is greater in air than in the dielectric medium, a converging lens is thicker in the middle than at the outer edges, just as in the optical case.

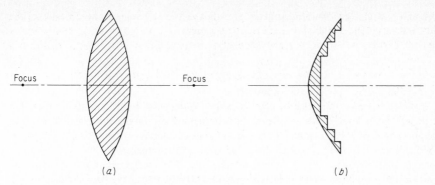

Figure 7.18 (*a*) Converging-lens antenna constructed of homogeneous solid dielectric. Direct microwave analogy of the optical lens. (*b*) Zoned dielectric lens.

One of the limitations of the solid homogeneous dielectric lens is its thick size and large weight. Both the thickness and the weight may be reduced considerably by stepping or zoning the lens (Fig. 7.18*b*). Zoning is based on the fact that a 360° change of phase at the aperture has no effect on the aperture phase distribution. Starting with zero thickness at the edge of the lens, the thickness of the dielectric is progressively increased toward the lens axis as in the design of a normal lens. However, when the path length introduced by the dielectric is equal to a wavelength, the path in the dielectric can be reduced to zero without altering the phase across the aperture. The thickness of the lens is again increased in the direction of the axis according to the lens design until the path length in the dielectric is once more 360°, at which time another step may be made. The optical path length through each of the zones is one wavelength less than the next outer zone.

Although zoning reduces the size and weight of a lens, it is not without disadvantages. Dielectric lenses are normally wideband; however, zoning results in a frequency-sensitive device. Another limitation is the loss in energy and increase in sidelobe level caused by the shadowing produced by the steps. The effect of the steps may be minimized by using a design with large *f/D*, on the order of 1 or more. Even with these limitations, a stepped lens is usually to be preferred because of the significant reduction in weight.

The larger the dielectric constant (or index of refraction) of a solid dielectric lens, the thinner it will be. However, the larger the dielectric constant, the greater will be the mismatch between the lens and free space and the greater the loss in energy due to reflections at the surface of the lens. Compromise values of the index of refraction lie between 1.5 and 1.6. Lens reflections may also be reduced with transition surfaces as in optics. These surfaces should be a quarter wave thick and have a dielectric constant which is the square root of the dielectric constant of the lens material.

Artificial dielectrics.[45-47] Instead of using ordinary dielectric materials for lens antennas, it is possible to construct them of artificial dielectrics. The ordinary dielectric consists of molecular particles of microscopic size, but the artificial dielectric consists of discrete metallic or dielectric particles of macroscopic size. The particles may be spheres, disks, strips, or rods imbedded in a material of low dielectric constant such as polystyrene foam. The particles are arranged in some particular configuration in a three-dimensional lattice. The dimension of the particles in the direction parallel to the electric field as well as the spacing between particles should be small compared with a wavelength. If these conditions are met, the lens will be insensitive to frequency.

When the particles are metallic spheres of radius a and spacing s between centers, the dielectric constant of the artificial dielectric is approximately

$$\epsilon \doteq 1 + \frac{4\pi a^3}{s^3} \tag{7.20}$$

assuming no interaction between the spheres.[47]

An artificial dielectric may also be constructed by using a solid dielectric material with a controlled pattern of voids. This is a form of Babinet inverse of the more usual artificial dielectric composed of particles imbedded in a low-dielectric-constant material.[48] The voids may be either spheres or cylinders, but the latter are easier to machine.

Lenses made from artificial dielectrics are generally of less weight than those from solid dielectrics. For this reason, artificial dielectrics are often preferred when the size of the antenna is large, as, for example, at the lower radar frequencies. Artificial-dielectric lenses may be designed in the same manner as other dielectric lenses.

Metal-plate lens.[49-52] An artificial dielectric may be constructed with parallel-plate wave-guides as shown in Fig. 7.19. The phase velocity in parallel-plate waveguide is greater than that in free space; hence the index of refraction is less than unity. This is opposite to the usual optical refracting medium. A converging metal-plate lens is therefore thinner at the center than at the edges, as opposed to a converging dielectric lens which is thinner at the edges. The metal-plate lens shown in Fig. 7.19 is an E-plane lens since the electric-field vector is parallel to the plates. Snell's law is obeyed in an E-plane lens, and the direction of the rays through the lens is governed by the usual optical laws involving the idex of refraction.

The surface contour of a metal-plate lens is, in general, not parabolic as in the case of the reflector.[7] For example, the surface closest to the feed is an ellipsoid of revolution if the surface at the opposite face of the lens is plane.

The spacing s between the plates of the metal-plate lens must lie between $\lambda/2$ and λ if only the dominant mode is to be propagated. The index of refraction for this type of metal-plate lens is

$$\eta = \left[1 - \left(\frac{\lambda}{2s} \right)^2 \right]^{1/2} \tag{7.21}$$

E ⊙ —— H →

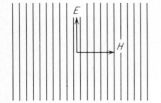

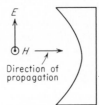

E ↑
⊙ H →
Direction of propagation

Figure 7.19 Plan, elevation, and end views of a converging lens antenna constructed from parallel-plate wave-guide. (E-plane metal-plate lens.)

where λ is the wavelength in air. Equation (7.21) is always less than unity. At the upper limit of spacing, $s = \lambda$, the index of refraction is equal to 0.866. The closer the spacing, the less will be the index of refraction and the thinner will be the lens. However, the spacing, and therefore the index of refraction, cannot be made arbitrarily small since the reflection from the interface between the lens and air will increase just as in the case of the solid-dielectric lenses. For a value of $s = \lambda/2$, the index of refraction is zero and the waveguide is beyond cutoff. The wave incident on the lens will be completely reflected. In practice, a compromise value of η between 0.5 and 0.6 is often selected, corresponding to plate spacings of 0.557λ and 0.625λ and to power reflections at normal incidence of 11 and 6.25 percent, respectively (Ref. 1, p. 410).

Even with an index of refraction in the vicinity of 0.5 to 0.6, the thickness of the metal-plate lens becomes large unless inconveniently long focal lengths are used. The thickness may be reduced by zoning just as with a dielectric lens. The bandwidth of a zoned metal-plate lens is larger than that of an unzoned lens, but the steps in the lens contour scatter the incident energy in undesired directions, reduce the gain, and increase the sidelobe level. An example of an X-band metal-plate zoned lens is shown in Fig. 7.20.

Another class of metal-plate lens is the constrained lens, or path-length lens, in which the rays are guided or constrained by the metal plates. In the H-plane metal-plate constrained lens, the electric field is perpendicular to the plates (H field parallel); thus the velocity of the wave which propagates through the plates is relatively unaffected provided the plate spacing is

Figure 7.20 X-band metal-plate zoned lens.

greater than $\lambda/2$. The direction of the rays is not affected by the refractive index, and Snell's law does not apply. Focusing action is obtained by constraining the waves to pass between the plates in such a manner that the path length can be increased above that in free space. In one type of cylindrical constrained lens with the E field parallel to the plates, a 1° beam could be scanned over a 100° sector by positioning the line feed.[52] The lens was 72 wavelengths in size, had an $f/D = 1.5$, and operated at a wavelength of 1.25 cm.

Luneburg lens. Workers in the field of optics have from time to time devised lenses in which the index of refraction varied in some prescribed manner within the lens. Although such lenses had interesting properties, they were only of academic interest since optical materials with the required variation of index of refraction were not practical. However, at microwave frequencies it is possible to control the index of refraction of materials (η is the square root of the dielectric constant ϵ), and lenses with a nonuniform index of refraction are practical.

One of the most important of the variable-index-of-refraction lenses in the field of radar is that due to Luneburg.[53] The Luneburg lens is spherically symmetric and has the property that a plane wave incident on the sphere is focused to a point on the surface at the diametrically opposite side. Likewise, a transmitting point source on the surface of the sphere is converted to a plane wave on passing through the lens (Fig. 7.21). Because of the spherical symmetry of the lens, the focusing property does not depend upon the direction of the incident wave. The beam may be scanned by positioning a single feed anywhere on the surface of the lens or by locating many feeds along the surface of the sphere and switching the radar transmitter or receiver from one horn to another. The Luneberg lens can also generate a number of fixed beams.

The index of refraction η or the dielectric constant ϵ varies with the radial distance in a Luneburg lens of radius r_0, according to the relationship

$$\eta = \epsilon^{1/2} = \left[2 - \left(\frac{r}{r_0}\right)^2\right]^{1/2} \tag{7.22}$$

The index of refraction is a maximum at the center, where it equals $\sqrt{2}$, and decreases to a value of 1 on the periphery.

Practical three-dimensional Luneburg lenses have been constructed of a large number of spherical shells, each of constant index of refraction. Discrete changes in index of refraction approximate a continuous variation. In one example of a Luneburg lens 10 concentric spherical shells are arranged one within the other.[54,55] The dielectric constant of the individual shells varies from 1.1 to 2.0 in increments of 0.1.

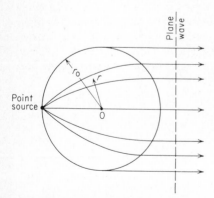

Figure 7.21 Luneburg-lens geometry showing rays from a point source radiated as a plane wave after passage through the lens.

The dielectric materials must not be too heavy, yet they must be strong enough to support their own weight without collapsing. They should have low dielectric loss and not be affected by the weather or by changes in temperature. They should be easily manufactured with uniform properties and must be homogeneous and isotropic if the performance characteristics are to be independent of position.

The antenna pattern of a Luneburg lens has a slightly narrower beamwidth than that of a paraboloidal reflector of the same circular cross section, but the sidelobe level is greater.[56-60] This is due to the fact that the paths followed by the rays in a Luneburg lens tend to concentrate energy toward the edge of the aperture. This makes it difficult to achieve extremely low sidelobes. In practice, the sidelobe level of a Luneburg lens seems to be in the vicinity of 20 to 22 dB.

When the full 4π radians of solid coverage is not required, a smaller portion of the lens can be used, with a saving in size and weight.[61,62] The Luneburg-lens principle can also be applied as a passive reflector in a manner analogous to a corner reflector.[61] If a reflecting cap is placed over a portion of the spherical lens, an incident wave emerges in the same direction from which it entered. The cap may be made to cover a sector as large as a hemisphere.

One of the limitations of a dielectric lens is the problem of removing heat dissipated within the interior. Dielectrics are generally poor conductors of heat. Unless materials are of low loss and can operate at elevated temperatures,[63] dielectric lenses such as the Luneburg lens are more suited to receiving, rather than transmitting, applications.

The Luneburg principle may also be applied to a two-dimensional lens which generates a fan beam in one plane. In the geodesic analog of a two-dimensional Luneburg lens the variation in dielectric constant is obtained by the increased path length for the RF energy traveling in the TEM mode between parallel plates.[64-67] The result is a dome-shaped parallel-plate region as shown in Fig. 7.22.

Other types of lenses based on the principle of nonuniform index of refraction have been described.[43,60,68] Homogeneous spherical lenses with uniform dielectric constant are also of interest[69-71] if the index of refraction is not too high and if the diameter is not greater than about 30λ. They can be competitive to the Luneburg lens, especially for high-power applications.[71]

Lens tolerances.[1,2] In general, the mechanical tolerances for a lens antenna are less severe than for a reflector. A given error in the contour of a mechanical reflector contributes twice to the error in the wavefront because of the two-way path on reflection. Mechanical errors in the lens contour, however, contribute but once to the phase-front error. Although the theoretical tolerance of a lens may be less than required of a reflector, in practice it might be more difficult to achieve a given level of performance in a lens. A reflector can be readily supported mechanically from the back. This is not available in a lens where the mechanical support is from the periphery of the lens and from the mechanical properties of the lens material itself. Another source of error in the lens not found in reflector antenna is the variation in the properties of the lens material. Both real and artificial dielectrics are not always perfectly uniform from sample to sample or even within the same sample.

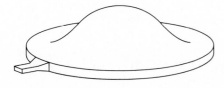

Figure 7.22 The "tin-hat" geodesic analog of a two-dimensional Luneburg lens.

Evaluation of lenses as antennas. One of the advantages of a lens over a reflector antenna is the absence of aperture blocking. Considerable equipment can be placed at the focus of the lens without interfering with the resultant antenna pattern. The first monopulse radars used lenses for this purpose, but with time the monopulse RF circuitry was reduced in size and the reflector antenna came to be preferred over the lens. The lens is capable of scanning the beam over a wide angle. Theoretically the Luneburg lens or the homogeneous sphere can cover 4π steradians. Constrained metal-plate lenses are capable of wide scan angles as compared to the limited scanning possible by moving the feed in a paraboloid reflector.

The lens is usually less efficient than comparable reflector antennas because of loss when propagating through the lens medium and the reflections from the two lens surfaces. In a zoned lens there will be additional, undesired scattering from the steps. Although it is dangerous to generalize, the additional losses from the sources in a stepped lens might be 1 or 2 dB.[72]

The lack of suitable solid or artificial dielectric materials has limited the development of lenses. The problem of dissipating heat from large dielectric lenses can sometimes restrict their use to moderate-power or to receiver applications. Conventional lenses are usually large and heavy, unless zoned. To reduce the loss caused by scattering from the steps in a zoned lens, the ratio of the focal length f to the antenna diameter D must be made large (of the order of unity). Lenses which must scan by positioning the feed should also have large f/D ratio. A large f/D requires a greater mechanical structure because the feeds are bigger and must be supported farther from the lens. The mechanical support of a lens is usually more of a problem than with a reflector.

The wide-angle scanning capability of a lens would be of interest in radar as a competitor for a phased array if there were available a practical means for electronically switching the transmitter and receiver among fixed feeds so as to achieve a rapidly scanning beam.

7.6 PATTERN SYNTHESIS

The problem of pattern synthesis in antenna design is to find the proper distribution of current across a finite-width aperture so as to produce a radiation pattern which approximates the desired pattern under some condition of optimization. Pattern-synthesis methods may be divided into two classes, depending upon whether the aperture is continuous or an array. The current distributions derived for continuous apertures may sometimes be used to approximate the array-aperture distributions, and vice versa, when the number of elements of the array-antenna is large. The discussion in this section applies, for the most part, to linear one-dimensional apertures or to rectangular apertures where the distribution is separable, that is, where $A(x, z) = A(x)A(z)$.

The synthesis techniques which apply to array antennas usually assume uniformly spaced isotropic elements. The element spacing is generally taken to be a half wave-length. If the elements were not isotropic but had a pattern $E_e(\theta)$, and if the desired overall pattern were denoted $E_d(\theta)$, the pattern to be found by synthesis using techniques derived for isotropic elements would be given by $E_d(\theta)/E_e(\theta)$.

Fourier-integral synthesis. The Fourier-integral relationship between the field-intensity pattern and the aperture distribution was discussed in Sec. 7.2. The distribution $A(z)$ across a continuous aperture was given by Eq. (7.14).

$$A(z) = \frac{1}{\lambda} \int_{-\infty}^{\infty} E(\phi) \exp\left(-j2\pi \frac{z}{\lambda} \sin \phi\right) d(\sin \phi) \qquad (7.14)$$

where z = distance along the aperture and $E(\phi)$ = field-intensity pattern. If only that portion of the aperture distribution which extends over the finite-aperture dimension d were used, the resulting antenna pattern would be

$$E_a(\phi) = \int_{-d/2}^{d/2} A(z) \exp\left(j2\pi \frac{z}{\lambda} \sin\phi\right) dz \tag{7.23}$$

Substituting Eq. (7.14) into the above and changing the variable of integration from ϕ to ξ to avoid confusion, the antenna pattern becomes

$$E_a(\phi) = \frac{1}{\lambda} \int_{-d/2}^{d/2} \int_{-\infty}^{\infty} E(\xi) \exp\left[j2\pi \frac{z}{\lambda}(\sin\phi - \sin\xi)\right] d(\sin\xi) \, dz \tag{7.24}$$

Interchanging the order of integration, the approximate antenna pattern is

$$E_a(\phi) = \frac{d}{\lambda} \int_{-\infty}^{\infty} E(\xi) \frac{\sin\left[\pi(d/\lambda)(\sin\phi - \sin\xi)\right]}{\pi(d/\lambda)(\sin\phi - \sin\xi)} d(\sin\xi) \tag{7.25}$$

where $E_a(\phi)$ is the Fourier-integral pattern which approximates the desired pattern $E(\phi)$ when $A(z)$ is restricted to a finite aperture of dimension d.

Ruze[73] has shown that the approximation to the antenna pattern derived on the basis of the Fourier integral for continuous antennas (or the Fourier-series method for discrete arrays) has the property that the mean-square deviation between the desired and the approximate patterns is a minimum. It is in this sense (least mean square) that the Fourier method is optimum. The larger the aperture (or the greater the number of elements in the array), the better will be the approximation.

The Fourier series may be used to synthesize the pattern of a discrete array, just as the Fourier integral may be used to synthesize the pattern of a continuous aperture.[74] Similar conclusions apply. The Fourier-series method is restricted in practice to arrays with element spacing in the vicinity of a half wavelength. Closer spacing results in supergrain arrays which are not practical.[75,76] Spacings larger than a wavelength produce undesired grating lobes.

Woodward-Levinson method. Another method of approximating the desired antenna pattern with a finite aperture distribution consists in reconstructing the antenna pattern from a finite number of sampled values. The principle is analogous to the sampling theorem of circuit theory in which a time waveform of limited bandwidth may be reconstructed from a finite number of samples. The antenna-synthesis technique based on sampled values was introduced by Levinson at the MIT Radiation Laboratory in the early forties and was apparently developed independently by Woodward in England.[7,77–79]

The classical sampling theorem of information theory states: If a function $f(t)$ contains no frequencies higher than W Hz, it is completely determined by giving its ordinates at a series of points spaced $1/2W$ seconds apart. The analogous sampling process applied to an antenna pattern is that the radiation pattern $E_a(\phi)$ from an antenna with a finite aperture d is completely determined by a series of values spaced λ/d radians apart, that is, by the sample values $E_s(n\lambda/d)$, where n is an integer.[73] In Fig. 7.23a is shown the pattern $E(\phi)$ and the sampled points spaced λ/d radians apart. The sampled values $E_s(n\lambda/d)$, which determine the antenna pattern, are shown in b.

The antenna pattern $E_a(\phi)$ can be constructed from the sample values $E_s(n\lambda/d)$ with a pattern of the form $(\sin\psi)/\psi$ about each of the sampled values, where $\psi = \pi(d/\lambda)\sin\phi$. The $(\sin\psi)/\psi$ function is called the *composing function* and is the same as that used in information

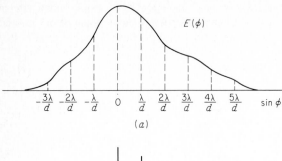

(a)

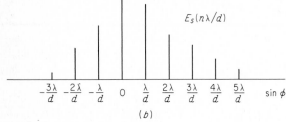

(b)

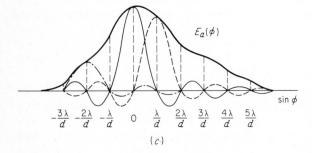

(c)

Figure 7.23 (a) Radiation pattern $E(\phi)$ with sampled values λ/d radians apart, where d = aperture dimension; (b) sampled values $E_s(n\lambda/d)$, which specify the antenna pattern of (a); (c) reconstructed pattern $E_a(\phi)$ using $(\sin \psi)/\psi$ composing function to approximate the desired radiation pattern $E(\phi)$.

theory to construct the time waveform from the sampled values. The antenna pattern is given by

$$E_a(\phi) = \sum_{n=-\infty}^{\infty} E_s\left(\frac{n\lambda}{d}\right) \frac{\sin\left[\pi(d/\lambda)(\sin\phi - n\lambda/d)\right]}{\pi(d/\lambda)(\sin\phi - n\lambda/d)} \qquad (7.26)$$

that is, the antenna pattern from a finite aperture is reconstructed from a sum of $(\sin \psi)/\psi$ composing functions spaced λ/d rad apart, each weighted according to the sample values $E_s(n\lambda/d)$, as illustrated in Fig. 7.23c.

The $(\sin \psi)/\psi$ composing function is well suited for reconstructing the pattern. Its value at a particular sample point is unity, but it is zero at all other sample points. In addition, the $(\sin \psi)/\psi$ function can be readily generated with a uniform aperture distribution. The Woodward-Levinson synthesis technique consists in determining the amplitude and phase of the uniform aperture distribution corresponding to each of the sample values and performing a summation to obtain the required overall aperture distribution.

The aperture distribution may be found by substituting the antenna pattern of Eq. (7.26) into the Fourier-transform relationship given by Eq. (7.14). The aperture distribution becomes

$$A_s(z) = \frac{1}{d} \sum_{n=-\infty}^{\infty} E_s\left(\frac{n\lambda}{d}\right) \exp\left(-\frac{j2\pi nz}{d}\right) \qquad (7.27)$$

Therefore the aperture distribution which generates the nth $(\sin \psi)/\psi$ composing pattern has uniform amplitude and is proportional to the sampled value $E_s(n\lambda/d)$. The phase across the aperture is such that the individual composing patterns are displaced from one another by a half a beamwidth (where the beamwidth is here defined as the distance between the two nulls which surround the main beam). The phase is given by the exponential term of Eq. (7.27) and represents a linear phase change of $n\pi$ radians across the aperture. The number of samples required to approximate the radiation pattern from a finite aperture of width d is $2d/\lambda$.

The essential difference between Fourier-integral synthesis and the Woodward-Levinson method is that the former gives a radiation pattern whose mean-square deviation from the desired pattern is a minimum, and the Woodward-Levinson method gives an antenna pattern which exactly fits the desired pattern at a finite number of points.

Dolph-Chebyshev arrays.[80–83] This pattern produces the narrowest beamwidth for a specified sidelobe level. The beamwidth is measured by the distance between the first nulls that straddle the main beam. The sidelobes are all of equal magnitude. Dolph[80] derived the aperture illumination with this property by forcing a correspondence between the Chebyshev polynomial and the polynomial describing the pattern of an array antenna. Although a radiation pattern with the narrowest beamwidth for a given sidelobe level seems a reasonable choice for radar, it is seldom employed since it cannot be readily achieved with practical antennas where high gain and low sidelobes are desired. As the antenna size increases, the currents at the end of the aperture become large compared with the currents along the rest of the aperture, and the radiation pattern becomes sensitive to the edge excitation. This sets a practical upper limit to the size of an antenna that can have a Dolph-Chebyshev pattern and therefore sets a lower limit to the width of the main beam which can be achieved.

Taylor aperture illumination. The Taylor aperture illumination is a realizable approximation to the Dolph-Chebyshev illumination.[84] It produces a pattern with uniform sidelobes of a specified value, but only in the vicinity of the main beam. Unlike the theoretical Dolph-Chebyshev pattern, the sidelobes of the Taylor pattern decrease outside a specified angular region. The sidelobe level is uniform within the region defined by $|(d/\lambda)\sin\phi| < \bar{n}$ and decreases with increasing angle ϕ for $|(d/\lambda)\sin\phi| > \bar{n}$, and where $\bar{n}$ is an integer, d is the antenna dimension and λ is the wavelength. Hence $\bar{n}$ divides the radiation pattern into a uniform sidelobe region straddling the main beam and a decreasing sidelobe region. The number of equal sidelobes on each side of the main beam is $\bar{n} - 1$.

The beamwidth of a Taylor pattern will be broader than that of the Dolph-Chebyshev. If the design sidelobe level is 25 dB, a Taylor pattern with $\bar{n} = 5$ gives a beamwidth 7.7 percent greater than the Dolph-Chebyshev, and with $\bar{n} = 8$ it is 5.5 percent greater.

The Taylor pattern is specified by two parameters: the design sidelobe level and $\bar{n}$, which defines the boundary between uniform sidelobes and decreasing sidelobes. The integer $\bar{n}$ cannot be too small. Taylor states that $\bar{n}$ must be at least 3 for a design-sidelobe ratio of 25 dB and at least 6 for a design-sidelobe ratio of 40 dB. The larger $\bar{n}$ is, the sharper will be the beam. However, if $\bar{n}$ is too large, the same difficulties as arise with the Dolph-Chebyshev pattern will occur. The aperture illuminations for high values of $\bar{n}$ are peaked at the center and at the edge of the aperture, and might be difficult to achieve in practice.

Care must be exercised in the selection of the sidelobe level of a Taylor pattern. Large antennas with narrow beamwidths can exhibit a severe degradation in gain because of the large energy contained within the sidelobes as compared with that within the main beam. The value of $\bar{n}$ must be properly chosen consistent with the beamwidth and sidelobe level.[85]

Although the Taylor pattern was developed as a realizable approximation to the Dolph-Chebyshev, it does not resemble the theoretical equal-sidelobe pattern. Values of $\bar{n}$ are not large so that a Taylor pattern exhibits decreasing sidelobes over most of its range. Decreasing sidelobes are not undesired in radar application. If the radar designer has a choice, it is preferred that the high sidelobes be near the main beam, where they are easier to recognize, rather than to have isolated high sidelobes elsewhere. This is one time nature is cooperative since it is natural for the sidelobes to be large in the vicinity of the main beam.

The Taylor aperture illumination has also been applied to synthesizing the patterns of circular, two-dimensional antennas.[86,87] It has been widely used as a guide for selecting antenna aperture illuminations.

When the difference pattern of a monopulse antenna can be selected independently of the sum pattern, as in a phased array, the criterion for a good difference pattern is to obtain maximum angle sensitivity commensurate with a given sidelobe level. Bayliss[88] has described a method for obtaining suitable monopulse difference patterns on this basis. It parallels Taylor's approach to the sum pattern.

7.7 COSECANT-SQUARED ANTENNA PATTERN

It was shown in Sec. 2.11 that a search radar with an antenna pattern proportional to $\csc^2\theta$, where θ is the elevation angle, produces a constant echo-signal power for a target flying at constant altitude, if certain assumptions are satisfied. Many fan-beam air-search radars employ this type of pattern. A constant echo signal with range, however, is probably not as important an application of the cosecant-squared pattern as is achieving the desired elevation coverage in an efficient manner. Shaping of the beam is desirable since the needed range at high angles is less that at low angles; hence, the antenna gain as a function of elevation angle can be tailored accordingly. Shaped patterns like the cosecant-squared pattern are also used in airborne radars that map the surface of the earth.[89]

Antenna design. The design of a cosecant-squared antenna pattern is an application of the synthesis techniques discussed in the preceding section. Examples of cosecant-squared-pattern synthesis are given in the literature.[7,73,77,90]

The cosecant-squared pattern may be approximated with a reflector antenna by shaping the surface or by using more than one feed. The pattern produced in this manner may not be as accurate as might be produced by a well-designed array antenna, but operationally, it is not necessary to approximate the cosecant-squared pattern very precisely. A common method of producing the cosecant-squared pattern is shown in Fig. 7.24. The upper half of the reflector is a parabola and reflects energy from the feed in a direction parallel to the axis, as in any other parabolic antenna. The lower half, however, is distorted from the parabolic contour so as to direct a portion of the energy in the upward direction.

A cosecant-squared antenna pattern can also be produced by feeding the parabolic reflector with two or more horns or with a linear array. If the horns are spaced and fed properly, the combination of the secondary beams will give a smooth cosecant-squared pattern over some range of angle. A reasonable approximation to the cosecant-squared pattern can be obtained with but two horns. A single horn, combined with a properly located ground plane, can also generate a cosecant-squared pattern with a parabolic reflector.[91] The feed horn, plus its image in the ground plane, has the same effect as two horns. The traveling-wave slot antenna[92] and the surface-wave antenna[93] can also be designed to produce a cosecant-squared antenna pattern.

The shaping of the beam is generally in one plane, with a narrow pattern of conventional

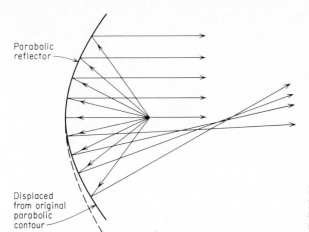

Parabolic
reflector

Displaced
from original
parabolic
contour

Figure 7.24 Cosecant-squared antenna produced by displacing the reflector surface from the original parabolic shape.

design in the orthogonal plane. The parabolic cylinder antenna fed from a line source is convenient for obtaining independent control of the patterns in the two orthogonal planes. This type of antenna, however, is generally bulkier and heavier than a reflector fed from a single point source. The line feed is more difficult to pressurize than a point feed. The antenna with a point-source feed requires a reflector surface with *double curvature*, as compared to the single curvature of the cylindrical antenna, in order to obtain a shaped beam. The double curvature reflector is designed to provide both the desired shaping of the beam in one plane and focusing in transverse planes.[1,94-96] The surface is formed by the envelope of a system of paraboloids whose axes all lie in the plane of the shaped beam, but at varying angles of inclination to each other and to a fixed line.

The antenna of the S-band Airport Surveillance Radar (ASR) of the AN/TPN-19 landing system is shown in Fig. 7.25. The 14 ft (4.3 m) by 8 ft (2.4 m) reflector is an offset paraboloid fed from a 12-element vertical line-source. The uppermost feedhorn is located at the focal point of the paraboloid. The azimuth beamwidth is 1.6° and the vertical beamwidth is 4° with cosecant-squared shaping from 6° to 30°. Being a transportable equipment, the antenna is self-erecting and is stowed inside the shelter through a roof hatch. The radar system provides coverage to 60 nmi and 40,000 feet with a 15 rpm rotation rate.

Loss in gain. An antenna with a cosecant-squared pattern will have less gain than a normal fan-beam pattern generated from the same aperture. To obtain an approximate estimate of the loss in gain incurred by beam shaping, the idealized patterns in Fig. 7.26 will be assumed. The normal antenna pattern is depicted in Fig. 7.26a as a square beam extending from $\theta = 0$ to $\theta = \theta_0$. The cosecant-squared pattern in Fig. 7.26b is shown as a uniform beam over the range $0 \leq \theta \leq \theta_0$ and decreases as $\csc^2 \theta / \csc^2 \theta_0$ over the range $\theta_0 < \theta \leq \theta_m$. The gain G of the square beam in Fig. 7.26a divided by the gain G_c of the cosecant-squared antenna beam in Fig. 7.26b is

$$\frac{G}{G_c} = \frac{\theta_0 + [1/(\csc^2 \theta_0)] \int_{\theta_0}^{\theta_m} \csc^2 \theta \, d\theta}{\theta_0} = \frac{\theta_0 + \sin^2 \theta_0 (\cot \theta_0 - \cot \theta_m)}{\theta_0} \qquad (7.28)$$

For small values of θ_0,

$$\frac{G}{G_c} \approx 2 - \theta_0 \cot \theta_m \qquad (7.29)$$

Figure 7.25 Cosecant-squared antenna of the AN/TPN-19 landing system. (*Courtesy Raytheon Company.*)

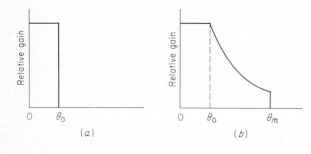

Figure 7.26 Idealized antenna pattern assumed in the computation of the loss in gain incurred with a cosecant-squared antenna pattern. (*a*) Normal antenna pattern; (*b*) cosecant-squared pattern.

where all angles in the above formulas are measured in radians. For example, if $\theta_0 = 6°$ and $\theta_m = 20°$, the gain is reduced by 2.2 dB compared with a fan beam 6° wide. If θ_m is made 40°, the loss is 2.75 dB. In the limit of large θ_m and small θ_0, the loss approaches a maximum of 3 dB.

Shaped beams and STC. A radar that can detect a 1 m² target at 200 nmi can detect a 10^{-4} m² target at 20 nmi because of the inverse fourth power variation of signal strength with range. Therefore, small nearby targets such as birds and insects can clutter the output of a radar. To avoid this, the receiver gain can be reduced at short range and increased during the period between pulses so that the received signal from a target of constant cross section remains unchanged with range. The programmed control of the receiver gain to maintain a constant echo signal strength is called *sensitivity time control (STC)*. It is an effective method for eliminating radar echoes due to unwanted birds and insects.

STC makes use of the inverse fourth power variation of signal strength with range. The cosecant-squared shaping of the antenna also utilizes the inverse fourth power variation with range (Sec. 2.11). Thus, when STC is used with a cosecant-squared antenna pattern, the high-angle coverage of the radar is reduced. Targets that are seen at long range and at a particular height with a cosecant-squared antenna, will be missed at shorter ranges when STC is used. To incorporate both beam shaping and STC, the pattern must have higher gain at higher-elevation angles than would a cosecant-squared pattern. Figure 7.27a illustrates the antenna elevation pattern for an air-search radar which is desired to compensate for STC and yet provide a signal independent of range, as does the cosecant-squared pattern.[97] Figure 7.27b shows the coverage of a long-range radar with such an antenna pattern.

Beam shaping may also be employed to increase the target-to-clutter ratio in some cases. A target at high elevation angles competes with surface clutter at low angles. Increasing the antenna gain at high angles but not at low angles will therefore improve the target echo with respect to the clutter.[98]

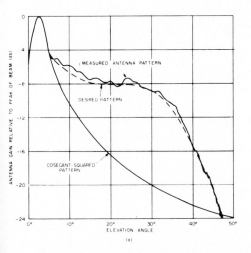

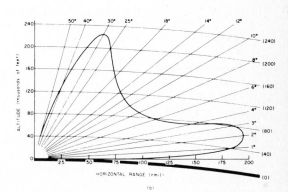

Figure 7.27 Antenna elevation pattern for a long-range air-search radar to achieve high-angle coverage when STC is employed. (a) Comparison with the cosecant-squared pattern; (b) free-space coverage diagram. (*From Shrader,*[97] *courtesy McGraw-Hill Book Company.*)

7.8 EFFECT OF ERRORS ON RADIATION PATTERNS[73,99-101]

Antenna-pattern synthesis techniques such as those discussed in Sec. 7.6 permit the antenna designer to compute the aperture illumination required to achieve a specified radiation pattern. However, when the antenna is constructed, it is usually found that the experimentally measured radiation pattern deviates from the theoretical one, especially in the region of the sidelobes. Generally, the fault lies not with the theory, but in the fact that it is not possible to reproduce precisely in practice the necessary aperture illumination specified by synthesis theory. Small, but ever-present, errors occur in the fabrication of an antenna. These contribute unavoidable perturbations to the aperture illumination and result in a pattern different in detail from the one anticipated.

Errors in the aperture illumination may be classed as either systematic or random. The former are predictable, but the latter are not and can only be described in statistical terms. Examples of systematic errors include (1) mutual coupling between the elements of an array, (2) aperture blocking in reflector antennas due to the feed and its supports, (3) diffraction at the steps in a zoned-lens antenna, and (4) periodicities included in the construction of the antenna. Random errors include (1) errors in the machining or manufacture of the antenna as a consequence of the finite precision of construction techniques, (2) RF measurement errors incurred in adjusting an array, (3) wall-spacing errors in metal-plate lenses, (4) random distortion of the antenna surface, and (5) mechanical or electrical phase variations caused by temperature or wind gradients across the antenna. Although random errors may be relatively small, their effect on the sidelobe radiation can be large. Systematic errors are usually the same from antenna to antenna in any particular design constructed by similar techniques. On the other hand, random errors differ from one antenna to the next even though they be of the same design and constructed similarly. Therefore the effect of random errors on the antenna pattern can be discussed only in terms of the average performance of many such antennas or in terms of statistics.

The effect of errors on the radiation pattern has long been recognized by the practical antenna designer. The usual rule-of-thumb criterion employed in antenna practice is that the phase of the actual wavefront must not differ from the phase of the desired wavefront by more than $\pm\lambda/16$ in order to ensure satisfactory performance. The application of this criterion to a reflector antenna requires the mechanical tolerance of the surface to be within $\pm\lambda/32$. It is possible, however, to obtain more precise criteria for specifying the maximum errors which may be tolerated in the aperture illumination.

Systematic errors. The effect of systematic errors on the radiation pattern may be found by properly modifying the aperture distribution to take account of the known errors. For example, a linear phase error across the antenna aperture causes the beam position to tilt in angle. A quadratic, or square-law, variation in phase is equivalent to defocusing the antenna. A periodic error with fundamental period p/λ, where p is measured in the same units as is the wavelength λ, will produce spurious beams displaced at angles ϕ_n from the origin, according to the relation $\sin \phi_n = n\lambda/p$, where n is an integer. The patterns of the spurious beams are of similar shape as the original pattern but are displaced in angle and reduced in amplitude.

Random errors in reflectors. The classical work on the effects of random errors on antenna radiation patterns is due to Ruze.[73,99,101] He pointed out that in a reflector antenna, only the phase error in the aperture distribution need be considered. Such a phase error, for example, might be caused by a deformation of the surface from its true value. Ruze[101] showed in a simple derivation that the gain of a circular aperture with arbitrary phase error is approximately

$$G = G_0(1 - \overline{\delta^2}) \qquad (7.30)$$

where G_0 is the gain of the antenna in the absence of errors, and δ is the phase error, in radians, calculated from the mean phase plane. This simple expression is valid for any aperture illumination and reflector deformation, provided that the latter is small compared to the wavelength. It indicates that the rms phase variation about the mean phase plane must be less than $\lambda/14$ for a one dB loss of gain. For shallow reflectors, the two-way path of propagation means that the surface error must be one-half this amount, or $\lambda/28$.

Errors do more than reduce the peak gain of an antenna. They affect the entire pattern. Using a model of an antenna in which the reflector is distorted by a large number of random gaussian-shaped bumps, Ruze showed that the radiation pattern can be expressed as

$$G(\theta, \phi) = G_0(\theta, \phi)e^{-\bar{\delta^2}} + (2\pi C/\lambda)^2 e^{-\bar{\delta^2}} \sum_{n=1}^{\infty} \frac{[\bar{\delta^2}]^n}{n!\, n} e^{-(\pi Cu/\lambda)^2/n} \tag{7.31}$$

where $G_0(\theta, \phi)$ is the no-error radiation pattern whose axial value (at $\theta = 0$, $\phi = 0$) is $\rho_a(\pi D/\lambda)^2$, D is the antenna diameter, ρ_a is the aperture efficiency, C is the correlation interval of the error, and u equals $\sin \theta$. The mean square phase error $\bar{\delta^2}$ is assumed to be gaussian. The angles θ, ϕ are those usually employed in classical antenna theory and are defined in Fig. 7.28. They are not to be confused with the usual elevation and azimuth angles. The antenna lies in the x-y plane of Fig. 7.28. The error current in one region of the antenna is assumed independent of the error currents in adjacent regions. The size of the regions in which the error currents cannot be considered independent is the correlation interval, C. The size of the correlation interval affects both the magnitude and the directional characteristic of the spurious radiation that results from the presence of errors.

The first term of Eq. (7.31) represents the no-error radiation pattern reduced by the factor, $\exp - \bar{\delta^2}$. The second term describes the disturbing pattern and represents a source of sidelobe energy which depends on the mean-square phase error and the square of the correlation interval. For small phase errors, when only the first term of the series ($n = 1$) need be considered, Eq. (7.31) becomes

$$G(\theta, \phi) = G_0(\theta, \phi)e^{-\bar{\delta^2}} + (2\pi C/\lambda)^2 \bar{\delta^2} e^{-(\pi Cu/\lambda)^2} \tag{7.32}$$

From Eq. (7.31), the reduction in gain on axis can be written as

$$G/G_0 = e^{-\bar{\delta^2}} + \frac{1}{\rho_a}\left(\frac{2C}{D}\right)^2 e^{-\bar{\delta^2}} \sum_{n=1}^{\infty} \frac{[\bar{\delta^2}]^n}{n!\, n} \tag{7.33}$$

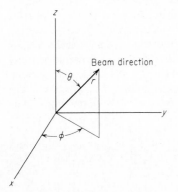

z

Beam direction

θ

r

y

ϕ

x

Figure 7.28 Coordinate system defining the angles θ, ϕ of Eq. (7.31).

When the second term can be neglected (as when the correlation interval is small compared to the diameter of the antenna and the phase error is not too large), the gain can be written

$$G = G_0 e^{-\delta^2} = \rho_a \left(\frac{\pi D}{\lambda}\right)^2 e^{-(4\pi\epsilon/\lambda)^2} \tag{7.34}$$

where ϵ is the effective reflector tolerance measured in the same units as λ; i.e., it is the rms surface tolerance of a shallow reflector (large focal-length-to-diameter ratio) which will produce the phase-front variance δ^2. For a given reflector size D, the gain increases as the square of the frequency until the exponential term becomes significant. Differentiation of Eq. (7.34) shows that the maximum gain corresponds to a wavelength

$$\lambda_m = 4\pi\epsilon \tag{7.35}$$

At this wavelength the gain will be 4.3 dB below what it would be in the absence of errors. The maximum gain is then

$$G_{\max} = \frac{\rho_a}{43} \left(\frac{D}{\epsilon}\right)^2 \tag{7.36}$$

The gain of an antenna is thus limited by the mechanical tolerance to which the surface can be constructed and maintained.[102] The most precise reflector antennas seem to be limited to a precision of not much greater than about one part in 20,000, which from Eq. (7.36) corresponds to a diameter of about 1600 wavelengths for maximum gain. The beamwidth of such an antenna would be about 0.04° with a gain of about 68 dB.

In practice, the construction tolerance of an antenna is often described by the "peak" error, rather than the rms error. The ratio of the peak to the rms error is found experimentally to be about 3 : 1. This truncation of errors occurs since large errors usually are corrected in manufacture.

The effect of errors in array antennas and further discussion of errors in continuous apertures is given in Sec. 8.8.

7.9 RADOMES[14,103,115,122–124]

Antennas for ground-based radars are often subjected to high winds, icing, and/or temperature extremes. They must be sheltered if they are to continue to survive and perform under adverse weather conditions. Antennas which must be operated in severe weather are usually enclosed for protection in a sheltering structure called a *radome*. Radomes must be mechanically strong if they are to provide the necessary protection, yet they must not interfere with the normal operation of the antenna. Antennas mounted on aircraft must also be housed within a radome to offer protection from large aerodynamic loads and to avoid disturbance to the control of the aircraft and minimize drag.

The design of radomes for antennas may be divided into two separate and relatively distinct classes, depending upon whether the antenna is for airborne or ground-based (or ship-based) application. The airborne radome is characterized by smaller size than ground-based radomes since the antennas that can be carried in an aircraft are generally smaller. The airborne radome must be strong enough to form a part of the aircraft structure and usually must be designed to conform to the aerodynamic shape of the aircraft, missile, or space vehicle in which it is to operate.

A properly designed radome should distort the antenna pattern as little as possible. The presence of a radome can affect the gain, beamwidth, sidelobe level, and the direction of the boresight (pointing direction), as well as change the VSWR and the antenna noise temperature. Sometimes in tracking radars, the rate of change of the boresight shift can be important. Generally, an antenna situated within a large ground-based radome sees the same approximate radome environment no matter where the beam points within its normal coverage. Radar antennas located in the nose of aircraft, however, generally require an ogive-shaped radome which does not present the same environment for all beam positions. When the antenna is directed forward (energy propagating parallel to the radome axis) the angle of incidence on the radome surface can be in excess of 80°. In other look directions the incidence angle might be zero degrees. Since the transmission properties of radome materials varies with angle of incidence and polarization, the design of an airborne radome to achieve uniform scanning properties might not be easy. The design is further complicated by the need for structural strength, lightning protection, and protection from erosion by rain, hail, and dust. It is not surprising therefore that the electrical performance of a radome must sometimes be sacrificed to accommodate these other factors.

A radome permits a ground-based radar antenna to operate in the presence of high winds. It also prevents ice formation on the antenna. Although it is possible to design antennas strong enough to survive extreme weather conditions and to provide sufficiently large motors to rotate them in high winds, it is often more economical to design lighter antennas with modest drive power and operate them inside radomes.

The shape of a radome for a ground-based antenna is usually a portion of a sphere. The sphere is a good mechanical structure and offers aerodynamic advantage in high winds. Precipitation particles blow around a sphere rather than impinge upon it. Hence snow or other frozen precipitation is not readily deposited.

The first large radomes (50-ft diameter or more) for ground-based radar antennas appeared shortly after World War II. They were constructed of a strong, flexible rubberized airtight material and were supported by air pressure from within.[103] Since the material of air-supported radomes can be relatively thin and uniform, they approximate the electrically ideal thin shell which provides good electrical properties. Such radomes can operate with high transmission efficiency at almost all radar frequencies. Materials include single-ply neoprene-coated terylene or nylon fabric, Hypalon-coated Dacron, and Teflon-coated fiberglass. Inflation pressures are in the vicinity of 0.5 lb/in² gauge. Air-supported radomes can be folded into a small package which make them suitable for transportable radars requiring mobility and quick erection times. Typically, a 50-ft radome can be installed at a prepared site in about one or two hours.[135] They are also of interest on static sites where wideband frequency operation is desired. One of the largest examples of an air-supported radome was the 210 ft diameter radome for the Bell Telephone System Telstar satellite communication antenna at Andover, Maine.

Air-supported radomes have a number of disadvantages. Their life is limited by exposure to ultraviolet light, surface erosion, and the constant flexing of the material in the wind. In high winds the material can be damaged by flying debris and the rotation of the antenna might have to cease to prevent the fabric from being blown against the antenna and torn. Maintenance of the internal pressure in high winds can sometimes be difficult. Another problem is that costly maintenance is required at frequent intervals.

The limitations of air-supported radomes are overcome by the use of rigid self-supporting radomes. The most common is the rigid space-frame, Fig. 7.29, which consists of a three-dimensional lattice of primary load-bearing members enclosed with thin dielectric panels. The panels can be very thin, even for large-diameter radomes, since they are not required to carry main loads or stresses. This type of construction, whereby a spherical structure is constructed from flat plastic panels of simple geometrical shapes, is sometimes called a *geodesic dome*. The

Figure 7.29 Rigid radome for ground-based antenna. (*From Davis and Cohen,*[125] *courtesy Electronics and MIT Lincoln Laboratory.*)

55-ft-diameter radome shown in Fig. 7.29 is designed so that the plastic flanges between the panels take the load while the plastic panels act as thin diaphragms which merely transmit wind-pressure loads to the framework to which they are attached. The supporting framework can also be of steel or aluminum members rather than plastic. Metal space-frame radomes can be of large size, 150-ft diameter being quite practical. Designs up to 500-ft diameter have been considered. The cross section of metallic members can be smaller than that of dielectric members of equivalent strength; hence, the effect of aperture blocking is less. Metal members are not only superior in electrical performance to the equivalent dielectric members but metal space-frame radomes are generally cheaper and easier to fabricate, transport, and assemble, and can be used for larger diameter configurations.

Aluminum structural members, which might be larger than steel of equivalent strength, are of light weight, noncorrosive, and require no maintenance. The load-bearing framework is covered with low-loss fiberglass-reinforced plastic panels. The panels should be non-erosive, water repellent, and designed to reject much of the incident solar radiation. In some radomes, the exterior surface is coated with a white radar-transparent paint, such as Hypalon, to reduce the interior temperature rise caused by solar radiation.

A metal space-frame radome might consist of individual triangular panels made up of a frame of aluminum extrusions encapsulating a low-loss dielectric reinforced plastics laminate membrane. Instead of the uniform panel sizes of Fig. 7.29, the space-frame radome can use a quasi-random selection of different panel sizes to minimize periodic errors in the aperture distribution which can give rise to spurious sidelobes. It also tends to make the radome insensitive to polarization. A typical metal space-frame radome of this type might have a transmission loss of 0.5 dB and cause the antenna sidelobes to increase an average of 1 dB at the −25 dB level. The boresight might be shifted less than 0.1 mrad and the antenna noise temperature might increase less than 5 K.[136] Approximate formulas are available for predicting the electrical effects (gain, beamwidth, sidelobes, and boresight error) of a metal space-frame radome, assuming a uniform or nearly uniform distribution of space-frame elements over the surface of the radome.[104]

The rubberized air-supported radome is an example of a *thin wall* radome in which the thickness of the wall is small compared to a wavelength. A thin-wall radome can also be constructed of plastic with low dielectric constant and low loss tangent. This type of radome approximates a thin shell where the loads and stresses are carried in the shell membrane itself. Unfortunately, the thin skin required for good electrical properties is not consistent with good mechanical properties. This limits the size and the frequency of operation of thin-wall rigid radomes.

Ground-based radomes may utilize foam materials, such as polyester polyurethane, of low dielectric constant and low loss tangent in a relatively thick-wall construction to meet structural requirements with excellent electrical performance over a wide frequency band. This is known as a *foam shell* radome. The individual panels may be joined together by application of an epoxy adhesive to the panel edges or by "welding" together with the same material to form a homogeneous shell.[105]

The structural limitation of the thin wall radome can be overcome by the *half-wave wall* radome which utilizes a homogeneous dielectric with an electrical thickness of half wavelength, or a multiple thereof, at the appropriate incidence angle. The half-wave thickness is nonreflecting if ohmic losses are negligible. The bandwidth of such a radome is limited, as is the range of incidence angles over which the energy is transmitted with minimal reflection.

The A sandwich is a three-layer wall consisting of a core of low-dielectric-constant material with a thickness of approximately one-quarter wavelength. This inner core is sandwiched between two thin outer layers, or skins, of a high-dielectric-constant material relative to that of the core. The skins might typically have a dielectric constant of about 4, and the core might have a value of about 1.2. The skins are thin compared to a wavelength. The core might be a honeycomb or fluted construction. The strength-to-weight ratio of the A sandwich is greater than that of a solid-wall radome. It is also capable of broader bandwidth. However, it is more sensitive to variations in polarization and angle of incidence. The electrical characteristics of the adhesive used for bonding the skins to the core must be taken into account in the design since it can make the skin look (electrically) thicker than its physical length. The structure must be properly sealed against moisture absorption.

The *B sandwich* is the "inverse" of the A sandwich. It is a three layer structure whose quarter-wave-thick skins have a dielectric constant lower than that of the core material. The B sandwich is the microwave analog of matching coatings used in optical lenses. Although it is superior electrically to the A sandwich, it is heavier and is not generally suited to the environmental conditions encountered in aircraft. It is not commonly used.

The *C sandwich* can be thought of as two back-to-back A sandwiches. It consists of five layers. There are two outer skins, a center skin, and two intermediate cores. It is used when the ordinary A sandwich does not provide sufficient strength.

When extreme structural rigidity or broadband capabilities are required with relatively light weight, multiple-layer sandwiches of seven, nine, eleven, or more layers may be considered.

Aircraft radomes, especially those used at supersonic speeds, are subject to mechanical stress and aerodynamic heating so severe that the electrical requirements of radomes made of dielectric materials must be sacrificed to obtain sufficient mechanical strength. The radome wall configurations described above are applicable to airborne radars, but another approach is based on the fact that a metal sheet with periodically spaced slots exhibits a bandpass characteristic. Thus thin metallic radomes, pierced with many openings (slots) to make it transparent to microwaves, offer the possibility of overcoming the mechanical limitations of dielectric radomes, yet result in good electrical properties.[106-108] A metallic structure not only has the potential for greater mechanical strength than dielectric radomes and to better distribute frictionally-induced heating, but it should be able to better withstand the stresses caused by rain, hail, dust, and lightning. Static buildup of charge and subsequent discharge to the airframe, encountered with dielectric radomes, can be eliminated with metallic radomes. In one experimental design of a conical-shape radome, periodic resonant slots were cut into the metallic surface so that approximately 90 percent of the radome surface was metal. Within its design band (8.8–9.0 GHz) the transmission properties were nearly ideal and accommodated scanning antennas transmitting arbitrary polarized signals. The reduced response outside the design band reduces the effects of out-of-band interference and can reduce the nose-on radar cross section of the aircraft at frequencies other than that for which the antenna is designed.

In conventional application, the radome is fixed and the antenna is scanned. It is sometimes of advantage to construct the antenna and radome to rotate together as a unit. This is called a *rotodome*. They have been used in ground-based systems as well as in airborne aircraft-surveillance radars such as the 24-ft-diameter radome in the U.S. Navy's E-2C (Fig. 7.30) or the 30-ft-diameter radome of AWACS. The E-2C radome weighs just over 2000 lb. The entire unit rotates at a 6 rpm rate. It houses a broadside array of Yagi-type endfire elements.

Figure 7.30 E-2C AEW aircraft with rotodome antenna.

Weather effects. A ground-based radar that operates without benefit of a radome must be designed to withstand wind loads.[109-113] Surveillance radars must not only be able to operate in strong winds, but must rotate uniformly. Rotating antennas outside of radomes cannot be operated at winds that are too great. They sometimes have to be shut down and securely fastened in strong gale winds. Even if extreme winds are not encountered, a radome-enclosed antenna has the advantage that it can be rotated with a much smaller motor than if it were outside the radome exposed to even normal winds.

Solid reflector surfaces require more drive power in wind than do lattice or tubular surfaces. If the exposed antenna is subject to icing that can close the holes of a mesh reflector, the mechanical design might have to be made on the basis of a solid surface anyway. Since the torque on an antenna depends on the wind direction, it will vary as the antenna rotates. It is often possible to reduce the wind torques by selecting the optimum position for the axis of rotation and by adding fins to the structure.[114] The torque on the fins can be made to be in the opposite direction to the torque on the antenna so as to cause a net reduction.

One of the attractions of the rigid radome is its ability to withstand the rigors of severe climate. Rime ice, the prevalent type of icing found in the Arctic region, has little or no effect on most radomes. Although it tends to collect on many types of structures and can obtain large thicknesses, both theory and experiment show a lack of rime-ice formation on a spherical radome.[115] The trajectories of water droplets in the air stream flowing around large spheres do not result in much impingement. Collection efficiencies might be only a few percent. The droplet sizes of freezing rain, however, are large and the collection efficiency can approach 100 percent. Dry snow does not stick to cold surfaces and is generally no problem. Snow might collect on the top portion of the radome but will be quickly removed by any following wind. Also, the effect of snow on electromagnetic propagation is less than with water. Wet snow can stick to the radome and its high liquid-water content can adversely affect propagation. The removal of snow by thermal means (heaters) is generally quite expensive. In some of the smaller rigid radomes, snow can be removed mechanically by tying a rope at the top and having someone walk the rope around the radome to knock off the snow.

Liquid water can collect on a radome due to condensation or rainfall. In some cases, the impingement of rain on a radome can have a more serious effect on the system performance than the rainfall along the propagation path. The losses due to water layers on radomes have been calculated by Blevis.[116,117]

Theory predicts that a 4-mm/h-rain will produce a layer of water 0.142 mm thick on a 55-ft diameter radome. The theoretical loss through such a water layer is 1.5 dB at 4.2 GHz and 4.4 dB at 9.36 GHz. Ruze[118] confirmed Blevis's calculations and noted that a nonuniformly wet radome can cause significant phase perturbations in the aperture illumination, as well as a transmission loss.

If the radome surface absorbs moisture it can seriously degrade its performance in rain. Cohen and Smolski[119] attribute most of the measured loss in the Bell Telephone air-inflated radome at Andover, Maine to this effect. In immersion tests, the Hypalon-coated Dacron fabric of the Andover air-inflated radome was found to absorb 13.2 percent water while the fibrous glass-reinforced laminate of rigid radomes absorbed less than 0.5 percent. Furthermore the fibrous-glass-reinforced laminate took about two minutes to dry but the air-inflated radome material required about 30 to 45 min.

The losses due to rain can be reduced significantly by treating the radome surface to make it nonwetting, or water repellent.[119-121] Experiments at 4.2 GHz with a 55-ft rigid metal space-frame radome at a simulated rain rate of 40 mm/h (a very high rate) had a measured loss of from 1.0 to 1.7 dB.[119] When treated to decrease its wettability the loss was less than 0.3 dB. The theoretical loss calculated by the method described by Blevis was 3.4 dB. The experimental values were less than predicted by theory since the water was not a uniform film, but formed many small streaks that rapidly ran off the radome in narrow rivulets.

The performance of a reflector without the protection of a radome can also be degraded by rain. The effect is small, however, at frequencies below X band. Water impinging on the antenna feed can have serious effects. The wetting of surfaces through which signals are transmitted should be avoided. Also, adequate drainage should be provided to prevent the accumulation of water in the reflector. Ice coating a reflector surface does not usually cause a problem. When ice is in the process of melting, however, the water surface can distort the pattern and in some instances can completely destroy the main beam.

7.10 STABILIZATION OF ANTENNAS[14]

If the radar platform is unsteady, as when it is located on a ship or an aircraft, the antenna pointing must be properly compensated for this undesired motion. *Stabilization* is the use of a servomechanism to control the angular position of an antenna so as to compensate automatically for changes in the angular position of the vehicle carrying the antenna. An antenna not compensated for the angular motion of the platform will have degraded coverage. On a ship, for example, an unstabilized antenna with its beam parallel to the deck plane will have its coverage shortened for surface targets since the beam will be looking into the water or up in the air as the ship rolls or pitches. In addition to degraded coverage, the measurement of the angular position with an unstabilized antenna can be in error unless the tilting of the platform is properly taken into account.

Stabilization of the antenna adds to the weight and size of the radar, but it is necessary in many applications of radar on ships and aircraft. The requirements for stabilization depend in part on the nature of the application. Requirements differ whether the radar is used for surface search, air search, tracking, or height finding.

There are three kinds of platform motion that can affect the location of the antenna beam. *Roll* is the side-to-side angular motion around the longitudinal (fore-and-aft) axis of the ship or aircraft. *Pitch* is the alternating motion of the fore and aft (bow and stern) parts of the vehicle. *Yaw* is the motion of the platform around the vertical axis.

Stabilization requires that the vehicle be equipped with a device, such as gyroscope, which is sensitive to directions in space and which measures the deviation of the platform from its level position. The gyroscope device used as the reference from which to stabilize the antenna is called a *stable element* or a *stable vertical*.

The direct approach to stabilization would be to provide a *stable base* that maintains the level no matter what the tilt of the ship or aircraft. The stable base compensates for roll and pitch. Yaw can usually be taken into account by a correction to the angle read-out.

In some applications, the pitch is small so that little harm results from omitting the stabilization of the pitch axis, leaving only the roll axis stabilized. This results in a simpler system compared to providing a stable base. This is called *roll stabilization*.

A different approach to the problem is *line-of-sight*, or *tilt*, stabilization. Instead of mounting the antenna on a level platform, the antenna is tilted about the elevation axis so as to automatically maintain the beam pointed at the horizontal. (The beam direction can also be maintained constant at whatever angle above or below the horizon is desired.) The line of sight is thereby stabilized.

The unsteady motions of the platform can cause errors in the angular measurement and distortion of the data on indicators like the PPI. Corrections can be applied in some cases to account for these distortions. This is called *data stabilization*. Correction of the data may still be required even with some form of mechanically stabilized mounts. Data stabilization is usually used with pencil-beam tracking antennas. A computer can readily calculate the angular corrections to the output data to account for platform tilt.

An example of the data stabilization is the correction applied for yaw. If a PPI were to display relative bearing (azimuth) the apparent bearing would change as the platform yawed. With the aid of a horizontal gyroscope to provide a north reference, a correction signal can be obtained which can be applied to the indicator to permit the display of the target in its true bearing rather than the bearing relative to the platform heading. Generally the top of the display corresponds to north. This true bearing display eliminates the blurring of a persistent screen display caused by the natural random changes in heading of the vehicle. It also eliminates the confusion sometimes found in a relative-bearing display caused by the entire display rotating as the course of the vehicle is altered.

Correction of the data is also needed with line-of-sight stabilization using a two-axis mount. Data stabilization is not necessary if a three-axis mount is used, such as e or f of Fig. 7.31.

Nine possible arrangements for mounting antennas are shown in Fig. 7.31. The one-axis mount is the simplest. Most commercial marine shipboard radars used for navigation employ such a mount. The elevation beamwidth is made broad enough (typically 20°) so that the surface of the sea remains illuminated during the roll of the ship. A correction can be applied to the indicator (data stabilization) to account for the error.

The two-axis mount of Fig. 7.31b, sometimes called an az-el mount, enables the beam to be maintained at the horizontal or at any angle above or below the horizon (line-of-sight stabilization). The beam can be directed to any point by the proper combination of azimuth (train) and elevation angles. For targets overhead, the angular accelerations required are quite high. It is not practical to use such a mount, therefore, for tracking targets through the zenith, as in the tracking of satellites. The rearrangement of the two axes as in (c) takes care of this problem. Tracking through the zenith is possible without encountering impractical drive accelerations. Such a mount, however, transfers the problem of excessive accelerations to some other direction; in this case, in the direction of the elevation axis. A three-axis mount avoids the problem of excessive acceleration.

The two-axis mount of (d) is similar to (c) except that it is arranged to provide roll stabilization. The roll data from a stable vertical can be applied to the basic roll axis without the need for computer orders as required in the other two-axis mounts. The antenna is then stabilized about the azimuth axis if the pitch angle is small.

The three-axis mount of (e) can provide full hemispheric coverage. The mount in (f) might be desired where antenna elevation is not required. The arrangement in (g) can be trained on target with minimum rate or acceleration requirements on all axes, irrespective of the position of the target or motion of the platform. In the three-axis arrangement of (h) a stable base is provided in which a roll axis lies in a fixed position parallel to the fore-and-aft line of the vehicle. This axis carries the pitch axis that supports the train axis. Pitch and roll signals from a stable vertical are applied as inputs to the corresponding axes. The result is that the azimuth (or train) axis is stabilized in the vertical and may receive a direct order of relative azimuth. Computations are not required as in other mounts. The antenna may be rotated at constant rates higher than would be practical with other mounts requiring application of a correction for a variable platform-tilt. This three-axis stable base antenna has no elevation angle and is limited to applications which scan the horizon or a fixed angle of elevation with respect to the horizon. If elevation scanning at high rates is required, as in height finders or 3-D radar, a fourth axis can be supplied as in (i). This configuration, although it requires no computer, is relatively massive and heavy because it has four servo systems for controlling

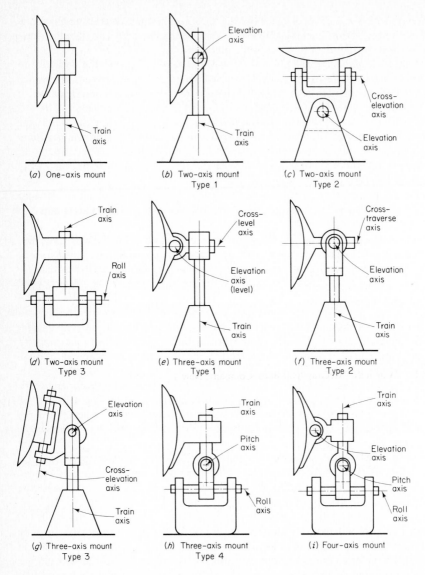

Figure 7.31 Arrangements of axes for stabilized antenna mounts. (*From Cady, Karelitz, and Turner,*[14] *chap. 4, courtesy McGraw-Hill Book Company.*)

rotation about each of the four axes. The three-axis mount of (*h*) is simpler, but also has similar disadvantages because of weight and size.

In the above, the stabilization of the mount was assumed to be by some form of mechanical compensation. Radars which electronically scan a beam in elevation (usually a pencil beam) by either phase shifters or frequency scan can stabilize the beam in elevation as a correction to the elevation scan orders, thus permitting a reduction in the size of the mount. The stabilization of a phased array antenna capable of electronically scanning the beam is accomplished in the computer that generates the steering orders.

Sometimes the terms *level* and *cross level* are used to refer to the angles in a stabilized system.[14] The *level angle* is the angle between the horizontal plane and the deck plane, measured in the vertical plane through the line of sight. The *cross-level angle* is the angle, measured about the line of sight, between the vertical plane through the line of sight and the plane perpendicular to the deck through the line of sight.

Antenna drives.[126,127] Mechanical tracking radars and nodding-beam height finders require variable drive power. The choice has been between electrical and hydraulic systems. Electrical drives are the Ward Leonard type or thyristor bridges driving dc motors. Hydraulic drives are often cheaper than electric drives and in the larger sizes (above 50 hp) they are more compact and often lighter. It has been said that hydraulics are less trouble to maintain. However, they tend to exhibit a slow but predictable deterioration with respect to internal leakage as a result of mechanical wear.

Both electric and hydraulic drives are used in surveillance radars. Surveillance antennas operate at constant rotation rate, but some variation can be tolerated so long as it is not sufficient to produce uneven illumination of the cathode-ray tube display or degradation of the MTI.

REFERENCES

1. Silver, S. (ed.): "Microwave Antenna Theory and Design," MIT Radiation Laboratory Series, vol. 12, McGraw-Hill Book Company, New York, 1949.
2. Kraus, J. D.: "Antennas," McGraw-Hill Book Company, New York, 1950.
3. Cutler, C. C., A. P. King, and W. E. Kock: Microwave Antenna Measurements, *Proc. IRE*, vol. 35, pp. 1462–1471, December, 1947.
4. Stratton, J. A.: "Electromagnetic Theory," McGraw-Hill Book Company, New York, 1941.
5. Sherman, J. W.: Aperture-antenna Analysis, chap. 9 of "Radar Handbook," M. I. Skolnik (ed.), McGraw-Hill Book Company, New York, 1970.
6. IEEE Test Procedures for Antennas, *IEEE Trans.*, vol. AP-13, pp. 437–466, May, 1965.
7. Fry, D. W., and F. K. Goward: "Aerials for Centimetre Wavelengths," Cambridge University Press, New York, 1950.
8. Sciambi, A. F.: The Effect of the Aperture Illumination on the Circular Aperture Antenna Pattern Characteristics, *Microwave J.*, vol. 8, pp. 79–84, August, 1965.
9. Cutler, C. C.: Parabolic-antenna Design for Microwaves, *Proc. IRE*, vol. 35, pp. 1284–1294, November, 1947.
10. Gray, C. L.: Estimating the Effect of Feed Support Member Blocking on Antenna Gain and Side Lobe Level, *Microwave J.*, vol. 7, pp. 88–91, March, 1964.
11. Ruze, J.: Feed Support Blockage Loss in Parabolic Antennas, *Microwave J.*, vol. 11, pp. 76–80, December, 1968.
12. Moore, R. L.: Passive ECM Applied to False Target Elimination, *Record of the IEEE 1975 International Radar Conference*, pp. 458–462, IEEE Publication 75 CHO 938-1 AES.
13. Foster, K.: Parabolic Cylinder Aerials, *Wireless Engr.*, vol. 33, pp. 59–65, March, 1956.
14. Cady, W. M., M. B. Karelitz, and L. A. Turner (eds.): "Radar Scanners and Radomes," MIT Radiation Laboratory Series, vol. 26, McGraw-Hill Book Company, New York, 1948.
15. Pugh, S., and D. E. Walker: Reflector Surfaces for Communications and Radar Aerials, "Design and Construction of Large Steerable Aerials," *Institution of Electrical Engineers (London) IEE Conference Publication* no. 21, pp. 369–373, 1966.
16. Richards, C. J., and M. Marchant: Structural Aspects: pt 1, Elementary Design Considerations, chap. 7 of "Mechanical Engineering in Radar and Communications," C. J. Richards (ed.), Van Nostrand Reinhold Co., New York, 1969.
17. Chu, T. S., and R. H. Turrin: Depolarization Properties of Offset Reflector Antennas, *IEEE Trans.*, vol. AP-21, pp. 339–345, May, 1973.

18. Paramonov, A. A.: Influence of Precipitation on the Electrical Properties of Wire Mesh Surfaces, *Radiotekhnika*, vol. 11, no. 9, pp. 12–20, 1956.
19. Hannan, P. W.: Microwave Antennas Derived from the Cassegrain Telescope, *IRE Trans.*, vol. AP-9, pp. 140–153, March, 1961.
20. Morgan, S. P.: Some Examples of Generalized Cassegrainian and Gregorian Antennas, *IEEE Trans.*, vol. AP-12, pp. 685–691, November, 1964.
21. Collins, G. W.: Shaping of Subreflectors in Cassegrain Antennas for Maximum Aperture Efficiency, *IEEE Trans.*, vol. AP-21, pp. 309–313, May, 1973.
22. Graham, R.: The Polarisation Characteristics of Offset Cassegrain Aerials, "Radar—Present and Future," Oct. 23–25, 1973, *IEE Conf. Publ. no. 105*, pp. 134–139.
23. Dahlsjö, O.: A Low Side Lobe Cassegrain Antenna, "Radar—Present and Future," Oct. 23–25, 1973, *IEE Conf. Publ. no. 105*, pp. 408–413.
24. Weiss, H. G.: The Haystack Microwave Research Facility, *Spectrum*, vol. 2, pp. 50–59, February, 1965.
25. Josefsson, L. G.: A Broad-band Twist Reflector, *IEEE Trans.*, vol. AP-19, pp. 552–554, July, 1971.
26. Lo, Y. T.: On the Beam Deviation Factor of a Parabolic Reflector, *IRE Trans.*, vol. AP-8, pp. 347–349, May, 1960.
27. Sandler, S. S.: Paraboloidal Reflector Patterns of Off-axis Feed, *IRE Trans.*, vol. AP-8, pp. 368–379, July, 1960.
28. Ruze, J.: Lateral Feed Displacement in a Paraboloid, *IEEE Trans.*, vol. AP-13, pp. 660–665, September, 1965.
29. Imbriale, W. A., P. G. Ingerson, and W. C. Wong: Large Lateral Feed Displacements in a Parabolic Reflector, *IEEE Trans.*, vol. AP-22, pp. 742–745, November, 1974.
30. Ashmead, J., and A. B. Pippard: The Use of Spherical Reflectors as Microwave Scanning Aerials, *J. IEEE (London)*, vol. 93, pt. IIIA, p. 627, 1946.
31. Ramsey, J. F., and J. A. C. Jackson: Wide-angle Scanning Performance of Mirror Antennas, *Marconi Rev.*, vol. 19, 3d qtr., pp. 119–140, 1956.
32. Li, T.: A Study of Spherical Reflectors as Wide-angle Scanning Antennas, *IRE Trans.*, vol. AP-7, pp. 223–226, July, 1959.
33. Spencer, R. C., and G. Hyde: Studies of the Focal Region of a Spherical Reflector: Geometric Optics, *IEEE Trans.*, vol. AP-16, pp. 317–324, May, 1968.
34. Love, A. W.: Scale Model Development of a High Efficiency Dual Polarized Feed for the Arecibo Spherical Reflector, *IEEE Trans.*, vol. AP-21, pp. 628–639, September, 1973.
35. Ronchi, L., and G. Toraldo di Francia: An Application of Parageometrical Optics to the Design of a Microwave Mirror, *IRE Trans.*, vol. AP-6, pp. 129–133, January, 1958.
36. Provencher, J. H.: Experimental Study of a Diffraction Reflector, *IRE Trans.*, vol. AP-8, pp. 331–336, May, 1960.
37. Kelleher, K. S.: A New Microwave Reflector, *IRE Natl. Conv. Record*, vol. 1, pt. 2, pp. 56–58, 1953.
38. Peeler, G. D. M., and D. H. Archer: A Toroidal Microwave Reflector, *IRE Natl. Conv. Record*, vol. 4, pt. 1, pp. 242–247, 1956.
39. Kelleher, K. S., and H. H. Hibbs: A New Microwave Reflector, *Naval Research Lab. Rept.* 4141, 1953.
40. Jackson, J., and E. Goodall: A 360° Scanning Microwave Reflector, *Marconi Rev.*, vol. 21, 1st qtr., pp. 30–38, 1958.
41. Barab, J. D., J. G. Marangoni, and W. G. Scott: The Parabolic Dome Antenna: A Large Aperture, 360 Degree, Rapid Scan Antenna, *IRE WESCON Conv. Record*, vol. 2, pt. 1, pp. 272–293, 1958.
42. Jasik, H.: "Antenna Engineering Handbook," McGraw-Hill Book Co., New York, 1961, sec. 15.6.
43. Johnson, R. C.: Optical Scanners, chap. 3 of "Microwave Scanning Antennas, vol. 1," R. C. Hansen (ed.), Academic Press, New York, 1964.
44. Foster, J. S.: A Microwave Antenna with Rapid Saw-Tooth Scan, *Canadian J. of Phys.*, vol. 36, pp. 1652–1660, 1958.
45. Martindale, J. P. A.: Lens Aerials at Centimetric Wavelengths, *J. Brit. IRE*, vol. 13, pp. 243–259, May, 1953.
46. Stuetzer, O. M.: Development of Artificial Microwave Optics in Germany, *Proc. IRE*, vol. 38, pp. 1053–1056, September, 1950.
47. Harvey, A. F.: Optical Techniques at Microwave Frequencies, *Proc. IEE*, vol. 106, pt. B, pp. 141–157, March, 1959. Contains an extensive bibliography.
48. Kelleher, K. S., and C. Goatley: Dielectric Lens for Microwaves, *Electronics*, vol. 28, pp. 142–145, August, 1955.
49. Kock, W. E.: Metallic Delay Lens, *Bell System Tech. J.*, vol. 27, pp. 58–82, January, 1948.
50. Kock, W. E.: Metal Lens Antennas, *Proc. IRE*, vol. 34, pp. 828–836, November, 1946.

51. Kock, W. E.: Path-length Microwave Lenses, *Proc. IRE*, vol. 37, pp. 852–855, August, 1949.
52. Ruze, J.: Wide-angle Metal-plate Optics, *Proc. IRE*, vol. 38, pp. 53–59, January, 1950.
53. Luneburg, R. K.: "Mathematical Theory of Optics" (mimeographed lecture notes), pp. 212–213, Brown University Graduate School, Providence, R.I., 1944. Published by University of California Press, Berkeley, 1964.
54. Peeler, G. D. M., and H. P. Coleman: Microwave Stepped-index Luneburg Lenses, *IRE Trans.*, vol. AP-6, pp. 202–207, April, 1958.
55. Buckley, E. F.: Stepped-index Luneburg Lenses, *Electronic Design*, vol. 8, pp. 86–89, Apr. 13, 1960.
56. Jasik, H.: The Electromagnetic Theory of the Luneburg Lens, *AFCRC Rept.* TR-54-121, November, 1954.
57. Braun, E. H.: Radiation Characteristics of the Spherical Luneburg Lens, *IRE Trans.*, vol. AP-4, pp. 132–138, April, 1956.
58. Morgan, S. P.: General Solution of the Luneburg Lens Problem, *J. Appl. Phys.*, vol. 29, pp. 1358–1368, September, 1958.
59. Webster, R. E.: Radiation Patterns of a Spherical Luneburg Lens with Simple Feeds, *IRE Trans.*, vol. AP-6, pp. 301–302, July, 1958.
60. Kelleher, K. S.: Designing Dielectric Microwave Lenses, *Electronics*, vol. 29, pp. 138–142, June, 1956.
61. Bohnert, J. I., and H. P. Coleman: Applications of the Luneburg Lens, *Naval Research Lab. Rept.* 4888, Mar. 7, 1957.
62. Peeler, G. D. M., K. S. Kelleher, and H. P. Coleman: Virtual Source Luneburg Lenses, *Symposium on Microwave Optics*, McGill University, Montreal, AFCRC-TR-59-118(I), ASTIA Document 211499, pp. 18–21, April, 1959.
63. Gunderson, L. C., and J. F. Kauffman: A High Temperature Luneburg Lens, *Proc. IEEE*, vol. 56, pp. 883–884, May, 1968.
64. Hollis, J. S., and M. W. Long: A Luneburg Lens Scanning System, *IRE Trans.*, vol. AP-5, pp. 21–25, January, 1957.
65. Rinehart, R. F.: A Solution to the Problem of Rapid Scanning for Radar Antennas, *J. Appl. Phys.*, vol. 19, pp. 860–862, September, 1948.
66. Johnson, R. C.: The Geodesic Luneburg Lens, *Microwave J.*, vol. 5, pp. 76–85, August, 1962.
67. Johnson, R. C., and R. M. Goodman: Geodesic Lenses for Radar Antennas, *EASCON Rec.*, 1968, pp. 64–69.
68. Huynen, J. R.: Theory and Design of a Class of Luneburg Lenses, *IRE WESCON Conv. Record*, vol. 2, pt. 1, pp. 219–230, 1958.
69. Bekefi, G., and G. W. Farnell: A Homogeneous Dielectric Sphere as a Microwave Lens, *Can. J. Phys.*, vol. 34, pp. 790–803, August, 1956.
70. ap Rhys, T. L.: The Design of Radially Symmetric Lenses, *IEEE Trans.*, vol. AP-18, pp. 497–506, July, 1970.
71. Free, W. R., F. L. Cain, C. E. Ryan, C. P. Burns, and E. M. Turner: High-Power Constant-Index Lens Antennas, *IEEE Trans.*, vol. AP-22, pp. 582–584, July, 1974.
72. Cheston, T. C.: Microwave Lenses, *Symposium on Microwave Optics*, McGill University, Montreal, AFCRC-TR-59-118(I), ASTIA Document 211499, pp. 8–17, April, 1959.
73. Ruze, J.: Physical Limitations on Antennas, *MIT Research Lab. Electronics Tech. Rept.* 248, Oct. 20, 1952.
74. Wolff, I.: Determination of the Radiating System Which Will Produce a Specified Directional Characteristic, *Proc. IRE*, vol. 25, pp. 630–643, May, 1937.
75. Block, A., R. C. Medhurst, and S. D. Pool: Superdirectivity, *Proc. IRE*, vol. 48, p. 1164, June, 1960.
76. Jordan, E. C.: "Electromagnetic Waves and Radiating Systems," sec. 12.17, Prentice-Hall, Inc., Englewood Cliffs, N. J., 1950.
77. Woodward, P. M.: A Method of Calculating the Field over a Plane Aperture Required to Produce a Given Polar Diagram, *J. IEE.* vol. 93, pt. IIIA, pp. 1554–1558, 1946.
78. Woodward, P. M., and J. D. Lawson: The Theoretical Precision with Which an Arbitrary Radiation Pattern May Be Obtained with a Source of Finite Size, *J. IEE*, vol. 95, pt. IIIA, pp. 362–370, September, 1948.
79. Rhodes, D. R.: "Synthesis of Planar Antenna Sources," Oxford University Press, London, 1974.
80. Dolph, C. L.: A Current Distribution for Broadside Arrays Which Optimizes the Relationship between Beamwidth and Side Lobe Level, *Proc. IRE*, vol. 34, pp. 335–348, June, 1946; also discussion by H. J. Riblet, vol. 35, pp. 489–492.
81. Drane, C. J.: Dolph-Chebyshev Excitation Coefficient Approximation, *IEEE Trans.*, vol. AP-12, pp. 781–782, November, 1964.
82. Stegen, R. J.: Gain of Tchebycheff Arrays, *IEEE Trans.*, vol. AP-8, pp. 629–631, November, 1960.

83. van der Maas, G. J.: A Simplified Calculation for Dolph-Tschebyscheff Arrays, *J. Appl. Phys.*, vol. 25, pp. 121–124, January, 1954.
84. Taylor, T. T.: Design of Line-source Antennas for Narrow Beamwidth and Low Side Lobes, *IRE Trans.*, vol. AP-3, pp. 16–28, January, 1955.
85. Hansen, R. C.: Gain Limitations of Large Antennas, *IRE Trans.*, vol. AP-8, pp. 490–495, September, 1960.
86. Taylor, T. T.: Design of Circular Apertures for Narrow Beamwidth and Low Sidelobes, *IRE Trans.*, vol. AP-8, pp. 17–22, January, 1960.
87. Hansen, R. C.: Tables of Taylor Distributions for Circular Aperture Antennas, *IRE Trans.*, vol. AP-8, pp. 23–26, January, 1960.
88. Bayliss, E. T.: Design of Monopulse Antenna Difference Patterns with Low Sidelobes, *Bell System Tech. J.*, vol. 47, pp. 623–650, May-June, 1968.
89. Levine, D.: "Radargrammetry," McGraw-Hill Book Co., New York, 1960.
90. Shanks, H. E.: A Geometrical Optics Method of Pattern Synthesis for Linear Arrays, *IRE Trans.*, vol. AP-8, pp. 485–490, September, 1960.
91. Hutchison, P. T.: The Image Method of Beam Shaping, *IRE Trans.*, vol. AP-4, pp. 604–609, October, 1956.
92. Hines, J. N., V. H. Rumsey, and C. H. Walter: Traveling-wave Slot Antennas, *Proc. IRE*, vol. 41, pp. 1624–1631, November, 1953.
93. Hougardy, R. W., and R. C. Hansen: Scanning Surface Wave Antennas: Oblique Surface Waves over a Corrugated Conductor, *IRE Trans.*, vol. AP-6, pp. 370–376, October, 1958.
94. Dunbar, A. S.: Calculation of Doubly Curved Reflectors for Shaped Beams, *Proc. IRE*, vol. 36, pp. 1289–1296, October, 1948.
95. Carberry, T. F.: Analysis Theory for the Shaped-Beam Doubly Curved Reflector Antenna, *IEEE Trans.*, vol. AP-17, pp. 131–138, March, 1969.
96. Brunner, A.: Possibilities of Dimensioning Doubly Curved Reflectors for Azimuth-Search Radar Antennas, *IEEE Trans.*, vol. AP-19, pp. 52–57, January, 1971.
97. Shrader, W. W.: MTI Radar, "Radar Handbook," M. I. Skolnik (ed.), McGraw-Hill Book Co., New York, 1970, sec. 17.20.
98. Shrader, W. W.: Antenna Considerations for Surveillance Radar Systems, *Seventh Annual East Coast Conference on Aeronautical and Navigational Electronics*, Baltimore, Md., October, 1960.
99. Ruze, J.: The Effect of Aperture Errors on the Antenna Radiation Pattern, *Suppl. al Nuovo cimento*, vol. 9, no. 3, pp. 364–380, 1952; also reprinted in *Proc. Symposium on Communication Theory and Antenna Design*, AFCRC Tech. Rept. 57–105, ASTIA Document 117067, January, 1957.
100. Bracewell, R. N.: Tolerance Theory of Large Antennas, *IRE Trans.*, vol. AP-9, pp. 49–58, January, 1961.
101. Ruze, J.: Antenna Tolerance Theory—A Review, *Proc. IEEE*, vol. 54, pp. 633–640, April, 1966.
102. Skolnik, M. I.: Large Antenna Systems, chap. 28 of "Antenna Theory, pt. 2," R. E. Collin and F. J. Zucker (eds.), McGraw-Hill Book Co., New York, 1969.
103. Simpson, P. J.: Mechanical Aspects of Ground-Based Radomes, Chap. 12 of "Mechanical Engineering in Radar and Communications," C. J. Richards (ed.), Van Nostrand Reinhold Co., New York, 1969.
104. Kay, A. F.: Electrical Design of Metal Space Frame Radomes, *IEEE Trans.*, vol. AP-13, pp. 188–202, March, 1965.
105. Larrench, W.: A Homogeneous, Rigid, Ground Radome, *Proc. IEE*, vol. 109B, pp. 445–446, November, 1962.
106. Oh, L. L., C. D. Lunden, and C. Chiou: Fenestrated Metal Radome, *Microwave J.*, vol. 7, pp. 62–65, April, 1964.
107. Oh, L. L., and C. D. Lunden: A Slotted Radome Cap for Rain, Hail, and Lightning Protection, *Microwave J.*, vol. 11, pp. 105–108, March, 1968.
108. Pelton, E. L., and B. A. Munk: A Streamlined Metallic Radome, *IEEE Trans.*, vol. AP-22, pp. 799–803, November, 1974.
109. Brown, J. S., and K. E. McKee: Wind Loads on Antenna Systems, *Microwave J.*, vol. 7, pp. 41–46, September, 1964.
110. Thom, H.: Distribution of Extreme Winds in the United States, *ASCE Trans.*, vol. 126, pt. II, paper 3191, 1961.
111. Wind Forces on Structures, *ASCE Trans.*, vol. 126, pt. II, paper 3269, 1961.
112. Hirst, H., and K. E. McKee: Wind Forces on Parabolic Antennas, *Microwave J.*, vol. 8, pp. 43–47, November, 1965.

113. Scruton, C., and P. Sachs: Wind Effects on Microwave Reflectors, "Design and Construction of Large Steerable Aerials," *IEEE (London) Conf. Publication* no. 21, pp. 29–38, 1966.
114. Webster, A. J.: Wind Torques on Rotating Radar Aerials, *Marconi Rev.*, vol. 28, pp. 147–170, 2d qtr., 1965.
115. Vitale, J. A.: Large Radomes, chap. 5 of " Microwave Scanning Antennas, vol. 1," R. C. Hansen (ed.), Academic Press, N.Y., 1964.
116. Blevis, B. C.: Rain Effects on Radomes and Antenna Reflectors, " Design and Construction of Large Steerable Aerials," *IEE (London) Conference Publication no. 21*, pp. 148–151, 1966.
117. Blevis, B. C.: Losses Due to Rain on Radomes and Antenna Reflecting Surfaces, *IEEE Trans.*, vol. AP-13, pp. 175–176, January, 1965.
118. Ruze, J. C.: More on Wet Radomes, *IEEE Trans.*, vol. AP-13, pp. 823–824, September, 1965.
119. Cohen, A., and A. Smolski: The Effect of Rain on Satellite Communications Earth Terminal Rigid Radomes, *Microwave J.*, vol. 9, pp. 111–121, September, 1966.
120. Weigand, R. M.: Performance of a Water-Repellent Radome Coating in an Airport Surveillance Radar, *Proc. IEEE*, vol. 61, pp. 1167–1168, August, 1973.
121. Anderson, I.: Measurements of 20-GHz Transmission Through a Radome in Rain, *IEEE Trans.*, vol. AP-23, pp. 619–622, September, 1975.
122. Kay, A. F.: Radomes and Absorbers, chap. 32 of "Antenna Engineering Handbook," H. Jasik (ed.), McGraw-Hill Book Company, New York, 1961.
123. Dicaudo, V. J.: Radomes, Chap. 14 of " Radar Handbook," M. I. Skolnik (ed.), McGraw-Hill Book Co., New York, 1970.
124. Walton, J. R., Jr.: " Radome Engineering Handbook," Marcel Dekker, Inc., New York, 1970.
125. Davis, P., and A. Cohen: Rigid Radome Design Considerations, *Electronics*, vol. 32, no. 16, pp. 66, 68–69, Apr. 17, 1959.
126. Hastings, K. N., and J. G. Chaplin: Drives for Steerable Antennas. Hydraulics versus Electro-Mechanics, " Design and Construction of Large Steerable Aerials," *IEE (London) Conference Publication no. 21*, pp. 314–318, 1966.
127. Selby, R. P.: Power Drives, chap. 9 of " Mechanical Engineering in Radar and Communications," C. J. Richards (ed.), Van Nostrand Reinhold Co., New York, 1969.
128. Kelleher, K. S., and H. H. Hibbs: Organ-pipe Radar Scanner, *Electronics*, vol. 25, pp. 126–127, May, 1952.
129. Peeler, G. D. M., K. S. Kelleher, and H. H. Hibbs: An Organ Pipe Scanner, *IRE Trans.*, no. PGAP-1, pp. 113–122, February, 1952.
130. Clarricoates, P. J. B., and G. T. Poulton: High-Efficiency Microwave Reflector Antennas—A Review, *Proc. IEEE*, vol. 65, pp. 1470–1504, October, 1977.
131. Green, T. J.: The Influence of Masts on Ship-Borne Radar Performance, RADAR-77, Oct. 25–28, 1977, London, IEE Conference Publication no. 105, pp. 405–408, available from IEEE, New York, 77CH1271-6 AES.
132. See Sec. 12.10 of ref. 1.
133. Mattingly, R. L.: Radar Antennas, chap. 25 of "Antenna Engineering Handbook," H. Jasik, (ed.), McGraw-Hill Book Company, New York, 1961, sec. 25.2.
134. Nicholls, R. B.: Advances in the Design of Foster Scanners, RADAR-77, Oct. 25–28, 1977, London, IEE Conference Publication no. 105, pp. 401–404, available from IEEE, New York, 77CH1271-6 AES.
135. Punnett, M. S.: Developments in Ground Mounted Air Supported Radomes, *IEEE 1977 Mechanical Engineering in Radar Symposium*, Nov. 8–10, 1977, Arlington, Va., pp. 40–45, IEEE Publication 77CH1250-0 AES.
136. Sangiolo, J. B., and A. B. Rohwer: Structural Design Improvements of ESSCO Radomes and Antennas, *IEEE 1977 Mechanical Engineering in Radar Symposium*, Nov. 8–10, 1977, Arlington, Va, pp. 46–51, IEEE Publication 77CH1250-0 AES.
137. Probert-Jones, J. R.: The Radar Equation in Meteorology, *Quart. J. Roy. Meteor. Soc.*, vol. 88, pp. 485–495, 1962.
138. Zhuk, M. S., and Yu. B. Molochkov: " Proektirovanie Linzovykx, Skaniruyushchikx, Shirokodiapa-zommykx Antenn i Fidernykx Ustroistv" (Design of Lenses, Scanning Antennas, Wideband Antennas, and Transmission Lines.) Energiya, Moscow, 1973, pp. 81–82.
139. Lewis, B. L.: 360° Azimuth Scanning Antenna Without Rotating RF Joints, U.S. Patent no. 3,916,416, Oct. 28, 1975.

EIGHT

THE ELECTRONICALLY STEERED PHASED ARRAY ANTENNA IN RADAR

8.1 INTRODUCTION

The phased array is a directive antenna made up of individual radiating antennas, or elements, which generate a radiation pattern whose shape and direction is determined by the relative phases and amplitudes of the currents at the individual elements. By properly varying the relative phases, it is possible to steer the direction of the radiation. The radiating elements might be dipoles, open-ended waveguides, slots cut in waveguide, or any other type of antenna. The inherent flexibility offered by the phased-array antenna in steering the beam by means of electronic control is what has made it of interest for radar. It has been considered in those radar applications where it is necessary to shift the beam rapidly from one position in space to another, or where it is required to obtain information about many targets at a flexible, rapid data rate. The full potential of a phased-array antenna requires the use of a computer that can determine in real time, on the basis of the actual operational situation, how best to use the capabilities offered by the array.

The concept of directive radiation from fixed (nonsteered) phased-array antennas was known during World War I.[1] The first use of the phased-array antenna in commercial broadcasting transmission was in the early thirties[2] and the first large steered directive array for the reception of transatlantic short-wave communication was developed and installed by the Bell Telephone Laboratories in the late thirties.[3] In World War II, the United States, Great Britain, and Germany all used radar with fixed phased-array antennas in which the beam was scanned by mechanically actuated phase shifters. In the United States, this was an azimuth scanning S-band fire control radar, the Mark 8, that was widely used on cruisers and battleships,[4] and the AN/APQ-7 (Eagle) high-resolution navigation and bombing radar at X band that scanned a 0.5° fan beam over a 60° sector in $1\frac{1}{3}$ seconds.[5] The British used the phased array in two height-finder radars, one at VHF and the other at S band.[6] The Germans employed VHF radars with fixed planar phased arrays in significant numbers.[7] One of these, called the Mammut, was 100 ft wide and 36 ft high and scanned a 10° beam over a 120° sector.

A major advance in phased-array technology was made in the early 1950s with the replacement of mechanically actuated phase shifters by electronic phase shifters. Frequency scanning in one angular coordinate was the first successful electronic scanning technique to be applied. In terms of numbers of operational radars, frequency scanning has probably seen more application than any other electronic scanning method. The first major electronically scanned phased arrays that performed beam steering without frequency scan employed the Huggins phase shifter (Sec. 8.4) which, in a sense, used the principle of frequency scan without the necessity of changing the radiated frequency. The introduction of digitally switched phase shifters employing either ferrites or diodes in the early 1960s made a significant improvement in the practicality of phased arrays that could be electronically steered in two orthogonal angular coordinates.

8.2 BASIC CONCEPTS

An array antenna consists of a number of individual radiating elements suitably spaced with respect to one another. The relative amplitude and phase of the signals applied to each of the elements are controlled to obtain the desired radiation pattern from the combined action of all the elements. Two common geometrical forms of array antennas of interest in radar are the linear array and the planar array. A *linear array* consists of elements arranged in a straight line in one dimension. A *planar array* is a two-dimensional configuration of elements arranged to lie in a plane. The planar array may be thought of as a linear array of linear arrays. A *broadside array* is one in which the direction of maximum radiation is perpendicular, or almost perpendicular, to the line (or plane) of the array. An endfire array has its maximum radiation parallel to the array.

The linear array generates a fan beam when the phase relationships are such that the radiation is perpendicular to the array. When the radiation is at some angle other than broadside, the radiation pattern is a conical-shaped beam. The broadside linear-array antenna may be used where broad coverage in one plane and narrow beamwidth in the orthogonal plane are desired. The linear array can also act as a feed for a parabolic-cylinder antenna. The combination of the linear-array feed and the parabolic cylinder generates a more controlled fan beam than is possible with either a simple linear array or with a section of a parabola. The combination of a linear array and parabolic cylinder can also generate a pencil beam.

The endfire array is a special case of the linear or the planar array when the beam is directed along the array. Endfire linear arrays have not been widely used in radar applications. They are usually limited to low or medium gains since an endfire linear antenna of high gain requires an excessively long array. Small endfire arrays are sometimes used as the radiating elements of a broadside array if directive elements are required. Linear arrays of endfire elements are also employed as low-silhouette antennas.

The two-dimensional planar array is probably the array of most interest in radar applications since it is fundamentally the most versatile of all radar antennas. A rectangular aperture can produce a fan-shaped beam. A square or a circular aperture produces a pencil beam. The array can be made to simultaneously generate many search and/or tracking beams with the same aperture.

Other types of array antennas are possible than the linear or the planar arrangements. For example, the elements might be arranged on the surface of a cylinder to obtain 360° coverage (360° coverage may also be obtained with a number of planar arrays). The radiating elements might also be mounted on the surface of a sphere, or indeed on an object of any shape, provided the phase at each element is that needed to give a plane wave when the radiation from all the elements is summed in space. An array whose elements are distributed on a nonplanar surface is called a *conformal array*.

An array in which the relative phase shift between elements is controlled by electronic devices is called an *electronically scanned array*. In an electronically scanned array the antenna elements, the transmitters, the receivers, and the data-processing portions of the radar are often designed as a unit. A given radar might work equally well with a mechanically positioned array, a lens, or a reflector antenna if they each had the same radiation pattern, but such a radar could not be converted efficiently to an electronically scanned array by simple replacement of the antenna alone because of the interdependence of the array and the other portions of the radar.

Radiation pattern.[8-11] Consider a linear array made up of N elements equally spaced a distance d apart (Fig. 8.1). The elements are assumed to be isotropic point sources radiating uniformly in all directions with equal amplitude and phase. Although isotropic elements are not realizable in practice, they are a useful concept in array theory, especially for the computation of radiation patterns. The effect of practical elements with nonisotropic patterns will be considered later. The array is shown as a receiving antenna for convenience, but because of the reciprocity principle, the results obtained apply equally well to a transmitting antenna. The outputs of all the elements are summed via lines of equal length to give a sum output voltage E_a. Element 1 will be taken as the reference signal with zero phase. The difference in the phase of the signals in adjacent elements is $\psi = 2\pi(d/\lambda) \sin \theta$, where θ is the direction of the incoming radiation. It is further assumed that the amplitudes and phases of the signals at each element are weighted uniformly. Therefore the amplitudes of the voltages in each element are the same and, for convenience, will be taken to be unity. The sum of all the voltages from the individual elements, when the phase difference between adjacent elements is ψ, can be written

$$E_a = \sin \omega t + \sin (\omega t + \psi) + \sin (\omega t + 2\psi) + \cdots + \sin [\omega t + (N - 1)\psi] \qquad (8.1)$$

where ω is the angular frequency of the signal. The sum can be written

$$E_a = \sin \left[\omega t + (N - 1)\frac{\psi}{2}\right] \frac{\sin (N\psi/2)}{\sin (\psi/2)} \qquad (8.2)$$

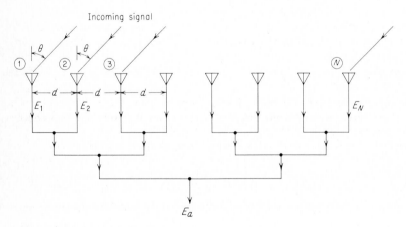

Figure 8.1 *N*-element linear array.

The first factor is a sine wave of frequency ω with a phase shift $(N - 1)\psi/2$ (if the phase reference were taken at the center of the array, the phase shift would be zero), while the second term represents an amplitude factor of the form $\sin (N\psi/2)/\sin (\psi/2)$. The field intensity pattern is the magnitude of Eq. (8.2), or

$$|E_a(\theta)| = \left| \frac{\sin [N\pi(d/\lambda) \sin \theta]}{\sin [\pi(d/\lambda) \sin \theta]} \right| \tag{8.3}$$

The pattern has nulls (zeros) when the numerator is zero. The latter occurs when $N\pi(d/\lambda) \sin \theta = 0, \pm\pi, \pm 2\pi, \ldots, \pm n\pi$, where $n =$ integer. The denominator, however, is zero when $\pi(d/\lambda) \sin \theta, = 0, \pm\pi, \pm 2\pi, \ldots, \pm n\pi$. Note that when the denominator is zero, the numerator is also zero. The value of the field intensity pattern is indeterminate when both the denominator and numerator are zero. However, by applying L'Hopital's rule (differentiating numerator and denominator separately) it is found that $|E_a|$ is a maximum whenever $\sin \theta = \pm n\lambda/d$. These maxima all have the same value and are equal to N. The maximum at $\sin \theta = 0$ defines the *main beam*. The other maxima are called *grating lobes*. They are generally undesirable and are to be avoided. If the spacing between elements is a half-wavelength ($d/\lambda = 0.5$), the first grating lobe ($n = \pm 1$) does not appear in real space since $\sin \theta > 1$, which cannot be. Grating lobes appear at $\pm 90°$ when $d = \lambda$. For a nonscanning array (which is what is considered here) this condition ($d = \lambda$) is usually satisfactory for the prevention of grating lobes. Equation 8.3 applies to isotropic radiating elements, but practical antenna elements that are designed to maximize the radiation at $\theta = 0°$, generally have negligible radiation in the direction $\theta = \pm 90°$. Thus the effect of a realistic element pattern is to suppress the grating lobes at $\pm 90°$. It is for this reason that an element spacing equal to one wavelength can be tolerated for a nonscanning array.

From Eq. (8.3), $E_a(\theta) = E_a(\pi - \theta)$. Therefore an antenna of isotropic elements has a similar pattern in the rear of the antenna as in the front. The same would be true for an array of dipoles. To avoid ambiguities, the backward radiation is usually eliminated by placing a reflecting screen behind the array. Thus only the radiation over the forward half of the antenna ($-90° \le \theta \le 90°$) need be considered.

The radiation pattern is equal to the normalized square of the amplitude, or

$$G_a(\theta) = \frac{|E_a|^2}{N^2} = \frac{\sin^2 [N\pi(d/\lambda) \sin \theta]}{N^2 \sin^2 [\pi(d/\lambda) \sin \theta]} \tag{8.4}$$

If the spacing between antenna elements is $\lambda/2$ and if the sine in the denominator of Eq. (8.4) is replaced by its argument, the half-power beamwidth is approximately equal to

$$\theta_B = \frac{102}{N} \tag{8.5}$$

The first sidelobe, for N sufficiently large, is 13.2 dB below the main beam. The pattern of a uniformly illuminated array with elements spaced $\lambda/2$ apart is similar to the pattern produced by a continuously illuminated uniform aperture [Eq. (7.16)].

When directive elements are used, the resultant array antenna radiation pattern is

$$G(\theta) = G_e(\theta) \frac{\sin^2 [N\pi(d/\lambda) \sin \theta]}{N^2 \sin^2 [\pi(d/\lambda) \sin \theta]} = G_e(\theta)G_a(\theta) \tag{8.6}$$

where $G_e(\theta)$ is the radiation pattern of an individual element. The resultant radiation pattern is the product of the *element factor* $G_e(\theta)$ and the *array factor* $G_a(\theta)$, the latter being the pattern of an array composed of isotropic elements. The array factor has also been called the *space factor.* Grating lobes caused by a widely spaced array may therefore be eliminated with directive elements which radiate little or no energy in the directions of the undesired lobes. For example, when the element spacing $d = 2\lambda$, grating lobes occur at $\theta = \pm 30°$ and $\pm 90°$ in addition to the main beam at $\theta = 0°$. If the individual elements have a beamwidth somewhat less than 60°, the grating lobes of the array factor will be suppressed.

Equation (8.6) is only an approximation, which may be seriously inadequate for many problems of array design. It should be used with caution. It ignores mutual coupling, and it does not take account of the scattering or diffraction of radiation by the adjacent array elements or of the outward-traveling-wave coupling.[12-14] These effects cause the element radiation pattern to be different when located within the array in the presence of the other elements than when isolated in free space.

In order to obtain an exact computation of the array radiation pattern, the pattern of each element must be measured in the presence of all the others. The array pattern may be found by summing the contributions of each element, taking into account the proper amplitude and phase.

In a two-dimensional, rectangular planar array, the radiation pattern may sometimes be written as the product of the radiation patterns in the two planes which contain the principal axes of the antenna. If the radiation patterns in the two principal planes are $G_1(\theta_e)$ and $G_2(\theta_a)$, the two-dimensional antenna pattern is

$$G(\theta_e, \theta_a) = G_1(\theta_e)G_2(\theta_a) \tag{8.7}$$

Note that the angles θ_e and θ_a are not necessarily the elevation and azimuth angles normally associated with radar.[15,16] The normalized radiation pattern of a uniformly illuminated rectangular array is

$$G(\theta_e, \theta_a) = \frac{\sin^2\left[N\pi(d/\lambda)\sin\theta_a\right]}{N^2\sin^2\left[\pi(d/\lambda)\sin\theta_a\right]} \frac{\sin^2\left[M\pi(d/\lambda)\sin\theta_e\right]}{M^2\sin^2\left[\pi(d/\lambda)\sin\theta_e\right]} \tag{8.8}$$

where N = number of radiating elements in θ_a dimension with spacing d and M = number in θ_e dimension.

Beam steering. The beam of an array antenna may be steered rapidly in space without moving large mechanical masses by properly varying the phase of the signals applied to each element. Consider an array of equally spaced elements. The spacing between adjacent elements is d, and the signals at each element are assumed of equal amplitude. If the same phase is applied to all elements, the *relative* phase difference between adjacent elements is zero and the position of the main beam will be broadside to the array at an angle $\theta = 0$. The main beam will point in a direction other than broadside if the relative phase difference between elements is other than zero. The direction of the main beam is at an angle θ_0 when the phase difference is $\phi = 2\pi(d/\lambda)\sin\theta_0$. The phase at each element is therefore $\phi_c + m\phi$, where $m = 0, 1, 2, \ldots,$ $(N - 1)$, and ϕ_c is any constant phase applied to all elements. The normalized radiation pattern of the array when the phase difference between adjacent elements is ϕ is given by

$$G(\theta) = \frac{\sin^2\left[N\pi(d/\lambda)(\sin\theta - \sin\theta_0)\right]}{N^2\sin^2\left[\pi(d/\lambda)(\sin\theta - \sin\theta_0)\right]} \tag{8.9}$$

The maximum of the radiation pattern occurs when $\sin\theta = \sin\theta_0$.

Equation (8.9) states that the main beam of the antenna pattern may be positioned to an angle θ_0 by the insertion of the proper phase shift ϕ at each element of the array. If variable, rather than fixed, phase shifters are used, the beam may be steered as the relative phase between elements is changed (Fig. 8.2).

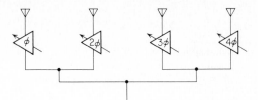

Figure 8.2 Steering of an antenna beam with variable phase shifters (parallel-fed array).

Using an argument similar to the nonscanning array described previously, grating lobes appear at an angle θ_g whenever the denominator is zero, or when

$$\pi \frac{d}{\lambda} (\sin \theta_g - \sin \theta_0) = \pm n\pi \tag{8.10a}$$

or

$$\left| \sin \theta_g - \sin \theta_0 \right| = \pm n \frac{\lambda}{d} \tag{8.10b}$$

If a grating lobe is permitted to appear at $-90°$ when the main beam is steered to $+90°$, it is found from the above that $d = \lambda/2$. Thus the element spacing must not be larger than a half wavelength if the beam is to be steered over a wide angle without having undesirable grating lobes appear. Practical array antennas do not scan $\pm 90°$. If the scan is limited to $\pm 60°$, Eq. (8.10) states that the element spacing should be less than 0.54λ. Note that antenna elements used in arrays are generally comparable to a half wavelength in physical size.

Change of beamwidth with steering angle. The half-power beamwidth in the plane of scan increases as the beam is scanned off the broadside direction. The beamwidth is approximately inversely proportional to $\cos \theta_0$, where θ_0 is the angle measured from the normal to the antenna. This may be proved by assuming that the sine in the denominator of Eq. (8.9) can be replaced by its argument, so that the radiation pattern is of the form $(\sin^2 u)/u^2$, where $u = N\pi(d/\lambda)(\sin \theta - \sin \theta_0)$. The $(\sin^2 u)/u^2$ antenna pattern is reduced to half its maximum value when $u = \pm 0.443\pi$. Denote by θ_+ the angle corresponding to the half-power point when $\theta > \theta_0$, and θ_-, the angle corresponding to the half-power point when $\theta < \theta_0$; that is, θ_+ corresponds to $u = +0.443\pi$ and θ_- to $u = -0.443\pi$. The $\sin \theta - \sin \theta_0$ term in the expression for u can be written[17]

$$\sin \theta - \sin \theta_0 = \sin (\theta - \theta_0) \cos \theta_0 - [1 - \cos (\theta - \theta_0)] \sin \theta_0 \tag{8.11}$$

The second term on the right-hand side of Eq. (8.11) can be neglected when θ_0 is small (beam is near broadside), so that

$$\sin \theta - \sin \theta_0 \approx \sin (\theta - \theta_0) \cos \theta_0 \tag{8.12}$$

Using the above approximation, the two angles corresponding to the 3-dB points of the antenna pattern are

$$\theta_+ - \theta_0 = \sin^{-1} \frac{0.443\lambda}{Nd \cos \theta_0} \approx \frac{0.443\lambda}{Nd \cos \theta_0}$$

$$\theta_- - \theta_0 = \sin^{-1} \frac{-0.443\lambda}{Nd \cos \theta_0} \approx \frac{-0.443\lambda}{Nd \cos \theta_0}$$

The half-power beamwidth is

$$\theta_B = \theta_+ - \theta_- \approx \frac{0.886\lambda}{Nd \cos \theta_0} \tag{8.13}$$

Therefore, when the beam is positioned an angle θ_0 off broadside, the beamwidth in the plane of scan increases as $(\cos \theta_0)^{-1}$. The change in beamwidth with angle θ_0 as derived above is not valid when the antenna beam is too far removed from broadside. It certainly does not apply when the energy is radiated in the endfire direction.

Equation 8.13 applies for a uniform aperture illumination. With a cosine-on-a-pedestal aperture illumination of the form $A_n = a_0 + 2a_1 \cos 2\pi n/N$, the beamwidth is[18]

$$\theta_B \simeq \frac{0.886\lambda}{Nd \cos \theta_0} [1 + 0.636(2a_1/a_0)^2] \tag{8.14}$$

The parameter n in the aperture illumination represents the position of the element. Since the illumination is assumed symmetrical about the center element, the parameter n takes on values of $n = 0, \pm 1, \pm 2, \dots, \pm (N-1)/2$. The range of interest is $0 \le 2a_1 \le a_0$ which covers the span from uniform illuminations to a taper so severe that the illumination drops to zero at the ends of the array. (The array is assumed to extend a distance $d/2$ beyond each end element.)

The above applies to a linear array. Similar results apply to a planar aperture;[17,18] that is, the beamwidth in the plane of the scan varies approximately inversely as $\cos \theta_0$, provided certain assumptions are fulfilled.

An interesting technique for graphically portraying the variation of the beam shape with scan angle has been described by Von Aulock,[15] an example of which is shown in Fig. 8.3. The antenna radiation pattern is plotted in spherical coordinates as a function of the two direction cosines, $\cos \alpha_x$ and $\cos \alpha_y$, of the radius vector specifying the point of observation. The angle ϕ is measured from the $\cos \alpha_x$ axis, and θ is measured from the axis perpendicular to the $\cos \alpha_x$ and $\cos \alpha_y$ axes. In Fig. 8.3, ϕ is taken to be a constant value of $90°$ and the beam is scanned in

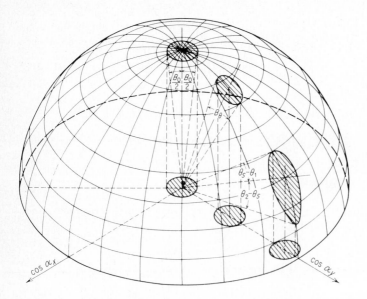

Figure 8.3 Beamwidth and eccentricity of the scanned beam. (*From Von Aulock*[15], *Courtesy Proc. IRE.*)

the θ coordinate. At $\theta = 0$ (beam broadside to the array) a symmetrical pencil beam of half-power width B_0 is assumed. The shape of the beam at the other angular positions is the projection of the circular beam shape on the surface of the unit sphere. It can be seen that as the beam is scanned in the θ direction, it broadens in that direction, but is constant in the ϕ direction. For $\theta \neq 0$, the beam shape is not symmetrical about the center of the beam, but is eccentric. Thus the beam direction is slightly different from that computed by standard formulas. In addition to the changes in the shape of the main beam, the sidelobes also change in appearance and position.

Series vs. parallel feeds. The relative phase shift between adjacent elements of the array must be $\phi = 2\pi(d/\lambda) \sin \theta_0$ in order to position the main beam of the radiation pattern at an angle θ_0. The necessary phase relationships between the elements may be obtained with either a *series-fed* or a *parallel-fed* arrangement. In the series-fed arrangement, the energy may be transmitted from one end of the line (Fig. 8.4a), or it may be fed from the center out to each end (Fig. 8.4b). The adjacent elements are connected by a phase shifter with phase shift ϕ. All the phase shifters are identical and introduce the same amount of phase shift, which is less than 2π radians. In the series arrangement of Fig. 8.4a where the signal is fed from one end, the position of the beam will vary with frequency (Sec. 8.4). Thus it will be more limited in bandwidth than most array feeds. The center-fed feed of Fig. 8.4b does not have this problem.

In the parallel-fed array of Fig. 8.2, the energy to be radiated is divided between the elements by a power splitter. When a series of power splitters are used to create a tree-like structure, as in Fig. 8.2, it is called a *corporate feed*, since it resembles (when turned upside down) the organization chart of a corporation. Equal lengths of line transmit the energy to each element so that no unwanted phase differences are introduced by the lines themselves. (If the lines are not of equal length, a compensation in the phase shift must be made.) The proper phase change for beam steering is introduced by the phase shifters in each of the lines feeding the elements. When the phase of the first element is taken as the reference, the phase shifts required in the succeeding elements are $\phi, 2\phi, 3\phi, \ldots, (N - 1)\phi$.

The maximum phase change required of each phase shifter in the parallel-fed array is many times 2π radians. Since phase shift is periodic with period 2π, it is possible in many applications to use a phase shifter with a maximum of 2π radians. However, if the pulse width is short compared with the antenna response time (if the signal bandwidth is large compared with the antenna bandwidth), the system response may be degraded. For example, if the energy

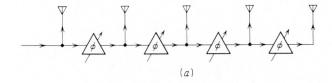

(a)

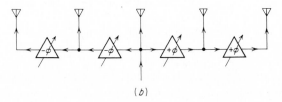

(b)

Figure 8.4 Series arrangements for applying phase relationships in an array. (a) fed from one end; (b) center-fed.

were to arrive in a direction other than broadside, the entire array would not be excited simultaneously. The combined outputs from the parallel-fed elements will fail to coincide or overlap, and the received pulse will be smeared. This situation may be relieved by replacing the 2π modulo phase shifters with delay lines.

A similar phenomenon occurs in the series-fed array when the energy is radiated or received at or near the broadside direction. If a short pulse is applied at one end of a series-fed transmitting array, radiation of energy by the first element might be completed before the remainder of the energy reaches the last element. On reception, the effect is to smear or distort the echo pulse. It is possible to compensate for the delay in the series-fed array and avoid distortion of the main beam when the signal spectrum is wide by the insertion of individual delay lines of the proper length in series with the radiating elements.[19]

In a series-fed array containing N phase shifters, the signal suffers the insertion loss of a single phase shifter N times. In a parallel-fed array the insertion loss of the phase shifter is introduced effectively but once. Hence the phase shifter in a series-fed array must be of lower loss compared with that in a parallel-fed array. If the series phase shifters are too lossy, amplifiers can be inserted in each element to compensate for the signal attenuation.

Since each phase shifter in the series-fed linear array of Fig. 8.4a has the same value of phase shift, only a single control signal is needed to steer the beam. The N-element parallel-fed linear array similar to that of Fig. 8.2 requires a separate control signal for each phase shifter or $N - 1$ total (one phase shifter is always zero). A two-dimensional parallel-fed array of MN elements requires $M + N - 2$ separate control signals. The two-dimensional series-fed array requires but two control signals.

8.3 PHASE SHIFTERS

The difference in phase ϕ experienced by an electromagnetic wave of frequency f propagating with a velocity v through a transmission line of length l is

$$\phi = 2\pi f l/v \tag{8.15}$$

The velocity v of an electromagnetic wave is a function of the permeability μ and the dielectric constant ϵ of the medium in which it propagates. Therefore, a change in phase can be had by a change in the frequency, length of line, velocity of propagation, permeability, or dielectric constant. The use of frequency to effect a change of phase is a relatively simple technique for electronically scanning a beam. It was one of the first practical examples of electronic scanning and has been widely employed. It is discussed separately in the next section. One of the more popular forms of phase shifters is one that varies the physical length of line to obtain a change in phase, especially when the lengths of line are quantized digitally. Varying the velocity of propagation by varying the permeability is the basis of ferrite phase shifters. Gas discharge and ferroelectric phase shifters are examples of devices that depend on changes in dielectric constant to vary the velocity of propagation, and hence the phase. The velocity of propagation may also be varied in a more direct manner as in the Eagle, or delta-a, scanner where the narrow wall of a waveguide is moved mechanically to produce a change in phase and a scanning beam.[20]

Many phase-shifting devices are *reciprocal* in that the phase change does not depend on the direction of propagation. Some important phase shifters, however, are *nonreciprocal*. These must have different control settings for reception and transmission. A phase shifter should be able to change its phase rapidly, be capable of handling high power, require control signals of little power, be of low loss, light weight, small size, have long life, and be of reasonable cost.

The various phase shifting techniques possess these properties in varying degree. No one device seems to be sufficiently universal to meet the requirements of all applications.

In this text a device for obtaining a change of phase is called a *phase shifter*, but they have also been known as *phasers*.

Digitally switched phase shifters. A change in phase can be obtained by utilizing one of a number of lengths of transmission line to approximate the desired value of phase. The various lengths of line are inserted and removed by high-speed electronic switching. Semiconductor diodes and ferrites are the devices commonly employed in digital phase shifters.

There are at least two methods for switching lengths of transmission lines. In one, the proper line length is selected from among many available lengths. This has been called a *parallel-line* configuration. In the other, the proper length of line is made up of the series combination of a relatively few selected lengths of line. This is a *series-line*, or a *cascaded*, configuration. Although the discrete nature of the digitally switched phase shifter means that the exact value of required phase shift cannot be achieved without a quantization error, the error can be made as small as desired. (Even analog phase shifters, which are continuously variable, cannot be conveniently set to a precise value of phase without special care in calibration over the desired range of temperature and frequency.)

Figure 8.5 illustrates the parallel-line configuration of the digitally switched phase shifter in which the desired length is obtained by means of a pair of one-by-N switches. Each of the boxes labeled S represents a SPST switch. The N ports of each one-by-N switch are connected to N lines of different lengths $l_1, l_2, \ldots, l_N$. The number of lines depends on the degree of phase quantization that can be tolerated. The number is limited by the quality of the switches, as measured by the difference between their impedance in the " off " and " on " positions. With many switches in parallel, the " off " impedance of each must be high if the combined impedance is to be large compared to the " on " impedance of a single switch. A parallel-line configuration with 16 lengths of line provides a phase quantization of 22.5° ($\pm 11.25°$), assuming the nth line is of length $n\lambda/16$. A suitable form of switch is the semiconductor diode. The diodes attached to the ends of the particular line selected are operated with forward bias to present a low impedance. The remaining diodes attached to the unwanted lines are operated with back-bias to present a high impedance. The switched lines can be any standard RF transmission line. Stripline has been used successfully, especially at the lower microwave frequencies. An advantage of the parallel-line configuration is that the signal passes through but two switches and, in principle, should have a lower insertion loss than the cascaded digitally switched phase shifter described below. A disadvantage is the relatively large number

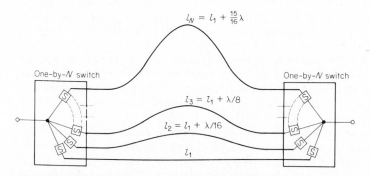

Figure 8.5 Digitally switched parallel-line phase shifter with N switchable lines.

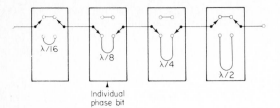

Individual
phase bit

Figure 8.6 Cascaded four-bit digitally switched phase shifter with $\lambda/16$ quantization. Particular arrangement shown gives 135° of phase shift ($\frac{3}{8}$ wavelength).

of lines and switches required when it is necessary to minimize the quantization error. The parallel-line configuration has also been used when phase shifts greater than 2π radians are needed, as in broadband devices which require true time delays rather than phase shift which is limited to 2π radians.

The cascaded digitally switched phase shifter (Fig. 8.6) has seen more application than the parallel-line configuration. A cascaded digitally switched phase shifter with four phase bits capable of switching in or out lengths of line equal to $\lambda/16$, $\lambda/8$, $\lambda/4$, and $\lambda/2$ yields a phase shift with a quantization of $\lambda/16$, just as the parallel line shifter with 16 lengths of line described above. The binary quantization of line lengths makes it convenient to apply digital techniques for actuating the shifter.

Figure 8.6 shows the schematic of a four-bit digital phase shifter consisting of four cascaded modules. Each module contains a switch that inserts either "zero" phase change or a phase change of $360/2^n$ degrees, where $n = 1, 2, 3, 4$. When the upper two switches are open, the lower two are closed, and vice versa. Note that in the "zero" phase state, the phase shift is generally not zero, but is some residual amount ϕ_0. Thus the two states provide a phase of ϕ_0 and $\phi_0 + \Delta\phi$. The difference $\Delta\phi$ between the two states is the desired phase shift required of the module.

The arrangements of Figs. 8.5 and 8.6 lend themselves to the use of semiconductor diodes. Ferrite phase shifters are also operated digitally, but in a slightly different manner, as described later.

Diode phase shifters.[22-26,155] The property of a semiconductor diode that is of interest in microwave phase shifters is that its impedance can be varied with a change in bias control voltage. This allows the diode to act as a switch. Phase shifters based on diode devices can be of relatively high power and low loss, and can be switched rapidly from one phase state to another. They are relatively insensitive to changes in temperature, they can operate with low control power, and are compact in size. They lend themselves well to microwave integrated circuit construction, and are capable of being used over the entire range of frequencies of interest to radar although their losses are generally less and their power handling is generally higher at the lower frequencies.

There are three basic methods for employing semiconductor diodes in digital phase shifters, depending on the circuit used to obtain an individual phase bit. These are: (1) the switched-line, (2) the hybrid-coupled, and (3) the loaded-line. The switched line was shown in Fig. 8.6. Each phase bit consists of two lengths of line that provide the differential phase shift, and two single-pole, double-throw switches utilizing four diodes.

The hybrid-coupled phase bit, as shown in Fig. 8.7, uses a 3-dB hybrid junction with balanced reflecting terminations connected to the coupled arms. Two switches (diodes) control the phase change. The 3-dB hybrid has the property that a signal input at port 1 is divided equally in power between ports 2 and 3. No energy appears at port 4. The diodes act to either pass or reflect the signals. When the impedance of the diodes is such as to pass the signals,

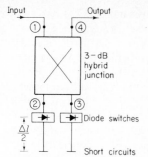

Figure 8.7 Hybrid-coupled phase bit.

the signals will be reflected by the short circuits located farther down the transmission lines. The signals at ports 2 and 3, after reflection from either the diode switches or the short circuits, combine at port 4. None of the reflected energy appears at port 1. The difference in path length with the diode switches open and closed is Δl. The two-way path Δl is chosen to correspond to the desired increment of digitized phase shift. An N-bit phase shifter can be obtained by cascading N such hybrids.

The loaded-line phase shifter, Fig. 8.8, consists of a transmission line periodically loaded with spaced, switched impedances, or susceptances. Diodes are used to switch between the two states of susceptance. The spacing between diodes is approximately one-quarter wavelength at the operating frequency. Adjacent quarter-wavelength-spaced loading-susceptances are equal and take either of two values. If the magnitude of the normalized susceptance is small, the reflection from any pair of symmetrical susceptances can be made to cancel so that matched transmission will result for either of the two susceptance conditions. Each pair of diodes spaced a quarter-wavelength apart produces an increment of the desired phase. The number of pairs that are cascaded determines the value of the transmission phase shift. To achieve high-power capacity, many such sections with small phase increments can be used so that there are many diodes to share the power. The ability to operate with high power is the advantage claimed for this type of diode phase shifter. If the largest practical phase shift per diode pair is $\lambda/16$, or $22.5°$, 32 diodes are required to shift the phase $360°$.

The hybrid-coupled phase shifter generally has less loss than the other two, and uses the least number of diodes. It can be made to operate over a wide band. The switched-line phase shifter uses more diodes than the other types and has an undesirable phase-frequency response which can be corrected at the expense of a higher insertion loss. This configuration is generally restricted to true-time-delay circuits and to low-power, miniaturized phase shifters where loss is not a major consideration. For a four-bit phase shifter, covering $360°$, the minimum number of diodes needed in the periodically loaded line is 32, the switched line requires 16, and the hybrid-coupled circuit needs only 8. The theoretical peak power capability of the switched line is twice that of the hybrid-coupled circuit since voltage doubling is produced by the reflection

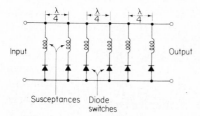

Figure 8.8 Periodically loaded-line phase shifter.

in the hybrid circuit. The switched-line phase shifter has the greatest insertion loss, but its loss does not vary with the amount of phase shift as it does in the other two types of circuits.

Diode phase shifters have been built in practically all transmission line media, including waveguide, coax and stripline. Microstrip is useful for medium-power devices because of the ease of manufacture, circuit reproducibility and its reduced size, weight, and production costs. The diode chips may be mounted directly on the substrate without the parasitic reactances of the diode packages. A multiple-bit diode phase shifter need not be constructed with all of the same types of phase bits. The loaded-line circuit is often preferred for small phase increments because of its compact size. It is not as suitable for large phase steps because it is difficult to match in both states under this condition. For example, in one particular design of a four-bit microstrip X-band phase shifter, using a sapphire substrate, it proved expedient to use the loaded-line construction for the 22.5 and 45° bits, and to use the hybrid coupled circuit for the 90 and 180° bits to obtain minimum insertion loss with reasonable bandwidth and power handling capability.[21] The switching times of phase shifters utilizing PIN diodes can be as low as 50 ns. Typically, switching of the order of one microsecond is suitable for most applications.

The PIN diode, Fig. 8.9, has been frequently used in high-power digital phase shifters. It consists of a thin slice of high-resistivity intrinsic semiconductor material sandwiched between heavily doped low-resistivity P^+ and N^+ regions. The intrinsic region acts as a slightly lossy dielectric at microwave frequencies and the heavily doped regions are good conductors. When biased in the reverse (nonconducting) state, the PIN diode resembles a low-loss capacitor. It is essentially an insulator situated between two conductors and exhibits the parallel-plate capacitance determined by the dielectric of the intrinsic region. At microwaves, the small-signal equivalent circuit of the PIN diode may be represented as a series RC circuit with the capacitance determined by the area, thickness, and dielectric constant of the intrinsic region. The capacitance is independent of the reverse-bias voltage. The series resistance is determined by the resistivity and geometry of the metallic-like P and N regions. In the forward-bias (conducting) state, when appreciable current is passed, the injection of holes and electrons from the P and N regions respectively, creates an electron-hole plasma in what was formerly the dielectric region. The slightly lossy dielectric is changed to a fairly good conductor with the application of forward bias. The capacitive component of the circuit disappears and the equivalent circuit becomes a small resistance which decreases in value with increasing forward current. The resistance can vary from thousands of ohms at zero bias to a fraction of

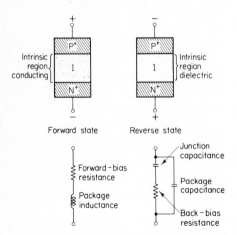

Figure 8.9 PIN diodes and simplified equivalent circuits for forward and reverse states.

an ohm with tens of milliamperes bias current. Thus with forward bias the diode resembles a low-value resistor. If the diode is housed in a package, the parasitic elements introduced by the package degrade the switching action and influence the voltage breakdown and thermal characteristics.

The variable capacitance semiconductor element, or *varactor* diode, also can be used as the switch in a diode phase shifter. It is capable of very rapid switching, of the order of a nanosecond, but can handle only low power. Since the capacitance of a varactor varies with the applied negative bias voltage, it may be employed as an analog (continuously variable) phase shifter using either the loaded line or the hybrid coupled configurations. Power levels are limited to a few milliwatts by the generation of harmonics and other nonlinear effects.

Ferrimagnetic phase shifters. A *ferrite* is a magnetizable metal-oxide insulator which contains magnetic ions arranged to produce spontaneous magnetization while maintaining good dielectric properties.[27-29] In contrast to ferromagnetic metals, ferrites are insulators and have a high resistivity which allows electromagnetic waves to penetrate the material so that the magnetic field component of the wave can interact with the magnetic moment of the ferrite. This interaction results in a change of the microwave permeability of the ferrite. The term *ferrimagnetism* was introduced to describe the novel magnetic properties of these materials that are now known as ferrites.

Ferrites have been derived principally from two basic metal-oxide families: the ferrimagnetic spinels and the garnets. Both have been used for microwave phase shifters. A typical spinel ferrite is $NiFe_2O_4$. In addition to nickel, compositions of magnesium-manganese and lithium have been used. A common garnet is $Y_3Fe_5O_{12}$, yttrium iron garnet (YIG). The characteristics of materials suitable for microwave devices are given by Ince and Temme.[30]

The physics of propagation of electromagnetic energy in ferrites is not easy to describe in simple terms.[31] The magnetic permeability of a ferrite is anisotropic (that is, it depends on the direction), and must be represented by a complex tensor rather than a scalar. For present purposes, it suffices to state that a change in the applied d-c magnetic field produces a change in the propagation properties because of a change in permeability, which results in a phase shift.

A ferrite phase shifter is a two-port device that permits variation of the phase shift between input and output by changing the magnetic properties of the ferrite. They may be analog or digital with either reciprocal or nonreciprocal characteristics. Several types of ferrite phase shifters have been developed, but those that have been of interest in applications are the Reggia-Spencer, toroidal latching, flux drive, and dual-mode phase shifters.

Reggia-Spencer phase shifter. This was one of the first ferrite phase shifters to be applied successfully to radar. It consists of a rod or a bar of ferrimagnetic material located at the center of a section of waveguide. A solenoid is wound around the waveguide to provide a longitudinal magnetic field. A change in phase is obtained by a change of the applied magnetic field that results from a change of the current passing through the coil. This is called an *analog* phase shifter since a continuous phase change is obtained as a function of the current through the solenoid. It is a *reciprocal* device in that the phase shift is the same for both directions of propagation. When it was introduced, the Reggia-Spencer phase shifter had a higher figure of merit (defined as the degrees of phase shift per dB of loss) than previous analog ferrite phase shifters, and was more compact. Since the rod of ferrite is located at the center of the guide out of contact with the walls, there is no means for dissipation of heat other than by radiation. This lack of a convenient thermal path to dissipate heat limits its power handling capability. However, in one design,[31,32] the heat transfer problem was overcome by having the axially located garnet bar directly cooled by a low-loss liquid dielectric that was allowed to flow along the surface of the garnet material. The flow was confined by completely encapsulating the garnet in a teflon jacket. Thus the cooling liquid was in direct contact with the garnet for efficient transfer of dissipated heat. Some of the highest power phase shifters have been obtained with this design. For example, a C-band phase shifter, operating over an 8 percent bandwidth was capable of handling 100 kW of peak power at an average power of 600 W. The insertion loss was 0.9 dB with a maximum VSWR of 1.25. However, it required 125 μs to switch the phase. At a 300-Hz switching rate, 16 W of switching power was needed. The phase shifter was 2.4 by 2.1 by 8.2 inches in size and weighed less than 1.5 lb.

The relatively long time to switch the phase of the Reggia-Spencer phase shifter from one value to another has limited its application. Switching times are measured in hundreds of microseconds rather than a few microseconds as is common with other microwave phase shifters. The long time is due to the large inductance of the solenoid and by the fact that the waveguide around which the solenoid is wrapped acts as a shorted turn. The resulting eddy currents generated by the shorted turn limit the speed with which the magnetic field can be changed. The shorted-turn effect may be minimized by a waveguide made of plastic which is coated with a thin copper plating. Sometimes a slit is cut in the side wall of the waveguide to reduce the eddy currents. Another factor that complicates operation is hysteresis which causes the value of permeability to differ for increasing and decreasing current. To eliminate hysteresis errors a current pulse must be applied to the phase shifter solenoid to drive the ferrite to

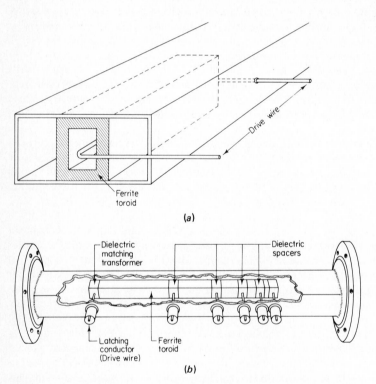

(a)

(b)

Figure 8.10 (a) Single bit of a latching ferrite phase-shifter mounted in waveguide; (b) sketch of five-bit latching ferrite phase-shifter. (*From Whicker and Jones,*[156] *courtesy IEEE.*)

saturation before a new value of phase shift can be applied. The phase shift of a ferrite device is especially sensitive to temperature. A change of 1°C might result in a 1° phase change. Hence, they usually must be operated in a temperature-controlled environment. Although the Reggia-Spencer phase shifter has been successfully applied in operational radar and was one of the first practical phase shifters suitable for radar, in its present form it has been largely superseded by other devices.

Latching ferrite phase shifters.[31,33] The use of a ferrite in the form of a toroid centered within a waveguide as in Fig. 8.10, results in a phase shifter with a fast switching time and with less drive power than required of a Reggia-Spencer device. Furthermore, it is not as temperature sensitive, and there is less of a problem caused by hysteresis in the ferrite. The toroid ferrite phase shifter, although not perfect by any means, has been in the past a popular choice for phased array application.

Consider the hysteresis loop, or B-H curve, of Fig. 8.11. This is a plot of the magnetization (gauss) as a function of the applied magnetic field (oersted) for a toroid-shaped section of ferrimagnetic material. The applied magnetic field is proportional to the current in the drive wire which can be considered a solenoid with one turn. When a sufficiently large current is passed through the drive wire threading the center of the toroid, the magnetization is driven to saturation. When the current is reduced to zero, there exists a remanent magnetization B_r. If a current of opposite polarity is passed through the drive wire, the ferrite is saturated with the opposite polarity of remanent magnetization. Thus a toroidal ferrite may take on two values of magnetization, $\pm B_r$, obtained by pulsing the drive wire with either a positive or negative current pulse. The difference in the two states of magnetization produces the differential phase

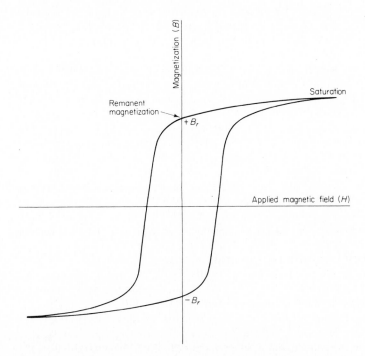

Figure 8.11 Hysteresis loop, or B-H curve, of a ferrite toroid.

shift. This is called a *latching phase shifter*. Thus a single, short-duration current pulse sets the phase. Continuous application of drive power is not needed as in other phase shifters. A "square" hysteresis loop is desired for such a phase shifter. The squareness is measured by the ratio of the remanent magnetization to the saturation magnetization, which with practical ferrite materials might typically be about 0.7.

The amount of differential phase shift depends on the ferrite material and the length of the toroid. A digital latching phase shifter is obtained by placing in cascade a number of separate toroids of varying length. The lengths of each toroid are selected to provide a differential phase shift of 180°, 90°, 45°, 22.5°, and so forth, depending on the number of bits required. A separate pair of drive wires is used for each section of toroid. Impedance matching is provided at the input and output toroids. By filling the center slot of the toroid with high dielectric-constant material a better figure of merit and lower switching power are obtained, but the peak power capability is decreased. The individual toroids are usually separated by thin dielectric spacers to avoid magnetic interaction (Fig. 8.10*b*). The drive wire is oriented for minimum RF coupling. It is also important to have proper mechanical contact between the toroid and the waveguide wall since an air gap can give rise to the generation of higher-order modes that can result in relatively large narrow-frequency-band, insertion-loss spikes. Careful attention must be paid to eliminating these gaps without excessive mechanical pressure that could cause magnetostriction which changes the magnetic properties of the material, especially if garnets are used. This type of latching phase shifter is also called a *twin-slab toroidal phase shifter* since the major action is due to the two vertical branches of the toroids. The function of the horizontal branches is to complete the magnetic circuit.

By introducing the applied magnetic field from within the waveguide via the single-turn drive wire, the shorted-turn effect that limits the switching speed of the Reggia-Spencer phase shifter is eliminated. This permits switching times of the order of microseconds. Whereas hysteresis was a nuisance to be tolerated in the Reggia-Spencer device, the latching phase shifter takes advantage of the hysteresis loop to produce the two discrete values of phase shift without the need for continuous drive power.

The toroid phase shifter is *nonreciprocal* in that the phase shift depends on the direction of propagation. In most applications this inconvenience is usually accepted in order to achieve the other advantages offered by the latching ferrite phase shifter. When used for both transmit and receive, the phase shift must be changed between the two modes of operation. With switching speeds of the order of microseconds, it is practical to reset the phase shifters just after transmission in order to receive. They are then reset just before transmitting the next pulse. The phase shift for proper operation when propagating in the reverse direction is obtained by simply reversing the polarity of the drive pulses. This reverses the direction of magnetization of the ferrite toroid, which is equivalent to reversing the direction of propagation. Nonreciprocal phase shifters cannot be used in reflectarrays (Sec. 8.6), since the electromagnetic energy must travel in both directions. Their use is also not practical in short-range radar or with high-duty cycle waveforms like those used in some pulse-doppler radars, since the switching between the two states is too slow.

The value of remanent magnetization is affected by temperature. Temperature changes may be due to microwave dissipation in the ferrite as well as ambient-temperature variations. Some materials are more temperature-sensitive than others, but the material cannot always be selected on this basis alone. Garnets are generally more temperature-stable and can handle higher power than other ferrimagnetic materials. A reduction in temperature sensitivity can be obtained with a composite magnetic circuit[30,31] in which the microwave ferrimagnetic material is internal to the waveguide, but a temperature-insensitive, magnetic-flux-limiting ferrimagnetic material is external to the waveguide. The latter limits and maintains the magnetic flux, and hence the phase shift, at an almost constant level over the stabilization range.

Flux drive.[34] The toroid, or twin-slab, ferrite phase shifter may be operated in an analog fashion by varying the current of the drive pulse so as to provide different values of remanent magnetization. This is called *flux drive*. It has the further advantage of having reduced temperature sensitivity. Instead of using a number of individual toroids of different lengths in cascade, a single long section of toroid is used that is capable of providing the total differential phase shift of 360°. The required digital phase increment is obtained by operating on a minor hysteresis loop, as in Fig. 8.12. If $B_r(1)$, for example, were the remanent magnetization needed to produce a phase change of 180° relative to the remanent magnetization $-B_r$, the amplitude and width of the driving pulse would be selected so as to rise up to point 1 on the hysteresis curve. When the current pulse decays to zero, the magnetization falls back to the remanent value $B_r(1)$ along the indicated curve. The difference in phase between $-B_r(1)$ and the value $B_r(1)$ determines the phase increment. With a different value of current pulse, a different value of remanent magnetization can be obtained and therefore a different phase shift. In this manner, the ferrite toroid is basically an analog device that can provide any phase increment. It acts as a digital phase shifter if the drive currents are digital. A digital-to-analog conversion is required to translate the digital control-signal generated by the array computer to a form that can set the flux-drive phase shifter. The length of the toroid is generally made 15 to 20 percent greater than normal to allow for some shrinkage of the total available increment of magnetization due to temperature changes. Analysis shows that when the drive output impedance is small, the effect of temperature-caused variations in the increment of magnetization will be small.

Dual-mode ferrite phase shifters. There are some applications which require a reciprocal phase shifter. As mentioned previously the Reggia-Spencer phase shifter is reciprocal but has limitations such as slow switching speed, temperature sensitivity, and a not particularly good figure

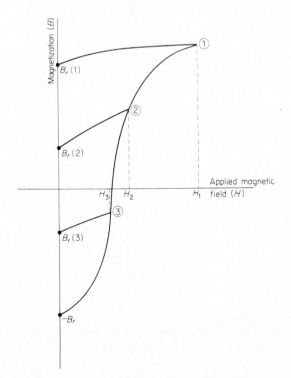

Figure 8.12 Hysteresis loop showing the principle of flux drive, where a single ferrite toroid is excited by discrete current pulses to produce digital phase-shift increments from what is basically an analog device.

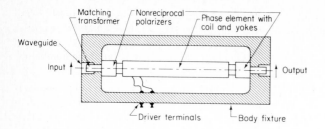

Figure 8.13 Dual-mode reciprocal phase shifter configuration. (*From Whicker and Young.*[162])

of merit. A more suitable reciprocal device that overcomes the limitations of previous reciprocal ferrite phase shifters in the *dual-mode* phase shifter based on the principle of the Faraday rotor.[25,33,35–37] It is competitive with the toroid phase shifter, especially at the higher microwave frequencies.

In the dual-mode phase shifter the linearly polarized signal in rectangular waveguide at the left-hand port in Fig. 8.13 is converted to circular polarization by a nonreciprocal circular polarizer (a ferrite quarter-wave plate). In the Faraday rotor portion, the applied longitudinal magnetic field rotates the circular polarization, imparting the desired phase shift. The circular polarization is converted to linear by a second nonreciprocal polarizer. A signal incident from the right is converted to circular polarization of the opposite sense by the nonreciprocal quarter-wave plate. However, since both the sense of polarization and the direction of propagation are reversed, the phase shift for a signal traveling from right to left is the same as that from left to right. The ferrite rod of the Faraday rotor is metallized to form a fully loaded waveguide and is accessible for heat sinking. The magnetic circuit is completed externally with a temperature-stable ferrite yoke. Flux drive is generally used to control the remanent magnetization.

The dual-mode phase shifter is of light weight and is capable of high average power. It also has a good figure of merit. Its chief limitation relative to the nonreciprocal toroid phase shifter is its longer switching time, being of the order of 10 to 20 μs. The switching speed is limited by the shorted-turn effect of the thin metallic film covering the ferrite rod.

Other ferrite phase shifters. There have been other kinds of ferrite phase shifters developed over the years based on different principles or on variations of those described above. Although the above phase shifters were discussed primarily as waveguide devices, some of them may be implemented in coax, helical line, and stripline.[33]

The ferrite Faraday rotor, which was mentioned above as being a part of the dual-mode phase shifter, also has been applied as an electronic phase shifter.[33,38] It is the equivalent of the mechanical Fox phase shifter.[39] Other examples of phase shifters are a helical slow-wave structure either surrounded by or surrounding a ferrite toroid,[40,41] a transversely magnetized slab of ferrite in rectangular waveguide,[42,43] a bucking rotator phase shifter,[44] circular polarization between quarter-wave plates,[44] a reciprocal, polarization-insensitive phase shifter for reflectarrays,[45] wideband phase shifters,[46] and phase shifters that operate at UHF.[47] Ferrites have also been used in continuous aperture scanning where a phase change is provided across an aperture without subdividing it into separate elements. In one experimental design,[48] ten discrete phase-shifter elements were replaced with a single ferrite-loaded aperture whose properties were externally controlled to scan the beam radiated by a five-wavelength, X-band antenna.

Other electronic phase shifters. In addition to the ferrite and diode devices, there have been other techniques suggested for electronically varying the phase shift. The traveling-wave tube can provide a fast, electronically controlled phase shift by variation of the helix voltage.[50] The phase shift is large enough to allow several taps on a single tube to control the phase of several antenna elements simultaneously.[51] Gaseous discharge, or plasma, phase shifters are based on the variation of the dielectric constant of the gaseous medium as a function of the number of free electrons, which depends on the current through the device.[51-53] Although they can handle about 1 kW of power and can be adapted to a wide range of frequencies, it is difficult to obtain stable operating characteristics with long life in sealed-off tubes. The switching properties of a ferrite circulator can also be used for constructing a digitally switched, nonreciprocal phase shifter.[54] It is not often used, however, because of its higher loss, lower peak power, and higher cost as compared with digitally switched diode phase shifters. Ferroelectric phase shifters, in which the dielectric of a ferroelectric material is a function of the applied electric field, are also possible.[49] They seem better suited for VHF and UHF than for higher frequencies.

Electromechanical phase-shifting devices. In addition to electronic phase shifters, electromechanical devices for changing phase have been used in phased-array radars, especially in the early models. Although electromechanical shifters are not now widely used, they are described here to illustrate the variety of devices that might be employed in array antennas.

One of the earliest and simplest forms of electromechanical phase shifters is a transmission line whose length is varied mechanically, as with a telescopic section. This is called a *line stretcher*. The telescoping section might be in the form of a " U ", and the length of the line changed in a manner similar to that of a slide trombone. The line stretcher is often implemented in coaxial line. A mechanical line stretcher that gives more phase shift for a given amount of motion than a conventional line stretcher is the helical-line phase shifter due to Stark.[51,55] The phase velocity on the helical transmission line is considerably less than the velocity of light. For this reason a given mechanical motion produces more phase change than would a line stretcher in conventional transmission line. Thus a shorter phase shifter can be had, which is especially advantageous at VHF or UHF frequencies. The reduction in length is essentially equal to the wind-up factor of the helix, which is the ratio of the circumference to the pitch. Wind-up factors may range from 10 to 20 in practical designs.

Both the coax and the helical line-stretchers are not well suited for the higher microwave frequencies. A waveguide device, suitable for the higher frequencies, that is analogous to the line stretcher is the hybrid junction such as the magic T, short-slot hybrid coupler (sometimes called the short-slot trombone phase shifter), or the equivalent in stripline.[51] A change in line length, with a corresponding change in phase, is obtained in the magic T with adjustable short circuits in the collinear arms.[56] The use of adjustable short circuits in the short-slot hybrid is somewhat more convenient to configure mechanically.

Another electromechanical phase shifter which has been used in array radar is the rotating-arm mechanical phase shifter.[11,57,58] It consists of a number of concentric transmission lines. Each line is a three-sided square trough with an insulated conductor passing down the middle. A moving arm makes contact with each circular assembly. The arms are rotated to produce a continuous and uniform variation of phase across the elements of the array. When the phase at one end of the concentric line is increasing, the phase at the other end is decreasing. Hence one line can supply the necessary phase variation to two elements, one on either side of array center. A total of $N/2$ concentric rings are required for a linear array of $N + 1$ elements.

There are several methods for generating phase shift that utilize the properties of circular polarization. One of the first devices to employ circularly polarized waves propagating in round waveguide was the Fox phase shifter.[39] This rotary-waveguide phase shifter was applied in World War II by the Bell Telephone Laboratories in the FH MUSA, or Mk 8, scanning radar.[4,59] This was the first US radar to use a phased-array antenna with phase shifters to steer the beam. The 42-element S-band array scanned a ± 9 degree sector at a rate of 10 per second. Related devices that obtain phase change by relative rotation of crossed dipoles in a circular guide or cavity are described by Kummer.[51]

A different form of mechanical beam steering is used in an array with spiral antenna elements.[60,61] The linearly polarized beam radiated by a flat, two-dimensional array of spirals may be scanned by rotating the individual spiral antenna elements. One degree of mechanical rotation corresponds to a phase change of one electrical degree. No additional phase-shifting devices are required. An array of spiral elements makes a simple scanning antenna. It is primarily useful in those applications where a broadband element is required and the power is not too high. The entire assembly, including the spiral radiators and feed networks, but possibly excluding the rotary joint, can be manufactured with printed circuit techniques. Helical radiating elements have also been used in arrays to obtain phase shifts by rotation of the elements.[62]

A change in phase in the waveguide transmission may be obtained by mechanically changing the dimensions of the guide.[20,59] One such device that has been applied to practical radars is the Eagle scanner, or delta-a scanner. The latter term is descriptive of the fact that the velocity of propagation, and therefore the phase, of a signal propagated through a waveguide depends on the width of the guide, or its " a " dimension. This phase shifting technique has been used for GCA (ground control approach) radars to mechanically scan fan beams in azimuth and elevation.

Mechanically actuated phase shifters are, of course, not capable of being actuated as rapidly as electronic devices, nor are they as flexible in being able to select any random value of phase. It is possible, however, with several electromechanical devices to scan a beam over its coverage at rates as fast as 10 times per second (0.1 s switching time), which is sufficiently rapid for many applications.

8.4 FREQUENCY-SCAN ARRAYS[63-65]

A change in frequency of an electromagnetic signal propagating along a transmission line produces a change in phase, as was indicated by Eq. (8.15). This provides a relatively simple means for obtaining electronic phase shift. Although parallel feeding of a frequency-scan array is possible, it is usually simpler to employ series feeding, as in Fig. 8.14. Since the attenuation in the transmission line connecting adjacent elements is small compared with that of conventional phase shifters, the series-fed arrangement can be used in frequency-scan arrays without excessive loss.

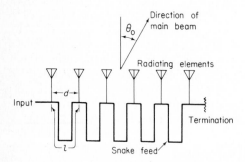

Figure 8.14 Series-fed, frequency-scan linear array.

The phase difference between two adjacent elements in the series fed array of Fig. 8.14 is

$$\phi = 2\pi f l/v = 2\pi l/\lambda \tag{8.16}$$

where f = frequency of the electromagnetic signal
l = length of line connecting adjacent elements (generally greater than the distance between elements)
v = velocity of propagation in the transmission line
λ = wavelength

For convenience of analysis, the velocity of propagation is taken to be equal to c, the velocity of light. This is applicable to coaxial lines or similar structures which propagate the TEM mode.

If the beam is to point in a direction θ_0, the phase difference between elements should be $2\pi(d/\lambda) \sin \theta_0$. In a frequency-scan array it is usually necessary to add an integral number of 2π radians of relative phase difference. This permits a given scan angle to be obtained with a smaller frequency change. Denote by the integer m the number of 2π radians added. By equating this phase difference to the phase shift obtained from a line of length l, as given by Eq. (8.16), we have

$$2\pi(d/\lambda) \sin \theta_0 + 2\pi m = 2\pi l/\lambda \tag{8.17a}$$

or
$$\sin \theta_0 = -\frac{m\lambda}{d} + \frac{l}{d} \tag{8.17b}$$

When the beam points broadside, $(\theta_0 = 0)$, Eq. (8.17b) yields $m = l/\lambda_0$, where λ_0 is defined as the wavelength corresponding to the beam position at broadside. The corresponding frequency is denoted f_0, and the direction of beam pointing becomes

$$\sin \theta_0 = \frac{l}{d}\left(1 - \frac{\lambda}{\lambda_0}\right) = \frac{l}{d}\left(1 - \frac{f_0}{f}\right) \tag{8.18}$$

If the beam is steered over the limits $\pm\theta_1$ the wavelength excursion $\Delta\lambda$ is given by

$$\sin \theta_1 = \frac{l}{2d}\frac{\Delta\lambda}{\lambda_0} \tag{8.19}$$

Thus there is a tradeoff between the wavelength excursion and the length of line connecting the elements. Figure 8.15 is a plot of the beam direction as a function of the frequency for various values of the *wrap-up factor* l/d, as given by Eq. (8.18). Note that the beam position is not symmetrical with frequency. To scan the beam $\pm45°$ about broadside requires a bandwidth of almost 30 percent with a wrap-up factor of 5, and 7 percent for a wrap-up factor of 20. (Percentage bandwidth = $100 \cdot f/f_0$.)

A frequency-scan radar requires a considerable portion of the available band of a radar to be devoted to beam steering. Although this is a simple method of electronic steering, it does not usually allow the frequency band to be used for other purposes, such as for high range-resolution or for frequency agility.

If too short a pulse (too wide a signal bandwidth) is used in the frequency-scan array, the pattern will be distorted. There are two equivalent methods for viewing this limitation. From the frequency domain point of view, each spectral component of frequency corresponds to a different pointing direction. If the signal contains widely spaced frequency components the beam will be spread over a considerable angular region rather than confined to a beamwidth as determined from diffraction theory. Alternatively, from the time delay point of view, a narrow pulse impressed at the input of the series-fed array of Fig. 8.14, requires a finite time to

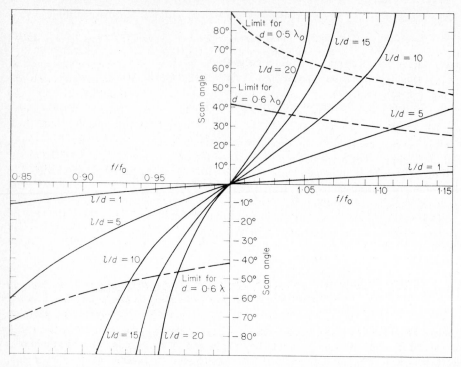

Figure 8.15 Scan angle θ_0 versus frequency. f_0 corresponds to $\theta_0 = 0$. Dashed curve gives onset of grating lobe for $d = 0.5\lambda_0$; dot-dash curve for $d = 0.6\lambda_0$.

travel down the transmission line. The finite time implies a finite bandwidth. The greater the wrap-up factor, the less the frequency band required to steer the beam over a given angle; however, the longer it will take to fill the array and the less will be the bandwidth that the array can support. For example, if the array antenna had an aperture of 5 m and if the wrap-up factor were 15, it would take 0.25 μs to fill the array. The pulse width should be long compared to this time if distortion is not to result.

A frequency-scan radar can radiate undesirable grating lobes, just as can any other array, if the electrical spacing between elements is too large. From Eq. (8.10b), the relationship between the angle θ_g at which the first grating lobe appears, and the angle θ_0 to which the main beam is steered, is given by

$$|\sin \theta_g - \sin \theta_0| = \lambda/d \qquad (8.20)$$

If we assume that the grating lobe can be tolerated if it is located at $\pm 90°$, then

$$|1 + \sin \theta_0| \leq \lambda/d \qquad (8.21)$$

The limiting scan angle for the appearance of grating lobes is shown by the dashed curves in Fig. 8.15 for the two cases $d = 0.5\lambda_0$ and $0.6\lambda_0$. The onset of the grating lobe can set a limit to the maximum angle of scan. The appearance of grating lobes is not symmetrical about $\theta_0 = 0$. If, for example, the element spacing were $0.5\lambda_0$ and $l/d = 15$, the beam can be scanned from $\theta_0 = 0$ to $\theta_0 = 62°$. (This assumes the lowest frequency is determined by the condition that the element spacing be not less than one-half wavelength.) With $d = 0.6\lambda_0$, the beam can be scanned from $-48°$ to $+36°$ without the appearance of grating lobes, for $l/d = 15$.

It is possible for a frequency-scan array to employ different frequencies to radiate over the same angular region, assuming that the antenna feed and elements are sufficiently broadband.[64] This can be seen from an examination of Eq. (8.17b). The factor m in this equation can take on any integer value. As the frequency is changed, one beam after another will appear and disappear, each beam corresponding to a different value of m. Only one beam at a time will be radiated if the grating lobe relation of Eq. (8.21) is satisfied; else multiple beams, or grating lobes, will appear. It can be shown that if an array radiates at a particular angle corresponding to a value m in Eq. (8.17b) when the frequency is f, then for some other value m_1 a beam will be radiated at the same angle when the frequency is $f_1 = (m_1/m)f$. Consider, for example an array with spacing $d = 0.6\lambda_0$ and $l/d = 15$. This corresponds to $m = l/\lambda_0 = 9$. From Fig. 8.15 it can be seen that the array will scan over the region $\pm 30°$ as the frequency is varied from $0.968f_0$ to $1.035f_0$, where f_0 is the frequency corresponding to the broadside position of the beam. As the frequency is increased, the factor $m = 10$ applies and the same angular region is scanned as the frequency varies from $1.075f_0$ to $1.149f_0$. For $m = 11$, the corresponding frequency range is from $1.183f_0$ to $1.264f_0$. The ability to radiate from the same array over more than one frequency region is of advantage when mutual interference between radars is of concern. It can also be of importance for military applications in which frequency diversity is desired.

Equation (8.18) assumed a transmission line whose velocity of propagation was that of the velocity of light. It is more usual, however, for transmission lines in a frequency-scan array to have a velocity of propagation which varies with frequency; i.e., they are *dispersive*. A waveguide is an example of a dispersive line. The velocity vs. frequency characteristic of such transmission lines can be used to advantage to provide more frequency sensitivity; that is, a smaller wrap-up factor can be obtained for a given scan angle and frequency excursion. A sketch of a folded waveguide for exciting an array of slotted waveguides is shown in Fig. 8.16. This type of feed is known as a snake feed, serpentine, or sinuous feed. Other slow-wave transmission lines can be used for the feed, including helical waveguide[69,70] and corrugated waveguide.[71] The configuration of Fig. 8.16 can be used to scan a pencil beam in elevation, with mechanical rotation providing the azimuth scan. The AN/SPS-48, Fig. 8.17, is such an example. It is a frequency-scan radar used on many U.S. Navy ships for the measurement of the elevation and azimuth of aircraft targets. It is sometimes called a *3D radar*, with range being the third coordinate, in addition to the two angles. The AN/SPS-48 radiates multiple frequencies so as to generate simultaneous multiple beams to permit a higher scan rate than would be possible with a single beam. Since the range to the target is less as the

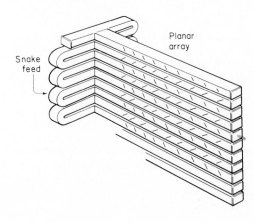

Figure 8.16 Planar-array frequency scan antenna consisting of a folded-wavelength delay-line feeding a set of linear waveguides that are so oriented that radiating slots in the narrow wall of the guide lie in the plane of the antenna.[63]

Figure 8.17 AN/SPS-48 frequency-scan radar antenna.

elevation angle increases, the transmitted power is decreased as the elevation angle is increased. (This is sometimes called *power programming.*) In a shipboard application, the elevation scanning control can be used to electronically stabilize the beam position to compensate for ship's motion.

In a normal frequency-scan radar, the beam dwells at each angular resolution cell (beamwidth) for one or more pulse repetition intervals. Another method for utilizing the frequency domain to scan an angular region is to radiate a single frequency-modulated pulse with a modulation-band wide enough to scan the beam over the entire angular region. Thus, over the duration of a single pulse, the antenna beam scans all angles. This is sometimes called *within-pulse* frequency scanning.[66,122] (The term within-pulse scanning has been applied to other methods of scanning, as described in Sec. 8.7.) The transmitted waveform is similar to the frequency-modulated pulse compression waveform (chirp) discussed in Sec. 11.5. The carrier frequencies of any echo pulses returned to the radar will be determined by the elevation angle of the targets. The receiver employs a bank of filters, each tuned to a different carrier frequency which in turn corresponds to a different angle. This is analogous to the output of a beamforming matrix, as in Sec. 8.7. The number of filters depends on the antenna beamwidth and the total angular coverage. The bandwidth Δf_B of these filters is determined by the frequency change needed to scan the antenna beam one beamwidth, that is

$$\Delta f_B = \frac{df}{d\theta_0} \theta_B = \frac{df}{d\theta_0} \frac{\lambda}{D} \tag{8.22}$$

where θ_B = beamwidth and D = aperture dimension. Substituting Eq. (8.18) into the above gives

$$\Delta f_B = \frac{f}{f_0} \frac{\cos \theta_0}{(l/d)(D/c)} \tag{8.23}$$

The time t_D for the signal to transit the snake feed from one end to the other is equal to $(l/d)(D/c)$. Thus, in the vicinity of broadside ($\theta_0 \simeq 0$), the bandwidth available for pulse compression is $\Delta f_B \simeq 1/t_D$. The frequency-modulated signal occupying a band equal to $1/t_D$ can be compressed in a pulse compression filter to produce a narrow pulse at the output of the filter. This then is one method for combining frequency scan with pulse compression, where the angular resolution depends on the antenna beamwidth and the range resolution depends on the group time delay t_D.

The frequency-scan technique is well suited to scanning a beam or a number of beams in a single angle coordinate. It is possible to use the frequency domain to produce a TV (raster) type of scan in two orthogonal angular coordinates by employing an array of slightly dispersive arrays fed from a single highly dispersive array. (The single snake line in this case is much longer electrically than the snake lines feeding a linear array.) The single snake line passes through several modes in traversing the frequency band. It has been said that a 90 by 20° sector might be possible using a 30 percent frequency band.[67] One of the disadvantages of this form of two-dimensional frequency-scan array is that it requires a wide band and is very limited in signal bandwidth, much more so than the array which steers in one dimension only.

Another method for employing frequency scan in a planar array is to use the frequency change to steer in one coordinate and phase shifters to steer in the other.[68] This is sometimes called a *phase-frequency array* (Fig. 8.18) in contrast to a *phase-phase* array which uses phase shifters to steer in both angular coordinates. The antenna may be considered as a number of frequency-scan arrays placed side by side. In an N by M element planar array there might be a separate snake feed for each of the N rows to obtain frequency steering in one coordinate, and one phase shifter for each of the M columns to achieve steering in the orthogonal plane.

Changing the frequency of a signal propagating through a length of transmission line is a convenient method for obtaining a phase shift, but it is not always desirable to operate a radar with a changing frequency. A phase shifting technique that uses a frequency change, but which then converts back to a constant frequency, is the Huggins phase shifter shown in Fig. 8.19. A signal of frequency f_0, whose phase is to be shifted an amount ϕ, is mixed with a signal from a variable frequency oscillator f_c in the first mixer. The output of the variable-frequency oscillator f_c is also passed through a delay line with a time delay τ. The output of the delay line is a signal of frequency f_c with a phase shift $\phi = 2\pi f_c \tau$. This phase-shifted signal and the output of the first mixer are then heterodyned in the second mixer. If the sum frequency is selected from the first mixer, the difference frequency is selected from the second mixer. The result is a signal with the same frequency as the input f_0, but with the phase shift ϕ. Because mixers are employed, this type of phase shifter operates at low power. It is convenient to implement at IF, although it is also possible to employ the principle at RF. A tapped delay

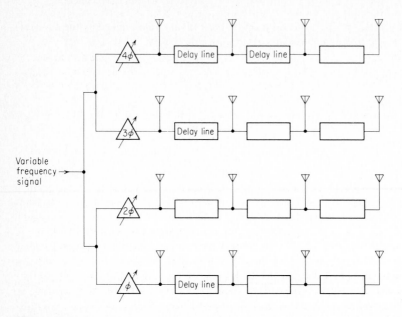

Figure 8.18 Volumetric scanning of a planar array using frequency scan in one coordinate and phase-shift scan in the other.

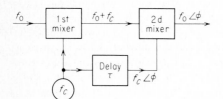

Figure 8.19 Schematic representation of the Huggins phase shifter.

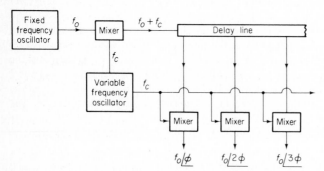

Figure 8.20 Huggins phase shifting applied to a linear array feed.

line, as illustrated in Fig. 8.20, can be utilized to obtain a series fed array. Two of these in a mixer-matrix array can provide steering in two angular coordinates. One delay line at frequency f_1 might be used to obtain the elevation steering and a second line at frequency f_2 might obtain the azimuth steering. The output of the mixer at each element is taken at the sum frequency, $f_1 + f_2$, to give the required phase shifts for two-dimensional beam steering.

8.5 ARRAY ELEMENTS[78,79]

Almost any type of radiating antenna element can be considered for use in an array antenna. Detailed descriptions of the various radiators used for arrays may be found in the standard texts on antennas and will not be discussed here. It should be cautioned, however, that the properties of a radiating element in an array can differ significantly from its properties when in free space. For example, the radiation resistance of an infinitely thin, half-wave dipole in free space is 73 ohms, but in an infinite array with half-wavelength element spacing and a back screen of quarter-wave separation it is 153 ohms when the beam is broadside. The impedance will also vary with scan angle. For a finite array the properties vary with location of the element within the array. In some arrays, dummy elements are placed on the periphery so as to provide the elements near the edge with an environment more like those located in the interior.

Although there have been many different kinds of radiators used in phased arrays, the dipole, the open-ended waveguide, and the slotted waveguide probably have been employed more than others. In addition to the conventional dipole, the dipole with its arms bent back (like an arrowhead) has been used for wide-angle coverage, the thick dipole has been used for reducing mutual coupling and for broad bandwidth (a dipole with a director rod also reduces mutual coupling), crossed dipoles are employed for operation with dual orthogonal polarization, and printed-circuit dipoles for simplification of fabrication. The dipole is always used over a reflecting ground plane, or its equivalent, in order to confine the main beam radiation to the forward direction.

Slots cut into the walls of a waveguide are similar in many respects to the dipole since the slot is the Babinet equivalent of the dipole. A slot array is generally easier to construct at the higher microwave frequencies than an array of dipoles. Although the slots may be in either the broad or the narrow wall of the waveguide, the narrow wall is generally preferred (edge slots) so that the waveguides may be stacked sufficiently close to obtain wide-angle scan without grating lobes. The waveguide slot array antenna is more suited for one-dimensional scanning than scanning in two coordinates. This type of construction is also suitable for mechanically rotating antennas. Figure 8.17 (AN/SPS-48) shows a stack of slotted waveguides for forming a beam that is frequency-scanned in elevation. The slotted guides forming the rows of the antenna are sometimes called *sticks*. The power coupled out of the guide by the slot is a function of the angle at which the slot is cut. When half-wavelength-spaced slots are fed in a series fashion, the field inside the guide changes phase by 180° between elements. The phases of every other slot therefore must be reversed to cause the radiated energy to be in phase. This is accomplished in a slotted waveguide by altering the direction of tilts of adjacent elements. In a dipole array, the phase is reversed by reversing every other dipole.

Open-ended waveguides are another popular form of array radiator. They are a natural extension of the waveguide sections in which the phase shifters are placed. Their performance can be calculated or measured in a simple phased-array simulator,[72,73] and good performance can be obtained with a well-designed radiator. The waveguide might be loaded with dielectric to reduce its physical size in order to fit the element within the required space. Ridge-loaded waveguide may be used for the same purpose. If wide-angle scan is not required, the open-ended waveguides may be flared to form a horn with greater directivity. An array of open-ended waveguides quite often will be covered with a thin, flat sheet of dielectric to serve as a means for better matching the array to free space, as well as to act as a radome to protect the array from the weather.

Other radiators that have been used in phased arrays include microstrip antennas,[160] polyrods, helices, spirals, and log-periodic elements. Radiating elements are generally of low gain, as is consistent with the broad-beamwidth needed for wide-angle scan and the narrow spacing to avoid grating lobes. More directive elements can be used when the scanning of the beam need not be over a wide angle. The more directive the element, the fewer that are needed to fill a given aperture.

Another important consideration in the selection of an antenna element and the design of an array, is the mutual coupling between the radiating elements. When the field intensity pattern of an array antenna was considered previously [Eq. (8.3)] it was assumed that the elemental radiators were independent of one another. In practice this is only an approximation. Electromagnetic radiation doesn't always respect such an assumption. Current in one element will affect the phase and amplitude of the currents in neighboring elements. In a planar array, the current at a particular element might be influenced significantly by a relatively large number of its neighbors (perhaps from 20 to 100). Since the phases and amplitudes of the currents at each of the elements change with the scan angle, the effects of mutual coupling also change with scan angle and complicate the problem of computing and compensating coupling effects. Mutual coupling, therefore, causes the actual radiation pattern to differ from that which would be predicted from an array of independent radiators. Thus to achieve the benefits from full control of the aperture illumination of a phased array, mutual coupling among the elements must be properly taken into account.

Most analyses of mutual coupling are concerned with the change in impedance at the input to the element.[74-77] It is important to know how the impedance changes in order to properly match the impedance of the transmitter and the receiver to the radiator. The effects of diffraction of energy by neighboring elements should also be considered as a mutual coupling problem. This is especially noted with end-fire elements (such as polyrods or log-periodic antennas) which are spaced close enough to couple or diffract energy. The effect of mutual coupling on the input impedance with such elements is usually small. However, the disturbance to the aperture illumination can be quite large and can result in significant differences from the pattern computed by ignoring such effects.

8.6 FEEDS FOR ARRAYS[20,81-84]

If a single transmitter and receiver are utilized in a phased array, there must be some form of network to connect the single port of the transmitter and/or the single port of the receiver to each of the antenna elements. The power divider used to connect the array elements to the single port is called an *array feed*. Several examples of such networks for linear arrays where shown in Figs. 8.2 and 8.4. For a planar array the problem is more complicated. The problem in combining the outputs of N ports into one port, or vice versa, is to do so with minimum loss. It is not always appreciated that loss in the array feed is equivalent to a loss in antenna power gain.

There are at least three basic concepts for feeding an array. The *constrained feed* utilizes waveguide or other microwave transmission lines along with couplers, junctions, or other power distribution devices. The *space feed* distributes the energy to a lens array or a

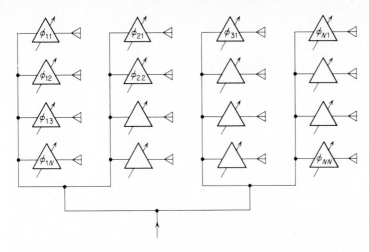

Figure 8.21 Planar array with phase-shift volumetric scan in two angular coordinates.

reflectarray in a manner analogous to a point-feed illuminating a lens or reflector antenna. The *parallel plate feed* uses the principles of microwave structures to provide efficient power division. It is, in some respects, a cross between the constrained feed and the space feed.

Constrained feed. Figure 8.21 shows a two-dimensional-scanning array which is sometimes known as a *parallel-series feed*. Each element has its own phase shifter. A separate command must be computed by the beam-steering computer and distributed to each phase shifter. The power distribution to the columns is by parallel feed. The power in each column is shown being distributed by a series feed to the vertical elements. If this were a parallel feed it would be called a *parallel-parallel feed*. (Series feeds are shown here so as not to overly complicate the figure. *Series-parallel* or *series-series* arrangements are also possible.) All the elements which lie in the same column utilize the same phase shift to steer the beam in azimuth. Likewise all the elements which lie in the same row utilize the same phase shift to steer the beam in elevation. The phase shift at the mnth element therefore is the sum of the phases required at the mth column for steering in azimuth and at the nth row for steering in elevation. This summation can be made at the computer and distributed to the MN elements of the array. Alternatively, $M + N$ control signals can be transmitted to the array if an adder is provided at each phase shifter to combine the azimuth and elevation phases.

In the *series-series feed* each of the M columns of an M by N array utilizes a series arrangement to produce steering in one coordinate (say elevation). A separate series-feed is used to provide the proper phase to each column to steer in the orthogonal coordinate (azimuth). A total of $M + 1$ series feeds are used. The computation of the phase-shifter commands is simplified since the azimuth and elevation phases are not added in the computer. Instead, the phase shifters of a single feed provide the azimuth steering while the M feeds at each column provide the elevation steering. In the series-series planar array all series phase shifters in the elevation plane take the same value and all series phase shifters in the azimuth plane take the same value. Thus only two control signals are required. The steering commands are simplified at the expense of $M - 1$ additional phase shifters and the added loss of a series arrangement.

When separate power amplifiers are used at each element in a transmitting array or separate receivers are at each element in a receiving array, series-feed arrays are attractive since their inherent loss is at low-power levels and is made up by the amplifiers. One method for producing low-power-level beam steering is to use a single series-fed array at frequency f_1 to provide the azimuth phases ϕ_n. At each element the signal is heterodyned in a mixer with the signal from a single series-fed array at frequency f_2 to provide the elevation phase ϕ_m. The sum signal at frequency $f_1 + f_2$ is used as the carrier frequency. It has the proper phase $\phi_m + \phi_n$ at each element to steer in the two directions. The mixer would be followed by a power amplifier if a transmitter, or it might be used to obtain the local oscillator frequency if a receiver. Only $M + N - 2$ phase shifters are required, but the mixer and power amplifier required at each element adds to the complication. This is sometimes called a *mixer-matrix feed*.

A convenient method for achieving two-dimensional scanning is to use frequency scan in one angular coordinate and phase shifters to scan in the orthogonal angular coordinate, as was diagrammed in Fig. 8.18. This is an example of a parallel-series feed. It may be considered as a number of frequency-scanned linear arrays placed side by side.

When the power splitters are four-port hybrid junctions, or the equivalent, the feed is said to be *matched*. Theoretically there are no spurious signals generated by internal reflections in a matched feed. It is not always convenient, however, to use four-port junctions. Three-port tee junctions are sometimes used for economic reasons to provide the power splitting, but the network is not theoretically matched. Internal reflections due to mismatch in the feed can appear as spurious sidelobes in the radiation pattern.[80]

Space feeds. There are two basic types of space feeds depending on whether they are analogous to a lens or to a reflector. The lens array, Fig. 8.22, is fed from a primary feed just as a lens antenna. An array of antenna elements collects the radiated energy and passes it through the phase shifters which provide a correction for the spherical wavefront, as well as a linear phase shift across the aperture to steer the beam in angle. Another set of elements on the opposite side of the structure radiate the beam into space. The primary pattern of the feed illuminating the space-fed array provides a natural amplitude taper. Spillover radiation from the feed, however, can result in higher sidelobes than from an array with a conventional constrained feed. The space-fed array can readily generate a cluster of multiple beams, as for monopulse angle measurement, by use of multiple horns or a multimode feed, rather than with a complicated feed network as in the conventional array.

There are two sets of radiators in the lens array requiring matching (the front and the back), thus increasing the matching problem and the potential for lower efficiency. The feed

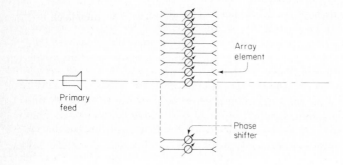

Figure 8.22 Principle of lens array.

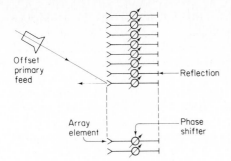

Figure 8.23 Reflectarray.

may be placed off-axis to avoid reflections from the back face of the lens, if desired. It is possible in a lens array to reduce the number of phase shifters by "thinning" the number of output radiators. This is accomplished by combining pairs of input elements and feeding the output of each pair to a single-phase shifter and radiating element. The thinned elements are near the outer portion of the antenna rather than at the center so as to produce a density taper (Sec. 8.10). This procedure, while reducing the number of phase shifters, generally results in lower gain and higher far-out sidelobes than would be produced by an amplitude taper.

A space-fed reflectarray with an offset feed is shown in Fig. 8.23. The energy enters the antenna elements, passes through the phase shifters, is reflected, and again passes back through the phase shifters to be radiated. Like the lens array, the phase shifters apply a linear phase distribution for beam steering and a correction for the curvature of the primary wavefront from the horn. Because the energy passes through the phase shifters twice, they need only half the phase-shift capability of a lens array or a conventional array; i.e., 180° of one-way phase shift is adequate, rather than 360°. The phase shifters, however, must be reciprocal. As with the lens array, multiple beams can be generated with additional feed horns.

The lens array allows more freedom than the reflectarray in designing the feed assembly since there is no aperture blocking, but the back surface of the reflectarray makes it easier to provide the phase shifter control and drive assemblies, structural members, and heat removal. Space-fed arrays are generally cheaper than conventional arrays because of the omission of the transmission-line feed networks and the use of a single transmitter and receiver rather than a distributed transmitter and receiver at each element. A space-fed array may be simpler than an array with a constrained feed, but a sacrifice is made in the control of the aperture illumination and in the maximum power capability of the array. Thus the ability to radiate large power by using a transmitter at each element is lost in this configuration.

Parallel-plate feeds. A folded pillbox antenna (Fig. 8.24), a parallel-plate horn, or other similar microwave device can be used to provide the power distribution to the antenna elements. These are *reactive* feed systems. They are basically used with a linear array and would have to be stacked to feed a planar array.

Subarrays. It is sometimes convenient to divide an array into subarrays. For example, the AN/SPY-1 AEGIS array utilizes 32 transmitting and 68 receiving subarrays of different sizes.[81] One reason for dividing the transmitting array into subarrays is to provide a distributed transmitter. In the AEGIS array a separate high-power amplifier feeds each of the 32 transmitting subarrays. It is also possible to give identical phase-steering commands to similar elements in each subarray, thus allowing simplification of the beam-steering unit and of the interface cabling between the array and the beam-steering unit. The term subarray has also been applied to the array feed networks for producing sum and difference radiation patterns.[82]

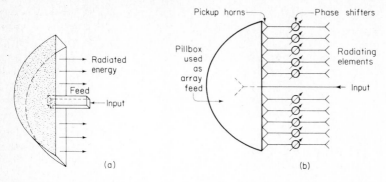

Figure 8.24 Example of a pillbox antenna (shown in (a)) used as a feed for an array (b). (*After Ricardi.*[161])

8.7 SIMULTANEOUS MULTIPLE BEAMS FROM ARRAY ANTENNAS

One of the properties of the phased array is the ability to generate multiple independent beams simultaneously from a single aperture. In principle, an N-element array can generate N independent beams. Multiple beams allow parallel operation and a higher data rate than can be achieved from a single beam. The multiple beams may be fixed in space, steered independently, or steered as a group (as in monopulse angle measurement). The multiple beams might be generated on transmit as well as receive. It is convenient in some applications to generate the multiple beams on receive only and transmit with a wide radiation pattern encompassing the total coverage of the multiple receiving beams. The ability to form many beams is usually easier on reception than transmission. This is not necessarily a disadvantage since it is a useful method of operating an array in many applications.

The simple linear array which generates a single beam can be converted to a multiple-beam antenna by attaching additional phase shifters to the output of each element. Each beam to be formed requires one additional phase shifter, as shown in Fig. 8.25. The simple array in this figure is shown with but three elements, each with three sets of phase shifters. One set of phase shifters produces a beam directed broadside to the array ($\theta = 0$). Another set of three phase shifters generates a beam in the $\theta = +\theta_0$ direction. The angle θ_0 is determined by the relationship $\theta_0 = \sin^{-1}(\Delta\phi\lambda/2\pi d)$, where $\Delta\phi$ is the phase difference inserted between adjacent elements. Amplifiers may be placed between the individual antenna elements and the beam-forming (phase-shifting) networks to amplify the incoming signal and compensate for any losses in the beam-forming networks. The output of each amplifier is subdivided into a number of independent signals which are individually processed as if they were from separate receivers.

Postamplification beam forming. When receiving beams are formed in networks placed after the RF amplifiers, as in Fig. 8.25, the antenna is sometimes called a *postamplification beam-forming array*, abbreviated PABFA. A separate transmitting antenna may be used to illuminate the volume covered by the multiple receiving beams or, alternatively, it is possible to transmit multiple beams identical to the multiple receiving beams, using the radiating elements of the same array antenna. Note that if the multiple transmitting beams are contiguous and at the same frequency, the composite transmitted pattern is similar to the pattern from a single beam encompassing the same angular region.

The receiving beam-forming network may be at IF or RF. Tapped delay lines have been a

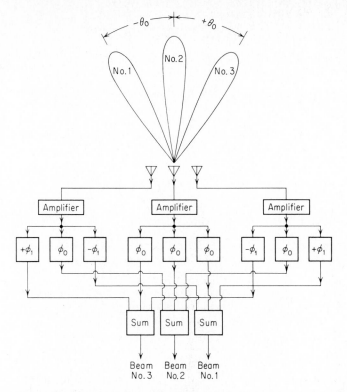

Figure 8.25 Simultaneous postamplifier beam formation. ϕ_0 = constant phase; $|\phi_1 - \phi_0| = |\Delta\phi| = |2\pi(d/\lambda)\sin\theta_0|$.

convenient method for obtaining multiple beams at IF. Beam-forming at IF is possible since phase is preserved during frequency translation from RF to IF (except for the constant phase shift introduced by the common local oscillator).

Blass beam-forming array.[85] The RF beam-forming principle shown in Fig. 8.26 has been used in the ASHR-1, a one-of-a-kind developmental height-finder radar built for the Federal Aviation Agency. Waveguide transmission lines were arranged to serve as the delay lines. Energy was tapped from each waveguide at the appropriate points by directional couplers to form beams at various elevation angles. Considerable waveguide was used in this design. To produce the 333 independent beams, the Blass height finder employed 30 miles of S-band waveguide.

Butler beam-forming array.[86–100] Another RF beam-forming device is the parallel network attributed to Butler, and independently discovered by Shelton. This is a lossless network which utilizes 3-dB directional couplers, or hybrid junctions, along with fixed phased shifters, to form N contiguous beams from an N-element array, where N is an integer that is expressed as some power of 2, that is, $N = 2^p$. The 3-dB directional coupler is a four-port junction which has the property that a signal fed into one port will divide equally (in power) between the other two ports and no power will appear in the fourth port. A 90° phase difference is introduced between the two equally divided signals. Similarly, a signal introduced into the fourth port will

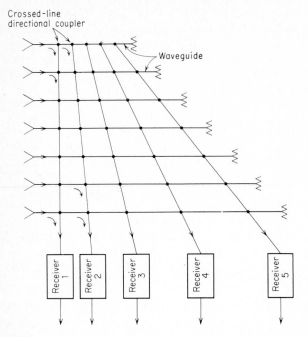

Figure 8.26 RF beam-forming using tapped transmission lines.

divide its power equally between the same two ports with a 90° relative phase difference, and no power will appear in the first port. The relative phase difference in this case is of opposite sign compared to the phase difference resulting from a signal introduced into the first port.

Consider a simple two-element array with half-wavelength spacing, connected to the two ports of a 3-dB directional coupler as shown in Fig. 8.27. If a signal is inserted in port No. 1, the 90° phase shift that results between the signals in ports 2 and 3 will produce a beam oriented in a direction 30° to the right of the array normal. A signal in port No. 4 results in

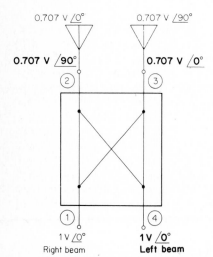

Figure 8.27 3-dB directional coupler generating two beams from a two-element array.

a phase distribution that produces a beam 30° to the left of the array normal. Thus this simple two-element array with a single 3-dB coupler produces two independent beams.

The two-element array is a trivial example of the Butler beam-forming antenna. Figure 8.28 illustrates the circuit of an eight-element array that generates eight independent beams. It utilizes 12 directional couplers and eight fixed phase shifters. The Butler matrix has 2^p inputs and 2^p outputs. Modifications of the Butler array to any number of elements have been suggested, but the resulting beam-forming network is not necessarily lossless.[88,97] The number of directional couplers or hybrids required for an N element array is equal to $(N/2) \log_2 N$, and the number of fixed phase shifters is $(N/2)(\log_2 N - 1)$.

The Butler beam-forming network is theoretically lossless; i.e., no power is intentionally dissipated in terminations. There will always be a finite insertion loss, however, due to the inherent losses in the directional couplers, phase shifters, and transmission lines that make up the network. A 16-element network at 900 MHz, for example, had an insertion loss of 0.74 dB, practically all of which was due to the strip transmission line used in the construction.[87] In a lossless, passive antenna radiating multiple beams from a common aperture it has been shown[88] that the radiation pattern and the crossover level of adjacent beams cannot be specified independently. With uniform illumination, as in the Butler array, the crossover level is 3.9 dB below the peak. This is independent of the beam position, element spacing, and wavelength.

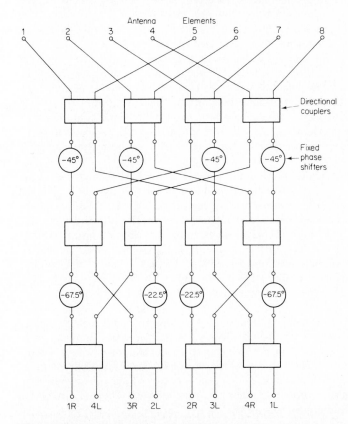

Figure 8.28 Eight-element Butler beam-forming matrix.

The low crossover level of the Butler array is one of its disadvantages. If a lossless network could be achieved with a cosine illumination so as to reduce the sidelobe levels compared to those obtained with a uniform illumination, the crossover level would be even worse (a level of -9.5 dB).

By combining the output beams of the networks with additional circuitry, the Butler beam-forming network can be modified to obtain aperture illuminations that result in lower sidelobes than available with uniform illumination. The beamwidth is widened, the gain lowered, and the network is no longer theoretically lossless. The addition of two adjacent beams of a Butler array, with the proper phase correction, results in an array with a cosine illumination. The crossover is lower than the lossless network, but the first sidelobe is -23 dB instead of -13.2 dB.

There is no theoretical limit to the bandwidth of a Butler array except for the bandwidth associated with the hardware making up the network, which can be greater than 30 percent. They have even been constructed with bandwidths of several octaves.[94] Operating over too wide a band, however, changes the beamwidth, shifts the location of the beams, and can introduce grating lobes just as with any other array antenna.

The complexity of the Butler beamforming network increases with the number of elements. A 64-element network, for example, requires 192 directional couplers and 160 fixed-phase shifters. The construction of a large Butler network requires a large number of cross-over connections in the transmission lines. These can present practical difficulties in the fabrication of the microwave printed circuits used to make up the device.[26] Many beams also require many parallel receivers, an added complexity. For these reasons, Butler beam-forming networks with large numbers of elements are not the general rule.

It is possible to construct planar arrays with Butler networks. A 2^p by 2^q element array (p, q are integers) requires $2^p + 2^q$ networks to achieve 2^{p+q} beams. Other methods of using Butler networks in planar arrays are possible. Shelton,[91] for example, describes a technique for generating multiple beams in hexagonal planar arrays with triangular spacing.

The Butler networks in this section were assumed to use 3-dB directional couplers with a $90°$ phase difference between the two equal outputs. Hybrid junctions can also be used. These produce a $180°$ phase difference between the two output signals and require a slightly different design procedure.[100]

It is of interest to note the relation of the Butler network to the Fast Fourier Transform (FFT). As stated previously, the radiated field of an antenna is related to the illumination across the aperture by the Fourier transform. Shelton[93] has pointed out that the flow diagram of the FFT is basically similar to the diagram of the Butler network. Thus the Butler network is a manifestation of the FFT. The antenna equivalent of the conventional Fourier transform is the Blass beam-forming network illustrated by Fig. 8.26. The Blass network required N^2 couplers for N inputs and N outputs, while the conventional Fourier transform also requires N^2 computations for an N-point transform. The Butler network utilizes $(N/2) \log_2 N$ junctions, just as the FFT uses $(N/2) \log_2 N$ computations for an N-point transform.

Within-pulse scanning.[101–104] If an antenna beam is scanned sequentially through its angular coverage, one position at a time, it illuminates all directions just as does a multiple-beam array. If the beam is scanned rapidly enough, however, it will have the effect of seeing "almost simultaneously" in all directions. The scan rate of the beam must be greater than the radar signal-bandwidth to preserve the information contained in the received signal. The entire scan is covered within a single pulse. Hence, the name *within-pulse scanning*. This is a method for achieving the equivalent of a multiple-beam array.

There have been at least two different variations of within-pulse scanning which have

been demonstrated experimentally. Both are more suited for electronic scanning in one angular coordinate than for two-angle-coordinate electronic scanning. Thus they are of interest for
determining elevation angle in 3D radar. In one method,[66] a frequency-scan beam is rapidly
swept through the angular region of interest, and the received signal is passed through a bank
of matched filters, each filter corresponding to a different angular direction. This was mentioned in Sec. 8.4.

The other is a receive-only method that uses a receiving array with mixers and local
oscillators (LOs), arranged so as to provide N separate receiving beams fixed in space
(Fig. 8.29). Each mixer is supplied with a different LO frequency. The transmitting antenna
provides a single beam illuminating the coverage of the N receiving beams. As described in
Sec. 8.2, in order to generate a beam at some angle θ_0, there must be a relative phase difference
between adjacent elements equal to $\phi = 2\pi(d/\lambda) \sin \theta_0$. If the left-hand element of an N-
element linear array is taken as the reference, then the phase shifters at the other elements must
have values of $\phi, 2\phi, \ldots, (N - 1)\phi$. If the beam is scanned as a function of time, these phase
shifts also change as a function of time. A constant rate of change of phase with time is
equivalent to a constant frequency. Thus a frequency difference at adjacent elements results in
a scanning beam. If the LOs in the mixers of Fig. 8.29 differ in frequency by f_s, the beam
repetitively scans its coverage at a rate of f_s. That is, the antenna beam occupies all possible
scan positions during the scan time $1/f_s$. Another way of looking at this is to note that if the
relative phase between adjacent elements of an array is changed by 2π radians, the radiation
pattern will assume all possible scan positions. If the 2π radians is changed in a time $1/f_s$, then the
rate of change of phase is $2\pi f_s$. Thus the linear phase change can be accomplished if
the frequency difference between the LOs at the mixers of adjacent elements is f_s.

By multiplying the summed output of all the element mixers with a periodic sampling
train of narrow gating pulses, a particular portion of space is observed. Changing the relative
time phase between the sampling pulse train and the array output results in observing a
different angular direction. Thus the beam can be steered by varying the time phase of the

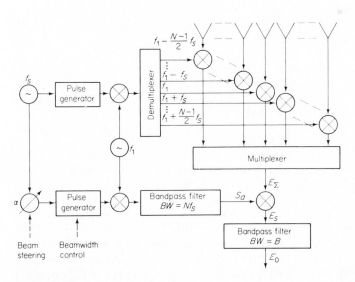

Figure 8.29 Within-pulse scanning using frequency-multiplexed linear array. This implementation has
been called MOSAR. (*From Johnson*[103] *Courtesy Proc. IEEE.*)

sampling signal. By providing more than one sampling signal, multiple simultaneous beams can be generated which can either be fixed or independently steered.

The local oscillator signals of Fig. 8.29 are derived coherently by mixing the two frequencies f_1 and f_s and filtering the desired mixer products $f_1 + nf_s$, $n = 1, 2, \ldots, N$. The frequency f_s, which is the rate at which the far field is sampled, must be greater than the total signal bandwidth if the modulation envelope of the radar pulses is to be preserved. The outputs from the elements must be limited to a bandwidth B before adding them all together, if noise overlap is to be avoided. The required bandwidth of the summed signal channel is Nf_s, which must be made greater than NB. This can be quite large (perhaps several gigahertz in some radars) and can represent a limitation on the implementation of the technique.

The various beam positions can be obtained by sampling the output at the proper time. In Fig. 8.29 one beam sampling channel is illustrated. The sampling signal S_a is a train of short pulses at a repetition rate f_s which can be time-phased to gate the scanning signal E_Σ according to the desired beam direction. The beam-steering phase is indicated by α in the figure. If α is variable, the beam is steerable. The generation of multiple fixed beams or a number of independent steerable beams requires a duplication of the sampling pulse generators and sampling mixers.

Other beam-forming methods. Several nonarray techniques using reflectors or lenses have been considered for the generation of multiple beams. The Luneburg lens and the torus reflector with multiple feeds as described in Chap. 7 are examples. A radar with a reflector antenna designed to generate a cluster of many beams has been sometimes called a *pincushion* radar.

In one example of a developmental radar that generated multiple beams, a spherical transmitting antenna was surrounded by three spherical Luneburg-lens receiving antennas, each covering a one-third sector of space.[105] The transmitting antenna was a spherical phased array of several thousand elements, with only a fraction of these energized at any one time to form a directive beam. Each energized array element was driven from a separate power amplifier, which obtained its properly phased input signal from a Luneburg lens. This lens functioned as a low-power RF analog beam-forming device for the spherical array. Such a lens is sometimes called a *computing lens*. The beam position of the spherical array was selected by switching the proper elements of the Luneburg beam-forming lens to the elements of the transmit array. Each of the three Luneburg-lens receive antennas in the radar system could generate a cluster of three beams pointing in the direction of the transmit beam. The angle of arrival was extracted by comparison of the amplitudes of the signals in the three-beam cluster. Beam switching was performed separately from beam forming on both transmit and receive. This example of a radar based on multiple-beam-forming is a system that is probably more complex than would be needed now to accomplish the same radar mission.

Another approach to multiple-beam antennas utilized a planar lens array fed by a constrained lens consisting of two connected hemispherical surfaces and a hemispherical feed surface.[82] The details of this antenna will not be described here. Its advantage is that it has true time delay, allowing a wide instantaneous bandwidth.

The Mubis[106] antenna, which uses a parallel-plate lens, and the Bootlace[107] antenna, which is a form of lens array, are also capable of RF beam forming. As mentioned, the Luneburg lens can be used as a beam-forming network to form multiple beams in conjunction with a circular or spherical array,[108] or the lens can be used directly to generate multiple beams. Since the Luneburg lens is not generally capable of high power, it is primarily a receiving antenna. For transmission, it can act at low power as an analog computer for the high-power spherical lens.

In principle, multiple receiving beams can be generated utilizing the Fast Fourier Transform (FFT) processor.[159] If the output of each receiving element is sampled at the Nyquist rate and if the sampled voltages are converted to a digital number, the FFT processor may be used to generate multiple beams digitally, just as the Butler beam-forming array generates them in analog fashion. Thus the antenna beams are generated by computation. Its practical implementation requires converting RF or IF signals at each element to digital numbers.

System considerations. One of the attractive features claimed for a multiple-beam-forming array is that it does away with phase shifters. These are replaced, however, by multiple receivers, one for each beam. This can represent an expensive trade. If instead of one receiver per beam, only one or a few receivers are time-shared over the total coverage, some form of switching is required. If the requirements for switching speed and flexibility are similar to those of a conventional electronically steered phased-array radar, the problem of switching a receiver between ports of the beam-forming array may be as difficult as that of providing the phase shifting in a conventional array radar.[105]

In principle, a surveillance radar with a fixed transmitting beam of width θ_t and a number N of fixed, narrow receiving beams of width θ_r covering the same volume ($N\theta_r = \theta_t$), has performance equivalent to a radar with a single scanning transmit-receive beam of width θ_r, provided the comparison is made on a similar basis and the received signals are processed in the optimum manner in each case. The transmit antenna gain in the multiple-beam system is $1/N$th that of the scanning single-beam system. The reduction of transmit antenna gain in a multiple-beam radar is compensated, however, in the ideal case by the increased number of hits available for integration. The gain of the transmitting antenna in the multiple-beam system is $\theta_r/\theta_t = 1/N$ that of the transmitting gain of the scanning-beam antenna. Thus the signal-to-noise ratio per pulse of the multiple-beam radar is less than the signal-to-noise ratio of the scanning-beam radar. If these N pulses are integrated without loss as in a perfect predetection integrator, the total signal-to-noise ratio in the multiple-beam radar just compensates for the lesser transmit gain. Thus the multiple-beam radar and the scanning-beam radar have equivalent detection capability, provided the data rates are the same and integration is without loss. Data rate is defined here as the revisit time in a scanning-beam radar, and in a multiple-beam radar it is the time over which the total number of pulses are integrated. In practice, postdetection integration is usually employed and there will be a finite integration loss. In such a case the scanning-beam system will have an advantage over the multiple-beam system.

In a multiple-beam system which uses a broad transmitting beam to illuminate the region covered by the contiguous receiving beams, the benefit of the two-way sidelobe levels that is characteristic of a conventional scanning radar antenna is not obtained. Thus it is usually desirable to suppress the sidelobes of the multiple receiving beams more than usual in order to reduce the likelihood of echoes from large targets being received via the one-way sidelobes.

Multiple-beam array antennas with a large number of simultaneous beams have not seen wide application, probably because of the complexity of such systems. They have, however, had application in 3D mechanically rotating air-surveillance radars which employ a small number of contiguous beams stacked in elevation to provide the elevation coordinate.

In some applications, the effect of multiple, independent beams can be obtained with a single-beam phased-array radar which is capable of flexible and rapid beam steering. For example, a sequential burst of pulses can be transmitted at the beginning of the transmission, with each pulse radiated in a different direction. This requires rapidly switching phase shifters to steer the beam between pulses. It also requires an application where a short minimum range is not important since reception cannot take place during the transmission of the burst of pulses. Thus this approach would be suitable for radars whose targets are at long range, such as satellite surveillance or BMD (Ballistic Missile Defense). On reception, a separate receive beam must be generated for each direction of transmission. In a surveillance application this could require the complication of some sort of beam-forming. In a tracking phased-array radar, however, a single time-shared receive beam can be used simultaneously to track many targets at different angular directions. Since the targets are under track, their approximate ranges are known so that a beam need be formed in the proper direction only during the time that a target echo is expected. In this manner, several targets can be held in track during the interpulse period, provided the phase shifters can switch sufficiently rapidly and a control computer is available to take advantage of the inherent flexibility of the array.

8.8 RANDOM ERRORS IN ARRAYS

In the analysis of the effects of reflector-antenna errors in Sec. 7.8 only the phase error was considered. In an array, however, other factors may enter to cause distortion of the radiation pattern. These include errors in the amplitude as well as the phase of the current at the individual elements of the array, missing or inoperative elements, rotation or translation of an element from its correct position, and variations in the individual element patterns. These errors can result in a decrease in gain, increase in the sidelobes, and a shift in the location of the main beam.

Since it is not always possible to know the exact nature of the errors that might be encountered in a specific antenna, the properties of the antenna must be described in statistical terms. That is, the average, or expected, value of the radiation pattern of an *ensemble* of antennas of similar type can be computed based on the statistics of the random errors. The statistical description of the antenna properties cannot be applied to any particular antenna, but applies to the collection of similar antennas whose errors are specified by the same statistical parameters.

The ensemble *average* power pattern of a uniform array of M by N isotropic elements arranged on a rectangular grid with equal spacing between elements can be expressed as[109]

$$|f(\theta, \phi)|^2 = P_e^2 e^{-\overline{\delta^2}} |f_0(\theta, \phi)|^2 + [(1 + \overline{\Delta^2})P_e - P_e^2 e^{-\overline{\delta^2}}] \sum_{m=1}^{M} \sum_{n=1}^{N} i_{mn}^2 \qquad (8.24)$$

where P_e = probability of an element being operative (or the fraction of the elements that remain working)

δ = phase error (described by a gaussian probability density function)

$|f_0(\theta, \phi)|^2$ = no-error power pattern

Δ = amplitude error

i_{mn} = no-error current at the mnth element

Thus the effect of random errors is to produce an average power pattern that is the superposition of two terms, similar to Eq. (7.31) for the continuous aperture. The first term represents the no-error power pattern multiplied by the square of the fraction of elements remaining and by a factor proportional to the phase error. The other term depends on both the amplitude error and the phase error as well as the fraction of elements remaining operative. It also depends on the aperture illumination, as given by the currents i_{mn}. Note that this second term is independent of the angular coordinates θ, ϕ. It can be thought of as a "statistical omnidirectional" pattern. It causes the far-out sidelobes to differ in the presence of error as compared to the no-error pattern. (The no-error pattern sidelobes generally drop off rapidly with increasing angle from broadside; therefore, beyond some angle the radiation pattern will be dominated by the error-produced sidelobes.) The shape of the main beam and the near-in sidelobes are relatively unaffected by errors, although their magnitudes are modified. Note that the factor P_e can also be used to evaluate the effect of random thinning of array antennas.

For $P_e = 1$ and small errors, the normalized pattern, obtained from Eq. (8.24) by dividing by the value of $|f_0(0, 0)|^2$, is

$$\overline{|f_n(\theta, \phi)|^2} = |f_{0n}(\theta, \phi)|^2 + (\overline{\Delta^2} + \overline{\delta^2}) \frac{\sum_m \sum_n i_{mn}^2}{\left(\sum_m \sum_n i_{mn}\right)^2} \tag{8.25}$$

The second term of this expression indicates that the larger the number of elements, the smaller will be the statistical sidelobe level. The main-beam intensity, being coherent, increases as the square of the number of elements, whereas the sidelobes due to errors, being incoherent, increases only directly with the number of elements. The gain of a broadside array of isotropic elements is approximately

$$G_0 = \frac{\left(\sum_m \sum_n i_{mn}\right)^2}{\sum_m \sum_n i_{mn}^2} \tag{8.26}$$

(Note that when $i_{mn} = $ constant, $G_0 = MN$.) Then the normalized pattern of Eq. (8.25) can be expressed as

$$\overline{|f_n(\theta, \phi)|^2} = |f_{0n}(\theta, \phi)|^2 + \frac{(\overline{\Delta^2} + \overline{\delta^2})}{G_0} \tag{8.27}$$

The greater the gain of the antenna, the less the relative effect of the errors on the sidelobes.

By substituting the radiation intensity of Eq. (8.24) into the definition of gain (or directivity) of Eq. (7.3) it can be shown that

$$G/G_0 = \frac{P_e}{(1 + \overline{\Delta^2}) \exp(-\overline{\delta^2})} \simeq \frac{P_e}{1 + \overline{\Delta^2} + \overline{\delta^2}} \tag{8.28}$$

Note that the relative reduction in gain is independent of the number of elements and depends only on the fraction of elements that are operative and the mean square value of the errors. When $P_e = 1$ and $\Delta = 0$, the expression is, for small-phase errors, similar to that of Eq. (7.30) for the continuous aperture.

In addition to raising the sidelobe level, random phase and amplitude errors in the aperture distribution cause an error in the position of the main beam. Rondinelli[110] has shown that for a uniform amplitude distribution across an M by M square array, the statistical rms beam pointing error is

$$\delta\theta_0 = (\overline{\theta_0^2})^{1/2} = \frac{\sqrt{3}\sigma}{(kd_e)M^2} \tag{8.29}$$

where $\sigma = $ rms value of normalized error current assuming Rayleigh distributed errors
$k = 2\pi/\lambda$
$d_e = $ element spacing
$M = $ number of elements along one dimension of square array

The phase angle is assumed uniformly distributed. Equation (8.29) indicates an error of 0.22×10^{-4} radian ($\sim 0.001°$) for a 100-by-100-element uniformly illuminated array with a beamwidth of approximately $1°$ when $\sigma = 0.4$.

Leichter's analysis[111] of beam-pointing errors was performed for a continuous line source, but may be applied to a linear array. Both uniform distributions and modified Taylor distributions were considered. The amplitude and phase distributions were described by the gaussian distribution and were assumed independent of one another. An example of Leichter's results for a uniform amplitude distribution is shown in Fig. 8.30.

Several conclusions may be derived from the various studies of errors described above and from Sec. 7.8. For array antennas the following seem to apply:

1. The larger the number of elements (MN) in the array, the smaller will be the spurious radiation for a given error tolerance and a given design sidelobe level. In other words, lower sidelobes are more likely to be achieved with larger antennas. This comes about because the intensity of the main beam increases as the square of the number of elements $(MN)^2$, while the spurious radiation increases only linearly since it represents the incoherent addition of many contributions.[112]
2. The rise in the sidelobe level due to random errors is independent of the beam scan angle.[110]
3. The lower the design sidelobe level, the greater will be the rise in the sidelobes, assuming a given antenna size and a given error tolerance.[113]
4. In a two-dimensional array, the most serious random error is in the translational position of the dipole elements. Of secondary importance are the errors in the currents applied to the elements. The angular position of the dipole elements is relatively unimportant.[113]

The following conclusions apply to the continuous antenna:

1. According to Ruze,[112] the spurious sidelobe radiation is proportional to the mean square error, just as in the discrete array, and in addition is proportional to the square of the correlation interval measured in wavelengths. Bates[114] defines his correlation interval differently from Ruze and obtains a first-power dependence for this reason.
2. If errors are unavoidable in a reflecting antenna surface, they should be kept small in extent; that is, for the same mechanical tolerance, the antenna with the smaller correlation interval

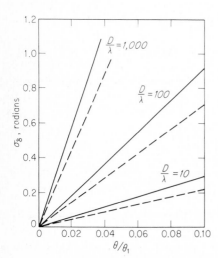

Figure 8.30 Plot of σ_δ versus θ/θ_1, where σ_δ is the rms phase error such that the pointing error will be in the interval $(-\theta, \theta)$ with a probability $p(\theta)$ for arrays with $\lambda/2$ spacing; solid curves apply for $p(\theta) = 0.95$; dashed curves apply for $p(\theta) = 0.99$; θ_1 = angle to the first null; D = antenna length. (*Courtesy Hughes Aircraft Co.*)

(rougher surface) will give lower sidelobes than an antenna with a larger correlation interval. An error stretching most of the length of the antenna is likely to have a worse effect than a localized bump or dent of much greater amplitude. Therefore small disturbances such as screws and rivets on the surface of the reflector will have little effect on the antenna radiation pattern.

3. An increase in frequency increases both the phase errors and the correlation interval in terms of wavelengths. Therefore the gain of a constant-area antenna does not increase as rapidly as the square of the frequency. For reflectors of equal gain (same diameter in wavelengths) Eq. (7.32) shows that the relative sidelobe level caused by errors will increase as the fourth power of the frequency, or 12 dB/octave.[112]

An important conclusion is that the details of the radiation pattern, especially in the region outside the main beam, are more likely to be determined by the accuracy with which the antenna is constructed than by the manner in which the aperture is illuminated. Thus the mechanical engineer and the skilled machinist and technician are just as important as the antenna designer in realizing the desired radiation pattern.

Effects of phase shifter quantization.[82,115] The discrete value of phase shift that results from the use of quantized phase shifters introduces an "error" in the desired aperture illumination. Phase quantization can cause a loss in antenna gain, an increase in the rms sidelobe level, the generation of spurious sidelobes, and a shift in the pointing of the main beam. The effect of quantized phase shifters on the radiation pattern is similar to the effect of random errors.

The gain of an array antenna with a mean square phase error $\overline{\delta^2}$ in its aperture illumination is approximately

$$G = G_0(1 - \overline{\delta^2}) \qquad (8.30)$$

where $G_0 =$ no-error gain. This follows from Eq. (8.28) with $P_e = 1$, $\overline{\Delta^2} = 0$, and $\overline{\delta^2}$ small. For a quantized phase shifter consisting of B bits, the phase error δ is described by a uniform probability density function extending over an interval $\pm\pi/2^B$. From Sec 2.4 the mean square phase error for the uniform probability density function is $\overline{\delta^2} = \pi^2/3(2^{2B})$. Substituting into Eq. (8.30), the loss of gain is found to be 1.0 dB for a two-bit phase shifter, 0.23 dB for a three-bit phase shifter, and 0.06 dB for four bits. On the basis of the loss in gain, a three- or four-bit phase shifter should be satisfactory for most applications.

Phase quantization errors also result in an increase in the rms sidelobe level and in the peak sidelobe level. With the assumptions that (1) the energy lost by the reduction in main beam gain shows up as an increase in the rms sidelobe level, (2) the element gain is the same for the main beam and the sidelobes (within the region of space scanned by the array), (3) an allowance of one dB for the reduction in gain due to the aperture illumination, and (4) one dB for scanning degradation, then the sidelobe level due to the quantization can be expressed as

$$\text{rms sidelobe level} \simeq \frac{5}{2^{2B}N} \qquad (8.31)$$

where $N =$ total number of elements in the array. For an array with four thousand elements, a three-bit phase shifter will give rms sidelobes better than 47 dB below the main beam, and a four-bit phase shifter gives 53 dB sidelobes. Thus three or four bits should be sufficient for most large arrays, except where very low sidelobes are desired.

Although the above assumed a random distribution of phase error across the aperture for computation of the rms sidelobe level, the actual phase distribution with quantized phase shifters is likely to be periodic. The periodic nature of the quantized phase will give rise to spurious *quantization lobes*, similar to grating lobes. The peak-quantization lobe relative to the main beam, when the phase error has a triangular repetitive distribution is

$$\text{Peak quantization lobe} = \frac{1}{2^{2B}} \qquad (8.32)$$

This applies when the main beam points close to broadside and there are many radiating elements within the quantized phase period. The position of the quantization lobe in this case is

$$\sin \theta_1 \simeq (1 - 2^B) \, \theta_0 \qquad (8.33)$$

where θ_0 = angle to which main beam is steered. Equation (8.32) is an optimistic estimate for the peak lobe. The greatest phase quantization lobe is said to occur when the element spacing is exactly one-half the phase quantization period or an exact odd multiple thereof.[82,115] With an element spacing of one-half wavelength, the quantization lobe will appear at $\sin \theta_1 \simeq \sin \theta_0 - 1$, with a value of

$$\text{Peak quantization lobe} \simeq \frac{\pi^2}{4} \frac{1}{2^{2B}} \frac{\cos \theta_1}{\cos \theta_0} \qquad (8.34)$$

The peak sidelobes due to the phase quantization can be significant, and attempts should be made to reduce them if their presence is objectionable. One method for reducing the peak lobe is to randomize the phase quantization. A constant phase shift can be inserted in the path to each element, with a value that differs from element to element by amounts that are unrelated to the bit size. This added phase shift is then subtracted in the command sent to the phase shifter. With an optical-fed array, such as the reflectarray or the lens array, decorrelation of the phase quantization is inherent in the array construction. In this case, the reduction in peak quantization lobes is said to be equivalent to adding one bit to the phase shifters in a 100-element array, 2 bits in a 1000-element array, and 3 bits in a 5000-element array.

The maximum pointing error due to quantization, is[115]

$$\Delta \theta_0 / \theta_B = \frac{\pi}{4} \frac{1}{2^B} \qquad (8.35)$$

where θ_B is the beamwidth. A four-bit phase shifter, for example would give $\Delta \theta_0 / \theta_B = 0.049$. Small steering increments are possible with quantized phase shifters. For example, a linear array of 100 elements can be steered in increments of about 0.01 beamwidth with 3-bit phase shifters.[82]

8.9 COMPUTER CONTROL OF PHASED-ARRAY RADAR[116-121]

A computer is a necessary part of a phased-array radar. It is vital in applications where flexible, multifunction operations are desired, as in satellite surveillance, air defense systems, ballistic missile defense, and multifunction airborne radar. The computer permits the inherent flexibility of an array to be exploited by efficiently controlling the radar and scheduling its operations. However, the cost of achieving this capability is not insignificant and has been one of the major factors in making the array radar expensive.

The computer is needed in a phased-array radar to provide *beam-steering commands* for the individual phase shifters; *signal management* by determining the type of waveform, the number of observations, data rate, power, and frequency; the corresponding *signal processing* and *data processing* in accordance with the mode of operation; outputs of *processed data to users*, including the generation of displays; "*housekeeping*" functions of performance monitoring, fault location, data recording and simulations; and the *executive management* of the radar by assigning priorities to the various tasks and how they should be performed so as to achieve a compromise between the required radar actions and the resources available in the radar and computer. In addition, the computer provides the means whereby an operator can manually interact with the radar.

Beam-steering computer. Although a single general-purpose computer can be used to perform all the computations and control required for a phased-array radar, it is often desirable to utilize a separate, special-purpose computer for beam steering. This can be an integral part of the radar hardware so as to minimize the problem of communicating the large number of phase-shifter orders. The general-purpose radar-control computer provides the beam-steering computer with the desired elevation angle and azimuth angle. The beam-steering computer translates this into commands necessary for each phase shifter. In some array designs with frequency-dependent phase shifters, the frequency also must be supplied to the beam-steering computer along with the two angles.

In some cases the generation and distribution of a large number of individual phase-shifter commands (one for each element) might be economically prohibitive. In devising computation algorithms and computer hardware, advantage can be taken of the fact that the phase shift ψ_{mn} required at the *mn*th element of a rectangular-spaced array can be separated by row and column since $\psi_{mn} = m\psi_y + n\psi_x$, where *m*, *n* are integers corresponding to the *m*th row and *n*th column, ψ_y is the phase difference needed between adjacent rows to steer the beam in elevation, and ψ_x is the phase difference between adjacent columns needed to steer in azimuth. This is sometimes called row/column steering. Baugh[121] describes a "non-time-critical" row/column beam-steering computer that represents a minimum equipment approach that takes from 10 to 20 ms to generate the phase-shifter commands. He also describes a "time-critical" design, which requires some form of adding device per element, that reduces the time to 50 to 100 μs. The better the hardware the less can be the computation time.

The use of subarrays usually simplifies the problem of the beam-steering computer. Instead of requiring a command for each of the elements of the array, the subarray steering requires only $p + q$ phase shifter commands per pointing angle, where *p* is the number of elements in the subarray and *q* is the number of subarrays in the array. However, with *q* identical subarrays of *p* elements each, the tolerance on the individual phase shifters must usually be better than with a similar conventional array of *pq* elements since with subarrays the errors across the entire aperture are no longer independent.

Array radar system functions. An array radar is sometimes required to perform more than one function with the same equipment. The problem with a multifunction array is to utilize effectively the resource of *time* and the resource of *radar energy*. The computer allows the radar to utilize its resources effectively by scheduling the execution of the various functions so as to perform the more important tasks first.

A multifunction array radar might be called upon to perform the following tasks:

Search of a specified volume of space at a specified rate, and the *detection* of targets.

Track initiation, or transition to track, after a new detection is established.

Track maintenance, or track update, by acquiring new data and consolidating it with existing tracks.

User services, whereby the specific information desired by the user is acquired, such as obtaining a satellite ephemeris or the solution of a fire-control problem.

In each of these, the proper radar waveform must be selected along with the corresponding receiver processing. This is known as *radar signal management* and is an important function of the computer. In the AN/FPS-85 satellite-surveillance radar, for example, seven different radar waveforms were available for performing the target detection and tracking mission as shown in Table 8.1.[118] An eighth waveform was used for alignment of the transmitter and receiver modules.

The *search and detection* function requires that the computer generate the angular coordinates of the region to be searched, the type of transmitted waveform to be used, the length of the dwell period, and the time assigned for the execution of the dwell by the radar. Since the radar is performing tracking and other functions as well as search, conflicts in scheduling the radar or the processor might arise. Tracking functions are often more critical of time than search, and would have a higher priority in the event of a scheduling conflict. When search must be interrupted for a higher-priority task, the computer program must be designed to do so with minimum disruption. In scheduling the various radar functions, the computer must insure that the average-power limits of the radar transmitter are not exceeded.

The flexibility of a phased-array radar allows more freedom in the selection of a detection criterion. For example, after initial crossing of the detection threshold, the radar beam can be returned to the same direction sooner than it would in normal search in order to verify that a target was indeed detected and that the threshold crossing was not a false alarm due to noise. This *verification pulse* can be of greater power and/or of longer duration to increase the probability of detection. The two-step process of (1) initial detection and (2) verification is sometimes called *sequential detection*. Because of the availability of a verification pulse, the power transmitted by the normal search-pulse can be less than if the detection decision had to be made on the basis of only a single observation.

There is always a limit to the information that can be processed by any general-purpose radar control computer. The process of initiating and maintaining track can be very demand-

Table 8.1 Waveforms used in AN/FPS-85 satellite surveillance radar[118]

Name	Description	Primary function
1. Search chirp pulse	250 μs, FM	Long-range search and track acquisition
2. Search simple pulse	10 μs	Short-range search and track acquisition
3. Track chirp pulse	250 μs, FM	Long-range track
4. Track simple pulse	1 μs	Short-range track and short-range SLBM search
5. Coherent extended range track	40-pulse burst of 125-μs pulses; burst duration 1 s	Extended-range track and signature
6. Coherent intermediate doppler	40-pulse burst of 25-μs pulses; burst duration 0.2 s	Intermediate doppler and signature
7. Coherent precision range	40-pulse burst of 5-μs pulses, stepped frequency; burst duration 1.2 ms	Precision-range tracking
8. Array calibration waveform	60 μs	Transmitter and receiver module calibration

ing on the computer resources, hence it is necessary to exclude information that is of no interest to the system. Such unwanted information may be from clutter, interference, or jamming. In principle, the computer can be programmed to recognize and reject these unwanted signals, but this is an inefficient method for eliminating them. It is more convenient to eliminate them in the radar signal-processor by either analog *bulk-processing* or its special-purpose digital equivalent. Such signal processing includes matched filtering, doppler (MTI) filtering, polarization filtering, and adaptive video thresholding.

Even if all unwanted detections are eliminated, there are two additional problems that can lead to a proliferation of detections and place a burden on the computer: (1) multiple detections of new targets and (2) redetections of old targets. If these are not recognized as such, the computer will attempt to correlate them with existing tracks and initiate new tracks, which can tie up the computer capacity needlessly. Multiple detections of the same target can occur in adjacent beam positions if the echo signal is of more than marginal strength. If the beam scans a uniform pattern, as it would if it were generated by a conventional mechanically scanned antenna, extraneous hits from adjacent beam positions can be readily recognized as such. An operator viewing a PPI would have no problem. However, in a flexible phased-array radar, the scanning of the beam positions might not be uniform. The likelihood of overdetection of a target will depend on how the volume is scanned. To avoid this problem the detection decision might have to be delayed until the neighboring beam positions have been scanned. After a detection decision is made it must be correlated against existing tracks to determine whether it is a new target or an existing target already in track. This is an important aspect of the detection process since the initiation of a track after a new target is detected requires a sequence of radar dwells that can consume the resources of the radar and the computer.

The *track-initiation* process after a new detection has been established is a demanding one. The radar must observe the target a significant number of times within a modest time interval to quickly establish the target's direction of travel and speed.

The *track-maintenance* function determines when new observations should be made on existing tracks in order to update the entries in the track file. Track maintenance not only establishes when the next radar observation must be made, but takes the steps necessary to obtain it. In performing this function it is often convenient to obtain the target position estimate in radar coordinates. Radar errors can be readily handled in this coordinate system. If the target is an aircraft, its position can then be converted into cartesian coordinates for smoothing and extrapolation of its trajectory. A constant-velocity target flying a straight line, nonradial course would have radial acceleration in the spherical coordinates of the radar, which is avoided with cartesian coordinates.

Figure 8.31, from Weinstock,[120] is an example of the timing of events that must occur in radar transmission and reception. The bottom figure shows the timing of the radar transmissions. While the radar is still receiving echoes from the $(N-1)$st dwell, a block of command words is communicated to the radar for control of the Nth dwell and the beam-steering computer calculates the phase-shifter orders needed for the next transmission. After the time of the maximum range return has elapsed, the phase shifters can be set for the next transmission. At this time, the signal processor can be set to accommodate the next dwell. If the array antenna employs nonreciprocal phase shifters, the complementary phase distribution needed for reception is set just after the transmission of the pulse. While the system is being set for the next transmission, the data received from the last transmission can be loaded into buffer storage for transfer to the radar control computer. Thus, during the Nth sequence of dwell executions, the returns from the $(N-1)$st sequence are processed and commands are generated for the $(N+1)$st sequence.

Time and radar power (or energy) are two resources that must be properly handled in a

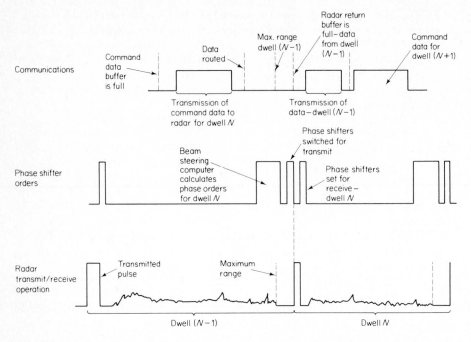

Figure 8.31 Representative command/response sequence. (*Courtesy RCA, Inc.*)

computer-controlled multifunction radar. The various functions to be performed must be efficiently prioritized so as not to exceed either resource. In search, the maximum range of interest strongly affects the time resource. In tracking, the target range does not usually limit the time resource since the range is known and the dwell interval can be adjusted accordingly. The power transmitted by the radar will depend on the type of function it is to perform as well as the target maximum range and possibly the past behavior of the target. Thus the radar resource of time is affected more by the long-range targets, and the radar power is more affected by the tactical function.

The data processor, as well as the radar, can cause the system to run out of available time. The time required to process each target dwell as well as the amount of computer memory needed per target is usually independent of the target range. Thus the saturation limit of the computer will be reached if it is given too many data points per unit time. This is more likely to occur with short-range targets since they generally require a higher data rate than long-range targets.

In order to maintain maximum effectiveness of the radar and its computer when overloading occurs, some sort of priority system is required. Tasks which are not as time critical should be deferred so that high-priority tasks can be accomplished. Table 8.2 is an example of a priority structure for a tactical air defense system in which the radar performs search and automatic tracking, as well as support missile engagements.[121] There are eight categories of priority, from the dedicated mode with the highest priority, to the standard testing and dummy operations (or time wasters) with the lowest priority. There is more than one similar function within each category which can be treated either on a first-in–first-out basis or by giving further priority ordering. Note that the subcategories of the "0" priority also appear in lower-priority categories.

Table 8.2 Example of priorities for a tactical system[121]

0	Dedicated mode	
	Burnthrough	Previously scheduled events
	Target definition	that capture radar for
	Special test	long periods of time
1	Engagements	
	Engaged hostile	
	Own missile	
	Preengaged hostile	
2	Time-critical	
	High-priority transition	
	High-priority confirmation	
	Horizon search	
3	Special request	
	Burnthrough	
	Target definition	
	Special scans	
	Target acquisition	
4	High-priority tracks	
	Confirmed hostile	
	Assumed hostile	
	Unevaluated	
	Controlled friendly	
	Confirmation	
	Track transition	
5	Low-priority track	
	Assumed friendly	
	Confirmed friendly	
6	Above-horizon search	
	All coverages	
	Special test	
7	Simulation, diagnostics, and dummies	

Table courtesy RCA, Inc.

The computer must be capable of responding to the immediate demands of high-priority tasks without throwing away the results of partially completed lower-priority tasks. The *executive program* must recognize the presence of a high-priority task and initiate it properly. There are two methods for interleaving high and low priorities. The first is *preemption* of control by a higher-priority task as soon as it arises. The interrupted task is put aside in such a way that it can be resumed when time permits. The second method, *time slicing*, breaks all lower-priority tasks into nonpreemptable segments. These segments must be short enough to avoid excessive delays in the transfer of control to the time-critical tasks.

The phased-array radar and its computer controller saturate differently. Long-range dwells are demanding of radar execution time, but since they generally involve a low data rate they are not particularly demanding of data processing time. On the other hand, high data rate tracking of short-range targets is not as demanding of radar execution time as for long-range targets, but the data-processing capability becomes the limiting factor. The handling of a large number of short-range targets in track can drive the processor to full utilization and require that search operations, which have lower priority than track, be reduced. Thus the mechanism by which the system saturates depends on the nature of the target situation.

The AN/FPS-85 is a large UHF phased-array radar located at Eglin Air Force Base, Florida which was designed for the detection and tracking of earth-orbiting objects (satellites).[118] It was one of the first major applications of a computer-controlled multifunction phased-array radar. The system computer consists of two IBM System 360/65I central processors, each with 131,072 thirty-two-bit words of core memory and a basic add time of 14 μs. There are digital devices that provide the interface between the system computer and the radar equipment, digital and analog signal processors, monitoring and controls, and equipment for space object identification. The programs for the computer total over 1,250,000 words of storage. About 60 percent of this large number of words consists of decision tables to locate faults when compared to expected outputs.

The cost of the computer hardware and software of a multifunction radar system can be a significant fraction of the total system cost. Computer system design must be integral with that of the radar hardware itself. The development of the software can be an especially demanding task that cannot be considered minor.

8.10 OTHER ARRAY TOPICS

Hemispherical coverage.[123-125] A single planar phased-array antenna might typically provide coverage of an angular sector $\pm 45°$ from the array normal; however, it is possible to achieve a scan coverage of $\pm 60°$ or greater. The amount of scan depends on the loss of gain, the amount of beam broadening, and the rise in sidelobes that can be tolerated. The frequency range over which the antenna must operate and the VSWR that can be accepted also determine how far the beam can be scanned. With a scan capability of $\pm 60°$, a minimum of three planar apertures, or faces, are required to cover the hemisphere. But when other factors are considered, more than three faces are usually desired. The greater the number of faces, the less will be the loss in gain, beam broadening, VSWR variation, polarization change, and the number of elements per face. Other factors which must be considered in selecting the optimum number of faces are the number of transmitters and receivers, the complexity of the control of the array, and the total cost.

Table 8.3 summarizes the properties of N array faces to give hemispherical coverage at a single frequency.[123] In the 5- and 6-face arrays, one face is normal to the zenith, which explains why the tilt angle is less for these two cases. Each array of a 4-face array, for example, would scan $\pm 55°$. To minimize the loss in gain, each face would be tilted back an angle of 35.3° from the vertical. Elements are arranged in an equilateral triangular pattern with element separa-

Table 8.3 Properties of hemispherical coverage arrays[123]

Number of faces	3	4	5	6
Maximum scan angle	63.4°	54.7°	47.1°	40.7°
Tilt angle	26.6°	35.3°	15.5°	20.4°
Element spacing (wavelengths)	0.628	0.691	0.679	0.700
Maximum power reflected	14%	7%	4%	2%
Maximum beam broadening	2.2	1.75	1.47	1.32
Maximum gain reduction (dB)	4.1	2.8	1.9	1.3
Relative total number of elements:				
1. Equal gain at max scan	1.22	1.00	1.05	1.04
2. Equal beamwidth at max scan	1.45	1.00	0.92	0.84
3. Equal average-gain over scan	1.05	1.00	1.17	1.24
4. Equal average-beamwidth over scan	1.15	1.00	1.09	1.11

tion of 0.691 wavelengths. Assuming an array matched at broadside, resistive mismatch off broadside results in 7 percent of the power reflected at the maximum scan angle. The beamwidth at the maximum scan angle is 1.75 times that of the broadside beamwidth. The reduction in gain is 2.8 dB, which includes the effect of 7 percent mismatch as well as the effect of beam broadening. The more faces that are used, the greater can be the element spacing before the onset of grating lobes. If it is required that the gains at the maximum scan angle be the same, it is seen in the table that the 4-face array requires the fewest elements. However, if the beamwidths at the maximum scan are to be the same, the 6-face array requires the fewest total number of elements. When the criterion is to maintain either the average gain or the average beamwidth the same over the scan, the 4-face array yields the minimum total number of elements.

These results apply to a ground-based array. If the array is on a ship, the lower limit of the elevation scan angle must be less than 0° to allow for the roll and pitch of the ship. (A value of −20° might be a typical requirement.) The optimum tilt angle of the array would be different than if the lower limit were 0°. In one example[125] a tilt angle of 27° is taken for a 4-face array covering from +90 to −20° in elevation, instead of the 35.3° of Table 8.3. The reduction in gain is 4.3 dB at the maximum scan angle instead of 2.8 dB.

Dome antenna.[126–129] A novel approach to obtaining hemispherical coverage is the dome antenna depicted in Fig. 8.32. A planar array of conventional design with variable phase shifters is shown situated below a hemispherical lens with fixed phase shifts. The lens, with its fixed phase shifts, alters the phase front of the field radiated by the planar array to cause a change in the direction of propagation. The dome acts as an RF analog to an optical prism that changes the scan angle of the planar array by a factor $K(\theta)$. For example, a constant value of $K = 1.5$ extends the coverage of a $\pm 60°$ planar array to $\pm 90°$. The loss of gain varies in this case from about 3.8 dB to 5.4 dB over the region of coverage. The dome need not be a hemisphere but can be shaped to favor particular regions of space. The dome can consist of dielectric-loaded circular waveguides inserted into holes drilled in the hemispherical shell. An experimental model at C band had a combined mismatch and insertion loss of about 0.8 dB. The dome might also be fabricated from radome-like material using a honeycomb skin with lightweight printed-circuit phase shifts glued to the inner surface.

In a conventional phased array, a linear phase gradient is applied across the aperture to steer the beam. However, a nonlinear phase distribution is required across the planar array of the dome antenna. The beamwidth can vary on the order of 4 to 1 over the entire scan range of the dome antenna. Furthermore, the beam might not be as narrow on the horizon as might be obtained with a conventional antenna configuration of equivalent aperture. As with any array

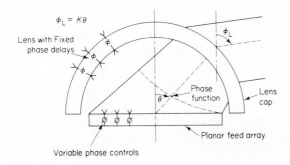

Figure 8.32 Dome antenna for achieving hemispherical coverage. (*Courtesy IEEE.*)

arranged on the surface of a sphere, the plane of polarization changes with the scan angle. With circular polarization, however, there is no change with scan.

A conventional array with four faces can track four times the number of targets as a dome antenna (which operates with a single planar array for hemispherical coverage), assuming the traffic is distributed uniformly. The average power required for the planar array of the dome antenna is approximately the sum of the average powers required for each of the individual arrays, assuming comparable performance.

It has been claimed that the dome antenna is of less cost than four arrays of conventional design covering the same volume. However, any comparison of this technique with conventional antennas must be done on the basis of the entire radar system and its intended application, not just the antenna. The required transmitter power, the loss in antenna gain, change of beamwidth, complexity of computer control, traffic handling, polarization change, and total system cost must all be considered when evaluating the relative merits of such antennas for a particular application.

Conformal arrays.[130-135] One of the desires of the radar systems engineer is to be able to place array elements anywhere on an arbitrary surface and achieve a directive beam, with good sidelobes and efficiency, which can be conveniently scanned electronically. An array of radiating elements arranged along the nose of an aircraft, on the wing, or the fuselage would thus be an attractive option for the systems designer. Such an array, which conforms to the geometry of a nonplanar surface, is called a *conformal array.*

In principle, the elements of an array located on any nonplanar surface can be made to radiate a beam in some given direction by applying the proper phase, amplitude, and polarization at each element. In practice, however, it is difficult to control the beamshape and obtain low sidelobes from an arbitrary surface, especially if the beam is electronically scanned. Furthermore, the mechanisms for feeding the elements and steering the beam of a conformal array, as well as the generation of the phase-shifter commands, are generally more complicated than those of a planar array. Although it is desired that conformal arrays be applicable to any surface, the complications that arise when dealing with a general nonplanar surface have restricted its consideration to relatively simple shapes, such as the cylinder, cone, ogive, and sphere. Even these shapes present difficulties in analysis and equipment implementation, and in realizing antennas competitive with the planar array.

The cylinder, which is probably the simplest form of nonplanar array, has a geometry suitable for antennas that scan 360° in azimuth. The scanning beam of a cylindrical antenna does not change its shape and broaden when steered, as does a scanning beam from a planar array. Beam steering in a circularly symmetric array can be accommodated by commutating or rotating the aperture distribution. Since the elements on the side of the cylindrical array opposite from the direction of propagation do not contribute energy in the desired direction, they are not excited. Unlike the planar array, the circular symmetry ensures that mutual coupling between elements is always the same as the beam scans in azimuth.

The truncated cone has similar properties to the cylindrical array, and might be utilized instead of the cylinder when the beam is to be scanned in elevation as well as azimuth. The conical surface as a radiating structure for a conformal array is also of interest since missiles are sometimes of this shape.

The ogive, which is the type of surface found on the nose of some aircraft, bears a resemblance to the cone. A conformal array arranged on the aircraft nose would allow wide coverage, provide a good aerodynamic shape, eliminate the distortion of the pattern of a mechanically scanned antenna caused by the conventional nose-radome, and would permit larger antenna apertures than the conventional nose-antenna configuration. The hemispherical array also attracts attention when hemispherical coverage is desired. In principle, the pattern is independent of beam direction.

Most of the work on conformal arrays has been with the cylinder. Even though it may be a relatively simple shape compared to the others mentioned above or to the generalized nonplanar surface, the properties of the cylindrical array are not as suitable in general as those of the planar array. In a cylindrical array, the radiation pattern cannot be separated into an element factor and an array factor as in planar array theory. Each element points in a different direction. Thus computations of patterns must include both element and array together. Another difficulty is that the phase as well as the magnitude of the element pattern varies with position. As a result of the cylindrical geometry, the radiation pattern is only partially separable in spatial coordinates. (It can be considered as the product of a ring-array pattern and a linear-array pattern.) The azimuth pattern of a vertical cylindrical array changes with both azimuth and elevation as the beam is scanned in elevation. The far-out sidelobes of a cylindrical array are generally large and broad in angle, as compared to those of a planar array. The design of efficient feed networks, the phase and amplitude control devices, distributed transmitter and/or receiver modules, and the control algorithms and logic are other problem areas.

Similar problems occur with other conformal-array shapes. In particular the conical shape can present a serious polarization problem since the polarization varies with the direction of propagation. Another problem is that the radiating elements can see different environments depending on their location on the cone. The pattern of the conical array cannot be separated into elevation and azimuth patterns.

Conformal arrays would have application for flush-mounted airborne radar antennas, especially on conical and ogive surfaces. The cylinder, truncated cone, and the hemisphere are possible antenna shapes for shipboard or surface applications. Such conformal arrays have been considered for radar but they do not appear suitable for general application. They have also been examined for IFF and air traffic control interrogation antennas.[132]

A serious competitor to the cylindrical array is a number of planar arrays arranged to approximate the cylinder. This is a much simpler system and is quite practical. (If necessary it can be surrounded by a cylindrical radome.) The change of beam shape with scan angle can be minimized, if desired, by radiating cooperatively from more than one planar face. As the beam is scanned off broadside, some of the transmitter power can be diverted to the face adjacent to the direction of scan to radiate energy in the same direction.[135] The broadening of the beam due to the foreshortening of the planar aperture is compensated by the cooperative use of the adjacent planar aperture.

Unequally spaced arrays.[136-139] Most array antennas have equal spacings between adjacent elements. Unequal spacings have sometimes been considered in order to obtain a given beamwidth with considerably fewer elements than an equally spaced array, or to approximate a desired pattern without the need for an amplitude-tapered aperture illumination. Since the minimum element spacing is about one-half wavelength, unequally spaced arrays generally have fewer elements than equally spaced arrays of the same size aperture. For this reason they are also called *thinned arrays*.

The design of a thinned array consists of selecting the diameter of the aperture to give the desired beamwidth, selecting the number of elements to give the desired gain (gain is directly proportional to the number of elements), and arranging the elements to achieve some desirable property of the sidelobes, such as minimizing the peak sidelobe or approximating some desired radiation pattern. There have been many methods proposed in the literature for determining the element spacings (or locations within the aperture) of such arrays. Only two will be mentioned here: dynamic programming and density tapering.

One approach is to try all possible element locations with the given number of elements and select that arrangement which produces the best result. This technique, called *total enumeration*, is generally impractical because of the large number of possible solutions that would need to be examined. An equivalent result, with less computation, is to apply *dynamic programming*, a method for determining the optimum solution to a multistage problem by optimizing each stage of the problem on the basis of the input to that stage. As applied to arrays, the various stages of dynamic programming involve the selection of the spacings of each element or pair of symmetrically placed elements. The application of dynamic programming has been successfully applied to the design of one dimensional (linear) unequally spaced array. Although the computations required are far less than for total enumeration, there is a practical limit to the number of elements in the array that can be treated by this method.

The other technique, *density taper*, is applicable to the design of either linear arrays or to large planar arrays. Consider a uniform grid of possible element locations with equal spacing. The aperture illumination function that would normally be considered for a conventional antenna is used as the model for determining the density of equal-amplitude elements. That is, the density of equal-amplitude elements is made to approximate the desired aperture illumination. The choice of element locations to achieve the desired density distribution may be made on either a statistical basis or deterministically. One possible application of density taper is in a transmitting array where it is desired to radiate equal power from each element, but where the high peak-sidelobe of the uniform illumination is not acceptable. Although lower peak-sidelobes can be obtained in this manner, the far-out sidelobes are generally considerably greater than if the same illumination function were used to establish an amplitude taper with an equally spaced array.

The unequally spaced array has seen only limited application, primarily because its advantages do not usually outweigh its disadvantages. The reduction in gain and the increases in sidelobes that are a consequence of thinning the elements are usually not desirable in most radar applications.

The purpose of unequal element spacing in a highly thinned array is to eliminate the grating lobes that would appear if equal element spacings of large value were used. Grating lobes are not necessarily undesirable, since the energy contained in the grating lobes can be used to detect targets. However, if this energy is distributed in the sidelobes by employing unequal spacings, it is wasted. The disadvantage with detecting targets in the grating lobes is that the angle measurement can be ambiguous. There are methods, however, for resolving these ambiguities.[140]

Instead of thinning elements, the number of phase shifters can be thinned.[141] That is, some of the phase shifters in the array can be used to adjust the phase of more than one element, so that the number of phase shifters is less than the number of elements. A 50 percent saving in the total number of phase shifters might be had. Thinning in this manner also gives rise to phase errors which cause a deterioration in the radiation pattern.

Adaptive antennas.[142–151] An adaptive antenna senses the received signals incident across its aperture and adjusts the phase and amplitude of the aperture illumination to achieve some desired performance, such as maximizing the received signal-to-noise ratio. The noise may be either internal receiver noise or external noise, as from jammers. Clutter echoes or interference from other electromagnetic radiations can also be minimized by adaptive antennas. Adaptive antenna techniques can automatically compensate for mechanical or electrical errors in an antenna by sensing the errors and applying corrective signals,[149] and can compensate for failed elements, radome effects, and blockage of the aperture from nearby structures. Much of the interest in adaptive antennas has been to reduce the effects of noise jamming in the antenna sidelobes. The ideal adaptive antenna acts automatically to adjust itself as a matched (spatial) filter by reducing the sidelobes in the direction of the unwanted signals. Adaptive antennas require some a priori knowledge of the desired signal, such as its direction, waveform, or statistical properties.

If all the elements of an adaptive phased-array antenna have a separate adaptive control loop, it is called a *fully adaptive array*. The adaption process is based on the application of some criterion such as the minimization of the mean-square error or the maximization of the signal-to-noise ratio. In arrays with a large number of adaptive elements the time required for the array to converge to the desired aperture illumination can be relatively long. The convergence time can be reduced at the cost of hardware complexity. The complexity of the adaptive mechanization and the speed of convergence limit the practical utility of the large adaptive array. The fewer the adjustable elements the more practical is the adaptive array.

One potential application of adaptive array antennas is for airborne surveillance radar. In addition to the reduction of jamming noise which enter the airborne surveillance radar, it is important to reduce the clutter that enters via the radar antenna sidelobes. The detection of targets from an airborne platform places severe demands on radar design in order to reduce or eliminate clutter that enters the radar receiver via the main beam. This is done in a conventional AMTI radar by signal processing (filtering) as described in Sec. 4.11. However, clutter that enters the radar receiver via the antenna sidelobes has doppler frequencies that cannot be rejected by conventional filtering, since targets of interest can have the same doppler as the sidelobe clutter. An adaptive array antenna can place nulls in the direction of sidelobe clutter by using the sidelobe clutter itself as the signal to be cancelled. This is accomplished by separating the sidelobe clutter from the main-beam clutter by doppler filtering. The filtered signal adaptively adjusts the aperture illumination to minimize the sidelobes in those directions from which clutter appears.

The *coherent sidelobe canceler* (Sec. 14.5) is a form of adaptive antenna that uses a small number of auxiliary elements to adaptively place nulls in the direction of external noise sources. It is an example of a successful application of the principles of the adaptive antenna that utilizes only a relatively few number of adaptive elements. The fully adaptive array of large size is, in theory, capable of nulling a larger region of space than a system with but a relatively few adaptive elements, as in the sidelobe canceler. However, the added complexity and longer convergence time that accompanies the greater degrees of freedom of the large, fully adaptive array has been a burden that is difficult to justify on a cost-effective basis for general application.

The objective of the usual adaptive antenna or the coherent sidelobe canceler is to adjust the sidelobe levels to minimize the effects of noise or other unwanted signals. If the radar application allows the use of antennas with extremely low sidelobes, the same result will be achieved as with the adaptive antenna. In general, the extremely low sidelobe antenna can be a better solution, if the desired low sidelobe levels can be achieved and maintained economically.

In principle, the automatic tracking radar discussed in Chap. 5 is another example of an adaptive antenna. The aperture illumination is sensed by a conical scan or monopulse feed and the antenna is repositioned to maintain the signal-to-noise ratio a maximum.

Triangular arrangement of elements.[152-154] If the elements of a planar array are arranged in a pattern of equilateral triangles rather than squares, a savings can be had in the total number of elements needed. This assumes that the number of elements in an array is determined by the requirement that no spurious beams (grating lobes) appear in the radiation pattern. The reduction in the number of elements depends on the solid angle over which the main beam is positioned. For example, if the beam is to be scanned anywhere within a cone defined by a half-angle of 45°, the number of elements required with triangular spacing is 13.4 percent less than with square spacing. In this case the altitude of the equilateral triangle in the triangularly arranged array is equal to the element spacing of the squarely arranged array. If instead, the main beam is positioned not within a cone, but over a "pyramidal" region, the reduction is less. The amount of reduction is only 9.2 percent in covering a ± 45 by $\pm 45°$ "square" angular region, and 7.7 percent for a ± 45 by ± 25 region. The smaller the angular region, the less the saving. One caution in the use of triangular spacings is that, compared to square spacings, it is more likely to produce high sidelobes in some portions of space. This is due to the phase quantization that results from digital phase shifters.[154] For most applications, however, these quantization lobes do not limit the system performance significantly.

Limited-scan arrays. The scan angle of a conventional array antenna might typically range from $\pm 45°$ to perhaps $\pm 60°$ in both angle coordinates. If the required scan angle can be limited to a much smaller angular region the number of radiating elements and the number of phase shifters can be reduced. This can result in a simpler and cheaper phased-array radar than the conventional array that must scan a wide angle in two orthogonal coordinates. Although the conventional phased-array radar has not seen extensive application because of its high cost and complexity, the limited-scan array has had significant application because it is more competitive in both performance and cost to mechanical scanning antennas. Applications for limited-scan arrays in radar have included the one-dimensional electronic scan of 3D radars (Sec. 14.4), aircraft landing or ground-control approach (GCA) radars,[157] and hostile-weapon-location radars.[158]

8.11 APPLICATIONS OF THE ARRAY IN RADAR

The phased-array antenna has been of considerable interest to the radar systems engineer because its properties are different from those of other microwave antennas. The array antenna takes several forms:

Mechanically scanned array. The array antenna in this configuration is used to form a fixed beam that is scanned by mechanical motion of the entire antenna. No electronic beam-steering is employed. This is an economical approach to air-surveillance radars at the lower radar frequencies, such as VHF. It is also employed at higher frequencies when a precise aperture illumination is required, as to obtain extremely low sidelobes. At the lower frequencies, the array might be a collection of dipoles or Yagis, and at the higher frequencies the array might consist of slotted waveguides.

Linear array with frequency scan. The frequency-scanned, linear array feeding a parabolic cylinder or a planar array of slotted waveguides has seen wide application as a 3D air-surveillance radar. In this application, a pencil beam is scanned in elevation by use of frequency and scanned in azimuth by mechanical rotation of the entire antenna.

Linear array with phase scan. Electronic phase steering, instead of frequency scanning, in the 3D air-surveillance radar is generally more expensive, but allows the use of the frequency domain for purposes other than beam steering. The linear array configuration is also used to generate multiple, contiguous fixed beams (stacked beams) for 3D radar. Another application is to use either phase- or frequency-steering in a stationary linear array to steer the beam in one angular coordinate, as for the GCA radar.

Phase-frequency planar array. A two-dimensional (planar) phased array can utilize frequency scanning to steer the beam in one angular coordinate and phase shifters to steer in the orthogonal coordinate. This approach is generally easier than using phase shifters to scan in both coordinates, but as with any frequency-scanned array the use of the frequency domain for other purposes is limited when frequency is employed for beam-steering.

Phase-phase planar array. The planar array which utilizes phase shifting to steer the beam in two orthogonal coordinates is the type of array that is of major interest for radar application because of its inherent versatility. Its application, however, has been limited by its relatively high cost. The phase-phase array is what is generally implied when the term *electronically steered phased array* is used.

The phased array antenna has seen application in radar for such purposes as aircraft surveillance from on board ship (AN/SPS-33), satellite surveillance (AN/FPS-85), ballistic missile defense (PAR, MSR), air defense (AN/SPY-1 and Patriot), aircraft landing systems (AN/TPN-19 and AN/TPS-32), mortar (AN/TPQ-36) and artillery (AN/TPQ-37) location, tracking of ballistic missiles (Cobra Dane), and airborne bomber radar (EAR).

There have been many developmental array radars built in the United States, including ESAR, ZMAR, MAR, Typhon, Hapdar, ADAR, MERA, RASSR, and others. Although much effort and funds have been expended, except for limited-scan arrays there has been no large serial production of such radars comparable to the serial production of radars with mechanically rotating reflector antennas.

8.12 ADVANTAGES AND LIMITATIONS

The array antenna has several unique characteristics that make it a candidate for consideration in radar application. However, the attractive features of the array antenna are sometimes nullified by several serious disadvantages.

The array antenna has the following desirable characteristics not generally enjoyed by other antenna types:

Inertialess, rapid beam-steering. The beam from an array can be scanned, or switched from one position to another, in a time limited only by the switching speed of the phase shifters. Typically, the beam can be switched in several microseconds, but it can be considerably shorter if desired.

Multiple, independent beams. A single aperture can generate many simultaneous independent beams. Alternatively, the same effect can be obtained by rapidly switching a single beam through a sequence of positions.

Potential for large peak and/or average power. If necessary, each element of the array can be fed by a separate high-power transmitter with the combining of the outputs made in "space" to obtain a total power greater than can be obtained from a single transmitter.

Control of the radiation pattern. A particular radiation pattern may be more readily obtained with the array than with other microwave antennas since the amplitude and phase of each array element may be individually controlled. Thus, radiation patterns with extremely low sidelobes or with a shaped main beam may be achieved. Separate monopulse sum and difference patterns, each with its own optimum shape, are also possible.

Graceful degradation. The distributed nature of the array means that it can fail gradually rather than all at once (catastrophically).

Convenient aperture shape. The shape of the array permits flush mounting and it can be hardened to resist blast.

Electronic beam stabilization. The ability to steer the beam electronically can be used to stabilize the beam direction when the radar is on a platform, such as a ship or aircraft, that is subject to roll, pitch, or yaw.

The above attributes of an array antenna offer the radar systems engineer additional flexibility in attempting to meet the requirements of radar applications. Some comments should be made, however, about the practical utility of these characteristics. These attributes are obtained for a price, so that they should only be considered when warranted. It is not obvious that they are always absolutely essential for the success of a particular application.

For example, it is certainly true that a mechanically scanned reflector antenna cannot switch a beam from one direction to another as fast as can a phased-array antenna. However, it is seldom, if ever, that the operational requirements do not permit the system designer to configure a system to do an *equivalent* job with a mechanically scanned antenna.

An *N*-element array can, in principle, generate *N* independent beams. However, in practice it is seldom required that a radar generate more than a few simultaneous beams (perhaps no more than a dozen), since the complexity of the array radar increases with increasing number of beams.

Although the array has the potential for radiating large power, it is seldom that an array is required to radiate more power than can be radiated by other antenna types or to utilize a total power which cannot possibly be generated by current high-power microwave tube technology that feeds a single transmission line.

Conventional microwave antennas cannot generate radiation patterns with sidelobes as low as can be obtained by an array antenna, especially a nonscanning array. However, when a planar array is electronically scanned, the change of mutual coupling that accompanies a change in beam position makes the maintenance of low sidelobes more difficult.

If an array has some margin in performance to permit graceful degradation, it is likely that this margin will be eliminated during the procurement process if the cost of the radar escalates. Even if the radar is delivered with margin for graceful degradation, it is likely that after some time in operation it will always be at the degraded level because of a desire to keep maintenance costs to a minimum. Another problem is to know when graceful degradation has gone too far and maintenance is needed.

The full testing of an array radar system is often more complicated than with conventional radar systems. Also, the incorporation of IFF can be more complicated than with rotating antennas.

Although the above are limitations to the phased array in radar, they are probably not sufficiently serious to restrict its greater use. However, the *major limitation* that has limited the widespread use of the conventional phased array in radar is its high cost, which is due in large part to its complexity. The software for the computer system that is needed to utilize the inherent flexibility of the array radar also contributes significantly to the system cost and complexity.

One of the factors that is often misleading is the usual picture of an array as a single radiating face of relatively modest size. A single array antenna can scan but a limited sector; $\pm 45°$ in each plane is perhaps typical. Four or more faces might therefore be necessary for hemispherical coverage. However, the usual photograph or drawing of a phased-array radar seldom reveals the amount of electronic equipment behind the array face that is required to make it a useful radar.

Another of the advantages sometimes claimed for an array radar is that it is capable of performing more than one function simultaneously; for example, it can do surveillance of a volume as well as track individual targets. The multifunction attribute of an array radar has been of considerable interest, but it can also be a serious liability in some applications. The preferred frequency for a ground-based aircraft surveillance radar is at the lower portion of the microwave region (UHF or L band). The major factors of importance that favor the lower frequencies for surveillance radar are the large average power and the large antenna aperture, both of which are easier to obtain at the lower frequencies. Also, a good MTI is easier to achieve and the effects of weather are less at these frequencies. The preferred frequency range for an aircraft-tracking radar is the upper portion of the microwave band (C or X bands) since it is easier to obtain the wide bandwidths and narrow beamwidths for precision target tracking than at the lower frequencies. Thus tracking radars differ significantly from surveillance radars. When a single radar is to perform both surveillance and tracking, a single frequency somewhere in the middle of the microwave band must be selected as a compromise (S band, or perhaps C or L bands). The compromise frequency might thus be higher than would be desired for search and lower than desired for tracking. The result is that a single-frequency, multifunction radar might not be as efficient as separate radars operating at separate frequencies, each optimized to fulfill a single, dedicated function. It is conceivable in some circumstances that two radars, one optimized for surveillance and the other for track, will perform better, be less costly, and take less total space than a single multifunction radar.

The caution regarding the drawbacks of a multifunction radar is especially true of aircraft-surveillance radar. However, if the particular application is such that the optimum frequency for search is the same as that for tracking, it is more likely that a multifunction array radar would have merit over separate radars. Such seems to be the case for satellite surveillance where the same frequency (UHF) seems desirable for both search and track, as in the AN/FPS-85.

The unique characteristics of an array antenna offer the radar systems designer capabilities not available with other techniques. As with any other device, the array will see major application when it can perform some radar function cheaper than any other antenna type or when it can do something not practical by other means.

REFERENCES

1. Southworth, G. C.: "Forty Years of Radio Research," Gordon and Breach, New York, 1962, pp. 98–100.
2. Wilmotte, R. M.: History of the Directional Antenna in the Standard Broadcast Band for the Purpose of Protecting Service Area of Distant Stations, *IRE Trans.*, vol. PGBTS-7, pp. 51–55, February, 1957.
3. Friis, H. T., and C. B. Feldman: A Multiple Unit Steerable Antenna for Short-Wave Reception, *Bell Sys. Tech. J.*, vol. 16, pp. 337–419, July, 1937.
4. Skolnik, M. I.: Survey of Phased Array Accomplishments and Requirements for Navy Ships, "Phased Array Antennas," A. A. Oliner and G. H. Knittel (eds.), Artech House, Inc., 1972, pp. 15–20.
5. Ridenour, L. N.: "Radar System Engineering," McGraw-Hill Book Co., New York, 1947, pp. 219–295.
6. Smith, R. A.: "Aerials for Metre and Decimetre Wave-Lengths," Cambridge University Press, London, 1949.
7. Price, A.: "Instruments of Darkness," Charles Scribner's Sons, New York, 1977.
8. Kraus, J. D.: "Antennas," McGraw-Hill Book Co., New York, 1950.
9. Schelkunoff, S. A.: A. Mathematical Theory of Linear Arrays, *Bell System Tech. J.*, vol. 22, pp. 80–107, January, 1943.
10. Schelkunoff, S. A., and H. T. Friis: "Antennas, Theory and Practice," John Wiley & Sons, Inc., New York, 1952.
11. Smith, R. A.: "Aerials for Metre and Decimetre Wave-lengths," Cambridge University Press, London, 1949.

12. Hansen, R. C.: " Microwave Scanning Antennas, Volume II," Academic Press, New York, 1966.
13. Hines, J. N., V. H. Rumsey, and T. E. Tice: On the Design of Arrays, *Proc. IRE*, vol. 42, pp. 1262–1267, August, 1954.
14. Amitay, N., V. Galindo, and C. P. Wu: "Theory and Analysis of Phased Array Antennas," Wiley-Interscience, New York, 1972.
15. Von Aulock, W. H.: Properties of Phased Arrays, *Proc. IRE*, vol. 48, pp. 1715–1727, October, 1960.
16. Ogg, F. C., Jr.: Steerable Array Radars, *IRE Trans.*, vol. MIL-5, pp. 80–94, April, 1961.
17. Bickmore, R. W.: A Note on the Effective Aperture of Electrically Scanned Arrays, *IRE Trans.*, vol. AP-6, pp. 194–196, April, 1958.
18. Elliott, R. S.: The Theory of Antenna Arrays, " Microwave Scanning Antennas, vol. II," R. C. Hansen (ed.), Academic Press, New York, N.Y., 1966, chap. 1.
19. Davies, D. E. N.: Radar Systems with Electronic Sector Scanning, *J. Brit. IRE*, vol. 18, pp. 709–713, December, 1958.
20. Kummer, W. H.: Feeding and Phase Scanning, " Microwave Scanning Antennas, vol. III," R. C. Hansen (ed.), Academic Press, New York, 1966, chap. 1.
21. Craig, C. W. and H. K. Hom: A Low-Loss Microstrip X-band Diode Phase Shifter, *1974 Government Microcircuit Applications Conference Digest of Papers*, pp. 58–59, June, 1974.
22. Hines, M. E.: Fundamental Limitations in RF Switching and Phase Shifting Using Semiconductor Diodes, *Proc. IEEE*, vol. 52, pp. 697–708, June, 1964.
23. White, J. F.: Review of Semiconductor Microwave Phase Shifters, *Proc. IEEE*, vol. 56, pp. 1924–1931, November, 1968.
24. Temme, D. H.: Diode and Ferrite Phaser Technology, " Phased Array Antennas," A. A. Oliner and G. H. Knittel (eds.), Artech House, Inc., Dedham, Mass., 1972, pp. 212–218.
25. Ince, W. J.: Recent Advances in Diode and Ferrite Phaser Technology for Phased Array Radars, pt. I, *Microwave J.*, vol. 15, pp. 36–46, Sept., 1972; pt. II, *Microwave J.*, vol. 15, pp. 31–36, October, 1972.
26. Stark, L.: Microwave Theory of Phased-Array Antennas—A Review, *Proc. IEEE*, vol. 62, pp. 1661–1701, December, 1974.
27. Dionne, G. F.: A Review of Ferrites for Microwave Application, *Proc. IEEE*, vol. 63, pp. 777–789, May, 1975.
28. Von Aulock, W. H., and C. E. Fay: "Linear Ferrite Devices for Microwave Application," Academic Press, New York, 1968.
29. Clarricoats, P. J. B.: " Microwave Ferrites," John Wiley and Sons, New York, 1961.
30. Ince, W. J., and D. H. Temme: Phasers and Time Delay Elements, *Advances in Microwaves*, vol. 4, Academic Press, New York, 1969.
31. Stark, L., R. W. Burns, and W. P. Clark: Phase Shifters for Arrays, chap. 12 of *Radar Handbook*, M. I. Skolnik (ed.), McGraw-Hill Book Co., New York, 1970.
32. Clark, W. P.: A High Power Phase Shifter for Phased-Array Systems, *IEEE Trans.*, vol. MTT-13, pp. 785–788, November, 1965.
33. Whicker, L. R. (ed.): "Ferrite Control Components, vol. 2, Ferrite Phasers and Ferrite MIC Components," Artech House, Inc., Dedham, Mass., 1974.
    (A collection of reprints of papers covering toroidal waveguide phase shifters, dual-mode phase shifters, Reggia-Spencer phase shifters, other novel ferrite phaser configurations, and ferrite microwave integrated components.)
34. DiBartolo, J., W. J. Ince, and D. H. Temme: A Solid State "Flux-Drive" Control Circuit for Latching-Ferrite-Phaser Applications, *Microwave J.*, vol. 15, pp. 59–64, September, 1972.
35. Whicker, L.: Selecting Ferrite Phasers for Phased Arrays, *Microwaves*, vol. 11, pp. 44–48, August, 1972.
36. Boyd, C. R., Jr.: A Dual-Mode Latching Reciprocal Ferrite Phase Shifter, *IEEE Trans.*, vol. MTT-18, pp. 1119–1124, December, 1970.
37. Boyd, C. R., Jr.: Comments on the Design and Manufacture of Dual-Mode Reciprocal Latching Ferrite Phase Shifters, *IEEE Trans.*, vol. MTT-22, pp. 593–601, June, 1974.
38. Fox, G. A., S. E. Miller, and M. T. Weiss: Behavior and Application of Ferrites in the Microwave Region, *Bell System Tech. J.*, vol. 34, pp. 5–103, January, 1955.
39. Fox, A. G.: An Adjustable Wave-guide Phase Changer, *Proc. IRE*, vol. 35, pp. 1489–1498, December, 1947.
40. Seckelmann, R.: Phase-Shift Characteristics of Helical Phase Shifters, *IEEE Trans*, vol. MTT-14, pp. 24–28, January, 1966.
41. Boxer, A. S., S. Hershenov, and E. F. Landry: A High-Power Coaxial Ferrite Phase Shifter, *IRE Trans*, vol. MTT-9, p. 577, November, 1961.

42. Sakiotis, N. G., and H. N. Chait: Ferrites at Microwaves, *Proc. IRE*, vol. 41, pp. 87–93, January, 1953.
43. Lax, B., K. J. Button, and L. M. Roth: Ferrite Phase Shifters in Rectangular Waveguide, *J. Applied Physics*, vol. 25, pp. 1413–1421, November, 1954.
44. Scharfman, H.: Three New Ferrite Phase Shifters, *Proc. IRE*, vol. 44, pp. 1456–1459, October, 1956.
45. Monaghan, S. R., and M. C. Mohr: Polarization Insensitive Phase Shifter for Use in Phased Array Antennas, *Microwave J.*, vol. 12, pp. 75–80, December, 1969.
46. Querido, H. J., J. Frank, and T. Cheston: Wide Band Phase Shifters, *IEEE Trans*, vol. AP-15, p. 300, March, 1967.
47. Johnson, C. M.: Ferrite Phase Shifter for the UHF Region, *IRE Trans*, vol. MTT-7, pp. 27–31, January, 1959.
48. Stern, E., and G. N. Tsandoulas: Ferroscan: Toward Continuous-Aperture Scanning, *IEEE Trans*, vol. AP-23, pp. 15–20, January, 1975.
49. Cohn, M., and A. F. Eikenberg: Ferroelectric Phase Shifters for VHF and UHF, *IRE Trans*, vol. MTT-10, pp. 536–548, November, 1962.
50. Senf, H. R.: Electronic Antenna Scanning, *Proc. Natl. Conf. on Aeronaut. Electronics* (Dayton, Ohio), 1958, pp. 407–411.
51. Kummer, W. H.: Feeding and Phase Scanning, chap. 1 in "Microwave Scanning Antennas, vol. III," R. C. Hansen (ed.), Academic Press, New York, 1966.
52. Alday, J. R.: Microwave Plasma Beam Sweeping Array, *IRE Trans*, vol. AP-10, pp. 96–97, January, 1962.
53. Pringle, D. H., and E. M. Bradley: Some New Microwave Control Valves Employing the Negative Glow Discharge, *J. Electron.*, (London), vol. 1, pp. 389–404, 1956.
54. Betts, F., D. H. Temme, and J. A. Weiss: A Switchable Circulator: *S*-band; Stripline; Remanent; 15 Kilowatts; 10 Microseconds; Temperature Stable, *IEEE Trans.*, vol. MTT-14, pp. 665–669, December, 1966.
55. Stark, L.: A Helical Line Scanner for Beam Steering a Linear Array, *IRE Trans.*, vol. AP-5, pp. 211–216, April, 1957.
56. Reed, R. H.: Modified Magic Tee Phase Shifter, *IRE Trans.*, no. PGAP-1, pp. 126–134, February, 1952.
57. Barrett, R. M.: Electronic Scanning: Past and Future, *Electronic Scanning Symposium*, Apr. 29–30, May 1, 1958, AFCRC-TR-58-145(I), ASTIA Document 152409.
58. Bacon, G. E.: Variable-elevation-beam Aerial Systems for $1\frac{1}{2}$ Metres, *J. IEE*, vol. 93, pt. IIIA, pp. 539–544, 1946.
59. Harvey, G. G.: Report of Conference on Rapid Scanning, *MIT Radiation Lab. Rept. 54-27*, June 15, 1943, ASTIA ATI 46616.
60. Kaiser, J. A.: Spiral Antennas Applied to Scanning Arrays, *Electronic Scanning Symposium*, Apr. 29–30, May 1, 1958, AFCRC-TR-58-145(I), ASTIA Document 152409.
61. Brown, R. M., Jr., and R. C. Dodson: Parasitic Spiral Arrays, *IRE Intern. Conv. Record*, vol. 8, pt. 1, pp. 51–66, 1960.
62. Waterman, A. T., Jr.: A Rapid Beam Swinging Experiment in Transhorizon Propagation, *IRE Trans.*, vol. AP-6, pp. 338–340, October, 1958.
63. Hammer, I. W.: Frequency-scanned Arrays, chap. 13 of "Radar Handbook," M. I. Skolnik (ed.), McGraw-Hill Book Co., N.Y., 1970.
64. Begovich, N. A.: Frequency Scanning, chap. 2 of "Microwave Scanning Antennas, Vol. III," R. C. Hansen (ed.), Academic Press, N.Y., 1966.
65. Radford, M. F.: Frequency Scanning Aerials, *Electronic Engineering*, vol. 36, pp. 222–226, April, 1964.
66. Milne, K.: The Combination of Pulse Compression with Frequency Scanning for Three-Dimensional Radars, *Radio Electronic Engr.*, vol. 28, pp. 89–106, Aug., 1964.
67. Croney, J.: Doubly Dispersive Frequency Scanning Antenna, *Microwave J.*, vol. 6, pp. 76–80, July, 1963.
68. Loughren, A. V.: System for Space-Scanning with a Radiated Beam of Wave Signals, U.S. Patent No. 2,409,944, issued Oct. 22, 1946.
69. Croney, J., and J. R. Mark: Design of a Volumetric Frequency Scanning Antenna, *Proceedings European Microwave Conference*, Sept. 8–12, 1969, IEE Conference Publication No. 58, pp. 148–151.
70. Okubo, G. H.: Helix Frequency Scanning Feed, *Microwave J.*, vol. 8, pp. 39–44, December, 1965.
71. Dewey, R.: Corrugated Waveguide Frequency Scanning Aerials, *Proceedings International Conference on Radar—Present and Future*, Oct. 23–25, 1973, IEE Conference Publication no. 105, pp. 100–105.

72. Wheeler, H. A.: A Systematic Approach to the Design of a Radiator Element for a Phased-Array Antenna, *Proc. IEEE*, vol. 56, pp. 1940–1951, November, 1968.
73. Wheeler, H. A.: A Survey of the Simulator Technique for Designing a Radiating Element in a Phased-Array Antenna, "Phased Array Antennas, Proceedings of the 1970 Phased Array Antenna Symposium," Artech House, Inc., Dedham, Massachusetts, 1972, pp. 132–148.
74. Carter, P. S., Jr.: Mutual Impedance Effects in Large Beam Scanning Arrays, *IRE Trans.*, vol. AP-8, pp. 276–285, May, 1960.
75. Edelberg, S., and A. A. Oliner: Mutual Coupling in Large Arrays: pt. I, Slot Arrays, *IRE Trans.*, vol. AP-8, pp. 286–297, May, 1960; pt. II, Compensation Effects, pp. 360–367, July, 1960.
76. Oliner, A. A., and R. G. Malech: Radiating Elements and Mutual Coupling, "Microwave Scanning Antennas, vol. II," R. C. Hansen (ed.), Academic Press, N.Y., 1966, chap. 2.
77. Oliner, A. A., and R. G. Malech: Mutual Coupling in Infinite Scanning Arrays, "Microwave Scanning Antennas, vol. II," R. C. Hansen (ed.), Academic Press, N.Y., 1966, chap. 3.
78. Stark, L.: Comparison of Array Element Types, "Phased Array Antennas, Proceedings of the 1970 Phased Array Antenna Symposium," Artech House, Inc., Dedham, Mass., 1972, pp. 51–67.
79. Knittel, G. H.: Design of Radiating Elements for Large Planar Arrays: Accomplishments and Remaining Challenges, *Microwave J.*, vol. 15, pp. 27–34, September, 1972.
80. Kurtz, L. A., and R. S. Elliott: Systematic Errors Caused by the Scanning of Antenna Arrays: Phase Shifters in the Branch Lines, *IRE Trans.*, vol. AP-4, pp. 619–627, October, 1956.
81. Scudder, R. M., and W. H. Sheppard: AN/SPY-1 Phased-Array Antenna, *Microwave J.*, vol. 17, pp. 51–55, May, 1974.
82. Cheston, T. C., and J. Frank: Array Antennas, chap. 11 of *Radar Handbook*, M. I. Skolnik (ed.), McGraw-Hill Book Co., New York, 1970.
83. Hill, R. T.: Phased Array Feed Systems, A Survey, " Phased Array Antennas, Proceedings of the 1970 Phased Array Antenna Symposium," A. A. Oliner and G. H. Knittel (eds.), Artech House, Dedham, Mass., 1972, pp. 197–211.
84. Odlum, W. J.: Selection of a Phased Array Antenna for Radar Applications, *IEEE Trans*, vol. AES-3, no. 6 (Supplement), pp. 226–235, November, 1967.
85. Blass, J.: Multidirectional Antenna—A New Approach to Stacked Beams, *1960 IRE International Convention Record*, pt. I, pp. 48–50.
86. Shelton, J. P., and K. S. Kelleher: Multiple Beams from Linear Arrays, *IRE Trans.*, vol. AP-9, pp. 154–161, March, 1961.
87. Delaney, W. P.: An RF Multiple Beam-Forming Technique, *IRE Trans.*, vol. MIL-6, pp. 179–186, April, 1962.
88. White, W. D.: Pattern Limitations in Multiple-Beam Antennas, *IRE Trans.*, vol. AP-10, pp. 430–436, July, 1962.
89. Moody, H. J.: The Systematic Design of the Butler Matrix, *IEEE Trans.*, vol. AP-11, pp. 786–788, Nov., 1964.
90. Butler, J. L.: Digital, Matrix, and Intermediate-Frequency Scanning, chap. 3 of " Microwave Scanning Antennas, vol. III," R. C. Hansen (ed.), Academic Press, New York, 1966, pp. 217–288.
91. Shelton, J. P.: Multibeam Planar Arrays, *Proc. IEEE*, vol. 56, pp. 1818–21, November, 1968.
92. Nester, W. H.: The Fast Fourier Transform and the Butler Matrix, *IEEE Trans.*, vol. AP-16, p. 360, May, 1968.
93. Shelton, J. W.: Fast Fourier Transform and Butler Matrices, *Proc. IEEE*, vol. 56, p. 350, March, 1968.
94. Withers, M. J.: Frequency Insensitive Phase-Shift Networks and their use in a Wide-Bandwidth Butler Matrix, *Electronics Letters*, vol. 5, No. 20, pp. 416–198, Oct. 2, 1969.
95. Chow, P. E. K., and D. E. N. Davies: Wide Bandwidth Butler Matrix, *Electronics Letters*, vol. 3, pp. 252–253, June, 1967.
96. Shelton, J. P.: Reduced Sidelobes for Butler-Matrix-Fed Linear Arrays, *IEEE Trans.*, vol. AP-17, pp. 645–47, September, 1969.
97. Foster, H. E., and R. E. Hiatt: Butler Network Extension to Any Number of Antenna Ports, *IEEE Trans.*, vol. AP-18, pp. 818–820, November, 1970.
98. Cheston, T. C., and J. Frank: Array Antennas, chap. 11 of " Radar Handbook," M. I. Skolnik (ed.), McGraw-Hill Book Co., New York, 1970, sec. 11.9.
99. Sheleg, B.: Butler Submatrix Feed Systems for Antenna Arrays, *IEEE Trans.*, vol. AP-21, pp. 228–9, March, 1973.
100. Hering, K. H.: The Design of Hybrid Multiple Beam Forming Networks, " Phased Array Antennas," A. A. Oliner and G. H. Knittel (eds.), Artech House, Inc., Dedham, Mass., 1972, pp. 240–242.
101. Cottony, H. V., and A. C. Wilson: A High-Resolution Rapid-Scan Antenna, *J. Res. Natl. Bur. Std.*, vol. 65D, pp. 101–110, 1961.

102. Davies, D. E. N.: A Fast Electronically Scanned Radar Receiving System, *J. Brit. Inst. Radio Engrs.*, vol. 22, pp. 305–318, April, 1961.
103. Johnson, M. A.: Phased-Array Beam Steering by Multiplex Sampling, *Proc. IEEE*, vol. 56, pp. 1801–1811, November, 1968.
104. Radford, M. F., and R. Greenwood: A Within-Pulse Scanning Height-Finder, *International Conference on Radar—Present and Future*, IEE Conference Publication no. 105, pp. 50–55, 1973.
105. Schrank, H. E., W. P. Hooper, and R. S. Davis: A Study of Array Beam Switching Techniques, *Rome Air Development Center Technical Documentary Report No. RADC-TDR-64-421*, December, 1964.
106. Earth-based Electronics, *Electronics*, vol. 34, no. 46, pp. 114, 115, 117, Nov. 17, 1961.
107. Gent, H.: The Bootlace Aerial, *RRE J.* (Royal Radar Establishment, Malvern, England), no. 40, pp. 47–58, October, 1957.
108. Shelton, J. P., Jr., and K. S. Kelleher: Recent Electronic Scanning Developments, *Conf. Proc., Fourth Natl. Conv. on Military Electronics*, pp. 30–34, 1960.
109. Skolnik, M. I.: Nonuniform Arrays, chap. 6 of "Antenna Theory, pt. I," R. E. Collin and F. J. Zucker (eds.), McGraw-Hill Book Co., New York, 1969, p. 231.
110. Rondinelli, L. A.: Effects of Random Errors on the Performance of Antenna Arrays of Many Elements, *IRE Natl. Conv. Record*, vol. 7, pt. 1, pp. 174–187, 1959.
111. Leichter, M.: Beam Pointing Errors of Long Line Sources, *Hughes Aircraft Co. Tech. Mem. 588*, April, 1959; also *IRE Trans.*, vol. AP-8, pp. 268–275, May, 1960.
112. Ruze, J.: Physical Limitations on Antennas, *MIT Research Lab. Electronics Tech. Rept. 248*, Oct. 30, 1952.
113. Elliot, R. S.: Mechanical and Electrical Tolerances for Two-dimensional Scanning Antenna Arrays, *IRE Trans.*, vol. AP-6, pp. 114–120, January, 1958.
114. Bates, R. H. T.: Random Errors in Aperture Distributions, *IRE Trans.*, vol. AP-7, pp. 369–372, October, 1959.
115. Miller, C. J.: Minimizing the Effects of Phase Quantization Errors in an Electronically Scanned Array, "Proceedings of Symposium on Electronically Scanned Array Techniques and Applications," *Rome Air Development Center Technical Documentary Report No. RADC-TDR-64-225*, vol. 1, pp. 17–38, July, 1964.
116. Champine, G. A.: Computer Control of Array Radar, *Sperry Engineering Review*, vol. 18, No. 4, pp. 18–23, Winter, 1965.
117. Scheff, B. H., and D. G. Hammel: Real-time Computer Control of Phased Array Radars, *Suppl. to IEEE Trans on Aerospace and Electronic Systems*, vol. AES-3, no. 6, pp. 198–206, Nov., 1967.
118. Reed, J. E.: The AN/FPS-85 Radar System, *Proc. IEEE*, vol. 57, pp. 324–335, March, 1969.
119. Johnson, C. M.: Ballistic-Missile Defense Radars, *IEEE Spectrum*, vol. 7, pp. 32–41, March, 1970.
120. Weinstock, W.: Computer Control of a Multifunction Radar, *RCA Engineer*, vol. 18, no. 1, June–July, 1972.
121. Baugh, R. A.: "Computer Control of Modern Radars," Privately published by RCA Inc., 1973.
122. Motkin, D. L.: Three-dimensional Air Surveillance Radar, *Systems Technology*, (Plessey Co, Iford, England), no. 21, pp. 29–33, June, 1975.
123. Knittel, G. H.: Choosing the Number of Faces of a Phased-Array Antenna for Hemisphere Scan Coverage, *IEEE Trans.*, vol. AP-13, pp. 878–882, Nov., 1965.
124. Kmetzo, J. L.: An Analytical Approach to the Coverage of a Hemisphere by N Planar Phased Arrays, *IEEE Trans.*, vol. AP-15, pp. 367–371, May, 1967.
125. Hering, K. H.: Optimization of Tilt Angle and Element Arrangement for Planar Arrays, *Microwave J.*, vol. 14, pp. 41–51, Jan., 1971.
126. Liebman, P. M., L. Schwartzman, and A. E. Hylas: Dome Radar—A New Phased Array System, *Record of the IEEE 1975 International Radar Conference*, pp. 349–353, Apr. 21–23, 1975, IEEE Publication 75 CHO 938-1 AES.
127. Bearse, S. V.: Planar Array Looks Through Lens to Provide Hemispherical Coverage, *Microwaves*, vol. 14, pp. 9–10, July, 1975.
128. Schwartzman, L., and J. Stangel: The Dome Antenna, *Microwave J.*, vol. 18, pp. 31–34, October, 1975.
129. Schwartzman, L., and P. M. Liebman: A Report on the Sperry Dome Radar, *Microwave J.*, vol. 22, pp. 65–69, March, 1979.
130. Hansen, R. C.: Conformal Arrays, *Microwave J.*, vol. 13, p. 39, April, 1970.
131. Sureau, J. C.: Conformal Arrays Come of Age, *Microwave J.*, vol. 16, pp. 23–26, October, 1973.
132. Giannini, R. J., J. H. Gutman, and P. W. Hannon: A Cylindrical Phased-Array Antenna for ATC Interrogation, *Microwave J.*, vol. 16, pp. 46–49, October, 1973.
133. Kummer, W. H. (ed.): Special Issue on Conformal Arrays, *IEEE Trans.*, vol. AP-22, no. 1, January, 1974.

134. Hansen, R. C.: Workshop on Conformal Arrays, *Microwave J.*, vol. 18, p. 28, October, 1975.
135. Hsiao, J. K., and A. G. Cha: Patterns and Polarizations of Simultaneously Excited Planar Arrays on a Conformal Surface, *IEEE Trans.*, vol. AP-22, pp. 81–84, January, 1974.
136. Skolnik, M. I.: Nonuniform Arrays, chap. 6 in "Antenna Theory, pt. I," R. E. Collin and F. J. Zucker (eds.), McGraw-Hill Book Co., New York, 1969.
137. Skolnik, M. I., G. Nemhauser, and J. W. Sherman: Dynamic Programming Applied to Unequally Spaced Arrays, *IEEE Trans.*, vol. AP-12, pp. 35–43, January, 1964.
138. Skolnik, M. I., J. W. Sherman, and F. C. Ogg: Statistically Designed Density-tapered Arrays, *IEEE Trans.*, vol. AP-12, pp. 408–417, July, 1964.
139. Steinberg, B. D.: The Peak Sidelobe of the Phased Array Having Randomly Located Elements, *IEEE Trans.*, vol. AP-20, pp. 129–136, March, 1972.
140. Skolnik, M. I.: Resolution of Angular Ambiguities in Radar Array Antennas with Widely-Spaced Elements and Grating Lobes, *IEEE Trans.*, vol. AP-10, pp. 351–352, May, 1962.
141. Goto, N., and D. K. Cheng: Phase Shifter Thinning and Sidelobe Reduction for Large Phased Arrays, *IEEE Trans.*, vol. AP-24, pp. 139–143, March, 1976.
142. Hansen, R. C. (ed.): Special Issue on Active and Adaptive Antennas, *IEEE Trans.*, vol. AP-12, no. 2, March, 1964.
143. Skolnik, M. I., and D. D. King: Self-Phasing Arrays, *IEEE Trans.*, vol. AP-12, pp. 142–49, March, 1964.
144. Widrow, B., P. E. Mantey, L. J. Griffiths, and B. B. Goode: Adaptive Antenna Systems, *Proc. IEEE*, vol. 55, pp. 2143–2159, December, 1967.
145. Gabriel, W. F.: Adaptive Array Antennas for AEW Radar, *Report of NRL Progress*, pp. 39–44, December, 1972.
146. Brennan, L. E., and I. S. Reed: Theory of Adaptive Radar, *IEEE Trans.*, vol. AES-9, pp. 237–252, March, 1973.
147. Zahm, C. L.: Application of Adaptive Arrays to Suppress Strong Jammers in the Presence of Weak Signals, *IEEE Trans.*, vol. AES-9, pp. 260–271, March, 1973.
148. Brennan, L. E., I. S. Reed, and P. Swerling: Adaptive Arrays, *Microwave J.*, vol. 17, pp. 43–46, p. 74, May, 1974.
149. O'Donovan, P. L., and A. W. Rudge: Adaptive Control of a Flexible Linear Array, *Electronics Letters*, vol. 9, pp. 121–122, Mar. 22, 1973.
150. Gabriel, W. F.: Adaptive Arrays—An Introduction, *Proc. IEEE*, vol. 64, pp. 239–272, February, 1976.
151. Gabriel, W. F. (ed.): Special Issue on Adaptive Antennas, *IEEE Trans.*, vol. AP-24, September, 1976.
152. Sharp, E. D.: A Triangular Arrangement of Planar-Array Elements that Reduces the Number Needed, *IRE Trans.*, vol. AP-9, pp. 126–129, March, 1961.
153. Cheng, D. H. S.: Characteristics of Triangular Lattice Arrays, *Proc. IEEE*, vol. 56, pp. 1811–1817, November, 1968.
154. Nelson, E. A.: Quantization Sidelobes of a Phased Array with a Triangular Element Arrangement, *IEEE Trans.*, vol. AP-17, pp. 363–365, May, 1969.
155. Garver, R. V.: "Microwave Diode Control Devices," Artech House, Inc., 1976.
156. Whicker, L. R., and R. R. Jones: A Digital Current Controlled Latching Ferrite Phase Shifter, *IEEE 1965 International Convention Record*, pt. V, pp. 217–223.
157. Ward, H. R., C. A. Fowler, and H. I. Lipson: GCA Radars: Their History and State of Development, *Proc. IEEE*, vol. 62, pp. 705–716, June, 1974.
158. Ethington, D. A.: The AN/TPQ-36 and AN/TPQ-37 Firefinder Radars, *International Conference RADAR-77*, Oct. 25–28, 1977, IEE (London) Publication no. 155, pp. 33–35.
159. Ruvin, A. E., and L. Weinberg: Digital Multiple Beamforming Techniques for Radar, *IEEE EASCON '78 Record*, pp. 152–163, Sept. 25–27, 1978, IEEE Publication 78 CH 1354-4 AES.
160. Malagisi, C. S.: Microstrip Disc Element Reflect Array, *IEEE EASCON '78 Record*, pp. 186–192, Sept. 25–27, 1978, IEEE Publication 78 CH 1352-4 AES.
161. Ricardi, L. J.: Array Beam-Forming Networks, *MIT Lincoln Laboratory Technical Note* 1965-12, 13 April 1965.
162. Whicker, L. R., and C. W. Young, Jr.: The Evolution of Ferrite Control Components, *Microwave J.*, vol. 21, pp. 33–37, November, 1978.

NINE

RECEIVERS, DISPLAYS, AND DUPLEXERS

9.1 THE RADAR RECEIVER

The function of the radar receiver is to detect desired echo signals in the presence of noise, interference, or clutter. It must separate wanted from unwanted signals, and amplify the wanted signals to a level where target information can be displayed to an operator or used in an automatic data processor. The design of the radar receiver will depend not only on the type of waveform to be detected, but on the nature of the noise, interference, and clutter echoes with which the desired echo signals must compete. In this chapter, the receiver design is considered mainly as a problem of extracting desired signals from *noise*. Chapter 13 considers the problem of radar design when the desired signals must compete with clutter. The current chapter also includes brief discussions of radar displays and duplexers.

Noise can enter the receiver via the antenna terminals along with the desired signals, or it might be generated within the receiver itself. At the microwave frequencies usually used for radar, the external noise which enters via the antenna is generally quite low so that the receiver sensitivity is usually set by the internal noise generated within the receiver. (External noise is discussed in Sec. 12.8.) The measure of receiver internal noise is the noise-figure (introduced previously in Sec. 2.3).

Good receiver design is based on maximizing the output signal-to-noise ratio. As described in Sec. 10.2, to maximize the output signal-to-noise ratio, the receiver must be designed as a *matched filter*, or its equivalent. The matched filter specifies the frequency response function of the IF part of the radar receiver. Obviously, the receiver should be designed to generate as little internal noise as possible, especially in the input stages where the desired signals are weakest. Although special attention must be paid to minimize the noise of the input stages, the lowest noise receivers are not always desired in many radar applications if other important receiver properties must be sacrificed.

Receiver design also must be concerned with achieving sufficient gain, phase, and amplitude stability, dynamic range, tuning, ruggedness, and simplicity. Protection must be provided against overload or saturation, and burnout from nearby interfering transmitters. Timing and reference signals are needed to properly extract target information. Specific applications such as MTI radar, tracking radar, or radars designed to minimize clutter place special demands on the receiver. Receivers that must operate with a transmitter whose frequency can drift need some means of automatic frequency control (AFC). Radars that encounter hostile counter-measures need receivers that can minimize the effects of such interference. Thus there can be many demands placed upon the receiver designer in meeting the requirements of modern high-quality radar systems. The receiver engineer has responded well to the challenge, and there exists a highly refined state of technology available for radar applications. Radar receiver design and implementation may not always be an easy task; but in tribute to the receiver designer, it has seldom been an obstacle preventing the radar systems engineer from eventually accomplishing the desired objectives.

Although the superregenerative, crystal video, and tuned radio frequency (TRF) receivers have been employed in radar systems, the superheterodyne has seen almost exclusive application because of its good sensitivity, high gain, selectivity, and reliability. No other receiver type has been competitive to the superheterodyne. A simple block diagram of a radar superheterodyne receiver was shown in Fig. 1.2. There are many factors that enter into the design of radar receivers; however, only the receiver noise-figure and the receiver front-end, as they determine receiver sensitivity, will be discussed here.

9.2 NOISE FIGURE

In Sec. 2.3 the noise figure of a receiver was described as a measure of the noise produced by a practical receiver as compared with the noise of an ideal receiver.[1,2] The noise figure F_n of a linear network may be defined as

$$F_n = \frac{S_{in}/N_{in}}{S_{out}/N_{out}} = \frac{N_{out}}{kT_0 B_n G} \tag{9.1}$$

where S_{in} = available input signal power
N_{in} = available input noise power (equal to $kT_0 B_n$)
S_{out} = available output signal power
N_{out} = available output noise power

"Available power" refers to the power which would be delivered to a matched load. The available gain G is equal to S_{out}/S_{in}, k = Boltzmann's constant = 1.38×10^{-23} J/deg, T_0 = standard temperature of 290 K (approximately room temperature), and B_n is the noise bandwidth defined by Eq. (2.3). The product $kT_0 \approx 4 \times 10^{-21}$ W/Hz. The purpose for defining a standard temperature is to refer any measurements to a common basis of comparison. Equation (9.1) permits two different but equivalent interpretations of noise figure. It may be considered as the degradation of the signal-to-noise ratio caused by the network (receiver), or it may be interpreted as the ratio of the actual available output noise power to the noise power which would be available if the network merely amplified the thermal noise. The noise figure may also be written

$$F_n = \frac{kT_0 B_n G + \Delta N}{kT_0 B_n G} = 1 + \frac{\Delta N}{kT_0 B_n G} \tag{9.2}$$

where ΔN is the additional noise introduced by the network itself.

The noise figure is commonly expressed in decibels, that is, 10 log F_n. The term *noise factor* is also used at times instead of noise figure. The two terms are now synonymous.

The definition of noise figure assumes the input and output of the network are matched. In some devices, less noise is obtained under mismatched, rather than matched, conditions. In spite of definitions, such networks would be operated so as to achieve the maximum output signal-to-noise ratio.

Noise figure of networks in cascade. Consider two networks in cascade, each with the same noise bandwidth B_n but with different noise figures and available gain (Fig. 9.1). Let F_1, G_1 be the noise figure and available gain, respectively, of the first network, and F_2, G_2 be similar parameters for the second network. The problem is to find F_o, the overall noise-figure of the two circuits in cascade. From the definition of noise figure [Eq. (9.1)] the output noise N_o of the two circuits in cascade is

$$N_o = F_o G_1 G_2 k T_0 B_n = \text{noise from network 1 at output of network 2}$$
$$+ \text{ noise } \Delta N_2 \text{ introduced by network 2} \quad (9.3a)$$

$$N_o = k T_0 B_n F_1 G_1 G_2 + \Delta N_2 = k T_0 B_n F_1 G_1 G_2 + (F_2 - 1) k T_0 B_n G_2 \quad (9.3b)$$

or

$$F_o = F_1 + \frac{F_2 - 1}{G_1} \quad (9.4)$$

The contribution of the second network to the overall noise-figure may be made negligible if the gain of the first network is large. This is of importance in the design of multistage receivers. It is not sufficient that only the first stage of a low-noise receiver have a small noise figure. The succeeding stage must also have a small noise figure, or else the gain of the first stage must be high enough to swamp the noise of the succeeding stage. If the first network is not an amplifier but is a network with loss (as in a crystal mixer), the gain G_1 should be interpreted as a number less than unity.

The noise figure of N networks in cascade may be shown to be

$$F_o = F_1 + \frac{F_2 - 1}{G_1} + \frac{F_3 - 1}{G_1 G_2} + \cdots + \frac{F_N - 1}{G_1 G_2 \cdots G_{N-1}} \quad (9.5)$$

Similar expressions may be derived when bandwidths and/or the temperature of the individual networks are not the same.[3]

Noise temperature. The noise introduced by a network may also be expressed as an *effective noise temperature*, T_e, defined as that (fictional) temperature at the input of the network which would account for the noise ΔN at the output. Therefore $\Delta N = k T_e B_n G$ and

$$F_n = 1 + \frac{T_e}{T_0} \quad (9.6)$$

$$T_e = (F_n - 1) T_0 \quad (9.7)$$

The *system noise temperature* T_s is defined as the effective noise temperature of the receiver system including the effects of antenna temperature T_a. (It is also sometimes called the system

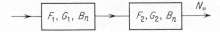

Figure 9.1 Two networks in cascade.

operating noise temperature.[60]) If the receiver effective noise temperature is T_e, then

$$T_s = T_a + T_e = T_0 F_s \tag{9.8}$$

where F_s is the *system noise-figure* including the effect of antenna temperature.

The effective noise temperature of a receiver consisting of a number of networks in cascade is

$$T_e = T_1 + \frac{T_2}{G_1} + \frac{T_3}{G_1 G_2} + \cdots \tag{9.9}$$

where T_i and G_i are the effective noise temperature and gain of the ith network.

The effective noise temperature and the noise figure both describe the same characteristic of a network. In general, the effective noise temperature has been preferred for describing low-noise devices, and the noise figure is preferred for conventional receivers. For radar receivers the noise figure is the more widely used term, and is what is used in this text.

Measurement of noise figure. The noise figure of a radar receiver can degrade in operation and cause reduced capability. Therefore some means for monitoring the noise figure should be provided in operating radars so that a worsening of receiver sensitivity can be detected and corrected. The monitoring of the noise figure can be accomplished either automatically or manually by the operator.

The receiver noise-figure can be measured with a broadband noise source of known intensity, such as a gas-discharge tube[22] or a solid-state noise source. The noise figure is determined by measuring (1) the noise power output N_1 of the receiver when a matched impedance at temperature $T_0 = 290$ K is connected to the receiver input and (2) the noise power output N_2 when a matched noise generator of temperature T_2 is connected to the receiver input.[2] The temperature T_2 is the equivalent noise temperature of the broadband noise generator. The noise figure can be shown to be

$$F_n = \frac{T_2/T_0 - 1}{Y - 1} \tag{9.10}$$

where $Y = N_2/N_1$.

The measurement of noise figure can be made during radar operation without degrading the receiver sensitivity by pulse-modulating the noise source in synchronism with the radar trigger and injecting the noise into the receiver during the "flyback" or "dead time" of the radar, just prior to the triggering of the next transmitter pulse. The measurement of the receiver output with the noise source on (N_2) and the noise source off (N_1) can be made on alternate pulse periods.

The receiver noise-figure or sensitivity can also be measured by use of a calibrated signal generator. With a matched resistance at the receiver input, the output power is due to receiver noise alone. The signal generator power is then applied to the receiver input and adjusted until the signal-plus-noise power is equal to twice the receiver noise power read with the matched resistance. The input signal under this condition is sometimes said to be the minimum discernible signal. It is also proportional to the receiver noise-figure.

The sensitivity of a radar may be visually displayed by using the measurement of receiver noise to display the normal range rings on the PPI only within the range at which the radar can detect targets reliably. This provides the operator with a continuous and immediate indication of radar sensitivity. When noise jamming is present, the appearance or nonappearance of the range rings can be made to be a function of azimuth as well as range.

In making a measurement of the receiver noise-figure, the noise source or signal generator is usually inserted by a directional coupler ahead of the duplexer and other RF components so that the overall noise-figure of the system is measured rather than that of the receiver alone.

9.3 MIXERS

Many radar superheterodyne receivers do not employ a low-noise RF amplifier. Instead, the first stage is simply the mixer. Although the noise figure of a mixer front-end may not be as low as other devices that can be used as receiver front-ends, it is acceptable for many radar applications when other factors besides low noise are important. The function of the mixer is to convert RF energy to IF energy with minimum loss and without spurious responses. Silicon point-contact and Schottky-barrier diodes[15,24] based on the nonlinear resistance characteristic of metal-to-semiconductor contacts have been used as the mixing element.[4,5] Schottky-barrier diodes are made of either silicon or GaAs, with GaAs preferred for the higher microwave frequencies. The Schottky-barrier diodes have had lower noise figures and lower flicker noise than conventional point-contact diodes, but the silicon point-contact diode has had better burnout properties. An integral part of the mixer is the local oscillator. The IF amplifier is also of importance in mixer design because of its influence on the overall noise-figure.

Conversion loss and noise-temperature ratio. The conversion loss of a mixer is defined as

$$L_c = \frac{\text{available RF power}}{\text{available IF power}} \qquad (9.11)$$

It is a measure of the efficiency of the mixer in converting RF signal power into IF. The conversion loss of typical microwave crystals in a conventional single-ended mixer configuration varies from about 5 to 6.5 dB. A crystal mixer is called "broadband" when the signal and image frequencies are both terminated in matched loads. A signal impressed in the RF signal channel of a broadband mixer is converted in equal portions to the IF signal and the RF image. Therefore the theoretical conversion loss can never be less than 3 dB with this configuration. (The image frequency is defined as that frequency which is displaced from the local oscillator frequency f_{LO} by the IF frequency, and which appears on the opposite side of the local oscillator frequency as the signal frequency f_{RF}. It is equal to $2f_{LO} - f_{RF}$.)

Short-circuiting or open-circuiting the image-frequency termination results in a "narrowband" mixer. The conversion loss is less in the narrowband than in the broadband mixer. In principle, it can be about 2 dB lower.[6] The design of a broadband mixer has been simpler to achieve and less critical than a narrowband mixer.

The *noise-temperature ratio* of a crystal mixer (not to be confused with effective noise temperature) is defined as

$$t_r = \frac{\text{actual available IF noise power}}{\text{available noise power from an equivalent resistance}} \qquad (9.12a)$$

or
$$t_r = \frac{F_c k T_0 B_n G_c}{k T_0 B_n} = F_c G_c = \frac{F_c}{L_c} \qquad (9.12b)$$

where F_c = crystal mixer noise figure and $L_c = 1/G_c$ = conversion loss. The noise temperature ratio of a crystal mixer varies approximately inversely with frequency from about 100 kHz (the exact value depends upon the diode[25]) down to a small fraction of a hertz. This is called *flicker noise*, or *1/f noise*. Above approximately 500 kHz, the noise-temperature ratio approaches a constant value. At a frequency of 30 MHz, a typical radar IF, it might range from 1.3 to 2.0.

From Eq. (9.12b) the noise figure of the mixer is $F_c = t_r L_c$. This, however, is not a complete measure of the sensitivity of a receiver with a mixer front-end. The overall noise-figure depends not only on the mixer stage, but also on the noise figure of the IF stage and the mixer conversion loss. It may be determined from the expression for the noise figure of two networks in cascade [Eq. (9.4)]. The first network is the mixer with noise figure $t_r L_c$ and gain $= 1/L_c$. The second network is the IF amplifier with a noise figure F_{IF}. The receiver noise-figure with a mixer front-end is then

$$F_o = F_1 + \frac{F_2 - 1}{G_1} = L_c(t_r + F_{IF} - 1) \tag{9.13}$$

(This does not include losses in the RF transmission line connecting the receiver to the antenna.) If, for example, the conversion loss of the mixer were 6.0 dB, the IF noise figure 1.5 dB, and the noise-temperature ratio 1.4, the receiver noise figure would be 8.6 dB. For low-noise-temperature-ratio diodes, the receiver noise figure is approximately equal to the conversion loss times the IF noise figure.

Balanced mixers. Noise that accompanies the local-oscillator (LO) signal can appear at the IF frequency because of the nonlinear action of the mixer. The LO noise must be removed if receiver sensitivity is to be maximized. One method for eliminating LO noise that interferes with the desired signal is to insert a narrow-bandpass RF filter between the local oscillator and the mixer. The center frequency of the filter is that of the local oscillator, and its bandwidth must be narrow so that LO noise at the signal and the image frequencies do not appear at the mixer. Since the receiver is tuned by changing the LO frequency, the narrowband filter must be tunable also.

A method of eliminating local-oscillator noise without the disadvantage of a narrow-bandwidth filter is the *balanced mixer* (Fig. 9.2). A balanced mixer uses a hybrid junction, a magic T, or an equivalent. These are four-port junctions. Figure 9.2 illustrates a magic T in which the LO and RF signals are applied to two ports. Diode mixers are in each of the remaining two arms of the magic T. At one of the diodes the sum of the RF and LO signals appears, and at the other diode the difference of the two is obtained. (In a magic T the LO would be applied to the H-plane arm, and the RF signal would be applied to the E-plane arm. The diode mixers would be mounted at equal distances in each of the collinear arms.) The two diode mixers should have identical characteristics and be well matched. The IF signal is

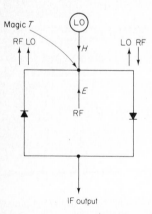

Figure 9.2 Principle of the balanced mixer based on the magic T.

recovered by subtracting the outputs of the two diode mixers. In Fig. 9.2 the balanced diodes are shown reversed so that the IF outputs can be added. Local-oscillator noise at the two diode mixers will be in phase and will be canceled at the output. It is only the AM noise of the local oscillator which is canceled. The FM noise inserted by the local oscillator is unaffected by the balanced mixer.[9]

In a single-ended mixer, the mixing action generates all harmonics of the RF and LO frequencies, and combinations thereof.[7] The output is designed to filter out the frequency of interest, usually the difference frequency. A balanced mixer suppresses the even harmonics of the LO signal. A *double-balanced mixer* is basically two single-ended mixers connected in parallel and 180° out of phase. It suppresses even harmonics of both the RF and the LO signals.[8,55]

Reactive image termination. If the image frequency of a mixer is presented with the proper reactive termination (such as an open or a short circuit), the conversion loss and the noise figure can be 1 to 2 dB less than with a "broadband" mixer in which the image frequency is terminated in a matched load.[6,10] The reactive termination causes energy converted to the image frequency to be reflected back into the mixer and reconverted to IF.[16,24]

Both the sum and the image frequencies can be reflected back to the diode mixer to achieve a lower conversion loss, but a number of adjustments are required for good results.[24] One method for terminating the image in a reactive load is to employ a narrow bandpass filter, or preselector, at the RF signal frequency. A limitation of this approach is that it is not suitable for very wide bandwidths. The filter has to be retuned if the mixer must operate at another frequency. Also, the high Q of the filter introduces a loss which will increase the system noise-figure.

A method for achieving a reactive termination without narrow-bandwidth components is the *image-recovery* mixer shown in Fig. 9.3. This has also been called an *image-enhanced mixer*,[17] or *product return mixer*.[56] (It is similar to the image-reject mixer[6,7,11,55] whose purpose is to reject the image response.) The RF hybrid junction on the left of the circuit produces a 90° phase difference between the LO inputs to the two mixers. The IF hybrid junction on the right imparts another 90° phase differential in such a manner that the images cancel, but the IF signals from the two mixers add in phase. The two mixers in Fig. 9.3 may be single-ended, balanced, or double-balanced mixers.[12] This mixer is capable of wide bandwidth, and is restricted only by the frequency sensitivity of the structure of the microwave circuit. The noise figure of an image-recovery circuit (Fig. 9.4) is competitive with other receiver front-ends. The image-recovery mixer is attractive as a receiver front-end because of its high dynamic range, low intermodulation products, less susceptibility to burnout, and less cost as compared to other front-ends.[13]

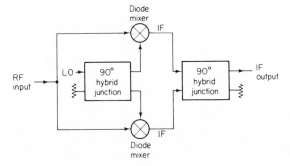

Figure 9.3 Image-recovery mixer.

Diode burnout. A crystal diode which is subjected to excessive RF power may suffer *burnout*. This is a rather loosely defined term which is applied to any irreversible deterioration in the detection or conversion properties of a crystal diode as the result of electrical overload. If excessive RF energy is applied to the diode the heat generated cannot be dissipated properly and the diode can be damaged. Excessive energy causes the diode to open-circuit or the semiconductor to puncture, resulting in failure of the device. As defined above, however, burnout of a diode can occur before the onset of physical destruction. An increase in the receiver noise due to the effects of excessive RF energy can be just as harmful as complete destruction; perhaps more so, for gradual deterioration of performance might not be noticed as readily as would catastrophic failure. It is for this reason that some means of automatic monitoring of receiver noise-figure is necessary if the radar is to be maintained in prime operating conditions.

A degradation in the noise figure of a predetermined amount usually is considered as the criterion for diode failure when defining burnout. Sometimes an increase in noise figure of 3 dB has been used as the criterion.[19] In other cases, a 1-dB increase has been used.[15] However, with Schottky and point-contact diodes, there is an increase in $1/f$ noise and a decrease in the breakdown voltage at lower power levels than would be indicated by the above criteria.[14]

One of the causes of diode burnout in radar receivers has been the increased RF leakage through a conventional duplexer due to aging of the TR tube. When the transmitter fires, the TR tube breaks down. A finite time, usually on the order of several nanoseconds, must elapse before breakdown is complete. During this time, RF energy leaks into the receiver. This is called the *spike-leakage* energy. From 1 to 10 ergs of spike-leakage energy might be required to burn out microwave crystal diodes. The amount of energy contained within the remainder of the pulse after the initial spike is usually small and is not as serious as spike leakage.

When a solid-state duplexer is used or when a solid-state limiter follows the TR switch, there need be no initial spike and burnout is not determined by the pulse energy. Burnout due to pulses without an initial spike, but greater than about 1 μs in duration, is determined primarily by the peak power.[16] (The burnout conditions for pulses 1 μs or greater is essentially the same as for CW.) Crystal diodes can withstand several watts or more of peak power under pulse conditions. For pulses shorter than 1 μs, the peak-power capability increases, but not at a sufficient rate for constant energy.[15]

In addition to leakage through the duplexer, diode burnout can result from the accidental reception of power from nearby radars or from the discharge of static electricity through the diode.

Noise figure due to RF losses. Any losses in the RF portion ahead of the receiver front-end result in an increase of the overall noise-figure. These losses, denoted L_{RF}, might be due to the receiver transmission line, duplexer, rotary joint, preselector filter, monitoring devices, or loss in the randome. The noise figure due to these RF losses may be derived from the definition of Eq. (9.1), which is

$$F_n = \frac{N_{out}}{k T_0 B_n G} \qquad (9.1)$$

The noise N_{out} from the lossy RF components is $k T_0 B_n$, and $G = 1/L_{RF}$. Therefore on substitution into Eq. (9.1), the noise figure F_L due to the RF losses is simply

$$F_L = L_{RF} \qquad (9.14)$$

The noise figure of a receiver with noise figure F_r, preceded by RF losses equal to L_{RF} is

$$F_0 = F_1 + \frac{F_2 - 1}{G_1} = L_{RF} + (F_r - 1)L_{RF} = F_r L_{RF} \qquad (9.15)$$

where $F_1 = L_{RF}$, $G_1 = 1/L_{RF}$, and $F_2 = F_r$.

9.4 LOW-NOISE FRONT-ENDS

Early microwave superheterodyne receivers did not use an RF amplifier as the first stage, or front-end, since the RF amplifiers at that time had a greater noise figure than when the mixer alone was employed as the receiver input stage. There are now a number of RF amplifiers that can provide a suitable noise figure. Figure 9.4 plots noise figure as a function of frequency for the several receiver front-ends used in radar applications. The parametric amplifier[10,18] has the lowest noise figure of those devices described here, especially at the higher microwave frequencies. However, it is generally more complex and expensive compared to the other front-ends.

The transistor amplifier can be applied over most of the entire range of frequencies of interest to radar.[10,19,20] The silicon bipolar-transistor has been used at the lower radar frequencies (below *L* band) and the galium arsenide field-effect transistor (GaAs FET) is preferred at the

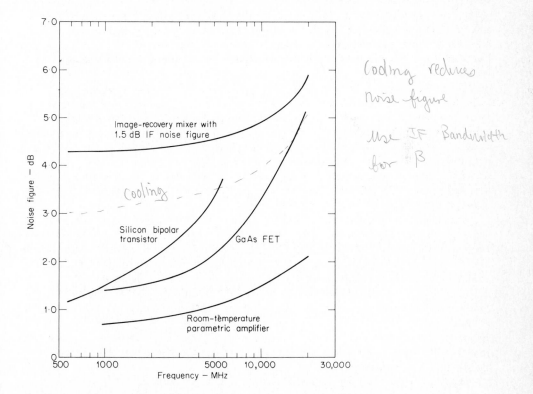

Figure 9.4 Noise figures of typical microwave receiver front-ends as a function of frequency—

higher frequencies. The transistor is generally used in a multistage configuration with a typical gain per stage decreasing from 12 dB at VHF to 6 dB at K_u band.[10] In the GaAs FET, the thermal noise contribution is greater than the shot noise. Cooling the device will therefore improve the noise figure.[21]

The tunnel-diode amplifier has been considered in the past as a low-noise front-end, with noise figures from 4 to 7 dB over the range 2 to 25 GHz.[10] It has been supplanted by the improvements made in the transistor amplifier. The traveling-wave-tube amplifier has also been considered as a low-noise front-end, but it has been overtaken by other devices. Cryogenic parametric amplifiers and masers produce the lowest noise figures, but the added complexity of operating at low temperatures has tempered their use in radar.

The noise figure of the ordinary "broadband" mixer whose image frequency is terminated in a matched load is not shown in Fig. 9.4. It would lie about 2 dB higher than the noise figure shown for the image-recovery mixer.

There are other factors beside the noise figure which can influence the selection of a receiver front-end. Cost, burnout, and dynamic range must also be considered. The selection of a particular type of receiver front-end might also be influenced by its instantaneous bandwidth, tuning range, phase and amplitude stability, and any special requirements for cooling. The image-recovery mixer represents a practical compromise which tends to balance its slightly greater noise figure by its lower cost, greater ruggedness, and greater dynamic range.[13]

Utility of low-noise front-ends. The lower the noise figure of the radar receiver, the less need be the transmitter power and/or the antenna aperture. Reductions in the size of the transmitter and the antenna are always desirable if there are no concomitant reductions in performance. A few decibels improvement in receiver noise-figure can be obtained at a relatively low cost as compared to the cost and complexity of adding the same few decibels to a high-power transmitter.

There are, however, limitations to the use of a low-noise front-end in some radar applications.[7] As mentioned above, the cost, burnout, and dynamic range of low-noise devices might not be acceptable in some applications. Even if the low-noise device itself is of large dynamic range, there can be a reduction of the dynamic range of the receiver as compared to a receiver with a mixer as its front-end. Dynamic range is usually defined as the ratio of the maximum signal that can be handled to the smallest signal capable of being detected. The smallest signal is the minimum detectable signal as determined by receiver noise, and the maximum signal is that which causes a specified degree of intermodulation or a specified deviation from linearity (usually 1 dB) of the output-vs.-input curve.

When an RF amplifier is inserted ahead of the mixer stage, and if no change is made in the remainder of the receiver, the minimum signal will be reduced because of the lower noise figure of the RF amplifier, but the maximum signal that can be handled by the receiver will also be reduced by an amount equal to the gain of the amplifier. Since the gain of the RF amplifier is usually high compared to the reduction in receiver noise-figure, the net result is a reduction in receiver dynamic range. However, this can be corrected and the original dynamic range recovered by reducing the gain of the IF amplifier to maintain a constant output (or constant noise level going into the display).[54] On the other hand, if the mixer rather than the IF amplifier is what limits the total dynamic range of the receiver, the introduction of the low-noise front-end will cause a sacrifice in the mixer dynamic range and, consequently, the dynamic range of the entire receiver.[7]

A low-noise receiver may also not be warranted if the RF losses preceding the receiver are high. From Eq. (9.15) the overall noise figure of a radar receiver with a noise figure F_r preceded by RF circuitry with a loss L_{RF}, is equal to $F_r L_{RF}$. In a radar the overall loss L_{RF} due to the transmission line, rotary joint, duplexer, receiver protector, and preselector filter might not be insignificant. In some nonradar applications, as for example radio astronomy, most of these lossy components are not necessary as they are in radar, so that a low noise figure front-end can be used effectively. However, an extremely low receiver noise-figure is not usually warranted in radar because of the unavoidable RF losses found in most radars. Even if the noise figure of the receiver were essentially 0 dB, the overall receiver noise-figure would still be equal to the losses in the RF portion of the system.

In a military radar, a low-noise receiver can make the radar more susceptible to the effects of deliberate electronic countermeasures (ECM). When practical, it may be preferred to deliberately employ a conventional receiver with modest sensitivity and to make up for the reduced sensitivity by larger transmitter power. This is not the most economical way to build a radar, but it does make the task of the hostile ECM designer more difficult.

A variety of low-noise radar receivers are available to the radar system designer. The well-recognized benefits of low-noise receivers, combined with their relative affordability, make them an attractive feature in modern radar design. However, low-noise receivers are sometimes accompanied by other less desirable properties that tend to result in a compromise in receiver performance. Thus a low-noise receiver might not always be the obvious selection, if properties other than sensitivity are important.

9.5 DISPLAYS

The purpose of the display is to visually present in a form suitable for operator interpretation and action the information contained in the radar echo signal. When the display is connected directly to the video output of the receiver, the information displayed is called *raw video*. This is the "traditional" type of radar presentation. When the receiver video output is first processed by an automatic detector or automatic detection and tracking processor (ADT), the output displayed is sometimes called *synthetic video*.

The cathode-ray tube (CRT) has been almost universally used as the radar display. There are two basic cathode-ray tube displays. One is the *deflection-modulated CRT*, such as the A-scope, in which a target is indicated by the deflection of the electron beam. The other is the *intensity-modulated CRT*, such as the PPI, in which a target is indicated by intensifying the electron beam and presenting a luminous spot on the face of the CRT. In general, deflection-modulated displays have the advantage of simpler circuits than those of intensity-modulated displays, and targets may be more readily discerned in the presence of noise or interference. On the other hand, intensity-modulated displays have the advantage of presenting data in a convenient and easily interpreted form. The deflection of the beam or the appearance of an intensity-modulated spot on a radar display caused by the presence of a target is commonly referred to as a *blip*.

The focusing and deflection of the electron beam may be accomplished electrostatically, electromagnetically, or by a combination of the two. Electrostatic deflection CRTs use an electric field applied to pairs of deflecting electrodes, or plates, to deflect the electron beam. Such tubes are usually longer than magnetic tubes, but the overall size, weight, and power dissipation are less. Electromagnetic deflection CRTs require magnetic coils, or deflection yokes, positioned around the neck of the tube. They are relatively lossy and require more drive power than electrostatic devices. Deflection-modulated CRTs, such as the A-scope, generally employ electrostatic deflection. Intensity-modulated CRTs, such as the PPI, generally employ electromagnetic deflection.

Magnetically focused tubes utilize either an electromagnet or a permanent magnet around the neck of the CRT to provide an axial magnetic field. Magnetic focus generally can provide better resolution, but the spot tends to defocus at the edge of the tube.

The CRT display is by no means ideal. It employs a relatively large vacuum tube and the entire display is often big and can be expensive. The cost is not simply the tube itself, which is usually modest, but the various circuits needed to display the desired information and provide the operator with flexibility. The amount of information that can be displayed is limited by the spot size, which in a high-performance display is less than 0.1 percent of the screen diameter.[27] In some high-range-resolution radars, however, the number of resolvable range cells available from the radar might be greater than the number of resolution cells available on the PPI screen. The result is a collapsing loss (Sec. 2.12). Increasing the CRT diameter does not necessarily help, since the spot diameter varies linearly with the screen diameter. Another limitation is the dynamic range, or contrast ratio, of an intensity modulated display which is of the order of 10 dB. This might cause blooming of the display by large targets so as to mask the blips from nearby smaller targets.

The decay of the visual information displayed on the CRT should be long enough to allow the operator not to miss target detections, yet short enough not to allow the information painted on one scan to interfere with the new information entered from the succeeding scan. However, there is usually not sufficient flexibility available to the CRT designer to always obtain the desired phosphor decay characteristics. The brilliance of the initial "flash" from the CRT phosphor may be high, but the afterglow is dim so that it is often necessary to carefully control both the color and the intensity of the ambient lighting to achieve optimum seeing conditions. The conventional CRT usually requires a darkened room or the use of a viewing hood by the operator. In spite of the limitations of the conventional CRT display, it is almost universally used for radar applications. Many of its limitations can be overcome, but sometimes with a sacrifice in some other property.

The ability of an operator to extract information efficiently from a CRT display will depend on such factors as the brightness of the display, density and character of the background noise, pulse repetition rate, scan rate of the antenna beam, signal clipping, decay time of the phosphor, length of time of blip exposure, blip size, viewing distance, ambient illumination, dark adaptation, display size, and operator fatigue. Empirical data derived from experimental testing of many of these factors are available.[28,29]

There has been much interest in applying solid-state technology as a radar display to replace the vacuum-tube CRT. Although the light-emitting diode has been too costly and complex to be a general replacement for the CRT, it might be applied when required to display limited information without the display occupying excessive space, such as in the Distance from Touchdown Indicator (DFTI) used in aircraft control towers to monitor landing aircraft.[27] Liquid crystal displays which can operate in high ambient lighting conditions have also been of interest for some special radar requirements, but they also are limited compared to the capability of a CRT. The plasma panel has also been considered as a bright radar display capable of incorporating alphanumeric labels.[59]

Types of display presentations. The various types of CRT displays which might be used for surveillance and tracking radars are defined as follows:

A-scope. A deflection-modulated display in which the vertical deflection is proportional to target echo strength and the horizontal coordinate is proportional to range.

B-scope. An intensity-modulated rectangular display with azimuth angle indicated by the horizontal coordinate and range by the vertical coordinate.

C-scope. An intensity-modulated rectangular display with azimuth angle indicated by the horizontal coordinate and elevation angle by the vertical coordinate.

D-scope. A C-scope in which the blips extend vertically to give a rough estimate of distance.

E-scope. An intensity-modulated rectangular display with distance indicated by the horizontal coordinate and elevation angle by the vertical coordinate. Similar to the RHI in which target height or altitude is the vertical coordinate.

F-Scope. A rectangular display in which a target appears as a centralized blip when the radar antenna is aimed at it. Horizontal and vertical aiming errors are respectively indicated by the horizontal and vertical displacement of the blip.

G-Scope. A rectangular display in which a target appears as a laterally centralized blip when the radar antenna is aimed at it in azimuth, and wings appear to grow on the pip as the distance to the target is diminished; horizontal and vertical aiming errors are respectively indicated by horizontal and vertical displacement of the blip.

H-scope. A B-scope modified to include indication of angle of elevation. The target appears as two closely spaced blips which approximate a short bright line, the slope of which is in proportion to the sine of the angle of target elevation.

I-scope. A display in which a target appears as a complete circle when the radar antenna is pointed at it and in which the radius of the circle is proportional to target distance; incorrect aiming of the antenna changes the circle to a segment whose arc length is inversely proportional to the magnitude of the pointing error, and the position of the segment indicates the reciprocal of the pointing direction of the antenna.

J-scope. A modified A-scope in which the time base is a circle and targets appear as radial deflections from the time base.

K-scope. A modified A-scope in which a target appears as a pair of vertical deflections. When the radar antenna is correctly pointed at the target, the two deflections are of equal height, and when not so pointed, the difference in deflection amplitude is an indication of the direction and magnitude of the pointing error.

L-scope. A display in which a target appears as two horizontal blips, one extending to the right from a central vertical time base and the other to the left; both blips are of equal amplitude when the radar is pointed directly at the target, any inequality representing relative pointing error, and distance upward along the baseline representing target distance.

M-scope. A type of A-scope in which the target distance is determined by moving an adjustable pedestal signal along the baseline until it coincides with the horizontal position of the target signal deflections; the control which moves the pedestal is calibrated in distance.

N-scope. A K-scope having an adjustable pedestal signal, as in the M-scope, for the measurement of distance.

O-scope. An A-scope modified by the inclusion of an adjustable notch for measuring distance.

PPI, or *Plan Position Indicator* (also called *P-scope*). An intensity-modulated circular display on which echo signals produced from reflecting objects are shown in plan position with range and azimuth angle displayed in polar (rho-theta) coordinates, forming a map-like display. An *offset,* or *offcenter,* PPI has the zero position of the time base at a position other than at the center of the display to provide the equivalent of a larger display for a selected portion of the service area. A *delayed PPI* is one in which the initiation of the time base is delayed.

R-scope. An A-scope with a segment of the time base expanded near the blip for greater accuracy in distance measurement.

RHI, or *Range-Height Indicator.* An intensity modulated display with height (altitude) as the vertical axis and range as the horizontal axis.

The above definitions are taken from the IEEE Standard Definitions,[23] with some modifications. The terms A-scope and A-display, B-scope and B-display, etc., are used interchangeably. These letter descriptions of radar displays date back to World War II. They are not all in current usage; however, the PPI, A-scope, B-scope, and RHI are among the more usual displays employed in radar. There are also other types of modern radar displays not included in the above listing which have not been given special letter designations. These include the various synthetic and raw video displays with alphanumeric characters and symbols for conveying additional information directly on the display.

CRT screens. A number of different cathode-ray-tube screens are used in radar applications. They differ primarily in their decay times and persistence. The properties of some of the phosphors which have been used in radar CRTs are listed in Table 9.1. The degree of image persistence required in a cathode-ray-tube screen depends upon the application. A long-persistence screen such as the P19 is appropriate for PPI presentations where the frame times are on the order of several seconds. On the other hand, where no persistence is needed, as when the frame time is less than the response time of the eye (0.1 s or less), a P1 phosphor might be used. The P1 phosphor is commonly found in most A-scope presentations.

In order to achieve long decay times a two-layer, or cascade, screen is sometimes used, as in the P7 or the P14. The first phosphor layer emits an intense light of short duration when excited by the electron beam. The light from the initial flash excites the second layer, emitting a persistent luminescence. The tandem action of the cascade screen results in greater efficiency than if the second, long-persistence layer had been excited by the electron beam directly, instead of by light emitted from the first layer. In the P14 screen, the initial flash of short duration from the first screen is blue in color and the persistent light radiated by the second screen is orange. An orange filter placed across the face of the tube eliminates the flash and leaves only the long-persistent trace. If a blue filter were used instead, the persistent afterglow of the second layer would be removed, leaving only the short-duration flash. Thus a single cascade screen can give either a short or a long persistence display, depending upon the filter employed.

In Table 9.1 the color emitted by the screen during both fluorescence (short duration) and phosphorescence (long duration) is given. In general terms, fluorescence may be thought of as luminescence which ceases almost immediately upon removal of the excitation, whereas in phosphorescence the luminescence persists. Cascaded screens produce different colors under fluorescence or phosphorescence.

Not all long-duration screens need be cascade. The P19 is a single-component phosphor with long persistence and no flash. If extremely long persistence is desired in a cathode-ray-tube display, a storage tube may be used. For all practical purposes, an image placed on a storage tube will remain indefinitely until erased.

Table 9.1 Radar CRT phosphor characteristics (*after Berg*[26])

Phosphor	Fluorescent color	Phosphorescent color	Persistence*
P1	Yellowish green	Yellowish green	Medium
P7	Blue	Yellowish green	Blue, medium short; yellow, long
P12	Orange	Orange	Long
P13	Reddish orange	Reddish orange	Medium
P14	Purplish blue	Yellowish orange	Blue, medium short; yellowish orange, medium
P17	Blue	Yellow	Blue, short; yellow, long
P19	Orange	Orange	Long
P21	Reddish orange	Reddish orange	Medium
P25	Orange	Orange	Medium
P26	Orange	Orange	Very long
P28	Yellowish green	Yellowish green	Long
P32	Purplish blue	Yellowish green	Long
P33	Orange	Orange	Very long
P34	Bluish green	Yellowish green	Very long
P38	Orange	Orange	Very long
P39	Yellowish green	Yellowish green	Long

* Persistence to 10 percent level: short = 1 to 10 μs; medium short = 10 μs to 1 ms; medium = 1 to 100 ms; long = 100 ms to 1 s; very long = > 1 s.

Resolution on the CRT is limited by the phosphor characteristics as well as the electron beam. A double-layer phosphor will have poorer resolution that a single-layer phosphor.

Color CRTs. The availability of color cathode-ray tubes provides another "dimension" for the display of target information. Target altitude, cross section, or identity might be color coded. The outputs from more than one radar or the outputs from each beam of a stacked-beam radar may be displayed on the same CRT in different color, as can alphanumeric data or symbols. Color can be an "attention-getter," and can improve the accuracy of recognition. In an MTI radar with a synthetic-video display that presents only moving targets, the clutter that is rejected (the raw video) can be superimposed on the display by showing it in a different color than the synthetic-video targets. (The display of weather clutter is of importance to the air traffic controller to avoid directing aircraft to areas of severe weather. The display of land clutter is also of importance in some applications so as to provide reference landmarks.) In principle, availability of color allows an intensity-modulated CRT, which is normally of limited dynamic range, to display target amplitude information by color coding the target blip according to the magnitude of the target cross section. Studies show that three or perhaps four colors is the optimum number for color display application.[57,58] A particular four-color display tested for air-traffic-control radar utilized yellow to present the aircraft data blocks, green for the area-sector map lines, red for navigation aids, and orange for showing precipitation.[61]

The tricolor shadow-mask cathode-ray tube commonly used for color television has a phosphor screen that consists of a series of three color dots. Although it may be well suited for commercial TV, it is not always a good display for radar since its resolution is limited by the size and spacing of the dots. Another color CRT of interest to radar is the *penetration color tube* using a multi-layer screen. The color is controlled by the anode voltage which determines the depth of penetration of the electron beam into a series of phosphor layers. Varying the anode voltage varies the depth of penetration and the particular phosphor layer that is excited. Typically, the difference between penetration of the layers is about 4 kV per layer.[30] A single-electron-gun, bicolor penetration-CRT, for example, might employ a layer of red phosphor and a layer of green separated by a layer of transparent dielectric material.[26,27] At low anode voltage (6 kV) the red phosphor is excited. At high voltage (12 kV), the green phosphor is excited. With intermediate voltages, yellow and orange can be obtained, giving a total of four different colors which can be distinguished. This method of color excitation permits better resolution to be obtained than with conventional TV color tubes. The range of colors available is not as great as with a color TV tube, but in many information-display applications it is sufficient.

A similar technique using anode-voltage control with two phosphors having the same color but different decay times can provide a CRT display with variable persistence. It is also possible to achieve a multicolor tube with variable persistence.

Although color displays may be pleasing and attractive to the operator, the applications in which they have proven superior to properly coded monochromatic displays apparently are fewer in number than might have been anticipated.[58]

Bright displays. There are applications where it is not possible or convenient to use the conventional CRT display that requires a darkened environment; such as in the cockpit of an aircraft or an airfield control tower. One form of bright display is the *direct-view storage tube.*[31] It not only operates under ambient light conditions, but it can provide a variable persistence. In this device, neither the brightness nor the persistence is directly affected by the short duration of the video signal, or by the characteristic of the viewing-screen phosphor.

The storage tube uses two electron beams generated by separate electron guns. One is a writing beam. The other is a flood beam that permits a bright, persistent display of the information carried by the writing beam. The writing beam does not impinge directly on the viewing screen as in a conventional CRT. It strikes a storage mesh, mounted just behind the screen, made up of a thin film of insulator material deposited on a metallic backing electrode. The insulator, which is the storage medium, is illuminated continuously by a "flood" beam. When illuminated by the writing beam, charge is stored on the insulator surface because of its secondary emission characteristics. The information written on the insulator storage surface forms a charge pattern that is made visible by the action of the flood beam. At those points where the action of the writing beam reduces the potential of the storage medium sufficiently, the electrons from the flood beam are allowed to pass through and reach the viewing screen. The display is continuous since the phosphor is continuously excited in the written areas by the flood beam.

A resolution of about 80 lines per inch with a screen brightness of 200 foot-lamberts can be obtained. The persistence can be varied from a fraction of a second to several minutes, or longer.[31] From five to seven gray-scale ranges are typical.[26] The direct-view storage tube can also be obtained with two colors and the shades in between.[30]

The direct-view storage tube generally is capable of less resolution than a conventional CRT. Also, the continuous buildup on the storage medium of strong, fixed clutter echoes when the persistence is long can result in a loss of picture quality due to a loss of resolution and a loss of contrast caused by the brightening of the dark area immediately surrounding the blip.

Another method for achieving a bright display is the *scan converter*. The output of a normal PPI display is stored in either an analog or digital storage medium, and the stored information is read out and displayed on a conventional TV monitor.[26,32,33] The conversion of the radial rho-theta raster of the PPI to the rectangular raster of the TV represents a "scan conversion," hence the name. The stored display may be read out repetitively to produce a bright flicker-free display. The display is bright not only because the stored information is displayed at a sufficiently high repetition rate to appear continuous, but also because TV monitors with typical brightness of 50 to 100 foot-lamberts are inherently brighter than the conventional CRT display.

One type of storage device that has been used in scan converters is a double-ended storage tube with two electron guns, one on each side of a charge-storage surface.[32] The radar output is written in PPI format by one electron gun on one side of the charge-storage surface. The second gun, on the other side of the storage surface, reads the stored charge pattern with a TV raster. The video outputs of several different radars, each with their own scan converter, can be combined on a single TV display. Use of the scan converter permits the convenient display of data from radars with different antenna rotation rates without the confusion produced by generating different rotation rates on the PPI. The complete displays of each radar or selected sectors from each radar can be combined as desired so that the TV display presents the available information in the best possible manner without including the sectors of those radars with poor or no data. Radar video clutter-maps or computer-generated information can also be combined on the TV display. A disadvantage of this form of scan converter is its low resolution.

A digital memory can be used as the storage device in a scan converter to overcome the limitations of the analog storage tube.

Rear Port. This is a flat plate-glass window in the cone of a cathode-ray tube aligned to be parallel to the tube faceplate.[34,57] It allows slide or film information to be projected onto the back of the tube phosphor, or it can be used to photograph the display without interfering with the operator. The CRT phosphor doubles as an optical and an electronic screen to give the operator a mixed presentation viewed in the usual manner from the front.

Synthetic-video Displays.[26,35,36] The use of a digital computer, as in an automatic detection and tracking processor, to extract target information results in a synthetic display in which target information is presented with standard symbols and accompanying alphanumerics. This is especially useful in an air-traffic-control display in which such information as target identity and altitude is desired to be displayed. The positions of targets on a predetermined number of previous scans might be shown on a synthetic display, or a line might be generated to indicate the direction of the target's trajectory and the target's speed. (The length of the line can be made proportional to target speed.) The use of a computer to generate the graphics and control the CRT display offers flexibility in the choice of such things as range scales, offcenter display, blow-up of selected areas, physical map outlines, grid displays, airport runways, stored clutter map, raw video, outlines of areas of weather blanked by the operator, display of stored flight plans, and time-compressed display of several successive radar scans. Also it can aid in the requesting of height-finder data by the operator, blanking of areas containing excessive interference or weather, providing training of operators by superimposing simulated targets, and other functions necessary for an air-traffic-control system or air-defense system. The operator can communicate in an interactive manner with the computer by such means as a keyboard, light pen, track ball, or even voice entry. Sometimes an auxiliary CRT display is mounted adjacent to the main radar display to provide tabular data and other information that would otherwise clutter the main display.

The characters and symbols that constitute the synthetic information may be inserted on a CRT during the radar dead time at the end of the sweep prior to the triggering of the next pulse. In some situations there might be too much information to be fully inserted during the available dead time. In a long-range air-traffic-control radar located in a busy area there might be more than a hundred targets for display extracted. For each target the necessary information to be presented might consist of perhaps ten alphanumeric characters and a number of line segments.[35] When the amount of data is large, writing time might have to be taken from the radar time by blanking one out of every n sweeps or writing the synthetic information during a random interruption of the sweep. It is also possible to utilize a second electron gun in the CRT to minimize the amount of time taken from the radar to insert the synthetic data. Another method for achieving the necessary time to display alphanumeric data is to store the video in a shift register and then read it out at a higher rate. The CCD, or charge coupled device, is one possibility as a shift register.

The data on the synthetic display must be refreshed at a sufficiently high rate to obtain a high brightness and to avoid flicker. When a number of displays are used with the output of a single radar, a dedicated minicomputer can be used at each display position for the refresh of data so as to reduce the data-transfer rate from the radar central processor.

9.6 DUPLEXERS AND RECEIVER PROTECTORS

The duplexer is the device that allows a single antenna to serve both the transmitter and the receiver. On transmission it must protect the receiver from burnout or damage, and on reception it must channel the echo signal to the receiver. Duplexers, especially for high-power applications, sometimes employ a form of gas-discharge device. Solid-state devices are also utilized. In a typical duplexer application the transmitter peak power might be a megawatt or more and the maximum safe power that can be tolerated at the receiver might be less than a watt. Therefore, the duplexer must provide, in this example, more than 60 dB of isolation between the transmitter and receiver with only negligible loss of the desired signal. In addition, during the interpulse period or when the radar is shut down, the receiver must be protected from high-power radiation, as from nearby radars, that might enter the radar antenna with less power than that needed to activate the duplexer, but with greater power than can be safely handled by the receiver.

There have been two basic methods employed that allow the use of a common antenna for both transmitting and receiving. The older method is represented by the branch-type duplexer and the balanced duplexer which utilize gas TR-tubes for accomplishing the necessary switching actions. The other method uses a ferrite circulator to separate the transmitter and receiver, and a receiver protector consisting of a gas TR-tube and diode limiter.

Branch-type duplexers. The branch-type duplexer, diagrammed in Fig. 9.5 was one of the earliest duplexer configurations employed. It consists of a TR (transmit-receive) switch and an ATR (anti-transmit receive) switch, both of which are gas-discharge tubes. When the transmitter is turned on, the TR and the ATR tubes ionize; that is, they break down, or fire. The TR in the fired condition acts as a short circuit to prevent transmitter power from entering the receiver. Since the TR is located a quarter wavelength from the main transmission line, it appears as a short circuit at the receiver but as an open circuit at the transmission line so that it does not impede the flow of transmitter power. Since the ATR is displaced a quarter wavelength from the main transmission line, the short circuit it produces during the fired condition appears as an open circuit on the transmission line and thus has no effect on transmission.

During reception, the transmitter is off and neither the TR nor the ATR is fired. The open circuit of the ATR, being a quarter wave from the transmission line, appears as a short circuit across the line. Since this short circuit is located a quarter wave from the receiver branch-line, the transmitter is effectively disconnected from the line and the echo signal power is directed to the receiver. The diagram of Fig. 9.5 is a parallel configuration. Series or series-parallel configurations are possible.[37]

The branch-type duplexer is of limited bandwidth and power-handling capability, and has generally been replaced by the balanced duplexer and other protection devices. It is used, in spite of these limitations, in some low-cost radars.

Balanced duplexers. The balanced duplexer, Fig. 9.6, is based on the short-slot hybrid junction which consists of two sections of waveguides joined along one of their narrow walls with a slot cut in the common narrow wall to provide coupling between the two.[38] The short-slot hybrid

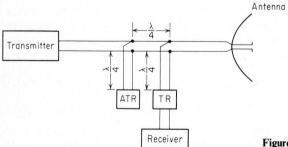

Figure 9.5 Principle of branch-type duplexer.

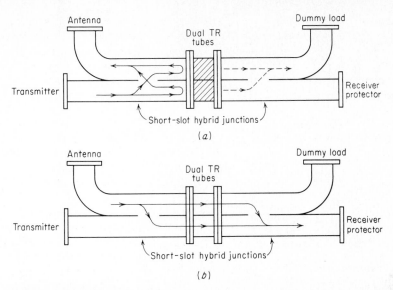

Figure 9.6 Balanced duplexer using dual TR tubes and two short-slot hybrid junctions. (a) Transmit condition; (b) receive condition.

may be considered as a broadband directional coupler with a coupling ratio of 3 dB. In the transmit condition (Fig. 9.6a) power is divided equally into each waveguide by the first short-slot hybrid junction. Both TR tubes break down and reflect the incident power out the antenna arm as shown. The short-slot hybrid has the property that each time the energy passes through the slot in either direction, its phase is advanced 90°. Therefore, the energy must travel as indicated by the solid lines. Any energy which leaks through the TR tubes (shown by the dashed lines) is directed to the arm with the matched dummy load and not to the receiver. In addition to the attenuation provided by the TR tubes, the hybrid junctions provide an additional 20 to 30 dB of isolation.

On reception the TR tubes are unfired and the echo signals pass through the duplexer and into the receiver as shown in Fig. 9.6b. The power splits equally at the first junction and because of the 90° phase advance on passing through the slot, the energy recombines in the receiving arm and not in the dummy-load arm.

The power-handling capability of the balanced duplexer is inherently greater than that of the branch-type duplexer and it has wide bandwidth, over ten percent with proper design. A *receiver protector*, to be described later, is usually inserted between the duplexer and the receiver for added protection.

Another form of balanced duplexer[39] uses four ATR tubes and two hybrid junctions (Fig. 9.7). During transmission (Fig. 9.7a) the ATR tubes located in a mount between the two short-slot hybrids ionize and allow high power to pass to the antenna. Dashed lines show the flow of power. During reception (Fig. 9.7b) the ATR tubes present a high impedance which results in the echo-signal power being reflected to the receiver. The ATR type of balanced duplexer has higher power-handling capability than that of Fig. 9.6, but it has less bandwidth.

TR tubes. The TR tube is a gas-discharge device designed to break down and ionize quickly at the onset of high RF power, and to deionize quickly once the power is removed. One common construction of a TR consists of a section of waveguide containing one or more resonant filters

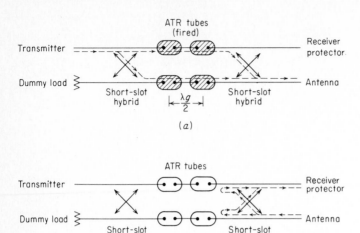

Figure 9.7 Balanced duplexer using ATR tubes. (a) Transmit condition; (b) receive condition.

and two glass-to-metal windows to seal in the gas at low pressure. A noble gas like argon in the TR tube has a low breakdown voltage, and offers good receiver protection and relatively long life. Pure-argon-filled tubes, however, have relatively long deionization times and are not suitable for short-range applications. The deionization process can be speeded up by the addition of water vapor or a halogen. The life of TR tubes filled with a mixture of a noble gas (argon) and a gas with high electron affinity (water vapor) is less than the life of tubes filled with a noble gas only.

To ensure reliable and rapid breakdown of the TR tube upon application of the RF pulse, an auxiliary source of electrons is supplied to the tube. This may be accomplished with a "keep-alive," which is a weak d-c discharge that generates electrons which diffuse into the TR where they act to trigger the breakdown once RF power is applied. The keep-alive generates noise just as any other gas-discharge device. The excess noise temperature is generally less than 50 K.[40]

Receiver protectors. Since the keep-alive in the TR is not usually energized when the radar is turned off, considerably more power is needed to break down the TR than when it is energized. Radiations from nearby transmitters may therefore damage the receiver without firing the TR. To protect the receiver under these conditions, a mechanical shutter can be used to short-circuit the input to the receiver whenever the radar is not operating. The shutter might be designed to attenuate a signal by 25 to 50 dB.

The TR is not a perfect switch; some transmitter power always leaks through to the receiver. The envelope of the RF leakage might be similar to that shown in Fig. 9.8. The short-duration, large-amplitude "spike" at the leading edge of the leakage pulse is the result of the finite time required for the TR to ionize or break down. Typically, this time is of the order of 10 nanoseconds. After the gas in the TR tube is ionized, the power leaking through the tube is considerably reduced from the peak value of the spike. This portion of the leakage pulse is termed the *flat*. Damage to the receiver front-end may result when either the *energy* contained within the spike or the power in the flat portion of the pulse is too large. "Typical" TR tubes might have a spike leakage of one erg or less and might provide an attenuation of the incident transmitter power of the order of 70 to 90 dB.

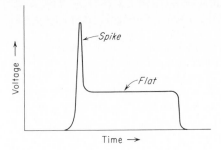

Figure 9.8 Leakage pulse through a TR tube.

A fraction of the transmitter power incident on the TR tube is absorbed by the discharge. This is called *arc loss*. Its magnitude depends upon the characteristics of the input window and the gas used. The arc loss might be $\frac{1}{2}$ to 1 dB in tubes filled with water vapor and 0.1 dB or less with argon filling. On reception, the TR tube introduces an insertion loss of about $\frac{1}{2}$ to 1 dB.

The life of a conventional TR is limited by the keep-alive, the amount of water vapor in the gas filling, and by disappearance, or "clean-up," of the gas due to the gas molecules becoming imbedded in the walls of the TR tubes.[41] The end of life of a TR tube is determined more by the amount of leakage power which it allows to pass than by physical destruction or wear.

The keep-alive can be replaced as a supplier of priming electrons in the TR by a radio-active source, such as tritium, which produces low-energy-level beta rays.[52] The tritium is in compound form as a tritide film. The replacement of the keep-alive by radioactive tritium eliminates the excess noise of the keep-alive and increases the life of the TR by an order of magnitude. Since a tritium TR needs no active voltages, a mechanical shutter is not required to protect the receiver from nearby transmissions when the radar is turned off. This also improves the reliability. The tritium-activated TR is usually followed by a diode limiter. The combination of the two is called a *passive TR-limiter* and is widely used as a *receiver protector*.

Solid-state limiters. Solid-state PN and PIN diodes can be made to act as RF limiters and are thus of interest as receiver protectors.[42-44] Ideally, a limiter passes low power without attenuation, but above some threshold it provides attenuation of the signal so as to maintain the output power constant. This property can be used for the protection of radar receivers in two different implementations depending on whether the diodes are operated unbiased (self-actuated) or with a d-c forward-bias current. Unbiased operation is also known as *passive*. It is used for low-power applications. Biased operation is also known as *active* and is capable of switching high power.[43,44,62] Biased PIN diodes can replace the gas-tube TRs of the balanced-duplexer configuration of Fig. 9.6 or they may be used as receiver protectors. For example,[44] a VHF balanced duplexer configuration using 32 PIN diodes mounted in $3\frac{1}{8}$ inch coaxial line (16 in each of two SPST shunt-type limiters) handled 150 kW peak power, 10 kW average power, with a pulse width of 200 μs. At the frequency of operation (200 to 225 MHz), the use of self-actuated diodes was practical. Recovery time was required to be less than 10 μs. The loss on transmit was 0.15 dB and on receive it was 2 dB. (The receive loss could be reduced to 1 dB with application of reverse bias to the diodes.) The device provided from 40 to 50 dB of protection. Another example[45] is an *L*-band receiver protector using 10 PIN diodes mounted in a coaxial line which was able to handle 1.6 MW peak power and 100 kW average power with a 50 μs pulse width over a 10 percent bandwidth. The loss was 0.6 dB, isolation 70 dB, turn-on time 2 μs and turn-off time 25 μs. Although the active diode-limiter offers many advantages for duplexer application, it does not protect the receiver when the bias is off. Therefore, it offers poor protection against nearby transmissions during the interpulse period or when the radar is shut down.

The passive diode-limiter without any bias current is operable at all times, but it is capable of handling much less power than the active limiter with bias. Thus its use as an all solid-state limiter is restricted to low power systems, perhaps handling less than 1.5 kW peak power.[45] When it can be used, the diode limiter has the advantage of fast recovery time (about 1 μs for 1 kW peak power, and 50 ns for 100 W);[43] no spike leakage; low flat leakage; reasonable loss, size and weight; and it can have a long life.[40]

Passive TR-limiter. The passive diode-limiter also finds important application when used with a radioactive-primed gas tube TR. The combination is known as a *passive TR-limiter.* The passive TR-limiter protects the receiver from nearby transmissions even when the radar is shut off since it requires no active voltages for either the radioactive-primed TR tube or the diode limiter. The presence of the diode limiter following the TR tube considerably reduces the spike leakage, increasing the life of the device itself as well as the receiver front-end it is supposed to protect. The passive TR-limiter, however, has a higher insertion loss than the TR tube but it generates no excess noise because it does not employ a keep-alive discharge. Its recovery time is superior to the conventional TR tube, but inferior to the diode limiter acting alone or the ferrite diode limiter.[40]

The passive TR-limiter has low leakage over a wide range of input power. As an example,[40] below 1 mW peak power the output power is linearly related to the input power. From 10 mW to 100 kW, the output flat-leakage increases by as little as 3 to 10 dB. Below 1-W input power, the diode alone provides the limiting action. Above 1-W, the TR tube ionizes to protect the diode limiter from burnout. More than one diode may be used to provide greater attenuation or larger bandwidth.

Before the introduction of the solid-state diode limiter, a low-power gas TR-tube, known as a protector TR, was often used as added protection for the receiver. It was placed between the receiver port of the duplexer and the receiver to safeguard against random pulses from nearby radar equipments which were too weak to fire the conventional TR tubes in the duplexer, but strong enough to damage the receiver. The passive TR-limiter or the diode limiter has replaced the protector TR. Such devices are known as *receiver protectors.* Receiver protectors also serve to protect against power reflected by mismatches at the antenna.

Ferrite limiter. The ferrite limiter[46] followed by a diode limiter, has also been employed as a solid-state receiver protector.[40,45] The diode is necessary to reduce the spike leakage to a safe level since the ferrite acting alone does not offer sufficient suppression of the spike leakage at high peak power. The ferrite diode limiter has fast recovery time (can be as low as several tens of nanoseconds), and if the power rating is not exceeded, the life should be essentially unlimited. The spike and flat leakage is low, but the insertion loss is usually higher and the package is generally longer and heavier than other receiver protectors. Except for the initial spike, the ferrite limiter is an absorptive device as compared to the gas-tube TR or the TR-limiter which are reflective. Therefore, the incident power is absorbed in the ferrite so that the average power capability can be a problem. Air or water cooling might be needed. The ferrite material may be biased, as for example a latching-ferrite circulator, to permit the handling of higher power.[45] Since the latching-ferrite receiver protector is undamaged by high power when in either state, protection from nearby transmitters can be had by inserting a passive TR-limiter following the ferrite.

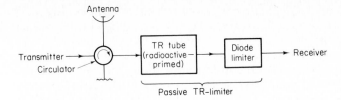

Figure 9.9 Circulator and receiver protector. A four-port circulator is shown with the fourth port terminated in a matched load to provide greater isolation between the transmitter and the receiver than provided by a three-port circulator.

Circulator and receiver protector. The ferrite circulator[47] is a three- or four-port device that can, in principle, offer separation of the transmitter and receiver without the need for the conventional duplexer configurations of Figs. 9.5 to 9.7. The circulator does not provide sufficient protection by itself and requires a receiver protector as in Fig. 9.9. The isolation between the transmitter and receiver ports of a circulator is seldom sufficient to protect the receiver from damage. However, it is not the isolation between transmitter and receiver ports that usually determines the amount of transmitter power at the receiver, but the impedance mismatch at the antenna which reflects transmitter power back into the receiver. The VSWR is a measure of the amount of power reflected by the antenna. For example, a VSWR of 1.5 means that about 4 percent of the transmitter power will be reflected by the antenna mismatch in the direction of the receiver, which corresponds to an isolation of only 14 dB. About 11 percent of the power is reflected when the VSWR is 2.0, corresponding to less than 10 dB of isolation. Thus, a receiver protector is almost always required. It also reduces to a safe level radiations from nearby transmitters. The receiver protector might use solid-state diodes for an all solid-state configuration,[48] or it might be a passive TR-limiter consisting of a radioactive primed TR-tube followed by a diode limiter.[49,50] The ferrite circulator with receiver protector is attractive for radar applications because of its long life, wide bandwidth, and compact design.

Other duplexer considerations. The duplexers described above using passive receiver protectors have recovery times from a fraction of a microsecond to several tens of microseconds. By employing the principle of multipacting, a recovery time as short as 5 ns is possible.[44,51] A multipactor is a vacuum tube which contains surfaces capable of a high yield of secondary emission upon impact by an electron. The secondary emission surfaces are biased with a d-c potential. The presence of RF energy causes electrons to make multiple impacts to generate by secondary emission a large electron cloud. The electron cloud moves in phase with the oscillation of the applied RF electric field to absorb energy from the RF field. The RF power is dissipated thermally at the secondary emission surfaces. The recovery time of the multipactor is extremely fast. It has low spike-leakage, and can handle high peak and average powers. The flat-leakage power passed by the multipactor is high enough to require a passive diode limiter following the multipactor. The multipactor has the disadvantage of being complex, and it offers no protection if the power is turned off.

With fast-rise-time, high-power RF sources, the receiver protector may be required to self-limit in less than one nanosecond. This can be achieved with fast-acting PN (varactor) diodes. A number of diode stages, preceded by plasma limiters, might be employed in such a receiver protector. In one design, an X-band passive receiver protector was capable of limiting a 1 ns risetime, multi-kilowatt RF pulses to 1-W spike levels.[53]

In many radar systems the STC (sensitivity time control), which is a time-dependent attenuator, is placed in the RF portion of the radar receiver rather than in the IF. When a diode limiter is used in the receiver protector, the STC function can be readily incorporated by inserting a variable bias into the limiter diodes to provide an attenuator that varies with time.[40,49] The PIN diodes serve the dual function of self-limiting during transmit, and act as a variable attenuator, or STC, during receive. One of the advantages of an integral STC using the diode limiter rather than a separate RF STC, is that no additional insertion loss is suffered for the STC other than that inherent in the TR-limiter itself.

In a solid-state limiter, it is also possible to couple a noise diode into the output to provide a convenient source for checking receiver noise-figure.[40]

Several hundred hours of life is considered typical for conventional TR tubes that use an active keep-alive discharge. However, a duplexer using a passive radioactive-primed TR and diode limiter is capable of perhaps five thousand hours of life,[50] with predictions of up to four years of continuous operation.[49]

REFERENCES

1. Friis, H. T.: Noise Figures of Radio Receivers, *Proc. IRE*, vol. 32, pp. 419–422, July, 1944.
2. Mumford, W. W., and E. Scheibe: "Noise: Performance Factors in Communication Systems," Horizon House–Microwave, Inc., Dedham, Mass. 1968.
3. Goldberg, H.: Some Notes on Noise Figures, *Proc. IRE*, vol. 36, pp. 1205–1214, October, 1948.
4. Van Voorhis, S. N. (ed.): "Microwave Receivers," MIT Radiation Laboratory Series, vol. 23, McGraw-Hill Book Company, New York, 1948.
5. Torrey, H. C., and C. A. Whitmer: "Crystal Rectifiers," MIT Radiation Laboratory Series, vol. 15, McGraw-Hill Book Company, New York, 1948.
6. Palamutcuoglu, O., J. G. Gardiner, and D. P. Howson: Image-Cancelling Mixers at 2 GHz, *Electronic Letters*, vol. 10, no. 7, pp. 104–106, Apr. 4, 1974.
7. Taylor, J. W., Jr., and J. Mattern: Receivers, chap. 5 of "Radar Handbook," M. I. Skolnik (ed.), McGraw-Hill Book Company, New York, 1970.
8. Neuf, D., D. Brown, and B. Jaracz: Multi-Octave Double-Balanced Mixer, *Microwave J.*, vol. 16, pp. 13–14, January, 1973.
9. Shigemoto, J.: Balanced Mixer Noise Considerations, *Microwave J.*, vol. 10, pp. 77–79, October, 1967.
10. Okean, H. C., and A. J. Kelly: Low-Noise Receiver Design Trends Using State-of-the-Art Building Blocks, *IEEE Trans.*, vol. MTT-25, pp. 254–267, April, 1977.
11. Cochrane, J. B., and F. A. Marki: Thin-Film Mixers Team Up To Block Out Image Noise, *Microwaves*, vol. 11, pp. 34–40, 84, March, 1977.
12. Neuf, D.: A Quiet Mixer, *Microwave J.*, vol. 16, pp. 29–32, May, 1973.
13. Neuf, D.: Radar Retrofit: Increase Receiver Sensitivity Cost Effectively, *Microwave System News*, vol. 5, pp. 55–60, April/May, 1976.
14. Anand, Y., and C. Howell: A Burnout Criterion for Schottky-Barrier Mixer Diodes, *Proc. IEEE*, vol. 56, p. 2098, November, 1968.
15. Anand, Y., and W. J. Moroney: Microwave Mixer and Detector Diodes, *Proc. IEEE*, vol. 59, pp. 1182–1190, August, 1971.
16. Torrey, H. C., and C. A. Whitmer: "Crystal Rectifiers," MIT Radiation Laboratory Series, vol. 15, McGraw-Hill Book Company, New York, 1948.
17. Dickens, L. E., and D. W. Maki: An Integrated-Circuit Balanced Mixer, Image and Sum Enhanced, *IEEE Trans.*, vol. MTT-23, pp. 276–281, March, 1975.
18. Chandler, H. M.: Microwave Low Noise Amplifiers For Use in Radar Systems, *The Marconi Review*, vol. 35, no. 185, pp. 94–120, 2d qtr., 1972.
19. Lindauer, J. T., and N. K. Osbrink: GaAs FET Amplifiers are Closing Fast on the Low-Noise Narrowband Leaders, *Microwave System News*, vol. 5, pp. 63–67, April/May, 1976.
20. Emery, F. E.: Low Noise GaAs FET Amplifiers, *Microwave J.*, vol. 20, pp. 79–82, June, 1977.
21. Miller, R. E., T. G. Phillips, D. E. Iglesias, and R. H. Kneer: Noise Performance of Microwave GaAs F.E.T. Amplifiers at Low Temperatures, *Electronics Letters*, vol. 13, no. 1, pp. 10–11, Jan. 6, 1977.
22. Mumford, W. W.: A Broadband Microwave Noise Source, *Bell System Tech. J.*, vol. 28, pp. 608–618, October, 1949.
23. IEEE Standard 686–1977: "IEEE Standard Radar Definitions," Nov. 9, 1977.

24. Watson, H. A., "Microwave Semiconductor Devices and Their Circuit Applications," McGraw-Hill Book Company, New York, 1969.

25. Howes, M. J., and D. V. Morgan (eds.): "Microwave Devices," John Wiley & Sons, New York, 1976.

26. Berg, A. A.: Radar Indicators and Displays, chap. 6 of "Radar Handbook," ed. by M. I. Skolnik, McGraw-Hill Book Company, New York, 1970.

27. Byatt, D. W. G., and J. Wild: Present and Future Radar Display Techniques, *International Conf. on Radar—Present and Future*, Oct. 23–25, 1973, pp. 201–206, IEE Conference Publication no. 105.

28. Lawson, J. L., and G. E. Uhlenbeck: "Threshold Signals," vol. 24, M.I.T. Radiation Laboratory Series, McGraw-Hill Book Company, New York, 1950.

29. Baker, C. H.: "Man and Radar Displays," Macmillan Company, New York, 1962.

30. Luxenberg, H. R., and R. L. Kuehn: "Display Systems Engineering," McGraw-Hill Book Company, New York, 1968.

31. Nairn, N. P.: Bright Radar Displays Employing Direct-View Storage Tubes, *Conference on Air Traffic Control Systems Engineering and Design*, London, Mar. 13–17, 1967, IEE Conference Publication no. 28, pp. 7–10.

32. Lyon, I. D.: Scan-Converted Bright Displays for Air Traffic Control Systems, *Conference on Air Traffic Control Systems Engineering and Design*, London, Mar. 13–17, 1967, IEE Conference Publication no. 28, pp. 41–44.

33. O'Neal, N. V., and P. M. Thiebaud: A Scan Converter Radar Display System, *Naval Research Laboratory Memorandum Report* 2979, Washington, D.C., January, 1975.

34. Robertson, J. W., and F. J. Savill: Rear Port Displays: Their Advantages, Limitations, and Design Problems for Radar Applications, *International Conference on Radar—Present and Future*, Oct. 23–25, 1973, London, IEE Conference Publication no. 105, pp. 207–212.

35. Odoardi, F., and C. Maggi: Modern Radar Data Display System: The Selenia IDM-7 Digital Display, *Rivista Tecnica Selenia* (Rome, Italy), vol. 2, no. 4, pp. 53–61, 1975.

36. Kernan, P.: Low-Cost Processing and Display System for Radar Tracking and Intercept, *IEEE EASCON '75 Convention Record*, pp. 108-A to 108G, 1975.

37. Reintjes, J. F., and G. T. Coate: "Principles of Radar," 3d ed., chap. 12, McGraw-Hill Book Company, New York, 1952.

38. Riblet, H. J.: The Short-Slot Hybrid Junction, *Proc, IRE*, vol. 40, pp. 180–184, February, 1952.

39. Jones, C. W.: Broad-Band Balanced Duplexers, *IRE Trans.*, vol. MTT-5, pp. 4–12, January, 1957.

40. Brown, N. J.: Modern Receiver Protection Capabilities with TR-Limiters, *Microwave J.*, vol. 17, pp. 61–64, February, 1974.

41. Maddix, H.: Clean up in TR Tubes, *IEEE Trans.*, vol. ED-15, pp. 98–104, February, 1968.

42. Leenov, D.: The Silicon PIN Diode as a Microwave Radar Protector at Megawatt Levels, *IEEE Trans.*, vol. ED-11, pp. 53–61, February, 1964.

43. Garver, R. V.: "Microwave Diode Control Devices," Artech House, Inc., Dedham, Mass., 1976.

44. White, J. F.: "Semiconductor Control," Artech House, Dedham, Mass., 1977.

45. Brown, N. J.: Control and Protection Devices, *IEEE NEREM 74 Record, pt 4: Radar Systems and Components*, Oct. 28–31, 1974, IEEE Catalog No. 74 CHO 9340 NEREM, Library of Congress Catalog no. 61–3748.

46. Kupke, W. F., T. S. Hartwick, and M. T. Weiss: Solid-State X-Band Power Limiter, *IRE Trans.*, vol. MTT-9, pp. 472–480, November, 1961.

47. Knerr, R. H.: An Annotated Bibliography of Microwave Circulators and Isolators: 1968–1975, *IEEE Trans.*, vol. MTT-23, pp. 818–825, October, 1975.

48. Gawronski, M. J., and H. Goldie: 200 W MIC L-Band Receiver Protector, *Microwave J.*, vol. 20, pp. 43–46, May, 1977.

49. Ratliff, P. C., W. Cherry, M. J. Gawronski, and H. Goldie: L-Band Receiver Protection, *Microwave J.*, vol. 19, pp. 57–60, January, 1976.

50. Goldie, H.: What's New With Receiver Protectors?, *Microwaves*, vol. 15, pp. 44–52, January, 1976.

51. Ferguson, P., and R. D. Dokken: For High-Power Protection . . . Try Multipacting, *Microwaves*, vol. 13, pp. 52–53, July, 1974.

52. Goldie, H.: Radioactive (Tritium) Ignitors for Plasma Limiters, *IEEE Trans.*, vol. ED-19, pp. 917–928, August, 1972.

53. Nelson, T. M., and H. Goldie: Fast Acting X-Band Receiver Protector Using Varactors, *IEEE MTT Symposium Digest*, pp. 176–177, 1974. Also Westinghouse Systems Development Division (Baltimore, Md.) report DSC-9731 by the same authors titled, Fast Acting Varactors for Sub-Nanosecond Power Limiting in Receiver Protectors (no date).

54. Hirsch, R. B.: Plain Talk on Log Amps, *Microwaves*, vol. 17, pp. 40–44, March, 1978. (The effect of IF gain on the receiver dynamic range was also indicated by Warren D. White, one of the reviewers of the revised manuscript of this book.)

55. Reynolds, J. F., and M. R. Rosenzweig: Learn the Language of Mixer Specification, *Microwaves*, vol. 17, pp. 72–80, May, 1978.
56. Hallford, B. R.: Low Conversion Loss X Band Mixer, *Microwave J.*, vol. 21, pp. 53–59, April, 1978.
57. Robertson, J. W., and F. J. Savill: Practical Experience with Rear Port Displays, *International Conference on Displays for Man-Machine Systems*, 4–7 April, 1977 IEE (London) Conference Publication no. 150, pp. 17–19.
58. Hopkin, V. D.: Colour Displays in Air Traffic Control, *International Conference on Displays for Man-Machine Systems*, Apr. 4–7, 1977, IEE (London) Conference Publication no. 150, pp. 46–49.
59. McLoughlan, S. D., M. B. Thomas, and G. Watkins: The Plasma Panel as a Potential Solution to the Bright Labelled Radar Display Problem, *International Conference on Displays for Man-Machine Systems*, Apr. 4–7, 1977, IEE (London) Conference Publication no. 150, pp. 14–16.
60. Jay, F. (ed-in-ch.): "IEEE Standard Dictionary of Electrical and Electronics Terms," 2d ed., Inst. of Electrical and Electronics Engineers, Inc., New York, 1977.
61. Anonymous: News item in *Aviation Week and Space Technology*, vol. 109, no. 13, p. 61, Sept. 25, 1978.
62. Tenenholtz, R.: Designing Megawatt Diode Duplexers, *Microwave J.*, vol. 21, pp. 30–34, Dec., 1978, and vol. 22, pp. 63–68, Jan., 1979.

DETECTION OF RADAR SIGNALS IN NOISE

10.1 INTRODUCTION

The two basic operations performed by radar are (1) *detection* of the presence of reflecting objects, and (2) *extraction* of information from the received waveform to obtain such target data as position, velocity, and perhaps size. The operations of detection and extraction may be performed separately and in either order, although a radar that is a good detection device is usually a good radar for extracting information, and vice versa. In this chapter some aspects of the problem of detecting radar signals in the presence of noise will be considered. Noise ultimately limits the capability of any radar. The problem of extracting information from the received waveform is described in the next chapter. The detection of signals in the presence of clutter is discussed in Chap. 13.

10.2 MATCHED-FILTER RECEIVER[1-8]

A network whose frequency-response function maximizes the output peak-signal–to–mean-noise (power) ratio is called a *matched filter*. This criterion, or its equivalent, is used for the design of almost all radar receivers.

The frequency-response function, denoted $H(f)$, expresses the relative amplitude and phase of the output of a network with respect to the input when the input is a pure sinusoid. The magnitude $|H(f)|$ of the frequency-response function is the receiver amplitude passband characteristic. If the bandwidth of the receiver passband is wide compared with that occupied by the signal energy, extraneous noise is introduced by the excess bandwidth which lowers the output signal-to-noise ratio. On the other hand, if the receiver bandwidth is narrower than the bandwidth occupied by the signal, the noise energy is reduced along with a considerable part of the signal energy. The net result is again a lowered signal-to-noise ratio. Thus there is an optimum bandwidth at which the signal-to-noise ratio is a maximum. This is well known to the radar receiver designer. The rule of thumb quoted in pulse radar practice is that the receiver bandwidth B should be approximately equal to the reciprocal of the pulse width τ. As we shall see later, this is a reasonable approximation for pulse radars with conventional superheterodyne receivers. It is not generally valid for other waveforms, however, and is mentioned to illustrate in a qualitative manner the effect of the receiver characteristic on signal-to-noise ratio. The exact specification of the optimum receiver characteristic involves the frequency-response function and the shape of the received waveform.

The receiver frequency-response function, for purposes of this discussion, is assumed to apply from the antenna terminals to the output of the IF amplifier. (The second detector and video portion of the well-designed radar superheterodyne receiver will have negligible effect on the output signal-to-noise ratio if the receiver is designed as a matched filter.) Narrowbanding is most conveniently accomplished in the IF. The bandwidths of the RF and mixer stages of the normal superheterodyne receiver are usually large compared with the IF bandwidth. Therefore the frequency-response function of the portion of the receiver included between the antenna terminals to the output of the IF amplifier is taken to be that of the IF amplifier alone. Thus we need only obtain the frequency-response function that maximizes the signal-to-noise ratio at the output of the IF. The IF amplifier may be considered as a filter with gain. The response of this filter as a function of frequency is the property of interest.

For a received waveform $s(t)$ with a given ratio of signal energy E to noise energy N_0 (or noise power per hertz of bandwidth), North[1] showed that the frequency-response function of the linear, time-invariant filter which maximizes the output peak-signal–to–mean-noise (power) ratio for a fixed input signal-to-noise (energy) ratio is

$$H(f) = G_a S^*(f) \exp\left(-j2\pi f t_1\right) \qquad (10.1)$$

where $S(f) = \displaystyle\int_{-\infty}^{\infty} s(t) \exp\left(-j2\pi ft\right) dt$ = voltage spectrum (Fourier transform) of input signal

$S^*(f)$ = complex conjugate of $S(f)$

t_1 = fixed value of time at which signal is observed to be maximum

G_a = constant equal to maximum filter gain (generally taken to be unity)

The noise that accompanies the signal is assumed to be stationary and to have a uniform spectrum (white noise). It need not be gaussian. If the noise is not white, Eq. (10.1) may be modified as discussed later in this section. The filter whose frequency-response function is given by Eq. (10.1) has been called the *North* filter, the *conjugate* filter, or more usually the *matched* filter. It has also been called the *Fourier transform criterion*. It should not be confused with the circuit-theory concept of impedance matching, which maximizes the power transfer rather than the signal-to-noise ratio.

The frequency-response function of the matched filter is the conjugate of the spectrum of the received waveform except for the phase shift $\exp\left(-j2\pi f t_1\right)$. This phase shift varies uniformly with frequency. Its effect is to cause a constant time delay. A time delay is necessary in the specification of the filter for reasons of physical realizability since there can be no output from the filter until the signal is applied. The frequency spectrum of the received signal may be written as an amplitude spectrum $|S(f)|$ and a phase spectrum $\exp\left[-j\phi_s(f)\right]$. The matched-filter frequency-response function may similarly be written in terms of its amplitude and phase spectra $|H(f)|$ and $\exp\left[-j\phi_m(f)\right]$. Ignoring the constant G_a, Eq. (10.1) for the matched filter may then be written as

$$|H(f)| \exp\left[-j\phi_m(f)\right] = |S(f)| \exp\left\{j[\phi_s(f) - 2\pi f t_1]\right\} \qquad (10.2)$$

or

$$|H(f)| = |S(f)| \qquad (10.3a)$$

and

$$\phi_m(f) = -\phi_s(f) + 2\pi f t_1 \qquad (10.3b)$$

Thus the amplitude spectrum of the matched filter is the same as the amplitude spectrum of the signal, but the phase spectrum of the matched filter is the negative of the phase spectrum of the signal plus a phase shift proportional to frequency.

The matched filter may also be specified by its impulse response $h(t)$, which is the inverse Fourier transform of the frequency-response function.

$$h(t) = \int_{-\infty}^{\infty} H(f) \exp{(j2\pi ft)} \, df \tag{10.4}$$

Physically, the impulse response is the output of the filter as a function of time when the input is an impulse (delta function). Substituting Eq. (10.1) into Eq. (10.4) gives

$$h(t) = G_a \int_{-\infty}^{\infty} S^*(f) \exp{[-j2\pi f(t_1 - t)]} \, df \tag{10.5}$$

Since $S^*(f) = S(-f)$, we have

$$h(t) = G_a \int_{-\infty}^{\infty} S(f) \exp{[j2\pi f(t_1 - t)]} \, df = G_a s(t_1 - t) \tag{10.6}$$

A rather interesting result is that the impulse response of the matched filter is the image of the received waveform; that is, it is the same as the received signal run backward in time starting from the fixed time t_1. Figure 10.1 shows a received waveform $s(t)$ and the impulse response $h(t)$ of its matched filter.

The impulse response of the filter, if it is to be realizable, is not defined for $t < 0$. (One cannot have any response before the impulse is applied.) Therefore we must always have $t < t_1$. This is equivalent to the condition placed on the transfer function $H(f)$ that there be a phase shift $\exp{(-j2\pi ft_1)}$. However, for the sake of convenience, the impulse response of the matched filter is sometimes written simply as $s(-t)$.

Derivation of the matched-filter characteristic. The frequency-response function of the matched filter has been derived by a number of authors using either the calculus of variations[1] or the Schwartz inequality.[9] In this section we shall derive the matched-filter frequency-response function using the Schwartz inequality.

$s(t)$

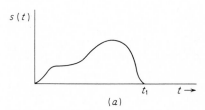

t_1 $t \rightarrow$

(a)

$h(t)$

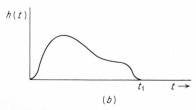

t_1 $t \rightarrow$

(b)

Figure 10.1 (a) Received waveform $s(t)$; (b) impulse response $h(t)$ of the matched filter.

We wish to show that the frequency-response function of the linear, time-invariant filter which maximizes the output peak-signal–to–mean-noise ratio is

$$H(f) = G_a S^*(f) \exp(-j2\pi f t_1)$$

when the input noise is stationary and white (uniform spectral density). The ratio we wish to maximize is

$$R_f = \frac{|s_o(t)|^2_{max}}{N} \tag{10.7}$$

where $|s_o(t)|_{max}$ = maximum value of output signal voltage and N = mean noise power at receiver output. The ratio R_f is not quite the same as the signal-to-noise ratio which has been considered previously in the radar equation. [Note that the peak power as used here is actually the peak *instantaneous* power, whereas the peak power referred to in the discussion of the radar equation in Chap. 2 was the average value of the power over the duration of a pulse of sine wave. The ratio R_f is *twice* the average signal-to-noise power ratio when the input signal $s(t)$ is a rectangular sine-wave pulse.] The output voltage of a filter with frequency-response function $H(f)$ is

$$|s_o(t)| = \left| \int_{-\infty}^{\infty} S(f)H(f) \exp(j2\pi f t) \, df \right| \tag{10.8}$$

where $S(f)$ is the Fourier transform of the input (received) signal. The mean output noise power is

$$N = \frac{N_o}{2} \int_{-\infty}^{\infty} |H(f)|^2 \, df \tag{10.9}$$

where N_0 is the input noise power per unit bandwidth. The factor $\frac{1}{2}$ appears before the integral because the limits extend from $-\infty$ to $+\infty$, whereas N_0 is defined as the noise power per cycle of bandwidth over positive values only. Substituting Eqs. (10.8) and (10.9) into (10.7) and assuming that the maximum value of $|s_o(t)|^2$ occurs at time $t = t_1$, the ratio R_f becomes

$$R_f = \frac{\left| \int_{-\infty}^{\infty} S(f)H(f) \exp(j2\pi f t_1) \, df \right|^2}{\frac{N_0}{2} \int_{-\infty}^{\infty} |H(f)|^2 \, df} \tag{10.10}$$

Schwartz's inequality states that if P and Q are two complex functions, then

$$\int P^*P \, dx \int Q^*Q \, dx \geq \left| \int P^*Q \, dx \right|^2 \tag{10.11}$$

The equality sign applies when $P = kQ$, where k is a constant. Letting

$$P^* = S(f) \exp(j2\pi f t_1) \quad \text{and} \quad Q = H(f)$$

and recalling that

$$\int P^*P \, dx = \int |P|^2 \, dx$$

we get, on applying the Schwartz inequality to the numerator of Eq. (10.10),

$$R_f \leq \frac{\int_{-\infty}^{\infty} |H(f)|^2 \, df \int_{-\infty}^{\infty} |S(f)|^2 \, df}{\frac{N_0}{2} \int_{-\infty}^{\infty} |H(f)|^2 \, df} = \frac{\int_{-\infty}^{\infty} |S(f)|^2 \, df}{\frac{N_0}{2}} \tag{10.12}$$

From Parseval's theorem,

$$\int_{-\infty}^{\infty} |S(f)|^2 \, df = \int_{-\infty}^{\infty} s^2(t) \, dt = \text{signal energy} = E \tag{10.13}$$

Therefore we have

$$R_f \leq \frac{2E}{N_0} \tag{10.14}$$

The frequency-response function which maximizes the peak-signal–to–mean-noise ratio R_f may be obtained by noting that the equality sign in Eq. (10.11) applies when $P = kQ$, or

$$H(f) = G_a S^*(f) \exp(-j2\pi f t_1) \tag{10.15}$$

where the constant k has been set equal to $1/G_a$.

The interesting property of the matched filter is that no matter what the shape of the input-signal waveform, the maximum ratio of the peak signal power to the mean noise power is simply twice the energy E contained in the signal divided by the noise power per hertz of bandwidth N_0. The noise power per hertz of bandwidth, N_0, is equal to $kT_0 F$ where k is the Boltzmann constant, T_0 is the standard temperature (290 K), and F is the noise figure.

The matched filter and the correlation function. The output of the matched filter is not a replica of the input signal. However, from the point of view of detecting signals in noise, preserving the shape of the signal is of no importance. If it is necessary to preserve the shape of the input pulse rather than maximize the output signal-to-noise ratio, some other criterion must be employed.[10]

The output of the matched filter may be shown to be proportional to the input signal cross-correlated with a replica of the transmitted signal, except for the time delay t_1. The cross-correlation function $R(t)$ of two signals $y(\lambda)$ and $s(\lambda)$, each of finite duration, is defined as

$$R(t) = \int_{-\infty}^{\infty} y(\lambda) s(\lambda - t) \, d\lambda \tag{10.16}$$

The output $y_0(t)$ of a filter with impulse response $h(t)$ when the input is $y_{in}(t) = s(t) + n(t)$ is

$$y_0(t) = \int_{-\infty}^{\infty} y_{in}(\lambda) h(t - \lambda) \, d\lambda \tag{10.17}$$

If the filter is a matched filter, then $h(\lambda) = s(t_1 - \lambda)$ and Eq. (10.17) becomes

$$y_0(t) = \int_{-\infty}^{\infty} y_{in}(\lambda) s(t_1 - t + \lambda) \, d\lambda = R(t - t_1) \tag{10.18}$$

Thus the matched filter forms the cross correlation between the received signal corrupted by noise and a replica of the transmitted signal. The replica of the transmitted signal is "built in" to the matched filter via the frequency-response function. If the input signal $y_{in}(t)$ were the

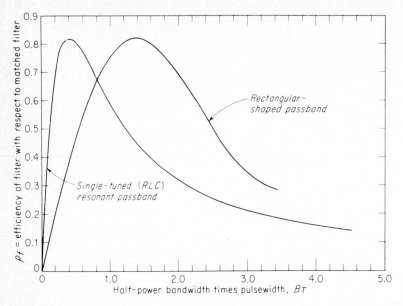

Figure 10.2 Efficiency, relative to a matched filter, of a single-tuned resonant filter and a rectangular shaped filter, when the input signal is a rectangular pulse of width τ. B = filter bandwidth.

same as the signal $s(t)$ for which the matched filter was designed (that is, the noise is assumed negligible), the output would be the autocorrelation function. The autocorrelation function of a rectangular pulse of width τ is a triangle whose base is of width 2τ.

Efficiency of nonmatched filters. In practice the matched filter cannot always be obtained exactly. It is appropriate, therefore, to examine the efficiency of nonmatched filters compared with the ideal matched filter. The measure of efficiency is taken as the peak signal-to-noise ratio from the nonmatched filter divided by the peak signal-to-noise ratio $(2E/N_0)$ from the matched filter. Figure 10.2 plots the efficiency for a single-tuned (RLC) resonant filter and a rectangular-shaped filter of half-power bandwidth B, when the input is a rectangular pulse of width τ. The maximum efficiency of the single-tuned filter occurs for $B\tau \approx 0.4$. The corresponding loss in signal-to-noise ratio is 0.88 dB as compared with a matched filter. Table 10.1 lists

Table 10.1 Efficiency of nonmatched filters compared with the matched filter

Input signal	Filter	Optimum $B\tau$	Loss in SNR compared with matched filter, dB
Rectangular pulse	Rectangular	1.37	0.85
Rectangular pulse	Gaussian	0.72	0.49
Gaussian pulse	Rectangular	0.72	0.49
Gaussian pulse	Gaussian	0.44	0 (matched)
Rectangular pulse	One-stage, single-tuned circuit	0.4	0.88
Rectangular pulse	2 cascaded single-tuned stages	0.613	0.56
Rectangular pulse	5 cascaded single-tuned stages	0.672	0.5

B bandwidth
τ pulse duration

the values of $B\tau$ which maximize the signal-to-noise ratio (SNR) for various combinations of filters and pulse shapes. It can be seen that the loss in SNR incurred by use of these non-matched filters is small.

Matched filter with nonwhite noise. In the derivation of the matched-filter characteristic [Eq. (10.15)], the spectrum of the noise accompanying the signal was assumed to be white; that is, it was independent of frequency. If this assumption were not true, the filter which maximizes the output signal-to-noise ratio would not be the same as the matched filter of Eq. (10.15). It has been shown[11-13] that if the input power spectrum of the interfering noise is given by $[N_i(f)]^2$, the frequency-response function of the filter which maximizes the output signal-to-noise ratio is

$$H(f) = \frac{G_a S^*(f) \exp(-j2\pi f t_1)}{[N_i(f)]^2} \tag{10.19}$$

When the noise is nonwhite, the filter which maximizes the output signal-to-noise ratio is called the NWN (nonwhite noise) matched filter. For white noise $[N_i(f)]^2 =$ constant and the NWN matched-filter frequency-response function of Eq. (10.19) reduces to that of Eq. (10.15). Equation (10.19) can be written as

$$H(f) = \frac{1}{N_i(f)} \times G_a\left(\frac{S(f)}{N_i(f)}\right)^* \exp(-j2\pi f t_1) \tag{10.20}$$

This indicates that the NWN matched filter can be considered as the cascade of two filters. The first filter, with frequency-response function $1/N_i(f)$, acts to make the noise spectrum uniform, or white. It is sometimes called the *whitening filter*. The second is the matched filter described by Eq. (10.15) when the input is white noise and a signal whose spectrum is $S(f)/N_i(f)$.

10.3 CORRELATION DETECTION[14-23]

Equation (10.18) describes the output of the matched filter as the cross correlation between the input signal and a delayed replica of the transmitted signal. This implies that the matched-filter receiver can be replaced by a cross-correlation receiver that performs the same mathematical operation, as shown in Fig. 10.3. The input signal $y(t)$ is multiplied by a delayed replica of the transmitted signal $s(t - T_r)$, and the product is passed through a low-pass filter to perform the integration. The cross-correlation receiver of Fig. 10.3 tests for the presence of a target at only a single time delay T_r. Targets at other time delays, or ranges, might be found by varying T_r. However, this requires a longer search time. The search time can be reduced by adding

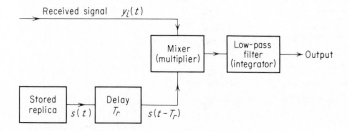

Figure 10.3 Block diagram of a cross-correlation receiver.

parallel channels, each containing a delay line corresponding to a particular value of T_r, as well as a multiplier and low-pass filter. In some applications it may be possible to record the signal on some storage medium, and at a higher playback speed perform the search sequentially with different values of T_r. That is, the playback speed is increased in proportion to the number of time-delay intervals T_r that are to be tested.

Since the cross-correlation receiver and the matched-filter receiver are equivalent mathematically, the choice as to which one to use in a particular radar application is determined by which is more practical to implement. The matched-filter receiver, or an approximation, has been generally preferred in the vast majority of applications.

10.4 DETECTION CRITERIA

The detection of weak signals in the presence of noise is equivalent to deciding whether the receiver output is due to noise alone or to signal-plus-noise. This is the type of decision probably made (subconsciously) by a human operator on the basis of the information present at the radar indicator. When the detection process is carried out automatically by electronic means without the aid of an operator, the detection criterion cannot be left to chance and must be carefully specified and built into the decision-making device by the radar designer.

In Chap. 2 the radar detection process was described in terms of threshold detection. Almost all radar detection decisions are based upon comparing the output of a receiver with some threshold level. If the envelope of the receiver output exceeds a pre-established threshold, a signal is said to be present. The purpose of the threshold is to divide the output into a region of no detection and a region of detection; or in other words, the threshold detector allows a choice between one of two hypotheses. One hypothesis is that the receiver output is due to noise alone; the other is that the output is due to signal-plus-noise. It was shown in Chap. 2 that the dividing line between these two regions depended upon the probability of a false alarm, which in turn is related to the average time between false alarms.

There are two types of errors that might be made in the decision process. These are unavoidable with observations of finite duration in the presence of noise. One kind of error is to mistake noise for signal when only noise is present. This occurs whenever the noise is large enough to exceed the threshold level. In statistical detection theory it is sometimes called a type I error. The radar engineer would call it a *false alarm*. A type II error is one in which the signal is erroneously considered to be noise when signal is actually present. This is a *missed detection* to the radar engineer. The setting of the threshold represents a compromise between these two types of errors. A relatively large threshold will reduce the probability of a false alarm, but there will be more missed detections. The nature of the radar application will influence to a large extent the relative importance of these two errors and, therefore, the setting of the threshold.

Neyman-Pearson observer. The threshold level was selected in Chap. 2 so as not to exceed a specified false-alarm probability; that is, the probability of detection was maximized for a fixed probability of false alarm. This is equivalent to fixing the probability of a type I error and minimizing the type II error. It is similar to the Neyman-Pearson test used in statistics for determining the validity of a specified statistical hypothesis.[24,25] Therefore this type of threshold detector is sometimes called a *Neyman-Pearson Observer*. In statistical terms it is claimed to be a uniformly most powerful test and is an optimum one, no matter what the a priori probabilities of signal and noise.[25] The Neyman-Pearson criterion is well suited for radar application and is usually used in practice, whether knowingly or not.

Likelihood-ratio receiver. The likelihood ratio is an important statistical tool and may be defined as the ratio of the probability-density function corresponding to signal-plus-noise, $p_{sn}(v)$, to the probability-density function of noise alone, $p_n(v)$.

$$L_r(v) = \frac{p_{sn}(v)}{p_n(v)} \tag{10.21}$$

It is a measure of how likely it is that the receiver envelope v is due to signal-plus-noise as compared with noise alone. It is a random variable and depends upon the receiver input. If the likelihood ratio $L_r(v)$ is sufficiently large, it would be reasonable to conclude that the signal was indeed present. Thus detection may be accomplished by establishing a threshold at the output of a receiver which computes the likelihood ratio. The selection of the proper threshold level will depend upon the statistical detection criterion used and by the probabilities of error desired and their relative importance.

The likelihood ratio is primarily useful in the analysis of the statistical-detection problem. It is difficult to conceive, however, of a receiver that computes the likelihood ratio directly as defined by Eq. (10.21). However, in certain cases the receiver which computes the likelihood ratio is equivalent to a receiver which computes the cross-correlation function or to one with a matched-filter characteristic, that is, one which maximizes the output signal-to-noise ratio.[24]

The Neyman-Pearson Observer is equivalent to examining the likelihood ratio and determining if $L_r(v) \geq K$, where K is a real, nonnegative number dependent upon the probability of false alarm selected.

Inverse probability receiver. A detection criterion that has been popular particularly for the theoretical analysis of statistical detection and for statistical parameter estimation is that based on *inverse probability*. The usual detection criteria employ the concept of *direct probability*, which describes the chance of an event happening on a given hypothesis. For example, the probability that a particular radar will detect a certain target under specified conditions is a direct probability. On the other hand, if an event actually happens, the problem of forming the best estimate of the cause of the event is a problem in *inverse probability*. For example, assume the event in question to be the output voltage v from a radar receiver. Upon obtaining this voltage, it is of interest to determine whether the output was caused by noise or by signal in the presence of noise. The probabilities of obtaining noise and signal-plus-noise before the event takes place are the a priori probabilities. They represent the initial state of knowledge concerning the event. The probability that the receiver output v was caused by noise or by signal-plus-noise is an a posteriori probability and represents the state of knowledge obtained as a result of observing the output.

The method of inverse probability involves the use of the a priori probabilities associated with each of the possible hypotheses which could explain the event. The a priori probabilities are used, along with a knowledge of the event, to compute the a posteriori probabilities. A separate a posteriori probability is computed for each hypothesis. That hypothesis which results in the largest a posteriori probability is selected as the most likely to explain the event.

This method has been applied by Woodward and Davies to the reception of signals in noise.[26-28] It is based upon the application of Bayes' rule for the probability of causes.[29] The joint probability of two events x and y is

$$p(x, y) = p(x)p(y|x) = p(y)p(x|y) \tag{10.22}$$

where $p(x)$ and $p(y)$ = probabilities of events x and y, respectively

$\quad$ $p(y|x)$ = conditional probability that event y will occur, given that event x has occurred

$\quad$ $p(x|y)$ = conditional probability of event x given that y has occurred

Let the event $x = SN$ represent signal-plus-noise, and let the event y be the receiver input, which may consist of either signal-plus-noise or noise alone. Equation (10.22) may be rewritten as

$$p(SN \mid y) = \frac{p(SN)p(y \mid SN)}{p(y)} \tag{10.23}$$

This is *Bayes' rule*. It expresses the (a posteriori) probability that the signal is present, given that the receiver input is y.

For a particular input y, the receiver can assess from Eq. (10.23) the probability that a particular signal was received. Since y will then be fixed, the denominator $p(y)$ will be constant and the a posteriori probability is

$$p(SN \mid y) = kp(SN)p(y \mid SN) \tag{10.24}$$

where the constant k is determined by the normalizing condition; that is, the integral of $p(SN \mid y)$ over all possible values must be unity. Therefore, if the a priori probability $p(SN)$ is known, the a posteriori probability may be found directly from Eq. (10.24) once $p(y \mid SN)$ has been evaluated. If the received waveform $y(t)$ as a function of time consists of the signal waveform $s_i(t)$ plus the white gaussian noise waveform $n(t) = y(t) - s_i(t)$, Woodward and Davies[27] show that

$$p(y \mid SN) = p_n[n(t)] = p_n[y(t) - s_i(t)] \propto \exp\left[-\frac{1}{N_0} \int n^2(t)\, dt\right] \tag{10.25}$$

where $p_n[n(t)] =$ probability-density function for noise waveform $n(t)$ and $N_0 =$ mean noise power per unit bandwidth (dimensions of energy). With this substitution, the a posteriori probability for the signal $s_i(t)$ becomes

$$p(SN \mid y) = kp(SN) \exp\left\{-\frac{1}{N_0} \int_0^{T_0} [y(t) - s_i(t)]^2\, dt\right\} \tag{10.26}$$

The integral in this expression is a definite one, with limits defined by the duration of the observation time $(0 \to T_0)$.

Equation (10.26) forms the basis of the technique used by Woodward and Davies for the analysis of radar reception problems when the interference is caused by white gaussian noise. Except for the a priori weighting factor $p(SN)$, Eq. (10.26) shows that the most probable waveform $s_i(t)$ is the one which has the least-mean-square deviation from the received waveform $y(t)$.

The computation of the a posteriori probability might be accomplished by computing the cross-correlation function between the actual waveform and the various possible waveforms that might be received. Expanding the integral in Eq. (10.26) we get

$$\int [y(t) - s_i(t)]^2\, dt = \int y^2(t)\, dt - 2 \int y(t)s_i(t)\, dt + \int s_i^2(t)\, dt \tag{10.27}$$

Upon reception, the waveform $y(t)$ is known, so that the first integral on the right-hand side of the equation is constant and can be absorbed in the constant k. The last integral is the energy E contained within the signal $s_i(t)$ and also is a constant. The second integral is not a constant. The a posteriori probability can be written

$$p(SN \mid y) = kp(SN) \exp\left[\frac{2}{N_0} \int_0^{T_0} y(t)s_i(t)\, dt\right] \tag{10.28}$$

where the first and third integrals of Eq. (10.27) are absorbed in the constant k. An a posteriori receiver, if it could be built, is one whose output is given by the above equation. If the receiver output (the a posteriori probability) is greater than a predetermined threshold, a target is said to be present. If the a priori probability $p(SN)$ can be considered constant, the computation of the a posteriori probability is equivalent to multiplying the received signal $y(t)$ by the signal waveform $s_i(t)$ and integrating with respect to time. This is the same process performed by the cross-correlation receiver (Sec. 10.3) and is equivalent to the operation of a matched filter (Sec. 10.2).

A limitation of the method of inverse probability based on the application of Bayes' rule is the difficulty of specifying the a priori probabilities. In most cases of practical interest, one is ignorant of the a priori probabilities. For example, it would be necessary to specify the a priori probability of finding a target at any particular range at any particular time. This is an almost impossible task. In the absence of better data, it might be assumed that all range intervals are equally probable a priori, and the a priori probability may be considered to be constant. However, such an assumption applied blindly to computations involving inverse probability can sometimes lead to erroneous and contradictory conclusions.[29] This difficulty in specifying the a priori probability was recognized by Woodward and Davies.[27] They suggest, however, that the a priori factor be omitted from the inverse-probability specification when it is doubtful, and in practice it may be supplied subjectively by the human observer. This merely begs the question, for it has not been proved that an operator can supply the necessary a priori probability, and in addition, there are many applications where no operator is involved in making the detection decision. Nevertheless, it may be stated that whenever the a priori probabilities are known, the inverse-probability method may be used with confidence. When the a priori probabilities are not known, the likelihood-ratio test is usually employed.

Relationship of inverse probability and likelihood-ratio receivers. The receiver input y may be either signal-plus-noise or noise alone. Therefore the probability of the event y may be expressed as

$$p(y) = p(y|SN)p(SN) + p(y|N)p(N) \tag{10.29}$$

The a posteriori probability of Eq. (10.23) becomes

$$p(SN|y) = \frac{p(SN)p(y|SN)}{p(y|SN)p(SN) + p(y|N)p(N)} \tag{10.30}$$

Or, in terms of the likelihood ratio[30]

$$L_r(y) = \frac{p(y|SN)}{p(y|N)}$$

$$p(SN|y) = \frac{L_r(y)p(SN)}{L_r(y)p(SN) + 1 - p(SN)} \tag{10.31}$$

Therefore, if a receiver can be built which computes the likelihood ratio and if the a priori probability $p(SN)$ is known, the a posteriori probability can be calculated. Since $p(SN|y)$ is a monotonic function of $L_r(y)$, the output of the likelihood receiver (or the matched-filter receiver) can be calibrated directly in terms of the a posteriori probability. The chief difference between the two representations is that the concept of inverse probability requires a knowledge of the a priori probabilities whereas the likelihood ratio does not. (The likelihood ratio follows from inverse probability if the assumption is made that the a priori probabilities are equally likely.) Both the a posteriori method and the likelihood method may be implemented by computing the cross-correlation function between the received signal and the signal $s_i(t)$.

Ideal observer. The criterion of Neyman-Pearson is not the only one which might be used for establishing a threshold level. One of the first mathematical criteria applied to the theory of radar detection was the *Idea Observer* as formulated by Siegert.[31] (The term "Ideal" does not necessarily imply that this criterion is *the* ideal criterion.) The criterion of the Ideal Observer maximizes the total probability of a correct decision (or minimizes the total probability of an error). Since the two ways in which an error might be made are (1) false alarm and (2) missed detection, the probability of an error is.

$$p(E) = p(N)p_{fa} + p(SN)p_m \qquad (10.32)$$

where $p(N)$ and $p(SN)$ = a priori probabilities of obtaining noise and signal-plus-noise, respectively $[p(N) + p(SN) = 1]$
p_{fa} = conditional probability of a false alarm, given that noise is present
p_m = conditional probability of a miss, given that signal is present (α, β are often used in mathematical references for p_{fa} and p_m)

The a priori probability is the probability of obtaining either noise alone or signal-plus-noise before the detection decision is made. These probabilities must be known beforehand in the Ideal Observer.

The false-alarm probability depends upon the miss probability in the case of the Ideal Observer. This is unlike the Neyman-Pearson Observer in which P_{fa} was fixed. The function of the Ideal Observer is to minimize the total probability of error. It accomplishes this by adjusting the threshold, which in turn affects both the false alarm and the miss probabilities. The dependence of the two errors upon one another is probably the chief limitation of the Ideal Observer as a radar detection criterion. Middleton[25] states that when the probability of false alarm is 0.05, the probability of detection is 0.90. A false-alarm probability of 0.05 is usually high for most radar applications. Radar false-alarm probabilities are rarely greater than 10^{-4}. Low values of false-alarm probability with the Ideal Observer require extremely large values of detection probabilities. For example, a false-alarm probability of 10^{-5} corresponds to a detection probability of 0.99998. Thus the privilege of minimizing the total error seems to result in overdetecting the target. Therefore, from a practical point of view, the Ideal Observer is less efficient than the Neyman-Pearson Observer for most radar applications.

The Ideal Observer applies equal weight to an error due to a false alarm and to an error due to a miss. However, there may be cases in which equal weight is not appropriate (Bayes criterion). If the errors are not of equal importance, the theory of the Ideal Observer may be modified to take this into account using the concepts of *statistical decision theory*.[24,32–34] Unfortunately, in most radar applications there is no realistic basis for assigning a value to a miss or to a false alarm; consequently, the usefulness of decision theory in radar is limited.

Sequential observer. Most radars utilize the equivalent of the Neyman-Pearson Observer and operate with a fixed number of pulses. However, the detection decision might very well be made on the basis of only a few observations or possibly a single observation, and it would not be necessary to record the later observations that occur once the threshold has been crossed. Hence there may be some advantage to using a flexible detection criterion which takes account of this fact. Such a detection criterion is the *Sequential Observer*.[30–43] The Sequential Observer makes an observation of the receiver output and, on the basis of this single observation, decides between one of three choices: (1) the receiver output is due to the presence of signal with noise; (2) the output is due to noise alone; or (3) the available evidence is not convincing enough to make a decision between choices 1 and 2. If the evidence is sufficient to allow a choice to be made between signal-plus-noise (1) and noise alone (2), the test is completed. But if choice 3 is indicated, no conclusive decision can be reached and another observation is made. The three choices are again examined on the basis of the combined observations. If no decision is reached as to the presence of signal or the presence of noise, another observation is made, and the process is repeated until the evidence is convincing enough to make a definite conclusion.

The Sequential Observer fixes the probability of errors beforehand but allows the integration time (or the number of observations) to be a variable. Two threshold levels are established with a gray region in between. If the output is definitely below the lower threshold, noise alone is said to be present. If the upper threshold is exceeded, the signal is declared to be present along with the noise. But if the output lies between the two thresholds, no decision is possible and another observation is made.

The Sequential Observer permits a significant reduction in the *average* number of samples (pulses) needed for making a decision. The improvement depends upon whether a signal is present or not. If a signal is present, the average number of samples (or observations) required for making a decision is significantly greater than when noise alone is present. The Sequential Observer makes a relatively prompt decision when only noise is present.

The average savings also depend upon whether detection is performed coherently or noncoherently. Assuming coherent detection (pulses integrated predetection) and $P_D = 0.90$ and $P_{fa} = 10^{-8}$, the Sequential Observer is able to determine, on the average, that noise alone is present with less than one-tenth the number of observations required for the Neyman-Pearson Observer.[90] In the presence of a threshold signal, the Sequential Observer requires, on the average, about one-half the number of observations of the equivalent fixed-sample observer. Again assuming $P_D = 0.90$ and $P_{fa} = 10^{-8}$, a noncoherent Sequential Observer (pulses integrated postdetection) requires only one-fourteenth the number of observations when only noise is present. When signal is present, approximately one-half the number of observations are needed as with an equivalent Neyman-Pearson Observer.

Unfortunately, the application of the Sequential Observer to radar is limited. At any particular antenna position a radar normally observes many range resolution intervals, perhaps several hundred. At each range interval within the beamwidth of the antenna a decision has to be made concerning the presence or absence of a target before the antenna beam can be positioned to a new direction. Therefore the dwell time in any direction is determined not by the average number of observations made by each sequential detector, but by that range interval which requires the largest number of observations. This can result in relatively long dwell times, and it is possible for the savings obtained with the Sequential Observer to be negated. The required dwell time might be even longer than that of the fixed sample-number Neyman-Pearson Observer.

If the Sequential Observer allows a savings in average power of from 8 to 10 dB when implemented for a single-range cell, the power advantage decreases to 3 to 4 dB for 200-range cells and can even be as little as 1 dB.[40] In addition, the Sequential Observer requires something more flexible than the usual rotating-antenna radar. Because of the variable dwell time the antenna beam-positioning system must usually be a phased-array antenna and the data processing must be digital. Thus if full benefit is to be had from the Sequential Observer in radar, only one or a few independent decisions per beam position should be made. It might be employed when only a single "guard-band" is desired for detecting targets within a selected range interval, a not too usual application. It has also been considered for use in a low-signal density environment when the criterion is that either no signal is present or one signal of fixed and specified level is present and equally likely to be in any range interval.[38,39]

The term *sequential detection* is sometimes used synonymously with Sequential Observer. Sequential detection has also been used to describe a two-step, or two-stage, detection process that can be employed with phaséd-array radar.[39,42] The radar transmits a pulse or a series of pulses in a particular direction as in an ordinary radar, except that the false alarm probability is slightly higher than would normally be used. If no threshold crossings are obtained, the antenna beam moves to the next angular position. However, if a threshold crossing is observed from any range cell, a second pulse or set of pulses is transmitted with higher energy, and with a different threshold. A target is declared to be present only if threshold crossings are observed from the same range cell on both transmissions. The second transmission might also be of greater resolution as well as higher power. When the two thresholds are optimized, it has been found that a second transmission is employed in about four percent of the beam positions.[42] It is claimed that a power saving of from 3 to 4 dB can be obtained as compared with uniform scanning.[42]

10.5 DETECTOR CHARACTERISTICS

No effect on receiver regardless of detector

The portion of the radar receiver which extracts the modulation from the carrier is called the *detector*. The use of this term implies somewhat more than simply a rectifying element. It includes that portion of the radar receiver from the output of the IF amplifier to the input of the indicator or data processor. We shall not be concerned about the problem of amplification, although it is an important aspect of receiver design. Instead, we shall be more interested in the effect of the detector on the desired signal and the noise.

One form of detector is the *envelope detector*, which recognizes the presence of the signal on the basis of the amplitude of the carrier envelope. All phase information is destroyed. It is also possible to design a detector which utilizes only phase information for recognizing targets. An example is one which counts the zero crossings of the received waveform. The *zero-crossings detector* destroys amplitude information. If the exact phase of the echo carrier were known, it would be possible to design a detector which makes optimum use of both the phase information and the amplitude information contained in the echo signal. It would perform more efficiently than a detector which used either amplitude information only or phase information only. The *coherent detector* is an example of one which uses both phase and amplitude. These three types of detectors—the envelope, the zero-crossings, and the coherent detectors—are considered in the present section.

Envelope detector—optimum-detector law. The function of the envelope detector is to extract the modulation and reject the carrier. By eliminating the carrier, all phase information is lost and the detection decision is based solely on the envelope amplitude. It will be recalled that most of the analysis of the radar equation in Chap. 2 was based on the envelope detector.

The envelope detector consists of a rectifying element and a low-pass filter to pass the modulation frequencies but to remove the carrier frequency. The rectifier characteristic relates the output signal to the input signal and is called the *detector law*. Most detector laws approximate either a linear or a square-law characteristic. In the linear detector the output signal is directly proportional to the input envelope. (Actually, the so-called linear detector is a nonlinear device, or else it would not be a detector.) Similarly, in the square-law detector, the output signal is proportional to the square of the input envelope. In some of the quoted mathematical results, the linear-detector law is assumed, while in others, the square law is assumed. In general, the difference between the two is small and the detector law in any analysis is usually chosen for mathematical convenience. When we speak of a detector law we really mean the combined law of the detector and video integrator, if one is used. If the detector were linear while the video integrator had a square-law characteristic, the combination of detector and integrator would be considered square law.

In the following we shall derive the optimum form of the second-detector law.[26,43] Assume that there are n independent pulses with envelope amplitudes $v_1, v_2, \ldots, v_n$ available from the radar receiver. The problem consists in determining whether these n pulses are due to signal-plus-noise or whether they are due to noise alone. The probability-density function for the envelope of n independent noise samples is the product of the probability-density function for each sample, or

$$p_n(n, v_i) = \prod_{i=1}^{n} p_n(v_i) \tag{10.33}$$

The probability-density function for ith noise pulse $p_n(v_i)$ is given by Eq. (2.21), rewritten

$$p_n(v_i) = v_i \exp\left(-\frac{v_i^2}{2}\right) \tag{2.21}$$

where v_i is the ratio of the envelope amplitude R to the rms noise voltage $\psi_0^{1/2}$. Likewise, the probability-density function for the envelope of n signal-plus-noise pulses is

$$p_s(n, v_i) = \prod_{i=1}^{n} p_s(v_i) \tag{10.34}$$

The probability-density function for signal-plus-noise, $p_s(v_i)$, is the Rice distribution of Eq. (2.27)

$$p_s(v_i) = v_i \exp\left[-\frac{(v_i^2 + a^2)}{2}\right] I_0(av_i) \tag{2.27}$$

where a = ratio of signal (sine-wave) amplitude to rms noise voltage and $I_0(x)$ = modified Bessel function of zero order. The detection process is equivalent to determining which of the two density functions [Eq. (10.33) or (10.34)] more closely describes the output of the receiver.

The ratio of the density function for signal-plus-noise to that for noise alone is the likelihood ratio. It may be used to decide whether or not the signal is present. The greater the likelihood ratio, the more probable it is that the receiver input is due to signal-plus-noise rather than noise alone. The likelihood ratio, or any other monotonic function of this ratio, must exceed a predetermined threshold value in order to declare that the signal is present. The selection of the threshold will depend upon the probability of false alarm. The likelihood ratio [Eq. (10.34) divided by (10.33)] is

$$L_r(v) = \exp\left(-n\frac{a^2}{2}\right) \prod_{i=1}^{n} I_0(av_i) \geq \lambda \tag{10.35}$$

where λ is the constant which depends on the false-alarm probability. Taking the logarithm gives

$$\sum_{i=1}^{n} \ln I_0(av_i) \geq \ln \lambda + n\frac{a^2}{2} \tag{10.36}$$

This states that for optimum processing one should take the pulses, each of amplitude v_i (where $i = 1, 2, \ldots, n$), and sum them according to the law $\sum_{i=1}^{n} \ln I_0(av_i)$. This sum is compared with a threshold as given by the right-hand side of Eq. (10.36). Therefore the combined detector and integrator must have a law given by

$$y = \ln I_0(av) \tag{10.37}$$

where y = output voltage of detector and integrator

       a = amplitude of sine-wave signal divided by rms noise voltage

       v = amplitude of IF voltage envelope divided by rms noise voltage

$I_0(x)$ = modified Bessel function of zero order

This equation specifies the form of the detector law which maximizes the likelihood ratio for a fixed probability of false alarm. A suitable approximation to Eq. (10.37) is[91]

$$y = \ln I_0(av) \approx \sqrt{(av)^2 + 4} - 2 \tag{10.38}$$

For large signal-to-noise ratios ($a \gg 1$), this is approximately

$$y \approx av \tag{10.39}$$

Thus the linear detector is a good approximation to the optimum $\ln I_0(av)$ detector law when the signal-to-noise ratio is large. For small signal-to-noise ratios the approximation of Eq. (10.38) can be written

$$y \approx \frac{(av)^2}{4} \tag{10.40}$$

which is the characteristic of a square-law detector.

Hence it may be concluded that for small signal-to-noise ratios the square-law detector may be a suitable approximation to the optimum detector, while for large signal-to-noise ratios the linear detector is more appropriate. In practice, it makes little difference which of the two detector laws is used. The difference between the square-law and the linear detectors was shown by Marcum[43] to produce less than 0.2 dB difference in the required signal-to-noise ratio. If one has a choice, the linear law might be preferred because of its linearity and, hence, its large dynamic range. The linear detector, like any detector law, approaches the square-law characteristic for small signal-to-noise ratios.

Logarithmic detector. If the output of the receiver is proportional to the logarithm of the input envelope, it is called a *logarithmic receiver*. It finds application where large variations of input signals are expected. It might be used to prevent receiver saturation or to reduce the effects of unwanted clutter targets in certain types of non-MTI radar receivers (Sec. 13.8).

The detection characteristics (probability of detection as a function of the probability of false alarm, signal-to-noise ratio, and the number of hits integrated) for the logarithmic receiver have been computed by Green,[44] following the methods of Marcum.[43] For 10 pulses integrated, the loss with the logarithmic receiver is about 0.5 dB, while for 100 pulses integrated, the loss is about 1.0 dB. As the number of pulses increases, the loss due to a logarithmic detector approaches a maximum value of 1.1 dB.[45]

Zero-crossings detector. The information contained in the zero crossings of the received signal can, in principle, be used for detecting the presence of signals in noise. The greater the signal-to-noise ratio the less will be the average number of zero crossings. The average number of zero-crossings per second at the output of a narrow-bandpass filter of rectangular shape when the input is a sine wave in gaussian noise is[46]

$$\bar{n}_0 = 2f_0 \left[\frac{S/N + 1 + (f_B^2/12f_0^2)}{S/N + 1} \right]^{1/2} \tag{10.42}$$

where f_0 = center frequency of filter

    f_B = filter bandwidth

  S/N = signal-to-noise (power) ratio

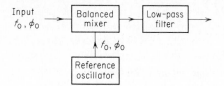

Figure 10.4 Basic configuration of a coherent detector.

If a suitable device is used to measure $\bar{n}_0$, a target signal is said to be present if this number is less than a predetermined (threshold) value and is said to be absent if the threshold is exceeded. (The value of $\bar{n}_0$ is a maximum when $S/N = 0$.)

Coherent detector. The coherent detector (Fig. 10.4) consists of a reference oscillator feeding a balanced mixer. The input to the mixer is a signal of known frequency f_0 and known phase ϕ_0 plus its accompanying noise. The reference-oscillator signal is assumed to have the same frequency and phase as the input signal to be detected. The output of the mixer is followed by a low-pass filter which allows only the d-c and the low-frequency modulation components to pass while rejecting the higher frequencies in the vicinity of the carrier. The coherent detector provides a translation of the carrier frequency to direct current. It does not extract the modulation envelope and is a truly linear detector, whereas the "linear" envelope detector was not linear in the same sense. Therefore the coherent detector will be a more efficient detector, especially when signal-to-noise ratios are low.

The coherent detector does not destroy phase information as does the envelope detector, nor does it destroy amplitude information as does the zero-crossings detector. Since it utilizes

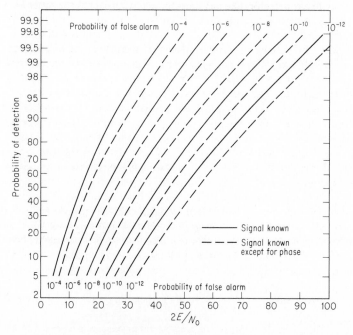

Figure 10.5 Comparison of detection probabilities for signal known completely (coherent detector) and for signal known except for phase (envelope detector).

more information than either the envelope or the zero-crossings detector, it is not surprising that the signal-to-noise ratio from the coherent detector is better than from the other two. The improvement in the signal-to-noise ratio might vary from 1 to 3 dB or more, over the range of signal-to-noise ratios of interest in most radar applications. A comparison is shown in Fig. 10.5 of the detection probabilities when the signal parameters are known completely (coherent detector) and when the signal is known except for phase (envelope detector).[24] The abscissa is plotted as $2E/N_0$ instead of signal-to-noise ratio, where E is the signal energy and N_0 is the noise power per hertz of bandwidth.

Although the coherent detector may be of superior sensitivity than other detectors it is seldom used in radar applications since the phase of the received signal is not usually known.

10.6 PERFORMANCE OF THE RADAR OPERATOR

The rate of information inherent in a typical radar signal is considerably greater than can be handled by a human operator. A one-megahertz-bandwidth signal, for example, is capable of conveying information at a rate of two megabits/s, but an operator can accept an information rate of only 10 to 20 bits/s. Thus there is a tremendous mismatch between the information content of a radar and the information-handling capability of an operator. The function of the radar display is to aid the operator to extract in an efficient manner the information contained in the radar signal that is important to the task. The usual type of radar display is a cathode-ray tube or its equivalent.

The ability of the operator to detect radar signals in the presence of noise or clutter cannot be determined with as great a reliability as can the performance of the electronic threshold detector described in Chap. 2. Human behavior is certainly less predictable than that of an electronic device. However, it appears that an operator's performance can be as good as that predicted for the ideal electronic threshold detector *if* the operator is well trained, motivated, alert, not fatigued, and the display is properly designed. Furthermore, the operator is probably better able to recognize and interpret patterns relating groups of associated echoes than can automatic devices.[47]

It has been shown experimentally that when an operator views a display in which the pulses received from successive sweeps are displayed side-by-side without loss of memory (fading of the recorded signals with time), the integration improvement achieved by an operator is equivalent to what would be expected from classical detection theory.[48,49] This is illustrated by the data of Fig. 10.6, which plots the experimentally observed signal-to-noise ratio necessary to achieve a probability of detection of 0.50 as a function of the number of pulses available on the display. Curve A applies to a chemical recorder which makes a permanent record, and curve B applies to an intensity modulated B-scope of long persistence. (The chemical recorder has been of more use in sonar than in radar.) The integration improvement was found to be from 2.2 to 2.5 dB per doubling of the number of pulses, which is consistent with the computations for the theoretical integration improvement obtained by Marcum[43] for ideal postdetection integration. The results for the chemical recorder (curve A) are slightly better than the B-scope (curve B) due to the imperfect memory of the cathode-ray tube B-scope as compared with the permanent memory of the chemical recorder. Similar results were noted for the improvement in a conventional PPI when the pulses were displayed side-by-side.

In other experiments with the PPI, the integration improvement has been reported to be proportional to 1.5 dB per doubling of the number of pulses, which corresponds to $n^{1/2}$.[50] Improvements as high as 1.8 dB per doubling were also observed, but this is less than the 2.2 to

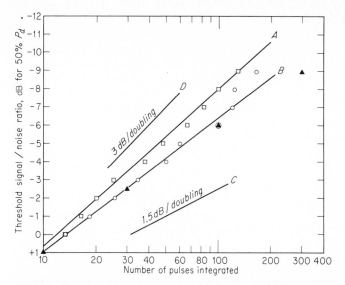

Figure 10.6 Improvement of operator detection threshold (signal-to-noise ratio per pulse) as a function of the number of pulses (sweeps) for side-by-side displays. Curve A, chemical recorder; curve B, cathode-ray-tube B-scope display; curve C, line of slope 1.5 dB/doubling ($\propto n^{1/2}$); curve D, line of slope 3 dB/doubling ($\propto n$) for perfect predetection integration. ▲'s are theoretical values for ideal postdetection-integration improvement as computed by Marcum[43] for a false-alarm probability of 10^{-3}. (*From J. Brit. IRE.*[49])

2.5 dB per doubling mentioned above. The applicability of these particular tests to the usual PPI display can be questioned since a 5-inch-diameter CRT was used with simulated echoes placed at an azimuth unknown to the operator at a distance of 1 inch from the center of the scope. The conclusion that might be drawn from such experiments is that somewhat less than theoretical integration improvement will be obtained if the pulses are not displayed properly; that is, the persistence of the display should be sufficient to prevent excessive "loss of memory," the dynamic range of the display must be large enough not to lose signal information, and the display resolution should be consistent with the radar resolution to avoid collapsing loss. Since the dynamic range of most displays is limited to a relatively small value, the pulses received from a target should be displayed side by side (as is usually the case on a PPI or B-scope) rather than piled up at a single point on the display. In such a case the integration is performed by the eye-brain combination of the operator.

Experimental measurements of the operator's ability to detect signals on an A-scope showed the integration-improvement factor to be $n^{1/2}$, or 1.5 dB per doubling of the number of pulses integrated.[31] This is less than predicted for an ideal postdetection integrator. It was also found that the operator's detection capability was maximum when $B\tau \approx 1$, where $B = $ IF bandwidth and $\tau = $ pulse width, as is consistent with the analysis of Sec. 10.2.

The detectability of a target echo on a CRT display depends on such factors as the size and brightness of the CRT spot, the noise background, pulse repetition frequency, antenna rotation rate, type of phosphor, and its decay characteristics, the size of the display, and the color and intensity of the ambient illumination. The effect of most of these factors has been determined empirically.[31,51,62] The name *pipology* is sometimes applied to the art of designing and operating the radar display so as to optimize the operator's ability to detect target pips. (A target *pip* is also called a *blip*.) It has also been found from experience that an operator should not, in general, be required to perform his duties for more than one-half hour at one sitting.[51]

10.7 AUTOMATIC DETECTION

The function of the radar operator viewing the ordinary radar display is to recognize the presence of targets and extract their location. When the function is performed by electronic decision circuitry without the intervention of an operator, the process is known as *automatic detection*. One of the chief reasons for employing automatic detection is to overcome the limitations of an operator due to fatigue, boredom, and overload. In addition, the use of automatic detection allows the radar output to be transmitted over telephone lines rather than by more expensive broadband microwave links, since only detected target information need be transmitted and not the full bandwidth signal (raw video). Automatic detection is also an important part of automatic detection and track (ADT) systems, as discussed in Sec. 5.10. The automatic detector has also been called *plot extractor* and *data extractor*.

Usually there are four basic aspects to automatic detection: (1) the integration of the pulses received from the target; (2) the detection decision; and the determination of the target location in (3) range and (4) azimuth. Sometimes there is also included in automatic-detection circuitry the means for maintaining a constant false alarm rate (CFAR). This is discussed separately in the next section.

Binary moving-window detector.[39,52-60] As a radar antenna scans by a target it will normally receive n echo pulses. If m of these expected n pulses exceed a predetermined value (threshold), a target may be declared to be present. The use of a criterion that requires m out of n echo pulses to be present is a form of integration. It is less efficient than ideal postdetection integration, but it has the advantage of simplicity. It is called the binary moving-window detector, but it has also been called double-threshold detector, *m*-out-of-*n* detector, coincidence detector, sliding-window detector, and binary integrator.

A block diagram of the binary moving-window detector is shown in Fig. 10.7. The radar video is passed through a threshold detector, which may be thought of as a bottom clipper. Only those signals whose amplitude exceeds the preset threshold are allowed to pass. This is the *first* of two thresholds, hence the name *double-threshold detector* which is sometimes used. The output of the first threshold is sampled by the quantizer at least once per range-resolution cell. A standard pulse is generated if the video waveform exceeds the first threshold, and nothing if it does not. These are designated by 1 or 0 respectively. Thus the output of the quantizer is a series of 1s and 0s. This is then separated into separate range cells by the range

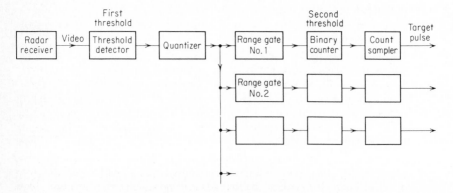

Figure 10.7 Block diagram of a binary moving window detector, or binary integrator.

gates, and the 1s and 0s from the last n sweeps are stored and counted in the binary counter. If there are at least m 1s within the last n sweeps, a target is said to be present. The number m is the *second* threshold to be passed in the double-threshold detector. The two thresholds must be selected jointly for best performance. The optimum value of m for a constant echo signal is shown in Fig. 10.8. (Similar results are available for fluctuating targets.[39,58,59]) This curve is approximate since there is some slight dependence upon the false-alarm probability, but it appears to be independent of the signal-to-noise ratio. The actual value of m can differ significantly from m_{opt} without a large penalty in signal-to-noise ratio. For example, the signal-to-noise ratio will be within 0.5 dB of optimum for m ranging over a value of 0.44 n for a nonfluctuating target or a Swerling case 1, and 0.34 n for a Swerling case 2.[59]

The quantization of signals into but two levels (zero or one) in the binary moving-window detector results in a loss of about 1.5 to 2 dB in signal-to-noise ratio as compared to the ideal post-detection integrator.[53–55,57] When the amplitude is quantized into more than two levels, the loss is less. For example, quantization into four levels (2 bits), reduces the loss to about one-third that experienced in two-level, or binary, quantization (1 bit).[60]

A corollary advantage of the binary moving-window detector is that it is less sensitive to the effects of a single large interference pulse that might exist along with the target echo pulses. In the usual integrator, the full energy of the interference pulse is added. In the binary moving-window detector, however, it contributes no more than would any other pulse that crosses the first threshold since a 1 is recorded no matter what the amplitude. Similarly, the double-threshold detector has some advantage over the usual integrator when the background interference is not receiver noise but is nongaussian, as is some sea clutter and land clutter.

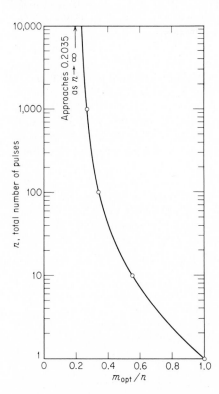

Figure 10.8 Optimum number of pulses m_{opt} (out of a maximum of n) for a binary moving window detector. A constant (nonfluctuating) target is assumed. (*After Swerling,*[52] *courtesy Rand Corporation.*)

Such clutter is characterized by having large "spiky" echoes. Although they may occur infrequently, they can be quite large and give undue weight in a conventional integrator. In the double-threshold detector they contribute the same as any threshold-crossing signal no matter what their amplitude. An analysis of the problem of detection of signals in clutter (or noise) that is nongaussian, indicates that a detector based on the *median* value crossing a threshold is significantly more efficient than the usual detection criterion based on the *mean* value.[61] The implementation of the binary moving window detector is similar to that of the median detector.

An estimate of the target's angular position may be made by locating the center of the group of n pulses.[56] This operation is called *beam splitting*. The moving-window detector has no prior knowledge of the target beginning. It must be sufficiently sensitive to quickly detect a region of increased density of 1s, yet it must not be so sensitive that it initiates false alarms due to noise. Once a target beginning is realized, the detector must be able to sense the end of the region of increased-density of 1s. Again, if it is too sensitive to change, the detector will tend to split targets.

Tapped-delay-line integrator. An integrator based on the tapped delay line is shown in Fig. 10.9. The time delay through the line is made equal to the total integration time, and the taps are spaced at intervals equal to the pulse-repetition period. The number of taps equals the number of pulses to be integrated. The outputs from each of the taps are tied together to form the sum of the previous n pulses. One of the advantages of this integrator is that any type of weighting may be applied to the individual pulses by simply inserting the proper attenuation at each tap of the delay line.

This form of integrator that sums the last n pulses is also known as the *moving-window detector* or the *analog moving-window detector*.[64] It, of course, can also be implemented in digital circuitry. It is similar to the binary moving-window detector discussed above, but it does not use a double threshold and it does not suppress the effects of large interference spikes as does the binary detector. It does not suffer the 1.5 to 2 dB loss of the binary detector and it can be employed to estimate the target's angle by beam splitting. The detection performance of the moving-window detector is about 0.5 dB worse than the optimal detector which weights the returned signals by the fourth power of the antenna voltage pattern.[63] The angular location of the target can be estimated from the output of this detector by taking the midpoint between the first and last crossings of the detection threshold, or by taking the maximum value of the running sum. After correcting for the bias, the accuracy of the location measurement is only about 20 percent worse than theoretical.[63]

Recirculating-delay-line integrator.[65-70] The delay-line integrator, or moving-window detector, described above requires that a number of pulses, equal to that expected from the target, be held in storage. A simplification can be had by recirculating the output through a single delay line (Fig. 10.10a) whose delay is equal to the pulse repetition period. The *recirculating-*

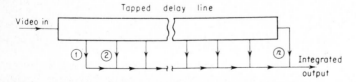

Figure 10.9 Pulse integrator using tapped delay line.

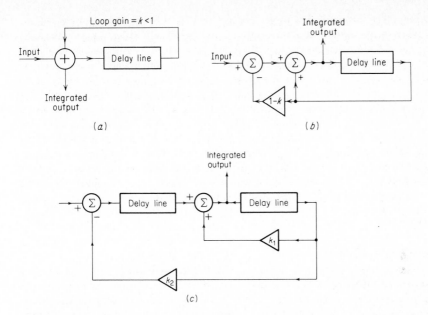

Figure 10.10 Recirculating-delay-line integrator, or feedback integrator, $k =$ loop gain < 1. (*a*) single delay loop; (*b*) double loop; (*c*) two-pole filter.

delay-line integrator, also called the *feedback integrator*, adds each new sweep to the *sum* of all the previous sweeps. To prevent unwanted oscillations, or "ringing," due to positive feedback, the sum must be attenuated by an amount k after each pass through the line. The factor k is the gain of the loop formed by the delay line and the feedback path. It must be less than unity for stable operation. The effect of $k < 1$ is that the integrator has imperfect "memory." The optimum value of k depends on the number of pulses received from the target (Sec. 2.6 and Ref. 65).

Analog delay lines that have been used in such devices include acoustic lines, lumped-parameter delay lines, electrostatic storage tubes, and magnetic drums or disks. They do not have, in general, sufficient stability to permit large values of k. When analog lines are used, loop gains are not much larger than 0.9 in the configuration of Fig. 10.10*a*, and 0.98 in the configuration of Fig. 10.10*b*. Since the effective number of pulses integrated is equal to $(1 - k)^{-1}$, a value of $k = 0.9$ corresponds to about 10 pulses integrated and $k = 0.98$ to about 50 pulses. However, when the recirculating-delay-line integrator is implemented with digital circuitry, values of k approaching unity can be achieved.[65]

The factor k in the recirculating-delay-line integrator of Fig. 10.10*a* is equivalent to an exponential weighting of the received pulses. It results in a loss of about 1.0 dB in signal-to-noise ratio as compared with the ideal postdetection integrator that weights the received pulses in direct proportion to the fourth power of the antenna beam pattern.[66] It is 0.5 dB less efficient than the moving window detector with uniform weights.[67] The double-loop integrator of Fig. 10.10*b* is a two-pole filter with a multiple pole. It is about 0.3 dB less efficient than the ideal integrator with optimum weights.[66] The double delay-line configuration of Fig. 10.10*c* is also a two-pole filter but unlike the double-loop integrator, the two poles need not be at the same location. Its detection performance is only 0.15 dB less efficient than the optimum.[66]

The recirculating delay-line integrators of Fig. 10.10 can be used to obtain an estimate of the angular location of the target. The target can be found, as in beam splitting, by taking the midpoint between the start and end of the threshold crossing. There will be a bias in the angle estimate that depends on the signal-to-noise ratio, but this bias can be estimated accurately.[67] The standard deviation of angular estimates obtained with the single delay-line is about 15 percent greater than the optimum estimates based on the Cramer-Rao lower bound.[66] If the maximum value of the output is used as an estimator of the target location, the bias is constant. However, the standard deviation of the estimates are 100 percent greater than the optimum.[66] Similarly the standard deviation of the double-loop integrator using the maximum value as the estimator produces a standard deviation 50 to 100 percent greater than the optimum. The two-pole filter of Fig. 10.10c, on the other hand, has a standard deviation 15 percent greater than optimum, and the estimator based on the maximum value has a constant bias.[66] Its relatively good angle-estimating accuracy, good detection performance, along with the relative simplicity of a feedback integrator makes the two-pole filter a good choice as an automatic detector for scanning radars.

10.8 CONSTANT-FALSE-ALARM-RATE (CFAR) RECEIVER

The threshold at the output of a radar receiver, as discussed in Sec. 2.5, is chosen so as to achieve a desired false-alarm probability. The false-alarm rate is quite sensitive to the threshold level. For example, a 1 dB change in the threshold can result in three orders of magnitude change in the false alarm probability (Fig. 2.5). It does not take much of a drift in the receiver gain, a change in receiver noise, or the presence of external noise or clutter echoes to inundate the radar display with extraneous responses.

If changes in the false-alarm rate are gradual, an operator viewing a display can compensate with a manual gain adjustment. It has been said[75] that the maximum increase in noise level that can be tolerated with a manual system using displays and operators is from 5 to 10 dB. But with an automatic detection and tracking (ADT) system, the tolerable increase is less than 1 dB. Excessive false alarms in an ADT system cause the computer to overload as it attempts to associate false alarms with established tracks or to generate new, but false, tracks. Manual control is too slow and imprecise for automatic systems. Some automatic, instantaneous means is required to maintain a constant false-alarm rate. Devices that accomplish this purpose are called *CFAR*.

A CFAR may be obtained by observing the noise or clutter background in the vicinity of the target and adjusting the threshold in accordance with the measured background. Figure 10.11 illustrates the *cell-averaging CFAR* which utilizes a tapped delay-line to sample the range cells to either side of the range cell of interest, or test cell. The output of the test cell is the radar output. The spacing between the taps is equal to the range resolution. The outputs from the delay line taps are summed. This sum, when multiplied by the appropriate constant, determines the threshold level for achieving the desired probability of false alarm. Thus the threshold varies continuously according to the noise or the clutter environment found within a range interval surrounding the range cell under observation.[76] This form of CFAR has sometimes been called *Adaptive Video Threshold*, or AVT. The particular CFAR shown in Fig. 10.11 does not sum all the range cells as described above, but sums the cells ahead of the test cell separately from the sum of the cells following the test cell. The threshold is determined by whichever of the two sums is the greater. This is done to minimize the generation of false alarms at the leading and trailing edges of abrupt clutter regions.[76] A small additional loss is incurred, however, compared to using all the taps to establish the threshold. (For example, with 32 reference cells, a 10^{-5} probability of false alarm, and 0.9 probability of detection, the

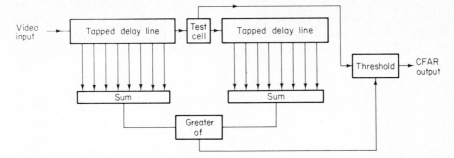

Figure 10.11 Cell averaging CFAR. In this version the greater of the outputs from the range cells ahead of or behind the cell of interest is used to set the threshold.

loss with a conventional cell-averaging CFAR is 0.8 dB. With the "greater-of" technique the loss is increased to 1.1 dB.[75])

Typically the number of taps used in a cell-averaging CFAR might vary from 16 to 20. The CFAR may be thought of as using the outputs of the sampled cells to estimate the unknown amplitude of the background noise or clutter. Because of the finite number of samples, the background is not completely known and a loss occurs compared to the ideal detector. For example, when only 10 independent samples are used, a loss of 3.5 dB is said to result for a probability of detection of 0.9 and probability of false alarm of 10^{-6}, when the background is broadband noise or clutter with a Rayleigh probability density.[74,75] With 20 independent samples the loss is 1.5 dB, and with 40 samples it is 0.7 dB. The above applies to single-hit detection. The loss decreases with increasing number of pulses integrated. With 10 pulses integrated, the loss with 10 independent samples decreases to 0.7 dB, and to 0.3 dB for 100 pulses integrated.[88]

If the target echo is large, energy can spill over into the adjacent range-resolution cells and affect the measurement of the average background. For this reason, the range cells surrounding the test cell are often omitted when averaging the background. As mentioned in Sec. 5.10, the resolution of targets in range is degraded by the adaptive threshold process.

The above description of the cell-averaging CFAR assumed that the background from which the threshold was set was determined by sampling in range. In those radars which extract the doppler frequency shift, as with a bank of doppler filters, the estimate of the background can be based on both the range and the doppler domains.[77] It is also possible to utilize data from the adjacent angle-resolution cells to establish the threshold.

A common assumption in the design of many CFARs is that the probability density function of the background noise amplitude is known (usually taken to be gaussian) except for a scale factor. Clutter, however, is often nonhomogeneous and thus nonstationary, as well as being of unknown probability density function in some cases. With such uncertainty in the background, a *nonparametric* method of detection must be used.[75,78-82] (A *nonparametric detector*, also called a *distribution-free detector*, in its most general form does not require prior knowledge of the probability density function of the noise or the signal.[39]) A nonparametric detector permits a constant false-alarm rate to be achieved for background noise that might be described by very broad classes of probability density functions. It has a greater loss than when the character of the noise is known and an optimum detector can be designed, but it does keep the false alarm rate fixed. One form of nonparametric detector is based on the "ranks" of observations in which the components of the observations are ranked in order of magnitude, and detection is based only on some function of these ranks.[39] This can be implemented with a *rank detector* which computes the ranks by pair-wise comparisons of the output from the range cell under test with each of the outputs from the neighboring range cells that sample the background noise.[78-81] After the detector ranks the sample under test with its neighboring samples, it integrates the ranks and a target is declared after testing against a fixed and an adaptive threshold.[80,81]

There are several other methods for achieving CFAR besides the use of cell averaging. One of the first CFAR receivers to be described in the literature used the output of a post-detection integrator (low-pass filter) to estimate the average noise. This was then applied as a feed-forward signal to control the threshold level to maintain the false-alarm rate constant.[71,72] The noise had to remain constant for a time corresponding to the total number of pulses returned from the target. Another CFAR technique is the hard limiter, sometimes called the *Dicke fix*.[72-74] This consists of a broadband IF filter followed by a hard limiter and a narrow-band matched filter. The hard limiter is set low enough to ensure that receiver noise is limited. Thus the output is unaffected by the level of the noise. A signal, however, causes the output of the matched filter to increase by a factor M, equal to the ratio of the bandwidth of the broadband filter to the bandwidth of the narrow-band (matched) filter. This form of CFAR may not be desirable with MTI since hard limiting reduces the improvement factor or the clutter attenuation (Sec. 4.8). Hard limiting also introduces an additional loss. The larger the ratio of the bandwidths M, the less the loss. For example, for a 0.5 probability of detection and a 10^{-3} probability of false alarm, the loss is 1.0 dB for $M = 100$, 2.2 dB for $M = 20$, and 7 dB for $M = 10$.[73] In essence the Dicke fix (that uses wideband limiting) samples the adjacent resolution cells in the frequency domain, as compared to the cell-averaging CFAR that samples the background in the adjacent resolution cells in the time domain.

When the product of the bandwidth B and the pulse width τ is greater than unity, as in a pulse compression radar, the loss is determined by the product $MB\tau$ rather than M. Thus, a loss of 1.0 dB corresponds to $MB\tau = 100$. The ratio of the bandwidths M therefore can be unity when $B\tau$ is large enough.[73] Both coded-pulse waveforms[85] and frequency-modulated waveforms have been considered for pulse compression radar with CFAR.[86]

The benefits of a large $B\tau$ for CFAR can also be obtained without transmitting a pulse-compression waveform.[87] A conventional pulse waveform can be transmitted, and on reception the received signal passed through a dispersive delay line (a pulse expansion network) that spreads the signal over a duration T. The output of the dispersive filter is hard limited and fed to a second dispersive delay line with a characteristic inverse to the first. The rms noise power at the output is smaller than the peak power from a point target by the factor BT. A detection threshold is set somewhere within this range of possible amplitude to allow point targets that are larger than the background to be detected. This dispersive CFAR may be placed either before or after the matched filter.

The log-FTC receiver described in Sec. 13.8 has CFAR properties when the background has a Rayleigh probability density function. The FTC, or fast time-constant, acts as a differentiating circuit, or high-pass filter, to remove the mean value of the clutter or noise. This function can be obtained with a more sophisticated filter consisting of a parallel combination of integrator and subtractor.[83] The integrator is a narrow-band filter that averages the order of ten range-resolution cells to establish the background level. A receiver implemented in this manner has been called a *log-CFAR*. The term *LOG/CFAR* has been applied to the cell-averaging CFAR which is preceded by a logarithmic detector.[84] The normalization of the threshold is accomplished in the LOG/CFAR by subtraction rather than by division as in the conventional cell-averaging CFAR. Also, the LOG/CFAR is capable of operating over a larger dynamic range of background noise levels, but it has poorer detectability for the same number of reference noise samples than the conventional cell-averaging CFAR.

CFAR is widely used to prevent clutter and noise interference from saturating the display of an ordinary radar and preventing targets from being obscured. It is also needed in ADT, or track-while-scan systems, to prevent the tracking computer from being overloaded by extraneous clutter targets or noise. CFAR, however, is not without its disadvantages and can be considered a necessary evil. It introduces an additional loss compared to optimum detection, and in some systems the number of pulses processed needs to be large to keep the loss low. More important, CFAR maintains the false-alarm rate constant at the expense of the probability of detection. Thus, it causes targets to be missed. Furthermore, the operator is usually given no indication that there may be missed detections. In radars with a simple CFAR, interference or hostile jamming can lower the sensitivity of the radar without the operator even being aware that it is happening since the jamming is not visible on the scope. Thus some means should be included to inform the operator when the detection probability has been lowered because of the CFAR action.

It is better to incorporate in the design of a radar means for eliminating unwanted signals *before* they enter the ADT than to depend on CFAR to eliminate them. Examples of signal processing techniques that eliminate unwanted signals without a severe penalty in detectability include MTI for clutter echoes, the low sidelobe antenna and/or sidelobe cancelers for jamming noise, and the sidelobe blanker or the prf filter for pulse interference.

REFERENCES

1. North, D. O.: An Analysis of the Factors Which Determine Signal/Noise Discrimination in Pulsed-carrier Systems, *RCA Tech. Rept.* PTR-6C, June 25, 1943 (ATI 14009). Reprinted in *Proc. IEEE*, vol. 51, pp. 1016–1027, July, 1963.

The following references (2 to 8) appear in *IRE Trans.*, vol. IT-6, no. 3, June, 1960, special issue on Matched Filters:

2. Turin, G. L.: An Introduction to Matched Filters, pp. 311–329.
3. Westerfield, E. C., R. H. Prager, and J. L. Stewart; Processing Gains against Reverberation (Clutter) Using Matched Filters, pp. 342–348.
4. Middleton, D.: On New Classes of Matched Filters and Generalizations of the Matched Filter Concept, pp. 349–360.
5. Sussman, S. M.: A Matched Filter Communications System for Multipath Channels, pp. 367–373.
6. Lerner, R. M.: A Matched Filter Detection System for Complicated Doppler Shifted Signals, pp. 373–385.
7. Cutrona, L. J., E. N. Leith, C. J. Palermo, and L. J. Porcello: Optical Data Processing and Filtering Systems, pp. 386–400.
8. Welte, G. R.: Quaternary Codes for Pulsed Radar, pp. 400–408.
9. Van Vleck, J. H., and D. Middleton: A Theoretical Comparison of Visual, Aural, and Meter Reception of Pulsed Signals in the Presence of Noise, *J. Appl. Phys.*, vol. 17, pp. 940–971, November, 1946.
10. Davenport, W. B., Jr., and W. L. Root: "Introduction to Random Signals and Noise," chap. 11, McGraw-Hill Book Company, Inc., New York, 1958.
11. Dwork, B. M.: Detection of a Pulse Superimposed on Fluctuation Noise, *Proc. IRE*, vol. 38, pp. 771–774, July, 1950.
12. Zadeh, L. A., and J. R. Ragazzini: Optimum Filters for the Detection of Signals in Noise, *Proc. IRE*, vol. 40, pp. 1223–1231, October, 1952.
13. Urkowitz, H.: Filters for the Detection of Small Radar Signals in Clutter, *J. Appl. Phys.*, vol. 24, pp. 1024–1031, August, 1953.
14. Raemer, H. R., and A. B. Reich: Correlation Devices Detect Weak Signals, *Electronics*, vol. 32, no. 21, pp. 58–60, May 22, 1959.
15. Lee, Y. W.: Application of Statistical Methods to Communication Problems, *MIT Research Lab. Electronics Tech. Rept.* 181, Sept. 1, 1950.
16. Lee, Y. W., T. P. Cheatham, Jr., and J. B. Wiesner: Application of Correlation Analysis to the Detection of Periodic Signals in Noise, *Proc. IRE*, vol. 38, pp. 1165–1171, October, 1950.

17. Lee, Y. W., and J. B. Wiesner: Correlation Functions and Communication Applications, *Electronics*, vol. 23, pp. 86–92, June, 1950.

18. Singleton, H. E.: A Digital Electronic Correlator, *Proc. IRE*, vol. 38, pp. 1422–1428, December, 1950.

19. Horton, B. M.: Noise-modulated Distance Measuring Systems, *Proc. IRE*, vol. 47, pp. 821–828, May, 1959.

20. Rudnick, P.: The Detection of Weak Signals by Correlation Methods, *J. Appl. Phys.*, vol. 24, pp. 128–131, February, 1953.

21. Fano, R. M.: Signal-to-noise Ratios in Correlation Detectors, *MIT Research Lab. Electronics Tech. Rept.* 186, Feb. 19, 1951.

22. George, S. F.: Effectiveness of Crosscorrelation Detectors, *Proc. Natl. Electronics Conf.* (Chicago), vol. 10, pp. 109–118, 1954.

23. Green, P. I., Jr.: The Output Signal-to-noise Ratio of Correlation Detectors, *IRE Trans.*, vol. IT-3, pp. 10–18, March, 1957.

24. Peterson, W. W., T. G. Birdsall, and W. C. Fox: The Theory of Signal Detectability, *IRE Trans.*, no. PGIT-4, pp. 171–212, September, 1954.

25. Middleton, D.: Statistical Criteria for the Detection of Pulsed Carriers in Noise, pts. I and II, *J. Appl. Phys.*, vol. 24, pp. 371–391, April, 1953; also letters to the editor by D. Middleton et al. in *J. Appl. Phys.*, January, 1954.

26. Woodward, P. M.: "Probability and Information Theory with Applications to Radar," McGraw-Hill Book Company, New York, 1953.

27. Woodward, P. M., and I. L. Davies: Information Theory and Inverse Probability in Telecommunications, *Proc. IEE*, vol. 99, pt. III, pp. 37–44, March, 1952.

28. Davies, I. L.: On Determining the Presence of Signals in Noise, *Proc. IEE*, pt. III, pp. 45–51, March, 1952.

29. Fuller, W.: "An Introduction to Probability Theory and Its Applications," 2d ed., vol. 1, p. 114, John Wiley & Sons, Inc., New York, 1957.

30. Peterson, W. W., and T. G. Birdsall: The Theory of Signal Detectability, *Univ. Mich., Dept. Elec. Eng. Tech. Rept.* 13, Contract DA-36-039 sc-15358, June, 1953.

31. Lawson, J. L., and G. E. Uhlenbeck (eds.): "Threshold Signals," MIT Radiation Laboratory Series, vol. 24, sec. 7.5, McGraw-Hill Book Company, New York, 1950.

32. Middleton, D.: Statistical Theory of Signal Detection, *IRE Trans.*, no. PGIT-3, pp. 26–51, March, 1954.

33. Van Meter, D., and D. Middleton: Modern Statistical Approaches to Reception in Communication Theory, *IRE Trans.*, no. PGIT-4, pp. 119–141, September, 1954.

34. Middleton, D., and D. Van Meter: Detection and Extraction of Signals in Noise from the Viewpoint of Statistical Decision Theory, *J. Soc. Ind. Appl. Math.*, vol. 3, pp. 192–253, December, 1955, and vol. 4, pp. 86–119, June, 1956.

35. Bussgang, J. J., and D. Middleton: Optimum Sequential Detection of Signals in Noise, *IRE Trans.*, vol. IT-1, pp. 5–18, December, 1955.

36. Blasbalg, H.: The Relationship of Sequential Filter Theory to Information Theory and Its Application to the Detection of Signals in Noise by Bernoulli Trials, *IRE Trans.*, vol. IT-3, pp. 122–131, June, 1957.

37. Wald, A.: "Sequential Analysis," John Wiley & Sons, Inc., New York, 1947.

38. Marcus, M. B., and P. Swerling: Sequential Detection in Radar with Multiple Resolution Elements, *IRE Trans.*, vol. IT-8, pp. 237–245, April, 1962.

39. Caspers, J. W.: Automatic-detection Theory, chap. 15 of "Radar Handbook," M. I. Skolnik (ed.), McGraw-Hill Book Company, Inc., New York, 1970.

40. Bussgang, J. J.: Sequential Methods in Radar Detection, *Proc. IEEE*, vol. 58, pp. 731–743, May, 1970.

41. Guarguaglini, P. F., and F. Marcoz: The DFTSD: A Sequential Suboptimum Processor for Multiple-range-bin Radar Systems, *IEEE Trans.*, vol. AES-10, pp. 193–203, March, 1974.

42. Brennan, L. E., and F. S. Hill, Jr.: A Two-step Sequential Procedure for Improving the Cumulative Probability of Detection in Radars, *IEEE Trans.*, vol. MIL-9, pp. 278–287, July–October, 1965.

43. Marcum, J. I.: A Statistical Theory of Target Detection by Pulsed Radar, Mathematical Appendix, *IRE Trans.*, vol. IT-6, pp. 145–267, April, 1960.

44. Green, B. A., Jr.: Radar Detection Probability with Logarithmic Detectors, *IRE Trans.*, vol. IT-4, pp. 50–52, March, 1958.

45. Hansen, V. G.: Postdetection Integration Loss for Logarithmic Detectors, *IEEE Trans.*, vol. AES-8, pp. 386–388, May, 1972. See correction, AES-10, p. 168, January, 1974.

46. Bendat, J. S.: "Principles and Applications of Random Noise Theory," chap. 10, John Wiley & Sons, Inc., New York, 1958.

47. Benjamin, R.: Man and Machine in the Extraction and Use of Radar Information, *J. Brit. IRE*, vol. 26, pp. 309–316, October, 1963.
48. Tucker, D. G.: Detection of Pulse Signals in Noise: Trace-to-Trace Correlation in Visual Displays, *J. Brit. IRE*, vol. 17, pp. 319–329, June, 1957.
49. Skolnik M. I., and D. G. Tucker: Discussion on " Detection of Pulse Signals in Noise: Trace-to-Trace Correlation in Visual Displays," *J. Brit. IRE*, vol. 17, pp. 705–706, December, 1957.
50. Payne-Scott, R.: The Visibility of Small Echoes on Radar PPI Displays, *Proc. IRE*, vol. 36, pp. 180–196, February, 1948.
51. Baker, C. H.: " Man and Radar Displays," The Macmillan Company, New York, 1962.
52. Swerling, P.: The " Double Threshold " Method of Detection, *Rand Corp. Rept.* RM-1008, Dec. 17, 1952, Santa Monica, Calif.
53. Harrington, J. V.: An Analysis of the Detection of Repeated Signals in Noise by Binary Integration, *IRE Trans.*, vol. IT-1, pp. 1–9, March, 1955.
54. Schwartz, M.: A Coincidence Procedure for Signal Detection, *IRE Trans.*, vol. IT-2, pp. 135–139, December, 1956.
55. Drukey, D. L., and L. C. Levitt: Radar Range Performance, *Hughes Aircraft Co. Tech. Mem.* 560, Aug. 1, 1957, Culver City, Calif.
56. Dinneen, G. P., and I. S. Reed: An Analysis of Signal Detection and Location by Digital Means, *IRE Trans.*, vol. IT-2, pp. 29–38, March, 1956.
57. Dillard, G. M.: A Moving-Window Detector for Binary Integration, *IEEE Trans.*, vol. IT-13, pp. 2–6, January, 1967.
58. Worley, R.: Optimum Thresholds for Binary Integration, *IEEE Trans.*, vol. IT-14, pp. 349–353, March, 1968.
59. Walker, J. F.: Performance Data for a Double-Threshold Detection Radar, *IEEE Trans.*, vol. AES-7, pp. 142–146, January, 1971. Comment by V. G. Hansen, p. 561, May, 1971.
60. Hansen, V. G.: Optimization and Performance of Multilevel Quantization in Automatic Detectors, *IEEE Trans.*, vol. AES-10, pp. 274–280, March, 1974.
61. Trunk, G. V., and S. F. George: Detection of Targets in Non-Gaussian Sea Clutter, *IEEE Trans.*, vol. AES-6, pp. 620–628, September, 1970.
62. Popov, G. P.: " Engineering Psychology in Radar," JPRS-55522, Joint Publications Research Service, Washington, D.C., 23 March, 1972.
63. Trunk, G. V.: Radar Signal Processing, "Advances in Electronics and Electron Physics," vol. 45, edited by L. Marton, Academic Press, Inc., New York, 1978, pp. 203–252. Also Trunk, G. V.: Survey of Radar Signal Processing, Naval Research Laboratory Report 8117, Washington, D.C., June 21, 1977.
64. Hansen, V. G.: Performance of the Analog Moving Window Detector, *IEEE Trans.*, vol. AES-6, pp. 173–179, March, 1970.
65. Trunk, G. V.: Detection Results for Scanning Radars Employing Feedback Integration, *IEEE Trans.*, vol. AES-6, pp. 522–527, July, 1970.
66. Cantrell, B. H., and G. V. Trunk: Angular Accuracy of a Scanning Radar Employing a Two-Pole Filter, *IEEE Trans.*, vol. AES-9, pp. 649–653, September, 1973.
67. Trunk, G. V.: Comparison of Two Scanning Radar Detectors: The Moving Window and the Feedback Integrator, *IEEE Trans.*, vol. AES-7, pp. 395–398, March, 1971.
68. Urkowitz, H.: Analysis and Synthesis of Delay Line Periodic Filters, *IRE Trans.*, vol. CT-4, pp. 41–53, June, 1957.
69. Cooper, D. C., and J. W. R. Griffiths: Video Integration in Radar and Sonar Systems, *J. Brit. IRE*, vol. 21, pp. 421–433, May, 1961.
70. Palmer, D. S., and D. C. Cooper: An Analysis of the Performance of Weighted Integrators, *IEEE Trans.*, vol. IT-10, pp. 296–302, October, 1964.
71. Siebert, W. M.: Some Applications of Detection Theory to Radar, *IRE Natl. Conv. Record*, vol. 6, pt. 4, pp. 5–14, 1958.
72. Hansen, V. G., and A. J. Zöttl: The Detection Performance of the Siebert and Dicke-Fix CFAR Radar Detectors, *IEEE Trans.*, vol. AES-7, pp. 706–709, July, 1971.
73. Carpentier, M. H.: " Radars: New Concepts," Gordon and Breach, New York, 1968, sec. 4.6.
74. Nathanson, F. E.: " Radar Design Principles," McGraw-Hill Book Company, New York, 1969, chap. 4.
75. Hansen, V. G.: Constant False Alarm Processing in Search Radars, *International Conference on Radar—Present and Future*, Oct. 23–25, 1973, pp. 325–332, IEE Publication No. 105.
76. Hubbard, J. V.: Digital Automatic Radar Data Extraction Equipment, *J. Brit. IRE*, vol. 26, pp. 397–405, November, 1963.

77. Finn, H. M., and R. S. Johnson: Adaptive Detection Mode with Threshold Control as a Function of Spatially Sampled Clutter-Level Estimates, *RCA Rev.*, vol. 29, pp. 414–464, September, 1968.

78. Dillard, G. M., and C. E. Antoniak: A Practical Distribution-Free Detection Procedure for Multiple-Range-Bin Radars, *IEEE Trans.*, vol. AES-6, pp. 629–635, November, 1970.

79. Hansen, V. G., and B. A. Olsen: Nonparametric Radar Extraction Using a Generalized Sign Test, *IEEE Trans.*, vol. AES-7, pp. 942–950, September, 1971.

80. Trunk, G. V., B. H. Cantrell, and F. D. Queen: Modified Generalized Sign Test Processor for 2-D Radar, *IEEE Trans.*, vol. AES-10, pp. 574–582, September, 1974.

81. Cantrell, B. H., G. V. Trunk, F. D. Queen, J. D. Wilson, and J. J. Alter: Automatic Detection and Integrated Tracking System, *Record of the IEEE 1975 International Radar Conference*, pp. 391–395, IEEE Publication 75 CHO 938-1 AES.

82. Dillard, G. M.: Mean-Level Detection Utilizing a Digital First-Order Recursive Filter, *IEEE Trans.*, vol. AES-12, pp. 793–798, November, 1976.

83. Taylor, J. W., Jr., and J. Mattern: Receivers, chap. 5 of "Radar Handbook," M. I. Skolnik (ed.), McGraw-Hill Book Co., Inc., New York, 1970.

84. Hanson, V. G., and H. R. Ward: Detection Performance of the Cell Averaging LOG/CFAR Receiver, *IEEE Trans.*, vol. AES-8, pp. 648–652, September, 1972.

85. Taylor, J. W., Jr., and G. Brunins: Long-Range Surveillance Radars for Automatic Control Systems, *Record of the IEEE 1975 International Radar Conference*, pp. 312–317.

86. Mortley, W. S., and S. N. Radcliffe: Pulse Compression and Signal Processing, *International Conference on Radar—Present and Future*, Oct. 23–25, 1973, pp. 292–296, IEE Publication no. 105.

87. Ward, H. R.: Dispersive Constant False Alarm Rate Receiver, *Proc. IEEE*, vol. 60, pp. 735–736, June, 1972.

88. Mitchell, R. L., and J. F. Walker: Recursive Methods for Computing Detection Probabilities, *IEEE Trans.*, vol. AES-7, pp. 671–676, July, 1971.

89. Gupta, D. V., J. F. Vetelino, T. J. Curry, and J. T. Francis: An Adaptive Threshold System for Nonstationary Noise Backgrounds, IEEE Trans., vol. AES-13, pp. 11–16, January, 1977.

90. Preston, G. W.: The Search Efficiency of the Probability Ratio Sequential Search Radar, *IRE Intern. Conv. Record*, vol. 8, pt. 4, pp. 116–124, 1960.

91. This expression was suggested by Warren D. White, one of the reviewers of the revised manuscript of this book.

ELEVEN

EXTRACTION OF INFORMATION AND WAVEFORM DESIGN

11.1 INTRODUCTION

In Chap. 10, which was concerned with detection, it was stated that the observation of radar signals in noise could be divided into (1) the *detection* of the presence of signals in noise and (2) the *extraction* of the target information contained in the signal. The analysis of the detection of signals in noise is based in large part on the mathematics of *statistical hypothesis testing*. Similarly, extraction of information from radar signals is a problem in *statistical parameter estimation*. The detection of signals and the extraction of information are not totally independent processes since either one without the other is meaningless.

   This chapter discusses some of the concepts involved in the extraction of information from a radar signal. The next section considers in a qualitative manner the type of target information available from a radar signal. The theoretical accuracies with which range, relative velocity (doppler frequency), and angle of arrival can be determined are derived in Sec. 11.3. This is followed by a treatment of the ambiguity function and its effect on waveform design. Pulse compression waveforms and processing are discussed in Sec. 11.5. Although pulse compression radar is of interest for other than the extraction of information, it is included in this chapter because of its close relation to the ambiguity function. The final section reviews several radar techniques that might be used to distinguish one class of target from another.

11.2 INFORMATION AVAILABLE FROM A RADAR

A radar obtains information about a target by comparing the received echo signal with the transmitted signal. The availability of an echo signal indicates the *presence* of a reflecting target; but knowing a target is present is of little use by itself. Something more must be known. Therefore, a radar provides the *location* of the target as well as its presence. It can also provide information about the type of target. This is known as *target classification*.

   The time delay between the transmission of the radar signal and the receipt of an echo is a measure of the distance, or *range*, to the target. The range measurement is usually the most significant a radar makes. No other sensor has been able to compete with radar for determining the range to a distant target. A typical radar might be able to measure range to an accuracy of several hundred meters, but accuracies better than a fraction of a meter are practical. Radar ranges might be as short as that of the police traffic-speed-meter, or as long as the distances to the nearby planets.

Almost all radars utilize directive antennas. A directive antenna not only provides the transmitting gain and receiving aperture needed for detecting weak signals, but its narrow beamwidth allows the target's *direction* to be determined. A typical radar might have a beam-width of perhaps one or two degrees. The angular resolution is determined by the beamwidth, but the angular accuracy can be considerably better than the beamwidth. A ten to one *beam splitting* would not be unusual for a typical radar. Some radars can measure angular accuracy considerably better than this. An rms error of 0.1 mrad is possible with the best tracking radars.

The echo from a moving target produces a frequency shift due to the doppler effect, which is a measure of the *relative velocity*. Relative velocity also can be determined from the rate of change of range. Tracking radars often measure relative velocity in this manner rather than use the doppler shift. However, radars for the surveillance and tracking of extraterrestrial targets, such as satellites and spacecraft, might employ the doppler shift to measure directly the relative velocity, but it is seldom used for this purpose in aircraft-surveillance radars. Instead, aircraft-surveillance radars use the doppler frequency shift to separate the desired moving targets from the undesired fixed clutter echoes, as in MTI radars.

If the target can be viewed from many directions, its *shape* can be determined. Space object identification (SOI) radars are an example of those that extract target shape information. The synthetic aperture radar (SAR) which maps the terrain is another example. Radars that determine the shape of a target are sometimes called *imaging* radars.

To obtain the target size or shape requires resolution in range and in angle. Good range resolution is generally easier to achieve than comparable resolution in angle. In some radar applications it is possible to utilize resolution in the doppler frequency shift as a substitute for resolution in angle, if there is relative motion between the distributed target and the radar. Resolution is possible since each element of the distributed target has a different relative velocity. This principle has been used in synthetic aperture radars for ground mapping, inverse SAR for SOI and the imaging of planets, and in the scatterometer for measuring the ground or sea echo as a function of incidence angle.

Internal motions of the target such as the rotation of aircraft engines, vibrations of vehi-cles, the spinning of a satellite, or the rotation of antennas can also provide information about the target. The different responses to different polarizations of the electromagnetic wave provide information on the target symmetry. It is this property that permits echoes from symmetrical raindrops to be discriminated against in favor of echoes from asymmetrical aircraft. Many microwave radars use circular polarization for this purpose. The nature of the surface roughness can be inferred from the radar echo, as can the dielectric properties of the scattering surface. The former has been applied to the measurement of sea state (from satel-lites), and the latter was used in early radar astronomy to probe the nature of the moon's surface.

11.3 THEORETICAL ACCURACY OF RADAR MEASUREMENTS

The ability of a radar to detect the presence of an echo signal is fundamentally limited by noise. Likewise, noise is the factor that limits the accuracy with which the radar signals may be estimated. The parameters usually of interest in radar applications are the range (time delay), the range rate (doppler velocity), and the angle of arrival. The amplitude of the echo signal might also be measured, but its precise value is usually not important except insofar as it influences the signal-to-noise ratio.

In this section the theoretical accuracies of radar measurements will be derived. To simplify the analysis, it is assumed that the signal is large compared with the noise. This is a reasonable assumption since the signal-to-noise ratio must be relatively large if the detection decision is to be reliable (Sec. 2.5). Furthermore, as will be evident later, large signal-to-noise ratios are necessary for accurate measurements. It is also assumed that the error associated with a measurement of a particular parameter is independent of the errors in any of the other parameters. The validity of this assumption depends upon the availability of a large signal-to-noise ratio. (Further information regarding radar measurements can be found in Ref. 7.)

Theoretical radar accuracies may be derived by a variety of methods including those based on (1) simple geometrical relationships between signal, noise, and the parameter to be measured, (2) inverse probability, (3) a suitably selected gating function preceded by a matched filter, and (4) the estimate of the variance using the likelihood function. The measure of the error is the root mean square of the difference between the measured value and the true value. The disturbance limiting the accuracy of the radar measurement is assumed to be the receiver noise. It is furthermore assumed that bias errors have been removed.

It will be shown that the rms error δM of a radar measurement M can be expressed as

$$\delta M = \frac{kM}{\sqrt{2E/N_0}} \tag{11.1}$$

where E is the received signal energy, N_0 is the noise power per unit bandwidth, and k is a constant whose value is of the order of unity. For a time-delay measurement, k depends on the shape of the frequency spectrum $S(f)$, and M is the rise time of the pulse; for a doppler frequency measurement, k depends on the shape of the time waveform $s(t)$ and M is the spectral resolution, or the reciprocal of the observation time; and for an angle measurement k depends on the shape of the aperture illumination $A(x)$, and M is the beamwidth.

Range-accuracy – leading-edge measurement. The measurement of range is the measurement of time delay $T_R = 2R/c$, where c is the velocity of light. One method of determining range with a pulsed waveform is to measure the time at which the leading edge of the pulse crosses some threshold[1] (Fig. 11.1). The pulse uncorrupted by noise is shown by the solid curve. The shape of the pulse is not perfectly rectangular; the rise and decay times are not zero, for this would require an infinite bandwidth. The effect of noise is to perturb the shape of the pulse and to shift the time of threshold crossing as shown by the dashed curve. The maximum slope (rate of rise) of the leading edge of a rectangular pulse of amplitude A at the output of a video filter is

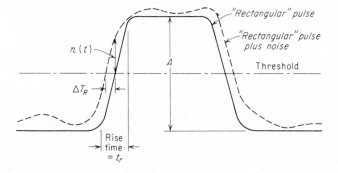

Figure 11.1 Measurement of time delay using the leading (or trailing) edge of the pulse. Solid curve represents echo pulse uncorrupted by noise. Dashed curve represents the effort of noise.

A/t_r, where t_r is the rise time. For large signal-to-noise ratios the slope of the pulse corrupted by noise is essentially the same as the slope of the uncorrupted pulse. From Fig. 11.1 the slope of the pulse in noise may be written as $n(t)/\Delta T_R$, where $n(t)$ is the noise voltage in the vicinity of the threshold crossing and ΔT_R is the error in the time-delay measurement. Equating the two expressions for the slope gives

$$\Delta T_R = \frac{n(t)}{A/t_r} \tag{11.2}$$

or

$$[\overline{(\Delta T_R)^2}]^{1/2} = \delta T_R = \frac{t_r}{(A^2/\overline{n^2})^{1/2}} = \frac{t_r}{(2S/N)^{1/2}} \tag{11.3}$$

where $A^2/\overline{n^2}$ is the video signal-to-noise (power) ratio. The last part of Eq. (11.3) follows from the fact that the video signal-to-noise power ratio is equal to twice the IF signal-to-noise power ratio (S/N), assuming a linear-detector law and a large signal-to-noise ratio.

If the rise time of the video pulse is limited by the bandwidth B of the IF amplifier, then $t_r \approx 1/B$. Letting $S = E/\tau$ and $N = N_0 B$, where E is the signal energy, N_0 the noise power per unit bandwidth, and τ the pulse width, the error in the time delay can be written

$$\delta T_R = \left(\frac{\tau}{2BE/N_0}\right)^{1/2} \tag{11.4}$$

If a similar independent time-delay measurement is made at the trailing edge of the pulse, the two combined measurements will be improved by $\sqrt{2}$, or

$$\delta T_R = \left(\frac{\tau}{4BE/N_0}\right)^{1/2} \qquad \text{rectangular pulse} \tag{11.5}$$

For constant pulse amplitude A, the rms time-delay error given by Eq. (11.3) is proportional to the rise time and is independent of the pulse width. An improvement in accuracy is obtained, therefore, by decreasing the rise time (increasing the bandwidth) or increasing the signal-to-noise ratio.

The actual estimate of the time delay obtained by determining when the leading or trailing edge of the pulse crosses a threshold will depend on the value of the threshold relative to the peak value of the pulse. This can result in an error in measurement even if no noise is present. It is possible to alleviate this situation, however, by using an adaptive threshold in which the level of the threshold is always a fixed fraction of the pulse amplitude.[2]

Range accuracy using gating signals and matched filters.[3] Consider the receiver block diagram shown in Fig. 11.2 consisting of a multiplier followed by a low-pass filter (or integrator). The two inputs to the multiplier are the received echo signal $y(t)$ and a *gating signal* $g(t - T_R)$. The time $T_R = 2R/c$ is the estimate of the true delay time T_0. The purpose of the gating signal is to aid in extracting an esimate of T_0. As before, it is assumed that the signal is large compared with noise; consequently there is no doubt as to the existence or the approximate position of the echo pulse.

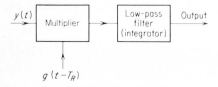

Figure 11.2 Receiver for measuring range (time delay) using a gating signal $g(t - T_R)$, where T_R is the estimate of the true time delay. Note the similarity to the cross-correlation receiver of Fig. 10.3.

The echo signal $y(t)$ is composed of signal and noise, $s(t - T_0) + n(t)$, where $s(t - T_0)$ is the echo signal in the absence of noise. The contributions due to signal s_0 and noise n_0 at the output of the low-pass filter may be expressed as

$$s_o(T_R - T_0) = \int g(t - T_R)s(t - T_0) \, dt \tag{11.6}$$

$$n_o(T_R) = \int g(t - T_R)n(t) \, dt \tag{11.7}$$

Defining $\Delta T_R = T_R - T_0$, the form of the output $s_o(\Delta T_R)$ with an optimum gating signal should be an odd function. Its value is zero at $\Delta T_R = 0$, and its even-order derivatives are zero. For small values of ΔT_R, the output will be directly proportional to ΔT_R. (Thus it is similar to the angle-error detector of a monopulse tracking radar.)

The ratio of the root mean-square noise voltage $(\overline{n_o^2})^{1/2}$ to the slope M of the output $s_o(\Delta T_R)$ evaluated at $\Delta T_R = 0$ will be taken as a measure of the rms error in time measurement, or

$$\delta T_R = (\overline{\Delta t^2})^{1/2} = \frac{(\overline{n_o^2})^{1/2}}{M} \tag{11.8}$$

where

$$M = \left[\frac{\partial s_o(\Delta T_R)}{\partial \, \Delta T_R} \right]_{\Delta T_R = 0}$$

The error is illustrated in Fig. 11.3. The receiver output characteristic with signal s_o only is represented by the solid curve. The effect of noise is shown by the dashed curve. Noise displaces the zero crossing by an amount Δt.

Using the calculus of variations, Mallinckrodt and Sollenberger[3] show that the Fourier transform $S_g(f)$ of the optimum gating function which minimizes the time-delay error [Eq. (11.8)] is given, except for an arbitrary constant factor, by

$$S_g(f) = \frac{j2\pi f S(f)}{|N_i(f)|^2} \tag{11.9}$$

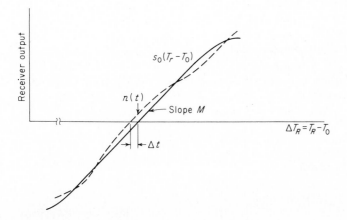

Figure 11.3 Effect of noise $n(t)$ in shifting the apparent zero crossing of the output $s_0(T_R - T_0)$ of the gating receiver of Fig. 11.2.

where $S(f)$ and $N_i(f)$ are the Fourier transforms of the input signal $s(t - T_0)$ and the input noise $n(t)$, respectively. The factor $j2\pi f$ corresponds to a differentiation. Therefore the optimum gating waveform appears as the time derivative of the received waveform, if the noise spectrum is constant. By applying the convolution theorem to Eq. (11.6), the Fourier transform of the output $S_o(f)$ is equal to $S_g(f)S^*(f)$, or

$$S_o(f) = \frac{j2\pi f \, |S(f)|^2}{|N_i(f)|^2} \tag{11.10}$$

This is a transform of an odd function.

The gating signal and the matched filter are related to one another. In Sec. 10.2 it was shown that the frequency-response function of the matched filter was given, except for a constant and a time delay, by

$$H(f) = \frac{S^*(f)}{|N_i(f)|^2} \tag{11.11}$$

where in the notation of Sec. 10.2, $S^*(f)$ is the complex conjugate of the Fourier transform of the input signal in the absence of noise and $N_i(f)$ is the spectrum of the input noise (the Fourier transform of the input noise voltage). Therefore, if the multiplier of Fig. 11.2 is preceded by a matched filter, the form of the optimum gating signal will be

$$S_g(f) = j2\pi f \tag{11.12}$$

This is the Fourier transform of the doublet impulse $u_2(t)$, or first derivative of the impulse function $\delta'(t)$. The gating waveform is therefore

$$g(t - T_R) = u_2(t - T_R) = \delta'(t - T_R) \tag{11.13}$$

The above analysis indicates that optimum range processing consists in passing the echo signal through a matched filter followed by a gating in time that samples the signal waveform at the instant before and the instant following the time T_R. The difference between these two samples is a measure of the difference between the estimated delay time T_R and the true delay time T_0. In some respects, the gating process is analogous to the split-range-gate technique for range-tracking radars described in Sec. 5.6, except that the sampling gates described here are of infinitesimal width whereas they are usually of the order of a pulse width in range tracking.

Substituting the optimum gating signal into the expression for accuracy gives (Ref. 3, Eq. 12)

$$\overline{\delta T_R^2} = \frac{1}{4 \int_0^\infty (2\pi f)^2 |S(f)|^2/|N_i(f)|^2 \, df} \tag{11.14}$$

If the noise spectrum is a constant with spectral energy equal to N_0 watts/Hz of bandwidth, the mean square error is

$$(\delta T_R)^2 = \frac{N_0}{4 \int_0^\infty (2\pi f)^2 |S(f)|^2 \, df} \tag{11.15}$$

An effective bandwidth β may be defined such that

$$\beta^2 = \frac{\int_{-\infty}^\infty (2\pi f)^2 |S(f)|^2 \, df}{\int_{-\infty}^\infty |S(f)|^2 \, df} = \frac{1}{E} \int_{-\infty}^\infty (2\pi f)^2 |S(f)|^2 \, df \tag{11.16}$$

where E is the signal energy. This definition of bandwidth is unlike those discussed previously in this book and is not simply related to either the half-power bandwidth or the noise bandwidth. In terms of the effective bandwidth, the time-delay error becomes

$$\delta T_R = \frac{1}{\beta(2E/N_0)^{1/2}} \tag{11.17}$$

and the range error is $\delta R = (c/2)\,\delta T_R$.

The effective bandwidth as defined by Eq. (11.16) was first introduced by Gabor[4] and has been used by Woodward[5] in his treatment of detection and accuracy by means of inverse probability. Both Gabor and Woodward define the effective bandwidth in terms of the complex-frequency representation, while the definition presented above is in terms of the real time waveforms. In essence, β^2 is the normalized second moment of the spectrum $|S(f)|^2$ about the mean (here taken to be at zero frequency). The larger the value of β^2, the more accurate will be the range measurement.

Examples of time-delay (range) accuracy. The computation of β^2 for a perfectly rectangular pulse—one with zero rise time and zero fall time—results in $\beta^2 = \infty$. This implies that the minimum rms range error for a perfectly rectangular pulse is zero and that the range measurement can be made with no error. In practice, however, pulses are not perfectly rectangular since a zero rise time or a zero fall time requires an infinite bandwidth. Finite bandwidths result in finite rise times and finite β^2.

In order to compute β^2 (and the range error) for a practical "rectangular" pulse, it will be assumed that the pulse spectrum is of the form $S(f) = (\sin \pi f\tau)/\pi f$, just as with the perfectly rectangular pulse of width τ, but that the bandwidth is limited to a finite value B as shown in Fig. 11.4.[6] This is equivalent to passing a perfectly rectangular pulse through a filter of width B. The time waveform of the output will be a pulse of width approximately τ, but the slopes of the leading and trailing edges are finite. An example of the shape of a bandwidth-limited rectangular pulse is shown in Fig. 11.5. The integrals in the expression for β^2 [Eq. 11.16] extend, in this instance, from $-B/2$ to $+B/2$ instead of from $-\infty$ to $+\infty$. Thus the effective bandwidth is

$$\beta^2 = \frac{(2\pi)^2 \displaystyle\int_{-B/2}^{B/2} f^2(\sin^2 \pi f\tau)/\pi^2 f^2 \, df}{\displaystyle\int_{-R/2}^{B/2} (\sin^2 \pi f\tau)/\pi^2 f^2 \, df} = \frac{1}{\tau^2}\,\frac{\pi B\tau - \sin \pi B\tau}{\mathrm{Si}\,(\pi B\tau) + (\cos \pi B\tau - 1)/\pi B\tau} \tag{11.18}$$

where $\mathrm{Si}\, x$ is the sine integral function defined by $\displaystyle\int_0^x (\sin u)/u \, du$. As $B\tau \to \infty$, the product $\beta^2\tau^2 \to 2B\tau$, or

$$\beta^2 \approx \frac{2B}{\tau} \qquad \text{for large } B\tau \tag{11.19}$$

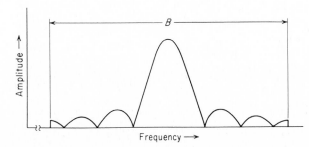

Figure 11.4 Spectrum of a bandwidth-limited "rectangular pulse."

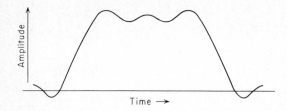

Figure 11.5 Shape of rectangular pulse after passing through a band-limited rectangular filter.

This is a reasonably good approximation for almost any value of $B\tau$. Therefore the rms error in the time-delay measurement for a "rectangular" pulse of width τ, limited to a bandwidth B, is approximately

$$\delta T_R \approx \left(\frac{\tau}{4BE/N_0}\right)^{1/2} \qquad \text{bandwidth-limited rectangular pulse} \qquad (11.20)$$

The pulse width τ in the above expression is that of the perfectly rectangular pulse before band limiting. It is a good approximation to the width of the band-limited pulse when $B\tau$ is large.

Equation (11.20) is the same as Eq. (11.5) but is derived in a totally different manner. This is a rather interesting result, for it seems to imply that the value of time delay obtained with a straightforward method like the leading-edge technique can be as accurate as the optimum processing technique described above. It was assumed in the analysis of a bandwidth-limited pulse that a matched filter was employed. If the spectral width of the "rectangular" pulse is changed, the matched filter must be changed also.

The rms time-delay error for a trapezoidal-shaped pulse of width $2T_1$ across the top, flat portion and with rise and fall times of T_2 may be shown from Eqs. (11.16) and (11.17) to be

$$\delta T_R = \left(\frac{T_2^2 + 3T_1 T_2}{6E/N_0}\right)^{1/2} \qquad \text{trapezoidal pulse} \qquad (11.21)$$

When the trapezoidal pulse approaches in shape the rectangular pulse, that is, when $T_1 \gg T_2$, Eq. (11.21) becomes

$$\delta T_R \approx \left(\frac{T_1 T_2}{2E/N_0}\right)^{1/2} \qquad (11.22)$$

The bandwidth B is approximately the reciprocal of the rise time T_2, and if $T_1 \approx \tau/2$, where τ is the pulse width, the rms error is

$$\delta T_R \approx \left(\frac{\tau}{4BE/N_0}\right)^{1/2}$$

which is the same as that derived previously for the bandwidth-limited rectangular pulse.

By letting $T_1 \to 0$ in Eq. (11.21), the time-delay error for a triangular-shaped pulse is obtained. Calling the width at the base of the triangle $\tau_B = 2T_2$, the rms error is

$$\delta T_R = \frac{\tau_B}{\sqrt{12}(2E/N_0)^{1/2}} \qquad \text{triangular pulse} \qquad (11.23)$$

Consider a pulse described by the gaussian function

$$s(t) = \exp\left(-\frac{1.384t^2}{\tau^2}\right) \qquad (11.24)$$

where τ is the half-power pulse width. The gaussian-shaped pulse is sometimes specified in those applications where interference with equipments operating at nearby frequencies is to be avoided. The gaussian pulse is well suited for this purpose since its spectrum decays rapidly on either side of the carrier frequency. Its rms range error is

$$\delta T_R = \frac{\tau}{1.18(2E/N_0)^{1/2}} = \frac{1.18}{\pi B(2E/N_0)^{1/2}} \qquad \text{gaussian pulse} \qquad (11.25)$$

where B is the half-power bandwidth of the gaussian-pulse spectrum.

The effective bandwidth of a waveform with a uniform frequency spectrum of width B is $\beta = \pi B/\sqrt{3}$. The waveform which gives rise to a uniform frequency spectrum is of the form $(\sin x)/x$, where $x = \pi B\tau$. The rms time-delay error is therefore

$$\delta T_R = \frac{\sqrt{3}}{\pi B(2E/N_0)^{1/2}} \qquad \frac{\sin x}{x} \text{ waveform} \qquad (11.26)$$

The radar waveform which yields the most accurate time-delay measurement, all other factors being equal, is the one with the largest value of effective bandwidth β. If the bandwidth is limited by external factors to a value B, the spectrum which produces the largest β, and hence the most accurate range measurement, would be one which crowded all its energy at the two ends of the band; that is,

$$S(f) = \delta(f - f_0 - B/2) + \delta(f - f_0 + B/2)$$

where f_0 is the carrier frequency and $\delta(x)$ is the delta function. The corresponding time waveform consists of two sine waves at frequencies $f_0 \pm B/2$. This is the two-frequency CW radar waveform discussed in Sec. 3.5. The two-frequency CW radar spectrum (and its corresponding waveform) are not always suitable in practice since they lead to ambiguous measurements if the frequency separation B between the two sine waves is greater than $c/2R_b$, where c is the velocity of propagation and R_b is the maximum unambiguous range. If the range measurement is to be unambiguous, the spectrum must be continuous over the bandwidth B.

Accuracy of frequency (doppler-velocity) measurement. Using the method of inverse probability, Manasse[8] showed that the minimum rms error in the measurement of frequency is

$$\delta f = \frac{1}{\alpha(2E/N_0)^{1/2}} \qquad (11.27)$$

where

$$\alpha^2 = \frac{\int_{-\infty}^{\infty} (2\pi t)^2 s^2(t)\, dt}{\int_{-\infty}^{\infty} s^2(t)\, dt} \qquad (11.28)$$

and $s(t)$ is the input signal as a function of time. Note the similarity between the definition of α and β, as defined in Eq. (11.16), as well as between the expressions for δf and δT_R. The parameter α is called the effective time duration of the signal, and $(\alpha/2\pi)^2$ is the normalized second moment of $s^2(t)$ about the mean epoch, taken to be $t = 0$. If the mean is not zero, but some other value t_0, the integrand in the numerator of (11.28) would be $(2\pi)^2(t - t_0)^2 s^2(t)$. In radar, the measurement error specified by Eq. (11.27) is that of the doppler frequency shift.

The value of α^2 for a perfectly rectangular pulse of width τ is $\pi^2\tau^2/3$; thus the rms frequency error is

$$\delta f = \frac{\sqrt{3}}{\pi\tau(2E/N_0)^{1/2}} \qquad \text{rectangular pulse} \qquad (11.29)$$

The longer the pulse width, the better the accuracy of the frequency measurement.

For a bandwidth-limited "rectangular" pulse as described by Fig. 11.5, the value of α^2 is

$$\alpha^2 = \frac{\pi^2 \tau^2}{3} \, \frac{\dfrac{\cos \pi B\tau - 3}{\pi B\tau} - \dfrac{8(\cos \pi B\tau - 1)}{(\pi B\tau)^3} - \dfrac{2 \sin \pi B\tau}{(\pi B\tau)^2} + \text{Si}\,(\pi B\tau)}{\text{Si}\,(\pi B\tau) + (\cos \pi B\tau - 1)/\pi B\tau} \tag{11.30}$$

In the limit as $B\tau \to \infty$, the value of α^2 approaches $\pi^2 \tau^2/3$, which is the same as that obtained for the perfectly rectangular pulse.

The value of α^2 for a perfectly rectangular pulse is finite even though its value of β^2 is infinite. The value of α^2 will be infinite, however, for a waveform with a perfectly rectangular frequency spectrum, corresponding to a $(\sin x)/x$ time waveform ($x = \pi B t$) of infinite duration. In practice, any waveform must be limited in time, and α^2 will therefore be finite. The frequency error (or α^2) for a waveform with rectangular spectrum may be found in a manner similar to that employed for computing β^2 for a bandwidth-limited rectangular pulse. The frequency error will be like that of Eq. (11.20) but with the roles of B and τ reversed.

The rms error in the measurement of doppler frequency with a trapezoidal pulse is

$$\delta f = \frac{(2T_2/3 + 2T_1)^{1/2}}{2\pi \left(\dfrac{2T_1^2 T_2}{3} + \dfrac{T_1 T_2^2}{3} + \dfrac{T_2^3}{15} + \dfrac{2T_1^3}{3} \right)^{1/2} \left(\dfrac{2E}{N_0} \right)^{1/2}} \qquad \text{trapezoidal pulse} \tag{11.31}$$

This reduces to the expression for a rectangular pulse [Eq. (11.29)] as the trapezoidal pulse becomes more rectangular, that is, when $T_1 = \tau/2 \gg T_2$.

For the triangular pulse, we set $T_1 = 0$ in Eq. (11.31) and let $2T_2 = \tau_B$. The rms error is

$$\delta f = \frac{(10)^{1/2}}{\pi \tau_B (2E/N_0)^{1/2}} \qquad \text{triangular pulse} \tag{11.32}$$

The rms frequency error for a gaussian-shaped pulse is

$$\delta f = \frac{1.18}{\pi \tau (2E/N_0)^{1/2}} = \frac{B}{1.18(2E/N_0)^{1/2}} \qquad \text{gaussian pulse} \tag{11.33}$$

The time-delay and frequency-error expressions obtained in this section apply for a single observation. When more than one independent measurement is made, the resultant error may be found by combining the errors in the usual manner for gaussian statistics; that is, the variance (square of δf or δT_R) of the combined observations is equal to $1/N$ of the variance of a single observation, where N is the number of independent observations. If β^2 or α^2 remains the same for each measurement, the expressions derived here still apply, but with E the *total* signal energy involved in all N observations.

Uncertainty relation. The so-called "uncertainty" relation of radar states that the product of the effective bandwidth β occupied by a signal waveform and the effective time duration α must be greater than or equal to π; that is,[4]

$$\beta\alpha \geq \pi \tag{11.34}$$

Equation (11.34), the radar "uncertainty" relationship, may be derived from the definitions of β and α given by Eqs. (11.16) and (11.28) and by applying the Schwartz inequality. It is a consequence of the Fourier-transform relationship between a time waveform and its spectrum and may be derived without recourse to noise considerations. The use of the word "uncertainty" is a misnomer, for there is nothing uncertain about the "uncertainty" relation of Eq. (11.34). It states the well-known mathematical fact that a narrow waveform yields a wide spectrum and a wide waveform yields a narrow spectrum and both the time waveform and the frequency spectrum cannot be made arbitrarily small simultaneously.

The relation of Eq. (11.34) is useful, however, as an indication of the accuracy with which time delay and frequency may be measured simultaneously. The product of the rms time-delay error [Eq. (11.17)] and the rms frequency error [Eq. (11.27)] is

$$\delta T_R \, \delta f = \frac{1}{\beta \alpha (2E/N_0)} \tag{11.35}$$

Substituting the inequality of Eq. (11.34) in the above gives

$$\delta T_R \, \delta f \leq \frac{1}{\pi (2E/N_0)} \tag{11.36}$$

This states that *the time delay and the frequency may be simultaneously measured to as small a theoretical error as one desires by designing the radar to yield a sufficiently large ratio of signal energy* (E) *to noise power per hertz* (N_0), *or for fixed* E/N_0, *to select a radar waveform which results in a large value of* $\beta \alpha$. Large $\beta \alpha$ products require waveforms long in duration and of wide bandwidth.

The poorest waveform for obtaining accurate time-delay and frequency measurements simultaneously is the one for which $\beta \alpha = \pi$. It may be shown that this corresponds to the gaussian-shaped pulse. The triangular-shaped pulse is little better, since its $\beta \alpha$ product is $\sqrt{\frac{6}{5}}\pi$.

The radar "uncertainty" relation seems to have the opposite interpretation of the uncertainty principle of quantum mechanics. The latter states that the position and the velocity of an electron or other atomic particles cannot be simultaneously determined to any degree of accuracy desired. Precise determination of one parameter can be had only at the expense of the other. This is not so in radar. Both position (range or time delay) and velocity (doppler frequency) may in theory be determined simultaneously if the $\beta \alpha$ product and/or the E/N_0 ratio are sufficiently large. The two uncertainty principles apply to different phenomena, and the radar principle based on classical concepts should not be confused with the physics principle that describes quantum-mechanical effects. In classical radar there is no theoretical limit to the minimum value of the $\delta T_R \, \delta f$ product since the radar systems designer is free to choose as large a $\beta \alpha$ product (by proper selection of the waveform) and E/N_0 ratio as he desires, or can afford. His limits are practical ones, such as power limitations or the inability to meet tolerances. In the quantum-mechanical case, on the other hand, the observer does not have control over his system as does the radar designer since the $\beta \alpha$ product of a quantum particle is fixed by nature and not by the observer.

Angular accuracy. The measurement of angular position is the measurement of the angle of arrival of the equiphase wavefront of the echo signal. The theoretical rms error of the angle measurement may be derived in a manner similar to the derivations of time (range) and frequency errors discussed above. The analogy between the angular error and the time-delay or frequency error comes about because the Fourier transform describes the relationship between the radiation pattern and the aperture distribution of an antenna in a manner similar to the relationship between the time waveform and its frequency spectrum.

For simplicity the angular error in one coordinate plane only will be considered. The analysis can be extended to include angular errors in both planes, if desired. It is assumed that the signal-to-noise ratios are large and that the noise can be described by the gaussian probability-density function.

Consider a linear in-phase receiving antenna of length D, or a rectangular receiving aperture of width D as shown in Fig. 11.6. The amplitude distribution across the aperture as a

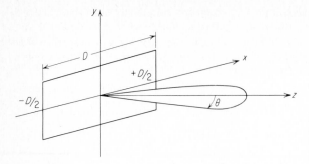

Figure 11.6 Rectangular receiving aperture of width D and amplitude distribution $A(x)$ giving rise to radiated pattern $G_v(\theta)$.

function of x is denoted $A(x)$. The (voltage) gain as a function of the angle θ (one-dimensional radiation pattern) in the xz plane is proportional to

$$G_v(\theta) = \int_{-D/2}^{D/2} A(x) \exp\left(j2\pi \frac{x}{\lambda} \sin\theta\right) dx \tag{11.37}$$

When the angle θ is small, $\sin\theta \approx \theta$, and Eq. (11.37) is recognized as an inverse Fourier transform

$$G_v(\theta) = \int_{-D/2}^{D/2} A(x) \exp \frac{j2\pi x\theta}{\lambda} dx \tag{11.38}$$

This is analogous to the inverse Fourier transform relating the frequency spectrum $S(f)$ and the time waveform $s(t)$, or

$$s(t) = \int_{-\infty}^{\infty} S(f) \exp\left(j2\pi ft\right) df \tag{11.39}$$

As the antenna scans at a uniform angular rate ω_s, the received signal voltage from a fixed point source will be proportional to $G_v(\theta) = G_v(\omega_s t)$ and may be considered a time waveform. If θ/λ [in Eq. (11.38)] is identified with t in Eq. (11.39), and if x is identified with f, the theoretical rms error can be obtained by analogy to the time-delay accuracy as given by Eq. (11.17), or

$$\delta\left(\frac{\theta}{\lambda}\right) = \frac{1}{\gamma(2E/N_0)^{1/2}} \tag{11.40}$$

where γ is the effective aperture width defined by

$$\gamma^2 = \frac{\int_{-\infty}^{\infty} (2\pi x)^2 |A(x)|^2 \, dx}{\int_{-\infty}^{\infty} |A(x)|^2 \, dx} \tag{11.41}$$

The theoretical angular error of an antenna with a rectangular amplitude distribution across the aperture is

$$\delta\theta = \frac{\sqrt{3}\lambda}{\pi D(2E/N_0)^{1/2}} \tag{11.42a}$$

The effective aperture width is 2π times the square root of the normalized second moment of $|A(x)|^2$ about the mean value of x, taken to be at $x = 0$. The half-power beamwidth θ_B of a rectangular distribution is $0.88\lambda/D$, where θ_B is in radians. Therefore

$$\delta\theta = \frac{0.628\theta_B}{(2E/N_0)^{1/2}} \tag{11.42b}$$

where the units of $\delta\theta$ are the same as those of θ_B. The relative error $(\delta\theta/\theta_B)$ is a function of E/N_0 only. The angular error can be many times less than the beamwidth, depending upon the value of E/N_0. The accuracy formulas derived previously for the time delay and the frequency may be readily applied to the determination of the angular error for various aperture distributions.

The effective aperture width γ for several aperture distributions which can be computed analytically are given below:

Parabolic distribution

$$A(x) = 1 - \frac{4(1 - \Delta)x^2}{D^2} \qquad |x| < \frac{D}{2}, \Delta < 1$$

For $\Delta = 0$ $\qquad\qquad\qquad\qquad \gamma^2 = 0.863D^2$

For $\Delta = 0.5$ $\qquad\qquad\qquad\qquad \gamma^2 = 1.88D^2$ $\qquad\qquad$ (11.43)

For $\Delta = 1.0$ $\qquad\qquad\qquad\qquad \gamma^2 = 3.287D^2$

Cosine distribution

$$A(x) = \cos\frac{\pi x}{D} \qquad |x| < \frac{D}{2}$$

$$\gamma^2 = 1.286D^2 \qquad\qquad (11.44)$$

Triangular distribution

$$A(x) = 1 - \frac{2}{D}|x| \qquad |x| < \frac{D}{2}$$

$$\gamma^2 = 0.986D^2 \qquad\qquad (11.45)$$

Inverse probability, likelihood ratio, and accuracy. The method of inverse probability as described by Woodward[5] can be used as a basis for determining the theoretical accuracies associated with radar measurements. The likelihood function also can be used for deriving measurement accuracy.[9] Both methods result in accuracy expressions like that of Eq. (11.17).

11.4 AMBIGUITY DIAGRAM[5,10–12]

The ambiguity diagram represents the response of the matched filter to the signal for which it is matched as well as to doppler-frequency-shifted (mismatched) signals. Although it is seldom used as a basis for practical radar system design, it provides an indication of the limitations and utility of particular classes of radar waveforms, and gives the radar designer general guidelines for the selection of suitable waveforms for various applications.

The output of the matched filter was shown in Sec. 10.2 to be equal to the cross correlation between the received signal and the transmitted signal [Eq. (10.18)]. When the received echo signal from the target is large compared to noise, this may be written as

$$\text{Output of the matched filter} = \int_{-\infty}^{\infty} s_r(t)s^*(t - T_R')\, dt \qquad (11.46)$$

where $s_r(t)$ is the received signal, $s(t)$ is the transmitted signal, $s^*(t)$ is its complex conjugate, and T_R' is the estimate of the time delay (considered a variable). Complex notation is assumed in Eq. (11.46). The transmitted signal expressed in complex form is $u(t)e^{j2\pi f_0 t}$, where $u(t)$ is the complex-modulation function whose magnitude $|u(t)|$ is the envelope of the real signal, and f_0 is the carrier frequency. The received echo signal is assumed to be the same as the transmitted signal except for the time delay T_0 and a doppler frequency shift f_d. Thus

$$s_r(t) = u(t - T_0)e^{j2\pi(f_0 + f_d)(t - T_0)} \tag{11.47}$$

(The change of amplitude of the echo signal is ignored here.) With these definitions the output of the matched filter is

$$\text{Output} = \int_{-\infty}^{\infty} u(t - T_0)e^{j2\pi(f_0 + f_d)(t - T_0)}[u(t - T_R')e^{j2\pi f_0(t - T_R')}]^* \, dt$$

$$= \int_{-\infty}^{\infty} u(t - T_0)u^*(t - T_R')e^{j2\pi(f_0 + f_d)(t - T_0)} \, e^{-j2\pi f_0(t - T_R')} \, dt \tag{11.48}$$

It is customary to set $T_0 = 0$ and $f_0 = 0$, and to define $T_0 - T_R' = -T_R' = T_R$. The output of the matched filter is then

$$\chi(T_R, f_d) = \int_{-\infty}^{\infty} u(t)u^*(t + T_R)e^{j2\pi f_d t} \, dt \tag{11.49}$$

In this form a positive T_R indicates a target beyond the reference delay T_0, and a positive f_d indicates an incoming target.[13] The squared magnitude $|\chi(T_R, f_d)|^2$ is called the *ambiguity function* and its plot is the *ambiguity diagram*.

The ambiguity diagram has been used to assess the properties of the transmitted waveform as regards its target resolution, measurement accuracy, ambiguity, and response to clutter.

Properties of the ambiguity diagram. The function $|\chi(T_R, f_d)|^2$ has the following properties:

$$\text{Maximum value of } |\chi(T_R, f_d)|^2 = |\chi(0, 0)|^2 = (2E)^2 \tag{11.50}$$

$$|\chi(-T_R, -f_d)|^2 = |\chi(T_R, f_d)|^2 \tag{11.51}$$

$$|\chi(T_R, 0)|^2 = \left| \int u(t)u^*(t + T_R) \, dt \right|^2 \tag{11.52}$$

$$|\chi(0, f_d)|^2 = \left| \int u^2(t)e^{j2\pi f_d t} \, dt \right|^2 \tag{11.53}$$

$$\iint |\chi(T_R, f_d)|^2 \, dT_R \, df_d = (2E)^2 \tag{11.54}$$

The first equation given above, Eq. (11.50), states that the maximum value of the ambiguity function occurs at the origin and its value is $(2E)^2$, where E is the energy contained in the echo signal. Equation (11.51) is a symmetry relation. Equations (11.52) and (11.53) describe the behavior of the ambiguity function on the time-delay axis and the frequency axis, respectively. Along the T_R axis the function $\chi(T_R, f_d)$ is the autocorrelation function of the modulation $u(t)$, and along the f_d axis it is proportional to the spectrum of $u^2(t)$. Equation (11.54) states that the total volume under the ambiguity function is a constant equal to $(2E)^2$.

Ideal ambiguity diagram. If there were no theoretical restrictions, the ideal ambiguity diagram would consist of a single peak of infinitesimal thickness at the origin and be zero everywhere else, as shown in Fig. 11.7. The single spike eliminates any ambiguities, and its infinitesimal

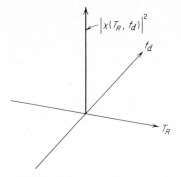

Figure 11.7 Ideal, but unattainable, ambiguity diagram.

thickness at the origin permits the frequency and the echo delay time to be determined simultaneously to as high a degree of accuracy as desired. It would also permit the resolution of two targets no matter how close together they were on the ambiguity diagram. Naturally, it is not surprising that such a desirable ambiguity diagram is not possible. The fundamental properties of the ambiguity function prohibit this type of idealized behavior. The two chief restrictions are that the maximum height of the $|\chi|^2$ function be $(2E)^2$ and that the volume under the surface be finite and equal $(2E)^2$. Therefore the peak at the origin is of fixed height and the function encloses a fixed volume. A reasonable approximation to the ideal ambiguity diagram might appear as in Fig. 11.8. This waveform does not result in ambiguities since there is only one peak, but the single peak might be too broad to satisfy the requirements of accuracy and resolution. The peak might be narrowed, but in order to conserve the volume under its surface, the function must be raised elsewhere. If the peak is made too narrow, the requirement for a constant volume might cause peaks to form at regions of the ambiguity diagram other than the origin and give rise to ambiguities. Thus the requirements for accuracy and ambiguity may not always be possible to satisfy simultaneously.

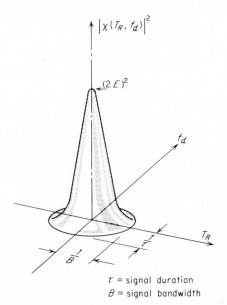

τ = signal duration
B = signal bandwidth

Figure 11.8 An approximation to the ideal ambiguity diagram, taking account the restrictions imposed by the requirement for a fixed value of $(2E)^2$ at the origin and a constant volume enclosed by the $|\chi|^2$ surface.

The ambiguity diagram in three dimensions may be likened to a box of sand. The total amount of sand in the box is fixed and corresponds to a fixed signal energy. No sand can be added, and none can be removed. The sand may be piled up at the center (origin) to as narrow a pile as one would like, but its height can be no greater than a fixed amount $(2E)^2$. If the sand in the center is in too narrow a pile, the sand which remains might find itself in one or more additional piles, perhaps as big as the one at the center.

The optimum waveform is one which has the desired ambiguity diagram for a given amount of "sand" (energy). The usual pulse radar or the usual CW radar, as we shall see, does not result in an ideal diagram. To produce an ambiguity diagram such as that shown in Fig. 11.8, the transmissions must be noiselike.

The synthesis of the waveform required to satisfy the requirements of accuracy, ambiguity, and resolution as determined by the ambiguity diagram is a difficult task. The usual design procedure is to compute the ambiguity diagram for the more common waveforms and to observe its behavior. Because of the limitations of synthesis, the ambiguity diagram has been more a measure of the suitability of a selected waveform than a means of finding the optimum waveform.

Single pulse of sine wave. The ambiguity diagram for a single rectangular pulse of sine wave is shown in Fig. 11.9. Contours for constant values of doppler frequency shift (velocity) are shown in Fig. 11.9a. The contour for zero velocity is triangular in shape and represents the autocorrelation function of a rectangular pulse such as would be predicted from Eq. (11.52). Contours for fixed values of time delay are shown in Fig. 11.9b. The center contour corresponding to $T_R = 0$ is the spectrum of a rectangular pulse [Eq. (11.53)]. The composite three-dimensional ambiguity surface is shown in Fig. 11.9c.

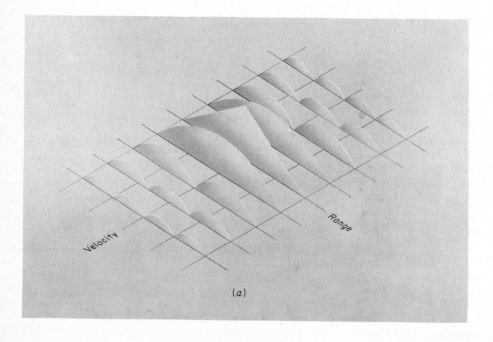

(a)

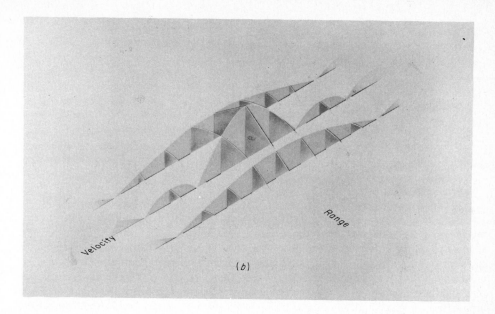

(b)

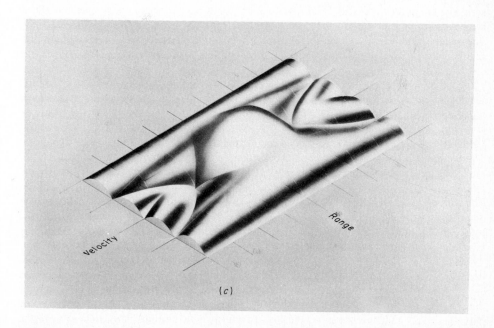

(c)

Figure 11.9 Three-dimensional plot of the ambiguity diagram for a single rectangular pulse. (a) Contours for constant dopppler frequency (velocity); (b) contours for constant time delay (range); (c) composite surface. (*Courtesy S. Applebaum and P. W. Howells, General Electric Co., Heavy Military Electronics Department, Syracuse, N.Y.*)

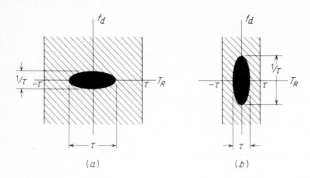

Figure 11.10 Two-dimensional ambiguity diagram for a single pulse of sine wave. (a) Long pulse; (b) short pulse.

It is usually inconvenient to draw a three-dimensional plot of the ambiguity diagram. For this reason a two-dimensional plot is often used to convey the salient features. Figure 11.10 is an example of the two-dimensional plot of the three-dimensional ambiguity diagram corresponding to the single pulse of Fig. 11.9c. Shading is used to give an indication of the regions in which $|\chi(T_R, f_d)|^2$ is large (completely shaded areas), regions where $|\chi|^2$ is small but not zero (lightly shaded areas), and regions where $|\chi|^2$ is zero (no shading). The plot for a single pulse shows a single elliptically shaped region in which $|\chi|^2$ is large. This is what would have been expected from our previous discussions since a single measurement does not result in ambiguity if the threshold is chosen properly. Range error is proportional to the pulse width τ, while doppler error is proportional to $1/\tau$. Shortening the pulse width improves the range accuracy, but at the expense of the doppler-velocity accuracy. Although the shape of the ellipse can be as thin or as broad as one likes in either axis, the opposite will be true for the other axis. The region in the vicinity of the origin cannot be made as small as we wish along both axes simultaneously without shifting some of the completely shaded region elsewhere in the diagram.

By letting τ become very large (essentially infinite), Fig. 11.10 may also be used to represent a CW radar. Similarly, by letting τ be very small (infinitesimal), the diagram applies to an impulse radar.

Periodic pulse train. Consider a sinusoid modulated by a train of five pulses, each of width τ. The pulse-repetition period is T_p, and the duration of the pulse train is T_d (Fig. 11.11a). The ambiguity diagram is represented in Fig. 11.11b. With a single pulse the time-delay- and frequency-measurement accuracies depend on one another and are linked by the pulse width τ. The periodic train of pulses, however, does not suffer this limitation. The time-delay error is determined by the pulse width τ as before, but the frequency accuracy is determined by the total duration of the pulse train. Thus the time- and frequency-measurement accuracies may be made independent of one another.

For the privilege of independently controlling the time and frequency accuracy with a periodic waveform, additional peaks occur in the ambiguity diagram. These peaks cause ambiguities. The total volume represented by the shaded areas of the ambiguity diagram for the periodic waveform approximates the total volume of the ambiguity diagram of the single pulse, assuming that the energy of the two waveforms are the same. This follows from the relationship expressed by Eq. (11.54). In practice, the radar designer attempts to select the pulse-repetition period T_p so that all targets of interest occur only in the vicinity of the central peak, all other peaks being far removed from the region occupied by the targets. The periodic-pulse waveform is a good one from the point of view of accuracy if the radar application is

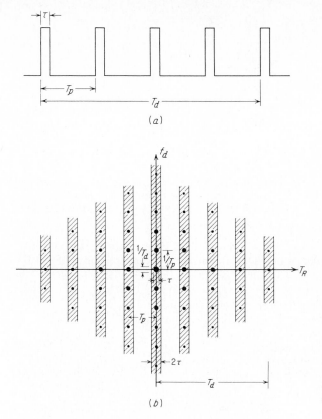

(a)

(b)

Figure 11.11 (a) Pulse train consisting of five pulses; (b) ambiguity diagram for (a).

such that it is possible to ignore or eliminate any ambiguities which arise. The fact that most practical radars employ this type of waveform attests to its usefulness far better than any theoretical analysis which might be presented here. It is encouraging, however, when theoretical considerations substantiate the qualitative, intuitive reasoning upon which most practical engineering decisions must usually be based, for lack of any better criterion.

Single frequency-modulated pulse. Ambiguities may be avoided with a single-pulse waveform rather than a periodic-pulse waveform. Although the accuracy of simultaneously measuring time and frequency with a simple pulse-modulated sinusoid was seen to be limited, it is possible to obtain simultaneous time and frequency measurements to as high a degree of accuracy as desired by transmitting a pulse long enough to satisfy the desired frequency accuracy and one with enough bandwidth to satisfy the time accuracy. In other words, the peak at the center of the ambiguity diagram may be narrowed by transmitting a pulse with a large bandwidth times pulse-width product (large $\beta\alpha$). One method of increasing the bandwidth of a pulse of duration T is to provide internal modulation. The ambiguity diagram for a frequency-modulated pulse is shown in Fig. 11.12. The waveform is a single pulse of sine wave whose frequency is decreased linearly from $f_0 + \Delta f/2$ to $f_0 - \Delta f/2$ over the duration of the pulse T, where f_0 is the carrier frequency and $\Delta f \approx B$ is the frequency excursion.

The ambiguity diagram is elliptical, as for the single pulse of unmodulated sine wave. However, the axis of the ellipse is tilted at an angle to both the time and frequency axes. This

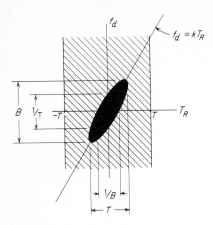

Figure 11.12 Ambiguity diagram for a single frequency-modulated pulse. (Also called the *chirp* pulse-compression waveform.)

particular waveform is not entirely satisfactory. The accuracy along either the time axis or the frequency axis can be made as good as desired. However, the accuracy along the ellipse major axis is relatively poor. This is a consequence of the fact that both the time delay (range) and the frequency (doppler) are both determined by measuring a frequency shift. Thus neither the range nor the velocity can be determined without knowledge of the other.

This limitation can be overcome by transmitting a second FM pulse whose slope on the ambiguity diagram is different from that of Fig. 11.12. The second modulation might be a linear frequency modulation which increases, rather than decreases, in frequency. This is analogous to the FM-CW radar of Chap. 3, in which the doppler frequency shift is extracted as well as the range. It will be recalled that the sawtooth frequency-modulated waveform of the FM-CW radar was capable of determining the range as long as there was no doppler frequency shift. By using a triangular waveform instead of the sawtooth waveform it was possible to measure both the range and the doppler frequency. The same technique can be used with the frequency-modulated pulse radar.

Classes of ambiguity diagrams. There are three general classes of ambiguity diagrams, Fig. 11.13. The knife edge, or ridge, is obtained with a single pulse of sine wave. Its orientation is along the time-delay axis for a long pulse, along the frequency axis for a short pulse, or it can be rotated to any direction in the T_R, f_d plane by the application of linear frequency modulation. The bed of spikes in Fig. 11.13b is obtained with a periodic train of pulses. The internal structure of each of the major components, illustrated figuratively by the simple arrows, depends on the waveform of the individual pulses. The thumbtack ambiguity diagram, Fig. 11.13c, is obtained with noise or pseudonoise waveforms. The width of the spike at the center can be made narrow along the time axis and along the frequency axis by increasing the bandwidth and pulse duration, respectively. However, the plateau which surrounds the spike is more complex than illustrated in the simple sketch. With real waveforms, the sidelobes in the plateau region can be higher than might be desired. Furthermore, the extent of the platform increases as the spike is made narrower since the total volume of the ambiguity function must be a constant, as was given by Eq. (11.54). There can be many variations of these three classes, as illustrated in Ref. 11.

Transmitted waveform and the ambiguity function. The particular waveform transmitted by a radar is chosen to satisfy the requirements for (1) detection, (2) measurement accuracy, (3) resolution, (4) ambiguity, and (5) clutter rejection. The ambiguity function and its plot, the

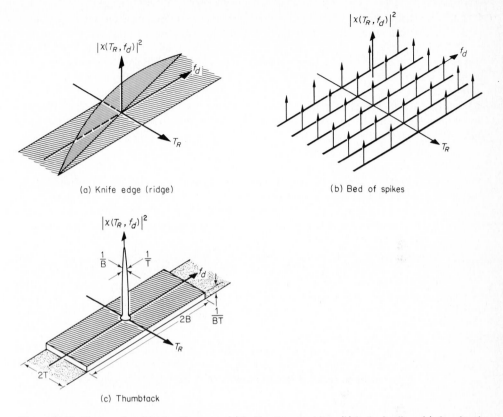

Figure 11.13 Classes of ambiguity diagrams: (*a*) knife edge, or ridge; (*b*) bed of spikes; (*c*) thumbtack. (*From G. W. Deley,*[12] *Courtesy McGraw-Hill Book Company.*)

ambiguity diagram, may be used to assess qualitatively how well a waveform can achieve these requirements. Each of these will be discussed briefly.

If the receiver is designed as a matched filter for the particular transmitted waveform, the probability of detection is independent of the shape of the waveform and depends only upon E/N_0, the ratio of the total energy E contained in the signal to the noise power per unit bandwidth. The requirements for *detection* do not place any demands on the *shape* of the transmitted waveform except (1) that it be possible to achieve with practical radar transmitters, and (2) that it is possible to construct the proper matched filter, or a reasonable approximation thereto. The maximum value of the ambiguity function occurs at $T_R = 0, f_d = 0$ and is equal to $(2E)^2$. Thus the value $|\chi(0, 0)|^2$ is an indication of the detection capabilities of the radar. Since the plot of the ambiguity function is often normalized so that $|\chi(0, 0)|^2 = 1$, the ambiguity diagram is seldom used to assess the detection capabilities of the waveform.

The *accuracy* with which the range and the velocity can be measured by a particular waveform depends on the width of the spike, centered at $|\chi(0, 0)|^2$, along the time and the frequency axes. The *resolution* is also related to the width of the central spike, but in order to resolve two closely spaced targets the central spike must be isolated. It cannot have any high peaks nearby that can mask another target close to the desired target. A waveform that yields good resolution will also yield good accuracy, but the reverse is not always so.

A continuous waveform (a single pulse) produces an ambiguity diagram with a single peak. A discontinuous waveform can result in peaks in the ambiguity diagram at other values of T_R, f_d. The pulse train (Fig. 11.11 or 11.13b) is an example. The presence of additional spikes can lead to *ambiguity* in the measurement of target parameters. An ambiguous measurement is one in which there is more than one choice available for the correct value of a parameter, but only one choice is appropriate. Thus the correct value is uncertain. The ambiguity diagram permits a visual indication of the ambiguities possible with a particular waveform. The ambiguity problem is characteristic of a single target, as is the detection and accuracy requirements of a waveform, whereas resolution is concerned with multiple targets.

The ambiguity diagram may be used to determine the ability of a waveform to reject clutter by superimposing on the T_R, f_d plane the regions where clutter is found. If the transmitted waveform is to have good clutter-rejection properties the ambiguity function should have little or no response in the regions of clutter.

The problem of synthesizing optimum waveforms based on a desired ambiguity diagram specified by operational requirements is a difficult one. The approach to selecting a waveform with a suitable ambiguity diagram is generally by trial and error rather than by synthesis.

In summary, this section has considered some of the factors which enter into the selection of the proper transmitted waveform. The problem of designing a waveform to achieve detection may be considered independently of the requirements of accuracy, ambiguity, resolution, and clutter rejection. A waveform satisfies the requirements of detection if its energy is sufficiently large and if the receiver is designed in an optimum manner, such as a matched-filter receiver. Waveform shape is important only as it affects the practical design of the matched filter. The ability of a particular waveform to satisfy the requirements of accuracy, ambiguity, resolution, and clutter rejection may be qualitatively determined from an examination of the ambiguity diagram. In general, periodic waveforms may be designed to satisfy the requirements of accuracy and resolution provided the resulting ambiguities can be tolerated. A waveform consisting of a single pulse of sinusoid avoids the ambiguity problem, but the time delay and frequency cannot simultaneously be measured to as great an accuracy as might be desired. However, it is possible to determine simultaneously both the frequency and the time delay to any degree of accuracy with a transmitted waveform containing a large bandwidth pulse-width product (large $\beta\alpha$ product). The problem of synthesizing optimum waveforms from an ambiguity diagram specified by operational requirements is a difficult one and is often approached by trial and error.

The name *ambiguity function* for $|\chi(T_R, f_d)|^2$ is somewhat misleading since this function describes more about the waveform than just its ambiguity properties. Woodward[5] coined the name to demonstrate that the total volume under this function is a constant equal to $(2E)^2$, independent of the shape of the transmitted waveform, [Eq. (11.54)]. Thus the total *area of ambiguity*, or uncertainty, is the same no matter how $|\chi(T_R, f_d)|^2$ is distributed over the T_R, f_d plane, as illustrated by the sandbox analogy mentioned earlier in this section. The reader is advised not to be distracted by trying to understand why this function is described by the ambiguous use of the term "ambiguity."

11.5 PULSE COMPRESSION

Pulse compression allows a radar to utilize a long pulse to achieve large radiated energy, but simultaneously to obtain the range resolution of a short pulse. It accomplishes this by employing frequency or phase modulation to widen the signal bandwidth. (Amplitude modulation is also possible, but is seldom used.) The received signal is processed in a matched filter that compresses the long pulse to a duration $1/B$, where B is the modulated-pulse spectral bandwidth. Pulse compression is attractive when the peak power required of a short-pulse radar cannot be achieved with practical transmitters.

Application of short pulse to radar. A conventional short-pulse radar may be desired for the following purposes:

Range resolution. It is usually easier to separate multiple targets in the range coordinate than in angle.

Range accuracy. If a radar is capable of good range-resolution it is also capable of good range accuracy.

Clutter reduction. A short pulse increases the target-to-clutter ratio by reducing the clutter contained within the resolution cell with which the target competes.

Glint reduction. In a tracking radar, the angle and range errors introduced by a finite size target are reduced with increased range-resolution since it permits individual scattering centers to be resolved.

Multipath resolution. Sufficient range-resolution permits the separation of the desired target echo from echoes that arrive via scattering from longer paths, or multipath.

Minimum range. A short pulse allows the radar to operate with a short minimum range.

Target classification. The characteristic echo signal from a distributed target when observed by a short pulse can be used to recognize one class of target from another.

ECCM. A short-pulse radar can negate the operation of certain electronic countermeasures such as range-gate stealers and repeater jammers, if the response time of the ECM is greater than the radar pulse duration. The wide bandwidth of the short-pulse radar also has some advantage against noise jammers.

Doppler tolerance. With a single short pulse, the doppler frequency shift will be small compared to the receiver bandwidth so that only one matched-filter is needed for detection, rather than a bank of matched filters with each filter tuned for a different doppler shift.

A short-pulse radar is not without its disadvantages. It requires large bandwidth with the possibility for interference to other users of the band. The shorter the pulse the more information there is available from the radar, and therefore the greater will be the demands on the information processing and display. In some high-resolution radars the number of resolution cells might be greater than the capability of conventional displays to present the information without overlap, so that a collapsing loss can result. The wide bandwidth might also mean less dynamic range in some radar designs. If the radar transmitter is peak-power limited, the shorter the pulse the less the energy transmitted. This has resulted in short-pulse radars usually being limited in range.

Pulse compression is a method for achieving most of the benefits of a short pulse while keeping within the practical constraints of the peak-power limitation. It is usually a suitable substitute for the short-pulse waveform except when a long minimum range might be a problem or when maximum immunity to repeater ECM is desired. Pulse compression radars, in addition to overcoming the peak-power limitations, have an EMC (Electromagnetic Compatability) advantage in that they can be made more tolerant to mutual interference. This is achieved by allowing each pulse-compression radar that operates within a given band to have its own characteristic modulation and its own particular matched filter. The pulse-compression radar, although widely used as a substitute for a short-pulse radar, has some limitations of its own, as will be discussed later in this section.

There are at least two ways of describing the concept of a pulse-compression radar. One is based on an approach similar to that of the ambiguity function of Sec. 11.4. A modulation of some form is applied to the transmitted waveform and its response after passing through the matched filter is examined. For example, the frequency-modulated pulse waveform whose ambiguity diagram was shown in Fig. 11.12 is the widely used *chirp* pulse-compression waveform. The other approach is to consider the modulation applied to a long pulse as providing distinctive marks over the duration of the pulse. For instance, the changing frequency of a linearly frequency-modulated pulse is distributed along the pulse and thus identifies each segment of the pulse. By passing this modulated pulse through a delay line whose delay time is a function of the frequency, each part of the pulse experiences a different time delay so that it is possible to have the trailing edge of the pulse speeded up and the leading edge slowed down so as to effect a time compression of the pulse.

The *pulse compression ratio* is a measure of the degree to which the pulse is compressed. It is defined as the ratio of the uncompressed pulse width to the compressed pulse width, or the product of the pulse spectral bandwidth B and the uncompressed pulse width T. Generally, $BT \gg 1$. The pulse compression ratio might be as small as 10 (13 is a more typical lower value) or as large as 10^5 or greater. Values from 100 to 300 might be considered as more typical. There are many types of modulations used for pulse compression, but two that have seen wide application are the *linear frequency modulation* and the *phase-coded pulse*.

Linear FM pulse compression. The basic concept of the linear frequency modulated (FM) pulse compression radar was described by R. H. Dicke, in a patent filed in 1945.[14] Figure 11.14, which is derived from Dicke's patent, shows the block diagram of such a radar. In this version of a pulse-compression radar the transmitter is frequency modulated and the receiver contains a pulse-compression filter (which is identical to a matched filter). The transmitted waveform consists of a rectangular pulse of constant amplitude A and of duration T (Fig. 11.15a). The frequency increases linearly from f_1 to f_2 over the duration of the pulse, Fig. 11.15b. (Alternatively, the frequency could be linearly decreased with time rather than increased.) The time waveform of the signal described by Fig. 11.15a and b is shown schematically in Fig. 11.15c. On reception, the frequency-modulated echo is passed through the pulse-compression filter, which is designed so that the velocity of propagation through the filter is proportional to frequency. When the pulse-compression filter is thought of as a dispersive delay line, its action can be described as speeding up the higher frequencies at the trailing edge of the pulse relative to the lower frequencies at the leading edge so as to compress the pulse to a width $1/B$, where $B = f_2 - f_1$ (Fig. 11.15d). When the pulse compression filter is considered as a matched filter, the output (neglecting noise) is the autocorrelation function of

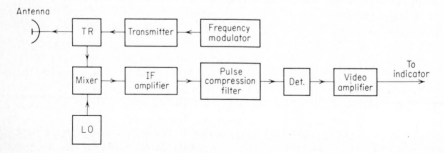

Figure 11.14 Block diagram of an FM pulse compression radar.

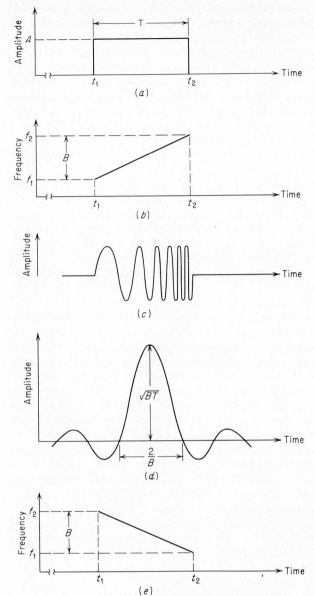

Figure 11.15 Linear FM pulse compression. (*a*) Transmitted waveform; (*b*) frequency of the transmitted waveform; (*c*) representation of the time waveform; (*d*) output of the pulse-compression filter; (*e*) same as (*b*) but with decreasing frequency modulation.

the modulated pulse, which is proportional to $(\sin \pi Bt)/\pi Bt$.[15] The peak power of the pulse is increased by the pulse compression ratio BT after passage through the filter.

The FM waveform in the block diagram of Fig. 11.14 is generated by directly modulating the high-power transmitter. It is not always convenient to directly modulate a transmitter in this manner. Alternatively, the waveform may be generated at a low-power level and amplified in a power amplifier. This is the more usual procedure. The waveform may be generated by a number of means including a voltage-controlled oscillator whose frequency is made to vary with an applied voltage; the so-called serrasoid modulator[16] which generates a quadratic waveform and compares this with a repetitive sawtooth wave to generate pulses that have a t^2 variation in spacing which are filtered to form a linear-FM waveform; a tapped delay-line with nonuniform tap spacings determined by the positive-going zero crossings of the desired linear-FM waveform, followed by a filter to form the output waveform; a synthesizer[17] which combines a staircase frequency waveform with the average desired slope, and a sawtooth frequency waveform which fills-in the steps with a short linear FM; or by digital means in which an algorithm performs a double integration to produce a phase at the output of the generator which varies as the square of time, as required for a linear FM.[18] The digital generator has the advantage of flexibility in the selection of bandwidth and time duration, as well as good stability and low residual generation errors. The above are sometimes called *active* methods for generating waveforms.

The linear FM waveform may also be generated by *passive* methods such as by exciting a dispersive delay line with an impulse. The frequency-response function of the dispersive delay line used for generating the transmitted waveform is the conjugate of that of the pulse-compression filter. In the special case of the linear FM waveform the same dispersive delay line that generates the transmitted waveform may be used as the receiver matched filter if the received waveform is mixed with an LO whose frequency is greater than that of the received signal. This results in a time inversion (it changes $s(t)$ to $s(-t)$) by converting an increasing FM to a decreasing FM, or vice versa.

Dispersive delay lines. There have been a number of devices used as dispersive delay lines, or pulse-compression filters, for linear FM waveforms.[16] They may be classed as ultrasonic, electromagnetic, or digital. Ultrasonic delay lines include those of aluminum or steel strip, piezoelectric materials such as quartz with propagation taking place through the bulk (or volume) of the material, piezoelectric materials with the propagation taking place along the surface (surface acoustic wave), and YIG (yttrium-iron-garnet) crystals. The all-pass, time delay using bridged-T networks with either lumped-constant circuit elements or stripline; the waveguide operated near its cutoff frequency; and the tapered folded-tape meander line are examples of electromagnetic dispersive delay lines suitable for linear-FM pulse compression. The charge-coupled device has also been considered for application in pulse compression.[19] Each of these delays lines has different characteristics and preferred regions of operation as regards bandwidth and pulse duration.[16] Only the surface acoustic wave (SAW) device will be described here as an illustration of a typical dispersive delay line suitable for radar application.[20]

The surface-acoustic-wave delay line, shown schematically in Fig. 11.16, consists of a piezoelectric substrate such as a thin slice of quartz or lithium niobate with input and output interdigital transducers (IDT) arranged on the surface. The design of the IDT determines the impulse response of the SAW delay line. Efficient electric-to-acoustic coupling occurs when the

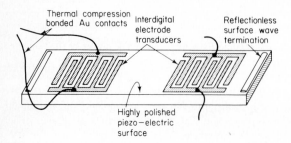

Thermal compression bonded Au contacts

Interdigital electrode transducers

Reflectionless surface wave termination

Highly polished piezo−electric surface

Figure 11.16 Schematic of a simple surface acoustic wave (SAW) delay line. (*From Bristol,*[22] *Courtesy Proc. IEEE.*)

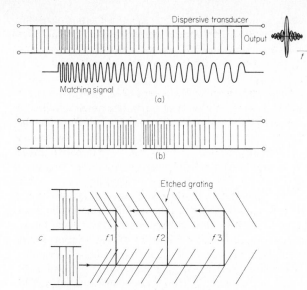

Figure 11.17 Three basic forms of SAW interdigital transducers for linear FM pulse compression. (*a*) Dispersive delay line with dispersion designed into one transducer; (*b*) dispersion in both transducers, and (*c*) a reflective array compressor (RAC). (*From Maines and Paige,*[21] *courtesy of Proc. IEEE.*)

comb fingers, or electrodes, of the IDT are spaced one-half the wavelength of the acoustic signal propagating along the SAW material.[22] Thus the frequency response of the delay line depends on the periodicity of the electrode spacings. A dispersive delay line for linear FM pulse compression is obtained with a variable electrode spacing, as illustrated in Fig. 11.17*a* or *b*. The duration of the resulting pulse is proportional to the length of the interdigital transducer. Amplitude shaping can be controlled by varying the overlap of the electrodes, as in Fig. 11.18. This is sometimes called *apodization*. Weighting of the amplitude, as discussed later in this section, is sometimes desired so as to reduce the time-sidelobes accompanying the compressed waveform.

The reflective-array compressor (RAC) form of SAW device, shown schematically in Fig. 11.17*c*, is a geometry that provides better performance for large pulse-compression ratios.[22] Shallow grooves etched in the delay path result in SAW reflections to form a delay that depends on the frequency. The structure is less sensitive to fabrication tolerances than conventional transducers.

A "typical" SAW dispersive delay line developed for linear FM pulse-compression radar had a bandwidth of 500 MHz and an uncompressed pulse width of 0.46 μs.[23] The center frequency was 1.3 GHz and the compressed pulse width was 3 ns. Amplitude weighting of the received signal reduced the highest sidelobe to -24 dB instead of -13.2 dB without weighting. (The compressed pulse was widened because of the weighting.) The filter package measured 0.5 by 1.5 by 2.25 in. The maximum insertion loss was 40 dB. Other designs have resulted in pulse widths as long as 100 μs and pulse-compression ratios as high as 10,000.[22]

The SAW dispersive delay line is one of the more important of the many devices that have been employed for pulse-compression radar. They have been claimed to be simple, low cost,

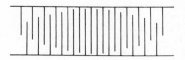

Figure 11.18 Interdigital transducer (nondispersive) showing overlap of comb fingers, or electrodes, to provide an amplitude weighting along the pulse.

small size, and highly reproducible in manufacture. The amplitude weighting to reduce sidelobes can be integrated directly into the interdigital transducer design. In addition to the linear FM, they can be designed to operate with nonlinear frequency modulations, phase-coded pulses, and the burst pulse.

Time sidelobes and weighting. The uniform amplitude of the linear FM waveform results in a compressed pulse shape of the form $(\sin \pi Bt)/\pi Bt$ after passage through the matched filter, or dispersive delay line, where B is the spectral bandwidth.[24] There are time, or range, sidelobes to either side of the peak response with the first, and largest, sidelobe -13.2 dB down from the peak. The large sidelobes are often objectionable since a large target might mask nearby, smaller targets. Also, near-in sidelobes might at times be mistaken for separate targets. These sidelobes can be reduced by amplitude weighting of the received-signal spectrum, just as the spatial sidelobes of an antenna radiation pattern can be reduced by amplitude weighting the illumination across the antenna aperture, as was described in Sec. 7.2. The same illumination functions used in antenna design to reduce spatial sidelobes can also be applied to the frequency domain to reduce the time sidelobes in pulse compression. A comparison of several types of spectral weighting functions is shown in Table 11.1.[16,24,25] The Taylor weighting with $\bar{n} = 8$ means the peaks of the first 7 sidelobes $(\bar{n} - 1)$ are designed to be equal, after which they fall off as $1/t$. The Dolph-Chebyshev weighting theoretically results in all sidelobes being equal. It is of academic interest only, since it is unrealizable. The Taylor is a practical approximation to the Dolph-Chebyshev. A suitable waveform might be the cosine-squared on a pedestal, as in the Hamming function, for example. The effect of weighting the received-signal spectrum to lower the sidelobes also widens the main lobe and reduces the peak signal-to-noise ratio compared to the unweighted linear FM pulse compression. This loss is due to the filter not being matched to the received waveform; that is, the filter is said to be *mismatched*. Thus to reduce the sidelobes to a level of -30 to -40 dB results in a loss in peak signal-to-noise ratio of from one to two dB. For many applications the beam broadening and the loss in peak signal-to-noise ratio due to mismatch are usually tolerated in order to achieve the benefits of the lower sidelobes.

Instead of weighting the received-signal spectrum to reduce the time sidelobes, it is possible, in principle, to achieve the same affect by amplitude-weighting either the envelope of transmitted FM signal or the received signal. In the case of linear FM with large pulse-compression ratio, the amplitude weighting applied to the time waveform is of the same form as the weighting applied to the frequency spectrum, for the same output response.[24] Amplitude-modulating the transmitted signal is not usually practical in high-power radar

Table 11.1 Properties of weighting functions

Weighting function	Peak sidelobe dB	Loss dB	Mainlobe width (relative)	Sidelobe decay function
Uniform	-13.2	0	1.0	$1/t$
$0.33 + 0.66 \cos^2 (\pi f/B)$	-25.7	0.55	1.23	$1/t$
$\cos^2 (\pi f/B)$	-31.7	1.76	1.65	$1/t^3$
Taylor $(\bar{n} = 8)$	-40	1.14	1.41	$1/t$
Dolph-Chebyshev	-40	.....	1.35	1
$0.08 + 0.92 \cos^2 (\pi f/B)$ (Hamming)	-42.8	1.34	1.50	$1/t$

$B =$ bandwidth

since most microwave tubes should be operated saturated; i.e., either full-on or off. Generally, weighting of the received frequency spectrum has been preferred over either time-weighting alternative.

If it were practical to amplitude-weight the transmitted waveform to reduce the time sidelobes, the receiver can be designed with the appropriate matched filter so that no theoretical loss in signal-to-noise will result. This implies that the transmitted signal energy is the same with or without weighting, as follows from the discussion in Sec. 10.2 of the output signal-to-noise ratio of a matched filter. However, when the transmitter is peak-power limited, it is preferable to use a constant-amplitude transmitted signal and perform the weighting in the receiver, in spite of the mismatched filter with its reduction in signal-to-noise ratio.[25] In one example,[24] transmitting a linear-FM waveform with a gaussian envelope and a matched-filter receiver to give -40 dB sidelobes resulted in 2.2 dB greater penalty in detection capability than when a uniform amplitude is transmitted with Hamming weighting on receive in a mismatched filter.

It is possible to achieve low time-sidelobes with uniform-amplitude transmitted waveforms and no theoretical loss in signal-to-noise ratio by means of nonlinear FM, as discussed later in this section.

Doppler-tolerant waveform. If an unknown doppler-frequency shift is experienced when a long pulse, a noise-modulated pulse, or a pulse train is reflected from a moving target, a receiver tuned to the transmitted signal will not accept the echo signal if the doppler shift places the echo frequency outside the band of the receiver matched filter. That is, the receiver may not be tuned to the correct frequency. To circumvent this potential loss of signal, a bank of contiguous matched filters must be used to cover the range of expected doppler-frequency shifts. It is possible, however, with a suitable transmitted waveform to employ a single matched filter that will accept doppler-shifted echoes with minimum degradation.

It has been shown that the waveform which allows a single pulse-compression filter to be matched for all doppler-frequency shifts (all target velocities) is[26-28]

$$s(t) = A(t) \cos \left[\frac{2\pi f_0^2\, T}{B} \ln \left(1 - \frac{Bt}{f_0\, T} \right) \right] \tag{11.55}$$

The amplitude $A(t)$ of Eq. (11.55) represents modulation by a rectangular pulse of width T. The band occupied by the signal is B, and the carrier frequency is f_0. This expression for the doppler-tolerant waveform is difficult to interpret as it stands, but if the natural-log factor is expanded in a series, Eq. (11.55) becomes

$$s(t) = A(t) \cos \left(2\pi f_0 t + \frac{\pi B t^2}{T} + \frac{2\pi B^2 t^3}{3 f_0\, T^2} + \cdots \right) \tag{11.56}$$

When terms greater than the first two can be neglected (which applies when $2\pi B^2 t^3 \ll 3 f_0\, T^2$), Eq. (11.56) reduces to the classical linear FM waveform. Thus the linear FM, or chirp, pulse-compression waveform is a practical approximation to the theoretical doppler-tolerant waveform. Differentiating, with respect to time, the argument of Eq. (11.55), the frequency of the doppler-tolerant waveform is found to be $2\pi f_0^2\, T/(f_0\, T - Bt)$. Inverting to obtain the period, it can be seen that the doppler-tolerant waveform is one with a *linear period modulation*.

The short-pulse waveform also can tolerate unknown shifts in the doppler frequency when using a single matched filter.

Phase-coded pulse compression.[16,29,30,38] In this form of pulse compression, a long pulse of duration T is divided into N subpulses each of width τ. The phase of each subpulse is chosen to be either 0 or π radians. If the selection of the 0, π phase is made at random, the waveform approximates a noise-modulated signal with a thumbtack ambiguity function, as in Fig. 11.13c. The output of the matched filter will be a spike of width τ with an amplitude N times greater than that of the long pulse. The pulse-compression ratio is $N = T/\tau = BT$, where $B = 1/\tau$ = bandwidth. The output waveform extends a distance T to either side of the peak response, or central spike. The portions of the output waveform other than the spike are called *time sidelobes*.

The binary choice of 0 or π phase for each subpulse may be made at random. However, some random selections may be better suited than others for radar application.[66] One criterion for the selection of a good "random" phase-coded waveform is that its autocorrelation function should have equal time-sidelobes. (Recall from Sec. 10.2 that the output of the matched filter is the autocorrelation of the input signal for which it is matched, if noise can be neglected.) The binary phase-coded sequence of 0, π values that result in equal sidelobes after passage through the matched filter is called a *Barker code*. An example is shown in Fig. 11.19a. This is a Barker code of length 13. The $(+)$ indicates 0 phase and $(-)$ indicates π radians phase. The autocorrelation function, or output of the matched filter, is shown in (b). There are six equal time-sidelobes to either side of the peak, each at a level -22.3 dB below the peak. In (c)

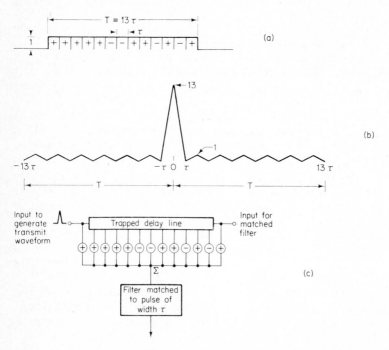

Figure 11.19 (a) Example of a phase-coded pulse with 13 equal subdivisions of either $0°(+)$ or $180°(-)$ phase. This is known as a Barker code of length 13. (b) Autocorrelation function of (a), which is an approximation to the output of the matched filter. (c) Block diagram of the filter for generating the transmitted waveform of (a) with the input on the left. The same tapped delay line can be used as the receiver matched filter by inserting the received echo at the opposite end (the right-hand side of the delay line in this illustration).

Table 11.2 Barker codes

Code length	Code elements	Sidelobe level, dB
2	+ −, + +	− 6.0
3	+ + −	− 9.5
4	+ + − +, + + + −	− 12.0
5	+ + + − +	− 14.0
7	+ + + − − + −	− 16.9
11	+ + + − − − + − − + −	− 20.8
13	+ + + + + − − + + − + − +	− 22.3

is illustrated schematically a tapped delay line that generates this Barker-coded waveform when an impulse is incident at the left-hand terminal. The same tapped delay line can be used as the receiver matched filter if the input is applied at the right-hand terminal.

The known Barker codes are shown in Table 11.2. The longest is of length 13. This is a relatively low value for a practical pulse-compression waveform. When a larger pulse-compression ratio is desired, some form of pseudorandom code is usually used. A popular technique is the generation of a linear recursive sequence using a shift register with feedback, as in Fig. 11.20. This device generates a binary pseudorandom code of zeros and ones of length $2^n - 1$, where n is the number of stages in the shift register. Feedback is provided by taking the output of the shift register and adding it, modulo two, to the output from one of the previous stages of the shift register. In the example of Fig. 11.20, the output of the 6th and 7th stage are combined. In modulo-two addition, the output is zero when the inputs are alike $[(0, 0)$ or $(1, 1)]$ and is one when the inputs are different. It is equivalent to ordinary base-two addition with only the least significant bit carried forward. A modulo-two adder is also called an *exclusive-or* gate. An n-stage shift register has a total of 2^n different possible states. However, it cannot have the state in which all the stages contain zeros since the shift register would remain in this state and produce all zeros thereafter. Thus an n-stage shift register can generate a binary sequence of length no greater than $2^n - 1$ before repeating. The actual sequence obtained depends on both the feedback connections and the initial loading of the register. When the output sequence of an n-stage shift register is of period $2^n - 1$, it is called a *maximal length sequence*, or *m-sequence*. These have also been called linear recursive sequences (LRS), pseudonoise (PN) sequences, and binary-shift-register sequences. Although the sequence is a series of zeros or ones, for application to the phase-coded pulse-compression radar the zeros can be considered as corresponding to zero phase and the ones to π radians phase.

Since an m-sequence takes on all possible states except the zero state, the initial state of the shift register does not affect the content of the sequence, but will define the starting point of the sequence. A different sequence is obtained with different feedback connections. (The number of stages connected modulo two must be even since an odd number of modulo-two additions taken from the all-ones state will produce a one for the next term and the continued generation of the all-ones state. This generates a sequence of length $2^n - 2$ and is not an m-sequence.[57]) Table 11.3 lists the number of possible m-sequences obtainable from an n-stage shift register.

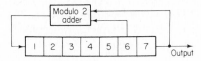

Figure 11.20 Seven-bit shift-register for generating a pseudorandom linear recursive sequence of length 127.

Table 11.3 Number of m-sequences obtainable from an n-stage shift register[16]

Number of stages, n	Length of maximal sequence $2^n - 1$	Number of maximal sequences	Example feedback stage connections
3	7	2	3, 2
4	15	2	4, 3
5	31	6	5, 3
6	63	6	6, 5
7	127	18	7, 6
8	255	16	8, 6, 5, 4
9	511	48	9, 5
10	1023	60	10, 7
11	2047	176	11, 9

For large $2^n - 1 = N$, the peak sidelobe is approximately $1/N$ that of the maximum response, measured in power. The actual values vary with the particular sequence. For example, with $N = 127$, the peak sidelobe is between -18 and -19.8 dB, instead of the -21 dB predicted on the basis of the code length. For $N = 255$, the peak sidelobe varies from -21.3 dB to -22.6 dB, instead of the -24 dB predicted.[29]

The binary codes generated in this manner fit many of the tests for randomness. (Randomness, however, is not necessarily a desirable property of a code used for pulse compression.) The number of ones in each sequence differs from the number of zeros by at most one (the *balance* property). Among the runs of ones and zeros in each sequence, one-half of the runs of each kind are of length one, one-fourth are of length two, one-eighth are of length three, and so on (the *run* property). If the sequence is compared term by term with any cyclic shift of itself, the number of agreements differs from the number of disagreements by at most one (the *correlation* property). These sequences are called *linear* since they obey the superposition theorem.

The peak sidelobe levels of the linear recursive sequences and of Barker codes greater than length 5 are lower than the -13.2 dB of the linear FM waveform. However, the sidelobes of the Barker codes can be further lowered by employing a mismatched filter and accepting a slight loss in the peak signal-to-noise ratio.[31]

Comparison of linear FM and phase-coded pulse compression. Both of these waveforms have their areas of application; but in the past, the linear FM, or chirp, pulse compression has probably been more widely used. The time sidelobes of the phase-coded pulse are of the order of $1/BT$. The peak sidelobe of the chirp waveform is generally higher, but at a slight sacrifice in signal-to-noise ratio it can be made low by means of weighting networks. The chirp waveform is doppler-tolerant in that a single pulse-compression filter can be used, but it cannot provide an independent range and doppler measurement. With moving targets, the phase-coded pulse might require a bank of contiguous matched filters covering the expected range of doppler frequencies. The sidelobes in the time-frequency plane of the ambiguity diagram are usually larger than desired for good doppler resolution without ambiguity. Although the ambiguity diagram of the phase-coded pulse has a narrow spike, the wide plateau means that there can be a large undesirable response from distributed clutter that extends in both the range and the doppler domain, and the resolution performance will be poor in a dense-target environment or when a small target is to be seen in the presence of much stronger echoes.

A different phase-coded sequence can be assigned to each radar so that a number of radars can share the same spectrum. In a military radar, the coding can be changed to help counter repeater jammers that attempt to simulate the waveform. The chirp waveform is more vulnerable to repeater jamming than is the phase-coded pulse. The chirp waveform is more likely to be used when a wide bandwidth, or very narrow compressed pulse, is required. The phase-coded pulse is more likely to be used when jamming or EMC is a problem, or when long-duration waveforms are desired. Generally, the implementation of the chirp pulse-compression has been less complex than that of the phase-coded pulse.

Although definite differences exist between these two basic waveforms, it would be difficult to provide precise guidelines describing where one is preferred to the other. Each application must be examined individually to determine the best form of pulse compression to use.

Other pulse-compression waveforms. Other pulse-compression methods include nonlinear FM, discrete frequency-shift, polyphase codes, compound Barker codes, code sequencing, complementary codes, pulse burst, and stretch. Each will be discussed briefly.

The *nonlinear-FM* waveform with constant-amplitude time envelope provides a compressed waveform with low time-sidelobes at the output of the receiver matched-filter without the 1- to 2-dB penalty obtained with the linear-FM waveform and mismatched filter.[16,24,32] The nonlinear variation of frequency with time has the same effect as amplitude-weighting the transmitted-signal spectrum, while maintaining the rectangular pulse-shape desired for efficient transmitter operation. If the nonlinear FM is symmetrical in time, the ambiguity diagram has a single peak rather than a ridge. (A symmetrical waveform means the frequency increases, or decreases, during the first half of the pulse and decreases, or increases, during the second half.) The nonlinear FM is thus more sensitive to doppler-frequency shifts and is not doppler-tolerant. The surface-acoustic-wave delay line is one method for generating the nonlinear FM waveform and for acting as the matched filter.

The *discrete frequency-shift*, or *time-frequency coded*, waveform is generated by dividing a long pulse into a series of contiguous subpulses and shifting the carrier frequency from subpulse to subpulse.[16,33] The frequency steps are separated by the reciprocal of the subpulse width. A linear stepping of the frequency gives an ambiguity diagram more like the ridge of the linear FM. When the frequencies are selected at random, the result is a thumbtack ambiguity diagram. The number of subpulses N required to achieve a thumbtack ambiguity diagram with random frequency stepping is far less than with the phase-coded pulse. To achieve a total bandwidth B, each subpulse need only have a bandwidth B/N. The width of each subpulse is thus N/B. A given pulse-compression ratio BT can be obtained with $N = T/(N/B)$ subpulses, or $N = \sqrt{BT}$, instead of the BT subpulses required for the phase-coded pulse. Pulse-compression ratios as high as 10^5 and bandwidths of several hundred megahertz have been obtained.[33] With the proper frequency-shift sequence, the sidelobe level on a power basis is $1/N^2$ down from the main response.[34] However on the doppler axis of the ambiguity function the resulting sidelobes are $1/N$ rather than $1/N^2$, which is the result of only $N = \sqrt{BT}$ subpulses.

According to Nathanson,[33] the discrete frequency-shift waveforms are preferable instead of linear FM when the pulse-compression ratio and the signal bandwidth are large. The resulting thumbtack ambiguity diagram means there will be no range-doppler coupling to cause erroneous measurements. The order in which the discrete frequencies are transmitted can be varied so that each radar can have its own code, and interference between radars will be reduced. The components in each channel need only have a bandwidth $1/N$ times the total processing bandwidth. Still another advantage of such systems is that a strong CW interference signal will only suppress the target signal in one of the N channels. However, in a linear FM system a strong CW signal anywhere in the total signal bandwidth can capture a limiting receiver and cause suppression of the target echo.

Instead of the 0, π binary phase shift, smaller increments of phase can be applied in the phase-coded pulse compression waveform. These are called *polyphase codes.*[16,24] The time sidelobes of a polyphase code can be lower than those of the binary phase-coded waveform of similar length.[63,64] However, the performance of polyphase codes deteriorates rapidly in the presence of a doppler-frequency shift and therefore they have been limited to situations where the doppler is negligible.

Other binary coding methods that have been considered for pulse compression include:[35] (1) *compound Barker codes* for obtaining longer codes from the Barker series by coding segments of one Barker code with another Barker code (a length 13 Barker code compounded within another length 13 code gives a pulse-compression ratio of 169 and a peak sidelobe of -22.3 dB); (2) *code sequencing on successive PRF periods,* in which a different code is used for each transmission to produce random sidelobes which when N pulses are added to produce a $\sqrt{N}$ improvement in mainlobe-to-sidelobe amplitude ratio, where N is the number of sequences employed; and (3) *complementary codes* in which a pair of equal-length codes have the property that the time sidelobes of one code are the negative of the other so that if two codes of a complementary pair are alternated on successive transmissions, the algebraic sum of the two autocorrelation functions is zero except for the cental peak.[36,37]

A *pulse burst* is a waveform in which a series of pulses are transmitted as a group before any of the echo signals are received. It is used to obtain simultaneous range and doppler-velocity resolution when the minimum range is relatively long; as for radars whose targets are extraterrestrial, such as satellites and ballistic missiles. Pulse compression might be applied to each of the individual pulses of the burst for better range resolution, and amplitude weighting of the pulse burst might be used to lower the doppler sidelobe level to improve the detection of small doppler-shifted target echoes close in frequency to large clutter echoes.

Stretch is a technique related to pulse compression that permits an exchange of signal-time duration for signal bandwidth.[58,59] Its advantage is that high range-resolution can be obtained with wideband transmitted signals, but without the usual wideband processing circuitry. However, only a portion of the range interval can be observed in this manner. If a signal occupies a time T and a bandwidth B, a change in time aT allows a change in bandwidth B/a. Thus if a signal is stretched in time, say by a factor $a = 10$, the bandwidth can be reduced by a factor of 10, and the signal can be processed with more practical narrow band circuitry. However, only $\frac{1}{10}$th the total range interval can be processed. This disadvantage may be bothersome in a surveillance radar, but it might be well suited to a tracking radar or to a high-range-resolution radar used for target classification. The technique uses elements similar to those of the linear-FM pulse-compression radar. A linear-FM waveform (chirp) of narrow bandwidth B_1 is mixed with a wideband chirp of bandwidth B_2. The radiated signal is a chirp of bandwidth $B_1 + B_2$. On receive, the signal is mixed with the same wideband chirp B_2 to give a chirp of narrow bandwidth B_1 which is then processed as a normal pulse-compression signal. The mixing operation results in a time expansion of $a = (B_1 + B_2)/B_1$; but the range-resolution possible is that of a signal of bandwidth $B_1 + B_2$, using processing circuitry of bandwidth B_1.

Compatibility with other processing. Pulse-compression systems are sometimes used in conjunction with MTI radar.[39] The increased range resolution afforded by pulse compression provides an increased target-to-clutter echo, as does the MTI processing. If there exists, however, inherent instabilities in the radar system there can result noiselike time-sidelobes accompanying the pulse-compression waveform. Such system instabilities might be caused by noise on local oscillators, noise on transmitter power supplies, transmitter time jitter, and transmitter tube noise. These noise sidelobes will not cancel in the MTI system and, if sufficiently large, they can result in uncancelled residue that will appear on the radar display. Thus when the system instabilities are high, the detection of targets in clutter can be seriously degraded. One approach for operating under such conditions is to use two limiters.[39] One limiter is placed before the pulse-compression filter and has an output dynamic range equal to the difference between the peak transmitter power and transmitter noise in the system bandwidth. The other limiter is between the pulse-compression filter and the MTI and has a dynamic range equal to the expected MTI improvement factor.

The technology of pulse compression has been applied to radar with conventional, unmodulated pulses to achieve CFAR (constant false alarm rate) performance better than that of the log-FTC. In one example,[40] a dispersive delay line with a linear time-delay vs. frequency characteristic is followed by a hard limiter and by a second dispersive delay line with a characteristic inverse to the first. (If the limiter were omitted, the circuit would operate simply as a linear nondispersive time delay.) It has been claimed that there was no discernible loss in detectability with this CFAR, but that the detectability loss for log-FTC is about 3 dB.

A variation of this technique has been applied to chirp pulse-compression radar to achieve CFAR in conjunction with pulse compression. Two dispersive delay lines are used with a hard limiter between them to provide CFAR. Both lines are of similar characteristics, and the total pulse-compression ratio is divided between the two. In one design,[41] the first delay line compressed a 5 μs pulse to 1.6 μs, and the second delay line following the limiter compressed the 1.6 μs pulse to 0.05 μs, for a total pulse-compression ratio of 100.

When a limiter is used preceding the pulse-compression filter in conventional pulse-compression radar for suppressing impulsive and other interference, the presence of multiple targets can cause degraded performance. If the uncompressed pulse has an amplitude less than the rms noise level, there is little degradation. This situation will apply in many cases when the pulse-compression ratio is large. However, if two signals are present simultaneously, and if one is much larger than the rms noise level at the limiter output, the weaker signal will be suppressed by the stronger over the time interval that they overlap. The result is that the probability of detecting the weaker signal is degraded.[42] The hard limiting of superimposed echoes in pulse-compression radar may also cause the generation of false targets in addition to small-signal suppression.[43]

Pulse compression has also been used in conjunction with frequency-scan radar to achieve improved range resolution.[44] This was discussed briefly in Sec. 8.4. The frequency-scan radar uses a linear-array antenna to scan a beam in one angular coordinate by changing the frequency. The use of the frequency domain for electronic beam scanning normally precludes the use of the frequency domain for range resolution. That is, pulse compression and frequency scan are often not compatible. However, if the entire angular region is swept within a single pulse by a frequency-scan antenna, the response from a point target will be frequency-modulated due to the finite beamwidth. A dispersive time delay filter can cause the received echo to be compressed so as to achieve better range resolution. The amount of range-resolution possible is limited by the finite time-delay of the signal propagating through the frequency-scan feed network. A bank of dispersive filters are needed to cover the angular sector, as described in Sec. 8.4. Each filter would be designed to accommodate the spread of frequencies expected from a particular elevation angle.

Limitations of pulse compression. Pulse compression is not without its disadvantages. It requires a transmitter that can be readily modulated and a receiver with a matched filter more sophisticated than that of a conventional pulse radar. Although it may be more complex than a conventional long-pulse radar, the equipment for a high-power pulse compression radar is more practical than would be required of a short-pulse radar with the same pulse energy. The time sidelobes accompanying the compressed pulse are objectionable since they can mask desired targets or create false targets. When limiting is employed, there can be small-target suppression and possibly spurious false-targets as well. The long uncompressed pulse can restrict the minimum range and the ability to detect close-in targets. A conventional short pulse at a different frequency might have to be generated at the end of the long pulse to provide coverage of the close-in range that is blanked by the long pulse. Since it only has to cover the range blanked by the long pulse, it need not be of large power. A separate receiver, or matched filter, might be needed for this short-range pulse. A pulse-compression waveform does not have the immunity to repeater jammers or range-gate stealers inherent in the short-pulse radar. By repeating a chirp signal with an offset frequency, the repeater can appear at the output of the pulse-compression filter ahead of the target skin-echo, thus making it harder to separate the true signal from the repeater signal. The frequency offset can compensate for the finite time-delay required for a repeater to respond. With a long coded-pulse signal, the repeater can also put energy into the radar ahead of the compressed pulse, or coincident with it, if sufficient power is used to overcome the mismatch of the repeater signal to the pulse-compression filter.

In spite of its limitations, pulse compression has been an important part of radar systems technology.

Spread spectrum. Spread spectrum communication systems[68] employ waveforms similar to those of pulse compression radar. The purpose of such waveforms in communications is to allow multiple simultaneous use of the same frequency spectrum. This is achieved by coding each signal differently from the others. In military applications, spread spectrum communications also has the capability of rejecting interference as well as reduce the probability of intercept by a hostile elint receiver. Sometimes pulse compression radars have been called spread spectrum radars. This terminology is misleading since pulse compression is used in radar for different reasons than spread spectrum is used in communications.

11.6 CLASSIFICATION OF TARGETS WITH RADAR

In most radar applications, the only properties of the target that are measured are its location in range and angle. Such radars are sometimes called *blob detectors* since they recognize targets only as "blobs" located somewhere in space. It is possible, however, to extract more information about the target. Radar may be able to recognize one type of target from another; that is, to determine that the target is a 747 aircraft and not a DC-10, or that a particular ship is a tanker and not a freighter. This capability is known as *target classification*. When the target is a spacecraft or satellite, the process is sometimes called SOI, or *Space Object Identification*. In this section, several possible radar techniques that might be used for target classification will be enumerated briefly. Generally, target classification by radar involves examining the detailed structure of the echo signal. It usually requires a larger signal-to-noise ratio than normally needed for detection. Thus the range at which target classification can be made is often less than the range at which the target can be first detected.

High-range-resolution. A short-pulse radar or a pulse-compression radar can provide sufficient range-resolution to obtain a profile of the target shape, as determined by the target's major scattering centers. From this range profile, an estimate of the target size can be made. The profile obtained by the radar is the projection in the direction of propagation. A complete "image" of the target would require multiple looks from different directions. If the trajectory of the target is known from the measurement of the target track, it is possible to infer the aspect of the radar projection. A radar should not be expected to provide the same target details as are seen visually. An electromagnetic sensor, whether the eye or a radar, responds to scattering from those details of the target which are comparable to the wavelength of observation. Since there is such a large difference in wavelength between microwave radar and visual sensors, the target details that are seen by radar can be quite different from what is seen visually. When attempting to measure target size with a high-range-resolution radar, an error can be incurred since the extremities of the target are not always good scatterers. Echoes from the forward and rear portions of the target might be obscured in the noise, if the radar is not sufficiently powerful.

High-range-resolution with monopulse. The inclusion of a monopulse angle-measurement to a high-range-resolution radar was mentioned briefly in Sec. 5.8. Resolution in range of the individual target scattering-centers permits an angle measurement of the scatterer without the errors introduced by the glint caused by multiple scatterers within the same resolution cell. This provides a three-dimensional "image" of the target and thus presents more target information from which to derive a classification than does a conventional high-range-resolution radar without monopulse. The implementation of this measurement capability is more complicated than a short-pulse radar without angle sensing. Also, it cannot be employed when the target angular extent is less than the sensitivity of the monopulse measurement.

Engine modulations. The radar echo from aircraft is modulated by the rotating propellers of piston engines, and by the rotating compressor and turbine blades of jet engines.[45] (The compressor would be seen by a radar looking into the forward part of the jet aircraft and the turbine when looking into the rear.) The characteristic modulations of the radar echoes from aircraft can, in some cases, be used to recognize one type of aircraft from another; or more correctly, one type of aircraft engine from another. Aircraft jet-engine modulations are likely to be of relatively high frequency (ten to twenty kilohertz perhaps) because of the high speed of the engine components that cause the modulated echo. The helicopter with its large rotating blades also provides a distinctive modulation of the radar echo that distinguishes it from the echoes of other aircraft. A ship is less likely to give distinctive modulations, unless it has large rotating radar antennas or rotating machinery within view of the radar.

Cross-section fluctuations. In the discussion of radar cross section in Sec. 2.8, it was stated that the angular pattern of the cross section of targets has a many-lobed structure whose spacings depend on the size and nature of the target. As the target moves relative to the radar, its aspect changes and the lobed pattern of the target cross section results in a fluctuating received signal. These amplitude fluctuations occur at a slow rate, compared with the much higher frequency fluctuations of the engine modulations mentioned above. The larger the size of the target the narrower will be the lobes and the greater will be the fluctuation frequency for a given angular rate of change of aspect. In principle, the cross-section fluctuations might be able to provide some information that can distinguish one type of target from another. If the target is rotating, as were some of the early unstabilized satellites, the amplitude fluctuations of the radar cross section with time can provide information from which the shape of the target can be determined.[60] It has also been suggested that the complex echo-signal fluctuations (amplitude *and* phase) can be employed for target classification.[65]

Synthetic aperture radar. The synthetic aperture radar, which is discussed in Sec. 14.1, is a radar in a moving vehicle that provides a high resolution image in both range and in a direction parallel to the vehicle motion. The latter dimension is sometimes called *cross range* when the radar uses a side-looking antenna directed perpendicular to the direction of motion. The range resolution is obtained with either a conventional short-pulse or pulse-compression waveform, and resolution in cross range is obtained by synthesizing the effect of a large antenna aperture. It is used chiefly for mapping of the ground and imaging of stationary objects on the ground.

Inverse synthetic aperture radar. In the ordinary synthetic aperture radar the target is stationary and the radar is in motion. The opposite will also permit target imaging; that is, the radar is stationary and the target is in motion. This is called *inverse synthetic aperture radar* or *delay-doppler mapping*. Although the signal processing required is similar to that of the conventional synthetic aperture radar, the inverse synthetic aperture process can also be viewed as an equivalent doppler filtering. Each part of a moving target has a slightly different relative velocity, or doppler-frequency shift. Filtering these various doppler frequencies resolves the different parts of the target to provide an image. Although the inverse synthetic aperture results from relative motion of the target, just as does the cross-section amplitude fluctuations mentioned previously, the processing of the inverse synthetic aperture radar signal requires a coherent system (one that preserves phase). In principle, inverse synthetic aperture radar can be used to image moving targets such as aircraft and ships. It has been applied in the past to imaging of the moon and to mapping the surface below the clouds surrounding the planet Venus.[61]

Polarization.[46,47] The polarization of the radar backscattered energy depends on the target properties and differs, in general, from the polarization of the energy incident on the target. This property can be used as a possible basis for discriminating one target from another. For example, a thin straight wire can be readily distinguished from a homogeneous sphere by observing the variation of the echo signal amplitude as the polarization is rotated. The echo from the sphere will be unmodulated, and the echo signal from the wire will vary between a maximum and a minimum at twice the rate at which the polarization is rotated. The use of circular polarization to reduce the radar echo from symmetrical raindrops relative to the echo from aircraft, as described in Sec. 13.8, takes advantage of the differences in target response to different incident polarizations.

For complete knowledge of the effect of polarization, the *polarization matrix* must be determined. If H stands for linear horizontal polarization, V for linear vertical, and if the first letter of a two-letter grouping denotes the transmitted polarization and the second letter denotes the polarization of the received signal, then the polarization matrix requires knowledge of the amplitudes and phase of the following components: HH, VV, HV and VH. HV and VH are sometimes called the *cross polarization* components. In general $HV = VH$ so that only one need be determined. Orthogonal circular-polarization components can also be used to describe the polarization matrix. By transmitting two orthogonal polarizations and measuring the amplitude and phase of the received echoes on each polarization, as well as the cross-polarization component, a means of target discrimination, or classification, can be provided. Some information about the target can be obtained from the amplitude only, and not the phase, of the polarization components. It has also been suggested that the cross-polarized component of the backscattered echo from simple axially symmetric objects (such as disks, cones, and cone-spheres) can provide an estimate of a transverse dimension of the body and give an indication of the severity of the edges, or "edginess."[47]

Nonlinear-contact effects (METRRA).[48] When metals come in contact with each other it is possible for their junctions to act as nonlinear diodes. The nonlinear properties of such junctions can be used to recognize metallic from nonmetallic reflectors when illuminated by radar. This technique has sometimes been called METRRA, which stands for Metal Reradiating Radar.[48] Most solid, mechanical, metal-to-metal bonds and properly made solder joints do not show nonlinear effects. However, if there is no molecular contact between the metals, and the space between them is small (of the order of 100 Å), then there can be discernible nonlinear effects. Such effects are noted with loose metal-to-metal contacts.

There are two basic approaches for taking advantage of the nonlinearity of these metal contacts. In one approach a single frequency is transmitted and a harmonic of the transmitted frequency is received. The nature of the nonlinearity of typical contacts is such that the third harmonic is usually the greatest. The other approach simultaneously transmits two frequencies f_1 and f_2, and the receiver is tuned to a strong cross-product such as $2f_1 \pm f_2$. When receiving at a frequency different from that transmitted, care must be exercised to ensure that the transmitter signal does not radiate a significant spectral component at the frequency to which the receiver is tuned.

The amount of signal returned from a nonlinear contact at a harmonic frequency is a nonlinear function of the incident field strength. Thus the nonlinear target cross section depends on the power, and the normal radar equation does not apply. In one system formulation the range dependence varied as the sixth power instead of the fourth power.[50] High peak power is more important with such targets than is high average power.

Similar techniques have also been proposed for cooperative targets by deliberately providing the target with a passive transponder employing microwave diodes.[49,50]

Another related target effect that might be utilized for target recognition is the random modulation of the scattered signal caused by the modification of the current distribution on a metal target that results from intermittent contacts on the target.[67] This modulation can be detected by examining the frequency spectrum in the vicinity of the received carrier. The acronym RADAM, which stands for *radar detection of agitated metals*, has sometimes been used to describe this effect.[69]

Inverse scattering. In principle, the size and shape of a target can be found by measuring the backscattered field, or radar cross section, at all frequencies and all aspects. It is not possible, of course, to obtain such complete information, but the process can be approximated by measuring the backscatter at a finite number of frequencies and aspects. The name *inverse scattering*[51] has been used to describe this method for obtaining the target size and shape, and thus provide a means for target classification. The use of high-range-resolution radar for profiling a target, and the inverse synthetic aperture radar mentioned above are two practical examples of target classification methods that might be called approximations of inverse scattering.

Instead of examining the radar echo as a function of frequency, the target response to an impulse can give the equivalent information since the Fourier transform of the impulse contains all frequencies. (In the above, an *impulse* is an infinitesimally short pulse, and the *target response* is the echo signal as a function of time.) This has been proposed as a means of target classification,[52] with the impulse being approximated by a short microwave pulse. The short pulse, with its high-frequency content, characterizes the fine detail of the target. Although this can be used as a means of target classification, it has been suggested that the usual short-pulse radar does not obtain important information about the target since its waveform does not contain the lower frequencies.[53-55,62] The lower frequencies that are suggested as being important are those corresponding to wavelengths from half the target size to wavelengths about ten times the target dimensions. These correspond to frequencies in the Rayleigh and the low-resonance regions. The use of such frequencies is said to provide overall dimensions, approximate shape, and material composition. Instead of the impulse response, the response of a target to a ramp function is sometimes more convenient to work with, especially when the longer wavelengths are used to obtain target classification.

The method of inverse scattering seems to require an examination of the target over a large frequency range, perhaps as much as 10 to 1. If the phase shift as well as the amplitude of the echo are measured, fewer frequencies might be utilized than when amplitude alone is obtained.[56] Experiments with as many as 12 frequencies have been carried out for simple scattering objects, but as few as four frequencies were said to be adequate for the discrimination of objects as complex as aircraft.[53]

Automatic target classification. For practical utilization, most of the target classification methods described above require automatic processing. Some method of signal recognition or pattern recognition must be applied to be able to correctly estimate the type of target. This can be just as important a part of target classification as the sensor itself.

Target track history. The above methods of target classification depend on extracting something about the target other than its location. However, the target track, which is obtained from location data alone, can provide significant information that can be used in the identification process, especially in a military situation. The speed, course, and maneuver of a target can indicate something about its intent; and therefore a form of target classification is possible. Automatic detection and tracking circuitry can aid in this form of discrimination since it provides a method for obtaining the target track. The long-range capability of HF over-the-horizon radar (Sec. 14.2) is also useful for this purpose since it can observe a large portion of the target track and might even be capable of observing where the target track originates (airport or seaport), which can be an important classification clue in some applications.

REFERENCES

1. Goldman, S.: "Frequency Analysis, Modulation, and Noise," p. 281, McGraw-Hill Book Company, New York, 1948.
2. Torrieri, D. J.: Arrival Time Estimation by Adaptive Thresholding, *IEEE Trans.*, vol. AES-10, pp. 178–184, March, 1974.
3. Mallinckrodt, A. J., and T. E. Sollenberger: Optimum Pulse-Time Determination, *IRE Trans.*, no. PGIT-3, pp. 151–159, March, 1954.
4. Gabor, D.: Theory of Communication, *J. IEE*, pt. III, vol. 93, pp. 429–441, 1946.
5. Woodward, P. M.: "Probability and Information Theory, with Applications to Radar," chap. 6, McGraw-Hill Book Co., New York, 1953.
6. Skolnik, M. I.: Theoretical Accuracy of Radar Measurements, *IRE Trans.*, vol. ANE-7, pp. 123–129, December, 1960.
7. Barton, D. K., and H. R. Ward: "Handbook of Radar Measurement," Prentice-Hall, Inc., N.J. 1969.
8. Manasse, R.: Range and Velocity Accuracy from Radar Measurements, unpublished internal report dated February, 1955, MIT Lincoln Laboratory, Lexington, Mass. (Not generally available.)
9. Slepian, D.: Estimation of Signal Parameters in the Presence of Noise, *IRE Trans.*, no. PGIT-3, pp. 68–89, March, 1954.
10. Siebert, W. McC.: A Radar Detection Philosophy, *IRE Trans.*, vol. IT-2, pp. 204–221, September, 1956.
11. Rihaczek, A. W.: "Principles of High-Resolution Radar," McGraw-Hill Book Company, New York, 1969.

12. Deley, G. W.: Waveform Design, chap. 3 of " Radar Handbook," M. I. Skolnik (ed.), McGraw-Hill Book Company, New York, 1970.
13. Sinsky, A. I., and C. P. Wang: Standardization of the Definition of the Radar Ambiguity Function, *IEEE Trans.*, Vol. AES-10, pp. 532–533, July, 1974.
14. Dicke, R. H.: Object Detection System, U.S. Patent no. 2,624,876, issued Jan. 6, 1953.
15. Cook, C. E.: Pulse Compression: Key to More Efficient Radar Transmission, *Proc. IRE*, vol. 48, pp. 310–315, March, 1960.
16. Farnett, E. C., T. B. Howard, and G. H. Stevens: Pulse Compression, chap. 20 of "Radar Handbook," M. I. Skolnik (ed.), McGraw-Hill Book Co., New York, 1970.
17. Peebles, P. Z., Jr., and G. H. Stevens: A Technique for the Generation of Highly Linear FM Pulse Radar Signals, *IEEE Trans.*, vol. MIL-9, pp. 32–38, January, 1965.
18. Eber, L. O., and H. H. Soule, Jr.: Digital Generation of Wideband LFM Waveforms, *IEEE 1975 International Radar Conference*, pp. 170–175, Apr. 21–23, 1975, IEEE Publication 75 CHO 938-1 AES.
19. Upton, L. O., and G. J. Mayer: Charge-coupled Devices and Radar Signal Processing, *RCA Engr.*, vol. 21, no. 1, pp. 30–34, June/July, 1975.
20. Holland, M. G., and L. T. Claiborne: Practical Surface Acoustic Wave Devices, *Proc. IEEE*, vol. 62, pp. 582–611, May, 1974.
21. Maines, J. D., and E. G. S. Paige: Surface-Acoustic-Wave Devices for Signal Processing Applications, *Proc. IEEE*, vol. 64, pp. 639–652, May, 1976.
22. Bristol, T. W.: Surface Acoustic Wave Devices—Technology and Applications, *IEEE 1976 WESCON Professional Program*, Los Angeles, Sept. 14–17, 1976, paper 24/1.
23. Lipka, M.: Pulse Compression Filter and Wideband Receiver Evaluation. *IEEE 1975 International Radar Conference*, pp. 283–287, Apr. 21–23, 1975, IEEE Publication 75 CHO 938-1 AES.
24. Cook, C. E., and M. Bernfeld: "Radar Signals," Academic Press, New York, 1967.
25. Temes, C. L.: Sidelobe Suppression in a Range-Channel Pulse-Compression Radar, *IRE Trans.*, vol. MIL-6, pp. 162–169, April, 1962.
26. Thor, R. C.: A Large Time-bandwidth Product Pulse-Compression Technique, *IEEE Trans.*, vol. MIL-6, pp. 169–173, April, 1962.
27. Kroszcynski, J. J.: Pulse Compression by Means of Linear-Period Modulation, *Proc. IEEE*, vol. 57, pp. 1260–1266, July, 1969.
28. Rowlands, R. O.: Detection of a Doppler-Invariant FM Signal by Means of a Tapped Delay Line, *J. Acoust. Soc. Am.*, vol. 37, pp. 608–615, April, 1965.
29. Taylor, S. A., and J. L. MacArthur: Digital Pulse Compression Radar Receiver, *APL Technical Digest*, vol. 6, pp. 2–10, March/April, 1967.
30. Golomb, S. W.: "Digital Communications with Space Applications," Prentice-Hall, Englewood Cliffs, N.J., 1964.
31. Ackroyd, M. H., and F. Ghani: Optimum Mismatch Filters for Sidelobe Suppression, *IEEE Trans.*, vol. AES-9, pp. 214–218, March, 1973.
32. Millett, R. E.: A Matched-Filter Pulse-Compression System Using a Nonlinear FM Waveform, *IEEE Trans.*, vol. AES-6, pp. 73–78, January, 1970.
33. Nathanson, F. E.: "Radar Design Principles," McGraw-Hill Book Co., New York, 1969.
34. Rihaczek, A. W.: Radar Waveform Selection—A Simplified Approach, *IEEE Trans.*, vol. AES-7, pp. 1078–1086, November, 1971.
35. Mutton, J. O.: Advanced Pulse Compression Techniques, *IEEE NAECON '75 Record*, pp. 141–148.
36. Welti, G. R.: Quaternary Codes for Pulsed Radar, *IRE Trans.*, vol. IT-6, pp. 400–408, June, 1960.
37. Golay, M. J. E.: Complementary Series, *IRE Trans.*, vol. IT-7, pp. 82–87, April, 1961.
38. MacWilliams, F. J., and N. J. A. Sloane: Pseudo-Random Sequences and Arrays, *Proc. IEEE*, vol. 64, pp. 1715–1729, December, 1976.
39. Shrader, W. W.: MTI Radar, chap. 17 of " Radar Handbook," M. I. Skolnik (ed.), McGraw-Hill Book Co., New York, 1970.
40. Ward, H. R.: Dispersive Constant False Alarm Rate Receiver, *Proc. IEEE*, vol. 60, pp. 735–736, June, 1972.
41. Mortley, W. S., and S. N. Radcliffe: Pulse Compression and Signal Processing, *IEEE Conf.* Publication No. 105, " Radar—Present and Future," Oct. 23–25, 1973, pp. 292–296.
42. Bogotch, S. E., and C. E. Cook: Effects of Limiting on the Detectability of Partially Time Coincident Pulse Compression Signals, *IEEE Trans.*, vol. MIL-9, pp. 17–24, January, 1965.
43. Woerrlein, H. H.: Spurious Target Generation Due to Hard Limiting in Pulse Compression Radars, *IEEE Trans.*, vol. AES-7, pp. 1170–1178, November, 1971.
44. Milne, K.: The Combination of Pulse Compression with Frequency Scanning for Three-dimensional Radars, *The Radio and Electronic Engr.*, vol. 28, pp. 89–106, August, 1964.

45. Hynes, R., and R. E. Gardner: Doppler Spectra of S Band and X Band Signals, *IEEE Trans. Suppl.*, vol. AES-3, no. 6, pp. 356–365, Nov., 1967. Also, *Report of NRL Progress*, pp. 1–10, January, 1968, PB 177017.
46. Copeland, J. R.: Radar Target Classification by Polarization Properties, *Proc. IRE*, vol. 48, pp. 1290–1296, July, 1960.
47. Knott, E. F., and T. B. A. Senior: Cross Polarization Diagnostics, *IEEE Trans.*, vol. AP-20, pp. 223–224, March, 1972.
48. Optiz, C. L.: Metal-detecting Radar Rejects Clutter Naturally, *Microwaves*, vol. 15, pp. 12–14, August, 1976.
49. Shefer, J., and R. J. Klensch: Harmonic Radar Helps Autos Avoid Collisions, *IEEE Spectrum*, vol. 10, pp. 38–45, May, 1973.
50. Davies, D. E. N., and H. Makridis: Two-Frequency Secondary Radar Incorporating Passive Transponders, *Electronics Letters*, vol. 9, no. 25, pp. 592–593, Dec. 13, 1973.
51. Lewis, R. M.: Physical Optics Inverse Diffraction, *IEEE Trans.*, vol. AP-17, pp. 308–314, May, 1969.
52. Kennaugh, E. M., and D. L. Moffatt: Transient and Impulse Response Approximations, *Proc. IEEE*, vol. 53, pp. 893–901, August, 1965.
53. Ksienski, A. A., Y. T. Lin, and L. J. White: Low-Frequency Approach to Target Identification, *Proc. IEEE*, vol. 63, pp. 1651–1660, December, 1975.
54. Moffatt, D. L., and R. K. Mains: Detection and Discrimination of Radar Targets, *IEEE Trans.*, vol. AP-23, pp. 358–367, May, 1975.
55. Chuang, C. W., and D. L. Moffatt: Natural Resonances of Radar Targets via Prony's Method and Target Discrimination, *IEEE Trans.*, vol. AES-12, pp. 583–589, November, 1976.
56. Goggins, W. B., Jr., P. Blacksmith, and C. J. Sletten: Phase Signature Radars, *IEEE Trans.*, vol. AP-22, pp. 774–780, November, 1974.
57. Berrie, D. W.: The Effects of Pseudo-Random Noise Upon Radar System Performance, *Aeronautical Systems Division, Wright-Patterson Air Force Base, Ohio*, ASD-TR-74-23, August, 1974. (Submitted as Ohio State University Dissertation), U.S. Gov. Printing Office: 1974-657-017/347.
58. Caputi, W. J., Jr.: Stretch: A Time-Transformation Technique, *IEEE Trans.*, vol. AES-7, pp. 269–278, March, 1971.
59. Hoft, D. J., and M. B. Fishwick: Analog Waveform Generation and Processing, *Electronic Progress*, vol. 17, no. 1, pp. 2–16, Spring, 1975. (Published by Raytheon Co., Lexington, Mass.)
60. Barton, D. K.: Sputnik II as Observed by C-Band Radar, *1959 IRE National Conv. Rec.*, vol. 7, pt 5, pp. 67–73.
61. Pettengill, G. H.: Radar Astronomy, chap. 33 of " Radar Handbook," M. I. Skolnik (ed.), McGraw-Hill Book Co., New York, 1970.
62. Lin, Y.-T., and A. A. Ksienski: Identification of Complex Geometrical Shapes by Means of Low-Frequency Radar Returns, *The Radio and Electronic Engineer*, vol. 46, pp. 472–486, October, 1976.
63. Frank, R. L.: Polyphase Codes with Good Nonperiodic Correlation Properties, *IEEE Trans.*, vol. IT-9, pp. 43–45, January, 1963.
64. Cantrell, B. H.: Sidelobe Reduction in Polyphase Codes, *Naval Research Laboratory Report* 8108, Washington, D.C., Apr. 13, 1977.
65. von Schlachta, K.: A Contribution to Radar Target Classification, *International Conference RADAR-77*, Oct. 25–28, 1977, pp. 135–139, IEE (London) Conference Publication no. 155.
66. Lindner, J.: Binary Sequences Up to Length 40 with Best Possible Autocorrelation Function, *Electronics Letters*, vol. 11, no. 21, p. 507, Oct. 10, 1975.
67. Bahr, A. J., and J. P. Petro: On the RF Frequency Dependence of the Scattered Spectral Energy Produced by Intermittent Contacts Among Elements of a Target, *IEEE Trans.*, vol. AP-26, pp. 618–621, July, 1978.
68. Dixon, R. C.: "Spread Spectrum Systems," Wiley Interscience, N.Y., 1976.
69. Newburgh, R. G.: Basic Investigations of the RADAM Effect, *Rome Air Development Center*, New York, Report RADC-TR-78-151, June, 1978, AD AO58099. (Approved for public release.)

CHAPTER
TWELVE

PROPAGATION OF RADAR WAVES

12.1 INTRODUCTION

The propagation of radar waves is affected by the earth's surface and its atmosphere. Complete analysis or prediction of radar performance must take into account propagation phenomena since most radars do not operate in "free space" as was assumed in the ideal formulation of the radar equation in Chaps. 1 and 2. Free-space radar performance is modified by *scattering* of electromagnetic energy from the surface of the earth, *refraction* caused by an inhomogeneous atmosphere, and *attenuation* by the gases constituting the atmosphere. Also included under the subject of propagation is the *external noise* environment in which the radar finds itself. The radar is also affected by the *reflection*, or backscatter, of energy from the earth's surface and from rain, snow, birds, and other clutter objects; but this subject is reserved for Chap. 13. The effects of propagation will modify the free-space performance of the radar, as well as introduce errors in the radar measurements.

It is usually convenient to distinguish between two different regions when considering radar propagation. One is the *optical*, or *interference*, region, which is within the line of sight (direct observation) of the radar. The other is the *diffraction* region, which lies beyond the line of sight, or beyond the horizon, of the radar. Radar energy found in this region is usually due to diffraction by the curvature of the earth or refraction by the earth's atmosphere.

Although the basic theory of radar wave propagation may be well understood, accurate quantitative predictions are not always easy to obtain because of the difficulty in acquiring the necessary knowledge of the environment in which the radar operates. In some respects, the prediction of propagation phenomena is like predicting the weather. Quite often the radar system designer must be content with only a qualitative knowledge of "average" propagation effects. Nevertheless, it is important to know and understand how propagation phenomena can influence radar performance since it can be a major factor in determining how well a radar performs in a particular application.

441

12.2 PROPAGATION OVER A PLANE EARTH

Although there are but few situations where accurate predictions of radar propagation effects can be made by assuming a flat rather than round earth, it is nevertheless instructive to examine this special case. The assumption of a flat earth simplifies the analysis and illustrates the type of changes introduced in radar coverage by a reflecting ground or sea surface.

Consider the earth to be a plane, flat, reflecting surface with the radar antenna located at height h_a. The target is at a height h_t and at a distance R from the radar. Figure 12.1 illustrates that energy radiated from the radar antenna arrives at the target via two separate paths. One is the direct path from radar to the target; the other is the path reflected from the surface of the earth. The echo signal reradiated by the target arrives back at the radar via the same two paths. Thus the received echo is composed of two components traveling separate paths. The magnitude of the resultant echo signal will depend upon the amplitudes and relative phase difference between the direct and the surface-reflected signals. Modification of the field strength (volts per meter) at the target caused by the presence of the surface may be expressed by the ratio

$$\eta = \frac{\text{field strength at target in presence of surface}}{\text{field strength at target if in free space}} \tag{12.1}$$

It is assumed in this analysis that the lengths of the direct and the reflected paths are almost (but not quite) equal so that the amplitudes of the two signals are approximately the same provided there is no loss suffered on reflection. Hence, if the amplitudes of the two waves differ from one another, it is assumed to be due to a surface-reflection coefficient less than unity. Although the two paths are comparable in length, they are not exactly equal. Any difference in the relative phase between the direct and the reflected waves can be attributed to the difference in the path length and the change in phase that occurs on reflection. The reflection coefficient of the surface may be considered as a complex quantity $\Gamma = \rho e^{j\psi}$. The real part ρ describes the change in amplitude, while the argument ψ describes the phase shift on reflection.

It is further assumed in the present example that the reflection coefficient $\Gamma = -1$. The reflected wave suffers no change in amplitude, but its phase is shifted 180°. A reflection coefficient of -1 applies at microwave frequencies to a smooth surface with good reflecting properties if the radiation is horizontally polarized and the angle of incidence is small.

The effective radiation pattern of the radar located at A in Fig. 12.1 may be found in a manner analogous to that of a two-element interferometer antenna formed by the radar antenna at A and its image mirrored by the ground at A'. The difference between the reflected path AMB and the direct path AB (or $A''MB$) is $\Delta = 2h_a \sin \xi$, when $R \gg h_a$. For ξ small, $\sin \xi$ may be replaced by $(h_a + h_t)/R$ so that $\Delta = 2h_a(h_a + h_t)/R \approx 2h_a h_t /R$. The latter expression

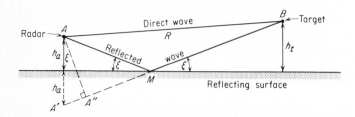

Figure 12.1 Propagation over a plane reflecting surface.

assumes that $h_t \gg h_a$. The phase difference corresponding to the path-length difference is

$$\psi_d = \frac{2\pi}{\lambda} \frac{2h_a h_t}{R} \text{ radians} \tag{12.2}$$

To this must be added the phase shift ψ_r resulting from the reflection of the wave at M, which is assumed to be π radians, or $180°$. The total phase difference between the direct and the ground-reflected signals as measured at the target is

$$\psi = \psi_d + \psi_r = \frac{2\pi}{\lambda} \frac{2h_a h_t}{R} + \pi \tag{12.3}$$

The resultant of two signals, each of unity amplitude but with phase difference ψ, is $[2(1 + \cos \psi)]^{1/2}$. Therefore the ratio of the power incident on the target at B to that which would be incident if the target were located in free space is

$$\eta^2 = 2\left(1 - \cos \frac{4\pi h_a h_t}{\lambda R}\right) = 4 \sin^2 \frac{2\pi h_a h_t}{\lambda R} \tag{12.4}$$

Because of reciprocity, the path from target to radar is the same as from radar to target. The power ratio at the radar is therefore

$$\eta^4 = 16 \sin^4 \frac{2\pi h_a h_t}{\lambda R} \tag{12.5}$$

The radar equation describing the received echo power must be modified by the propagation factor η^4 of Eq. (12.5). Since the sine varies in magnitude from 0 to 1, the factor η^4 varies from 0 to 16. The received signal strength also varies from 0 to 16; hence the fourth-power relation between range and echo signal results in a variation of radar range from 0 to 2 times the range of the same radar in free space.

The field strength is a maximum when the argument of the sine term in Eq. (12.5) is equal to $\pi/2, 3\pi/2, \ldots, (2n + 1)\pi/2$, where $n = 0, 1, 2, \ldots$. The maxima are therefore defined by

$$\frac{4h_a h_t}{\lambda R} = 2n + 1 \qquad \text{maxima} \tag{12.6}$$

The minima, or nulls, occur when the sine term is zero, or when

$$\frac{2h_a h_t}{\lambda R} = n \qquad \text{minima} \tag{12.7}$$

Thus the presence of a plane reflecting surface causes the continuous elevation coverage to break up into a lobed structure as indicated in Fig. 12.2. A target located at the maximum of a

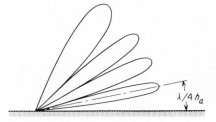

$\lambda/4 h_a$

Figure 12.2 Vertical lobe structure caused by the presence of a plane reflecting surface.

particular lobe will be detected at a range twice that of the same radar located in free space. However, at other angles, the radar detection range can be less than the free-space range. When the target lies in the nulls, no echo signal is received.

The angle (in radians) of the first (lowest) lobe is approximately equal to $\lambda/4h_a$. If low-angle coverage is desired, the radar antenna height should be high and the wavelength of the radiated energy small. The antenna pattern lobes caused by the presence of the ground might sometimes be of advantage when the longest possible detection range is desired against low-altitude targets and where continuous coverage is not required.

The simple form of the radar equation [Eq. (1.9)] may be written with the propagation factor η included to illustrate the effect of the plane earth:

$$P_r = \frac{P_t G^2 \lambda^2 \sigma}{(4\pi)^3 R^4} 16 \sin^4 \frac{2\pi h_a h_t}{\lambda R} \approx \frac{4\pi P_t G^2 \sigma (h_a h_t)^4}{\lambda^2 R^8} \qquad \text{for small angles} \qquad (12.8)$$

The received signal power for targets at low angles (on the lower side of the first lobe) varies as the eighth power of the range instead of the more usual fourth power. This phenomenon has been experimentally verified for ship targets at short ranges where the plane-earth approximation has some validity.[1] Another difference between Eq. (12.8) and the normal radar equation is the factor G/λ, which appears in place of the factor $G\lambda$.

The analysis in this section is based on many simplifying assumptions; therefore care should be exercised in adapting the results and conclusions to more realistic situations. The assumption of a perfectly reflecting plane earth applies in but a few cases. Also assumed was an omnidirectional antenna pattern in the elevation plane. Since radars utilize directive antennas, the idealized antenna lobe structure as given by Eq. (12.4) must be appropriately modified to account for the actual antenna radiation pattern. Theoretically, the nulls in the lobe structure are at zero field strength since the direct and reflected signals are assumed to be of equal amplitude. In practice, the nulls are "filled in" and the lobe maxima are reduced because of nonperfect reflecting surfaces with reflection coefficients less than unity. The nulls will also be filled in if broadband signals are radiated by the radar.

In the above example, the reflection coefficient of the ground was taken to be -1, which applies for horizontal polarization and a smooth reflecting surface. The magnitude and phase of the reflection coefficients of vertically polarized energy behave differently from waves with horizontal polarization (Fig. 12.3a and b). The calculated amplitude and phase of the reflection coefficient are plotted for a smooth sea surface at 100 and 3,000 MHz. It is seen that the reflection coefficient for vertically polarized energy is less than that for horizontally polarized energy. The angle corresponding to the minimum reflection coefficient is called *Brewster's angle*.

The different reflection coefficients with the two polarizations result in different coverage patterns. The nulls are not as deep with vertical polarization, nor are the lobe maxima as great. Vertical polarization might be preferred when complete vertical coverage is required. Horizontal polarization might be specified when enhanced range capability is desired and complete vertical coverage is not necessary. The theoretical curves of Fig. 12.3 assume a smooth reflecting surface. In practice, this condition is seldom met, and the difference in the coverage diagrams obtained with each of the two types of polarization is often not as pronounced as might be expected on the basis of theoretical computations. Roughness is more important than the electrical properties in determining whether reflection is specular or not. Measurements have shown that the reflection coefficient for normal (nonsmooth) ground terrain is in the range 0.2 to 0.4 and is seldom greater than 0.5 at frequencies above 1500 MHz except for low angles of incidence.[2,3]

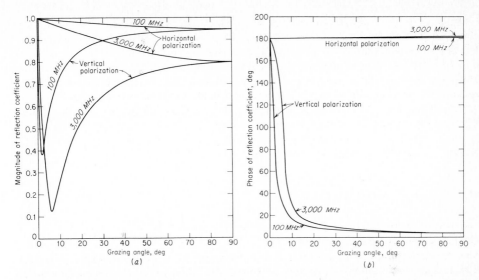

Figure 12.3 (*a*) Magnitude of the reflection coefficient as a function of the grazing angle (for seawater). (*b*) Phase of the reflection coefficient as a function of the grazing angle (for seawater). Phase of the reflected wave lags the phase of the incident wave. (*From Burrows and Attwood,*[5] *courtesy Academic Press, Inc.*)

For a rough, perfectly conducting surface, such as the sea, Ament[70] derived the reflection coefficient as

$$\rho_0 \exp\left[-2k^2\overline{h^2} \sin^2 \psi\right]$$

where ρ_0 = complex reflection coefficient for a smooth surface
$k = 2\pi/\lambda$
λ = wavelength
$\overline{h^2}$ = mean-square surface height (mean value of $h = 0$)
ψ = grazing angle

Experimental data taken over the sea fit this expression well; except for large values of the roughness parameter $[(h \sin \psi)/\lambda \approx h\psi/\lambda > 0.11]$, where the theory underestimates the experimental observations.[71,72]

The reflection of electromagnetic waves from the sea may be separated into a *coherent* and an *incoherent* component.[71-73] The coherent component has a reflection coefficient whose amplitude and phase are fixed by a given geometry and sea state. The term "reflection coefficient" as usually found in the literature is generally that of the coherent component. The incoherent, or fluctuating, component of a surface-scattered signal is characterized by a random phase and amplitude. The incoherent component of the forward-scattered signal from a rough surface behaves differently from the coherent component as a function of roughness. It increases linearly with increasing surface-roughness parameter $(h\psi/\lambda)$, levels off to a maximum and then decreases as the inverse square root of the roughness parameter.[74]

The presence of a reflecting surface affects the measurement of low elevation angles as well as the radar coverage. When the target elevation angle is less than a beamwidth, the radar receives both the direct and the ground-reflected waves, and the effective antenna pattern is altered. The radar sees the image as well as the target. Height-finding radars which measure elevation angle of arrival by comparing the amplitudes of the signals received at two different beam elevation angles (lobe comparison) can give erroneous and ambiguous measurements at low angles.[4,75] Elevation errors near the ground may be considerably reduced in magnitude and the ambiguities eliminated by surrounding the radar with a metallic fence to remove the ground-reflected wave. The fence replaces the ground-reflected wave with a diffracted wave of lesser importance. The diffraction fence also removes objectionable ground clutter.

Tracking radars also are affected at low elevation angles by the ground-reflected wave. The tracker might give an erroneous angle measurement just as in the case of the height finder, or it might track the "image" in the ground instead of the true target. The overall effect on the tracker is somewhat analogous to "glint" (Sec. 5.5).

12.3 THE ROUND EARTH

In general, the curvature of the earth cannot be neglected when predicting radar coverage. This is especially true for coverage at low elevation angles near the horizon. The two regions of interest in radar propagation are the *interference* region and the *diffraction* region. The interference, or optical, region is located within line of sight of the radar. The direct and reflected waves interfere to produce a lobed radiation pattern similar to that described for the plane

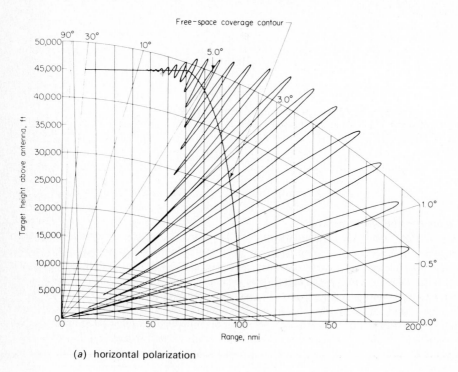

(a) horizontal polarization

Figure 12.4 Example of a vertical-plane coverage diagram for (a) horizontal polarization and (b) vertical polarization. Frequency = 1300 MHz, antenna height = 50 ft, antenna vertical beamwidth = 12° with beam maximum pointing on the horizon, sea surface with 4 ft wave-height and a free space range = 100 nmi.

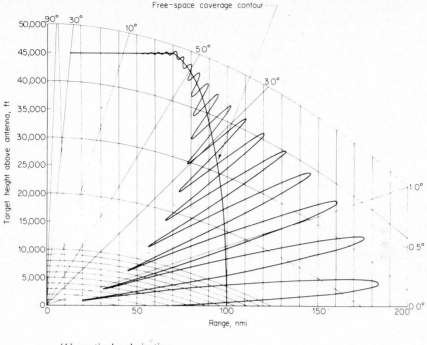

(*b*) vertical polarization

Figure 12.4 (*continued*)

earth. Lobing, however, is not as pronounced in the case of the round earth: the minima are not as deep nor are the maxima as great since a wave reflected from a curved surface is more divergent than one reflected from a plane. The other region of interest is that which lies just beyond the interference region below the radar line of sight and is the diffraction, or the shadow, region. Here radar signals are rapidly attenuated. Very few microwave radars have the capability of penetrating the diffraction region to any great extent because of the severe losses.

The effect of the round earth on radar coverage can be predicted by analytical means for the idealized case of a "smooth" earth of known, uniform properties. There exist in the literature the necessary graphs and nomographs which simplify the computation of radar coverage.[5-9] Computers may be used to perform the calculations and to automatically plot the coverage patterns by computer-controlled plotting machines.[10] Figure 12.4 illustrates, for a particular case, the computed vertical-plane radar coverage patterns for (*a*) horizontal polarization and (*b*) vertical polarization.

12.4 REFRACTION

Radio and radar waves travel in straight lines in free space. However, electromagnetic waves propagating within the earth's atmosphere do not travel in straight lines but are generally bent or refracted. One effect of refraction is to extend the distance to the horizon, thus increasing

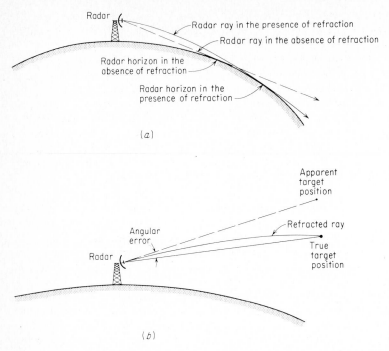

Figure 12.5 (a) Extension of the radar horizon due to refraction of radar waves by the atmosphere; (b) angular error caused by refraction.

the radar coverage (Fig. 12.5a). Another effect is the introduction of errors in the measurement of elevation angle (Fig. 12.5b). Bending, or refraction, of radar waves in the atmosphere is caused by the variation with altitude of the velocity of propagation, or the *index of refraction*, defined as the velocity of propagation in free space to that in the medium in question.

At microwave frequencies, the index of refraction n for air which contains water vapor is[11,12]

$$(n - 1)10^6 = N = \frac{77.6p}{T} + \frac{3.73 \times 10^5 e}{T^2} \tag{12.9}$$

where p = barometric pressure, mbar (1 mm Hg = 1.3332 mbar)

e = partial pressure of water vapor, mbar

T = absolute temperature, K

The parameter $N = (n - 1)10^6$ is the "scaled-up" index of refraction and is called *refractivity*. It is often used in propagation work instead of n because it is a more convenient unit. The prime difference between optical and microwave refraction is that water vapor has a negligible effect on the former; consequently the second term of Eq. (12.9) may be neglected at optical frequencies. Since the barometric pressure p and the water-vapor content e decrease rapidly with height, while the temperature T decreases slowly with height, the index of refraction normally decreases with increasing altitude. A typical value of the index of refraction near the surface of the earth is 1.0003. In a standard atmosphere the index decreases at the rate of about 4×10^{-8} m^{-1} of altitude.

The decrease in refractive index with altitude means that the velocity of propagation increases with altitude, causing radio waves to bend downward. The result is an increase in the effective radar range as was illustrated in Fig. 12.5a. (Variations of the refractive index in the horizontal plane may also exist, but they do not materially alter the bending.)

Refraction of radar waves in the atmosphere is analogous to bending of light rays by an optical prism. The path of the radar waves through the atmosphere may be plotted using ray-tracing techniques, provided the variation of refractive index is known.

The classical method of accounting for atmospheric refraction in computations is by replacing the actual earth of radius a ($a = 3440$ nautical miles) by an equivalent earth of radius ka and by replacing the actual atmosphere by a homogeneous atmosphere in which electromagnetic waves propagate in straight lines rather than curved lines (Fig. 12.6). It may be shown from Snell's law in spherical geometry that the value of the factor k by which the earth's radius must be multiplied in order to plot the ray paths as straight lines is

$$k = \frac{1}{1 + a(dn/dh)} \tag{12.10}$$

where dn/dh is the rate of change of refractive index n with height.[5] The vertical gradient of the refractive index dn/dh is normally negative. If it is assumed that this gradient is constant with height and equal to a value of 39×10^{-9} per meter, the value of k is $\frac{4}{3}$. The use of the $\frac{4}{3}$ effective earth's radius to account for the refraction of radio waves predates radar and, because of its convenience, has been widely used in radio communications, propagation work, and radar.[13] It is only an approximation, however, and may not yield correct results if precise radar measurements are desired, as, for example, in a long-range height finder. The term *standard refraction* is applied when the index of refraction decreases uniformly with altitude in such a manner that $k = \frac{4}{3}$ (Ref. 14).

The distance d to the horizon from a radar at height h may be shown from simple geometrical considerations to be approximately

$$d = \sqrt{2kah} \tag{12.11a}$$

where ka is the effective radius of the earth and h is assumed small compared with a. For $k = \frac{4}{3}$, Eq. (12.11a) reduces to a particularly convenient relationship if d and h are measured in statute miles and feet, respectively.

$$d \text{ (statute miles)} = \sqrt{2h(\text{ft})} \tag{12.11b}$$

or,
$$d \text{ (nautical miles)} = 1.23 \sqrt{h(\text{ft})} \tag{12.11c}$$

or,
$$d \text{ (km)} = 130 \sqrt{h(\text{km})} \tag{12.11d}$$

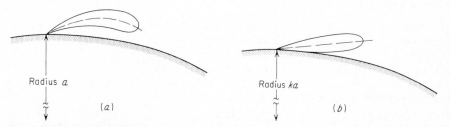

Figure 12.6 (a) Bending of antenna beam due to refraction by the earth's atmosphere; (b) shape of beam in equivalent-earth representation with radius ka.

Equations (12.11) have been used at times as a measure of the line-of-sight coverage of a ground-based radar viewing a target at a height h. This may lead to incorrect results in some cases since the optical line of sight does not necessarily correspond to the radar line of sight, as explained in Sec. 12.6.

The four-thirds earth approximation has several limitations. It is only an average value and should not be used for other than general computational purposes. The correct value of k depends upon meteorological conditions. Bean[15,16] found that the average value of k measured at an altitude of 1 km varies from 1.25 to 1.45 over the continental United States during the month of February and from 1.25 to 1.90 during August. In general, the higher values of k occur in the southern part of the country. Burrows and Attwood[5] state that k lies between $\frac{6}{5}$ and $\frac{4}{3}$ in arctic climates.

The use of an effective earth's radius implies that dn/dh is constant with height, or in other words, that n decreases linearly with height. This assumption is in disagreement with the experimentally observed refractive-index structure of the atmosphere at heights above 1 km.[17] The variation of refractivity with altitude is found to be described more nearly by an exponential function of height rather than the linear variation assumed by the $\frac{4}{3}$ earth model or any model of constant effective earth's radius. A more appropriate refractivity model is one in which the refractivity varies exponentially with height,[12,17]

$$N = N_s \exp\left[-c_e(h - h_s)\right] \tag{12.12}$$

where
N_s = refractivity at surface of earth
h = altitude of target
h_s = altitude of radar
$c_e = \ln(N_s/N_1)$ = a constant which depends upon value of N_s and N_1, the latter being refractively at altitude of 1 km

It is found that the exponential model gives a more accurate determination of the effects of atmospheric refraction than does a linear model.

The use of the correct atmospheric model is quite important in a height-finder radar, especially for targets at long ranges.[12,17,18] Refraction causes the radar rays to bend, resulting in an apparent elevation angle different from the true one. In certain radar applications, corrections must be made to the radar data to obtain a better estimate of elevation angle, range, or height.[19] Surface observations of refractivity often suffice for ascertaining the effects of refraction.[20-22]

Refraction is troublesome primarily at low angles of elevation, especially at or near the horizon. It can usually be neglected at angles greater than 3 to 5° in most radar applications.

Although more refined models of atmospheric refraction must be considered where precise radar measurements are important, the simplicity of the usual $\frac{4}{3}$ earth approximation makes it attractive for rough predictions.

The above discussion of refraction has been directed primarily to the aircraft target located within the lower portion of the atmosphere called the troposphere. Targets such as satellites and ballistic missiles operate above both the troposphere and the ionosphere. The effects of the entire atmosphere must be considered in such cases.[23,24]

12.5 ANOMALOUS PROPAGATION

A decrease in atmospheric index of refraction with increasing altitude, as described in the previous section, bends the radar rays so as to extend the coverage beyond that expected with a uniform atmosphere. If the gradient of the index of refraction is strong enough for the rays to have the same curvature as the earth itself, it would be possible for initially horizontal rays to bend around the surface of the earth. From Eq. 12.10, a gradient $dn/dh = -1.57 \times 10^{-7} \text{ m}^{-1}$ will make the effective earth's radius infinite which allows initially horizontal rays to follow the curvature of the earth. Under such conditions, the radar range is significantly increased and detection beyond the radar horizon can result.

The abnormal propagation of electromagnetic waves is called *superrefraction, trapping, ducting,* or *anomalous propagation.* According to one definition,[25] *normal atmospheric refraction* corresponds to the gradient dn/dh varying from 0 to $-0.787 \times 10^{-7} \text{ m}^{-1}$, or when the effective earth's radius (Eq. 12.10) varies from $k = 1$ to $k = 2$. *Superrefraction* corresponds to dn/dh from -0.787×10^{-7} to $-1.57 \times 10^{-7} \text{ m}^{-1}$, or $k = 2$ to $k = \infty$. *Trapping,* or *ducting,* occurs when dn/dh is greater than $-1.57 \times 10^{-7} \text{ m}^{-1}$. The radar ranges with ducted propagation are greatly extended. Holes can also appear in the coverage. If the atmospheric index of refraction were to increase, rather than decrease, with increasing height, the rays would curve upward and the radar range would be reduced as compared to normal conditions. This is called *subrefraction.* Its occurrence is rare. The term *anomalous,* or *nonstandard, propagation* applies to any of the above propagation conditions other than normal; but it is almost always used to describe those conditions where the radar range is extended well beyond normal.

A superrefracting duct which lies close to the ground is called a *ground-based duct,* or a *surface duct.* Over water the surface duct is also called the *evaporation duct,* since it is the result of water vapor evaporated from the sea. A duct which lies above the surface is called an *elevated duct.* Surface ducts apparently are more usual than elevated ducts. To propagate energy within the duct, the angle the radar ray makes with the duct should be small, usually less than one degree.[26,27] Only those radar rays launched nearly parallel to the duct are trapped. With a surface-based radar, ducting is limited to low angles of elevation so that the chief effect is to extend the surface coverage.

Since the angle between the radar beam and the duct direction (as defined by the levels of constant index of refraction) cannot be greater than about 1° if power is to be coupled into an elevated duct, the radar must usually be at an altitude that permits the required shallow coupling angles to be achieved. The radar in this case will be more sensitive to targets within the duct than those outside it.

The extension of the radar range within the duct results in a reduction of coverage in other directions. The regions with reduced coverage are called *radar,* or *radio, holes.* If, for example, the radar range is extended against surface targets by the presence of a surface duct, air targets just above the duct that would normally be detected might be missed.

Atmospheric ducts are generally of the order of 10 or 20 meters in height, never more than perhaps 150 to 200 meters. Propagation within a surface duct is similar to propagation within a waveguide with a "leaky top wall." A duct supports only certain modes of propagation and does not readily support propagation below a critical wavelength. A simplified approximate model of propagation in atmospheric ducts gives the maximum wavelength that can be propagated in a surface duct of depth d as[28]

$$\lambda_{\max} = 2.5 \left(-\frac{\Delta n}{\Delta h} \right)^{1/2} d^{3/2} \tag{12.13}$$

where $\lambda_{\max}$, Δh, and d are in the same units. For example, at X band ($\lambda_{\max} = 3$ cm), Eq. (12.13) predicts that the duct must be at least 10 m thick; at S band ($\lambda = 10$ cm) it must be 22 m; and at UHF ($\lambda = 70$ cm), the duct thickness must be at least 80 m thick. (The value of $-\Delta n/\Delta h$ was taken as $1.57 \times 10^{-7} \text{ m}^{-1}$.) Since atmospheric ducts are not usually deep, extended range propagation is more likely to be experienced at the higher microwave frequencies than at the lower frequencies. Unlike standard waveguide propagation, the cutoff wavelength for ducting does not sharply divide the regions of propagation and no propagation, so that radiation whose wavelength is several times the "cutoff wavelength" may be affected by the duct.

A duct is produced when the index of refraction decreases with altitude at a rapid rate. If the index of refraction [Eq. (12.9)] is to decrease with height, the temperature must increase and/or the humidity (water-vapor content) must decrease with height. An increase of temperature with height is called a *temperature inversion* and occurs when the temperature of the sea or land surface is appreciably less than that of the air. A temperature inversion, by itself, must be very pronounced in order to produce superrefraction. Water-vapor gradients are more effective than temperature gradients alone; thus superrefraction is usually more prominent over oceans, especially in warm climates.

Ducting occurs when the upper air is exceptionally warm and dry in comparison with the air at the surface. There are several meteorological conditions which may lead to the formation of superrefracting ducts.[31,32] Over land, ducting is usually caused by radiation of heat from the earth on clear nights, especially in the summer when the ground is moist. The earth loses heat, and its surface temperature falls, but there is little or no change in the temperature of the upper atmosphere. This leads to conditions favorable to ducting, that is, a temperature inversion at the ground and a sharp decrease in the moisture with height. Therefore, over land masses ducting is most noticeable at night and usually disappears during the warmest part of the day.

Another common cause of ducting is the movement of warm dry air, from land, over cooler bodies of water. Warm dry air blown out over the cooler sea is cooled at the lowest layers and produces a temperature inversion. At the same time, moisture is added from the sea to produce a moisture gradient. This form of anomalous propagation over the sea tends to be more prominent on the leeward side of land masses. Ducting occurs during either the day or night and can last for long periods of time. It is most likely to occur, however, in the late afternoon and evening when the warm afternoon air drifts out over the sea.[33]

Thus the character of ducting is likely to differ over land and sea. Land masses change temperature much more quickly than does the sea. As a result, there is much more of a diurnal variation of ducting over land than over sea, where it is likely to be more continuous and widespread.[32]

Superrefracting ground ducts may also be produced by the diverging downdraft under a thunderstorm.[31] The relatively cool air which spreads out from the base of a thunderstorm results in a temperature inversion in the lowest few thousand feet. The moisture gradient is also appropriate for the formation of a duct. Duct formation by thunderstorms may not be as frequent as other ducting mechanisms, but it is of importance since it may be used as a means of detecting the presence of a storm. An operator carefully watching a radar display can detect the presence of a storm by the sudden increase in the number and range of ground targets. The conditions appropriate to the formation of a thunderstorm duct are short-lived and have a time duration of the order of perhaps 30 min to 1 h.

With the exception of thunderstorms, ducting is essentially a fine-weather phenomenon. As tropical (but not equatorial) climates are noted for their fine weather, it is not surprising to find the most intense ducting occurring in such regions.[32] In temperate climates ducting is more common in summer than in winter. It does not occur when the atmosphere is well mixed, a condition generally accompanying poor weather. When it is cold, rough, stormy, rainy, or cloudy, the lower atmosphere is well stirred up and propagation is likely to be normal. Both rough terrain and high winds tend to increase atmospheric mixing, consequently reducing the occurrence of ducting. Radar propagation is normal whenever the upper air is unusually cold in comparison with the earth's surface.

Elevated ducts. An example of propagation in elevated ducts is found in the "tradewind region" between the midocean, high-pressure cells and the equatorial doldrums. Two such tradewind areas that have been studied lie between Brazil and the Ascension Islands[29] and between Southern California and Hawaii.[30] The tradewind temperature inversion is usually found over the eastern portions of the tropical oceans. In these regions, there is a slow-sinking of high-altitude air (large-scale subsidence) which meets low-level maritime air flowing toward the equator. The general sinking of air from high altitudes results in adiabatic heating due to compression, and a decrease in moisture content. Thus this warmer, drier air lies above the cooler, moist air to produce a temperature inversion; i.e., an increase in temperature with height. This results in a strong duct along the interface of the temperature inversion. In some cases, a stratus cloud layer will form at the base of the temperature inversion.[27] Thus the duct altitude can be identified by the height of the cloud tops that are suppressed by the tempera-ture inversion. When the temperature inversion occurs below the altitude at which clouds are formed, a haze layer in the air below the temperature inversion can be observed. The altitude of the tradewind duct varies from hundreds of meters at the eastern part of the tropical oceans to thousands of meters at the western end. Thus, the height gradually rises in going from east to west. There is also a general decrease of the refractivity gradients to the west. Off the southern California coast, in the vicinity of San Diego, the boundary between moist and dry air is relatively stable and exists at an average altitude of 300 to 500 meters. Since the elevated duct is due to meteorological effects there are seasonal, as well as diurnal, variations. The optimum season for duct formation in the tradewind region between Brazil and Ascension Islands occurs in November.[29] It has been said, however, that elevated ducts giving rise to strong persistent anomalous propagation occur throughout most of the year over at least one-third of the oceans.[27]

To take maximum advantage of propagation in an elevated duct, the radar and target should be at an altitude near that of the duct. This is found from both theory and measure-ments made by aircraft flying at various altitudes. It was noted,[30] however, that the propaga-tion of electromagnetic energy from antennas well below the duct is greater than would be predicted from classical ray theory. One explanation postulated for explaining this discrepancy is that energy can be scattered into and out of the elevated duct by irregularities in the index-of-refraction profile.

Evaporation duct. The evaporation duct that lies just above the surface of the sea is a result of the water vapor, or humidity, evaporated from the sea. The air in contact with the sea is saturated with water vapor, with a saturation vapor pressure appropriate to the temperature of the sea surface.[34] The air several meters above the sea is not usually saturated so there will be a gradual decrease in water vapor pressure from the surface value to the ambient value well above the surface. The decrease in water vapor pressure is approximately logarithmic, which contributes a logarithmic decreasing term to the index of refraction.

The evaporation duct exists over the ocean, to some degree, almost all of the time. The height of the duct, which is usually within the range from 6 to 30 m, varies with the geographic location, season, time of day, and the wind speed. In the North Sea the mean duct thickness is about 6 m while in tropical regions it is of the order 10 to 15 m.[34] In the North Atlantic at Weather Ship *D* (44°N, 41°W) the median value of duct thickness (half the ducts have thickness of lesser value) is 10 m in the summer and 30 m in the winter.[35] In the area offshore of San Diego, the medium value of the duct thickness as calculated from 5 years of meteorological data is 6 m, and a similar calculation for the Mediterranean Sea is 10 m.[36] Measurements made in the Atlantic tradewind area off of Antigua showed ducts averaging 6 to 15 m in height.[37] The height and strength of the duct was found to vary with wind speed. Stronger winds generally resulted in stronger signals and lower attenuation rates. Wind speeds from 8 to 15 knots produced a low, moderately strong duct, and winds from 20 to 30 knots produced a higher, but weaker duct. Passing squalls and rain showers did not wipe out the duct or decrease the propagated signals.

The thicker the duct, the lower the frequency that can be propagated, as can be seen from Eq. (12.13). Thus the lower microwave frequencies (below L band) are not usually strongly affected by the evaporation duct. The upper frequency limit for ducting is determined by the increased attenuation due to roughness of the sea and by absorption of electromagnetic energy in the vicinity of the water vapor resonance at 22 GHz. Thus there is probably an optimum frequency for the utilization of ducted propagation. In a duct with *complete* trapping of the electromagnetic energy, the cylindrical spreading of the energy radiated by a point source results in the power density decreasing as R^{-1}, where R = range, instead of R^{-2} as with free space propagation. The energy is seldom completely trapped, however, and there is attenuation due to the leaking of energy from the upper surface of the duct as well as by scattering and absorption. Energy is also scattered out of the duct because of a rough sea surface.[76] It has been found experimentally[37] that S-band radiation (10 cm wavelength) is only partially trapped by the evaporation duct, with attenuation rates averaging about 0.85 dB/nmi. Generally, there is less attenuation at S band the higher the radar antenna or the target altitude.[36,37] At X band (3 cm wavelength), the coupling to the duct is stronger and the average (one-way) attenuation was reported as 0.45 dB/nmi.[37] Unlike S band, the attenuation at X band was less at low antenna and target heights (3 to 5 m), giving stronger signals and greater ranges than antennas and targets at heights up to 30 m; which is a further indication of effective trapping at the higher frequencies.

It was observed that with a large enough duct more than one mode can be propagated and there can be more than one antenna height suitable for low-loss propagation. For example,[37] one set of X-band data showed the minimum attenuation to occur with an antenna height of 2 m. Increasing the antenna height increased the loss to a maximum at about 10 m height. Further increase of height decreased the attenuation until a secondary minimum was obtained at 20 m height, after which the attenuation again increased (at least to a height of 30 m). On a ship it might not be practical to site a radar antenna 2 m over the sea. Instead it might be sited at 20 m height, with the slightly greater loss being a price to pay for the convenience of the higher antenna location. The above values apply to a particular experiment in a particular location. Hence, the optimum antenna heights might vary.

Theoretical models of propagation in evaporation ducts confirm the general experimental observations presented above.[77] When multiple modes of propagation are trapped in the duct, theory predicts considerable signal fades due to interference between the several modes. Fades can be of the order of 20 dB. In one example, it was shown that the fading is strongly dependent on the radar and the target heights, and that for centimeter wavelengths the fading occurs at intervals of from two to three miles.

The height of the evaporation duct, from which the propagation conditions can be inferred, can be readily calculated from measurements of the surface water temperature and, at some convenient height, the air temperature, relative humidity, and wind speed.[78] These four measurements have been found sufficient for the description of ducting conditions.

The above has been concerned with propagation beyond the normal horizon. Within the horizon, the refractive effects of the duct can lead to a modification of the normal lobing pattern caused by the interference of the direct ray and the surface-reflected ray. The relative phase between the direct and reflected rays can be different in the presence of the duct, and focusing can change the relative amplitudes of the two components. The focusing effect can even cause the amplitude of the surface-reflected ray to sometimes exceed that of the direct ray.[38] The effect of the duct on the line-of-sight propagation is to reduce the angle of the lowest lobe, bringing it closer to the surface.

Consequences of ducted propagation. Although both the elevated and the surface (evaporation) duct can result in extended radar ranges, the consequences of their presence are often bad rather than good. The presence of extended ranges cannot always be predicted in advance. Furthermore, they cannot be depended upon. The extension of range along some propagation paths means a decrease of range along others, or radio holes. These holes in the coverage can seriously affect airborne surveillance radars as well as ground-based and ship-borne radars. The extended ranges during ducting conditions mean that ground clutter is likely to be present at longer ranges. This can put a severe burden on some MTI radars that are designed on the assumption that clutter will not appear beyond a certain range. This will be particularly bad with radars which cannot cancel second-time-around clutter, such as those with pulse-to-pulse staggered pulse repetition frequencies or those which use magnetron transmitters with random starting phase pulse to pulse. The accuracy of precision range measuring systems that utilize phase measurements, as in the Tellurometer (Sec. 3.5), also can be affected by the evaporation duct.[38] On the positive side, the evaporation duct, when used with a properly sited antenna, can provide extended range against surface targets or low-flying aircraft considerably beyond that which would be expected from a uniform atmosphere.

Prediction of refractive effects. The effect of atmospheric refraction on electromagnetic propagation can be determined from a knowledge of the variation of index of refraction with altitude over the path of propagation. From profiles of index of refraction, classical ray tracing[28] can be applied to determine how the rays propagate. Generally, it is reasonable to assume that the properties of the atmosphere vary only with height. This assumption is made to simplify the computations. One of the most accurate methods for obtaining the atmospheric profile of the index of refraction is with an airborne microwave refractometer.[40,41] In one version,[39] two precision microwave transmission cavities are employed, one of which is open to collect a sample of the atmosphere. The other is hermetically sealed and acts as a reference. The two cavities are fed by the same microwave source which is swept in frequency. The measured difference in the resonant frequencies of the two cavities is due to the different dielectric properties (or index of refraction) of the gases within the two cavities. This frequency difference can be calibrated in terms of the index of refraction of the atmosphere in the sampling cavity. Although the airborne refractometer may give excellent data on the variation of the index of refraction, it is not always suitable for many applications since it requires an aircraft or helicopter.

The index of refraction profile can also be determined by measuring the pressure, temperature, and humidity as a function of altitude and using Eq. (12.9) to compute the refractivity. The radiosonde is a balloon-borne package of instruments for obtaining such measurements.[40] The data is telemetered back to the ground. One drawback of the radiosonde is that it is generally a slow-response instrument, and is not always able to detect with sufficient accuracy the significant refractive index gradients needed for obtaining an accurate description of the ray paths.

In some applications a complete ray trace is not necessary; instead, only the amount by which the rays are bent might be desired, or the error in elevation angle which occurs. A correction to the measured elevation angle can be made using only the value of the index of refraction at the surface,[20,21] obtained with either a microwave refractometer or from meteorological measurements of temperature, pressure, and humidity. An improvement to the elevation angle correction can be had by making a measurement of the brightness temperature of the atmosphere with a microwave radiometer, along with a surface refractive-index measurement.[22] Since refractive bending and radiometric brightness-temperature measurements both depend on the atmospheric profile, the use of the radiometer provides information that aids in the correction of elevation angle. It has been estimated that the brightness-temperature measurement made with a radiometer operating near the water vapor absorption line of 22 GHz, along with a measurement of surface refractivity, can result in an improvement in angle accuracy of almost 50 percent as compared with a surface refractivity measurement alone.[22] The ground-based microwave radiometer can also be used to provide a correction for the error in time delay introduced by the atmosphere, by making brightness-temperature measurements near the 22-GHz water vapor absorption line and the 60-GHz absorption line of oxygen.[42]

If no other information is available, a correction for the elevation angle can be had on the basis of the yearly statistics of the meteorological data that enters into Eq. (12.9) for determining the index of refraction.[22]

It is possible to automate, by means of a computer, the necessary calculations for determining the refractive effects of the atmosphere on the coverage of the radar. Such computations could be made at the radar site to provide the operator with the information needed to know how a radar is affected by the natural environment. One implementation for naval application, known as IREPS (Integrated Refraction Effects Prediction System), uses a mini-computer and interactive graphic display terminal to provide (1) plots of refractivity as a function of altitude (which indicate the location of ducts); (2) a plain language narrative description of the propagation effects that can be expected over surface-to-surface, surface-to-air, and air-to-air paths; (3) vertical coverage diagrams for specific equipments; (4) display of path loss with range; and (5) a ray-trace used primarily to assess the performance of airborne systems.[79,80] Input data for IREPS can be obtained from balloon-borne radiosonde measurements of upper air temperature, pressure, and humidity. Alternatively, an aircraft-borne microwave refractometer can be used. Inputs are obtained from surface meteorological measurements made on board the ship on which the IREPS is located. When no other input data is available, the IREPS utilizes a stored library of historic refractivity and climatology statistics as a function of the latitude, longitude, season, and time of day. When historic data is used the output is a prediction of propagation performance in probabilistic terms. Also stored are the necessary system parameters for the various electromagnetic systems whose predicted propagation performance is desired.

12.6 DIFFRACTION

In free space, electromagnetic waves travel in straight lines. In the earth's atmosphere, radar waves can propagate beyond the geometrical horizon by refraction. Another mechanism that permits radar coverage to be extended beyond the geometrical horizon is *diffraction*. Radar waves are diffracted around the curved earth in the same manner that light is diffracted by a straight edge. The ability of electromagnetic waves to propagate around the earth's curvature by diffraction depends upon the frequency, or more precisely, upon the size of the object compared with the wavelength. The lower the frequency, the more the wave is diffracted. The mechanism of diffraction is especially important at very low frequencies (VLF) where it provides world-wide communications. However, at radar frequencies the wavelength is small compared with the earth's dimensions and little energy is diffracted. Thus radar coverage cannot be extended much beyond the line of sight by this mechanism.

Figure 12.7 is a plot of the electric field strength (relative to free space) at the target as a function of the distance from the transmitting antenna. Both the radar antenna and the target are assumed to be at a fixed height (100 m in this example). The computed curves apply to propagation over an idealized smooth earth in the absence of an atmosphere. The line of sight is the straight-line distance between radar and target that is just tangent to the surface of the earth. The distance between radar and target along the line of sight is

$$d_0 = \sqrt{2kah_1} + \sqrt{2kah_2} \qquad (12.14)$$

where h_1, h_2 = heights of radar antenna and target, respectively

a = earth's radius

k = factor discussed in Sec. 12.4, accounting for refraction due to a uniform gradient of refractivity

The point of tangency of the line of sight with the earth is the *geometrical*, or *optical, horizon*.

At optical frequencies ($\lambda \approx 0$) the field strength within the interference region (that is, between the radar and the geometrical horizon) is essentially the same as in free space. The field does not penetrate beyond the horizon. Thus, for optical frequencies or very short radar

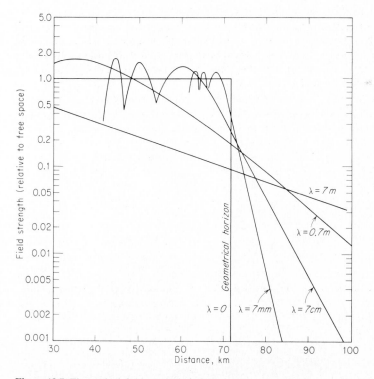

Figure 12.7 Theoretical field strength (relative to free-space field strength) as a function of the distance from the transmitting antenna. Vertical polarization, $h_a = h_t = 100$ m, $k = 1$, ground conductivity = 10^{-2} mho/m, dielectric constant = 4. (*After Burrows and Attwood,*[5] *courtesy Academic Press, Inc.*)

wavelengths, the geometrical horizon represents the approximate boundary between the regions of propagation and no propagation. As the frequency decreases (increasing wavelength), Fig. 12.7 indicates that more and more energy propagates beyond the geometrical horizon. However, the field strength at, and just within, the geometrical horizon decreases with decreasing frequency.

It is concluded that if low-altitude radar coverage is desired beyond the geometrical horizon in the diffraction region, the frequency should be as low as possible. If, on the other hand, low-altitude coverage is to be optimized within the interference region and if there is no concern for coverage beyond the horizon, the radar frequency should be as high as possible. (This assumes the absence of ducting.)

The formula for the distance along the line of sight [Eq. (12.14)] should not be used as a measure of the radar coverage without some reservation. Figure 12.7 shows that a target located at the geometrical horizon is *not* in free space but is definitely within the diffraction region of the radar. The field strength for a target on the radar line of sight might vary from 10 to 30 dB below that in free space.[43] The loss of signal strength in the diffraction region can be quite high. At a frequency of 500 MHz, the one-way propagation loss is roughly 1 dB/mi at low altitudes. It can be even greater at higher frequencies. Therefore, to penetrate 10 miles within the diffraction region, the radar power at 500 MHz must be increased by at least 20 dB over that required for free-space propagation.

The decrease in radar coverage due to the attenuation of electromagnetic energy in the diffraction region is illustrated by Fig. 12.8 for a radar operating at a frequency of 500 MHz. These curves are theoretical contours of constant radar coverage. The radar height is assumed to be 200 ft above the curved earth. Curve 1 represents the locus of the geometrical line of sight as defined by Eq. (12.14) for $k = \frac{4}{3}$. Curve 2 is the constant signal contour in the diffraction region for a signal strength equal to the free-space signal that would be received from a range of approximately 220 nautical miles; that is, if the radar is to detect a target that lies along this contour, it must be capable of detecting the same target in free space at a range of 220 nautical miles. If the target were at an altitude of 200 ft, the maximum detection range would be reduced from 220 to about 35 nautical miles. Curves 3, 4, and 5 are similar to curve 2, except that they apply to a free-space signal of 110, 55, and 27.5 nautical miles, respectively. Curve 6

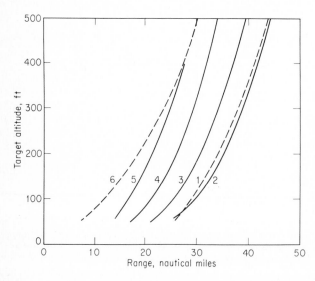

Figure 12.8 Contours of "radar coverage" for radar height of 200 ft above curved earth. (1) Geometrical line-of-sight contour for $k = \frac{4}{3}$; (2) constant-radar-signal contour in the diffraction region, assuming a radar capable of a free-space range of 220 nmi, vertical polarization, seawater, $k = \frac{4}{3}$, $f = 500$ MHz; (3) same as (2), but for 110 nmi free-space range; (4) same as (2), but for 55 nmi free-space range; (5) same as (2), but for 27.5 nmi free-space range; (6) contour defining start of diffraction region. (*Courtesy Proc. IRE.*)

represents the approximate boundary between the interference region and diffraction region. Any target to the right of curve 6 may be considered to be within the diffraction region. This illustrates why most radars operating at microwave or UHF frequencies are limited to coverage within the geometrical line of sight.

12.7 ATTENUATION BY ATMOSPHERIC GASES[5,12,28]

The attenuation of radar energy in a clear atmosphere in the absence of precipitation is due to the presence of oxygen and water vapor. Attenuation results when a portion of the energy incident on the molecules of these atmospheric gases is absorbed as heat and is lost. The reduction in radar signal power when propagating over a distance R and back (two-way path) may be expressed as $\exp(-2\alpha R)$, where α is the (one-way) attenuation coefficient measured in units of (distance)$^{-1}$. Instead of plotting α, it is more usual to plot the one-way attenuation in decibels per unit distance. This is equivalent to plotting the quantity 4.34α, where the constant accounts for the conversion from the natural logarithm to the base 10 logarithm.

Attenuation by oxygen and water vapor is shown in Fig. 12.9.[44,45] Resonance peaks for water vapor occur at 22.24 GHz (1.35 cm wavelength) and at about 184 GHz, while the

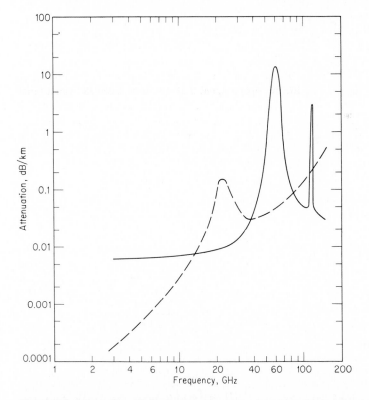

Figure 12.9 Attenuation of electromagnetic energy by atmospheric gases in an atmosphere at 76 cm pressure. Dashed curve is absorption due to water vapor in an atmosphere containing 1 percent water vapor molecules (7.5 g water/m^3). The solid curve is the absorption due to oxygen. (*From Burrows and Attwood*[5] *and Straiton and Tolbert.*[46])

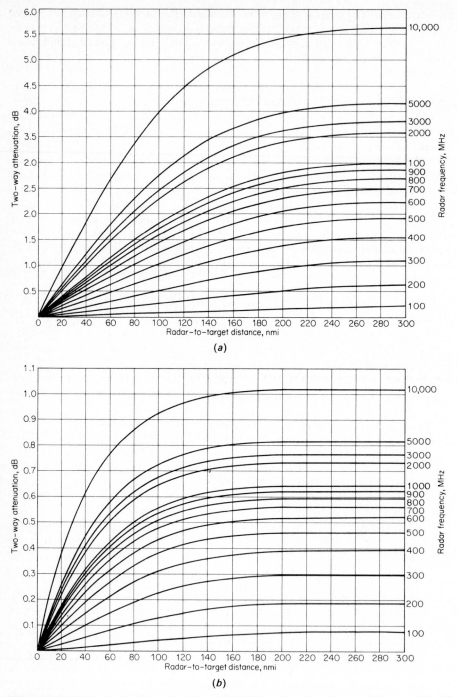

Figure 12.10 Attenuation for two-way, radar propagation as a function of range and frequency for (a) zero elevation angle and (b) 5° elevation angle. (*From Blake.*[47])

oxygen molecule has resonances at 60 GHz (0.5 cm wavelength) and 118 GHz.[46] At frequencies below about 1 GHz (*L* band) the effect of atmospheric attenuation is negligible. Above 10 GHz, it becomes increasingly important. The large attenuations experienced at millimeter wavelengths is one of the chief reasons long-range, ground-based radars are seldom found above 35 GHz (K_a band).

The attenuation of the atmospheric gases decreases with increasing altitude. Thus the attenuation experienced by a radar will depend on the altitude of the target as well as the range. With a ground-based radar the attenuation is greatest when the antenna points to the horizon, and it is least when it points to the zenith. Figure 12.10 gives examples of the two-way attenuation for elevation angles of 0 and 5°.

12.8 ENVIRONMENTAL NOISE

At microwave frequencies the normal noise level is relatively low and the sensitivity of conventional radar receivers is usually determined by internal noise. However, radar receivers with very low noise input stages sometimes can be affected by the ambient noise from the natural environment. This is more likely to occur at both the upper and lower extremes of the microwave frequency region. Several sources of ambient, or external, noise are described in this section.

Cosmic noise[48,49] There is a continuous background of noiselike electromagnetic radiation which arrives from such extraterrestrial sources as our own galaxy (the Milky Way), extragalactic sources, and "radio stars." In general, cosmic noise decreases with increasing frequency and can usually be neglected at frequencies above *L* band. It can be a serious limitation, however, to those radars operating at VHF or lower. The magnitude of cosmic noise depends upon the portion of the celestial sphere in which the antenna points. It is a maximum when looking toward the center of our own galaxy, and it is a minimum when observing along the pole about which the galaxy revolves. A plot of the maximum and minimum cosmic-noise *brightness temperature* as a function of frequency is shown by the dashed curves of Fig. 12.11. The brightness temperature of an extended source of radiation is the temperature of a blackbody which yields the same noise power at the receiver. A highly directive antenna viewing a distributed source of radiation under "ideal" conditions would receive a noise power equal to $kT_B B$, where k = Boltzmann's constant, T_B = brightness temperature of the source, and B = bandwidth of the receiver. By "ideal" conditions is meant an antenna with negligible sidelobes and negligible resistive losses, and which looks at a distributed source of brightness temperature (the cosmic noise) in the absence of the earth's atmosphere or any other source of noise. For a practical antenna the *antenna temperature* is defined as the integral of the brightness temperature over all angles, weighted by the antenna pattern.

Atmospheric absorption noise. It is known from the theory of blackbody radiation that any body which absorbs energy radiates the same amount of energy that it absorbs, else certain portions would increase in temperature and the temperature of other portions would decrease.[56] Therefore a lossy transmission line absorbs a certain amount of energy and reradiates it as noise. The same is true of the atmosphere since it also attenuates or absorbs microwave energy. The radiation arising in the atmosphere (or any other absorbing body) must just compensate for the partial absorption of the blackbody radiation.

Consider an absorbing atmosphere at an ambient temperature T_a surrounded by an imaginary blackbody at the same temperature. The loss L is the factor by which energy is

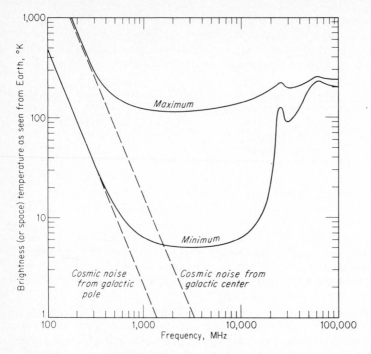

Figure 12.11 Maximum and minimum brightness temperatures of the sky as seen by an ideal single-polarization antenna on earth. (*After Greene and Lebenbaum,*[50] *Microwave J.*)

attenuated in passing through the atmosphere. The noise power available over a bandwidth B_n from the imaginary blackbody is kT_aB_n. The noise power after passing through the atmosphere is kT_aB_n/L. Thus the amount of power absorbed by the atmosphere is $kT_aB_n(1 - 1/L)$ and is equal to the noise power ΔN radiated by the atmosphere itself. From the definition of effective noise temperature and the fact that $1/L$ is the " gain " (less than unity), the following equation is obtained:

$$\Delta N = kT_e B_n G = \frac{kT_e B_n}{L} = kT_a B_n\left(1 - \frac{1}{L}\right)$$

Hence
$$T_e = T_a(L - 1) \tag{12.15}$$

If the atmospheric loss were 1 dB at a temperature of 260 K, the effective noise temperature would be 68 K; a 3-dB loss results in $T_e = 260$ K, while a 10-dB loss gives $T_e = 1340$ K.

A plot of the single-polarization brightness temperature or space temperature due to both cosmic noise and atmospheric absorption is shown by the solid curves of Fig. 12.11. An ambient temperature of 260 K is assumed in the computation of atmospheric-absorption noise. At the higher frequencies (X band or above) atmospheric absorption is the predominant contributor to the brightness temperature, while at the lower frequencies (L band or lower), the cosmic noise predominates. There exists a broad minimum in the brightness temperature extending from about 1,000 to 10,000 MHz. It is in this region that it is advantageous to operate low-noise receivers to achieve maximum system sensitivity. The minimum atmospheric absorption occurs when the antenna is vertical (pointed at the zenith), while the maximum occurs when the antenna is directed along the horizon. The noise is greater along the horizon than at the zenith since the antenna "sees" more atmosphere. The antenna beams must be oriented at elevation angles greater than about 5° to avoid excessive atmospheric-absorption noise in the main beam.

The computed brightness temperatures of Fig. 12.11 do not agree in detail with those presented by others.[47,51,52] The fact that such computations made by different authors do not always agree precisely need not be a limitation to the radar systems engineer. Disagreements often result from the oversimplifying nature of the assumptions or in the model used in formulating the calculations. A good radar design is one which is not overly sensitive to small variations in the model or the assumptions.

Atmospheric and urban noise. A single lightning stroke radiates considerable RF noise power. At any one moment there are an average of 1800 thunderstorms in progress in different parts of the world. From all these storms about 100 lightning flashes take place every second.[53] The combined effect of all the lightning strokes gives rise to a noise spectrum which is especially large at broadcast and short-wave radio frequencies. Noise that arises from lightning-stroke radiation is called *atmospheric noise* (not to be confused with noise produced by atmospheric absorption as described previously). The spectrum of atmospheric noise falls of rapidly with increasing frequency and is usually of little consequence above 50 MHz.[54] Hence atmospheric noise is seldom an important consideration in radar design, except, perhaps, for radars in the lower VHF region.

Another source of noise predominant at the lower radar frequencies is *urban noise*, also known as man-made noise. Noise from automobile ignition, electric razors, power tools, and fluorescent lights are examples. It is of little concern at UHF or higher frequencies.[55]

Solar noise. The sun is a strong emitter of electromagnetic radiation, the intensity of which varies with time. The minimum level of solar noise is due to blackbody radiation at a temperature of about 6000 K.[56,57] The solar noise is unlike most other noise mechanisms in that its power increases with increasing frequency. Solar storms (sunspots and flares) can increase the solar-noise level several orders of magnitude over that of the "quiet," or undisturbed sun. The solar noise can sometimes be of significant magnitude to affect the sensitivity of low-noise radar receivers from energy received in the antenna sidelobes. The sun also can be used as a source to calibrate the beam-pointing (boresight) of large antennas.[58] Discrete radio sources, called radio stars, are too weak at radar frequencies to be a serious source of interference, but they have been used in conjunction with sensitive receivers to determine pointing and focusing corrections for large antennas.[59]

System noise temperature. Figure 12.12 illustrates some of the sources of noise which generally must be considered when computing the system effective noise temperature. The antenna sees the cosmic noise at a temperature T_c with an intervening absorbing atmosphere at a temperature T_{at} and a loss L_{at}. The atmosphere may be characterized for present purposes by a single temperature and a single loss, but it can be subdivided, if desired, into an ionospheric component, an oxygen component, and a water-vapor component. The combined temperature of cosmic noise and atmospheric noise $[T_c + (L_{at} - 1)T_{at}]$ is called the *space temperature*, the *brightness temperature*, or the *antenna temperature of an ideal antenna*. The RF losses L_{rf} indicated in the figure are meant to include the antenna, radome, and duplexer losses, as well as transmission-line loss. The effective noise temperature of the receiver is denoted T_{re}.

If it is assumed that the noise contributions enter the receiver via the main beam only, the

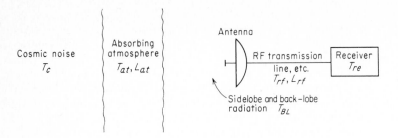

Figure 12.12 Contributions to the total system effective noise temperature.

effect of the sidelobes may be neglected, and the total system effective noise temperature T_e may be found from a straightforward application of Eq. (9.10); therefore

$$T_e = T_c + (L_{at} - 1)T_{at} + (L_{rf} - 1)T_{rf}L_{at} + T_{re}L_{rf}L_{at} \qquad (12.16)$$

Note that T_c, T_{at}, and T_{rf} are actual temperatures while T_e and T_{re} are effective noise temperatures.

In general, the contributions to the total effective system noise temperature may be divided into three categories: (1) the effective space noise temperature, (2) the effective noise-temperature contributions due to RF lossy components, and (3) the effective noise temperature of the receiver itself.

Equation (12.16) applies to an ideal antenna with no sidelobes. In a practical antenna the noise which appears at the antenna terminals enters via the sidelobe radiation as well as from the main beam. In many cases the total noise power due to the sidelobes can be greater than the noise power in the main beam. This is especially true when the main beam views the relatively "cool" sky but the sidelobes view the "hot" earth. A portion of the main beam might also view the relatively "hot" earth if pointed at or near the horizon.

The amount of noise which enters the antenna depends upon the entire antenna radiation pattern, including the sidelobes and the type of objects they illuminate. Land is almost a complete absorber; hence those portions of the radiation pattern which illuminate the ground see a noise source at the ambient temperature. Perfectly reflecting sources, such as a smooth sea or a road, act as a mirror to reflect the radiation from the sky or other objects. Thus the sea or a metallic object may appear very cold if it is oriented to reflect radiation from the sky to the antenna. The choice of polarization also influences the amount of sea or land absorption. Vertical polarization is absorbed more than horizontal.

The total antenna temperature can be found by integrating the temperature "seen" by the antenna, weighted by the antenna gain over the entire sphere.[60]

$$T_a = \frac{\int T_B(\theta, \phi)G(\theta, \phi) \, d\Omega}{\int G(\theta, \phi) \, d\Omega} \qquad (12.17)$$

where $d\Omega$ = solid angle given by $\sin \theta \, d\theta \, d\phi$. The brightness temperature $T_B(\theta, \phi)$ is often a complicated function, and T_a must be approximated by numerical means. The antenna temperature is an average value of the brightness, or space, temperature in the field of the antenna pattern.

Figure 12.13 gives the antenna noise temperature computed for a typical 10-ft-diameter parabolic reflector operating at 1000 MHz as a function of the antenna elevation angle.[61] These data assume (1) vertical polarization, (2) the antenna located on a seacoast and looking over

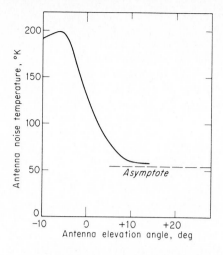

Figure 12.13 Computed antenna noise-temperature as a function of elevation angle for a 10-ft-diameter (3 m) parabolic-reflector antenna operating at a frequency of 1 GHz. (*After Greene,*[61] *courtesy Airborne Instruments Laboratory.*)

the sea, (3) the antenna always pointing at the galactic center (4) no intense radio stars in the antenna pattern, (5) resistive losses in the reflector, feed, and transmission line absorbing 2 percent of the incident power, and (6) the feed producing a parabolic illumination taper.

12.9 MICROWAVE-RADIATION HAZARDS

High power microwave energy can produce spectacular effects. It has been reported for example, that 50 kW of UHF power radiating from the open end of a 6- by 15-in. waveguide will cause ordinary light bulbs to explode, fluorescent lamps many feet away to light up, and a piece of steel wool to explode into arcs.[62] It is not surprising, therefore, that microwave energy, if of sufficient intensity, is a health hazard and can produce biological damage in humans.

Heating is the chief effect of microwave radiation on living tissue. In controlled dosages, radiation heating is beneficial and forms the basis of diathermy, a therapeutic heating of the tissue beneath the skin. Frequencies ranging from HF to microwaves have been used for diathermy.[63] The heating effects of microwave radiation have also been applied commercially in the form of microwave ovens, used for cooking food rapidly.

Harmful effects of excessive microwave radiation result from either a general rise in the total body temperature or from selective heating of sensitive parts of the body. Exposure of the whole body will cause the internal temperature to rise and produce fever. An increase in the total body temperature of 1°C is considered excessive,[64] and prolonged exposure or too high a temperature rise can be fatal. Discomfort resulting from a general rise in body temperature can be perceived by the victim and serve as a warning.

The danger of localized heating depends upon whether compensating cooling mechanisms exist to dissipate the heat generated at the radiated part of the body. For example, localized heating is least serious in muscle tissue which is well equipped with blood vessels capable of dissipating heat. Heating is more dangerous in the brain, the testes, the hollow viscera, and the eyes, where there is little opportunity for the exchange of heat with the surrounding tissue.

Many instances have been reported where cataracts have been deliberately formed in the eyes of animals by exposure to microwave radiation. The viscous material of the eyeball is

466 INTRODUCTION TO RADAR SYSTEMS

affected by heat in much the same manner as the white of an egg. It is transparent at room temperature but becomes opaque if its temperature is raised excessively. The process is an irreversible one. Although the testes are apparently more sensitive to heat than the eye, testicular damage is often temporary and reversible.[63,65]

At frequencies below 400 and above 3000 MHz, the body absorbs less than half the incident energy. Lower frequencies pass through, and higher frequencies are reflected at the skin's surface. Between 1000 and 3000 MHz the percentage of radiation absorbed can approach 100 percent, depending on the thickness of skin and subcutaneous layers of fat.[65]

If the whole body is immersed in microwave radiation, a rise in temperature or a sensation of warmth serves as a warning before damage to localized parts of the body becomes severe. However, if only parts of the body are exposed, there may or may not be a sensation of warmth, depending upon the frequency. Heating caused by frequencies penetrating the interior of the body can be of concern because the sparsity of sensory nerves may make it imperceptible. Higher frequencies absorbed at or close to the surface of the body are more likely to be perceived than interior heating.

There have been few authenticated incidents where radiations have been the cause of biological damage in humans. However, as radar powers increase, the likelihood of biological damage becomes greater, and if serious harm is to be avoided, proper safety precautions must be observed. The United States armed services have established[63,67] the maximum safe continuous exposure level to be an average power density of 10 mW/cm². There has been substantial basis for setting this limit. However, when working under conditions of moderate to severe heat stress the maximum exposure level should be reduced appropriately.[66]

The criterion of 10 mW/cm² is based on the experimental observation that thermal effects are dominant. However, there is evidence indicating that nonthermal biological effects also occur from exposure to microwave radiation.[68] Pulsed power can produce biological change not obtained with CW power of the same average value. Therefore, the possibility of dangerous effects with excessively high peak powers should not be overlooked even if the average power is less than the safe threshold. There has been, however, no direct link made between biological effects attributed to nonthermal microwave origin and an adverse effect in humans. The assumption of hazardous nonthermal effects at low densities of radiation is a fear that has not been substantiated.[69] The demonstration of a biological effect due to microwaves does not of necessity demonstrate a peril. Confusion also can result at times between microwaves and other more harmful radiations. Microwave radiation is nonionizing and is vastly different from ionizing radiations such as X rays.

As a safety precaution, areas of high power density should be fenced off, locked, or otherwise made inaccessible when transmitting. Personnel should never look into an open waveguide or antenna feed horn connected to energized transmitters. When personnel must work in areas where the power density is at a dangerous level, they should be protected with screened enclosures or with protective apparel made from reflective material.

Another potential safety hazard in working with high power is the generation of X rays when high voltages are used to operate RF power tubes. Tubes must be properly shielded with lead, and X-ray safety badges worn by operating personnel to warn of excessive dosage.

REFERENCES

1. Ridenour, L. N.: "Radar System Engineering," MIT Radiation Laboratory Series, vol. 1, sec. 2.12, McGraw-Hill Book Company, New York, 1947.
2. Bachynski, M. P.: Microwave Propagation over Rough Surfaces, RCA Rev., vol. 20, pp. 308–335, June, 1959.

3. Sherwood, E. M., and E. L. Ginzton: Reflection Coefficients of Irregular Terrain at 10 Cm, *Proc. IRE*, vol. 43, pp. 877–878, July, 1955.
4. Hey, J. S., and S. J. Parsons: The Radar Measurement of Low Angles of Elevation, *Proc. Phys. Soc.*, ser. B, vol. 69, pp. 321–328, 1956.
5. Burrows, C. R., and S. S. Attwood: " Radio Wave Propagation," Academic Press, Inc., New York, 1949.
6. Reed, H. R., and C. M. Russell: " Ultra High Frequency Propagation," John Wiley & Sons, Inc., New York, 1953.
7. Norton, K. A.: The Calculation of Ground-wave Field Intensity over a Finitely Conducting Spherical Earth, *Proc. IRE*, vol. 29, pp. 623–639, Dec., 1941.
8. Domb, C., and M. H. L. Pryce: The Calculation of Field Strengths over a Spherical Earth, *J. IEE*, vol. 94, pt. III, pp. 325–339, 1947.
9. Senior, T. B. A.: Radio Propagation over a Discontinuity in the Earth's Electrical Properties, *Proc. IEE*, pt. C, vol. 104, pp. 43–53, 139–147, Mar., 1957.
10. Blake, L. V.: Machine Plotting of Radio/Radar Vertical-Plane Coverage Diagrams, *Naval Research Laboratory, Washington, D.C., Report* 7098, June 25, 1970 (AD 709897).
11. Smith, E. K., and S. Weintraub: The Constants in the Equation for Atmospheric Refractive Index at Radio Frequencies, *Proc. IRE*, vol. 41, pp. 1035–1037, August, 1953.
12. Bean, B. R., E. J. Dutton, and B. D. Warner: Weather Effects on Radar, chap. 24 of " Radar Handbook," M. I. Skolnik (ed.), McGraw-Hill Book Company, New York, 1970.
13. Schelling, J. C., C. R. Burrows, and E. B. Ferrell: Ultra-short Wave Propagation, *Proc. IRE*, vol. 21, pp. 427–463, March, 1933.
14. Jay, F. (editor in chief): "IEEE Standard Dictionary of Electrical and Electronic Terms," 2d ed. John Wiley & Sons, Inc., New York, 1977.
15. Bean, B. R.: The Geographical and Height Distribution of the Gradient of Refractive Index, *Proc. IRE*, vol. 41, pp. 549–550, April, 1953.
16. Bean, B. R., B. A. Cahoon, C. A. Samson, and G. D. Thayer: "A World Atlas of Atmospheric Radio Refractivity," U.S. Dept. of Commerce, *ESSA Monograph* 1, 1966.
17. Bean, B. R., and G. D. Thayer: Models of Atmospheric Radio Refractive Index, *Proc. IRE*, vol. 47, pp. 740–755, May, 1959.
18. Fannin, B. M., and K. H. Jehn: A Study of Radar Elevation-angle Errors Due to Atmospheric Refraction, *IRE Trans.*, vol. AP-5, pp. 71–77, January, 1957.
19. Blake, L. V.: Ray Height Computation for a Continuous Nonlinear Atmospheric Refractive-Index Profile, *Radio Science*, vol. 3, pp. 85–92, January, 1968.
20. Bean, B. R., and B. A. Cahoon: The Use of Surface Weather Observations to Predict the Total Atmospheric Bending of Radio Rays at Small Elevation Angles, *Proc. IRE*, vol. 45, pp. 1545–1546, November, 1957.
21. Bean, B. R., G. D. Thayer, and B. A. Cahoon: Methods of Predicting the Atmospheric Bending of Radio Rays, *Natl. Bur. Standards (U.S.) J. Research*, vol. 64D, pp. 487–492, September/October, 1960.
22. Gallop, M. A., Jr., and L. E. Telford: Estimation of Tropospheric Refractive Bending from Atmospheric Emission Measurements, *Radio Science*, vol. 8, pp. 819–827, October, 1973.
23. Millman, G. H.: Atmospheric Effects on VHF and UHF Propagation, *Proc. IRE*, vol. 46, pp. 1492–1501, August, 1958.
24. Crane, R. K.: Ionospheric Scintillation, *Proc. IEEE*, vol. 65, pp. 180–199, February, 1977.
25. Meteorological Aspects of Radio-Radar Propagation, *Navy Weather Research Facility*, Norfolk, VA, NWRF 31-0660-035, 1960.
26. Battan, L. J.: " Radar Meteorology," University of Chicago Press, Chicago, 1959.
27. Purves, C. G.: Geophysical Aspects of Atmospheric Refraction, *Naval Research Laboratory*, Washington, D.C. Report 7725, June 7, 1974.
28. Kerr, D. E. (ed.): " Propagation of Short Radio Waves," MIT Radiation Laboratory Series, vol. 13, McGraw-Hill Book Company, New York, 1951.
29. Ringwalt, D. L., and F. C. Macdonald: Elevated Duct Propagation in the Tradewinds, *IRE Trans.*, vol. AP-9, pp. 377–383, July, 1961.
30. Guinard, N. W., J. Ransone, D. Randall, C. Purves, and P. Watkins: Propagation Through an Elevated Duct: Tradewinds III, *IEEE Trans.*, vol. AP-12, pp. 479–490, July, 1964.
31. Battan, L. J.: " Radar Observation of the Atmosphere," University of Chicago Press, Chicago, 1973.
32. Booker, H. G.: Elements of Radio Meteorology: How Weather and Climate Cause Unorthodox Radar Vision beyond the Geometrical Horizon, *J. IEE*, vol. 93, pt. IIIA, pp. 69–78, 1946.
33. Saxton, J. A.: The Influence of Atmospheric Conditions on Radar Performance, *J. Inst. Navigation (London)*, vol. 11, pp. 290–303, 1958.

34. Rotheram, S.: Radiowave Propagation in the Evaporation Duct, *The Marconi Review*, vol. 37, pp. 18–40, First Quarter, 1974.
35. Schneider, A.: Oversea Radar Propagation Within a Surface Duct, *IEEE Trans.*, vol. AP-17, pp. 254–255, March, 1969.
36. Hitney, H. V.: Radar Detection Range Under Atmospheric Ducting Conditions, *IEEE 1975 International Radar Conference*, Arlington, Va., pp. 241–243, IEEE Publication 75 CHO 938-1 AES.
37. Katzin, M., R. W. Bauchman, and W. Binnian: 3- and 9-Centimeter Propagation in Low Ocean Ducts, *Proc. IRE*, vol. 35, pp. 891–905, September, 1947.
38. Früchtenicht, H. W.: Notes on Duct Influences on Line-of-Sight Propagation, *IEEE Trans.*, vol. AP-22, pp. 295–302, March, 1974.
39. Stewart, C. H., and G. J. Vincent: Radar-Tracking Accuracy Increased, *Electronics*, vol. 37, pp. 73–75, May 4, 1964.
40. Bean, B. R., and E. J. Dutton: " Radio Meteorology," National Bureau of Standards Monograph 92, Mar. 1, 1966.
41. Crain, C. M.: Survey of Airborne Microwave Refractometer Measurements, *Proc. IRE*, vol. 43, pp. 1405–1411, October, 1955.
42. Schaper, L. W., Jr., D. H. Staelin, and J. W. Waters: The Estimation of Tropospheric Electrical Path Length by Microwave Radiometry, *Proc. IEEE*, vol. 58, pp. 272–273, February, 1970.
43. Skolnik, M. I.: Radar Horizon and Propagation Loss, *Proc. IRE*, vol. 45, pp. 697–698, May, 1957.
44. Van Vleck, J. H.: The Absorption of Microwaves by Oxygen, *Phys. Rev.*, vol. 71, pp. 413–424, Apr. 1, 1947.
45. Van Vleck, J. H.: The Absorption of Microwaves by Uncondensed Water Vapor, *Phys. Rev.*, vol. 71, pp. 425–433, Apr. 1, 1947.
46. Straiton, A. W., and C. W. Tolbert: Anomalies in the Absorption of Radio Waves by Atmospheric Gases, *Proc. IRE*, vol. 48, pp. 898–903, May, 1960.
47. Blake, L. V.: Prediction of Radar Range, chap. 2 of " Radar Handbook," M. I. Skolnik (ed.), McGraw-Hill Book Company, New York, 1970.
48. Ko, H. C.: The Distribution of Cosmic Radio Background Radiation, *Proc. IRE*, vol. 46, pp. 208–215, January, 1958.
49. Menzel, D. H.: " The Radio Noise Spectrum," Harvard University Press, Cambridge, Mass., 1960.
50. Greene, J. C., and M. T. Lebenbaum: Letter in *Microwave J.*, vol. 2, pp. 13–14, October, 1959.
51. Hogg, D. C.: Effective Antenna Temperature Due to Oxygen and Water Vapor in the Atmosphere, *J. Appl. Phys.*, vol. 30, pp. 1417–1419, September, 1959.
52. Hogg, D. C., and W. W. Mumford: The Effective Noise Temperature of the Sky, *Microwave J* , vol. 3, pp. 80–84, March, 1960.
53. Schonland, B. F. J.: " The Flight of Thunderbolts," 2d ed., Oxford, Clarendon Press, London, 1964.
54. Crichlow, W. Q., D. F. Smith, R. N. Morton, and W. R. Corliss: Worldwide Radio Noise Levels Expected in the Frequency Band 10 Kilocycles to 100 Megacycles, *Natl. Bur. Standards (U.S.) Circ.* 557, Aug. 25, 1955.
55. Spaulding, A. D.: The Electromagnetic Interference Environment: Man-Made Noise, pt. 1: Estimation of Business, Residential and Rural Areas, *Office of Telecommunications, U.S. Department of Commerce*, 1974.
56. Pawsey, J. L., and R. N. Bracewell: " Radio Astronomy," Oxford University Press, London, 1955.
57. Wild, J. P.: Observational Radio Astronomy, *Advances in Electronics and Electron Physics*, vol. 7, 1955.
58. Graf, W., R. N. Bracewell, J. H. Deuter, and J. S. Rutherford: The Sun as a Test Source for Boresight Calibration of Microwave Antennas, *IEEE Trans.*, vol. AP-19, pp. 606–612, September, 1971.
59. Baars, J. W. M.: The Measurement of Large Antennas with Cosmic Radio Sources," *IEEE Trans.*, vol. AP-21, pp. 461–474, July, 1973.
60. Staelin, D. H.: Passive Remote Sensing at Microwave Wavelengths, *Proc. IEEE*, vol. 57, pp. 427–439, April, 1969.
61. Greene, J.: Antenna Noise Temperature, AIL Advertisement, *Proc. IRE*, vol. 46, p. 2a, May, 1958.
62. Salisbury, W. W.: The Resnatron, *Electronics*, vol. 19, pp. 92–97, February, 1946.
63. Mumford, W. W.: Some Technical Aspects of Microwave Radiation Hazards, *Proc. IRE*, vol. 49, pp. 427–447, February, 1961.
64. Schwan, H. P., and K. Li: Hazards Due to Total Body Irradiation by Radar, *Proc. IRE*, vol. 44, pp. 1572–1581, November, 1956.
65. Leary, F.: Researching Microwave Health Hazards, *Electronics*, vol. 32, no. 7, pp. 49–53, Feb. 20, 1959.

66. Mumford, W. W.: Heat Stress Due to RF Radiation, *Proc. IEEE*, vol. 57, pp. 171–178, February, 1969.
67. Setter, L. R., D. R. Snavely, D. L. Solem, and R. F. Van Wye: " Regulations, Standards, and Guides for Microwaves, Ultraviolet Radiation, and Radiation from Lasers and Television Receivers—An Annotated Bibliography," U.S. Dept. of Health, Education, and Welfare, Public Health Service Publication no. 999-RH-35, April, 1969.
68. Special Issue on Biological Effects of Microwaves, *IEEE Trans.*, vol. MTT-19, February, 1971.
69. Justesen, D. R.: Diathermy Versus the Microwaves and Other Radio-frequency Radiations: A Rose by Another Name is a Cabbage, *Radio Science*, vol. 12, pp. 355–364, May–June, 1977.
70. Ament, W. S.: Toward a Theory of Reflection by a Rough Surface, *Proc. IRE*, vol. 41, pp. 142–146, January, 1953.
71. Beard, C. I., I. Katz, and L. M. Spetner: Phenomenological Vector Model of Microwave Reflection from the Ocean, *IRE Trans.*, vol. AP-4, pp. 162–167, April, 1956.
72. Beard, C. I.: Coherent and Incoherent Scattering of Microwaves from the Ocean, *IRE Trans.*, vol. AP-9, pp. 470–483, Sept., 1961.
73. Beard, C. I., D. L. Drake, and C. M. Morrow: Measurements of Microwave Forward Scattering from the Ocean at L-Band, *IEEE NAECON '74 Record*, pp. 269–274, 1974.
74. Spetner, L. M.: Incoherent Scattering of Microwaves from the Ocean, *Radio Science*, vol. 10, pp. 585–587, June, 1975.
75. Brown, B. P.: Radar Height Finding, chap. 22 of " Radar Handbook," M. I. Skolnik (ed.), McGraw-Hill Book Company, New York, 1970.
76. Kanevskii, M. B.: The Propagation of Millimeter and Centimeter Waves in Tropospheric Waveguides Close to the Sea, *Soviet Radiophysics*, vol. 9, no. 5, pp. 867–875, 1966.
77. Joseph, R. I., and G. D. Smith: Propagation in an Evaporation Duct: Results in Some Simple Analytic Models, *Radio Science*, vol. 7, pp. 433–441, April, 1972.
78. Richter, J. H., and H. V. Hitney: The Effect of the Evaporation Duct on Microwave Propagation, *Naval Electronics Laboratory Center*, San Diego, California, Technical Report 1949, Apr., 17 1975.
79. Hitney, H. V., and J. H. Richter: Integrated Refractive Effects Prediction System (IREPS), *Naval Engineers Journal*, vol. 2, pp. 257–262, April, 1976.
80. Hitney, H. V., and J. H. Richter: Integrated Refractive Effects Prediction System (IREPS), URSI Commission F Open Symposium, Propagation in Non-Ionized Media, La Baule, France, Apr. 28 to May 6, 1977.

THIRTEEN

RADAR CLUTTER

13.1 INTRODUCTION TO RADAR CLUTTER

Clutter may be defined as any unwanted radar echo. Its name is descriptive of the fact that such echoes can "clutter" the radar output and make difficult the detection of wanted targets. Examples of unwanted echoes, or clutter, in a radar designed to detect aircraft include the reflections from land, sea, rain, birds, insects, and chaff. Unwanted echoes might also be obtained from clear-air turbulence and other atmospheric effects, as well as from ionized media such as the aurora and meteor trails. Clutter is generally distributed in spatial extent, in that it is usually much larger in physical size than the radar resolution cell. There are also "point" clutter echoes, such as towers, poles, and similar objects. The echo from a single bird is also an example of point clutter. When clutter echoes are sufficiently intense and extensive, they can limit the sensitivity of a radar receiver, and thus determine the range performance. In such circumstances, the optimum radar waveform and receiver design can be quite different than when receiver noise alone is the dominant effect.

Radar echoes from land, sea, rain, birds, and other such objects are not always undesired. Reflections from storm clouds, for example, can be a bother to a radar that must see aircraft, but storm clouds are what the radar meteorologist wants to see in order to measure rainfall rate over a large area. The backscatter echoes from land can degrade the performance of many radars; but it is the target of interest for a ground-mapping radar, for remote sensing of the earth resources, and for most synthetic-aperture radars. Thus the same object might be the desired target in one application, and the undesired clutter echo in another.

In this chapter, the echoes from land, sea, and weather will be considered only for their harmful effects; that is, as clutter to be avoided or eliminated. The characteristics of clutter will be described as well as the various methods for reducing their harmful effects when they interfere with the detection of desired targets.

Echoes from the land or the sea are known as *surface* clutter, and echoes from rain or other atmospheric phenomena are known as *volume* clutter. Because of its distributed nature, the measure of the backscattering echo from such clutter is generally given in terms of a radar-cross-section *density* rather than the radar cross section as was described for conventional targets in Sec. 2.7. For surface clutter a cross section per unit area is defined as

$$\sigma^0 = \frac{\sigma_c}{A_c} \qquad (13.1)$$

where σ_c is the radar cross section from the area A_c. The symbol σ^0 is spoken, and sometimes written, as *sigma zero*. The advantage of using an expression such as Eq. (13.1) to describe distributed surface clutter is that it is usually independent of the area A_c. For volume distributed clutter a cross section per unit volume, or reflectivity, is defined as

$$\eta = \frac{\sigma_c}{V_c} \qquad (13.2)$$

where σ_c in this case is the radar cross section from the volume V_c.

13.2 SURFACE-CLUTTER RADAR EQUATIONS

Consider the geometry of Fig. 13.1 which depicts a radar illuminating the surface at a grazing angle ϕ. It is assumed that the width of the area A_c is determined by the azimuth beamwidth θ_B, but that the dimension in the range dimension is determined by the radar pulse width τ rather than the elevation beamwidth. Using the simple radar equation of Sec. 1.2, the power C received from the clutter is

$$C = \frac{P_t G A_e \sigma_c}{(4\pi)^2 R^4} \qquad (13.3)$$

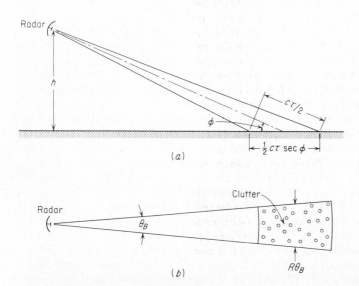

Figure 13.1 Geometry of radar clutter. (*a*) Elevation view showing the extent of the surface intercepted by the radar pulse, (*b*) plan view showing clutter patch consisting of individual, independent scatterers.

where P_t = transmitter power

  G = antenna gain

  A_e = antenna effective aperture

  R = range

  σ_c = clutter cross-section, which is equal to

$$\sigma_c = \sigma^0 A_c = \sigma^0 R\theta_B(c\tau/2) \sec \phi \tag{13.4}$$

where c = velocity of propagation. With this substitution the radar equation for surface clutter is

$$C = \frac{P_t G A_e \sigma^0 \theta_B(c\tau/2) \sec \phi}{(4\pi)^2 R^3} \tag{13.5}$$

Thus the echo from surface clutter varies inversely as the cube of the range, rather than inversely as the fourth power as is the case for point targets.

The signal power S returned from a target with cross section σ_t is

$$S = \frac{P_t G A_e \sigma_t}{(4\pi)^2 R^4} \tag{13.6}$$

Combining Eqs. (13.5) and (13.6), the signal-to-clutter ratio for a target in a background of surface clutter at low grazing angle is

$$\frac{S}{C} = \frac{\sigma_t}{\sigma^0 R\theta_B(c\tau/2) \sec \phi} \tag{13.7}$$

If the maximum range R_{max} corresponds to the minimum discernible signal-to-clutter ratio $(S/C)_{min}$, then the radar equation can be written

$$R_{max} = \frac{\sigma_t}{(S/C)_{min} \, \sigma^0 \theta_B(c\tau/2) \sec \phi} \tag{13.8}$$

In this equation, the clutter power C is assumed large compared to receiver noise power. This is an entirely different form of the radar equation than when the target detection is dominated by receiver noise alone. The range in Eq. (13.8) appears as the first power rather than as the fourth power in the usual radar equation of Eq. (13.6). This means there is likely to be greater variation in the maximum range of a clutter-dominated radar than a noise-dominated radar. For example, if the target cross section in Eq. (13.8) were to vary by a factor of two, the maximum range would also vary by a factor of two. However, the same variation in target cross section would only cause a variation in range of a factor of 1.2 when the radar performance is determined by receiver noise.

There are other significant differences in Eq. (13.8) that should be noted. The transmitter power does not appear explicitly. Increasing the transmitter power will indeed increase the target signal, but it will also cause a corresponding increase in clutter. Thus there is no net gain in the detectability of desired targets. The only demand on the transmitter power is that it be great enough to cause the clutter power at the radar receiver to be large compared to receiver noise. If otherwise, Eq. (13.8) would not apply.

The antenna gain does not enter, except as it is affected by the azimuth beamwidth θ_B. The narrower the pulse width the greater the range. This is just opposite to the case of conventional radar detection of targets in noise. A long pulse is desired when the radar is limited by noise in order to increase the signal-to-noise ratio. When clutter dominates noise, a long pulse decreases the signal-to-clutter ratio. (When pulse compression is used, the pulse width τ in Eq. (13.8) is that of the compressed pulse.) If the statistics of the clutter echoes are similar to the statistics of receiver noise, then the signal-to-clutter ratio in Eq. (13.8) can be selected similar to that for signal-to-noise ratio as described in Chap. 2. (It will be seen in Sec. 13.3 that a decrease in pulse width can change the nature of the clutter statistics in some cases and might negate the benefits of the greater signal-to-clutter ratio obtained with the short pulse.) The improvement in range due to the integration of n pulses is not indicated in this equation. There can be a considerable difference in the integration improvement when clutter-limited from when noise-limited. Clutter echoes, unlike receiver noise, might be correlated pulse to pulse, especially if the clutter is stationary relative to the radar. Receiver noise is usually decorrelated in a time equal to $1/B$, where B = receiver (IF) bandwidth. The decorrelation time of clutter is usually much greater than this.

Next consider the case where the radar observes surface clutter near perpendicular incidence. (At perpendicular incidence the grazing angle ϕ is 90°.) The clutter area viewed by the radar will be determined by the antenna beamwidths θ_B and ϕ_B in the two principal planes. The area A_c in Eq. (13.1) is $(\pi/4)R\theta_B R\phi_B \sin\phi$, where ϕ = grazing angle. The factor $\pi/4$ accounts for the elliptical shape of the area. Substituting A_c into Eq. (13.3) and taking $G = \pi^2/\theta_B \phi_B$,[69] the clutter radar equation is

$$C = \frac{\pi P_t A_e \sigma^0}{64R^2 \sin\phi} \tag{13.9}$$

The clutter power is seen to vary inversely as the square of the range. Equation (13.9) applies, for example, to the signal received from the ground by a radar altimeter. An equation for detecting a target in this type of clutter background could be derived, but it is a situation not often found in practice.

In describing the geometry of surface clutter, the incidence angle and the depression angle are sometimes used instead of the grazing angle. These are shown in Fig. 13.2. The *incidence angle* is defined relative to the normal to the surface; the *grazing angle* is defined with respect to the tangent to the surface, and the *depression angle* is defined with respect to the local horizontal at the radar. When the earth's surface can be considered flat, the depression angle and the grazing angle are the same. When the earth's curvature must be considered, these two angles are not equal. The incidence angle is preferred when considering earth backscatter effects from near perpendicular incidence, as in the case of the altimeter. The grazing angle is the preferred measure in most of the other radar applications, and will be the angle used in this chapter.

Descriptions presented in this chapter of radar scattering from the land and the sea are by no means complete. Perhaps the chief difficulty in trying to understand the nature of clutter is the lack of adequate quantitative descriptions of the nature of the scattering objects. In this respect, the information regarding radar scattering from the sea is probably better understood than radar scattering from the land. The state of the sea is determined primarily by the

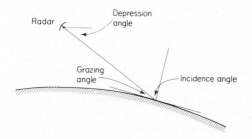

Figure 13.2 Angles used in describing geometry of the radar and surface clutter.

strength of the wind and the distance and time over which the wind has been blowing. Thus the sea in the North Atlantic is not that much different from the sea in the South Pacific if the wind conditions are the same. Land clutter, on the other hand, is highly dependent on the nature of the terrain and the local conditions. Urban areas, cultivated fields, forests, mountains, desert, and tundra all can produce different radar echoes. Furthermore, radar scattering from land is affected by rain, snow cover, the type of vegetation or crops, the time of year, the presence of streams and lakes, and man-made objects interspersed among the terrain. There can be considerable variability in both land and sea clutter. This inherent variability must be understood and properly accounted for in radar system design.

13.3 SEA CLUTTER

The echo from the surface of the sea is dependent upon the wave height, wind speed, the length of time and the distance (fetch) over which the wind has been blowing, direction of the waves relative to that of the radar beam, whether the sea is building up or is decreasing, the presence of swell as well as sea waves, and the presence of contaminants that might affect the surface tension. The sea echo also depends on such radar parameters as frequency, polarization, grazing angle, and, to some extent, the size of the area under observation. Although there is much that is known about the nature of sea clutter, the quantitative, and sometimes even the qualitative, effects of many of the above factors are not known to the degree often desired. The relative uniformity of the sea over the oceans of the world makes it easier to deal with sea clutter than with land clutter, but it is difficult to collect data at sea under the controlled and reproducible conditions necessary to establish quantitative causal relationships. In spite of the limitations, there does exist a body of information regarding the radar echo from the sea, some of which will be reviewed briefly in this section.

Average value of σ^0. A composite average of sea clutter data is shown in Fig. 13.3.[1] This is a plot of the mean cross section per unit area, σ^0, as a function of grazing angle for various frequencies and polarizations. It does not correspond to any particular set of experimental data, but it represents what might be typical of "average" conditions. It was derived from a body of data that extended from 10- to 20-knot wind speeds. (The uncertainty of the data ought to have been indicated in this figure by making the vertical thickness of each curve at least ± 3 dB wide. This was not done so as to avoid the confusion that would be caused by the wide overlap of the curves.)

The sea state is a measure of the wave height, as shown in Table 13.1. Although the sea state is commonly used to describe radar sea-clutter measurements, it is not a complete measure in itself since, as mentioned above, the radar sea echo depends on many other factors. Sea clutter is also sometimes described by the wind speed, but likewise it is not a suitable measure by itself. Unfortunately, the reporting of radar sea echo measurements is seldom accompanied by a complete description of the sea and wind conditions.

Variation with grazing angle. There are usually identified three scattering regions according to the grazing angle. At near vertical incidence ($\phi \approx 90°$) the radar echo is relatively large. This is called the *quasi-specular region* and seems to be the result of specular scatter from facetlike surfaces oriented perpendicular to the direction of the radar. At the other extreme, when the grazing angles are low, of the order of several degrees or less, σ^0 decreases very rapidly with decreasing angle. This is called the *interference region* since the direct wave and

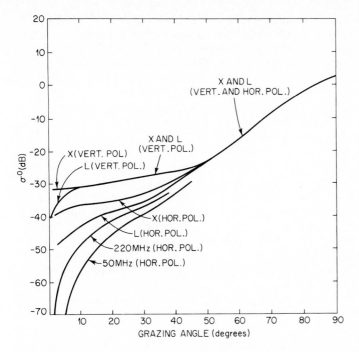

Figure 13.3 Composite of σ^0 data for average conditions with wind speeds ranging from 10 to 20 knots.

the wave scattered from the sea surface are out of phase and produce destructive interference. (The interference region is not well defined at the higher microwave frequencies plotted in Fig. 13.3.) The angular region between the interference region and the quasi-specular region is the *plateau*, or *diffuse*, *region*. The chief scatterers are those components of the sea that are of a dimension comparable to the radar wavelength (Bragg scatter). Scattering in this region is similar to that from a rough surface.

Table 13.1 World Meteorological Organization sea state

Sea state	Wave height		Descriptive term
	Feet	Meters	
0	0	0	Calm, glassy
1	$0-\frac{1}{3}$	0–0.1	Calm, rippled
2	$\frac{1}{3}-1\frac{2}{3}$	0.1–0.5	Smooth, wavelets
3	2–4	0.6–1.2	Slight
4	4–8	1.2–2.4	Moderate
5	8–13	2.4–4.0	Rough
6	13–20	4.0–6.0	Very rough
7	20–30	6.0–9.0	High
8	30–45	9.0–14	Very high
9	over 45	over 14	Phenomenal

Variation with frequency and polarization. From Fig. 13.3, the value of σ^0 is seen to be essentially independent of frequency for vertical polarization, except at low grazing angles. With horizontal polarization, the clutter echo decreases with decreasing frequency. This is most pronounced at the lower grazing angles. The effect of polarization on σ^0 also depends on the wind, as mentioned below.

Variation with wind.[1-4] Except at grazing angles near normal incidence, the value of σ^0 increases with increasing wind speed. The backscatter is quite low with winds less than about 5 knots, increases rapidly with wind speeds from about 5 to about 20 knots, and increases less rapidly at higher wind speeds.[1] An example is shown in Fig. 13.4.[3] This is a composite plot of experimental X-band data obtained at three different locations and at three different times for a grazing angle of 10°. (The data represented by the solid circles were obtained in the Bermuda area; the x's in the North Atlantic; and the triangles, off Puerto Rico.) The dashed curve in each figure is drawn as if all three sets of data were from a single population. This curve is of the form $\exp(-c/U)$ where c is an arbitrary constant and U is the wind speed. When all three sets of data are considered as one, the dashed curve shows a saturation of σ^0 with wind speed above about 20 knots. However, conclusions drawn from viewing the three sets of data as a whole may not be fully valid because of the accuracy with

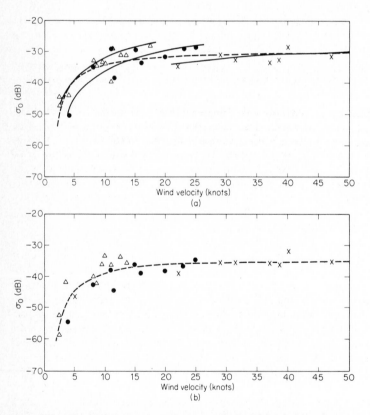

Figure 13.4 Median value of σ^0 as a function of wind speed, for X band and 10° grazing angle. (a) Vertical polarization, (b) horizontal polarization. (*After Daley, et al.*[3])

which such measurements can be made. The solid curves of Fig. 13.4*a*, therefore, are drawn for each separate set of data. (A similar set could be drawn for (*b*).) There is also a saturation effect evident when the three sets of data are considered separately, but it is not as pronounced as when the three sets of data are considered as one. Over the range from 10 to 20 knots, σ^0 increases at a rate of about 0.5 dB/kn. Above 30 knots the increase is about 0.1 dB/kn. Similar behavior is exhibited at 30° grazing angle. The effect of wind at C band is like that at X band. At lower frequencies, however, the variation of σ^0 with wind is less pronounced.

The energy backscattered from the sea will depend on the direction of the wind relative to the direction of the radar antenna beam. It is generally higher when the radar beam looks into the wind than when looking downward or crosswind. There might be from 5 to 10 dB variation in σ^0 as the antenna scans 360° in azimuth. It has been found[4] that the backscatter is more sensitive to wind direction at the higher radar frequencies than at the lower frequencies, that horizontal polarization is more sensitive to wind direction than is vertical polarization, that the ratio of σ^0 measured upwind to that measured downwind decreases with increasing grazing angle and sea state, and that at UHF the backscatter is practically insensitive to wind direction at grazing angles greater than ten degrees.

When viewing the sea at or near vertical incidence the backscatter is greatest with a calm sea and decreases with increasing wind. One set of data[5] gives the value of σ^0 at short wavelengths to be proportional to $U^{-0.6}$, where U is the wind speed.

The wind has been assumed here to be a driving force in determining the radar backscatter from the sea. It is indeed, but the relation between the wind and the sea conditions is complex; hence, so will be the relation between the wind and the radar backscatter. The capillary waves and the short gravity waves build up quickly with the wind so that the value of σ^0 measured with radars at the higher microwave frequencies should be quite responsive to wind conditions. The longer gravity waves require a finite time to build up and reach saturation; hence, the wind alone might not be a true measure of sea conditions unless account is taken of the length of time and the distance (fetch) over which the wind has been blowing. Thus radars at the lower frequencies which are more responsive to the longer gravity waves might experience a value of σ^0 that does not seem responsive to the immediate value of wind speed. Another complicating factor is the breaking of waves and the generation of spray, foam, or whitecaps which have an effect on the radar backscatter. The presence of swell waves along with the wind-driven sea can also affect the radar backscatter. Still another important factor when attempting to understand the effect of the wind on sea clutter is the difficulty in accurately estimating wind speed at sea.[6]

Effect of pulse width. In principle, the value of σ^0, the sea backscatter per unit area, is independent of the illuminated area. In practice, the area illuminated by the radar does affect the nature of the backscatter. This is especially so with short-pulse radars. When the radar resolution is sufficiently small it is found that the sea echo is not spatially uniform, but is composed of "occasional" individual targetlike echoes of short duration that result in a spiky appearance on a radar display. On an A-scope display it is found that sea clutter appears more spiky for horizontal than for vertical polarization, when the radar looks up or downwind rather than crosswind, at low radar frequencies rather than for high, and for low grazing angles rather than for high grazing angles.[7] To resolve the individual spikes the radar pulse width should be several tens of nanoseconds rather than microseconds. An individual spike might have a radar cross section at X band of the order of magnitude of one square meter (more or less), and can often last for several seconds.[8-10] Spikes can give rise to excessive false alarms in some radar applications. These spikes seem to be associated with breaking waves.

According to Long,[9] about half the time the whitecap forms either simultaneously with the appearance of a radar spike or a fraction of a second thereafter, leading to the conclusion that the whitecap occurs after the radar spike develops. The greater the duration of the spike, the greater the size and duration of the whitecap. A spike echo can appear without the presence of a whitecap, but no whitecap is observed in the absence of a spike. The spike echo at X band is amplitude modulated with frequencies of the order of 50 Hz and with a high modulation index. The spiky nature of sea echo indicates that the breaking of water waves plays an important role in producing backscatter.[119]

In addition to changes in clutter characteristics with narrow pulse widths, changes are also noted with narrow beamwidths. It has been reported[11] that at X band, small ships stand out much better with a 2° horizontal beamwidth antenna than when illuminated by a 0.85° beamwidth antenna. This is contrary to what would normally be expected if the clutter were uniform.

Thus, as the resolution cell size is decreased the nature of the sea clutter changes. The changes can have a profound effect on radar performance. The individual resolved spikes can cause false alarms. Raising the receiver threshold level to reduce the false alarms can result in a significant loss in sensitivity to desired targets. This negates the benefit of the increased signal-to-clutter ratio achieved by high resolution as expressed by Eq. (13.8) for the detection of targets in clutter. (As will be discussed in Sec. 13.4 there are techniques for receiver detector-design that can avoid much of this loss.) A corollary consequence of high resolution is a change in the probability density function of the sea echo, as described next.

Probability density function of sea clutter. The amplitude of sea clutter is characterized by statistical fluctuations which must be described in terms of the probability density function. When the echo from the sea can be modeled as that from a number of independent, random scatterers, with no one individual scatterer producing an echo of magnitude commensurate with the resultant echo from all scatterers, then the amplitude fluctuations of the sea echo is described by the gaussian probability density function, [Eq. (2.15)]. The statistical fluctuations of the envelope v of such an echo at the output of the envelope detector in the radar receiver is given by the Rayleigh probability density function (pdf), which is

$$p(v) = \frac{2v}{\sigma^2} \exp\left(-\frac{v^2}{\sigma^2}\right) \tag{13.10}$$

where σ is the standard deviation of the envelope v, which for the Rayleigh pdf is proportional to the mean value. (The median value is $\sqrt{\ln 2}\,\sigma$.) Clutter which conforms to this model is called *Rayleigh clutter*. The Rayleigh pdf applies to sea clutter when the resolution cell, or area illuminated by the radar, is relatively large. Experimental data show, however, that this distribution does not apply with a high-resolution radar which illuminates only a small area of the sea. The probability density function deviates from Rayleigh when the area illuminated by the radar has a dimension (usually the dimension determined by the pulse width) which is comparable to, or smaller than, the water wavelength.[13] There does not seem to exist any single analytical form of probability density function that fits all the observed data. The actual form of the pdf will depend on the size of the radar resolution cell and the sea state. An example of sea clutter statistics is shown in Fig. 13.5, which plots the probability that the clutter will exceed the abscissa, rather than the probability density function. (The two are related, however.) It is seen that actual clutter has a greater probability of a large value of sea clutter than does Rayleigh clutter. Stated in other terms, the actual distribution has higher "tails" than the Rayleigh. This means that clutter seen by a high-resolution radar will have a greater likelihood of false alarm than Rayleigh clutter if the receiver is designed in the

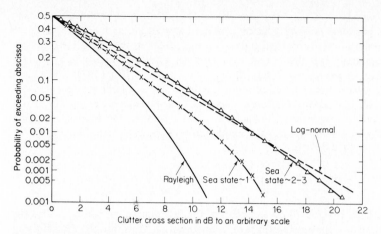

Figure 13.5 Experimental statistics of vertical polarization, X-band sea clutter for two sea states. Pulse width = 20 ns, beamwidth = 0.5°, grazing angle = 4.7°, range = 2 nmi. The Rayleigh distribution is shown for comparison. Dashed curve shows attempt to fit the sea state 2–3 data to log-normal distribution. (*After Trunk and George.*[12])

customary way based on gaussian noise. To avoid false alarms, the threshold detector at the output of such a receiver must be set to a higher value than when the clutter is Rayleigh. The increased threshold needed to avoid false alarms is significant (10 to 20 dB, or more, in some cases).

The log-normal probability density function has been proposed to model the clutter echo with very high-resolution radar and at the higher sea states. If the clutter cross section is log normal, the probability density function describing its statistics is

$$p(\sigma_c) = \frac{1}{\sqrt{2\pi}\,\sigma\sigma_c} \exp\left[-\frac{1}{2\sigma^2}\left(\ln \frac{\sigma_c}{\sigma_m}\right)^2 \right] \qquad \sigma_c > 0 \tag{13.11a}$$

where σ_c = clutter cross section

σ_m = median value of σ_c

σ = standard deviation of $\ln \sigma_c$ (natural logarithm)

The probability density function that describes the statistics of the voltage amplitude at the output of the envelope detector when Eq. (13.11a) is the clutter input, is given by

$$p(v) = \frac{2}{\sqrt{2\pi}\,\sigma v} \exp\left[-\frac{1}{2\sigma^2}\left(2\ln \frac{v}{v_m}\right)^2 \right] \tag{13.11b}$$

where v_m is the median value of v and σ remains the standard deviation of $\ln \sigma_c$.[114] A fit of the log-normal probability density function with $\sigma = 6$ dB to actual data for sea state 2 to 3 is shown in Fig. 13.5.[12] The median value of the theoretical distribution in this case was equated to the median of the actual data and the σ was selected to minimize the maximum difference in dB between the theoretical curve and the actual data.

With many radars that operate in the presence of sea clutter the Rayleigh pdf generally underestimates the range of values obtained from real clutter, and the log-normal pdf tends to overestimate the range of variation. Several other analytical probability density functions that lie between these two extremes have been suggested for modeling the actual statistical variations received from clutter. One of these is the contaminated-normal pdf which consists of the sum of two gaussian pdfs of different standard deviations and different relative weightings.[12] This pdf can be fitted to the data of Fig. 13.5 for sea state 1 by proper selection of the parameters. Other pdf's that have been suggested for approximating non-Rayleigh sea clutter include the Ricean[13] [Eq. (2.27)], the chi-square[13] [Eq. (2.40)], and a class of pdf's called the "K-distributions" based on modeling the clutter as a finite two-dimensional random walk.[29] The Weibull pdf has also been examined for modeling sea clutter.[14,15] It is a two-parameter pdf, like the log normal, but is intermediate between the Rayleigh and the log normal. In the Weibull clutter model the amplitude probability density function of the voltage v out of the envelope detector is

$$p(v) = \alpha \ln 2 \left(\frac{v}{v_m}\right)^{\alpha-1} \exp\left[-\ln 2\left(\frac{v}{v_m}\right)^{\alpha}\right] \tag{13.12}$$

where α is a parameter that relates to the skewness of the distribution (sometimes called the Weibull parameter), and v_m is the median value of the distribution. By appropriately adjusting its parameters, the Weibull can be made to approach either the Rayleigh or the log-normal distributions. The Rayleigh is actually a member of the Weibull family, and results when $\alpha = 2$. The exponential pdf is also a member of the Weibull family when $\alpha = 1$. (Further discussion of the Weibull pdf is given in Sec. 13.5.)

A knowledge of the statistics of the clutter is important in order to properly design a CFAR (constant false-alarm rate) receiver[115] and to avoid the loss associated with a receiver detector designed on the basis of the improper statistical model of clutter.

Theory of sea clutter. In the past, attempts were made to explain the mechanism of sea clutter by reflection from a corrugated surface,[16] by backscatter from the droplets of spray thrown by the wind into the air above the sea surface,[16,17] or by backscattering from small facets, or patches, on the sea surface.[18] None of these models were able to explain adequately the experimental observations. The *composite surface model*, in which the sea is recognized as composed of a number of waves of different wavelengths and amplitudes, has been more successful than previous models for describing the mechanism of sea clutter.[19,20,118] It is based on the application of classical scattering theory to a "slightly rough surface" which is described by the mean-squared-height spectrum of the sea surface. (By a slightly rough surface is meant one whose height variations are small compared to the radar wavelength and whose slopes are small with respect to unity.) The resultant radar backscatter can be interpreted as that obtained from the component of the sea spectrum which is "resonant" with the radar wavelength. That is, the radar wavelength λ_r is related to the water wavelength λ_w by the relation

$$\lambda_r = 2\lambda_w \cos \phi \tag{13.13}$$

where ϕ = grazing angle. This is sometimes called the *Bragg backscattering resonance condition* because of its similarity to the X-ray scattering in crystals as observed by Bragg.[2] Thus the radar is responsive only to those water waves which satisfy Eq. (13.13). An X-band radar ($\lambda = 3$ cm) at zero degrees grazing angle would backscatter from water waves of length 1.5 cm. These short water waves are known as *capillary waves*. An X-band radar when viewing large swell waves at low grazing angles in the absence of wind would receive little or no backscatter since there would be no capillary waves present, even though the swell might be large. As soon as wind appears, capillaries are formed, and the X-band radar will see backscatter since water waves are now present that are resonant to the X-band wavelength. A

Figure 13.6 Composite rough-surface model of the sea with the small resonant water waves riding on top of the larger water waves, resulting in a tilting of the resonant waves from the horizontal.

UHF radar ($\lambda = 70$ cm) at zero degrees grazing angle responds to water waves of length 35 cm. These are known as *gravity waves.*

The effect of the large water waves is to cause a tilting of the scattering surface of the smaller water waves, as in Fig. 13.6. The radar wavelengths are usually too short to be responsive to these long water waves, but the effect of such waves is noted by the tilting of the smaller water waves which are resonant with the incident radar energy. The tilting effect of the large waves is especially important at low grazing angles since the grazing angle is determined not only by the angle the radar energy makes with the mean sea surface, but also by the angle of the resonant waves that ride on the large wave structure.

The composite-model theory of sea clutter seems to account for the experimental results obtained with vertical polarization better than with horizontal polarization.[19] Also, it is more applicable to the midrange of grazing angles than to very low grazing angles or to near normal incidence.

There are other effects that should probably be part of any complete theory of sea clutter. Such factors include: the shadowing of the waves when the surface is viewed at low grazing angle; the specular reflection from plane facets when viewed at near normal incidence; the effect of breaking waves, whitecaps, and foam, especially when the winds are greater than about 15 to 20 knots; and the presence of refraction effects, such as that due to the evaporation duct over the surface of the sea, which enhance the propagation of radar waves near zero grazing angle so that sea clutter is increased at the longer ranges compared to normal propagation conditions.

Backscatter from sea ice. At low grazing angles, as might occur with a shipboard radar, there is little backscattered energy from smooth, flat ice. On a PPI or similar display the areas of ice will be dark except perhaps at the edges. If areas of water are present along with the ice, the sea clutter seen on the display will be relatively bright in contrast to the ice. Backscatter from rough ice, such as floes and pack ice, can produce an effect on the radar display similar to sea clutter. However, the backscatter from rough ice can be distinguished from sea clutter since its pattern on the display will remain stationary from scan to scan but the sea clutter pattern will change with time.[33] Measurements made over ice fields in the region of Thule, Greenland,[47] indicated the radar backscatter (σ^0) to vary linearly with frequency over a range of grazing angles from 1 to 10°. It was also found that σ^0 was proportional to the elevation angle over grazing angles from 2° to 10°, but was inversely proportional to grazing angle over the region from 1° to 2°. (The increase of σ^0 with decreasing grazing angle is similar to the behavior of land clutter but not sea clutter.) The scatterometer, a radar which measures σ^0 as a function of incidence angle in the region near vertical incidence, has been used as an ice sensor and to differentiate between first-year and multi-year ice.[65]

Radar can also detect large icebergs, especially if they have faces that are nearly perpendicular to the radar direction of propagation.[33] Icebergs with sloping faces can have a small echo even though they may be large in size. Growlers, which are small icebergs large enough to be of danger to ships, are poor radar targets because of their small size and shape. They seem to be more detectable with radar having a moderate resolution rather than a high resolution.[66]

The backscatter from ice at high grazing angles is complicated by the relative ease of penetration of radar energy into the ice. The reflection from the air-ice interface may be accompanied by a reflection from the underlying ice-water or ice-land interface. These reflections from the two interfaces, as well as multiple internal reflections, can result in radar backscatter that varies with the ice thickness, water content of the ice, and radar wavelength. The penetration properties of radar energy in ice and snow, especially at UHF or lower frequencies, can result in serious errors in the height measurement of radar altimeters.[67] Under some conditions, there might be no echo from the top surface of the ice, and if the ice is thick enough there might be no reflection from the bottom. When there is a recognizable reflection from the bottom, the distance measured by an altimeter calibrated on the assumption that propagation is in air, can be erroneous because the velocity of propagation in ice is 0.535 times that in air, and in snow it is less by a factor of 0.8.[67] Altimeters at the higher microwave frequencies are less affected by penetration into the ice and are more likely to indicate the air-ice interface than UHF altimeters.

Oil slicks. Oil on the surface of the sea has a smoothing effect on breaking waves, and damps out the capillary and short gravity waves normally generated by the wind. Thus oil slicks appear dark on a radar PPI compared to the surrounding sea and are readily detectable. High-resolution X-band radar, for example, can sometimes detect the oil leaking from the engine of a small boat at anchor. Airborne synthetic-aperture radar has been demonstrated to locate and map oil slicks over a large area. It is found that vertical polarization produces higher contrast than horizontal polarization. At low sea-state and wind it is possible to image an oil spill of less than 400 liters during the period of spillage.[68]

13.4 DETECTION OF TARGETS IN SEA CLUTTER

The ability of a radar to detect targets on or above the sea can be limited by sea clutter as well as by receiver noise. The magnitude and extent of the sea clutter seen by the radar will depend not only on the sea state and wind, but on the radar height. Radar is used to detect at least three different classes of targets over the sea: aircraft, ships, and small objects such as buoys. (Navigation radars must also see land-sea boundaries in addition to buoys.) The detection of such targets in the presence of sea clutter generally requires a different design approach for each, although there are similarities among them. In brief, a radar required to detect aircraft in the presence of sea clutter would likely employ MTI (moving target indication); a radar for the detection of buoys and other small objects on the sea would generally be of high resolution and employ techniques to prevent receiver saturation by large clutter; and a radar designed to detect ships could be of moderate resolution since the radar cross section of ships is relatively large. In the remainder of the section, the various techniques for detecting targets in clutter will be reviewed.

MTI. As mentioned, MTI can be used to separate the moving aircraft targets from the relatively stationary sea clutter. Except for some naval applications for the detection of very low-flying aircraft, the normal detection of aircraft over the sea does not usually represent a severe MTI design problem. Sea clutter is not as strong and does not usually extend to as great a range as does land clutter. Thus high-flying aircraft over the sea can be detected at long range with relatively conventional radar with only a minimal MTI, or even with no MTI in some cases. A more severe problem occurs when a seaborne radar must operate near land. Land clutter can be so large that clutter echoes might enter the radar receiver via the antenna sidelobes. In this case, when strong land clutter is a problem, a good MTI or pulse-doppler radar is required. MTI or pulse-doppler radar is also necessary in radar carried by high-flying aircraft that must detect other aircraft masked by sea clutter.

Although ships are moving targets that might be candidates for MTI radar, it is not often that MTI is needed for this application. The large cross section of ships generally results in suitable target-to-clutter ratios with conventional radars.

Frequency. The plot of σ^0 vs. grazing angle for various frequencies (Fig. 13.3) indicates that the lower the frequency, the less the sea clutter. This applies at the lower grazing angles and with horizontal polarization. Thus, to reduce clutter, the radar should be at a low frequency. There are some reservations, however, that would tend to favor the higher frequencies in many applications. The lower the frequency, the more difficult it is to direct the radar energy at low angles. This is illustrated in Sec. 12.2, which gives the elevation angle of the lowest interference lobe as $\lambda/4h_a$, where λ = wavelength and h_a = antenna height. Thus if targets low on the water are to be seen, higher frequencies may be preferred even if the clutter is greater. Another reason for preferring the higher frequencies in some applications, in spite of the larger clutter, is that greater range-resolution and azimuth-resolution (shorter pulse width and narrower beamwidths) are easier to obtain than at lower frequencies. The higher resolution usually results in greater target-to-clutter ratio. Civil marine radars are available at both S band (10 cm wavelength) and X band (3 cm). Although both bands have distinctive advantages for ship application, when only a single radar is employed it is usually at X band for the reasons given above.[21] Thus the fact that sea clutter is less at the lower frequencies is but one of several factors to be considered in the selection of the optimum frequency for a particular radar application.

Polarization. Figure 13.3 indicates that horizontal polarization results in less sea clutter than vertical, for the sea conditions which apply to that data. The majority of radars that operate over the sea usually employ horizontal polarization. Air-surveillance radars with horizontal polarization achieve greater range than with vertical polarization because of the higher reflection coefficient of horizontal polarization (Sec. 12.2). A firm conclusion in favor of one polarization or the other is difficult to make for a civil marine radar, but most sets use horizontal polarization.[21] At high sea states the difference between horizontal and vertical polarization is said to disappear.[22]

Horizontal polarization, as was mentioned in Sec. 13.3, has a probability density function which deviates from Rayleigh more than does vertical polarization when the pulse widths are short. The non-Rayleigh nature of the clutter must be properly considered in the detector design if large increases in the minimum detectable signal-to-clutter ratios are to be avoided.

Pulse width. The radar equation that applies to the detection of targets in surface clutter at low grazing angles (Eq. 13.8) shows that, all other factors remaining constant, the shorter the pulse width the greater will be the maximum range. Decreasing the pulse width decreases the amount of clutter with which the target must compete. There are three cautions to remember, however, when considering the use of Eq. (13.8). *First*, the clutter echo power must be large compared to the receiver noise power for Eq. (13.8) to be valid. Thus the energy within the pulse must not be too low. If the peak power is limited, a short pulse might not have enough energy to make clutter, rather than receiver noise, dominate. Pulse compression (Sec. 11.5) can overcome this limitation of a short pulse since it can achieve a resolution equivalent to that of a short pulse, but with the energy of a long pulse. *Second*, the required signal-to-clutter ratio depends on the pulse width because of the change in clutter statistics (probability density function) with short pulse widths, as discussed in Sec. 13.3. The changes in clutter statistics must be accounted for in the detector design, else the large signal-to-clutter ratios required of a conventional threshold detector might negate any benefit of the shortened pulse. As mentioned previously, a radar with a very short pulse (or narrow beamwidth) might prove to be worse than one with a longer pulse (or a wider beamwidth) because of the change in clutter statistics. (Receiver detection-criteria for compensating the effects of a change in clutter statistics with high resolution are discussed later in this section.) *Third*, if the pulse is too short, it might encompass only part of a large target such as a ship, and the target cross section might be reduced. However, if the echoes from the various parts of a large target are displayed without collapsing loss and if the operator knows the general shape of the target, there will be little, if any, loss in detectability.

Antenna beamwidth. The discussion above regarding the benefits and limitations of the short pulse also apply to the azimuth beamwidth. The use of high resolution in angle and/or range is the principal method available for detecting a fixed target in clutter, assuming the target "cross section per unit area" is greater than that of clutter σ^0. Too narrow a beamwidth, just as too narrow a pulse width, is not desired since it can also alter the clutter statistics in an unfavorable manner.[31]

Antenna scan rate. The sea clutter background does not change significantly during the time the usual civil-marine radar antenna (with a 20 rpm rotation rate) scans by a particular clutter patch. That is, the sea can be considered "frozen" during the observation time. For a medium-resolution X-band radar at low grazing angles, the time required for the sea clutter echo to decorrelate is about 10 ms.[16,21] Any pulses received from sea clutter during this time will be correlated and no improvement in signal-to-clutter ratio will be obtained by integration. (This is unlike receiver noise since integration of signal pulses and noise can provide significant improvement in signal-to-noise ratio, as described in Sec. 2.6.) Although the clutter received during a single scan past a target will be correlated, the sea surface will usually change by the time of the next antenna scan so that the clutter will be decorrelated from scan to scan. Thus scan-to-scan integration can usually result in an improvement of target-to-clutter ratio. To make available more independent clutter echoes, the antenna rotation rate can be increased.[21,23] If, for example, the radar antenna has a 2° beamwidth, a pulse repetition frequency of 3600 Hz, and a 20 rpm rotation rate, there will be 60 pulses returned from a target on each scan of the antenna. There is little integration improvement obtained from these 60 pulses since they occur within a time ($\frac{1}{60}$ s) comparable to the decorrelation time for X-band sea clutter. If the antenna rotation rate is speeded up to 600 rpm, there will be 2 pulses received during a scan past a target. (There should be at least 2 pulses per scan rather than one, in order to avoid a large scanning loss.) On the next scan, one-tenth second later, two additional pulses are received that are decorrelated from the previous two pulses so that an integration improvement can be obtained. The number of pulses processed by the slow-speed and the high-speed antennas are the same. The 20 rpm antenna receives 60 pulses per scan with a 3-second rotation period. In the same 3 seconds, the 600 rpm antenna scans by the target 30 times, receiving two pulses on each scan. A scan-to-scan integration improvement can be achieved with the high-speed antenna if the storage properties of the CRT screen permit efficient integration (as with a storage tube) or if some form of automatic integration is used.

Experimental measurements using civil-marine radar with a 1° beamwidth and 5000 Hz pulse repetition frequency show that increasing the antenna rotation rate from 20 to 420 rpm improves the target-to-clutter ratio from about 4 to 8 dB depending on the sea state.[24] (The smaller improvement corresponds to the higher sea state.) An attempt was also made to decorrelate the two pulses received from a target with the high-speed rotation rate by transmitting each of the two pulses at a different frequency. Unexpectedly, no discernible improvement was noted when the two frequencies were used instead of a single frequency, even though the two frequencies were 60 MHz apart. This would indicate that there is a slowly fading component of the clutter echo which cannot be decorrelated and is probably similar to the clutter spikes discussed in Sec. 13.3.

The decorrelation time of the clutter is approximated by the reciprocal of the width of the clutter doppler spectrum. It is found that the width of the frequency spectrum decreases with decreasing frequency so that the decorrelation time is greater at the lower frequencies than at the high frequencies. Normally it would be expected that the decorrelation time would be linearly proportional to the frequency, since the doppler-frequency shift is also linearly related to the frequency. However, experimental measurements show that the ratio of the decorrelation time at L band to that at X band is less than would be expected if the linear relation applied. For example, if the decorrelation time at X band were 10 ms, then it would be about 60 ms at L band, instead of the 70 to 75 ms predicted on the basis of a linear law. Therefore, when the clutter spectra are given in units of velocity rather than frequency, the velocity spectra are slightly broader at the lower frequencies, an unexpected result.[25] The spectra are broader for horizontal than for vertical polarization, and increase in width as the square root of the wave height for both polarizations.

Observation time. In the above, a rapid antenna-rotation rate takes advantage of the fact that the clutter echo changes with time, but the target echo does not. With a high-resolution radar (nanoseconds pulse width) the individual sea-clutter spikes, as mentioned in Sec. 13.3, can persist for several seconds, which is much greater than the 10 ms decorrelation time quoted for X-band sea clutter. The persistence of these sea-clutter spikes makes difficult the detection of small targets, such as buoys and small boats. A single "snapshot" of a PPI would not likely differentiate the echo of a small target from that of the sea spikes. The individual sea spikes, however, will disappear with time and new spikes will appear at other locations. If it is possible to observe the radar display for a sufficient period of time, the small targets can be recognized since they will remain relatively fixed in amplitude while the sea spikes come and go. The penalty paid for this procedure is a long observation time.

Frequency agility. If the frequency of a pulse of length τ is changed by at least $1/\tau$, the echo from uniformly distributed clutter will be decorrelated. Thus pulse-to-pulse frequency changes (frequency agility) of $1/\tau$ or greater will decorrelate the clutter and permit an increase in target-to-clutter ratio when the decorrelated pulses are integrated. With a high-resolution radar in which the persistent sea-clutter spikes appear, the benefit of frequency agility is decreased since the sea spikes are correlated over a relatively long period of time and appear similar to target echoes. Thus frequency agility will be of less value the more non-Rayleigh the clutter statistics. (If the radar is on a rapidly moving platform, the clutter might also be decorrelated as the radar resolution cell views a different patch of clutter.)

Receiver detection criteria. In Chap. 2 the receiver detection criterion was based on the methods of Marcum and Swerling. Relationships between threshold levels, probabilities of detection and false alarm, signal-to-noise ratio, and number of pulses integrated were obtained using as a model a receiver with the following characteristics: (1) the envelope of the receiver noise at the output of the IF filter was described by the Rayleigh probability density function (pdf); (2) the envelope detector had either a linear or a square-law characteristic; (3) the integrator consisted of a linear addition of the pulses in the video portion of the receiver; and

(4) the detection decision was determined by whether or not the receiver output crossed a threshold that depended on the desired probability of false alarm. When the noise, or clutter, is not described by a Rayleigh pdf, the receiver model of Chap. 2 is not optimum and can cause degraded performance. The higher "tails" of the probability density function of non-Rayleigh clutter mean that there will be a greater number of false alarms in the conventional receiver designed on the basis of Rayleigh statistics. The false alarms with non-Rayleigh clutter can be reduced to any desired level by raising the receiver threshold. Raising the threshold, however, can require a significantly larger signal-to-noise, or signal-to-clutter, ratio for a given probability of detection.

Trunk and George[12] were among the first to note that something other than the conventional receiver design should be used when the clutter pdf is not Rayleigh. They showed that the *median detector* gave better performance in non-Rayleigh clutter than the conventional receiver (which can be called a *mean detector*). In this detector, the median value of the n received pulses is found and compared against a threshold. This requires that the amplitudes of the n individual samples (pulses) be stored and then arranged (ranked) in order of amplitude to find the median. An alternate approach, which is usually easier to implement, is to employ the m-out-of-n detector (the double threshold or binary integrator of Sec. 10.7) with the second threshold set at $m = (n - 1)/2$. The median detector is "robust," in that the threshold values and the detection probabilities do not depend on the detailed shape of the clutter pdf, but only on the median value.

The conventional receiver may be considered as comparing the *mean* value of the n pulses to a predetermined threshold. Trunk has shown that the trimmed-mean detector can be better than either the mean or the median detectors.[26,27] If there are n pulses available from which to make a detection decision, the trimmed-mean detector discards the smallest and the largest of the n pulses before taking the mean. Although its performance may be good, it is difficult to implement in practice because the n pulses must be rank-ordered to find the smallest and the largest.

Although the median detector is superior to the mean detector (conventional receiver), it is inferior to both the m-out-of-n detector and the logarithmic receiver when clutter can be described by the log-normal[28] or the Weibull[14] models. The m-out-of-n detector, or binary integrator, was originally employed in radar as an automatic integrator, but it has other properties of interest. The optimum number m (or the second threshold) will depend on the nature of the pdf. When the optimum m is used with non-Rayleigh-clutter, the m-out-of-n detector seems to be better than the other receiver criteria.[14,28]

The logarithmic receiver has an output voltage whose amplitude is proportional to the logarithm of the input envelope. A logarithmic characteristic provides a constant false alarm rate (CFAR) when the input clutter or noise has a Rayleigh pdf. The performance of the logarithmic receiver in non-Rayleigh clutter is almost as good as the m-out-of-n detector and is probably easier to implement.[14,28]

A comparison of the performance of the various receivers for a specific case is shown in Fig. 13.7.

Log-FTC and log-log receivers. The techniques discussed thus far attempt to improve the detection of targets in clutter by increasing the target-to-clutter ratio. There are also several techniques found in radars for the detection of targets in clutter whose purpose is to prevent the clutter from saturating the display or the receiver, or to maintain a constant false-alarm rate (CFAR). They have no subclutter visibility in that they offer no ability to see targets masked by clutter, but they do permit targets large compared to the clutter to be detected that might otherwise be lost because of the limited dynamic range of the receiving system. An

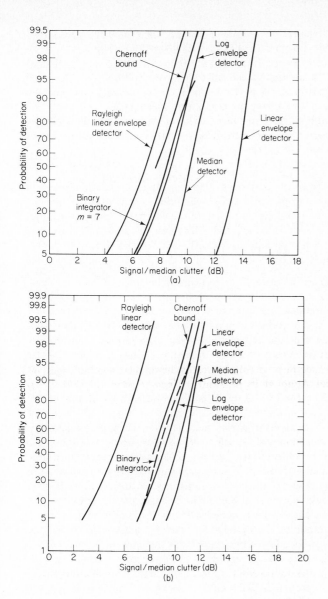

Figure 13.7 Comparison of various detectors in (*a*) log-normal clutter with a mean-to-median ratio of 1 dB, $\sigma = 0.693$ and (*b*) Weibull clutter with $\alpha = 1.2$ in Eq. (13.12). Probability of false alarm $= 10^{-6}$, 10 hits integrated. The curve labeled "Rayleigh linear detector" applies to the conventional receiver with Rayleigh clutter or noise. The curve labeled "Chernoff bound" provides an upper bound to the optimum detector for the particular clutter model indicated.

(*After* Schleher,[14,28] *courtesy IEEE.*)

example of one such technique is the log-FTC receiver.[32] This is a receiver with a logarithmic input-output characteristic followed by a high-pass filter (one with a fast time-constant, or FTC). The log-FTC has the property that when the input clutter or noise is described by a Rayleigh pdf, the output clutter or noise is a constant independent of the input amplitude. Thus the log-FTC receiver has a constant false-alarm rate. As with any CFAR, the false-alarm rate is maintained constant by a sacrifice in the probability of detection. (A further description as to why the log-FTC receiver is CFAR when the input is Rayleigh clutter is given in Sec. 13.8.)

Sea clutter has a Rayleigh pdf when the resolution is low. (A low-resolution radar in this case might have a 5° beamwidth and 1-μs pulse width.[30]) With better resolution the pdf of sea clutter deviates from Rayleigh and the log-FTC receiver is no longer CFAR. A variant sometimes useful for reducing the false-alarm rate in non-Rayleigh clutter is the *log-log receiver*. This is a logarithmic receiver in which the slope of the logarithmic characteristic progressively declines by a factor of 2 to 1 over the range from noise level to +80 dB.[30,31] (This assumes that the normal log-FTC has a logarithmic characteristic from −20 dB below the receiver noise to +80 dB above noise.) In the log-log receiver the higher clutter values (tails of the distribution) are subjected to greater suppression.

Adaptive video threshold. Other receiver techniques than the log-FTC were discussed in Sec. 10.7 for achieving CFAR. One example suitable for a clutter environment is the *adaptive video threshold* (AVT) which sets the threshold according to the amount of clutter (or noise) in a number of range cells ahead of and behind the particular range cell of interest. This form of CFAR is popular in automatic detection and track (ADT) systems since it is not strongly dependent on the type of clutter and can provide CFAR with land and weather clutter as well as sea clutter.

Other methods to avoid clutter saturation. The dynamic range of a radar display, whether PPI or A-scope, is far less than the range of echo-signal amplitudes that can be expected from clutter. It is necessary, therefore, to keep the large clutter echoes from saturating the display and preventing the detection of wanted targets. Echoes from the various sources of clutter experienced by a civil-marine radar might be 80 dB greater than receiver noise, yet the receiver might only be able to display without saturation signals that are less than about 20 to 25 dB above noise.[22] *Sensitivity time control* (STC), or *swept gain*, is widely used to reduce the large echoes from close-in clutter. STC is a time variation of the receiver gain. At the end of the transmission of the radar pulse the receiver gain is made low so that large signals from nearby clutter are attenuated. Echoes from nearby targets will also be attenuated; but because of the inverse fourth-power variation of signal power with range, they will usually be large enough to exceed the threshold and be detected. The receiver gain increases with time until maximum sensitivity is obtained at ranges beyond which clutter echoes are expected. Ideally, STC should make the average clutter power always equal to the noise power. Thus its range variation should be identical to the clutter-power range variation. It has been suggested that STC for a civil-marine radar have a gain proportional to R^{-3} out to a range of $8h_a h_w/3\lambda$, after which it is proportional to R^{-7}, where R = range, h_a = radar antenna height, h_w = wave height from peak to trough, and λ = radar wavelength.[34] STC may be implemented in the IF of the receiver, but it can also be incorporated at RF by inserting variable-attenuation microwave diodes ahead of the receiver. A particular STC characteristic can be limited by changes in clutter due to changes in ambient conditions (such as wind speed and direction, or by anomalous propagation), and by the nonuniformity of clutter with azimuth. Clutter-CFAR techniques that are "adaptive" do not have this limitation. (STC is also sometimes employed in radar not bothered by clutter to make more uniform the target echo power with range.)

Another technique for reducing saturation is *instantaneous automatic gain control* (IAGC) based on negative feedback controlling the gain of the IF amplifier. The response time of the IAGC is adjusted so that echo pulses from point targets pass with little attenuation, but longer pulses such as those from extended clutter are attenuated. That is, the automatic gain control is made to act within the time of a few pulse widths. The IAGC acts something like a pulse-width filter, permitting target pulses to pass and attenuating the longer pulses from clutter. Other means for attenuating long pulses from distributed clutter, yet pass the echoes from point targets, include the high-pass filter and various forms of pulse-width discriminators that inhibit echo signals not of the correct pulse width.

13.5 LAND CLUTTER

The clutter from land is generally more of a problem than the clutter from sea, both in theory and in practice. The backscatter from land is significantly greater than from sea except in the vicinity of near vertical incidence. Therefore, over the grazing angles of usual interest, a radar which must detect targets over land has a more difficult task than one which must detect targets over the sea. Even though a radar at sea might not be bothered by sea clutter, nearby land clutter can be so large that it can enter the radar via the antenna sidelobes and degrade performance. At vertical incidence there is less backscatter from land than from sea, but this is usually undesirable since it reduces the range of radar altimeters over land.

Land clutter is difficult to quantify and classify. The radar echo from land depends on the type of terrain, as described by its roughness and dielectric properties. Desert, forests, vegetation, bare soil, cultivated fields, mountains, swamps, cities, roads, and lakes all have different scattering characteristics. Furthermore, the radar echo will depend on the moisture content of the surface scatterers, snow cover, and the stage of growth of any vegetation. Buildings, towers, and other structures give more intense echo signals than forests or vegetation because of the presence of flat reflecting surfaces and "corner reflectors." Bodies of water, roads, and airport runways backscatter little energy but are recognizable on radar PPI displays as black areas amid the brightness of the surrounding ground echoes. A hill will appear to stand out in high relief on a PPI. The near side of the hill will give a large return, while the far side, which is relatively hidden from the view of the radar, will give little or no return. The radar cross section of a farmer's field will differ before and after ploughing, as well as before and after harvesting. It will also depend on the direction of the radar beam relative to the direction of the ploughed furrows. The echo from forests differs depending on the season. By contrast, sea echo is more uniform over the oceans of the world, providing the wind conditions are the same. Although knowledge of sea clutter is far from complete, it is better understood than is the knowledge of land clutter.

Information about the radar backscatter from land is required for several different applications, each of which has its own special needs. These applications include:

The detection of aircraft over land, where the clutter echoes might be as much as 50 to 60 dB greater than aircraft echoes. MTI or pulse-doppler radar is commonly used for this application to remove the background clutter.

The detection of surface targets over land, where moving vehicles or personnel can be separated from clutter by means of MTI. Fixed targets require high resolution for their detection.

Altimeters which measure the height of aircraft or spacecraft. Large clutter energy is desired since the "clutter" is the target.

The detection of terrain features such as hills and mountains ahead of an aircraft to warn of approaching high ground (*terrain avoidance*) or to allow the aircraft to follow the contour of the land (*terrain following*).

Mapping, or imaging, radars that utilize high resolution. Ground objects are recognized by their shape and contrast with surroundings.

Remote sensing with imaging radars, altimeters, or scatterometers to obtain specific information about the nature of the surface characteristics.

Examples of land clutter. There exist many measurements of land clutter, but there has been relatively little codified and condensed information. References 9, 35, and 36 summarize much of what has been published regarding land clutter. Not only is there a lack of published data for the various types of land clutter, but there is not always agreement among similar data taken by different investigators. Long[9] points out that "two flights over apparently the same type of terrain may at times differ by as much as 10 dB in σ^0." Such variation and uncertainty in the value of land clutter must be accounted for by conservative radar design.

The data for land clutter is usually reported in terms of σ^0, the cross section per unit area, just as for sea clutter. However, it is sometimes given by a parameter γ which equals $\sigma^0/\sin \phi$, where ϕ is the grazing angle. For ideal rough terrain, γ is approximately independent of the angle ϕ, except at low grazing angles and near perpendicular incidence.

An example of clutter σ^0 for several broad classes of terrain is shown in Fig. 13.8. This applies to X-band clutter. The boundaries of the various regions are wide to indicate the wide variation of the data within the classes of terrain. Figure 13.9 illustrates airborne data at X and L bands obtained by the Naval Research Laboratory.[37] The azimuth beamwidth was 5° and the pulse width was 0.5 μs at each frequency. The lack of smoothness of the data is due, in part, to the fact that the data was not all taken at the same time. For a particular grazing angle the two frequencies had to be obtained by reflying the aircraft along the same flight path. Different grazing angles also required reflying the aircraft over the same area. Each point on the curve is an average over 1 to 2 miles of ground track.

The curve of Fig. 13.10 illustrates the behavior of land clutter at low grazing angles.[38] The value of σ^0 decreases with decreasing grazing angle until, in this case, at about 3.5° the value of σ^0 actually increases as the grazing angle is lowered. This behavior is said to be due to the stronger backscatter from vertically oriented structures (trees, buildings, crops) as the grazing angle approaches zero. Other reported measurements at low grazing angles of heavily wooded, rolling hills typical of New England show σ^0 at these angles to be unaffected by changes in season or weather.[39]

There also exists much experimental data of the σ^0 of crops such as corn, soybeans, milo, alfalfa, and sorghum.[48-53] The radar backscatter depends not only on the type of crop, and frequency, but also on the state of its growth, the moisture content of the soil, and the time of day.

There is wide variability in the clutter measurements made by different experimenters under supposedly similar conditions. Some of these differences are due to lack of adequate ground-truth data, accurate system calibration, or accurate data processing.[54] There is usually a practical limit as to how well the significant ground-truth can be measured and how accurate a system can be calibrated.

The backscatter from snow-covered terrain depends on the nature of the underlying surface as well as the depth and wetness of the snow. Radar energy can propagate in snow with but moderate attenuation so that the backscatter will be determined by the energy reflected from the snow's surface as well as the energy reflected from the surface beneath the snow. The amount of attenuation in propagating through the snow will also be important. The backscatter from snow-covered terrain is dependent on the wetness, or amount of liquid and solid water in the snow, more so than the thickness of the snow. It was found,[46] for example, that 15 cm of dry powderlike snow had no effect on the measured backscatter over the frequency range 1 to 8 GHz. Any backscatter contribution by the snow was completely dominated by the contribution of the underlying soil. With 12 cm of wet snow, σ^0 decreased approximately 5 to 10 dB at grazing angles between 30 and 80°. At 30° grazing angle, for example, the variation of σ^0 as a function of the water content of the snow is in the range -0.3 to -0.6 dB/0.1 g/cm^2 over the frequency range from L to C bands. Other measurements indicate that the presence of

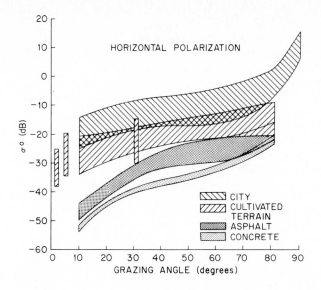

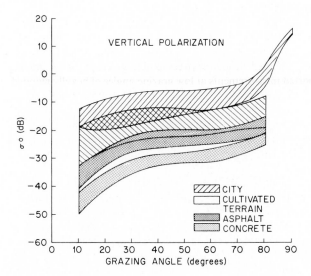

Figure 13.8 Boundaries of measured radar return at X band from various land clutter.[35] (*Courtesy of I. Katz.*)

a soft, wet snowcover on vegetation tends to lower the return by as much as 6 dB.[40] Measurements made over snowfields near Thule, Greenland, showed the backscatter to be proportional to the square of the frequency over the frequency range from UHF to X band and at grazing angles from 5 to 20°.[47]

Frequency dependence. Figure 13.11 shows an example of the frequency dependence of land clutter for a particular grazing angle.[37] Clutter from urban areas in these measurements is independent of frequency; rural terrain shows no frequency dependence from L to X band;

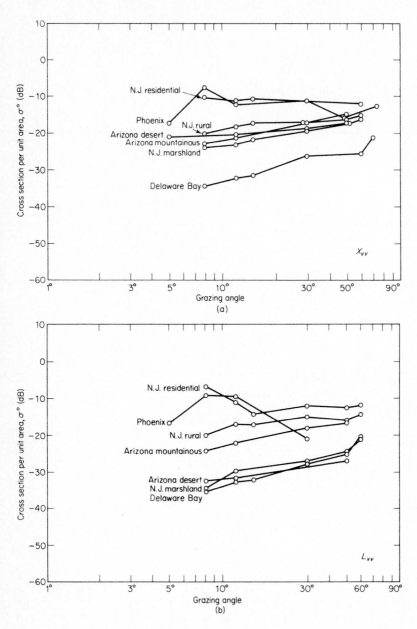

Figure 13.9 Median values of σ^0 for various terrain: (*a*) *X* band, vertical polarization, (*b*) *L* band, vertical polarization, (*c*) *X* band, horizontal polarization, (*d*) *L* band, horizontal polarization. (*From Daley, et al.*[37])

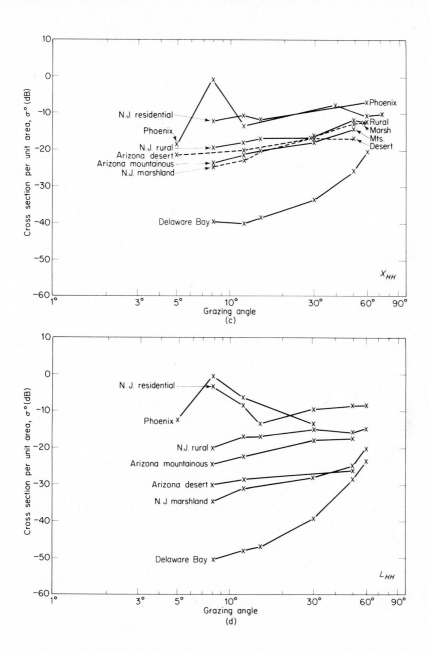

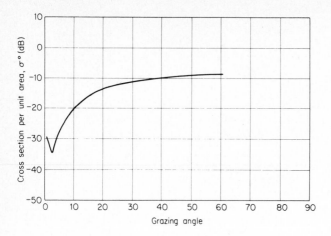

Figure 13.10 Cross-section per unit area of cultivated terrain illustrating the increase of σ^0 with decreasing grazing angle for small values of grazing angle. X band, horizontal polarization. (*From Katz and Spetner*,[38] *Courtesy J. Res. Nat. Bur. Stds.*)

mountainous terrain also shows no frequency dependence from L to X band but is considerably lower at UHF; rough hills, desert, and cultivated farmland show a linear dependence with frequency; and the backscatter from marsh varies as the $\frac{3}{2}$ power of frequency. The frequency dependence indicated by the data of Ref. 37 is independent of the polarization and of the angle of incidence (at least over the range from 8 to 30°).

Measurements of crops[40] and dry trees[41] indicate that above X band, σ^0 varies approximately linear with frequency, even up to 3.2-mm wavelength. The frequency dependence for concrete, asphalt, and cinder and gravel roads varies approximately as the square of the frequency.[40]

Effect of resolution. The use of σ^0, the cross section per unit area, to describe the radar scattering from clutter, implies that the clutter is uniform and independent of the radar-resolution-cell area. In reality, land clutter is not uniform. Radars with high resolution might actually be capable of better performance than predicted on the basis of a Rayleigh clutter model and a σ^0 independent of radar resolution. A radar with a narrow pulse width or narrow beamwidth not only sees less land clutter than a radar with larger pulse width or beamwidth, but it can see targets in those areas where clutter is less than the average, if these areas can be resolved. In practice, there are regions of land where the clutter might be considerably greater than the average and regions where it is considerably less than average. If the resolution of the radar is great enough to resolve the areas of lower clutter from the areas of greater clutter, targets within the clear areas can often be detected and tracked even though it might be predicted on the basis of the average clutter σ^0 that it would not be possible.

The ability of some radars to resolve the strong clutter regions into discrete areas, between which targets may be detected, is called *interclutter visibility*. It is difficult to establish a quantitative measure of this effect; but it has been suggested[42] that the improvement in target detectability can be approximated by the ratio of the average clutter level to the median clutter, which can be as much as 20 dB for a medium-resolution radar (for example,[64] one with a 2 μs pulse width and a 1.5° beamwidth). Thus a medium resolution radar might have a 20 dB advantage over low-resolution radars in the detection of targets in land clutter. The

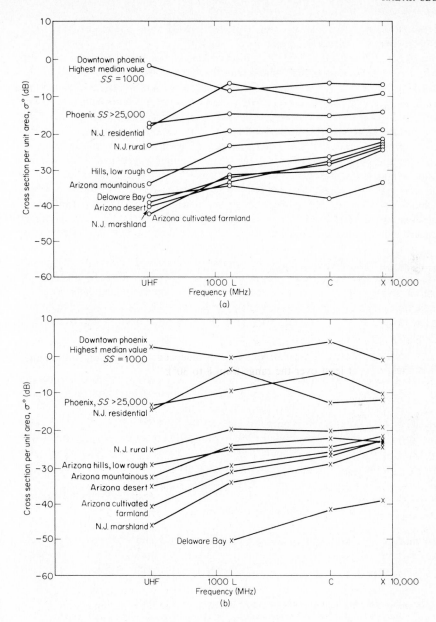

Figure 13.11 Median values of σ^0 as a function of frequency. Grazing angle $= 8°$. (*a*) Vertical polarization, (*b*) horizontal polarization. Two curves are shown for Phoenix: one is for a sample size of 1000 and corresponds to the downtown area, the other is for a sample size of 25,000 which represents a larger area and includes a smaller density of point targets than the smaller sample. (*From Daley, et al.*[37])

effectiveness of interclutter visibility has been exploited in MTI radar and in non-MTI radar with log-FTC receiver characteristics.

As an example of the effects of high resolution on land clutter, measurements have been reported with a horizontally polarized C-band radar having a 15-ns pulse width (3 m resolution) and a 1.5° beamwidth.[43] A rural region in England consisting of woods, fields, buildings, villages, small towns, and structures such as pylons, was examined at ranges from 7 to 11 km. It was found that 65 percent of the clutter exceeded 0.1 m², 18 percent exceeded 1 m², and less than 1 percent exceeded 10 m². The clutter with radar cross sections greater than 10 m² was limited to 6 m in length and was found to be associated with man-made objects such as electricity pylons and buildings. The separations between clutter larger than 10 m² was between 135 and 675 m. For clutter which was greater than 0.1 m², the patch sizes varied from 2 m to well over 300 m in length. The majority of these clutter-patch separations were less than 30 m but a few exceeded 110 m. For this terrain, it was concluded that if a minimum clutter patch separation of 75 m (0.5 μs) were necessary for target tracking, a 10 m² target could be tracked 99 percent of the time, but a 1 m² target could be tracked only 55 percent of the time.

In the above example of C-band data, the echoes from man-made objects were "point" targets of radar cross sections greater than 10 m². At the lower microwave frequencies, the echoes from the strong point-scatterers can be several orders of magnitude greater. They are often so large that they might not be completely removed by MTI. Thus it is not uncommon for MTI radar to have at its output many fixed point-scatterers that must be recognized so as not to be confused with desired targets.

Probability density function for land clutter. Much of the discussion regarding the statistics of sea clutter applies as well to land clutter, *except* that the physical mechanisms may be different. The statistical variation of σ^0 from homogeneous terrain such as deserts and some types of farmland can sometimes be described by the gaussian probability density function (pdf), so that the envelope of the radar receiver output is given by the Rayleigh pdf (Eq. 13.10).[44] Some urban areas, rural areas with buildings and silos, and mountainous terrain approximate the log-normal pdf (Eq. 13.11).[44,45] In general, neither the Rayleigh nor the log-normal pdf's accurately describe the statistics of real clutter, which usually lies between the two. The Weibull pdf (Eq. 13.12), which is intermediate between the Rayleigh and the log normal, has been suggested as applicable to land clutter. Table 13.2 lists examples of the Weibull parameter α (in Eq. 13.12) for several types of clutter and sea state.

Land clutter models. The wide variety and complexity of terrain and what grows or is built upon it, makes it difficult to formulate a satisfactory theoretical model to describe the backscatter from land. Adding to the difficulty are the effects of weather, time of day, and season, as

Table 13.2 Weibull clutter parameters[14]

Terrain/Sea state	Frequency	Beamwidth (deg)	Pulse width (μs)	Grazing angle	Weibull parameter
Rocky mountains	S	1.5	2	....	0.52
Wooded hills	L	1.7	3	~0.5	0.63
Forest	X	1.4	0.17	0.7	0.51–0.53
Cultivated land	X	1.4	0.17	0.7–5.0	0.61–2.0
Sea state 1	X	0.5	0.02	4.7	1.45
Sea state 3	K_u	5	0.1	1.0–30.0	1.16–1.78

well as the fact that radar waves can penetrate the surface and can be scattered from discontinuities underneath the surface.

Land clutter has been modeled as a Lambert surface with σ^0 varying as $\sin^2 \phi$, where ϕ = grazing angle[55]; as assemblies of spheres, hemispheres, hemicylinders[55,56]; as specular reflection from small flat-plate segments, or facets, which backscatter radar energy only when the facets are oriented perpendicular to the radar line of sight[57]; as a slightly rough surface (similar to the model used to describe sea clutter) to describe scattering from concrete and asphalt[40]; as an application of the Kirchhoff-Huygens principle which makes the assumption that the current flowing at each point in a locally curved surface is the same as would flow in the surface if it were flat and oriented tangent to the actual surface[58]; and as a model consisting of long, thin, lossy cylinders randomly distributed to describe certain types of vegetation such as grass, weeds, and flags.[40] None of these have been too successful as a general model; however, they do provide some guidance for the understanding of specific land clutter.[35]

13.6 DETECTION OF TARGETS IN LAND CLUTTER

Many of the techniques discussed in Sec. 13.4 for the detection of targets in sea clutter have application for land clutter as well. MTI and high resolution are two important examples. High-resolution results in a probability density function that is non-Rayleigh for both land and sea clutter, but the spatial distribution of land clutter provides interclutter visibility that permits the detection of targets that lie outside the separated clutter patches of large σ^0. The non-Rayleigh pdf, however, limits the utility of the log-FTC receiver and frequency agility in land clutter.

There are two clutter reduction techniques, peculiar to land clutter, which will be mentioned briefly in this section. These are the clutter fence and the Kalmus clutter filter.

Kalmus clutter filter.[59] This is a technique for extracting moving targets, such as a walking person or a slowly moving vehicle, from land clutter whose doppler-frequency spectrum masks that of the moving target. A moving target will produce a doppler-frequency shift that is either at a lower or a higher frequency than the transmitted signal. Vegetation or trees will have a back-and-forth motion due to the wind; hence, the doppler spectrum from such clutter will be distributed on both sides of the transmitted frequency, especially if observed over a sufficient period of time. By splitting the received spectrum into two halves about the transmitted carrier and subtracting the lower part from the upper part, the symmetrical clutter spectrum will cancel and the asymmetrical target spectrum will not. This method of suppressing the clutter relative to the target has been called the *Kalmus clutter filter*. The separation of the receding doppler signals from the approaching doppler signals may be accomplished with a technique similar to that shown in Fig. 3.8. Quantitative measurements of the target-to-clutter enhancement of this technique are not known, but it has been said[59] that "small targets moving in one direction could be easily detected in the presence of clutter signals exceeding the target return by many orders of magnitude." The effectiveness of this technique is limited by the degree of symmetry of the clutter spectrum, the need for a sufficient averaging time, and by the assumption that the clutter fluctuations are large compared to other factors that might affect the symmetry of the receiver spectrum.

Clutter fence. Reflections from nearby mountains and other large clutter can sometimes be of such a magnitude that it is not practical to completely suppress their undesirable effects by either MTI or range gating. One technique for reducing the magnitude of such large clutter seen by a fixed-site radar is to erect an electromagnetically opaque fence around the radar (or simply between the radar and the clutter source) to prevent the radar from viewing the clutter directly. The two-way isolation provided by a typical fence with a straight edge might be about 40 dB, where the isolation is given by the ratio of the clutter signal in the absence of a fence to that in the presence of the fence. The isolation is limited by the diffraction of the electromagnetic energy behind the fence. Greater isolation than that provided by a straight-edged fence can be had by incorporating two continuous slots near to, and parallel with, the upper edge of the fence to cancel a portion of the energy diffracted by the fence. The double-slot fence can increase the two-way isolation by 20 dB or more.[62]

A fence can suppress the clutter seen by the radar, but it produces other effects not always desirable. For example, it will limit the accuracy of elevation-angle measurement because of the blockage of the fence and the error caused by the energy diffracted by the fence. Energy diffracted by the fence also interferes with the direct path from the radar to cause multipath lobing of the radiation pattern in the angular region just above the fence. Radar energy backscattered from the fence can sometimes be large enough to damage the receiver front-end. In one design, the fence was tilted 15° away from the radar to prevent this from happening.[60]

Fences also can serve to prevent high power densities from existing in areas where personnel might be located.[61] A fence can also suppress the ground-reflected multipath signal that causes radiation-pattern lobing and elevation-angle errors, but the design of fences for this purpose is different from the design of a clutter-suppression fence.[61,63]

13.7 EFFECTS OF WEATHER ON RADAR

On the first page of the first chapter it was stated that radar could see through weather effects such as fog, rain, or snow. This is not strictly true in all cases and must be qualified, as the performance of some radars can be stongly affected by the presence of meteorological particles (hydrometeors). In general, radars at the lower frequencies are not bothered by meteorological or weather effects, but at the higher frequencies, weather echoes may be quite strong and mask the desired target signals just as any other unwanted clutter signal.

Whether the radar detection of meteorological particles such as rain, snow, or hail is a blessing or a curse depends upon one's point of view. Weather echoes are a nuisance to the radar operator whose job is to detect aircraft or ship targets. Echoes from a storm, for example, might mask or confuse the echoes from targets located at the same range and azimuth. On the other hand, radar return from rain, snow, or hail is of considerable importance in meteorological research and weather prediction. Radar may be used to give an up-to-date pattern of precipitation in the area around the radar. It is a simple and inexpensive gauge for measuring the precipitation over relatively large expanses. As a rain gauge it is quite useful to the hydrologist in determining the amount of water falling into a watershed during a given period of time. Radar has been used extensively for the study of thunderstorms, squall lines, tornadoes, hurricanes, and in cloud-physics research. Not only is radar useful as a means of studying the basic properties of these phenomena, but it may also be used for gathering the information needed for predicting the course of the weather. Hurricane tracking and tornado warning are examples of applications in which radar has proved its worth in the saving of life and property. Another important application of radar designed for the detection of weather echoes is in airborne weather-avoidance radars, whose function is to indicate to the aircraft pilot the dangerous storm areas to be avoided.

Radar equation for meteorological echoes.[69-71] The simple radar equation is

$$P_r = \frac{P_t G^2 \lambda^2 \sigma}{(4\pi)^3 R^4} \tag{13.14}$$

The symbols were defined in Sec. 1.2. In extending the radar equation to meteorological targets, it is assumed that rain, snow, hail, or other hydrometeors may be represented as a large number of independent scatterers of cross section σ_i located within the radar resolution cell. Let $\sum \sigma_i$ denote the average total backscatter cross section of the particles per unit of volume. The indicated summation of σ_i is carried out over the unit volume. The radar cross section may be expressed as $\sigma = V_m \sum \sigma_i$, where V_m is the volume of the radar resolution cell. The volume V_m occupied by a radar beam of vertical beamwidth ϕ_B, horizontal beamwidth θ_B, and a pulse of duration τ is approximately

$$V_m \approx \frac{\pi}{4}(R\theta_B)(R\phi_B)(c\tau/2) \tag{13.15}$$

where c = velocity of propagation. In the radar-meteorology literature, the radar pulse-extent h (in units of length) is often used instead of $c\tau$ in this equation. The factor $\pi/4$ is included to account for the elliptical shape of the beam area. In some instances this factor is omitted for convenience; however, radar meteorologists almost always include it since they are concerned with accurate measurement of rainfall rate using the radar equation. In the interest of even further accuracy, a correction is usually made to Eq. (13.15) to account for the fact that the effective volume of uniform rain illuminated by the two-way radar antenna pattern is less than that indicated when the half-power beamwidths are used to define the volume. Assuming a gaussian-shaped antenna pattern, the volume given by Eq. (13.15) must be reduced by a factor of 2 ln 2 to describe the equivalent volume that accounts for the echo power received by the two-way antenna pattern from distributed clutter.[69] Thus the radar equation of Eq. (13.14) can be written

$$\bar{P}_r = \frac{P_t G^2 \lambda^2 \theta_B \phi_B c\tau}{1024(\ln 2)\pi^2 R^2}\sum_i \sigma_i = \frac{P_t G\lambda^2 c\tau}{1024(\ln 2)R^2}\sum_i \sigma_i \tag{13.16}$$

In the above, the relationship $G = \pi^2/\theta_B \phi_B$ for a gaussian beamshape[69] was substituted. The bar over P_r denotes that the received power is averaged over many independent radar sweeps to smooth the signal fluctuations. This equation assumes that the volume of the antenna resolution cell is completely filled with uniform precipitation. If not, a correction must be made by introducing a dimensionless beam-filling factor ψ which is the fraction of the cross-sectional area of the beam intercepted by the region of scattering particles. It is difficult to estimate this correction. The resolution cell is not likely to be completely filled at long range or when the beam is viewing the edge of a precipitation cell. If the "bright band" (to be described later) is within the radar resolution cell, the precipitation also will not be uniform.

When the radar wavelength is large compared with the circumference of a scattering particle of diameter D (Rayleigh scattering region), the radar cross section is

$$\sigma_i = \frac{\pi^5 D^6}{\lambda^4}|K|^2 \tag{13.17}$$

where $|K|^2 = (\epsilon - 1)/(\epsilon + 2)$, and ϵ = dielectric constant of the scattering particles. The value of $|K|^2$ for water varies with temperature and wavelength. At 10°C and 10 cm wavelength, it is approximately 0.93. Its value for ice at all temperatures is about 0.197 and is independent of frequency in the centimeter-wavelength region. Substituting Eq. (13.17) into (13.16) yields

$$\bar{P}_r = \frac{\pi^5 P_t G c \tau}{1024(\ln 2)R^2\lambda^2} \, |K|^2 \sum_i D^6 \tag{13.18}$$

Since the particle diameter D appears as the sixth power, in any distribution of precipitation particles the small number of large drops will contribute most to the echo power.

Equation (13.18) does not include the attenuation of the radar energy by precipitation, which can be significant at the higher microwave frequencies and when accurate measurements are required. The two-way attenuation of the radar signal in traversing the range R and back is $\exp(-2\alpha R)$, where α is the one-way attenuation coefficient. If α is not a constant over the path R, the total attenuation must be expressed as the integrated value over the two-way path.

Scattering from rain. Equation (13.18), which applies for Rayleigh scattering, may be used as a basis for measuring with radar the sum of the sixth power of the raindrop diameters in a unit volume. The Rayleigh approximation is generally applicable below C band (5 cm wavelength) and, except for the heaviest rains, is a good approximation at X band (3 cm). Rayleigh scattering usually does not apply above X band. Another complication at frequencies above X band is that the attenuation due to precipitation precludes the making of quantitative measurements conveniently.

The sum of the sixth power of the diameters per unit volume in Eq. (13.18) is called Z, the *radar reflectivity factor*, or

$$Z = \sum_i D^6 \tag{13.19}$$

In this form Z has little significance for practical application. Experimental measurements, however, show that Z is related to the rainfall rate r by

$$Z = ar^b \tag{13.20}$$

where a and b are empirically determined constants. With this relationship the received echo power can be related to rainfall rate. A number of experimenters have attempted to determine the constants in Eq. (13.20), but considerable variability exists among the reported results.[70] Part of this is probably due to the difficulty in obtaining quantitative measurements and the variability of rain with time and from one location to another. One form of Eq. (13.20) that has been widely accepted is

$$Z = 200r^{1.6} \tag{13.21}$$

where Z is in mm^6/m^3 and r is in mm/h. This has been said to apply to stratiform rain. For orographic rain $Z = 31r^{1.71}$ and for thunderstorm rain $Z = 486r^{1.37}$. Thus a single expression need not be used, and the choice of a Z-r relationship can be made on the basis of the type of rain.[70] Substituting Eq. (13.21) into (13.18) with $|K|^2 = 0.93$ yields

$$\bar{P}_r = \frac{2.4P_t G\tau r^{1.6}}{R^2\lambda^2} \times 10^{-8} \tag{13.22}$$

where r is in mm/h, R and λ in meters, τ in seconds and P_t in watts. This indicates how the radar output can be made to measure rainfall.

The backscatter cross section per unit volume as a function of wavelength and rainfall

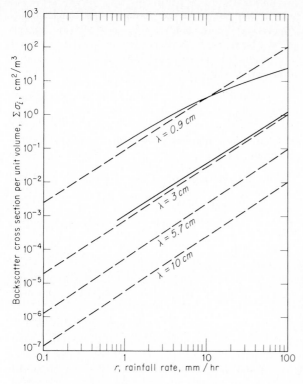

Figure 13.12 Exact (solid curves) and approximate (dashed curves) backscattering cross section per unit volume of rain at a temperature of 18°C. Exact computations obtained from F. T. Haddock, approximate curves based on the Rayleigh approximation. (*From Gunn and East,*[71] *Quart. J. Roy. Meteor. Soc.*)

rate is shown plotted in Fig. 13.12. The dashed lines are plotted by summing the Rayleigh cross section of Eq. (13.17) over unit volume and substituting Eq. (13.21) to give

$$\eta = \sum_i \sigma_i = 7f^4 r^{1.6} \times 10^{-12} \text{ m}^2/\text{m}^3 \qquad (13.23)$$

where f is the radar frequency in GHz and r the rainfall rate in mm/h. The solid curves are exact values computed by Haddock.[71,72] The Rayleigh scattering approximation is seen to be satisfactory over most of the frequency range of interest to radar.

The reflectivity factor Z of Eq. (13.19), which was defined as the sum of the sixth power of the particles' diameter per unit volume, was based on the assumption of Rayleigh scattering. When the scattering is not Rayleigh, a quantity similar to Z is defined, which is called the *equivalent radar reflectivity factor* Z_e given by[70,73]

$$Z_e = \lambda^4 \eta / (\pi^5 |K|^2) \qquad (13.24)$$

where η is the actual radar reflectivity, or backscatter cross section per unit volume, and $|K|^2$ is taken as 0.93.

Instead of the rainfall rate r, the intensity of precipitation is sometimes stated in terms of the dB reflectivity factor $Z = 200r^{1.6}$, or dBz = 10 log Z. A rainfall rate of 1 mm/h equals 23 dBz, 4 mm/h equals 33 dBz, and 16 mm/h equals 42 dBz. (This may be an incorrect usage of the precise definition of decibels as a power ratio, but it is the jargon used by the radar meteorologist.)

Scattering from snow. Dry snow particles are essentially ice crystals, either single or aggregated. The relationship between Z and snowfall rate r is as given by Eq. (13.20) for rain, but with different constants a and b. There have been less measurements of the Z-r relationship for snow than for rain, and there have been several different values proposed for the constants a and b. The following two expressions have been suggested

$$Z = 2000r^2 \qquad\qquad (13.25a)^{74}$$

$$Z = 1780r^{2.21} \qquad\qquad (13.25b)^{75}$$

Measurements show a correlation between surface temperature and the coefficient a of the $Z = ar^b$ relationship, which suggest the following[76]

$$Z = 1050r^2 \text{ for dry snow (ave temp.} < 0°\text{C)} \qquad (13.26a)$$

$$Z = 1600r^2 \text{ for wet snow (ave temp.} > 0°\text{C)} \qquad (13.26b)$$

A lower surface temperature results in a lower value of the coefficient a. Still another value that has been suggested is[77]

$$Z = 1000r^{1.6} \qquad\qquad (13.27)$$

There does not seem to be any agreed-upon value; the reader can take his pick. In all of the above, the snowfall rate r at the ground is in millimeters per hour of water measured when the snow is melted.

A radar is usually less affected by snow and ice than by rain because the factor $|K|^2$ in Eq. (13.18) is less for ice than for rain, and snowfall rates are generally less than rainfall rates.

Scattering from water-coated ice spheres. Moisture in the atmosphere at altitudes where the temperature is below freezing takes the form of ice crystals, snow, or hail. As these particles fall to the ground they melt and change to rain in the warmer environment of the lower altitudes. When this occurs, there is an increase in the radar backscatter since water particles reflect more strongly than ice. As the ice particles, snow, or hail begin to melt, they first become water-coated ice spheroids. At radar wavelengths, scattering and attenuation by water-coated ice spheroids the size of wet snowflakes is similar in magnitude to that of spheroidal water drops of the same size and shape. Even for comparatively thin coatings of water, the composite particle scatters nearly as well as a similar all-water particle.

Radar observations of light precipitation show a horizontal " bright band " at an altitude at which the temperature is just above 0°C. The measured reflectivity in the center of the bright band is typically about 12 to 15 dB greater than the reflectivity from the snow above it and about 6 to 10 dB greater than the rain below.[70] The center of the bright band is generally from about 100 to 400 m below the 0°C isotherm. Although the bright band is relatively thin, considerable attenuation can occur when radar observations are made through it at low elevations.

The bright band is due to changes in snow falling through the freezing level.[71] At the onset of melting the snow changes from flat or needle-shaped particles which scatter feebly to similarly shaped particles which, owing to a water coating, scatter relatively strongly. As melting progresses, the particles lose their extreme shapes, and their velocity of fall increases causing a decrease in the number of particles per unit volume and a reduction in the backscatter.

Scattering from clouds. Most cloud droplets do not exceed 100 μm in diameter (1 μm = 10^{-6} m); consequently Rayleigh scattering may be applied at radar frequencies for the prediction of cloud echoes. In Rayleigh scattering, the backscatter is proportional to the sixth power of the diameter, Eq. (13.17). Since the diameter of cloud droplets is about one-hundredth the diameter of raindrops, the echoes from fair-weather clouds are usually of little concern.

It is also possible to obtain weak echoes from a deep, intense fog at millimeter wavelengths, but at wavelengths of 3 cm and longer, echoes due to fog may generally be regarded as insignificant.

Attenuation by precipitation. In the frequency range for which Rayleigh scattering applies (particles small in size compared with the wavelength) the attenuation due to absorption is given by

$$\text{Attenuation (dB/km)} = 0.434 \left[\frac{\pi^2}{\lambda} \left(\sum_i D^3 \right) \text{Im} \left(-K \right) \right] \qquad (13.28)$$

where the summation is over 1 m³, D is the particle diameter in centimeters, λ is the wavelength in centimeters, Im $(-K)$ is the imaginary part of $-K$, and K is a factor which depends upon the dielectric constant of the particle. At a temperature of 10°C, the value of Im $(-K)$ for water is 0.00688 when the wavelength is 10 cm (S band) and 0.0247 for 3.2-cm wavelength (X band).[71] Equation (13.28) is a good approximation for rain attenuation at S-band or longer wavelengths. Since rain attenuation is usually small and unimportant at the longer wavelengths where this expression is valid, the simplicity offered by the Rayleigh scattering approximation is of limited use for predicting the attenuation through rain.

The computation of rain attenuation must therefore be based on the exact formulation for spheres as developed by Mie.[16] The results of such computations are shown in Fig. 13.13 as a function of the wavelength and the rainfall rate.

The attenuation produced by ice particles in the atmosphere, whether occurring as hail, snow, or ice-crystal clouds, is much less than that caused by rain of an equivalent rate of precipitation.[78] Gunn and East[71] state that the attenuation in snow is

$$\text{Attenuation at 0°C (dB/km)} = \frac{0.00349 r^{1.6}}{\lambda^4} + \frac{0.00224 r}{\lambda} \qquad (13.29)$$

where r = snowfall rate (mm/h of melted water content), and λ = wavelength, cm.

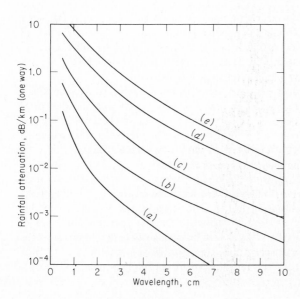

Figure 13.13 One-way attenuation (dB/km) in rain at a temperature of 18°C. (*a*) Drizzle—0.25 mm/h; (*b*) light rain—1 mm/h; (*c*) moderate rain—4 mm/h; (*d*) heavy rain—16 mm/h; (*e*) excessive rain—40 mm/h. In Washington, D.C., a rainfall rate of 0.25 mm/h is exceeded 450 h/yr, 1 mm/h is exceeded 200 h/yr, 4 mm/h is exceeded 60 h/yr, 16 mm/h is exceeded 8 h/yr and 40 mm/h is exceeded 2.2 h/yr.

13.8 DETECTION OF TARGETS IN PRECIPITATION

The chief effect of weather on radar performance is the backscatter, or clutter, from precipitation within the radar resolution cell. Attenuation by precipitation usually has little effect on the detection of targets except at frequencies above X band. Weather clutter, however, can be a serious factor in limiting the performance of microwave radars even at frequencies as low as L band. Good radar design must include methods for maintaining performance in the presence of precipitation.

The most effective method for reducing the effects of precipitation is to operate at a low frequency to take advantage of the significant decrease in backscatter from precipitation with decrease in frequency. The backscatter cross section of a particle in the Rayleigh region, as shown by Eq. (13.17) varies as the fourth power of the frequency. A radar operating at L band (1.3 GHz), for example, might experience about 34 dB less precipitation clutter than a radar at X band (9 GHz). Thus when precipitation is of concern, the lower frequencies are to be preferred.

At the higher microwave frequencies, precipitation clutter can be reduced by means of high resolution in range and in angle. Since rain is relatively uniform, the statistics of the backscatter are described by the Rayleigh probability density function, even when the radar is of high resolution. Therefore the problem of non-Rayleigh statistics with its deleterious effect on the false-alarm probability as occurs with high-resolution radar viewing sea clutter and land clutter (Secs. 13.3 and 13.5), does not usually occur with precipitation clutter.

Moving target indication (MTI) radar can provide some improvement in the detection of moving targets in precipitation *if* the MTI is properly designed to attenuate the doppler-shifted precipitation echoes. Unlike land clutter, storm clouds are not stationary and usually have a nonzero relative velocity that results in a doppler-frequency shift. The internal motions of the storm can also widen the spectrum of the precipitation clutter. At the higher microwave frequencies, where precipitation can be a bother, it is difficult to achieve effective MTI because of the reduced blind speeds (Sec. 4.2) and the increased spread of the clutter spectrum. The Moving Target Detector, described in Sec. 4.7, is an example of an MTI radar designed specifically to cope with the special problems of precipitation clutter.

Polarization. Raindrops are spherical, or nearly so, but aircraft are complex targets. Thus the backscattered energy from rain and aircraft will be affected differently by the polarization of the incident radar energy. Advantage can be taken of this difference to enhance the target-to-clutter ratio when the clutter background is precipitation. This is accomplished by utilizing a radar with circular polarization or with crossed linear polarization.

A circularly polarized wave incident on a spherical scatterer is reflected as a circularly polarized wave with the opposite sense of rotation and is rejected by the antenna that originally transmitted it. With a complex target such as an aircraft the reflected energy is more or less equally divided between the two senses of rotation so that some target-echo energy is accepted by the same radar antenna that transmitted the circularly polarized signal. This is the basis for target-to-clutter enhancement using circular polarization.

A circularly polarized wave is one in which the electric field vector rotates with constant amplitude about the axis of propagation at the radar frequency. To an observer looking in the direction of propagation, a clockwise-rotating electric field is known as *right-hand* circular polarization, and a counterclockwise rotation is known as *left-hand* circular polarization. Right-hand and left-hand circular polarization are said to be *orthogonal* polarizations since an antenna capable of accepting one will not accept the other. Similarly, horizontal and vertical linear polarizations are orthogonal. If the radar radiates one sense of circular polarized energy, it cannot accept the backscattered echo signal from a sphere or a plane sheet, since the direction of the polarization is reversed on reflection; that is, if right-hand circular polarization is transmitted, spherical raindrops reflect the energy as left-hand circular polarization. If the same antenna is used for both transmitting and receiving, the antenna is not responsive to the opposite sense of rotation and the echo energy will not appear at the receiver. A target such as an aircraft, however, will return some energy with the correct polarization as well as energy with the incorrect polarization. Energy incident on the aircraft may be returned after one "bounce," as from a plane sheet or a spherical surface; or it might make two or more bounces between various portions of the target (similar to a corner reflector) before being returned to the radar. On each bounce the direction of polarization rotation is reversed. Signals which make single reflections (or any odd number) will be rejected by the circularly polarized antenna that transmitted it, but those which make two reflections (or any even number) will be accepted. In addition to the mechanism of the double bounce, depolarization (or generation of the orthogonal component) can occur when the objects causing the scattering are not symmetrical. With most targets of interest, the scatterers are asymmetrical and backscatter energy appears in both orthogonal polarization components.

Raindrops may not always be perfect spheres. Their deviation from the symmetrical shape of a sphere will result in the reflected signal containing some energy in that polarization component accepted by the antenna. This limits the ability of circular polarization to reject precipitation clutter.

The rejection of rain echoes by a circularly polarized radar depends on the purity with which circular polarization can be generated by a practical antenna, as well as the deviation of the precipitation particles from a spherical shape.[117] The cancellation of the orthogonal polarization by an exceptionally well-designed, well-maintained antenna might be limited to about 40 dB.[79] To achieve 40 dB of cancellation the voltage ellipticity ratio of the antenna (ratio of the minor axis to the major axis of the polarization ellipse) must be 0.99, a difficult value to achieve. For 24 dB cancellation, the ellipticity ratio must be 0.94. Cancellations in excess of 30 dB have been achieved in light rain and in dry snow.[79] However, cancellations of only 15 dB or less are obtained from nonspherical precipitation such as heavy rain, from the melting layer, and from large wet snowflakes.[79,80,82] In some heavy rains and thunderstorms, the cancellation might be only 5 dB.[82]

It has been found that better cancellation of precipitation is obtained when the polarization is not quite circular. That is, there is a particular elliptical polarization which produces the best cancellation. The optimum elliptical polarization depends on the nature of the rain. This effect is due to the nonspherical raindrops which cause the phase shift and attenuation of the radar energy to be different depending on the direction of polarization. As the radar energy propagates through the rain, the differential phase shift and the differential attenuation results in the circular polarization being converted to elliptical polarization. By selecting the optimum elliptical polarization, it has been said that the cancellation in some regions of "heavy rain" might be increased by as much as 12 dB over that obtained with circular polarization.[81] However, the polarization that is optimum for one particular region might actually prove to be worse than the cancellation obtained with circular polarization in some other region. The optimum polarization thus depends on the distance traveled in rain, so that the antenna polarization needs to be continuously adjusted.[79] The greater the penetration into a rainstorm, the more elliptical will become the polarization of an originally circular polarized wave. Thus to maintain the improvement in rain cancellation with elliptical polarization, the polarization on receive should be made variable with range (time). With such adaptive polarization it has been suggested that an improvement in cancellation of 6 to 9 dB might be obtained consistently.[82]

Still another factor which reduces the effectiveness of circular polarization in rain clutter is the different reflection coefficients experienced by the horizontal and vertical polarization components on reflection from the surface of the land or the sea. This results in a change of polarization and a degradation of the rain cancellation.[82-86] It will be recalled from Sec. 12.2 that it is possible for a portion of the transmitted energy to arrive at the target via a surface-reflected path as well as by the direct path. Similarly, precipitation clutter can result from a surface-reflected path as well as the direct path. On reflection from the surface the vertically polarized component is attenuated more than the horizontal component, and it experiences a different phase shift (Fig. 12.3). The result is that the horizontally polarized component is cancelled to a greater extent than the vertical component so that the original circular polarization is converted to elliptical polarization with the vertical component predominating.[82] Assuming a 3.2° vertical beamwidth with cosecant-squared shaping up to 10°, theoretical calculations give the following maximum cancellation ratios as limited by surface reflections: 20.2 dB for sea, 23.6 dB for marsh, 27.2 dB for average land, and 34.1 dB for desert.[85] Other calculations for ground-mapping radars, such as those used for airport surface-traffic control, indicate significantly worse cancellation limitations. A 3° vertical beamwidth, pointed 0.7° below the horizon, with cosecant-squared shaping up to 30° will result in the cancellation ratio being limited to about 13 dB over water and 17 dB over moist soil.[86] To minimize the problem, narrow vertical beamwidths should be employed.

The radar cross section of aircraft targets is, in general, less with circular polarization than with linear polarization. Experimental measurements[87] indicate that when an aircraft is illuminated with one sense of circular polarization, the echo power on a statistical basis is divided more or less equally between right-hand and left-hand circular polarization. With linear polarization, the amount of energy converted to the orthogonal polarization is about 0.5 dB. Thus the net loss of aircraft-echo power on changing from linear to circular polarization is about 2.5 dB. In some cases, however, a greater value of radar cross section may be obtained with circular polarization rather than with linear.[79]

An alternative to circular polarization is to transmit any linear polarization and receive on the orthogonal linear polarization.[79,81] Spherical raindrops will give no return in the orthogonal channel if no cross-polarization distortion takes place in the radar or the propagation medium. There is some evidence to indicate that crossed linear polarization may give better rain rejection than conventional circular polarization.

Log-FTC receiver. A receiver with a logarithmic input-output characteristic followed by a high-pass filter (fast-time-constant, or FTC, circuit), will provide a constant false-alarm rate (CFAR) at the output when the input clutter or noise is described by the Rayleigh probability density function. CFAR is important since, without it, clutter can excite the radar display and obscure the presence of desired targets even if the targets are of greater strength than the clutter. CFAR prevents obscuration of detectable targets by weaker clutter by maintaining the clutter output from the receiver at a constant value well below the saturation level of the display. This is especially important in track-while-scan, or automatic detection and tracking (ADT) systems, where even small increases in background clutter could result in excessive false alarms that overload the capacity of the tracking computer. As with any CFAR, the constant false-alarm rate is maintained at the expense of the probability of detection. With log-FTC there is no improvement in target-to-clutter ratio and there is no subclutter visibility. The log-FTC, or similar CFAR, is necessary, however, in order to extract the target information contained in the radar signal that otherwise would be undetected if displayed along with the clutter, or which would be lost because of computer overload in an automatic system.

Although one of the first applications considered for the log-FTC receiver was the detection of targets in sea clutter,[88,89] it is probably more useful for operation in precipitation clutter. Such clutter is more likely to be Rayleigh than sea clutter (Sec. 13.3) or land clutter (Sec. 13.5) that often exhibit non-Rayleigh characteristics, especially with high-resolution radar. The log-FTC has sometimes been called *weather fix*.

The Rayleigh probability density function as was given in Sec. 13.3 is

$$p(v) = \frac{2v}{\sigma^2} \exp\left(-\frac{v^2}{\sigma^2}\right) \qquad v > 0 \qquad (13.30)$$

where σ^2 is the mean square value of v. The Rayleigh pdf has the property that the rms amplitude of the fluctuations about the mean (denoted by δv_{in}) is proportional to the mean $\bar{v}_{in}$, or $\delta v_{in} = k\bar{v}_{in}$. A logarithmic receiver has the characteristic

$$v_{out} = a \log b v_{in} \qquad (13.31)$$

The slope of the logarithmic receiver characteristic at $\bar{v}_{in}$ is

$$\frac{\Delta v_{out}}{\Delta v_{in}} = \frac{a}{\bar{v}_{in}} \qquad (13.32)$$

If the input clutter fluctuations δv_{in} are small compared to the total range of the logarithmic characteristic, the output fluctuations δv_{out} are approximately

$$\delta v_{out} = \text{slope} \times \delta v_{in} = \frac{a}{\bar{v}_{in}} \delta v_{in} = ak \qquad (13.33)$$

Thus the output fluctuations are constant, independent of the input mean.

Although the output fluctuations about the mean are constant, the output mean is not. A high-pass filter removes the mean value of the output, leaving the fluctuation of the clutter at a constant level on the display. The high-pass filter is equivalent to a differentiation, or to a circuit with a fast time-constant (FTC). The noise or clutter fluctuations that appear at the output of a logarithmic receiver are not symmetrical since the large amplitudes are suppressed due to the nature of the logarithmic characteristic. To make the output more like that of a linear receiver, the log-FTC may be followed by an amplifier with the inverse of the logarithmic characteristic (antilog). This restores the contrast of the display and eliminates the loss in detectability associated with the logarithmic characteristic.

A true logarithmic characteristic cannot be maintained down to zero input since $v_{out} \rightarrow -\infty$ as $v_{in} \rightarrow 0$. At some point the receiver characteristic must deviate from logarithmic and go through the origin. The practical logarithmic receiver will have a law given by $v_{out} = a \log (1 + b v_{in})$. The receiver characteristic is linear at low signal levels and logarithmic at large signals. This is called a *lin-log receiver*. The logarithmic characteristic must be maintained to about 20 dB below the rms noise level.[88] A variation of the log-FTC is the log-CFAR receiver in which the order of 10 resolution cells containing clutter are averaged by a narrow-band video filter with a symmetrical impulse response.[90]

Although the logarithmic receiver acts similarly to an automatic STC, it does not suppress properly, as does STC, the nearby clutter echoes which enter via the sidelobes. STC turns down the gain at close ranges, thereby reducing sidelobe clutter signals. Clutter large enough to appear in the sidelobes of a logarithmic receiver might not be suppressed and may confuse the radar display. For this reason STC and logarithmic receivers are sometimes used together. The STC action should be in the RF portion of the receiver rather than in the logarithmic amplifier.

13.9 ANGEL ECHOES

Radar echoes can be obtained from regions of the atmosphere where no apparent reflecting sources seem to exist. These have been called by various names, but they are commonly called *ghosts* or *angels*. They can take several different forms and have been attributed to various causes. There are two general classes of angel echoes: *dot* angels, which are point targets due to birds and insects, and *distributed* angels, which have substantial horizontal or vertical extent and are due to inhomogeneities of the refractive index of the atmosphere. Birds and insects in substantial number can also appear as distributed angels, and can have a degrading effect on radar. Since they are moving clutter to an MTI radar they are difficult to remove by doppler filtering. Sensitivity time control (STC) has proven a satisfactory method, in many cases, for reducing the effect of such clutter. Operation at UHF can reduce the backscatter from insects and, to some extent, birds because of the fourth-power relationship between cross section and frequency of a scatterer in the Rayleigh region (Eq. 13.17). Inhomogeneities of the atmospheric index of refraction generally do not produce strong enough backscatter to be a serious source of clutter to most radars.

Birds. Probably the most prominent source of angels is birds. Although the radar cross section of a single bird is small compared with that of an ordinary aircraft, the backscatter echo from a bird can be readily detected by many radars, especially at the shorter ranges, because of the inverse-fourth-power variation of echo signal with range. If, for example, the cross section of a bird the size of a sea gull were 0.01 m^2, it would produce as large an echo signal at a range of 10 nmi as would a 100 m^2 radar cross-section target at 100 nmi. When birds travel in flocks, the total cross section can be significantly greater than that of a single bird. Because the radar screen collapses a relatively large volume of space onto a small radar screen, the display can appear cluttered with bird echoes even though only a few birds can be seen by visual examination of the surrounding area. If there were, on the average, only one bird per square mile, more than 300 echoes would be displayed on the PPI within a 10-mile radius from the radar. This represents a significant amount of clutter. It has been said that as few as eight birds per square mile can completely blank a PPI screen. Increased echoes from birds are to be expected during migratory periods (spring and fall) and during those times of day when bird activity is large (sunrise and sunset).

Birds typically fly at speeds of from 15 to 40 knots,[92] and some can fly 50 knots or greater. These speeds are usually too high to be completely rejected by most microwave MTI radars. Most birds fly at altitudes below about 2500 m, with peak numbers between 300 and 1200 m, or even lower.[95,96]

Table 13.3 gives some examples of the radar cross sections of birds taken at three frequencies with vertical polarization.[91] The largest values occur at S band. Other examples of cross section are given in Fig. 13.14, which plots the average radar cross section as a function of the weight of the bird. The solid circles are the averages over a $\pm 20°$ sector around the broadside aspect.[92] The x's are the average of the $\pm 20°$ head-on and $\pm 20°$ tail-on aspects.[92,93] (Note that the values given for the pigeon in Fig. 13.10 differ from those given in Table 13.3. The two sets of values are from different sources.)

The radar cross section of birds does not show a simple wavelength or size dependence.[91] There are resonant effects, as illustrated by the measured cross section of a $2\frac{1}{2}$-lb duck at UHF being nearly twice that of a $4\frac{1}{4}$-lb duck.[94] (The median value of the $4\frac{1}{4}$-lb duck head-on was 600 cm^2, and 24 cm^2 tail-on.) The backscatter from birds fluctuates over quite large values with the maximum and minimum differing at times by more than two orders of magnitude.[91] Thus it is difficult to describe the radar cross section by a single value. It should properly be

Table 13.3 Radar cross sections of birds[91]

Bird	Frequency band	Mean radar cross section (cm^2)	Median radar cross section (cm^2)
Grackle	X	16	6.9
	S	25	12
	UHF	0.57	0.45
Sparrow	X	1.6	0.8
	S	14	11
	UHF	0.02	0.02
Pigeon	X	15	6.4
	S	80	32
	UHF	11	8.0

described statistically. There is some evidence to indicate that the probability density function for the radar cross section of (or received power from) a single bird in flight is log-normal. The mean-to-median ratio of the cross section, which is a measure of the amount of fluctuation in the cross section, is found to be independent of the magnitude of the radar cross section but is a function of the physical size of the bird relative to the radar wavelength.[91] Thus measurements of the mean-to-median ratio might be used to determine the size of the bird being observed. The fluctuations in the radar cross section have been attributed to the relative motions between the various parts of the bird and to changes in aspect, as well as to the wing-beat frequency. For example, spectral measurements of a small nocturnal migrant (a pipit) showed a wing-beat frequency of 15.8 Hz; and a migrating rapit (a honey buzzard) gave a frequency of 3.2 Hz.[92] A mallard produced a frequency of 6.5 Hz, plus harmonics at 13 and

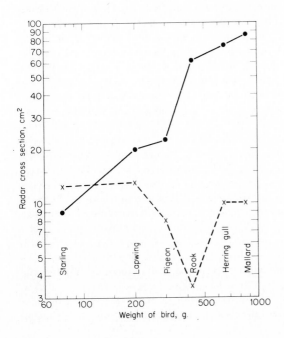

Figure 13.14 Radar cross section of birds (with closed wings) at S band. Vertical polarization. Solid circles apply to average values $\pm20°$ around broadside, x's apply to the average of the $\pm20°$ sector about the head and the $\pm20°$ sector about the tail. (*After Houghton and Smith.*[93])

19.5 Hz.[97] The wing-beat frequency f in hertz and the length l of the wing in millimeters are found[98] to be related by $fl^{0.827} = 572$. The term *bird activity modulation* (BAM) has been applied to the distinctive waveforms obtained from birds.[92] The spectral components of the BAM pattern of a bird in flight are said to be remarkably stable[97] and suited for determining identity.

Insects.[120] Even though they are small, insects are readily detected by radar, and in sufficient numbers can clutter the display and reduce the capability of a radar to detect desired targets. A radar cross section of 0.1 cm^2, which might correspond to an insect the size of a housefly at K_a band (8.6 cm wavelength),[99] can be detected at a range of about 11 nmi by a radar capable of seeing a 1 m^2 target at 200 nmi. Even modest insect concentrations (one insect in 10^4 m^3) can cause angel activity which can be classed as moderate.[100]

At X band, radar cross section measurements of a variety of insects range from 0.02 to 9.6 cm^2 with longitudinal polarization, and from 0.01 to 0.96 cm^2 for transverse polarization.[101] A desert locust[102] or a honeybee[101] might have a cross section of about 1 cm^2 at X band. At S band the cross section of a cabbage looper moth is about 2×10^{-3} cm^2, and for an adult field cricket it is 0.1 cm^2.[103] The cross section of insects below X band is approximately proportional to the fourth-power of the frequency.[104,105] Appreciable echoes are obtained only when insect body lengths are greater than a third of the radar wavelength. Insects observed broadside have echoes 10 to 1000 times greater than when viewed end-on.[116]

Heavy angel activity can be readily produced by insect concentrations that would scarcely cause visual awareness. Insects can be carried by the wind; therefore angels due to insects might be expected to have the velocity of the wind. Insect echoes are more likely to be found at the lower altitudes, near dawn and twilight. The majority of insects are incapable of flight at temperatures below 40°F (4.5°C) or above 90°F (32°C); consequently, large concentrations of insect angel echoes would not be expected outside this temperature range. As with clutter due to birds, sensitivity time control (STC) can reduce the adverse effects of clutter due to insects.

Clear-air turbulence. Some types of angel echoes which are nonpoint targets are attributed to atmospheric effects rather than to birds or insects. Early attempts to account for such echoes assumed specular reflection from gradients in the refractive index (dielectric constant) of the atmosphere. Theoretical calculations of the necessary gradients required to account for the observed angel reflections were excessively large compared to what is found in the real atmosphere. Instead, it is believed that the observed backscattering can be explained as reflections from atmospheric turbulence associated with inhomogeneities in the refractive index. These inhomogeneities might be caused by differences in the water vapor, temperature, and pressure. At low altitude, variations in the water-vapor pressure (humidity) are probably the dominant effect. At high altitude, there is little water vapor, and changes in temperature have the most effect on the refractive index.

Reflections from clear-air turbulence are thus a potential source of radar angel echoes. Turbulent motion is characterized by a variable velocity field and the presence of nonuniformities, or eddies, that produce mixing. The atmosphere can be assumed to be turbulent everywhere, but its intensity varies widely both in space and time. It is only when turbulence is concentrated into regions of greater or lesser intensity than its surroundings that it is of interest as an electromagnetic scatterer. There are at least two types of turbulent atmospheric formations that can result in angel activity. One is the *convective cell* or *plume* that occurs in the lower part of the atmosphere. The other is the *atmospheric horizontal layer* that can occur at any altitude. The latter is the form of turbulent effect that aircraft encounter at the higher altitudes. The convective cell is generally of greater turbulent intensity than the layer, but it does not extend over as wide an area of space and is not present as often as layers seem to be.

Convective cells, which are also called *thermals*, are the mechanism by which birds and gliders soar. When the surface of the earth is heated sufficiently so that the air becomes hotter than its surroundings, the buoyancy of the heated air will cause it to rise. If the source of heat is fixed, the rise of buoyant air is called a thermal plume. When the rising buoyant air is separated from the ground it may break away and form a freely floating thermal (convective cell), especially if there is a wind. Near the ground the convective cells might be a few tens of meters in diameter. As they rise, they grow in size and can reach a diameter of 1 to 3 km. They rise until they lose buoyancy or until they reach a level where condensation takes place. The tops of the cells might be at altitudes from 1 to 2 km. An individual cell might have a life of from 15 to 30 minutes. The cells can drift with the wind and align themselves in "streets." Much of the knowledge of the behavior of convective cells has been obtained from radar measurements.[70,105-108]

An atmospheric layer is a stratum within which the mean vertical gradient and/or the variance of refractive index are much greater than elsewhere. Layers may be from a few meters to more than a hundred meters in vertical thickness and might extend in the horizontal from about one kilometer to several tens of kilometers. Layers have been observed from altitudes of about 0.3 to more than 22 km, with the greatest number appearing in the vicinity of 1 to 2 km.[107,109] A typical layer associated with a subsidence inversion might exhibit a decrease of about $20N$ units in a thickness of 30 to 50 m[110] [where $N = (n - 1)10^6$, n = refractive index].

The echoes from clear-air turbulence are quite weak and are seen only by high-power radar. The basic theory for scattering from homogeneous and isotropic turbulent media was first given by Tatarski.[111,112] The scattering mechanism from turbulent media is similar to Bragg scatter in that a radar of wavelength λ scatters from that particular component of the turbulence with eddy sizes equal to $\lambda/2$. The volume reflectivity, or radar cross section (m^2) per cubic volume, from a turbulent medium is

$$\eta = 0.38 C_n^2 \lambda^{-1/3} \tag{13.34}$$

where C_n^2, the structure constant, represents a measure of the intensity of the refractive-index fluctuations, and λ is the radar wavelength. At altitudes of several hundred meters, values of C_n^2 are between 10^{-9} and 10^{-11} m$^{-2/3}$, which correspond to a volume reflectivity of about 0.82×10^{-9} to 0.82×10^{-11} m^{-1} at S band ($\lambda = 10$ cm). This is quite low as can be seen by comparison of the volume reflectivity for rain in Fig. 13.12. At 10 km altitude C_n^2 is approximately 10^{-14} m$^{-2/3}$. An S-band radar with one degree beamwidth and 1 μs pulsewidth viewing a turbulent medium with $\eta = 10^{-10}$ m^{-1} yields a radar cross section at 10 km of about 3×10^{-4} m^2. Thus, angel echoes from clear-air turbulence are not likely to bother most radars.

Other angel echoes. It has been suggested that other forms of meteorological angels are the mantle-shaped echoes (inverted U- or V-shape) associated with the upper surfaces of cumulus clouds, and echoes believed to be produced by the boundary surfaces between differentially moistened surface air over adjacent cold and warm water.[113] Radar also can detect the passage of an invisible sea breeze as it moves toward the shore.[70,100] Occasionally radar echoes may be obtained from large mineral or organic particles carried into the air by heavy winds or thunderstorms.[100] Echoes have also been received from the vicinity of forest fires and from the smoke plumes of dump fires.[100] The reflectivity of smoke particles is too small to account for these reflections, but the echoes might be due to the numerous large particles and debris sometimes present in the air above the fires. The heat from the fire also might cause atmospheric turbulence which is detectable by radar. " Ring echoes " have been observed on PPI displays that start at a point and form a rapidly expanding ring.[100] After one ring grows to a diameter of several miles, a second ring forms. Other rings can form similarly. They expand at velocities ranging from 20 to 30 knots and can attain diameters of 30 km or more. These ring-angels are associated with birds flying away from roosting areas. Angels can also be caused by second-time-around echoes or large signals that enter the radar via the antenna sidelobes.

REFERENCES

1. Skolnik, M. I.: Sea Echo, chap. 26 of "Radar Handbook," M. I. Skolnik (ed.) McGraw-Hill Book Company, New York, 1970.
2. Guinard, N. W., J. T. Ransone, Jr., and J. C. Daley: Variation of the NRCS of the Sea with Increasing Roughness, *J. Geophys. Res.*, vol. 76, pp. 1525–1538, Feb. 20, 1971.
3. Daley, J. C., J. T. Ransone, Jr., and J. A. Burkett: Radar Sea Return—JOSS I, *Naval Research Laboratory Report* 7268, Washington, D.C., May 11, 1971.
4. Daley, J. C., J. T. Ransone, Jr., J. A. Burkett, and J. R. Duncan: Upwind-Downwind-Crosswind Sea-Clutter Measurements, *Naval Research Laboratory Report* 6881, Washington, D.C., Apr. 14, 1969.
5. Daley, J. C., J. T. Ransone, Jr., and W. T. Davis: Radar Sea Return—JOSS II, *Naval Research Laboratory Report* 7534, Washington, D.C., Feb. 21, 1973.
6. Moskowitz, L. I.: The Wave Spectrum and Windspeed as Descriptors of the Ocean Surface, *Naval Research Laboratory Report* 7626, Washington, D.C., Oct. 30, 1973.
7. Macdonald, F. C.: Characteristics of Radar Sea Clutter, pt 1—Persistent Target-Like Echoes in Sea Clutter, *Naval Research Laboratory Report* 4902, Washington, D.C., Mar. 19, 1957.
8. Lewis, B. L., and I. D. Olin: Some Recent Observations of Sea Spikes, *International Conference RADAR-77*, pp. 115–119, Oct. 25–28, 1977, IEE (London) Conference Publication no. 155.
9. Long, M. W.: "Radar Reflectivity of Land and Sea," Lexington Books, D. C. Heath and Co., Lexington, Mass., 1975.
10. Lewis, B. L., J. P. Hansen, I. D. Olin, and V. Cavaleri: High-Resolution Radar Scattering Characteristics of a Disturbed Sea Surface and Floating Debris, *Naval Resarch Laboratory* Report 8131, Washington, D.C., July 29, 1977.
11. Williams, P. D. L.: Limitations of Radar Techniques for the Detection of Small Surface Targets in Clutter, *The Radio and Electronic Engineer*, vol. 45, pp. 379–389, August, 1975.
12. Trunk, G. V., and S. F. George: Detection of Targets in Non-Gaussian Sea Clutter, *IEEE Trans.*, vol. AES-6, pp. 620–628, September, 1970.
13. Trunk, G. V.: Radar Properties of Non-Rayleigh Sea Clutter, *IEEE Trans.*, vol. AES-8, pp. 196–204, March, 1972.
14. Schleher, D. C.: Radar Detection in Weibull Clutter, *IEEE Trans.*, vol. AES-12, pp. 736–743, November, 1976.
15. Fay, F. A., J. Clarke, and R. S. Peters: Weibull Distribution Applies to Sea Clutter, *International Conference RADAR-77*, pp. 101–104, IEE Conference Publication no. 155.
16. Kerr, D. E. (ed.): "Propagation of Short Radio Waves," MIT Radiation Laboratory Series, vol. 13, McGraw-Hill Book Company, New York, 1951.
17. Goldstein, H.: Frequency Dependence of the Properties of Sea Echo, *Phys. Rev.*, vol. 70, pp. 938–946, Dec. 1 and 15, 1946.
18. Katzin, M.: On the Mechanisms of Radar Sea Clutter, *Proc. IRE*, vol. 45, pp. 44–54, January, 1957.
19. Wright, J. W.: A New Model for Sea Clutter, *IEEE Trans.*, vol. AP-16, pp. 217–223, March, 1968.
20. Guinard, N. W., and J. C. Daley: An Experimental Study of a Sea Clutter Model, *Proc. IEEE*, vol. 58, pp. 543–550, April, 1970.
21. Croney, J.: Civil Marine Radar, chap. 31 of "Radar Handbook," M. I. Skolnik (ed.), McGraw-Hill Book Company, New York, 1970.
22. Paoli, L.: Sea Clutter Cancellation & Attenuation in Modern Marine Radars, *Rivista Tecnia Selenia*, (Rome, Italy), vol. 3, no. 4, pp. 1–9, 1976.
23. Croney, J.: Improved Radar Visibility of Small Targets in Sea Clutter, *The Radio and Electronic Engineer*, vol. 32, pp. 135–148, September, 1966.

24. Croney, J., and A. Woroncow: Dependence of Sea Clutter Decorrelation Improvements Upon Wave Height, *IEE Int. Conf. on Advances in Marine Navigational Aids*, July 25–27, 1972, IEE (London) Conference Publication no. 87, pp. 53–59.
25. Valenzuela, G. R., and M. B. Laing: Study of Doppler Spectra of Radar Sea Echo, *J. Geophys. Res.*, vol. 75, pp. 551–563, Jan. 20, 1970.
26. Trunk, G. V.: Trimmed-Mean Detector for Noncoherent Distributions, *Naval Research Laboratory Report* 6997, Washington, D.C., Dec. 11, 1969.
27. Trunk, G. V.: Further Results on the Detection of Targets in Non-Gaussian Sea Clutter, *IEEE Trans.*, vol. AES-7, pp. 553–556, May, 1971.
28. Schleher, D. C.: Radar Detection in Log-Normal Clutter, *IEEE 1975 International Radar Conference*, Arlington, Va., Apr. 21–23, 1975, pp. 262–267.
29. Jakeman, E., and P. N. Pusey: A Model for Non-Rayleigh Sea Echo, *IEEE Trans.*, vol. AP-24, pp. 806–814, November, 1976.
30. Croney, J.: Clutter and Its Reduction on Shipborne Radars, *International Conference on Radar— Present and Future*, Oct. 23–25, 1973, IEE (London) Conference Publication no. 105, pp. 213–220.
31. Croney, J., A. Woroncow, and B. R. Gladman: Further Observation on the Detection of Small Targets in Sea Clutter, *The Radio and Electronic Engineer*, vol. 45, pp. 105–115, March, 1975.
32. Croney, J.: Clutter on Radar Displays, *Wireless Engr.*, vol. 33, pp. 83–96, April, 1956.
33. Wylie, F. J.: "The Use of Radar at Sea," Hollis and Carter, Ltd., London, 1968.
34. Harrison, A.: Marine Radar Today—A Review, *The Radio and Electronic Engineer*, vol. 47, pp. 177–183, April, 1977.
35. Moore, R. K.: Ground Echo, chap. 25 of "Radar Handbook," M. I. Skolnik (ed.), McGraw-Hill Book Co., New York, 1970.
36. Barton, D. K.: "Radars, vol. 5, Radar Clutter," Artech House, Inc., Dedham, Mass., 1975. (A collection of 38 reprints on sea, land, and atmospheric clutter.)
37. Daley, J. C., W. T. Davis, J. R. Duncan, and M. B. Laing: NRL Terrain Clutter Study, Phase II, *Naval Research Laboratory Report* 6749, Washington, D.C., Oct. 21, 1968.
38. Katz, I., and L. M. Spetner: Polarization and Depression-Angle Dependence of Radar Terrain Return, *J. Res. Nat. Bur. Stds.*, vol. 64D, pp. 483–486, September/October, 1960.
39. Krason, H., and G. Randig: Terrain Backscattering Characteristics at Low Grazing Angles for *X*- and *S*-Band, *Proc. IEEE*, vol. 54, pp. 1964–1965, December, 1966.
40. Cosgriff, R. L., W. H. Peake, and R. C. Taylor: Terrain Scattering Properties for Sensor System Design (Terrain Handbook II), *Engineering Experiment Station Bulletin no. 181*, The Ohio State University, Columbus, Ohio, May, 1960. (Reprinted in ref. 36.)
41. Dyer, F. B., N. C. Currie, and M. S. Applegate: Radar Backscatter from Land, Sea, Rain, and Snow at Millimeter Wavelengths, *International Conference RADAR-77*, pp. 559–563, 25–28 Oct., 1977, IEE (London) Conference Publication no. 155.
42. Barton, D. K., and Shrader, W. W.: Interclutter Visibility in MTI Systems, *IEEE Eascon Record*, 1969, pp. 294–297.
43. Rigden, C. J.: High Resolution Land Clutter Characteristics, IEE Conf. Publ. No. 105, *Radar— Present and Future*, London, Oct. 23–25, 1973, pp. 227–232.
44. Valenzuela, G. R., and M. B. Laing: Point-Scatterer Formulation of Terrain Clutter Statistics, *Naval Research Laboratory Report* 7459, Sept. 27, 1972, Washington, D.C.
45. Warden, M. P.: An Experimental Study of Some Clutter Characteristics, *AGARD Conf. Proc. no. 66 on Advanced Radar Systems*, November, 1970.
46. Ulaby, F. T., W. H. Stiles, L. F. Dellwig, and B. C. Hanson: Experiments on the Radar Backscatter of Snow, *IEEE Trans.*, vol. GE-15, pp. 185–189, October, 1977.
47. Ringwalt, D. L., and F. C. MacDonald: Terrain Clutter Measurements in the Far North, *Report of NRL Progress*, Naval Research Laboratory, Washington, D.C., pp. 9–14, December, 1956.
48. Ulaby, F. T.: Radar Response to Vegetation, *IEEE Trans.*, vol. AP-23, pp. 36–45, January, 1975.
49. Ulaby, F. T., T. F. Bush, and P. P. Batlivala: Radar Response to Vegetation II: 8–18 GHz Band, *IEEE Trans.*, vol. AP-23, pp. 608–618, September, 1975.
50. Ulaby, F. T., and P. P. Batlivala: Diurnal Variations of Radar Backscatter from a Vegetation Canopy, *IEEE Trans.*, vol. AP-24, pp. 11–17, January, 1976.
51. Bush, T. F., and F. T. Ulaby: Radar Return from a Continuous Vegetation Canopy, *IEEE Trans.*, vol. AP-24, pp. 269–276, May, 1976.
52. Ulaby, F. T., and T. F. Bush: Corn Growth as Monitored by Radar, *IEEE Trans.*, vol. AP-24, pp. 819–828, November, 1976.
53. deLoor, G. P., A. A. Jurriens, and H. Gravestein: The Radar Backscatter from Selected Agricultural Crops, *IEEE Trans.*, vol. GE-12, pp. 70–77, April, 1974.

54. Bush, T. F., F. T. Ulaby, and W. H. Peake: Variability in the Measurement of Radar Backscatter, *IEEE Trans.*, vol. AP-24, pp. 896–899, November, 1976.
55. Clapp, R. E.: A Theoretical and Experimental Study of Radar Ground Return, *MIT Radiation Laboratory Rept.* 1024, April, 1946.
56. Twersky, V.: On Scattering and Reflection of Electromagnetic Waves by Rough Surfaces, *IRE Trans.*, vol. AP-5, pp. 81–90, January, 1957.
57. Katz, I., and L. M. Spetner: Two Statistical Models of Radar Return, *IRE Trans.*, vol. AP-8, pp. 242–246, May, 1960.
58. Beckmann, P., and A. Spizzichino: "The Scattering of Electromagnetic Waves from Rough Surfaces," The Macmillan Company, New York, 1963.
59. Kalmus, H. P.: Doppler Wave Recognition with High Clutter Rejection, *IEEE Trans.*, vol. AES-3 Supplement, no. 6, pp. 334–339, November, 1967.
60. Ruze, J., F. I. Sheftman, and D. A. Cahlander: Radar Ground-Clutter Shields, *Proc. IEEE*, vol. 54, pp. 1171–1183, September, 1966.
61. Becker, J. E., and J.-C. Sureau: Control of Radar Site Environment by Use of Fences, *IEEE Trans.*, vol. AP-14, pp. 768–773, November, 1966.
62. Becker, J. E., and R. E. Millett: A Double-Slot Radar Fence for Increased Clutter Suppression, *IEEE Trans.*, vol. AP-16, pp. 103–108, January, 1968.
63. Albersheim, W. J.: Elevation Tracking Through Clutter Fences, *IEEE Trans.*, vol. AES-3, no. 6 (EASTCON Suppl.), pp. 366–373, November, 1967.
64. Shrader, W. W.: Moving Target Indication Radar, *IEEE NEREM 74 Record, pt 4: Radar Systems and Components*, Oct. 28–31, 1974, pp. 18–26, IEEE Catalog no. 74 CHO 934–0 NEREM.
65. Parashar, S. K., R. M. Haralick, R. K. Moore, and A. W. Biggs: Radar Scatterometer Discrimination of Sea-Ice Types, *IEEE Trans.*, vol. GE-15, pp. 83–87, April, 1977.
66. Williams, P. D. L.: Detection of Sea Ice Growlers by Radar, *IEE Conf. Publ. no. 105, Radar—Present and Future*, London, Oct. 23–25, 1973, pp. 239–244.
67. Waite, A. H., and S. J. Schmidt: Gross Errors in Height Indication from Pulsed Radar Altimeters Operating over Thick Ice or Snow, *Proc. IRE*, vol. 50, pp. 1515–1520, June, 1962.
68. Pilon, R. O., and C. G. Purves: Radar Imagery of Oil Slicks, *IEEE Trans.*, vol. AES-9, pp. 630–636, September, 1973.
69. Probert-Jones, J. R.: The Radar Equation in Meteorology, *Quart. J. Roy. Meteor. Soc.*, vol. 88, pp. 485–495, 1962.
70. Battan, L. J.: "Radar Observation of the Atmosphere," *Univ. of Chicago Press*, Chicago, Ill., 1973.
71. Gunn, K. L. S., and T. W. R. East: The Microwave Properties of Precipitation Particles, *Quart. J. Roy. Meteor. Soc.*, vol. 80, pp. 522–545, October, 1954.
72. Haddock, F. T.: Scattering and Attenuation of Microwave Radiation Through Rain, *Naval Research Laboratory*, Washington, D.C. (unpublished manuscript), 1948.
73. Smith, P. L., Jr., K. R. Hardy, and K. M. Glover: Applications of Radar to Meteorological Operations and Research, *Proc. IEEE*, vol. 62, pp. 724–745, June, 1974.
74. Gunn, K. L. S., and J. S. Marshall: The Distribution with Size of Aggregate Snowflakes, *J. Meteor.*, vol. 15, pp. 452–466, 1958.
75. Sekhon, R. S., and R. C. Srivastava: Snow Size Spectra and Radar Reflectivity, *J. Atmos. Sci.*, vol. 27, pp. 299–307, 1970.
76. Puhakka, T.: On the Dependence of the Z-R Relation on the Temperature in Snowfall, *Preprints 16th Radar Meteorology Conference*, Am. Meteor. Soc., Apr. 22–24, 1975, Houston, Texas, pp. 504–507.
77. Austin, P. M.: Radar Measurements of the Distribution of Precipitation in New England Storms, *Proc. 10th Weather Radar Conf.*, pp. 247–254, 1963.
78. Saxton, J. A.: The Influence of Atmospheric Conditions on Radar Performance, *J. Inst. Navigation (London)*, vol. 11, pp. 290–303, 1958.
79. Schneider, A. B., and P. D. L. Williams: Circular Polarization in Radars, *Radio and Electronic Engineer*, vol. 47, no. 1/2, pp. 11–29, January/February, 1976.
80. Offutt, W. B.: A Review of Circular Polarization as a Means of Precipitation Clutter Suppression and Examples, *Proc. Natl. Electronics Conf.* (Chicago), vol. 11, pp. 94–100, 1955.
81. Hendry, A., and G. C. McCormick: Deterioration of Circular-polarization Clutter Cancellation in Anisotropic Precipitation Media, *Electronics Letters*, vol. 10, no. 10, pp. 165–166, May 16, 1974.
82. Nathanson, F. E.: Adaptive Circular Polarization, *IEEE 1975 International Radar Conference*, Apr. 21–23, 1975, pp. 221–225.
83. White, W. D.: Circular Polarization Cuts Rain Clutter, *Electronics*, vol. 27, pp. 158–160, March, 1954.
84. McFee, R., and T. M. Maher: Effect of Surface Reflections on Rain Cancellation of Circularly Polarized Radars, *IRE Trans.*, vol. AP-7, pp. 199–201, April, 1959.

85. Beasley, E. W.: Effect of Surface Reflections on Rain Cancellation in Radars Using Circular Polarization, *Proc. IEEE*, vol. 54, pp. 2000–2001, December, 1966.

86. Kalafus, R. M.: Rain Cancellation Deterioration Due to Surface Reflections in Ground-Mapping Radars Using Circular Polarization, *IEEE Trans.*, vol. AP-23, pp. 269–271, March, 1975.

87. Gent, H., I. M. Hunter, and N. P. Robinson: Polarization of Radar Echoes, Including Aircraft, Precipitation, and Terrain, *Proc. IEE*, vol. 110, pp. 2139–2148, December, 1963.

88. Croney, J.: The Reduction of Sea and Rain Clutter in Marine Radars, *J. Inst. Navigation (London)*, vol. 7, pp. 175–180, 190–192, 1954.

89. Croney, J.: Clutter on Radar Displays, *Wireless Engr.*, vol. 33, pp. 83–96, April, 1956.

90. Taylor, J. W., Jr., and J. Mattern: Receivers, chap. 5 of *Radar Handbook*, M. I. Skolnik (ed.), McGraw-Hill Book Company, New York, 1970.

91. Konrad, T. G., J. J. Hicks, and E. B. Dobson: Radar Characteristics of Birds in Flight, *Science*, vol. 159, pp. 274–280, Jan. 19, 1968.

92. Houghton, E. W., F. Blackwell, and T. A. Wilmot: Bird Strike and the Radar Properties of Birds, *International Conf. on Radar—Present and Future*, Oct. 23–25, 1973, pp. 257–262, IEE Conference Pub. no. 105.

93. Houghton, E. W., and N. J. Smith: Radar Echoing Areas of Birds, *Royal Radar Establishment Memorandum* no. 2557, July, 1969, Malvern, England.

94. Blacksmith, P., Jr., and R. B. Mack: On Measuring the Radar Cross Sections of Ducks and Chickens, *Proc. IEEE*, vol. 53, p. 1125, August, 1965.

95. Flock, W. L.: Monitoring Bird Movements by Radar, *IEEE Spectrum*, vol. 5, pp. 62–66, June, 1968.

96. Eastwood, E.: " Radar Ornithology," Methuen & Co., Ltd., London, 1967.

97. Blackwell, F., and E. W. Houghton: Radar Tracking and Identification of Wild Duck During the Autumn Migration, *Proc. World Conf. on Bird Hazards to Aircraft, Canada*, pp. 359–376, 1969.

98. Flock, W. L., and J. L. Green: The Detection and Identification of Birds in Flight, Using Coherent and Noncoherent Radars, *Proc. IEEE*, vol. 62, pp. 745–753, June, 1974.

99. Tolbert, C. W., A. W. Straiton, and C. O. Britt: Phantom Radar Targets at Millimeter Radio Wavelengths, *IRE Trans.*, vol. AP-6, pp. 380–384, October, 1958.

100. Plank, V. G.: A Meteorological Study of Radar Angels, *U.S.A.F. Cambridge Research Center Geophys. Research Papers*, no. 52, July, 1956, AFCRC-TR-56-211, AD 98752.

101. Hajovsky, R. G., A. P. Deam, and A. H. LaGrone: Radar Reflections from Insects in the Lower Atmosphere, *IEEE Trans.*, vol. AP-14, pp. 224–227, March, 1966.

102. Riley, J. R.: Angular and Temporal Variations in the Radar Cross-Sections of Insects, *Proc. IEE*, vol. 120, pp. 1229–1232, October, 1973.

103. Richter, J. H., and D. R. Jensen: Radar Cross-Section Measurements of Insects, *Proc. IEEE*, vol. 61, pp. 143–144, January, 1973.

104. Glover, K. M., K. R. Hardy, T. G. Konrad, W. N. Sullivan, and A. S. Michaels: Radar Observations of Insects in Free Flight, *Science*, vol. 154, pp. 967–972, 1966.

105. Hardy, K. R., and I. Katz: Probing the Clear Atmosphere with High Power, High Resolution Radars, *Proc. IEEE*, vol. 57, pp. 468–480, April, 1969.

106. Konrad, T. G.: The Dynamics of the Convective Process in Clear Air as Seen by Radar, *J. Atm. Sci.*, vol. 27, pp. 1138–1147, November, 1970.

107. Gossard, E. E., D. R. Jensen, and J. H. Richter: An Analytical Study of Tropospheric Structure as Seen by High-Resolution Radar, *J. Atm. Sci.*, vol. 28, pp. 794–807, July, 1971.

108. Gossard, E. E., and W. H. Hooke: " Waves in the Atmosphere," Elsevier Scientific Publishing Co., New York, 1975.

109. Saxton, J. A., J. A. Lane, R. W. Meadows, and P. A. Matthews: Layer Structure of the Troposphere, *Proc. IEE*, vol. 111, pp. 275–283, February, 1964.

110. Lane, J. A.: Radar Echoes from Clear Air in Relation to Refractive-Index Variations in the Troposphere, *Proc. IEE*, vol. 116, pp. 1956–1660, October, 1969.

111. Tatarski, V. I.: " Wave Propagation in a Turbulent Medium," McGraw-Hill Book Co., New York, 1961.

112. Tatarski, V. I.: The Effects of the Turbulent Atmosphere on Wave Propagation, TT68-50464, National Technical Information Service, Springfield, VA, 1971.

113. Atlas, D.: Meteorological "Angel" Echoes, *J. Meteorol.*, vol. 16, pp. 6–11, February, 1959.

114. George, S. F.: The Detection of Nonfluctuating Targets in Log-Normal Clutter, *Naval Research Laboratory Report* 6796, Washington, D.C., Oct. 4, 1968.

115. Goldstein, G. B.: False-Alarm Regulation in Log-Normal and Weibull Clutter, *IEEE Trans.*, vol. AES-9, pp. 84–92, January, 1973.

116. Schaefer, G. W.: Radar Observations of Insect Flight, chap. 8, of " Insect Flight," R. C. Rainey (ed.), Blackwell Scientific Publications, London, 1976.

117. Peebles, P. Z., Jr.: Radar Rain Clutter Cancellation Bounds Using Circular Polarization, *IEEE 1975 International Radar Conference*, Apr. 21–23, 1975, Arlington, Va, pp. 210–214, IEEE Publication 75 CHO 938-1 AES.

118. Bass, F. G., I. M. Fuks, A. I. Kalmykov, I. E. Ostrovsky, and A. D. Rosenberg: Very High Frequency Radiowave Scattering by a Disturbed Sea Surace, *IEEE Trans.*, vol. AP-16, pp. 554–568, September, 1968.

119. Kalmykov, A. I., and V. V. Pustovoytenko: On Polarization Features of Radio Signals Scattered From the Sea Surface at Small Grazing Angles, *J. Geophys. Res.*, vol. 81, no. 12, pp. 1960–1964, Apr. 20, 1976.

120. Greneker, E. F., and M. A. Corbin: Radar Reflectivity of Airborne Insects, A Literature Survey, *Georgia Institute of Technology Radar and Instrumentation Laboratory*, Atlanta, Ga., sponsored by Western Cotton Research Laboratory, U.S. Dept. of Agriculture.

FOURTEEN

OTHER RADAR TOPICS

14.1 SYNTHETIC APERTURE RADAR[1-3]

A synthetic aperture radar (SAR) achieves high resolution in the cross-range dimension by taking advantage of the motion of the vehicle carrying the radar to synthesize the effect of a large antenna aperture. The imaging of the earth's surface by SAR to provide a maplike display can be applied to military reconnaissance, measurement of sea state and ocean wave conditions, geological and mineral explorations, and other remote sensing applications.[98]

The resolution in the cross-range dimension of a conventional antenna is

$$\delta_{cr} = R\theta_B \tag{14.1}$$

where R is the range and θ_B is the beamwidth. The narrower the beamwidth, the better the resolution (smaller δ_{cr}). There is a limit, however, to the minimum beamwidth of a practical microwave antenna because of the difficulty of achieving and maintaining the necessary mechanical and electrical tolerances (Secs. 7.8 and 8.8). If the antenna beamwidth were as small as $0.2°$, the resolution δ_{cr} at a range of 100 km would be about 350 m, which is far larger than the fraction of a meter resolution possible in the range coordinate with the use of pulse-compression radar. SAR permits the attainment of high resolution by using the motion of the vehicle to generate the antenna aperture *sequentially* rather than *simultaneously* as with a conventional array antenna. In this section, the radar may be thought of as carried by an aircraft, but similar arguments apply for satellites or other moving vehicles.

Figure 14.1 shows an aircraft traveling with a constant velocity v along a straight path. Its radar antenna is mounted so as to radiate in the direction perpendicular to the direction of motion. Such a radar is known as a *sidelooking radar*, or SLR. The x's in the figure represent the position of the radar antenna each time a pulse is transmitted. If the echo received at each position is stored and if the last n pulses are combined (added together), the effect will be similar to a linear-array antenna whose length is the distance traveled during the transmission of the n pulses. The "element" spacing of the synthesized antenna is equal to the distance traveled by the aircraft between pulse transmissions, or $d_c = vT_p = v/f_p$, where T_p is the pulse-repetition period and f_p is the pulse-repetition frequency.

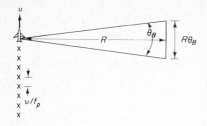

Figure 14.1 Geometry of the synthetic aperture radar traveling with a velocity v. The radar transmits a pulse at each position marked by an x. f_p = pulse repetition frequency.

Resolution of the SAR. The beamwidth of a conventional antenna of width D at a wavelength λ is

$$\theta_B = k\lambda/D \tag{14.2}$$

where k is a constant that depends on the shape of the current distribution across the aperture. (The constant k might vary from 0.9 to 1.3 or greater.) For convenience of analysis, take $k = 1$. Substituting Eq. (14.2) into (14.1) gives the cross-range resolution as

$$\delta_{cr} = R\lambda/D \tag{14.3}$$

The beamwidth of a synthetic aperture antenna of effective length L_e is similarly

$$\theta_s = k\lambda/2L_e \tag{14.4}$$

The factor 2 appears in the denominator because of the two-way propagation path from the antenna "element" to the target and back as compared with the one-way path of a conventional antenna. As a consequence of the two-way path the phase difference between the equally spaced elements of a synthetic array is twice that of a conventional array with the same spacing. (The terms synthetic array and synthetic aperture are used interchangeably here.) Again the factor k will be taken to be unity.

There are two fundamental limits to the maximum effective length of the synthetic aperture. One limit is determined by the width of the region illuminated at the range R by the real antenna. The length of the effective aperture L_e can be no greater than the width of the illuminated region as given by Eq. (14.3). Thus $L_e \le R\theta_B$. Note that the maximum effective length varies directly as the range. The other limit is determined by the far field of the synthetic aperture; i.e., by the need to restrict the aperture size so that the phase front can be considered as a plane wave. When this condition applies, the SAR is called *unfocused*. Figure 14.2 defines the maximum aperture of an unfocused SAR such that the difference between the minimum and maximum (two-way) paths is a quarter wavelength. From this geometry it can be derived that the effective length L_e of the synthesized antenna is $\sqrt{R\lambda}$, so that the cross-range resolution for the unfocused synthetic antenna is

$$\delta_{cr} = \sqrt{R\lambda}/2 \tag{14.5}$$

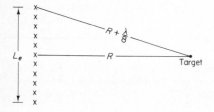

Figure 14.2 Geometry of the unfocused SAR.

The resolution of the unfocused SAR does not depend on the size of the real antenna. (In a sidelooking SAR the cross-range resolution is also called the *along-track* or *azimuth* resolution.)

The limit on resolution due to operation in the far field can be overcome by correcting the received signal for the curvature of the spherical wavefront experienced when the target is within the Fresnel region of the synthetic array. At each "element" of the synthetic array antenna a phase correction $\Delta\phi = 2\pi x^2/\lambda R$ is applied, where x is the distance from the center of the synthetic aperture. Note that a different correction must be applied for each range R. When this correction is applied at each element of the synthetic array, the antenna is said to be *focused* at a distance R and all the received echo signals from a target at that range are in phase. The angular resolution when the antenna is focused in the Fresnel region is equivalent to that in the far field. Hence, the cross-range resolution of the focused synthetic aperture using Eq. (14.4) with $L_e = R\lambda/D$ is

$$\delta_{cr} = R\theta_s = \frac{D}{2} \tag{14.6}$$

The resolution of the focused SAR is independent of the range and the wavelength, and depends solely on the dimension D of the real antenna.

An example comparing the resolution of the conventional antenna and the two types of synthetic aperture antennas is shown in Fig. 14.3.

It was mentioned that the factor of 2 in the denominator of Eq. (14.4) was a consequence of the relative phase shift between elements of the synthetic array antenna being due to the two-way propagation path, instead of the one-way propagation path as in the conventional array antenna. This results in the synthetic aperture radar having a two-way antenna pattern

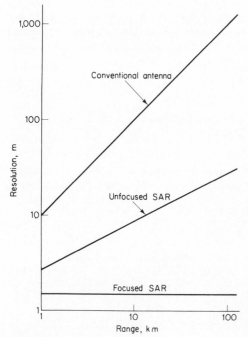

Figure 14.3 Comparison of the resolution of a synthetic aperture radar and a radar with a conventional antenna, assuming an X-band antenna with dimension $D = 3$ m.

equal to the one-way pattern of the conventional antenna of the same length, but with one-half the beamwidth. The two-way patterns with uniform weighting, assuming small angles, are approximately

$$\text{SAR} \rightarrow \frac{\sin\left[2\pi(L_e/\lambda)\sin\theta\right]}{2\pi(L_e/\lambda)\sin\theta} \qquad \text{real array} \rightarrow \frac{\sin^2\left[\pi(L/\lambda)\sin\theta\right]}{\left[\pi(L/\lambda)\sin\theta\right]^2}$$

Thus the two-way beamwidth of the SAR antenna is narrower than a real radar antenna of the same aperture size (if it could be built), but the sidelobes are not as low and do not drop off with increasing angle as fast. Two-way sidelobes of 13.2 dB in the SAR are not satisfactory for most purposes. Weighting of the received signals, similar to the weighting of the aperture illumination of a real antenna, is often applied to reduce the sidelobe levels.

Constraint on resolution and swath. Ambiguities can arise when signals are sampled, rather than continuous. As discussed in Sec. 2.10, the radar pulse repetition frequency (prf) must be low enough to avoid range ambiguities and multiple-time-around echoes. An additional constraint on the prf occurs in a synthetic aperture radar. The prf must be high enough to avoid angle ambiguities and image-foldover that results from grating lobes produced when the spacing between the elements of the synthetic array is too large. These two conflicting requirements on the prf of a SAR mean that the resolution and the coverage (swath) cannot be selected independently.

To avoid grating lobes in a phased-array antenna of isotropic radiating elements (with the main beam perpendicular to the aperture), the element spacing must be less than the wavelength λ (Sec. 8.2). A similar condition for a synthetic aperture antenna is that the distance traveled by the radar between pulse transmissions should be less than $\lambda/2$. (The factor of $\frac{1}{2}$ appears here for the same reason it was included in Eq. (14.3).) When a "directive" element-pattern is used the spacing necessary between elements to avoid grating lobes can be much greater than $\lambda/2$. In the SAR, the directive pattern of the real antenna can be considered as the element pattern of the synthetic array. By making the angular location of the first grating lobe coincide with the first null of the element pattern (that of the real antenna), grating lobes can be attenuated to a small value. The position of the first grating-lobe maximum of the synthetic array is

$$\theta_g = \frac{\lambda}{2d_e} = \frac{\lambda f_p}{2v} \tag{14.7}$$

where $d_e = v/f_p = $ spacing between elements of the synthetic array
$\quad v = $ velocity of vehicle carrying the radar
$\quad f_p = $ pulse repetition frequency

The position of the first null of the real antenna is approximately $\theta_n = \lambda/D$, where $D = $ width of the antenna. Since θ_g must be greater than or equal to θ_n to avoid grating lobes, the following condition is obtained

$$f_p \geq \frac{2v}{D} = \frac{v}{\delta_{cr}} \tag{14.8}$$

The right-hand portion of the equation applies for a focused SAR, since $\delta_{cr} = D/2$.

Combining the restriction on prf due to unambiguous range R_u with that of Eq. (14.8) yields

$$\frac{v}{\delta_{cr}} \leq f_p \leq \frac{c}{2R_u} \tag{14.9}$$

This leads to the condition

$$\frac{R_u}{\delta_{cr}} \leq \frac{c}{2v} \tag{14.10}$$

Thus the unambiguous range and the resolution cannot be selected independently of one another.

The condition given by Eq. (14.10) is optimistic in that there are at least two factors which result in the right-hand portion being reduced. One such condition results from the use of actual antenna patterns rather than idealized patterns. The unambiguous regions must be smaller than assumed in the above to allow for the fact that antenna patterns are not zero outside the mapped region.[4,16] For a uniform aperture illumination of the real antenna, Eq. (14.9) becomes[4]

$$1.53 \frac{v}{\delta_{cr}} \leq f_p \leq \frac{c}{1.53 \times 2R_u} \tag{14.11a}$$

or

$$\frac{R_u}{\delta_{cr}} \leq \frac{c}{4.7v} \tag{14.11b}$$

Essentially the same result is obtained for a cosine-weight aperture illumination. It has been suggested[4] that a practical procedure for estimating the required aperture and prf is to size the antenna such that the required swath and processed doppler bandwidth are within the 6-dB two-way antenna pattern, and the prf is such that the unambiguous time-delay and doppler intervals are within the 16-dB response.

There is another factor which reduces the right-hand side of Eqs. (14.10) or (14.11b) when optical processing or similar processing is used. As will be explained later, this increases the right-hand side of Eq. (14.8) by 2 so that Eq. (14.10) becomes

$$\frac{R_u}{\delta_{cr}} \leq \frac{c}{4v} \tag{14.12}$$

Similarly, the right-hand side of Eq. (14.11b) is also reduced by a factor of 2.

When a synthetic-aperture radar images the ground from an elevated platform, the unambiguous range can correspond to the distance between the forward edge and the far edge of the region to be mapped. This requires that the elevation beamwidth be tailored to illuminate only the swath S_w that is to be imaged by the radar. The swath S_w is often much smaller than the maximum range so that the prf can be increased to allow the umambiguous range R_u to encompass the distance $S_w \cos \psi$, where ψ is the grazing angle. Equation (14.12) becomes

$$\frac{S_w}{\delta_{cr}} \leq \frac{c}{4v \cos \psi} \tag{14.13}$$

Equation 14.11b would also be modified accordingly. The condition given by Eq. (14.13), and similar relations, represent an upper bound on the capabilities of an SAR to achieve a resolution δ_{cr} over a swath S_w.

Radar equation for SAR. Each particular application has its own form of the radar equation. The SAR has a particularly interesting form, especially when the relationships derived above are taken into account. In the literature on SAR, the radar equation is usually written with the signal-to-noise ratio (S/N) on the left-hand side. We start with the following

$$S/N = \frac{P_t A_e^2 \sigma n}{4\pi \lambda^2 k T_0 B F_n R^4} \tag{14.14}$$

where P_t = peak transmitted power
$\quad A_e$ = effective aperture of the real antenna
$\quad \sigma$ = target cross section
$\quad n$ = number of pulses integrated (coherently)
$\quad \lambda$ = wavelength
$\quad kT_0 = 4 \times 10^{-21}$ W/Hz
$\quad B$ = receiver bandwidth
$\quad F_n$ = receiver noise figure
$\quad R$ = range

This equation is modified by substituting $P_t = P_{av}/\tau f_p$, where P_{av} = average power, τ = pulse width $\approx 1/B$, and f_p = pulse repetition frequency. Also $n = f_p t_0$, where $t_0 = L_e/v$ = time required to generate the synthetic aperture whose length is given by $L_e = R\lambda/D$, v = velocity of the vehicle carrying the radar, and D = real-antenna dimension. For low grazing angle $\sigma = \sigma^0 \delta_{cr} \delta_r \sec \psi$ where σ^0 = radar cross section of the ground per unit area illuminated, δ_{cr} = along-track, or cross-range, resolution (equal to $D/2$ for a focused SAR), δ_r = range resolution, and ψ = grazing angle. Substituting the above relations into Eq. (14.14) yields

$$S/N = \frac{P_{av} A_e^2 \sigma^0 \delta_r \sec \psi}{8\pi \lambda k T_0 F_n v R^3} \tag{14.15}$$

This is essentially the same equation as given by Cutrona[2] and Harger.[3]

Equation (14.15) does not include the ambiguity constraints described previously. To use the equality of Eq. (14.13) in Eq. (14.15), write $A_e = \rho_a A$, where ρ_a = antenna aperture efficiency, and A = physical aperture area assumed to be rectangular of width D and height H. The height H must produce a vertical beamwidth which illuminates only the swath S_w that is to be covered. With such constraints, Eq. (14.15) becomes

$$S/N = \frac{2P_{av} \rho_a^2 \sigma^0 \delta_{cr} \delta_r}{\pi f k T_0 F_n R S_w \sin^2 \psi} \tag{14.16}$$

where f = radar frequency. This equation applies for pulse compression if δ_r is the resolution of the compressed pulse. It is seen from Eq. (14.16) that for a fixed signal-to-noise ratio, the average power must be increased if the resolution in range or azimuth are decreased, or if the range or swath are increased.

Equipment considerations. The synthetic-aperture radar requires a coherent reference signal and means for storing and processing the radar echoes. The coherent reference is necessary since an angle measurement is a measurement of phase from spatially separate positions. The effective length of the synthetic aperture can be limited by the stability of the transmitter or the receiver.

The heart of the SAR is the processor which must provide the proper amplitude and phase weights to the stored pulses, and sum them to obtain the image of the scene. Optical processors and digital processors have both been used.

Angular motions in yaw, roll, and pitch of the aircraft carrying the radar will cause the real antenna beam to point incorrectly. To avoid degradation of the SAR due to angular motions, the antenna must be stabilized.[1,18] Roll and pitch angles can be stabilized by means of gyroscopes. Yaw angle can be compensated by reference to a gyroscope or by " clutter-lock " in which the antenna position is adjusted so as to maintain a symmetrical doppler spectrum about zero frequency.

Atmospheric turbulence as well as deliberate maneuvers result in the aircraft trajectory deviating from a straight line. These deviations must be sensed and proper compensation applied to the received-signal phase so as to "straighten out" the synthetic antenna.[19] The required phase correction is a function of range with the more rapid corrections required at steep depression angles. Thus both motion compensation and antenna stabilization are necessary to achieve the resolution inherent in an SAR.

Optical processing.[1-3,92] In a synthetic aperture radar the coherent echo signals S_n from each range interval must be stored, without loss of phase, over the time interval required to form the synthetic aperture. An amplitude weighting W_n is usually applied to each of the signals to taper the "illumination" of the synthetic aperture so as to achieve lower sidelobes than given by a uniform illumination. In a focused SAR, a phase weighting $\exp(j\phi_n)$ also must be applied to compensate for the spherical wavefront when the scene to be imaged lies within the Fresnel region of the synthesized aperture. Then the signals at each range interval must be summed, as expressed by the operation $\sum_n S_n W_n \exp(j\phi_n)$. The processing is complicated by the fact that the number of pulses to be summed is proportional to the range. Furthermore, the phase correction in a focused SAR also depends on the range. Adding to the complication of processing is the large bandwidth and high information content of synthetic-aperture radar. Optical processing has proved to be well suited to the needs of SAR, except that it is seldom done in real time. The radar output is usually stored photographically and processed later on the ground.

In optical processing, the electrical signals at the radar output are converted to optical images on film. The weighting, filtering, and summation of signals are accomplished with the proper optical lenses and transparencies. Optical processing is basically two dimensional so that processing in the range coordinate is possible with the same apparatus. The output of the optical processor is a maplike photographic film of the terrain, as in Fig. 14.4. The SAR produces images without the slant-range distortion that occurs with the photographic camera.

The radar output is stored on film as depicted in Fig. 14.5. The output of the radar receiver is applied to the "z axis" of a cathode-ray tube so as to produce an intensity-modulated vertical sweep. The vertical deflection of the CRT is in synchronism with the range sweep of the radar. A lens focuses the CRT output on a moving strip of film whose rate of travel is proportional to the speed of the vehicle carrying the radar. The vertical dimension on the film represents the range and the horizontal dimension represents the cross-range, or along-track, coordinate. Returns from a single point-target fall on a horizontal line. Since film records light intensity, the sign of the signal amplitude will be lost if the signal alone is recorded on the film. To retain the sign of the amplitude, a dc bias equal to or greater than the most negative signal excursion is added to the signal. (It is this bias, plus the offset frequency described below, which causes the factor of 2 to be inserted in Eq. (14.12).)

The signal received from a single scatterer will appear as a varying doppler-frequency due to the relative motion as the antenna scans by. The doppler frequency is initially of high value, decreases to zero beat, and then increases again. When this signal is recorded on film the resulting intensity modulation is a one-dimensional hologram, or Fresnel zone-plate, and has focusing properties. If the film were illuminated by a coherent, collimated beam of light like that from a laser, a focused target-image will be obtained. Figure 14.6 shows the basic arrangement of an optical processor. The film contains the superposition of the holographic signals from all the individual scatterers illuminated by the radar. It is sometimes called the *phase history* or *signal history*. The image that results by illuminating the film is focused on a tilted plane since the curvature of the wavefront is proportional to the range. (This is somewhat like an optical analog of the radar itself.) Since it is undesirable to operate with the image focused

(a)

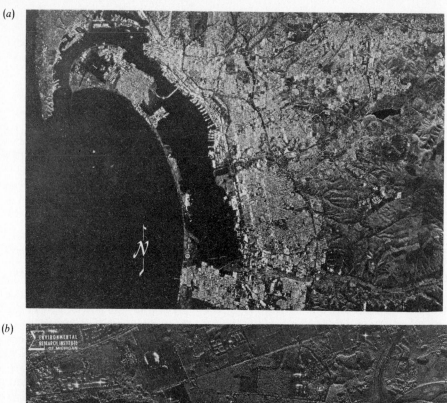

(b)

Figure 14.4 (a) Synthetic aperture radar image of San Diego, California obtained with the X-band GEMS (Goodyear Electronic Mapping System) radar. Resolution = 12 m; aircraft altitude = 40,000 ft (12 km). This is a mosaic covering approximately 19 by 32 km. (*Courtesy Goodyear Aerospace and Litton Aero Service.*) (b) A bend in the Huron river (center) just east of Ann Arbor, Michigan. Note football stadium (with lights) top left center, and large orchard in upper right center. South is in the up direction. Resolution is 5 ft (1.5 m) in azimuth and 7 ft (2.1 m) in range. (*Courtesy Environmental Research Institute of Michigan.*)

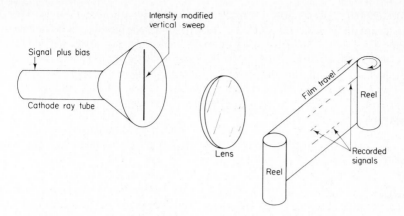

Figure 14.5 Recording of SAR signal on photographic film with the aid of a CRT.

on a tilted plane, the different phase-front curvatures are corrected and the image is erected by insertion of a conical lens placed at the phase-history film located in plane P_1. This erect image has its focus at infinity. The one-dimensional hologramlike signals recorded along the horizontal dimension on the phase-history film provide focusing in the horizontal plane. A cylindrical lens is used to focus the image in the vertical plane. The front focal-plane of the cylindrical lens coincides with that of the phase-history film. (A lens has the property that a Fourier transform relation exists between the amplitude distribution of the illumination at the front and back focal planes.[5])

With the conical and cylindrical lens inserted, the image is erect and is simultaneously focused in the range and cross-range dimensions, but it lies at infinity. To bring the image to some conveniently located plane, a spherical lens is inserted. A vertical slit in an opaque screen placed in the focal plane of the spherical lens displays all ranges in a particular direction. The slit is offset from the origin because of the dc bias that had to be added to the signal at the input to the CRT in order to record on film. If the slit were placed on the optical axis, the presence of the dc bias will cause a degraded image because the energy from the bias overlaps the desired image. This is avoided by placing the input signal on a residual carrier

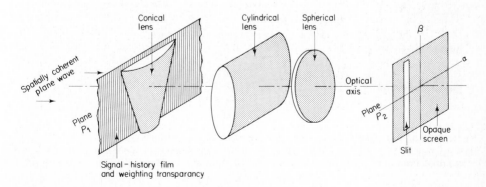

Figure 14.6 Optical processor for synthetic aperture radar.

frequency, called the *offset frequency*, along with the dc bias. (The offset frequency is equal to the doppler frequency shift associated with a scatterer located at the edge of the real antenna beam.) The result, which is similar to reconstructing the image of a hologram, is that the desired real image can then be separated from the energy associated with the bias and the virtual image. The slit is displaced from the optical axis at the place where the offset frequency focuses the real image. A recording film in the output plane P_2, when moved with a speed proportional to that of the vehicle carrying the radar, produces a map of the scene originally viewed by the radar.

If amplitude weighting of the synthetic aperture is desired to reduce the sidelobes, a shaded transparency with uniform phase thickness can be inserted adjacent to the data film in plane P_1.

Digital processing. The recirculating delay-line integrator, range-gated filter bank, and storage-tube integrator all have been used in the past for the electronic processing of unfocused SAR.[1] The optical processor was used for focused systems. Digital processing is also practical and has the advantage of real-time operation as compared with optical processing. The digital processor has I and Q channels with a high-speed A/D convertor in each channel. The sampling rate of the A/D is determined by the radar signal bandwidth, and the number of bits is set by the dynamic range desired. Motion compensation for azimuth and range slip (range walk) can be applied, as well as phase corrections for focusing. Semiconductor devices are used for memory and arithmetic. When the swath is significantly less than the unambiguous range, a buffer can be inserted after the A/D convertors to read the data out at a slower speed so as to process at a lower data rate than if the entire unambiguous range had to be imaged. Digital processing also allows the use of a nonlinear sweep to convert slant range to ground distance so as to make distances correct on the image.

Doppler-frequency model. The synthetic aperture radar may be considered as a vector summation of synthetic-array elements (which is the model generally taken in this section), or it may be considered in terms of doppler filtering. In fact, it was originally conceived by Carl Wiley of Goodyear Aircraft Corporation in 1951 as a doppler-filtering process rather than as a synthetic antenna. The two models are sometimes used interchangeably, depending upon which describes more clearly a particular effect. The basic difference between the two is that the coordinate system moves with the radar in the doppler description, while in the synthetic-aperture model the coordinate system is fixed to the ground.[9]

Consider the geometry of Fig. 14.1 in which an aircraft with a sidelooking SAR travels at a velocity v. (The effect of the elevation angle is neglected in this simple analysis but must be included in more precise considerations.) When a point scatterer just enters the forward edge of the beam, it has a doppler frequency $2(v/\lambda) \cos (\theta_B/2)$. For small beamwidths, the doppler frequency decreases linearly as the point-scatterer position changes relative to the radar and goes to zero when the scatterer is at the center of the beam, after which the frequency increases. The time-varying doppler frequency can be shown to be approximately[1]

$$f_d(t) = \frac{2v^2 |t - t_0|}{\lambda R} \tag{14.17}$$

where t_0 is the time when the point scatterer is at the center of the beam, and R is the range to the scatterer at that time. The total width Δf_d of this doppler signal as the scatterer progresses through the beam is $2v^2 T_d/\lambda R$, where T_d is the time during which the scatterer is within the beam. By analogy to the linear-FM pulse-compression waveform, the linearly varying doppler signal can be passed through a matched filter to produce a pulse of duration $1/\Delta f_d$. This corresponds to a cross-range, or along-track, resolution of $v/\Delta f_d$, or

$$\delta_{cr} = \frac{\lambda R}{2v T_d} = \frac{\lambda R}{2L_s} = D/2 \tag{14.18}$$

which is what was obtained in Eq. (14.6) from the synthetic aperture model. (In the above, the relations $v T_d = L_s = R\lambda/D$ were employed.)

Range resolution. In mapping or imaging, it is usually desired to have the range resolution equal to the along-track, or cross-range resolution. When optical processing is employed, the linear FM, or chirp, pulse-compression waveform readily lends itself to the same type of processing as described for the cross-range signal.[7] The form of the linear FM signal in the range dimension is similar to the form of the signal in the cross-range dimension. The linear FM signal has a phase that varies as t^2 (t = time), and the cross-range signal has a phase that varies as x^2 (x = distance along the ground track). Just as the phase history of the cross-range signal records on film as a Fresnel zone-plate, which then has a focusing action when illuminated by coherent light, so does the signal of the linear FM waveform also record as a Fresnel zone-plate. The range and the cross-range recorded signals may be treated as orthogonal aspects of a single two-dimensional filtering operation. Although the usual SAR optical processing provides dechirping (matched filtering) of the linear FM pulse-compression waveform, the chirp modulation of the transmitter is accomplished as in other pulse-compression radars.

Other aspects of SAR. The SAR has been described as a *side-looking radar* with the antenna beam directed perpendicular to the path of the vehicle such that the doppler frequency of scatterers illuminated by the center of the antenna beam is zero. It is possible, however, to operate with the antenna beam pointed either forward or aft of broadside. This is called the *squint mode.* The signal processor must be modified to account for the average doppler frequency not being zero. Recorders and displays must be designed to account for the geometry of the offset beam. Compensation might also be necessary for " range walk " which is the result of the target " walking" through one or more range-resolution cells during the time of observation. The achievable cross-range, or along-track, resolution worsens as the squint angle increases from broadside ($\delta_{cr} \approx 1/\sin \theta$, where θ = angle between aircraft heading and squinted antenna beam direction).

The squint mode produces a strip map just as does the sidelooking SAR. The *doppler beam-sharpening mode* is used with a circularly scanning antenna and a normal PPI display. As the squint angle of the doppler beam-sharpening mode varies, the integration time changes to keep the resolution constant.[1] The along-track resolution for either the squint mode or the doppler beam-sharpening mode is

$$\delta_a = \frac{\lambda R}{2v T \sin \theta} \tag{14.19}$$

where the symbols have been defined previously. As θ decreases in the doppler beam-sharpening mode, T is made to increase. As θ approaches 0 degrees, the required integration time T becomes too large to provide beam sharpening so that improved resolution is not obtained in a finite angular sector about the direction of the vehicle velocity.

In the *telescope,* or *spotlight, mode* very high resolution of a particular patch on the ground is obtained by steering the real antenna aperture to dwell longer than is possible with a fixed antenna. Theoretically, resolutions better than $D/2$ can be obtained. Another benefit of this mode is that the scene to be imaged is observed over a range of incidence angles which averages the speckle and makes a smoother image.

Since the synthetic-aperture radar is coherent, the image produced will have speckle; i.e., there will be constructive and destructive interference which results in a "breakup" of distributed scatterers. To reduce the effect of speckle and make a more "filled-in" image, the same scene can be viewed from different aspects or at different frequencies, or both, and the several images superimposed. The multiple looks can be obtained as in the searchlight mode by dwelling with a positionable antenna on the same area. Another approach, applicable with a fixed antenna, is to not use the full synthetic antenna length L_e to achieve a resolution δ_{cr}, but to break up the synthetic length into m subsections and look at the scene from slightly different aspects each with a resolution $m\,\delta_{cr}$. The m independent images are then combined noncoherently into a single image. It has been suggested that the noncoherent combining of images of lesser resolution produces a better image with less speckle than a single image of greater resolution.[11,17]

If the SAR is mapping a scene in which there are moving objects such as cars or trains, the resulting image will be smeared in range and shifted in the along-track dimension due to the radial motion of the object. The image also will be defocused in the along-track dimension because of radial acceleration or cross-range velocity of the object.[8] These effects can cause a distortion and displacement of the moving-target image. A reduction in signal strength will also occur if the target doppler shift is sufficiently large to be outside the passband of the along-track-dimension processor. When the SAR is located on a satellite, the effect of earth rotation must be properly compensated since it results in the earth's surface appearing as a moving target.[10,91]

The difference in signal characteristics from a moving target as compared to that from a stationary target can be used as a basis for combining AMTI (airborne moving target indication) with SAR.[8] The moving-target doppler shift can be detected when it exceeds the clutter doppler-spectrum width. It is also possible to use a dual sidelooking antenna pattern and detect the phase difference between the two mutually coherent observations of a target separated in time. The phase-detection AMTI method allows slowly moving targets to be detected when the aircraft velocity is large.

Ground relief, such as due to hills or mountains, stands out on SAR imagery because of the shadows they form, especially when viewed at low grazing angles. Shadows help emphasize topographic and geologic features that permit a trained interpreter to learn much about the nature of the terrain. (Optical photographs are generally taken from large grazing angles and do not show the same shadowing effects produced by radar.) Stereoscopic techniques can also be applied to SAR to provide a three-dimensional image of the terrain. This is accomplished by viewing the same terrain from different aspects, such as by flying two separate flight paths to produce images from two different elevation angles.[12] It is also possible to generate a pair of stereo images on a single pass using two vertical fan beams at different azimuth angles, or one fan beam and a conical beam.[13] The stereo processing of SAR images has been successfully used for mineral exploration.

At short ranges it is possible to use conventional noncoherent sidelooking radar with a large antenna aperture and pulse compression to obtain high-resolution terrain imaging. The images obtained with conventional noncoherent radar are generally different than those of coherent SAR in that they are less speckled.

Inverse SAR. Instead of moving a radar relative to a stationary object, it is possible to generate an image by moving the object relative to a stationary radar. Imaging with a stationary radar and moving target is called *inverse SAR* or *range-doppler imaging*. It has been applied

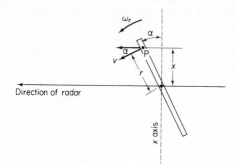

Figure 14.7 Inverse SAR, or range-doppler imaging, of a rotating object.

to the imaging of rotating targets,[14] especially the moon and planets.[15] Figure 14.7 shows a rigid body rotating at an angular speed of ω_r radians per second, with the axis of rotation normal to the paper. The doppler frequency associated with a point P on the body located a distance r from the axis of rotation is

$$f_d = \frac{2v \cos \alpha}{\lambda} = \frac{2\omega_r r \cos \alpha}{\lambda} = \frac{2\omega_r x}{\lambda} \qquad (14.20)$$

where v = velocity of point P. From this equation, the resolution in the x dimension can be written

$$\Delta x = \Delta f_d \frac{\lambda}{2\omega_r} = \frac{1}{T} \frac{\lambda}{2(\Delta\theta/T)} = \frac{\lambda}{2\,\Delta\theta} \qquad (14.21)$$

where T is the coherent integration time equal to $1/(\Delta f_d)$, and $\Delta\theta$ is the angle through which the body rotates during the time T. It does not take much of an angular rotation to produce good resolution at microwave frequencies. The contours of constant doppler are perpendicular to the x axis, and contours of constant range lie parallel to the x axis.

If the angle of target rotation over which the doppler is observed is too short, the doppler spectrum is broad and the resolution is low. Increasing the observation time narrows the spectrum and the resolution will increase. However, if the time of observation is too long, the doppler frequency of a point P on the rotating target will not be constant and the doppler spectrum will broaden with a consequent reduction in resolution. That is, there is an apparent target acceleration which limits the resolution. The result is that there will be an optimum time of observation, or aspect angle change, when attempting to image a target with the inverse SAR technique.[101]

14.2 HF OVER-THE-HORIZON RADAR[20]

Frequencies at VHF or lower are seldom used for conventional radar applications because of their narrow bandwidths, wide beamwidths, high ambient noise levels, and the potential interference from other users of the crowded electromagnetic spectrum. In spite of these limitations, the HF region of the spectrum is of special interest for radar because of its unique property of allowing propagation to long distances beyond the curvature of the earth by means of refraction from the ionosphere. A single refraction allows radar ranges to be extended to almost 4000 km. The targets of interest to HF over-the-horizon (OTH) radar are the same as those of interest to microwave radar and include aircraft, missiles, and ships. In addition, the long wavelengths characteristic of HF radar also provide distinctive information regarding the sea, as well as aurora, meteors, and land features. (Although the HF band is officially defined as extending from 3 to 30 MHz, for radar usage the lower frequency limit might lie just above the broadcast band, and the upper limit can extend to 40 MHz or more.)

The ability to see a target at long range by means of ionospheric refraction depends on the nature of the ionosphere (the density of electron concentration) and the radar frequency, as well as the normal parameters that enter into the radar range equation. Unlike conventional microwave radar, the specific frequency to be used by an OTH radar is a function of the range that is desired and the character of the ionosphere. Since the ionosphere varies with time of day, season, and solar activity, the optimum radar frequency will vary widely. Such radars must therefore be capable of operating over a wide portion of the HF band, as much as three octaves (4 to 32 MHz for example).[21] The ionosphere often consists of more than one refracting region. The highest region, denoted F_2, and the most important for HF propagation, is at altitudes of from 230 to 400 km. It provides the greatest ranges for a single refraction and can support the highest usable frequencies. The F_1 region, from about 180 to 240 km, is observed only during the day and is more pronounced during the summer than the winter. The E region, which lies between 100 and 140 km, can also support refraction. At these heights there can appear at times patches of high-density ionization called sporadic E which, when available, can be quite effective in providing stable propagation. The multiple refracting regions give rise to multipath propagation which can result in degraded performance because of the simultaneous arrival of radar energy at the target via more than one propagation path, each with different time delays. The effects of multipath can be reduced by the proper selection of frequency and by use of narrow elevation beamwidths which allow the energy to travel to the target via only a single path. The presence of the various refracting regions with different ionization densities at different altitudes requires good frequency management if an OTH radar is to operate with reliability.

The minimum range to which HF energy can be propagated by ionospheric refraction is determined by the lowest frequency at which the radar can operate. A nominal value for the minimum range (or skip distance) is about 1000 km.

The backscatter from the earth's surface is generally many orders of magnitude larger than the echo from desired moving targets. Thus HF radar must employ some form of doppler processing such as MTI, pulse-doppler, FM-CW, or CW radar to separate desired moving targets from clutter. The equivalent of a high-pass filter must be used to detect moving aircraft and missiles and reject stationary surface clutter. The detection of ships requires more sophisticated processing since the relatively low velocity of ships produces doppler-frequency shifts comparable to those of the sea (which is also a moving target). Even though the radar cross section of ships is often greater than that of aircraft, longer observation times are required to provide sufficient resolution in doppler frequency.

Character of OTH radar. The factors affecting the design of an HF OTH radar are slightly different than those affecting conventional microwave radar. This is illustrated by the simple radar equation commonly used in OTH radar analysis, which is

$$R^4 = \frac{P_{av} \, G_t \, G_r \, \lambda^2 \sigma F_p^2 \, T_c}{(4\pi)^3 N_0 (S/N) L_s} \tag{14.22}$$

where R = range

$\quad P_{av}$ = average power

$\quad G_t$ = transmitting antenna gain

G_r = receiving antenna gain
λ = wavelength
σ = target cross section
F_p = factor to account for the one-way propagation effects
T_c = coherent processing time
N_0 = receiver noise power per unit bandwidth
(S/N) = signal-to-noise (power) ratio
L_s = system losses

The transmitting and receiving antenna gains are shown separately in Eq. (14.22) since it is sometimes convenient in OTH radar to have separate antennas for these two functions. If narrow beamwidths are to be achieved, the radar antenna must be a physically large phased array. A one-degree beamwidth, for example, requires an aperture of about 1200 m at a frequency of 15 MHz. Since the transmitting antenna must handle high power, it is more costly to obtain than large receiving apertures. In one type of OTH radar design, a (relatively) small transmitting antenna with broad azimuth beamwidth is used along with a large receiving aperture consisting of a number of narrow contiguous beams covering the angle illuminated by the wide transmitting beam. Thus the complexity of a large transmitting antenna is traded for a number of parallel receiving channels. It is also possible to utilize a common aperture for both transmit and receive, with equal transmit and receive beamwidths. Duplexers would be required as in a microwave phased-array radar. The antenna might also support several simultaneous, independent radar beams; or multiple beams can be generated sequentially (a pulse burst). A narrow, steerable beam in the elevation plane is desired for OTH radar in order to avoid multipath propagation and to concentrate the energy at the desired range. However, the large vertical apertures to achieve such a capability are quite costly, so that HF OTH radars seldom have as large a vertical aperture, or as narrow an elevation beamwidth, as might be desired.

The propagation factor (F_p), receiver noise (N_0), and coherent integration time (T_c) of Eq. (14.22) represent major differences between the HF OTH radar and the conventional microwave radar. The factor F_p includes the energy loss along the ionospheric path, the mismatch loss due to a change in polarization caused by the ionosphere, ionospheric focusing gain or loss, and losses due to the dynamic nature of the path.[22,23] The receiver noise N_0 includes the ambient noise radiated by natural sources (chiefly lightning discharges from around the world) as well as the combined interference from the many users of the HF band. It is the latter which generally determines system sensitivity at HF. The processing time T_c is included to emphasize that an OTH radar is usually a doppler-processing radar that requires a dwell time of T_c seconds if a frequency resolution of $1/T_c$ hertz is to be achieved.

The radar cross section σ of targets at HF is often different than at microwaves. Since many targets have dimensions comparable to the HF wavelength, or component structures with dimensions comparable to the wavelength, the target is often in the resonance region where the radar cross sections are generally larger than at microwaves. At the lower HF frequencies where the wavelength is large compared to the target dimensions, the cross section will be in the Rayleigh region where σ decreases rapidly with decreasing frequency. If the frequency is sufficiently low, the cross section of small aircraft or missiles might be smaller than their microwave values.

It was mentioned that external sources of interference from other HF users can limit the sensitivity of the HF radar receiver. Thus the design of the radar for maximum performance will differ from that when receiver thermal noise sets the limit. In addition to satisfying the needs for an adequate signal-to-noise ratio, there must be adequate signal-to-clutter ratio.

Signal-to-clutter ratio usually can be increased with narrow beamwidths and narrow pulsewidths.

An OTH radar designed for the detection of aircraft at ranges out to 4000 km might have, for example, an average power of several hundreds of kilowatts or more, antenna gains from about 20 to 30 dB, and operating frequencies from several megahertz to several tens of megahertz. Antennas must be large in order to obtain reasonably narrow beamwidths. The antenna horizontal length might be 300 m or greater. The transmitted waveform can be pulse, CW, FM-CW, FM (chirp) pulse, or other pulse-compression coded waveforms. Pulse compression is used for the same reason as in microwave radars. Spectral bandwidths of from approximately 5 kHz to 100 kHz might be used, corresponding to effective pulse widths of 200 μs and 10μs, respectively. (Actual pulse widths can be much longer, especially if pulse compression is used.) The lower bandwidth limit of 5 kHz is set by the desire to be able to operate in the quieter " holes " of the HF spectrum. The wider the spectral width the less likely that regions of the spectrum can be found without significant interference. Even if interference is not a problem, the upper limit of spectral width, about 100 kHz, is set by the dispersive nature of the ionosphere. Doppler filter bandwidths for separating targets from clutter in an OTH radar might range from 1 Hz down to 0.05 Hz or less, depending on the target's characteristics and the stability of the propagation path. The pulse repetition frequency (prf) of an OTH radar is generally low to avoid range ambiguities. A prf of 30 Hz, for example, corresponds to an unambiguous range of about 5000 km.

The ionosphere is not a benign propagation medium. As the amount of ionization changes, the optimum frequency for propagating energy to a particular distance must be changed accordingly. The losses in propagating through the ionosphere change with time, and the transmitter must have sufficient excess power to overcome the maximum loss expected. Fading of the HF signals can occur due to the rotation of the plane of polarization over the ionized region of the propagation path. Multipath interference from more than one refracting region, as well as dynamic irregularities in the ionospheric propagation path, are two other sources of fading.

The specific region on the earth's surface illuminated by an HF radar depends on ionospheric conditions. A " typical " patch of the ground illuminated by a single frequency might be about 1000 km in the range coordinate. The region from 1000 to 4000 km might therefore require the radiation of three different frequencies for full coverage. On the other hand, there might be times when ionospheric conditions allow this region to be covered by only a single frequency, while at other times five or six frequencies might be required. This illustrates the necessity for flexible management of the radar. The ionosphere must be sensed and the parameters of the radar adjusted to optimum operation. With proper radar design and management it should be possible to achieve a propagation path reliability comparable to that of a microwave radar. This requires a large frequency range of operation, high power to overcome propagation losses, and diagnostics of the ionospheric conditions to determine proper radar parameters.

Example capabilities. The following describe the nominal performance capabilities that might be achieved with an over-the-horizon radar operating in the HF band:[20]

Range coverage—1000–4000 km; longer ranges are possible with multihop propagation, but with degraded performance.
Angle coverage—can be 360° in azimuth from a single site, if desired; 60 to 120° is more typical.
Targets—aircraft, missiles, and ships; also nuclear explosions, prominent surface features (such as mountains, cities, and islands), sea, aurora, meteors, and satellites.

Range resolution—could be as low as 2 km, but is more typically 20 to 40 km.

Relative range accuracy—typically 2 to 4 km for a target location relative to a known location observed by the same radar.

Absolute range accuracy—10 to 20 km, assuming real-time analysis of the propagation path is made.

Angle resolution—determined by the beamwidth; it can be less than 1° which corresponds to 50 km at a distance of 3000 km.

Angle accuracy—"beam splitting" of one part in 10 should be possible if the signal-to-noise ratio is sufficient; ionospheric effects might limit the angle measurement accuracy to some fraction of a degree.

Doppler resolution—resolution of targets whose doppler frequencies differ by 0.1 Hz or less is generally possible; at a radar frequency of 20 MHz, 0.1 Hz corresponds to a difference in relative velocity of about 1.5 knots.

Although the resolution of an HF radar is poorer in range and angle than that of microwave radar, its resolution in the doppler-frequency domain is quite good. Targets not resolvable in range or angle can be readily resolved in doppler. After targets are resolved in doppler, measurements in range and angle can be made to a greater accuracy than given by the nominal resolution in those coordinates.

Application to air traffic control. The order of magnitude increase in range possible with an HF OTH radar as compared with conventional microwave radar makes it attractive for coverage of those geographical areas where it is not convenient to locate line-of-sight radars. The observation of targets over large areas of the sea is an example where HF OTH radar can find effective applications. The unique properties of these radars also make them of interest for military applications. To illustrate the types of applications well suited for HF OTH radar, two examples will be mentioned: air traffic control over the sea, and remote observation of sea conditions.

In the continental United States and similar land areas of the world with large air traffic, long-range microwave air-surveillance radars can keep track of aircraft for the purpose of providing safe and efficient air travel. Such coverage over the ocean is not practical because of the unavailability of suitable sites for microwave radar. A shore-based HF OTH radar can cover large areas of the ocean and detect and track aircraft so as to provide air-traffic control. For example, an OTH radar with 120° angle coverage and a range interval extending from 1000 to 4000 km can survey an area of almost sixteen million square kilometers. Aircraft at any altitude within this region can be detected, located, and tracked so as to provide a cost-effective over-ocean air-traffic-control capability. Target height is not obtained with this OTH radar. (Nor is height obtained with the usual microwave air-traffic-control radars.) It is possible to utilize modified HF communications equipment as transponders on each aircraft which can relay back to the radar the height of the aircraft as determined by the onboard altimeter, as well as the identity of the aircraft. Limited communications also can be effected by this means. The cost of an HF OTH radar for the detection of aircraft might be expected to be high relative to the cost of a conventional microwave radar, but on the basis of cost per unit area of coverage it is quite competitive. Its chief advantage is that it can cover areas not feasible with conventional radars.

Application to measurement of sea conditions. The distinctive nature of the doppler frequency shift from the sea allows information to be extracted regarding the sea conditions and the winds driving the sea.[24,25] The major portion of the doppler-frequency spectrum from the sea

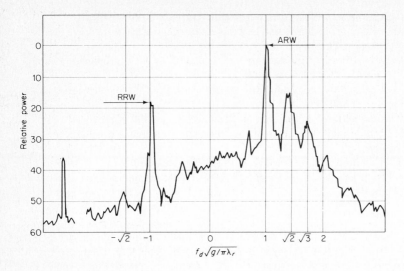

Figure 14.8 Spectrum of the radar echo from the sea obtained from an area about 9.5 by 7.5 km via groundwave propagation. The sea was developed by a 25-knot approaching wind. Radar frequency was 13.4 MHz producing a resonant doppler frequency shift of 0.37 Hz. The doppler frequency scale has been normalized so that the dominant components are at ±1. ARW = approach resonant wave; RRW = recede resonant wave; calibration signal is at left of figure.

is not found at zero frequency as is the case for land backscatter, but is centered at two discrete frequencies symmetrically spaced around zero, Fig. 14.8. The sea may be thought of as composed of a large number of individual wave trains, each with a different wavelength and amplitude and traveling in different directions. This collection of wave trains is described by a two-dimensional spectrum (wave amplitude as a function of water-wavelength and direction of travel). At grazing incidence the radar responds chiefly to the two wave trains which are traveling toward and away from the radar, each with a water-wavelength λ_w equal to half the radar-wavelength λ_r. For a grazing angle ϕ, this condition becomes $\lambda_w = (\lambda_r/2) \cos \phi$. The scattering from these two "resonant" wave components is similar to that from a diffraction grating. The term Bragg scatter is sometimes used to describe this form of scattering, by analogy to the Bragg-scatter mode for the X-ray diffraction by crystals.

The velocity of a water wave (a gravity wave in deep water) is $v = (g\lambda_w/2\pi)^{1/2}$, where g is the acceleration of gravity. Substituting this velocity into the classical formula for doppler frequency shift [Eq. (3.2)], and using the resonant condition $\lambda_w = \lambda_r/2$, gives the doppler shift for the resonant sea waves at grazing incidence as

$$f_d = \pm\sqrt{g/\pi\lambda_r} \qquad (14.23)$$

The plus sign applies to the approaching resonant wave, and the minus sign corresponds to the receding resonant wave. When the wind is blowing toward the radar the approaching-wave spectral line is the larger of the two. When the wind is blowing away from the radar, the receding-wave spectral line is the larger. When the wind is blowing perpendicular to the direction of the radar beam, the two spectral lines are equal. Thus the relative magnitude of the two major components of the doppler spectrum can provide a measure of the direction of the wind. (More precisely, it determines the direction of the waves driven by the wind since the wave and wind directions are not exactly coincident.) Note that in Eq. (14.23) the doppler frequency shift is proportional to the square root of the carrier frequency instead of the linear dependence normally characteristic of a doppler shift.

As the wind speed increases, higher-order components appear at $\sqrt{2}f_d$, $\sqrt{3}f_d$, etc. The appearance of higher-order components can be used as a measure of the wind speed, but a more useful measure is the magnitude of the continuum about zero doppler relative to the magnitude of the larger resonant spectral line. (The continumum between the two spectral lines is determined by second-order scattering from sets of waves that form corner reflectors.[24]) If land clutter is present in the radar resolution cell along with sea clutter, or if it enters the radar via the antenna sidelobes, there will be a spectral component at zero frequency which can degrade the usefulness of this measurement. When the sea is saturated, which is not unusual for those water wavelengths resonant with HF wavelengths, theory predicts, and experiments verify, that the cross section per unit area σ^0 corresponding to the larger resonant component is -29 dB, independent of sea state and frequency.[25,26] This allows that component to be used as a calibration standard.

Thus an examination of the doppler spectrum of the sea can give the sea roughness and direction, from which can be inferred something about the winds driving the sea. Swell waves or ship echoes can be recognized as distinct components. Surface currents can be noted by the asymmetrical placement of the two resonant components about zero frequency. With sufficient radar measurements the two-dimensional sea spectrum can be derived. The HF radar represents a unique tool for measuring sea conditions at a distance. The cost of a radar to measure sea conditions can be considerably less than the cost of an HF radar to detect aircraft. It is interesting to note that the development of such a capability was originally not as an attempt to learn more about the sea, but was a byproduct of attempts to improve the detection of targets in a sea-clutter background.

HF radar equation. The simple OTH radar equation of Eq. (14.22) does not account for the fact that a surveillance radar must cover a specified angular sector in a specified time. The radar equation for a microwave surveillance radar was given by Eq. (2.57). This must be modified for an HF radar since the coherent integration time T_c is fixed in an HF radar by the doppler processing requirements. Also in an HF radar, the elevation beamwidth θ_e is often not available as a design parameter because of the large cost of high antenna structures. Substituting into Eq. (14.22) $G_t = G_r = G = \pi^2/\theta_a\theta_e$, where θ_a = azimuth beamwidth; and $T_c = t_s\theta_a/\theta_T$, t_s = revisit time (scan time) and θ_T = total azimuth angle coverage, we get

$$R^4 = \frac{\pi P_{av} \lambda^2 \sigma F_p t_s}{64 N_0 \theta_a \theta_e^2 (S/N) L_s \theta_T} \tag{14.24}$$

Lumping into a single constant K the constants in this equation and the parameters assumed to be given (including θ_e) or not under the control of the radar designer, we get

$$R^4 = K \frac{P_{av}}{\theta_a} \tag{14.25}$$

Thus a measure of the performance of an HF surveillance radar is the ratio of the average power divided by the azimuth beamwidth. When comparing the performance of two radars with different antenna heights as well as different widths, a measure of relative performance is

$$\text{Measure of performance} = P_{av} w h^2 \tag{14.26}$$

where w = antenna width and h = antenna height. If separate antennas are used for transmit and receive, w is the width of the receiving antenna and h^2 is replaced by $h_t h_r$ where h_t = transmitting antenna height and h_r = receiving antenna height.

A "figure of merit" that has been used in the past for comparison of OTH radars is the DBJ value which is defined as the product of $P_{av} G_t G_r T_c$, expressed in dB. The units are joules (energy), or dB relative to a joule, hence the name DBJ. It is quite different than the measure of performance given above. This figure of merit is more representative of a "searchlighting" radar and not a surveillance radar.

Ground-wave OTH radar. The type of OTH radar described in the above that propagates via refraction from the ionosphere is sometimes called a sky-wave radar. It is also possible at HF to propagate energy around the curvature of the earth by diffraction. This is commonly called ground-wave propagation. A ground-wave radar can detect the same kind of targets as can a sky-wave radar. Detection is somewhat easier than with sky-wave propagation since ionospheric effects are not present and clutter returns from aurora generally can be eliminated by time gating. The ground-wave radar has a far shorter range than can be obtained via sky wave because of the propagation loss which increases exponentially with range. A ground-wave radar of a size and frequency comparable to the sky-wave radar discussed in the above might have a range against low-altitude aircraft targets of perhaps 200 to 400 km.

The microwave radar that uses the over-ocean evaporative duct (Sec. 12.5) to obtain extended propagation to detect low-altitude or surface targets beyond the normal line of sight is also sometimes called an over-the-horizon radar. It should not be confused with the HF radars described in this section that operate at much lower frequencies and at much longer ranges.

14.3 AIR-SURVEILLANCE RADAR[27-29,46-49,103]

The first successful application of radar was for the detection and tracking of aircraft. Air surveillance continues to be one of the more important radar applications for both civilian (air traffic control) and military purposes. The military employ such radars for general surveillance of the airspace, for providing acquisition information to anti-air-warfare systems, and for directing aircraft to an interception. There are two different types of civilian air-surveillance radars used by the Federal Aviation Administration for the control of air traffic in the United States. One is the S-band 60 nmi airport-surveillance radar (ASR), Fig. 14.9, which provides information on aircraft in the vicinity of airports. The other is the L-band, 200 nmi air-route surveillance radar (ARSR), Fig. 14.10, that provides coverage between airports. Table 14.1 lists the major characteristics of these two radars.

There is no simple formula or step-by-step guide for the design of an air-surveillance radar. One place to start, however, is with the proper form of the radar equation, along with a clear understanding of the task to be accomplished by the radar and knowledge of the constraints imposed by the environment. Quite often there is more than one factor affecting the selection of a particular radar characteristic, and these factors are sometimes conflicting. The designer must then make subjective judgments or even arbitrary choices in order to arrive at a compromise set of radar characteristics. Logical design of an entire radar system by computer is therefore difficult and dangerous. Experience and educated guesses are still necessary in many cases, just as is true of other engineering disciplines in which reasonable design decisions must be made on the basis of incomplete information or conflicting constraints. It is for these reasons that different design groups, in meeting the same user's requirements, often produce different radar designs that bear little outward resemblance to one another, yet accomplish the same objectives.

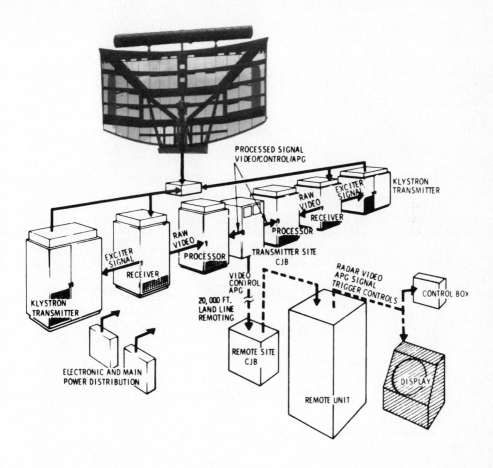

Figure 14.9 Block diagram of the ASR-8 airport surveillance radar. APG is the azimuth pulse generator which provides antenna timing information, and CJB is the cable junction box. (*Courtesy of Texas Instruments, Inc.*)

User's requirements. Since the design of a radar is strongly influenced by the task it is to accomplish, the radar designer should start with the user's requirements. For an air-traffic-control radar these might include the following: purpose the radar is to serve; types of aircraft it must detect; maximum range, with a stated probability of detection and average time between false alarms; coverage; number of aircraft expected within the radar coverage; minimum spacing of targets, which then determines the required resolution; accuracy of target location, or the accuracy of the target trajectory if a track-while-scan radar; weather and the environment in which the radar must operate, including restrictions imposed by the site or the

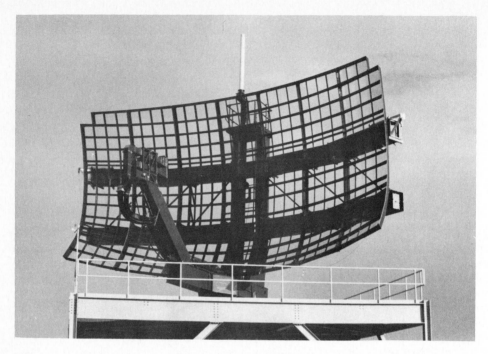

Figure 14.10 Photograph of the ARSR-3, air route surveillance radar. (*Courtesy of Westinghouse, Inc.*)

Table 14.1 Characteristics of two air traffic control radars[28,46,56]

	ARSR-3	ASR-8
Frequency band	*L*	*S*
Frequency	1250–1350 MHz	2700–2900 MHz
Instrumented range	200 nmi (370 km)	60 nmi (111 km)
Peak power	5 MW	1.4 MW
Average power	3.6 kW	875 W
Noise figure	4 dB	4 dB
Pulse width	2 μs	0.6 μs
Pulse repetition frequency	310–365 Hz	700–1200 Hz
		1040 Hz (ave)
Antenna rotation rate	5 rpm	12.8 rpm
Antenna size	12.8 m by 6.9 m	4.9 m by 2.7 m
Azimuth beamwidth	1.25°	1.35°
Elevation coverage	40°	30°
Antenna gain	34 dB	33 dB
Polarization	hor, vert, or circular	vert or circular
Blind speed	1200 knots	800 knots
MTI improvement factor	39 dB	34 dB

vehicle on which the radar is carried; reliability (mean time between failures), availability (fraction of time the radar works properly), and maintainability (sometimes these last three are called RAM); electromagnetic compatibility (EMC) requirements; restrictions on size, weight, cost, and delivery; type of prime-power available; restrictions on warm-up time and shut-down procedures; and the form in which the output information from the radar is desired. The user (or buyer) should attempt to limit the requirements to statements of performance rather than give them in terms of specific radar characteristics which restrict the radar designer in the choices available.

Characteristics of a long-range air-surveillance radar. To illustrate the nature of a long-range air-surveillance radar, the ARSR-3 (Table 14.1) will be described. No claim is made that the rationale given for each characteristic was that which influenced the original specifications for this radar. The brief discussion here is simply meant to convey some of the general philosophy that might enter into radar design.

The ARSR-3 has an instrumented range of 200 nmi, and is required to detect a 2 m² cross-section target with a single-scan probability of detection of 0.8 and a false alarm proba-bility of 10^{-6}. The greater the radar range the fewer the number of radars required to cover a specified area. On the other hand, the radars cannot be placed too far apart since the curvature of the earth will limit the minimum altitude at which targets can be seen. For example, at 200 nmi, all targets below 8 km altitude are beneath the radar line of sight and normally would not be detected. Most commercial aircraft have a cross section greater than the 2 m² specified for this radar. This low value is necessary, however, if small general-aviation aircraft are also to be detected.

The 5 rpm rotation rate of the antenna corresponds to a 12-second interval between target observations (scan time). This is about the longest interval that can usually be tolerated between observations of aircraft targets. A high rotation rate is needed for good target track-ing. A low rotation rate is desired for more hits on target, which reduces the requirements for large transmitter power or antenna aperture. For a surveillance radar the product of average power, antenna aperture, and scan time is a constant, Eq. (2.57). Reducing scan time (increas-ing rotation rate) requires an increase in the average power and/or antenna size. Some long-range military radars have rotation rates of 15 rpm (4-s scan time) because of the likelihood of target maneuver. A rapid change of course is less likely for civilian aircraft; hence, the slower rotation rate of 5 rpm for the ARSR-3.

The pulse width of 2 μs corresponds to a range resolution of about 300 m. (In practice, the resolution is said to be about 500 m.[46]) If limitations on peak power require a considerably longer pulse width in order to achieve the necessary energy within the pulse, some form of pulse compression could be used. (The ARSR-3 does not have pulse compression, however.) FM (chirp) is a common choice of pulse-compression waveform.

The 12.8 m (42 ft) wide by 6.9 m (22.6 ft) high antenna reflector produces a 1.25° azimuth beamwidth and a shaped elevation beam extending beyond 40° so as to provide coverage to an altitude of 18.6 km (61 ft). The upper corners of the antenna aperture have a square rather than rounded outline. This causes the underside of the elevation beam to have a sharp drop-off which minimizes the ground-reflected energy that causes a lobed elevation pattern and a degradation of the rain-rejection capability of circular polarization. The azimuth resolution and accuracy obtained by the 1.25° beamwidth is said to be 2° and 0.2°, respectively.[46] The antenna can be housed in a 17.4 m (57 ft) diameter rigid geodesic-radome.

The ARSR-3 antenna generates two beams slightly displaced in elevation for the purpose of reducing the echoes from high-speed surface clutter, such as from automobiles and trucks. These clutter echoes can be large enough and have a sufficiently high doppler frequency shift to not be completely eliminated by MTI doppler processors. Transmission takes place only on the lower of the two beams. At short range, reception is on the upper beam only. This beam is tilted to minimize illumination of the ground. As the transmitted pulse travels beyond the ground clutter range (typically 50 miles or greater) the receiver is switched to the lower beam for long-range reception. In the ARSR-3 the ratio of the lower-beam gain in the direction of the horizon to that of the upper-beam gain in the same direction is 16 dB.[46] (The lower beam has its half-power point on the horizon.) The antenna elevation pattern is shaped to have at the higher angles a greater gain than would be normal with a cosecant-squared pattern. This permits sensitivity time control (STC) to be used with the radar without a loss of coverage at high altitudes and short range (Fig. 7.27). The STC is employed to reduce near-in clutter, especially from birds and insects. In this radar the STC is applied in the RF ahead of the low-noise amplifier. The advantage of STC in the RF as compared to the IF is that it prevents saturation of the RF amplifier, as well as the IF amplifier, by close-in clutter. The noise level in the receiver is but little affected by the STC, a condition that is desired for operation of the log-CFAR receiver. Four PIN diodes provide an STC range of 63 dB with an insertion loss of less than 0.6 dB.

The ARSR-3 is really two separate radars, each at a different frequency, operating into a single antenna. Either system can be used separately (simplex operation) or both can be used simultaneously (diplex operation). The chief reason for the dual channels in this radar is to provide greater availability of the radar. A failure in one channel does not require the radar to shut down. The radar operates with a single channel while repairs are being made. The availability of dual-channel radars with built-in-test equipment and fault isolation has been demonstrated to be considerably greater than 99 percent. High availability also requires that spare parts be at hand when needed and that maintenance personnel be experienced and motivated. In diplex operation the two frequencies are radiated on orthogonal polarizations. Frequency and polarization diversity can provide an improved signal-to-noise ratio by converting a Swerling type 1 target (Sec. 2.8) with scan-to-scan fluctuations to a Swerling type 2 with pulse-to-pulse fluctuation. From 4 to 7 dB improvement in signal-to-noise ratio might be achieved in the diplex mode as compared with the simplex mode.[46]

The ARSR-3 utilizes a klystron amplifier to achieve 5 MW peak power and 3.6 kW average power. The receiver front-end is a low-noise transistor with 4 dB noise figure. The logarithmic receiver is followed by a CFAR which divides the amplitude of each range sample in the logarithmic output by the average amplitude of the returns within $\frac{5}{8}$ nmi of that sample. A 4-pulse digital canceller is employed with variable interpulse periods to produce a calculated MTI improvement factor of 39 dB. The pulse repetition frequencies range from 310 to 360 Hz.

The ARSR-3 contains provision for switching in or out various processing features. A range-azimuth generator (RAG) permits the selection to be made on the basis of both range and azimuth. Among the features that might be selected by range and azimuth are included the choice of MTI or normal (log-CFAR) video, two STC control curves (one optimized for terrain clutter, the other for sea clutter), the crossover range in switching from the upper to the lower beam, fixed pulse-repetition period (for eliminating second-time-around clutter) or variable pulse periods, different RF receiver gains, and sectors for transmitter blanking to avoid RFI.

The digital target extractor provides the radar output in a form suitable for transmission over narrowband telephone lines rather than require wideband microwave data links. Sliding-window detectors determine the range and azimuth centroid of the radar returns using the *m*-out-of-*n* detection criterion.

Circular polarization can be selected to reduce weather clutter. MTI and the log-CFAR receivers also reduce weather clutter at the radar output. However, in air-traffic-control operations, it is important that the controller know the locations of the bad weather so as not to vector aircraft through it. In the ARSR-3, the weather information is available at the orthogonal port of the circular-polarization diplexer and can be viewed by switching this output to the display.

Frequency considerations. The ARSR-3 described above operates at L band (1250 to 1350 MHz). This is a good compromise frequency for a long-range air-surveillance radar. Higher frequencies can provide the same angular beamwidths with smaller antennas, but the smaller apertures must be compensated by greater power if the maximum ranges are the same. Although high-power transmitters are available at S band, the peak power that can be transmitted is less because of the smaller waveguide sizes. Pulse compression can be used to compensate for the reduced peak power, but at the price of additional complexity. At the higher frequencies the backscatter from rain is greater, and the MTI blind speeds are more of a problem. Because of the smaller apertures and the greater effect of rain clutter, S-band radars are likely to have less range capability than radars at lower frequencies. Frequencies higher than S band are seldom used for long-range air surveillance.

High-power, large antenna apertures, and good MTI are easier to achieve at frequencies below L band. Also, weather clutter is much less. The lower the frequency the easier it is to obtain long range. However, at the lower frequencies, the azimuth beamwidths are broader and the available bandwidths are narrower than might be desired. Below UHF, the external noise increases with decreasing frequency and can limit receiver sensitivity. The lobing of the elevation pattern due to ground reflections results in wider elevation nulls at the lower frequencies and can cause long fade times of the target signals. The elevation angle of the first multipath lobe varies inversely with frequency (Sec. 12.2), so that low-altitude coverage is more difficult to achieve at the lower frequencies. In spite of limitations, the lower frequencies are attractive when low cost and long range are more important than other factors.

14.4 HEIGHT-FINDER AND 3D RADARS

In many applications a knowledge of target height might not be needed. An obvious example is where the target is known to lie on the surface of the earth, and its location is given by the range and azimuth coordinates. However, there are other instances in which the target's position in three dimensions must be known. In this section, radar methods for obtaining the elevation angle or height of a target will be discussed briefly. The height of a target above the surface of the earth may be derived from the measurement of elevation angle and range. The use of height as the third target-coordinate is more desirable than the elevation angle in applications where the height is apt to be constant. This is usually true for commercial aircraft and for satellites with nearly circular orbits. A radar whose purpose is the measurement of elevation angle, and which usually does not measure the azimuth angle, is called a *height-finder* radar. A radar which measures the elevation angle along with the azimuth angle (and range) is called a *3D radar*. One which measures the range and one angle coordinate (usually azimuth) is called a *2D radar*.

Nodding-beam height finder. Just as the range-azimuth coordinates of a target can be obtained with a vertical fan-beam antenna, the elevation coordinate can be obtained with a horizontal fan beam. (The three coordinates can thus be obtained with two 2D-radars.) Such a height finder would be directed by the 2D air-surveillance radar to the azimuth of the target. It then scans its horizontal fan beam in elevation to make an elevation angle measurement of the target found at the range designated by the air-surveillance radar. The *nodding-beam height finder* scans its beam in elevation by mechanically rocking the entire antenna. It is possible to mechanically slew a nodding-beam height finder a full 180° in a relatively short time (within two seconds, for example). With an operator making a height measurement manually, from 2 to 4 measurements per minute might be made. In one nodding-beam height finder, up to 22 target-heights per minute can be obtained when the slewing is controlled automatically for maximum data rate by the computer of the associated data-handling system.[29] The absolute height accuracy of a nodding-beam height finder can be ± 1500 ft (460 m) at 150 nmi (280 km) range.

The nodding-beam height finder is one of the oldest techniques for measuring the elevation angle of aircraft targets. It is also one of the best. Its accuracy is probably as good or better than any other technique. Even though a separate radar is employed to measure target height, the combined cost of the 2D air-surveillance radar and the nodding-beam height finder can be less than the cost of a comparable single 3D radar. (It is not always true that two radars cost more, are more complex, or occupy more volume than a "single" radar designed to do the same job.) Another advantage of the separate nodding-beam height finder in military applications is that it generally operates at a higher frequency (*S* or *C* bands are common choices) than does the 2D air-surveillance radar. This increases the ECCM capability of the system since a jammer must radiate in both radar bands simultaneously to deny the location of aircraft targets. The use of a higher frequency for the height finder is appropriate since it can be of shorter range than the 2D air-surveillance radar, and the antenna aperture can be of smaller size for a given beamwidth. The height finder can also be made to provide good range resolution for target counting by utilizing a narrower pulse than might be desired for a long-range 2D air-surveillance radar operating at a lower frequency.

Thus there are many reasons for the nodding-beam height finder being a good choice for obtaining the third coordinate on aircraft.

Instead of mechanically rocking the entire antenna structure, the horizontal fan beam of a height finder can be scanned in elevation by electromechanical means, such as with the Robinson scanner, organ-pipe scanner, or delta-a (or Eagle) scanner. It is also possible to scan the beam with electronic phase shifters. With a scanning phased-array antenna, however, the radiation pattern, which is a fan beam at broadside, becomes a conical beam when scanned off broadside. This can cause an error in the elevation measurement if the target is off the center of the beam. For this reason, height finders which use electronic scan usually radiate pencil beams rather than broad fan beams.

V-beam radar. This radar generates two fan beams: one vertical and the other slanted at some angle to the vertical (perhaps at 30 to 45°). The time separation between the echoes received in the vertical and slant beams is a measure of the target height. A short time-separation signifies low altitude, while longer separations occur with high altitude targets. Two separate reflectors may be used to generate the two fan beams,[30] the two reflectors might be back to back, a single reflector with two feeds can be used,[31,36] or the two beams can be generated with a single phased array antenna. The *V*-beam radar is a satisfactory method for obtaining three-coordinate target information if the number of targets is not excessive. The larger the number of targets, the more difficult is the problem of correlating the echoes from each of the two beams. The closer the beams, the easier it is to correlate the echoes, but the less the accuracy. Another useful modification is to separate the beams at zero elevation so that the data from the vertical and the slant beam do not appear close in time.

Monopulse. An elevation angle measurement can be made similar to that of the monopulse radar described in Sec. 5.4. Two fan beams are displaced ("squinted") in elevation angle and the sum and difference patterns are obtained. The measurement of angle is similar to that of an amplitude-comparison monopulse tracking radar except that it is made open loop; i.e., the output voltage from the angle-error detector is calibrated to read elevation angle. Two displaced horns with combining circuitry, or a single multimode horn, is the principle modification required of the antenna. Although monopulse tracking radars are capable of excellent accuracy, when monopulse is applied to a 2D surveillance radar the angle accuracy is usually poor. One reason is the broad elevation beamwidths. (The rms error of an angle measurement is proportional to the beamwidth.) Another limitation is the effect of multipath due to reflection from the earth's surface, also a consequence of the broad fan beams. The effect of multipath is similar to the effect of glint discussed in Sec. 5.5 for tracking radars. The multipath error can be quite severe and make the angle measurement in many cases almost useless, as discussed in Sec. 22.3 of Ref. 32. This effect will generally be less severe over land than over water because of the smaller reflection coefficient of land. If the antenna beam can be tilted up so as to reduce the illumination of the surface, or if the antenna elevation pattern can be designed to have a sharp cutoff on the underside of the beam, the effect of surface reflections is reduced and the monopulse angle measurement is improved. However, this reduces the coverage of targets at low elevation angles.

Since the problem of multipath errors occurs for antenna beams which illuminate the surface, one approach is to use three elevation beams. The lowest beam is made as narrow as practical in elevation. It illuminates the low angles. No monopulse angle measurement would be attempted with this beam. The other two beams cover the higher angles with monopulse processing. Since these fan beams do not illuminate the surface they do not suffer from multipath problems.

Phase in space. This technique employs a single reflector with a rather unique type of feed. The effect of multiple beams is generated with this feed, but only two receivers plus a phase detector are needed to determine angular location.[33-35] In order to obtain wide coverage in elevation angle a parabolic-torus reflector is employed rather than a section of a paraboloid. A curved piece of waveguide with radiating slots acts as the feed. The transmitter is fed into one end of the waveguide feed whose slots are designed to produce a number of contiguous beams in elevation with amplitudes controlled so as to produce a cosecant-squared radiation pattern. On reception, a receiver is inserted at each end of the feed. The echo signal from a particular direction is received at one of the slots and is divided in the waveguide feed. The two signals travel in opposite directions and are received at each end of the feed. The phase difference between the two signals (or the difference in travel time) depends on from which slot they originated, which in turn depends on the elevation angle of arrival. Thus a measurement of phase difference results in a measurement of elevation angle.

A detailed analysis of the accuracy of the phase-in-space technique is not available, but in one implementation it has been said[34] to have a potential elevation accuracy of about 0.2°. Multipath reflections can also degrade the accuracy of this method, but since the phase-in-space technique employs multiple beams to shape its elevation pattern, it might be possible to control the radiation so as to illuminate the surface with less energy than the conventional fan-beam antenna.

Multiple elevation (stacked) beams. The use of contiguous beams stacked in elevation has been employed for 3D radar. It is sometimes called a *stacked-beam radar*. It is a good technique from a fundamental point of view since it uses simultaneous pencil-beam radiation patterns from a single aperture to cover the elevation angles of interest. Each beam can be considered a separate radar. It is, however, costly and complex. The transmitter radiates a fan beam from the summation of all overlapping pencil beams to give the desired elevation coverage. A separate receiver is provided each pencil beam and some means of interpolation between the beams is used to refine the angle measurement. If automatic detection, sidelobe cancellation, or MTI is needed in the radar, these must be separately employed in each receiving channel, thus adding to the cost and complexity of the radar. The individual pencil beams, however, limit the volume of space observed, which can be an advantage in rain clutter or chaff. The MTI in the lower beams can be optimized for surface clutter and MTI in the upper beams, if used at all, can be optimized for rain and chaff. The individual pencil beams have a higher gain than a fan-beam antenna, and can provide a larger number of hits at a higher data rate than can a 3D radar with a single scanning beam in elevation.

Scanning pencil beam. Three-coordinate, or 3D, information can be obtained on each rotation of the radar by electronically scanning a single pencil beam in elevation while mechanically rotating the antenna in azimuth. The beam is rapidly scanned through the entire elevation coverage in the time the antenna rotates one azimuth beamwidth. The 3D information is obtained by the sequential scanning of a single beam as compared with the simultaneous multiple beams described in the above. A relatively simple form of electronic scan in this application is frequency scan (Sec. 8.4) in which a change in frequency results in a change of elevation angle.[37,38] The beam can also be scanned with phase shifters.[39]

One of the limitations of a long-range 3D radar with a single scanning beam is that the angular coverage consists of a large number of individual beam positions so that the dwell time in any one resolution cell is small. The minimum dwell time in each position is determined by the time it takes for the radar energy to travel to maximum range and back. The number of beam positions multiplied by the dwell time of each equals the scan time. Long dwell times and short scan time (high data rate) are not compatible. For example, assume that a 3D radar with a one-degree pencil beam must cover 20 elevation beam positions and 360 azimuth beam positions (a total of 7200 beam positions) every four seconds. If there is but one pulse per beam position, the pulse repetition frequency can be no less than 1800 Hz, which corresponds to an unambiguous range of 83 km. This is a rather short unambiguous range for an air-surveillance radar. A 3D radar that obtains only one pulse per beam position can suffer a relatively large beam-shape loss. Generally 2 or 3 pulses are required as a minimum to reduce this loss. Also, one pulse per beam position does not allow good MTI processing and it makes difficult the accurate measurement of elevation angle. To obtain more than one pulse per beam position and/or increase the unambiguous range requires some compromises or an increase in complexity. The compromises require a fatter beamwidth or a slower scan rate. Some benefit can be obtained by using a pulse repetition frequency (prf) that varies with elevation angle. At low angles where the ranges are long, a low prf is used. At the higher angles where the ranges are short, a high prf can be used. If an increase in complexity can be tolerated, a number of multiple elevation beams can be scanned simultaneously to increase the data rate, unambiguous range, and/or number of pulses per beam position. A separate receiver is required for each of the simultaneous beams.

The elevation angle accuracy of single-scanning-beam 3D radars is usually less than that of other systems. This is due to the relatively few pulses (sometimes only one pulse) per beam position, as well as the error introduced by the target echo amplitude fluctuations when sequential rather than simultaneous angle measurements are made. Changes in echo amplitude cause errors since the sequential measurements are necessarily made at different instants of time. This is similar to the effect of amplitude fluctuations on tracking radar, as discussed in Sec. 5.5. If frequency scan is used, there is even more likelihood of echo fluctuations because of the change in target echo with frequency.

Another method for obtaining 3D coverage with a scanning pencil beam is to employ helical scan, as in the renowned SCR-584 of World War II. A single pencil beam is rotated in azimuth with its elevation angle increased one beamwidth per revolution so as to trace a helical pattern. It is a simple technique, but is not of high accuracy since it is difficult to interpolate between adjacent elevation beam positions.

Within-pulse scanning.[40,41] In the frequency-scan 3D radar discussed above, a single pencil beam is step-scanned in elevation. Each pulse is at a constant frequency but the frequency is changed every pulse or every few pulses to position the beam at different elevation angles. Another method of utilizing a frequency-scan antenna is to sweep the frequency over the entire frequency range on *each* pulse so that energy is radiated throughout the entire elevation coverage for the duration of a single pulse. The frequency of the received echo signal will depend on the elevation angle of the target. A bank of contiguous receivers, each tuned to a different frequency, provides coverage of the elevation sector with the equivalent of parallel receiving beams. The elevation angle of the target is determined by which receiver is excited. This technique has been called *within-pulse* scanning.

The height resolution depends on the antenna beamwidth, as in a conventional frequency-scan antenna; but the range resolution is determined by the frequency spectral components, received from a given elevation angle, that correspond to one beamwidth. As stated in Sec. 8.4, the echo from a target is frequency modulated when illuminated by such a frequency-scanned beam. Pulse-compression filtering (similar to chirp) can be used to achieve a range resolution equal to the reciprocal of the one-way group delay along the dispersive delay line (snake feed) feeding the frequency-scan array.

There is another array antenna technique called *within-pulse scanning* that is different from the frequency scan technique described above. This other technique, which is discussed in Sec. 8.7, is a method for generating multiple receiving beams at a single frequency and can be considered more like the stacked-beam radar than a frequency-scan height finder.

Interferometer.[42-44] An interferometer consists of two individual antennas spaced so as to obtain a narrow beamwidth for accurate angle measurement. The phase difference between the signals of the two antenna elements of the interferometer provides the elevation angle, as given by Eq. (5.1). This type of angle measurement is similar to the phase-comparison monopulse radar, except that the size of the individual antennas is small compared to the spacing between them. Grating lobes (also called interferometer lobes) result from the wide spacing between the antennas. These grating lobes can cause ambiguities in the measurement. The ambiguities can be resolved by use of more than two antennas with unequal spacings. The outer two antennas provide accurate, but ambiguous, measurement of elevation angle. The function of the one or more inner antennas is to resolve the ambiguities.

In one implementation, an interferometer was attached to the rotating antenna of a 2D S-band air-surveillance radar.[42] The 2D radar provided the range and azimuth as well as the transmitted energy for the receiving interferometer. The interferometer consisted of four linear horizontally oriented arrays, or sticks, with one stick mounted below the 2D radar antenna and the other three sticks mounted above.

The elevation beamwidths of each of the individual antennas of the interferometer are generally broad so that multipath due to ground reflections can cause errors. Shaping the underside of the 2D transmitting antenna to minimize the energy illuminating the surface can reduce the angular error. The effective receiving area of the interferometer is usually less than that of the conventional radar reflector antenna with which it is used. Thus, the range capability of the interferometer angle measurement might be less than that of the 2D radar. This reduced range capability is acceptable in many situations since the height measurement is usually of more significance at shorter rather than longer ranges.

Lobe recognition. A coarse indication of height sometimes can be obtained by recognizing the fading of the echo signal as an aircraft target flies through the multipath lobes of the antenna pattern. Measuring the range at which the target is first seen on the bottom lobe, or the range where it disappears because of the first null of the lobing pattern, can be related to the target height. This effect is more applicable at the lower microwave frequencies and for radars sited over water. At the higher frequencies where the lobes are narrow, a count of the lobes per unit distance can be related to the target height. Although the technique of lobe recognition is relatively simple and has been reported to have been used in World War II,[45] its poor reliability, low target capacity, and poor accuracy, make it unattractive.

Time-difference height finding. The several height-finding techniques included in this section were described in terms of a ground-based radar. Most can be modified to apply to an airborne radar for the measurement of height of other aircraft or the depression angle if a ground target. The time-difference height-finding technique to be described is applicable, in principle, to almost any radar situation, but in practice it is more applicable to an elevated radar such as in an aircraft, especially one which operates over water. It derives the height of a target from the time difference between the radar signal reflected directly from the target and the multipath signal that arrives via reflection from the surface of the earth.

Consider the geometry of Fig. 12.1 for propagation over a plane reflecting surface. There are four possible paths by which the energy travels to the target and back. These are:

1. From radar to target and return by the same path (AB-BA).
2. From radar to target and return via reflection from the surface (AB-BMA).
3. From radar to target via reflection from the surface and return directly to the radar (AMB-BA). (This is the opposite of No. 2.)
4. From radar to target via reflection from the surface and return by the same path (AMB-BMA).

Path 1 is the shortest, path 4 is the longest and paths 2 and 3 are equal. Thus, there can be three separate responses from a target, as illustrated in the idealized sketch of Fig. 14.11. The relative amplitudes of the three echoes are in the ratio of 1, 2ρ and ρ^2, where ρ = surface reflection coefficient. A measurement of the time delay t_D along with the range R to the target and knowledge of the height h_a of the radar antenna yields the height of the target. This technique works best over surfaces with large reflection coefficient, and when the pulse widths

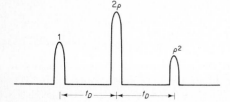

Figure 14.11 Idealized echo response of a point target located above a reflecting surface, for the case where the pulse width is less than the time separation t_D between the direct and surface-reflected signals. The surface reflection coefficient is ρ.

are short compared to the time difference t_D. For a flat earth, the height of the target is approximately

$$h_t = \frac{cRt_D}{2h_a} \tag{14.27}$$

where c = velocity of propagation. Usually the round-earth geometry must be considered, which results in a more complex formulation than the simple expression given above.

14.5 ELECTRONIC COUNTER-COUNTERMEASURES

All radars, civilian and military, must be able to operate in the crowded electromagnetic environment that results from the transmissions of other radiating sources, both in the radar's own band as well as from outside the band.[50] The radar must be designed to reject these unwanted radiations and to minimize the likelihood of its own transmissions causing trouble to other users of the spectrum. This is the subject of EMC, or *electromagnetic compatibility*. It is important if the electromagnetic spectrum, considered as a natural resource, is to be effectively and efficiently utilized for the benefit of all. Military radars, however, must also operate in a hostile environment where they may be subjected to deliberate interference designed to degrade their performance. The various methods for interfering electronically with radar are called *electronic countermeasures*, or ECM. *Active ECM* is sometimes referred to as *jamming*. There is also passive ECM, such as chaff, which reflects radar energy to create clutter and false targets. The methods employed to combat ECM are called *electronic counter-countermeasures*, or ECCM.

The several forms of ECM directed against radar may be categorized as noise jamming, deception jamming, chaff, and decoys. Intercept receivers and direction finders (which are called *electronic support measures*, or ESM), as well as antiradiation missiles (ARM) are also aspects of *electronic warfare* (EW) that must be of concern to the military radar systems designer.

If a determined adversary is willing to pay the resultant price, sufficient ECM can be brought to bear against any *single* radar to significantly reduce its effectiveness. This should not evoke pessimism on the part of the radar designer since it can be said that any military objective can be accomplished by a determined force if the force is skillful and large enough, and if cost is of no consequence. The goal of ECCM is to raise the cost of ECM to the point where it is prohibitive. The effect of ECM on a single radar can seldom be considered in isolation since it is seldom that a single radar acts as an entity in itself. A military radar is almost always in support of a weapon system. The question is not whether a single radar can operate without degradation, but whether the weapon system of which the radar is a part can fulfill its mission in spite of hostile ECM. It is easier to ensure the accomplishment of the weapon system's mission than to guarantee that operation of a single radar will not be degraded. This larger and more important goal, that of fulfilling a military mission, is what should guide the military systems planner and the radar systems designer. This broader concept, however, is not appropriate for discussion here. Instead, a brief review will be given of the various radar-ECCM options that can make the task of ECM more difficult.

As a general rule, good radar design practice can reduce vulnerability to electronic countermeasures. Good design is based on maximizing the ratio of the signal energy to noise power per hertz (E/N_0), as well as employing techniques to reduce mutual interference. It is important to avoid receiver saturation, or overloading. A wide dynamic range is desired, and linear rather than square-law detectors are preferred. The radar should be designed conservatively with larger power and larger antenna aperture than the minimum required for marginal detection.

One of the better measures against many forms of ECM is an alert, highly motivated, well trained, and experienced operator.

Noise jamming. Receiver noise generally limits the sensitivity of most microwave radars. Raising the noise level by external means, as with a jammer, further degrades the sensitivity of the radar. Noise is a fundamental limitation to radar performance and therefore can be an effective countermeasure. The ECCM designer must minimize the amount of noise a jammer can introduce into the radar receiver. It is difficult, however, to keep the noise out when the jammer is being illuminated by the main beam of the radar antenna. When this occurs, the narrow sector in the direction of the jammer will appear as a radial strobe on the PPI display. The direction to the jammer can be determined, but its range and the ranges of any targets masked by the noise strobe is not likely to be known. If noise enters the radar via the antenna sidelobes, the entire display can be obliterated and no target information obtained. Thus it is essential that noise be prevented from entering the receiver via the antenna sidelobes.

A jammer whose noise energy is concentrated within the radar receiver bandwidth is called a *spot jammer*. The spot jammer can be a potent threat to the radar if it is allowed to concentrate large power entirely within the radar bandwidth. The radar systems designer must prevent this by forcing the jammer to spread its power over a much wider band. This can be accomplished by changing the radar frequency from pulse to pulse in an unpredictable fashion over the entire tuning band available to the radar. A radar capable of changing its frequency from pulse to pulse is said to possess *frequency agility*. (Even in the absence of ECM, frequency agility has advantages in filling in the nulls of the elevation radiation pattern and in decorrelating target echoes so as to increase the probability of detection.) A jammer can also be forced to widen its jamming band if there are many radars operating within the same geographic area, each at a different frequency distributed over the available radar tuning range. A jammer which radiates over a wide band of frequencies is called a *barrage jammer*.

To take full advantage of frequency agility, the radar should employ a *prelook receiver* to examine the jammer's spectrum and select a frequency for the next radar transmission where the noise is a minimum. The jamming power is seldom uniform over the band. Prelook sampling of the environment can take place during the radar interpulse period just prior to each transmission, so as to select on a pulse-to-pulse basis that frequency which offers the least jamming interference.[51]

Pulse compression is sometimes credited as causing the jammer to spread its energy over a wider band than that of a normal spot jammer. However, pulse compression is seldom deliberately employed as a prime ECCM technique. It is almost always used as a means to achieve good range resolution with a long pulse. Furthermore, the jammer usually must be forced to spread its power over a much wider band than the spectral width of most pulse-compression radars in order to significantly reduce its effectiveness. Thus pulse compression should not be given too large a credit as a major ECCM technique, although it is certainly a positive factor.

Forcing the jammer to spread power over the entire band available to a radar is generally not sufficient in itself. The jammer also must be forced to spread its available power over more than one radar band. This can be accomplished with *frequency diversity* by using two or more radars. A 2D air-surveillance radar in one band used with a height-finder radar in another band is a good method for achieving frequency diversity. Both radar bands have to be jammed simultaneously if target location is to be denied.

In general, higher-frequency radars usually are less vunerable to jamming than are lower-frequency radars. One reason is that the bandwidth over which the higher-frequency radar can operate is greater, thus causing the jammer to spread its available power over a greater number of megahertz. Also, the antenna gain can be greater at the higher frequencies, a factor often favoring the radar when confronted with jamming, as seen by Eqs. (14.28) and (14.30). Furthermore, at the higher frequencies the antenna sidelobe levels can be lower, making it more difficult for sidelobe jamming. However, the advantages of operating against jammers at the higher frequencies are balanced in part by the disadvantages of the higher frequencies, especially above L band, for long-range air-surveillance radar.

The noise that enters the radar via the antenna sidelobes can be reduced by coherent sidelobe cancelers. This consists of one or more omnidirectional antennas and cancelation circuitry used in conjunction with the signal from the main radar antenna. Jamming noise in the omnidirectional antennas is made to cancel the jamming noise entering the sidelobes of the main antenna.[52] An antenna can also be designed to have very low sidelobe levels to reduce the effect of sidelobe jamming. Low sidelobe antennas require unobstructed siting if reflections from nearby objects are not to degrade the sidelobe levels.

By employing some or all of the above techniques, the effect of the sidelobe noise jammer can be significantly reduced. Some of the above techniques can also reduce the jamming that enters via the main beam. The effects of main-beam jamming can be further reduced by employing a narrow beamwidth to limit the region over which the jamming appears. If the main beam cannot be made narrow because of constraints on the antenna size, an auxiliary antenna can be employed to create a notch in the main-beam radiation pattern in the direction of the jammer. With adaptive circuitry similar to that of the sidelobe canceler, this *main-beam notch* can be automatically adjusted to be maintained in the direction of the jammer.

Multiple radars viewing the same coverage in a coordinated manner can provide some benefit against a jammer since it is unlikely that jamming power can be distributed uniformly in space. With netting of radars, those radars with less jamming can provide target data for radars with more jamming. Multiple radar sites also allow a jammer to be located by *triangulation* when main-beam jamming denies a direct measurement of range. This may be a satisfactory tactic for one or a few jammers, but triangulation cannot be used with a large number of jammers because of the generation of ghost targets. N jammers can produce $N^2 - N$ ghost targets when two radars attempt triangulation.

Locating the jammer might not be as important, however, as locating the attacking missiles or aircraft screened by the jammer, especially when the jamming aircraft is beyond the range of defensive weapons. A jammer that operates outside the range of normal defenses is known as a *stand-off jammer*. An important ECCM tactic is to engage hostile jammers with *home-on-jam* (HOJ) missile guidance. If the jammer is radiating, HOJ is generally a better guidance technique than is radar guidance. Stand-off jammers are also subject to direct attack by interceptor aircraft.

Increasing the radar energy in the direction of the jammer in the hope of increasing the radar echo power above the jamming noise is called *burnthrough*. This may be accomplished with reserve transmitter power or by dwelling longer in the direction of the jammer. Dwelling longer on a target reduces the data rate, and thus can degrade the overall radar performance. A significant reduction in data rate generally is not desirable. Dependence on burnthrough as a major ECCM tactic is questionable. It should be used where cost-effective and where the reduction in data rate is tolerable.

Impulsive noise, that can shock-excite the "narrow-band" radar receiver and cause it to ring, can be reduced with the Lamb noise-silencing circuit,[53] or Dicke fix.[54] This consists of a wideband IF filter in cascade with a limiter, followed by the normal IF matched filter. The wideband filter is designed to include most of the spectrum of the interfering signal. Its purpose is to preserve the short duration of the narrow impulsive spikes. These spikes are then clipped by the limiter to remove a considerable portion of their energy. If the large noise spikes are not limited and are allowed to pass they would shock-excite the narrowband IF amplifier and produce an output pulse much wider in duration than the input pulse. Therefore the interference would be in the receiver for a much longer time and at a higher energy level than when limited before narrowbanding. Desired signals which appear simultaneously with the noise spike might not be detected, but the circuit does not allow the noise to influence the receiver for a time longer than the duration of a noise spike. This device depends on the use of a limiter. Limiters, however, can generate undesired spurious responses and small-signal suppression, and reduce the improvement factor that can be achieved in MTI processors. It should therefore be used with caution as an ECCM device. If incorporated in a radar, provision should be included for switching it out of the receiver when it does more harm than good.

A *constant false-alarm rate* (CFAR) receiver is used with automatic detection systems to keep the false-alarm rate constant as the noise level at the receiver varies (Sec. 10.7). Without CFAR the computer in automatic systems can quickly become overloaded and cease to function. CFAR is sometimes classed as an ECCM, but in reality it is not in the same category as other ECCM. CFAR does not give immunity to jamming; it merely makes operation in the presence of jamming more convenient by automatically reducing the effective sensitivity of the receiver. If the jamming is severe enough, CFAR can produce almost the same effect as turning off the receiver. In a jamming environment, it can lull the radar operator into a false sense of security since the ideal CFAR, by maintaining the output noise level constant no matter what the level of the input noise, provides no indication that jamming is present. Additional means must be incorporated into a CFAR to provide the operator with warning of the presence of noise jamming.

A sensitive receiver with a low noise figure may be desirable in many civilian applications of radar, but in military applications it is not always an asset since it makes the receiver more vulnerable to jamming. Most low-noise receivers also have less dynamic range than one with a mixer front-end.

Since the jammer power received at the radar varies inversely as the square of the distance between radar and jammer (one-way propagation), while the radar echo power varies with distance inversely as the fourth power (two-way propagation), there will be some distance below which the radar echo will exceed the jammer signal. This is called the *self-screening range*, or the *crossover range*, and is approximately

$$R_{ss}^2 = \frac{P_{tr}}{P_{tj}} \frac{G_r}{G_j} \frac{\sigma}{4\pi} \frac{B_j}{B_r} \frac{J}{S} \tag{14.28}$$

where P_{tr} = radar transmitter power
  P_{tj} = jammer transmitter power
  G_r = radar antenna gain
  G_j = jammer antenna gain
  σ = target cross section
  B_j = jammer bandwidth
  B_r = radar signal bandwidth
  J/S = jammer-to-radar signal (power) ratio at the output of the IF required to mask the radar signal

A jammer located on a target of cross section σ will overpower the radar if the jammer is at a range R_{ss} or greater.

A radar equation for detecting a target in the presence of jamming noise large compared to receiver noise can be derived by substituting the jammer noise power per hertz received at the radar for the receiver noise power per hertz, or $kT_0 F_n$. The jammer noise power per hertz at the radar is

$$N_{0j} = \frac{P_{tj} G_j A_e}{4\pi R^2 B_j} \tag{14.29}$$

where A_e = effective receiving aperture of the radar antenna and R = range of jammer from radar. Substituting Eq. (14.29) for $kT_0 F_n$ in Eq. (2.54), the radar equation for a searchlighting or tracking radar, yields

$$R_{max}^2 = \frac{P_{av} G_t t_0 E_i(n)}{4\pi} \frac{\sigma}{(S/N)_1} \frac{B_j}{P_{tj} G_j} \tag{14.30}$$

where $t_0 = n/f_p$ = integration time. The system losses have been omitted. The remaining symbols have been defined above or in Sec. 2.14. Making the same substitution of Eq. (14.29) into the surveillance radar equation, Eq. (2.57), gives

$$R_{max}^2 = \frac{P_{av} E_i(n)\sigma}{(S/N)_1} \frac{t_s}{\Omega} \frac{B_j}{P_{tj} G_j} \tag{14.31}$$

For jamming which enters via the antenna sidelobes rather than via the main beam, the right-hand side of the above surveillance radar equation would be multiplied by the ratio of maximum antenna gain to the gain of the sidelobe in the direction of the jammer.

Repeater jamming. A target under observation by a radar can generate false echoes by delaying the received radar signals and retransmitting at a slightly later time. This is accomplished in a *repeater jammer*. Delaying the retransmission causes the repeated signals to appear at a range and/or azimuth different from that of the jammer. Thus the signal from the repeater generates a false target echo at the output of the radar receiver which, in principle, cannot be distinguished from a real target. A *true repeater* is one which retransmits the same signal that a target would reradiate.

A *transponder* repeater plays back a stored replica of the radar signal after it is triggered by the radar. The transmitted signal is made to resemble the radar signal as closely as practicable. The transponder might also radiate a noise pulse. It can be programmed to remain silent when illuminated by the main radar beam and to transmit only when illuminated by the sidelobes, creating spurious targets on the radar display at directions other than that of the true target.

A *range-gate stealer* is a repeater jammer whose function is to cause a tracking radar to "break lock" on the target. It will be recalled from Sec. 5.6 that a tracking radar can track a target in range by generating a pair of range gates within the radar receiver and adjusting these gates to center on the target. The tracking radar is said to be "locked on" the target when the echo is maintained between the two range gates. As the target moves, the range gates automatically follow. The range-gate stealer operates by initially transmitting a single pulse in synchronism with each pulse received from the radar, thereby strengthening the target echo. The repeater slowly shifts the timing of its own pulse transmissions to cause an apparent change in the target range. If the jamming signal is larger than the echo signal, the radar tracking circuits will follow the false signal from the jammer and ignore the weaker echo from the target. In this manner, the repeater "steals" the radar tracking circuits from the target. The delay between the true echo and the false echo can be lengthened or shortened to such an extent that the range servo limits in the radar receiver are exceeded, or else the repeater can be turned off, leaving the tracker without a target and forcing it to revert to the search mode.

CW radars sometimes employ a tunable filter called a velocity gate to track the doppler frequency shift. A *velocity-gate stealer* transmits a signal which falsifies the target's speed or pretends that it is stationary.

Repeater jammers may also be designed to break conical-scan angle track by transmitting a signal at the conical-scan frequency. This will either confuse the operation of the radar antenna servo or prevent it from following the target.

A repeater can be very effective against an unprepared radar system. Repeater jamming, however, is generally easier to counter than is noise jamming. Against a properly prepared radar, repeater jamming can be made to have only minimal effect. It is difficult to design a repeater jammer that will play back an exact reproduction of the radar signal. The radar echo from a target can be considered as being linear, but the signal from a repeater is generally not linearly related to that which was transmitted by the radar. Thus one strategy for defeating repeater jamming is to utilize radar transmissions with a form of identification difficult to mimic by the "nonlinear" repeater. Repeater "false echoes" can be unmasked as such by utilizing in an unpredictable manner different pulse repetition frequencies, rf frequencies, pulse widths, internal pulse modulations, or polarizations. Sidelobe blankers can prevent repeater signals from entering the radar via the antenna sidelobes and appearing on the display in directions different from that of the jammer. In a tracking radar, circuitry can be devised to prevent range-gate stealers from capturing the range gate. Monopulse tracking radar can be employed to avoid the vulnerability of conical-scan tracking radars to modulations that cause it to break lock in angle.

Passive ECM. The noise jammer and the repeater jammer were examples of active ECM. They internally generate or amplify electromagnetic energy, which is then radiated. Passive electronic countermeasures do not generate or amplify electromagnetic radiation. They act in a passive manner to change the energy reflected back to the radar. Examples of passive ECM are *chaff*, *decoys*, and *radar cross-section reduction*.

One of the earliest forms of countermeasures used against radar was *chaff*. Chaff consists of a large number of dipole reflectors, usually in the form of metallic foil strips packaged as a bundle. The many foil strips constituting the chaff bundle, on being released from an aircraft, are scattered by the wind and blossom to form a highly reflecting cloud. A relatively small bundle of chaff can form a cloud with a radar cross section comparable to that of a large aircraft.

Chaff is used either to deceive or to confuse. *Spot chaff* is the name usually associated with the deception role, while *corridor chaff* is a confusion countermeasure. Spot chaff is dropped as individual bundles which appear as additional targets on the radar in an effort to deceive the operator as to their true nature.

A chaff corridor is produced by aircraft continuously releasing chaff to form a long corridorlike cloud through which following aircraft can fly undetected. The effect is to mask the aircraft, much like a smoke screen.

Chaff is a relatively slow moving target compared with aircraft. Its vertical descent is determined by gravity and by the drag characteristics of the individual foil strips. Its horizontal component of velocity depends on the wind. Chaff may be distinguished from moving aircraft targets on the basis of its slower velocity. Discrimination is performed either by the radar operator or automatically with MTI.

Chaff dropped from an aircraft can also be used to "break lock" on tracking radars; that is, if a tracking radar is "locked on" and following a particular target, the dropping of chaff might cause the tracker to follow the chaff and not the target.

Chaff was a very effective countermeasure when used with the relatively slow bomber aircraft of World War II. With modern high-speed aircraft, a bundle of chaff quickly separates from the dispensing aircraft and makes the job of discriminating between target and chaff easier. However, chaff need not be simply dropped from the target. It can be dispensed from aerial rockets and fired ahead, behind, above, or below the target aircraft. Forward-shot chaff can deceive the range and velocity tracking gates of tracking radars.

The effects of chaff can be reduced by a properly designed MTI. The design of an MTI to detect moving targets in chaff is different from the design of an MTI for surface clutter since the chaff is generally in motion. The MTI must be capable of coping with moving, rather than stationary, clutter. In this regard the MTI design is more like that for detecting targets in weather clutter. The MTI design is further complicated if the moving chaff appears in the same resolution cell as stationary surface clutter. The MTI must then be made to cancel simultaneously both stationary surface clutter and moving volume clutter. High range and angle resolution is another technique effective in reducing the amount of clutter with which the target must compete.

A decoy is a small aircraftlike vehicle made to appear to the radar as a realistic target. If the decoy and the aircraft are made indistinguishable to the radar, the radar operator may be deceived into thinking the decoy is hostile and commit a weapon to attack. If sufficient decoys are present, the defense system could be overloaded.

A small aircraft or missile decoy can be designed to have a radar cross section comparable with that of a large aircraft by fitting it with radar signal enhancement devices such as corner reflectors, Luneburg reflectors, or active repeaters. The decoy could also be outfitted with a small jammer to mimic jammers on the target aircraft in order to make the two appear identical.

Decoys might be carried on board attacking bomber aircraft and launched outside the normal radar detection ranges. Since decoys can be made to closely resemble real targets, and since a decoy might conceivably carry a bomb, one defensive strategy would be to destroy all unfriendly targets, including known decoys.

Reducing the radar cross section of the target by proper shaping is another possible passive countermeasure. A target with doubly curved surfaces (curvature in two dimensions) will have small cross section. It should not have any flat, cylindrical, or conical surfaces which might be illuminated by the incident radar wave from a direction along the normal to the surface. The cone-sphere, described in Sec. (2.7), is a good example of a target shape that yields a low value of cross section over a wide range of angles.

The radar cross section of a target can also be reduced by electromagnetic absorbent materials.[55] One type of electromagnetic absorber is based on destructive interference. It is a quarter wavelength thick and designed so that energy reflected from the front surface cancels the energy which enters the material and is reflected from the inner surface. A destructive interference absorber is analogous to the antireflection coatings applied to optical lenses. It is inherently narrowband.

Another type of absorber is one which internally dissipates the energy incident upon it. It is usually much thicker than the destructive interference absorber but has the advantage of being broadband. Relatively thin absorbers, however, can be obtained with magnetic materials having appropriate dielectric properties.[99,100,104]

14.6 BISTATIC RADAR[57,58]

Throughout this book, it has been assumed that a common antenna is used for both transmitting and receiving. Such a radar is called *monostatic*. A *bistatic* radar is one in which the transmitting and receiving antennas are separated by a considerable distance (Fig. 14.12). If

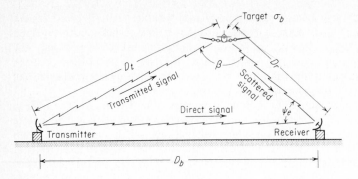

Figure 14.12 Geometry of a bistatic radar.

many separated receivers are employed with one transmitter, the radar system is called *multistatic*. Some of the characteristics, capabilities, and limitations of the bistatic radar will be described in this section and compared with the more usual monostatic radar.

Any radar which employs separate antennas for transmitting and receiving might be called bistatic, but for purposes of this discussion a bistatic radar is assumed to be one in which the separation between transmitter and receiver is comparable with the target distance. For aircraft targets, the separations might be of the order of a few miles to as much as several hundred miles. For satellite targets the separation might be hundreds or even thousands of miles. A distinction is made between radars with closely spaced antennas and radars with widely spaced antennas because the former resemble the conventional monostatic radar more than the type of bistatic radar to be discussed here.

Description. The bistatic radar is not a new concept. Its principle was known and demonstrated many years before the development of practical monostatic radar. In Sec. 1.5 it was mentioned that the first "radar" observations in both the United States and in Great Britain were made with separated CW transmitters and receivers. These early radars were known as *wave-interference* equipments but were the same as what would now be called bistatic radar. Taylor and Young of the Naval Research Laboratory first demonstrated bistatic radar for the detection of ships in 1922. Their work was disclosed in a patent issued in 1934.[59] The early experiments with wave-interference (bistatic) radar led to the development of monostatic radar in the late 1930s in both this country and abroad. Further development was put aside after the demonstration of the more versatile monostatic-radar principle. Bistatic radar lay dormant for about fifteen years until it was "reinvented" in the early 1950s and received new interest.[60]

Separating the transmitter and receiver in the bistatic radar results in considerably different radar characteristics than those obtained with the monostatic radar. The physical configuration of a bistatic radar is closer to that of a point-to-point microwave communications system than to the usual scanning monostatic radar. In fact, bistatic-radar detection of aircraft with point-to-point communications systems has often been reported in the literature. Perhaps the most common manifestation of the principle of the bistatic radar is the rhythmic flickering observed in a TV picture when an aircraft passes overhead, especially if the television receiver is tuned to a weak channel.

The inherent geometry of the bistatic radar is more suited to a fixed (nonscanning) fencelike coverage as in Fig. 14.12, as is obtained with fixed antennas generating fan beams.

The fence coverage of the bistatic radar is seen to be quite different from the hemispherical coverage of the monostatic radar. Similar coverage can also be obtained with the monostatic radar by operating with fixed, rather than scanning, antennas.

The radiated signal from the bistatic transmitter as shown in Fig. 14.12 arrives at the receiver via two separate paths, one being the *direct* path from transmitter to receiver, the other being the *scattered* path which includes the target. The measurements which can be made at the bistatic receiver are:

1. The total path length $(D_t + D_r)$, or transit time, of the scattered signal.
2. The angle of arrival of the scattered signal.
3. The frequency of the direct and the scattered signals. These will be different if the target is in motion (doppler effect).

A knowledge of the transmitted signal is necessary at the receiver site if the maximum information is to be extracted from the scattered signal. The transmitted frequency is needed to determine the doppler frequency shift. A time or phase reference is also required if the total scattered path length $(D_t + D_r)$ is to be measured. The frequency reference can be obtained from the direct signal. The time reference also can be obtained from the direct signal provided the distance D_b between transmitter and receiver is known. If the separation between transmitter and receiver is sufficiently large, the direct signal will be highly attenuated by propagation losses and might be too weak to be detected at the receiver. (The signal scattered by the target will not be highly attenuated if the target lies above the radar line of sight, but the direct signal must overcome the losses due to its over-the-horizon path.) When the direct signal is not available at the receiver, its function may be performed by a stable clock or reference oscillator synchronized to the transmitter. It will be assumed, therefore, that a knowledge of the transmitted signal is always available at the receiver, whether from the direct signal or from a suitable reference clock.

The bistatic radar can be operated with either a pulse modulation or CW, just as can a monostatic radar. The simplicity of CW or modulated CW has an advantage in the bistatic radar, not usually enjoyed by monostatic radar. A CW radar requires considerable isolation between transmitter and receiver to prevent the transmitted signal from leaking into the receiver. Isolation is obtained in the bistatic radar because of the inherent separation between transmitter and receiver.

Information available from the bistatic-radar signal. Radar, whether bistatic or monostatic, is capable of determining (1) the *presence* of a target (of sufficient size) within the coverage of the radar, (2) the *location* of the target position in space, and (3) a *component of velocity* (doppler) relative to the radar.

The method of locating the target position is similar in either radar. Both require the measurement of a distance and the angle of arrival in two orthogonal angular coordinates. The distance measured by the bistatic radar is the sum $S = D_t + D_r$, the total scattered path. The sum $D_t + D_r$ locates the target somewhere on the surface of a prolate spheroid (an ellipse rotated about its major axis) whose two foci are at the location of the transmitter and receiver. To further localize the target position the scattered-signal angle of arrival is required at the receiver. The intersection of the ray defined by the angle of arrival and the surface of the prolate spheroid determines the position of the target in space.

Applying the law of cosines to the geometry of Fig. 14.12 gives

$$D_t^2 = D_r^2 + D_b^2 - 2D_r D_b \cos \psi_e \tag{14.32}$$

where ψ_e is the angle of arrival measured in the plane of the fence, here assumed to be vertical. The bistatic radar can measure ψ_e and $S = D_t + D_r$. The separation D_b between transmitter and receiver is assumed known. Equation (14.32) may be written

$$D_r = \frac{S^2 - D_b^2}{2(S - D_b \cos \psi_e)} \tag{14.33}$$

The above equation locates the target in the plane of the angle ψ_e. The location of the target in the third dimension is found from the measurement of the orthogonal angle coordinate ψ_a (not shown in Fig. 14.12).

When the sum $S = D_t + D_r$ approaches the base-line distance D_b, the prolate spheroid degenerates into a straight line joining the two foci. Under these conditions, the location of the target position is indeterminate other than that it lies somewhere along the line joining the transmitter and receiver.

Locating a target with bistatic radar is not unlike locating a target with monostatic radar. The latter measures the total path length from radar to target to receiver just as does the bistatic radar. Since the two parts of the path are equal, the distance to the target is one-half the total path length. The distance or range measurement in the monostatic radar locates the target on the surface of a sphere. (The sphere is the limiting case of the prolate spheroid when the separation between the two foci becomes zero.) Hence the target position is found with monostatic radar as the intersection of a ray (defined by the angle of arrival) and the surface of a sphere.

The doppler beat frequency f_d between the scattered and direct signals in the bistatic radar is proportional to the time rate of change of the total path length of the scattered signal,

$$f_d = \frac{1}{\lambda} \left| \frac{d}{dt} (D_t + D_r) \right| \tag{14.34}$$

where λ is the wavelength of the transmitted signal. The doppler frequency shift provides a means for discriminating stationary objects from moving targets, but it is *not* a measure of the radial velocity as with the monostatic radar. In principle, it is possible to determine the trajectory of the target from doppler measurements only.[57]

Bistatic radar measurements. The measurements made with a bistatic radar are similar to those of the conventional monostatic radar except they are usually more complicated and difficult to accomplish. The measurement of distance relative to one of the sites requires a knowledge of both the angle of arrival at the receiving site and the distance between transmitter and receiver, as was indicated by Eq. (14.33). Any of the methods for measuring range with a monostatic radar can be applied to bistatic radar, including pulse and FM-CW modulations. If the direct signal from the transmitter is available at the receiving site, it can be used as a reference signal to extract the doppler frequency shift. However, the direct signal can seriously interfere with the measurement of the angle of arrival, just as does a multipath signal. To extract the angular location of the target, the direct signal must be separated in some manner from the target signals. This is not an easy task since the angles of arrival of the direct and scattered signals are generally close to one another when the bistatic radar is arranged to provide fence coverage.

Bistatic Radar equation. The simple form of the radar equation for monostatic radar is given by the familiar expression

$$P_r = \frac{P_t G^2 \lambda^2 \sigma_m}{(4\pi)^3 R^4 L_p^2 L_s} \qquad \text{monostatic} \tag{14.35}$$

where P_r = received signal power, W
  P_t = transmitted power, W
  G = antenna gain
  λ = wavelength, m
  σ_m = monostatic cross section (backscatter), m^2
  R = range to target, m
  L_p = one-way propagation loss
  L_s = system losses

The corresponding equation for the bistatic radar is

$$P_r = \frac{P_t G_t G_r \lambda^2 \sigma_b}{(4\pi)^3 D_t^2 D_r^2 L_p(t) L_p(r) L_s} \qquad \text{bistatic} \qquad (14.36)$$

where G_t = transmitting antenna gain in direction of target
  G_r = receiving antenna gain in direction of target
  σ_b = bistatic cross section, m^2
  D_t = transmitter-to-target distance, m
  D_r = receiver-to-target distance, m
  $L_p(t)$ = propagation loss over transmitter-to-target path
  $L_p(r)$ = propagation loss over receiver-to-target path

Equations (14.35) and (14.36) represent but one of the several forms in which the radar equation may be written. They are not meant to be complete descriptions of the performance of radar systems since they do not explicitly include many important factors, but they are suitable if only relative comparisons are to be made.

Target cross section. The radar cross section σ_b of a target illuminated by a bistatic radar is a measure of the energy scattered in the direction of the receiver. Bistatic radar cross sections for various-shaped objects have been reported in the literature.[60-68] Two cases of bistatic radar cross section will be considered. In one the scattering angle β (defined in Fig. 14.12) is exactly equal to 180°. In the other, β can take any value except 180°.

Consider first the case where $\beta \neq 180°$, for which the following theorem applies: " In the limit of vanishing wavelength the bistatic cross section for the transmitter direction $\hat{k}$ and receiver direction $\hat{n}_0$ is equal to the monostatic cross section for the transmitter-receiver direction $\hat{k} + \hat{n}_0$ with $\hat{k} \neq \hat{n}_0$ for bodies which are sufficiently smooth." In the preceding, $\hat{k}$ is the unit vector directed from the transmitter to the target and $\hat{n}_0$ is the unit vector directed from the receiver to the target. The target is assumed to be located at the origin of the coordinate system. This theorem permits bistatic cross sections to be determined from monostatic cross sections provided the conditions under which the theorem is valid are met.

It may be concluded from the above theorem that as long as $\beta \neq 180°$ the *range* of values of bistatic cross section for a particular target will be comparable with the *range* of values of monostatic cross section. This does not necessarily imply that a monostatic radar and a bistatic radar viewing the same target will see the same cross section. In some cases the monostatic cross section will be greater; in others, the bistatic cross section will be greater. But on the average, the two will vary over comparable values.

The case where $\beta = 180°$ (forward scatter) is not covered by the above theorem. The forward-scatter cross section can be many times the monostatic (backscatter) cross section. Siegel has shown that the forward-scatter cross section of a target with projected area A is $\sigma_f = 4\pi A^2/\lambda^2$, where λ, the wavelength of the radiation, is assumed small compared with the target dimension.[60] If a bistatic radar can be designed to take advantage of the large forward-scatter cross section, a significant improvement in detection capability can be had, or for the same detection capability as a radar with $\beta \neq 180°$, less power need be radiated. However, the radar applications in which advantage can be taken of the large forward-scatter signal are limited. The scattering angle β must be exactly, or reasonably close to, 180° in order to obtain forward scatter. Therefore the target must lie along the line joining the transmitter and receiver. Thus the transmitting and receiving antennas must be within line of sight of each other or nearly so. (The forward-scatter beamwidth from a sphere of radius a is approximately $2\lambda/\pi a$, when $a/\lambda \gg 1.$[60])

Another consequence of a bistatic radar designed to take advantage of the large forward-scatter cross section is the loss of doppler and target-position information. When $\beta = 180°$, the doppler frequency is zero; therefore moving targets cannot be discriminated on the basis of

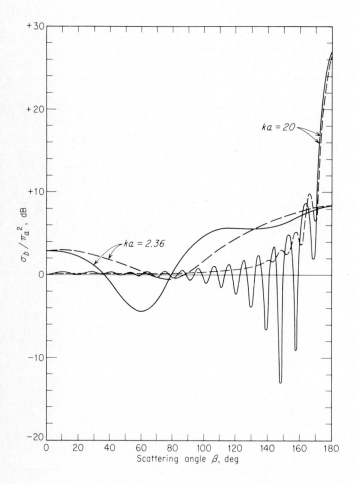

Figure 14.13 Bistatic cross section σ_b of a sphere as a function of the scattering angle β and two values of $ka = 2\pi a/\lambda$, where a is the sphere radius and λ is the wavelength. Solid curves are for the E plane (β measured in the plane of the E vector); dashed curves are for the H plane (β measured in the plane of the H vector, perpendicular to the E vector).[65,69]

frequency alone and the radar has no MTI capability. Also, the location of the target position from the measurement of the total scattered path length S is indeterminate, as mentioned previously.

It is concluded that the conditions for obtaining the large forward-scatter signal are too restrictive to be applicable in most radar situations. In general, the bistatic radar does not possess an exploitable advantage over the monostatic radar because of any cross-section enhancement. When conditions permit the utilization of the enhanced forward-scatter cross section, it is obtained at the expense of position location and moving-target discrimination.

Examples of the theoretical bistatic cross section of a sphere are shown in Fig. 14.13.

Comparison of bistatic and monostatic radars. It is difficult to make a precise comparison of bistatic and monostatic radars because of the dissimilarity in their geometries. The coverage of a monostatic radar is basically hemispherical, while the bistatic radar coverage is more or less planar. The monostatic radar is the more versatile of the two because of its ability to scan a hemispherical volume in space and because of the relative ease with which usable target information can be extracted from the received signal. Another advantage of monostatic radar is that only one site is required as compared with the two sites of the bistatic radar. Thus a bistatic-radar system might be more expensive than a monostatic radar of comparable detection ability since the cost of developing the additional site (building roads, sleeping quarters, mess facilities, etc.) can be a significant fraction of the total.

Although the bistatic radar cannot readily imitate the hemispherical coverage of monostatic radar, it is possible for the monostatic radar to give fence coverage by using fixed, rather than rotating, antennas. In order to compare the two on the basis of similar coverage, it will be assumed that a nonscanning monostatic radar is operated at each end of a radar fence. The monostatic radar requires two transmitters, two receivers, and two antennas to generate the fence. The bistatic radar also needs two antennas, but only one transmitter and one receiver. If similar equipment is used in the two types of radars, that is, the same antenna, same transmitter, etc., and if for the sake of analysis it is assumed that $\sigma_b = \sigma_m$, Eqs. (14.35) and (14.36) show that the echo signals from the monostatic and the bistatic radars will be equal when the target is at the mid-point of the fence ($R = D_t = D_r$). For targets at locations other than midway, the detection capability of both radars improves, within the limits of the antenna coverage. The monostatic-radar signal increases quite rapidly as the target approaches the radar because of the inverse relationship between the echo signal P_r and R^4 [Eq. (14.35)]. The bistatic radar signal also increases as either end of the fence is approached since the echo signal is inversely proportional to $D_t^2 D_r^2$ [Eq. (14.36)]. However, the total variation (dynamic range) of the received signal with target position is not as pronounced with bistatic radar as with monostatic radar. In bistatic radar, as either D_t or D_r decreases, the other increases. Thus the bistatic radar does not " overdetect " at short ranges as does monostatic radar.

The monostatic radar is well suited to volumetric coverage. If the bistatic radar were to provide volumetric coverage with directive antenna beams, the transmitting and receiving beams would have to be scanned in synchronism. If the receiving beam is always to observe the volume of space illuminated by the transmitted pulse packet, the receiving beam must be scanned rapidly. Since the pulse packet travels at the speed of light, the receiving beam sometimes must be scanned more rapidly than possible by mechanical means. This is called *pulse chasing.* If the receiving antenna uses a broad fixed beam, the problem of pulse chasing is alleviated. However, the smaller effective antenna aperture associated with the broad receiving beam results in less echo signal. This must be compensated by greater transmitter power. In short, the use of bistatic radar to obtain other than fence coverage usually results in more complicated and less efficient systems.

Although the bistatic radar has many interesting attributes, it cannot compete with monostatic radar in most radar system applications. The history of radar substantiates this conclusion. The monostatic radar is the more versatile of the two because of its ability to scan a large volume in space and because of the relative ease with which usable information concerning the target's position and relative velocity can be extracted from the received signal. The superiority of the monostatic radar even seems to extend to the application of the radar fence. Although less equipment is needed with the bistatic-radar fence, this advantage is offset to a large extent by the difficulties involved in extracting the target's position.

The bistatic radar deserves credit for its historical role in the early days of radar in leading to the development of monostatic radar. It should be given consideration, along with other possible radar techniques, in those applications where some inherent characteristic may be a desirable attribute or when the application does not require complete target information. But as a means for the general detection and location of targets, it is overshadowed by its offspring, the monostatic radar.

14.7 MILLIMETER WAVES AND BEYOND

Radar has been applied primarily in the microwave portion of the electromagnetic spectrum, with K_a band (35 GHz) representing the nominal upper limit for traditional radar applications. There has been continual interest, however, in the extension of radar to higher frequencies in the millimeter wave portion of the spectrum. Figure 14.14 illustrates the electromagnetic spectrum above the microwave region, showing the place of millimeter waves in relation to the infrared and visible regions. The millimeter region is defined from 40 GHz (7.5 mm wavelength) to 300 GHz (1.0 mm wavelength).[70] Although K_a band (35 GHz) corresponds to a wavelength of 8.6 mm, and radars at that band might qualify by some definition as belonging to the millimeter wave region, the technology at K_a band is basically that of the microwave region. At frequencies above K_a band most microwave techniques are inadequate, so that new technology is required. Thus K_a band should not usually be included as part of the millimeter-wave region.

Advantages of millimeter waves. Interest in millimeter-wave applications stems from the special properties exhibited by radar at these frequencies, as well as from the challenge of exploiting a region of the spectrum not widely used. The major attributes of the millimeter-wave region of interest to radar are the large bandwidth, small antenna size, and the characteristic wavelength. Large bandwidth means that high range-resolution can be achieved. It also reduces the likelihood of mutual interference between equipments, and makes more difficult

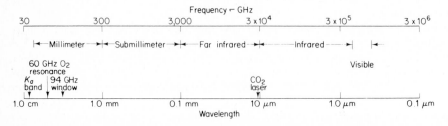

Figure 14.14 The electromagnetic spectrum of frequencies above the microwave region.

the effective application of electronic countermeasures. The short wavelengths allow narrow beamwidths of high directivity with physically small antennas. Narrow beamwidths are important for high-resolution imaging radar and to avoid multipath effects when tracking low-altitude targets. The short wavelengths of millimeter waves are also useful when exploring scattering objects whose dimensions are comparable to the millimeter wavelengths, such as insects and cloud particles. These are examples of scatterers whose radar cross sections are greater at millimeter wavelengths than at microwaves since they are generally in the resonance region (Fig. 2.9) at millimeter wavelengths, but in the Rayleigh region at microwaves (where the cross section varies as the fourth power of the frequency). Another advantage of the short wavelengths is that a doppler-frequency measurement of fixed accuracy gives a more accurate velocity measurement than at lower frequencies. The large attenuation of millimeter waves propagating in the clear atmosphere sometimes can be employed to advantage in those special cases where it is desired to reduce mutual interference or to minimize the probability of the radar being intercepted by a hostile intercept (elint) receiver at long range.

The above attributes of the millimeter-wave region suggest as potential applications low-angle tracking, interference-free radar, ECCM, cloud-physics radar, high-resolution radar, fuzes, and missile guidance. These, as well as other applications, have not been widely employed because of the concomitant limitations that occur with operation at millimeter waves.

Limitations of millimeter waves. The application of millimeter wavelengths has been limited by the availability of suitable components. Transmitter powers are low, receiver noise figures are high, transmission lines lossy, and equipment generally costly and unreliable. There is reason to believe, however, that this is not fundamental and that equipment shortcomings eventually can be overcome with sufficient effort. There exists adequate technology to either side of the millimeter wave spectrum: at the longer wavelengths (microwaves) and at the shorter wavelengths (IR and visual). Also, there are component developments at millimeter wavelengths which indicate that technology is not basically limited. For example, high power has been demonstrated with the gyrotron, or electron cyclotron maser, comparable to that achieved with microwave devices. Kilowatts of CW power can be obtained in the vicinity of 1-mm wavelength and several tens of kilowatts at 3-mm wavelength.[72] Over a megawatt of pulse power is claimed at 3-mm wavelength. Millimeter-wave gyrotrons require extremely high voltages (the electrons travel at relativistic velocities) and superconducting magnets. Receivers with mixer front-ends using Schottky-barrier diodes at room temperature have demonstrated respectable noise figures. Double-sideband noise figures of 6.5 dB can be obtained at 94 GHz, 8.5 dB at 140 GHz, and 11 dB at 325 GHz. (It has been suggested that a 7-dB single-sideband noise figure is achievable over these wavelengths.)[73] Even lower noise figures are possible with supercooled Josephson junctions operating at liquid-helium temperatures. There has also been significant development of high-gain antennas at millimeter wavelengths, especially for radio astronomy.[74] Unconventional transmission lines have also been developed.[95-97]

Thus the advances in components made in the past as well as technology evident to either side of the millimeter region give reason to believe that Nature has not ruled out attainment of adequate capabilities at millimeter wavelengths.

Unfortunately, the limited availability of suitable millimeter wave components is not the major reason for the lack of significant applications. The problem with millimeter wavelengths is that several of its favorable characteristics are also factors that limit its performance; in particular, the small antenna size and the consequences of its characteristic wavelengths.

The physically small antenna sizes at millimeter wavelengths result in high gain, but the small area means that less of the echo energy will be collected by the antenna. A large antenna aperture is important for long-range surveillance radars, as was shown by Eq. (2.57). The capability of a surveillance radar is proportional to the product of its average power and antenna aperture. Thus a small antenna requires a large transmitter power. Since large power is not easy to achieve at millimeter wavelengths, it is not likely that radars in this portion of the spectrum will find wide application for long-range surveillance.

The characteristic wavelengths of the millimeter region are responsive to individual scatterers of size comparable to the wavelength. Hence, information might be obtained about some targets that cannot be as readily obtained at lower frequencies (long wavelengths). Also the cross sections will be greater for those scatterers which are comparable in size to the radar wavelength. In the millimeter region, however, resonances of the atmospheric constituents are excited (in particular oxygen and water vapor) which result in attenuation that can be relatively high, even in the clear atmosphere. High attenuation might be a desirable attribute for a short-range radar that is required to avoid mutual interference with distant radars or to avoid being detected at long range (if a military radar). However, the relatively high attenuation in this part of the spectrum is a serious obstacle to achieving long-range radar.

The attenuation of electromagnetic energy in the portion of the electromagnetic spectrum above 10 GHz is shown in Fig. 14.15. The absorption due to oxygen at 60 GHz (5-cm wavelength) in the clear atmosphere is quite large (about 16 dB/km). Operation at that frequency is almost useless for radar applications within the earth's atmosphere. The same

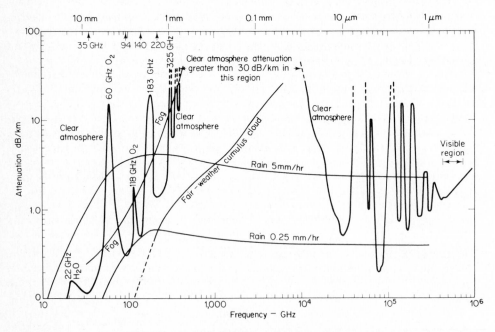

Figure 14.15 One-way attenuation of electromagnetic energy propagating in the clear atmosphere[87] (oxygen and water vapor at 7.5 g/m³, 1 atmosphere pressure, and absolute temperature equal 293 K), rain[88] at 293 K, the total attenuation of fog with a visibility of 300 m and at 293 K (includes attenuation through a clear standard atmosphere), and fair weather cumulus cloud with a mode radius of 4 μm, a water content of 0.063 g/m³, and a drop density of 100/cm³.[89,90]

applies for the region around 183 GHz. The relative minimum at 94 GHz with an attenuation of 0.3 dB/km is a likely frequency for millimeter radar applications. Note, however, that the 94-GHz "window" has even greater attenuation than the 22-GHz water-vapor absorption line (0.16 dB/km). Rain adds to the absorption and eliminates the regions of relatively low attenuation. For example, a rain of about 5 mm/h increases the attenuation of the 94-GHz window by almost an order of magnitude. Figure 14.15 shows that a rain of 0.25 mm/h would not significantly increase the clear-air attenuation. (At microwave frequencies the backscatter from rain has far greater effect on radar performance than the attenuation. The opposite is true in the millimeter-wave region.) The effect of a fog with 300-m visibility is seen to increase slightly the clear-air attenuation. A fair-weather cumulus cloud has low attenuation at millimeter wavelengths.

A one-way attenuation of 0.3 dB/km at 94 GHz might not be significant for short-range radar, but it can be overwhelming for long-range radar. For example, a 5-km radar will experience a two-way attenuation of 3 dB, which is tolerable in most situations. A 50-km radar, however, will have 30-dB attenuation, which is usually intolerable. Thus long-range radar is not likely at millimeter wavelengths under normal circumstances. Although the attenuation at millimeter wavelengths is far greater than that at microwave wavelengths, the losses incurred in propagating millimeter wavelengths through fog, haze, and smoke is less than at infrared or visible wavelengths.

Another example where a particular property of millimeter-wave radar has both favorable and unfavorable aspects is that of the doppler frequency shift. It was shown in Sec. 3.1 that the doppler frequency shift was proportional to the carrier (rf) frequency. This results in more accurate relative-velocity measurements with millimeter wavelengths than at lower frequencies. The large doppler shifts at millimeter wavelengths, however, can sometimes result in the echo signal being outside the receiver bandwidth, which complicates the receiver design. Also, the large doppler frequencies at millimeter wavelengths cause the blind speeds of MTI radars to appear at lower velocities than at microwaves, an undesirable property.

Thus, a frustrating paradox of the millimeter-wavelength region is that some of its claimed good points are also its weaknesses.

Clutter at millimeter wavelengths. In chap. 13 it was implied that knowledge of radar clutter characteristics at microwaves was less than desired. There is even less information on clutter at frequencies above 10 GHz. Details will not be given here, but the general trends will be summarized for the several types of clutter that have been measured.

The average clutter cross section per unit area for one set of measurements of trees and vegetation at millimeter wavelengths is given by[75,102]

$$\sigma^0(\text{dB}) = -20 + 10 \log{(\theta/25)} - 15 \log{\lambda} \qquad (14.37)$$

where θ is the grazing angle in degrees and λ is in centimeters. Thus, σ^0 increases with increasing frequency. Vertical polarization produced echos 3 to 4 dB greater than horizontal polarization. These measurements were made in Georgia during the summer and fall months and included deciduous trees and pine trees. The grazing angles were from 2 to 25°. The pulse width was 100 ns and the antenna beamwidth was in the vicinity of one degree.

The value of σ^0 for sea clutter at low grazing angle and low winds has been reported[76] to "not increase with frequency much above X band," but also[77] the "trend of increasing reflectivity with decreasing wavelength also holds for sea clutter although the effect is not nearly as pronounced as for land clutter." However, both references state that for higher sea states (the maximum wind speed was 14 knots) the value of σ^0 at 94 GHz is less than that at X band. Another observation is that vertical polarization seems to produce lower values of σ^0 at

94 GHz by about 5 dB (on average) than horizontal polarization, which is the opposite to the experience at microwaves. (It should be cautioned that the few measurements available were taken at low grazing angle and over a limited range of wind speeds.)

As with trees and grass, the value of σ^0 for snow increases with frequency and grazing angle.[77,78,102] Crusted snow (that which has melted and refrozen), produces higher values of σ^0 at 35 GHz and 94 GHz than wet snow, grass, or trees.

At microwaves the backscatter from rain can be explained according to the Rayleigh scattering model with the reflectivity varying as the fourth power of the frequency (Sec. 13.7). In the millimeter-wave region, the drop sizes of the raindrops span both the Rayleigh and the Mie scattering regions so that a difference in behavior can be expected. Experimental measurements indicated that, on the average, the radar backscatter at 9.4 and 35 GHz appeared to follow the Rayleigh scattering law.[79,102] Deviations from Rayleigh were found at 70 and 95 GHz, and the backscatter at 95 GHz was observed to be *less* than at 70 GHz. The backscatter at millimeter waves was found to depend on the drop size distribution of the pricipitation.[102] Even though the rain rates were the same, the backscatter from rain containing drops 2 to 6 mm in diameter was approximately 10 dB greater than rain containing drops 1 mm in diameter or less. The amplitude fluctuations of the rain backscatter were approximated by a log-normal probability distribution over the range from 9.4 to 94 GHz and at rain rates from 1 mm/h to 60 mm/h. The variance (and standard deviation) of the log-normal distribution was independent of rain rate and decreased with increasing frequency. The decorrelation times at 95 GHz varied from about 3.4 ms at 5 mm/h rain rate to 1.4 ms at 100 mm/h. At X band the decorrelation times were longer and varied from 14 ms at 5 mm/h to 5.4 ms at 100 mm/h.

Utility of millimeter waves. Experimentation with millimeter wavelength radiation and microwave radiation both date back to the end of the nineteenth century. The first microwave experiments were those of Hertz in 1886. Experiments at 6-mm wavelength occurred in 1895.[80] In the early 1920s, millimeter-wave research was reported in the United States, Germany, and Russia, with wavelengths as short as 0.22 mm.[81] Although many important applications of the microwave region have been developed, there has been almost no comparable activity in the millimeter-wave region. There are many possible reasons for this lack of activity, including lack of adequate millimeter-wave components, small antenna sizes, and difficult MTI. But the most restricting of all has been the relatively large attenuations experienced when propagating through the clear atmosphere, as well as the added attenuation during rain. The large attenuations are likely to limit millimeter radar to short-range applications where the total attenuations are tolerable or to applications where the atmosphere is absent, such as in space or at very high altitudes. Millimeter wavelengths might also find application when the propagation path does not traverse a large part of the earth's atmosphere, as when a ground-based radar directs its energy at or near the zenith. (At 94 GHz the two-way loss in transiting the entire atmosphere is about 1.7 dB at zenith, 3.5 dB at 60° from the zenith, and 10 dB at 80° from the zenith.)[82]

Submillimeter wavelengths.[83,84] The advantages of high resolution, wide bandwidth, and small antenna aperture are even more prevalent in the submillimeter proportion of the electromagnetic spectrum than at millimeters. The attenuation in the clear atmosphere, however, is far worse than at millimeters and varies from about 10 dB/km at 1-mm wavelength to more than 1000 dB/km at 0.1-mm wavelength.

Laser radar. The large attenuations experienced in clear air in the millimeter (60 GHz) and submillimeter regions are not present in the infrared and visible regions of the spectrum

(Fig. 14.15). (The attenuation in rain, however, is still large.) Thus laser radars operating in the infrared and visible regions achieve the advantages of high angular resolution, wide bandwidth, and doppler frequency sensitivity without the accompanying disadvantage of high attenuation as experienced in the submillimeter region. Because of its small physical aperture a laser radar is not suited for most surveillance applications. It is, however, well suited for precision measurement and target imaging. The design of a laser radar follows the same general principles as other radars, but with some exceptions.[85,93,94,105]

In addition to using a transmitter unlike those found at microwave frequencies, the laser radar exhibits other differences that must be accounted for when considering radar design. For example, the receiver sensitivity is not determined by thermal noise as at microwaves, but by quantum effects. The noise power per unit bandwidth at most laser frequencies is given by

$$N_0 = hf \tag{14.38}$$

where h = Planck's constant = 6.626×10^{-34} J·s, and f = frequency. The familiar expression for thermal noise power at microwave frequencies ($N_0 = kT$, where k = Boltzmann's constant and T = absolute temperature) does not apply when $kT \ll hf$. It is the coarseness of the laser signal itself, due to its quantized nature, that ultimately sets the limit to sensitivity as given by Eq. (14.38). By setting $kT_e = hf$, an "equivalent noise temperature" can be obtained for a laser receiver. At the CO_2 laser wavelength of 10.6 μm, $T_e = 1360$ K and at 1.0 μm wavelength $T_e = 14,400$ K, which indicates that laser radar receivers are generally of greater effective temperature (or noise figure) than low-noise microwave receivers.

The laser radar can employ the equivalent of either the microwave video receiver or the superheterodyne receiver. The former is called an incoherent (envelope) receiver or direct photodetection. The latter is called a coherent (heterodyne) receiver or photomixer. When the background noise is low and for short-pulse modulation the laser envelope detector can operate as a quantum-limited device and give essentially the same *detectivity* (inverse of the noise equivalent power) as heterodyne detectors.[85] (This is unlike microwaves where the video detector is far less sensitive.) The heterodyne, or photomixing, receiver can have a narrow passband and significantly reduce the effects of background noise. Its sensitivity also can approach that of an ideal quantum detector. The photomixing receiver is more complicated than the direct photodetection receiver and it requires a stable transmitter and local oscillator. When the target is in motion relative to the radar a large doppler frequency shift occurs which can place the echo signal outside the receiver passband. For example, with a relative velocity of 5 m/s (10 kt) the doppler frequency shift at 1-μm wavelength is 10 MHz. A rapidly tuning laser local oscillator or a large bank of IF filters are necessary to compensate for the large doppler frequency shift.

When the target is in the far field of the laser antenna, and if the antenna beam is larger than the target, the usual form of the simple radar equation of Sec. 1.2 applies. However, the beamwidth rather than the antenna gain is usually measured at laser frequencies so that $G_t = \pi^2/\theta_B^2$ may be substituted into the radar equation. The minimum detectable signal for quantum-limited detection is[85]

$$S_{\min} = \frac{n_p hf}{\eta \tau} \tag{14.39}$$

where n_p = required number of signal photoelectrons, η = quantum efficiency of the detector, and τ = observation time. With these substitutions, and with $2B = 1/\tau$ (assuming a video receiver) the radar equation can be written

$$R_{\max}^4 = \frac{P_t A_e \eta \sigma}{32\theta_B^2 n_p hfB} \tag{14.40}$$

If the target is much larger than the laser beam, as can happen in some situations, the radar equation becomes

$$R_{max}^2 = \frac{\pi P_t A_e \eta \rho \cos \phi}{32 n_p hfB} \tag{14.41}$$

In the above the surface is assumed to be a diffuse (Lambert) scatterer with cross section $\sigma = 4\rho(\pi/4)R^2\theta_B^2 \cos \phi$, where $\rho =$ surface reflectivity and $\phi =$ angle between the surface normal and the incident radar energy. These are only approximate equations. Losses should be included for propagation through the atmosphere and in the system optics.

If a quantum-limited laser must search a given solid angle Ω in a time t_s, the following relationship can be obtained[85]

$$P_{av} A_e \lambda = \text{const } \frac{\Omega}{t_s} \tag{14.42}$$

Thus the higher the frequency (shorter the wavelength) the more difficult it is for a laser to search a large solid angle in a short time. The limited search capability of a laser means that some sort of other cueing sensor, either electro-optical and/or radar, may be needed in order for the laser radar to acquire a target.[86]

The fundamental measurement capabilities of a properly designed laser radar cannot be equalled with a microwave radar. Thus laser radars are used for those special applications where its exceptional measurement capabilities are required, and where the need to search a large solid-angle and all-weather operation are not important.

REFERENCES

1. Kovaly, J. J.: "Synthetic Aperture Radar," Artech House, Inc., Dedham, Mass., 1976. (A collection of 33 reprints covering the development, theory, performance, effect of errors, motion compensation, processing, and application of SAR.)
2. Cutrona, L. J.: Synthetic Aperture Radar, chap. 23 of "Radar Handbook," M. I. Skolnik (ed.), McGraw-Hill Book Company, New York, 1970.
3. Harger, R. O.: "Synthetic Aperture Radar Systems: Theory and Design," Academic Press, New York, 1970.
4. Bayma, R. W., and P. A. McInnes: Aperture Size and Ambiguity Constraints for a Synthetic Aperture Radar, *IEEE 1975 International Radar Conference*, pp. 499–504, IEEE Publication 75 CHO 938-1 AES. (Also available in ref. 1.)
5. Cutrona, L. J.: Optical Computing Techniques, *IEEE Spectrum*, vol. 1, pp. 101–108, October, 1964.
6. Kirk, J. C., Jr.: A Discussion of Digital Processing in Synthetic Aperture Radar, *IEEE Trans.*, pp. 326–337, May, 1975. (Also available in ref. 1.)
7. Leith, E. N.: Optical Processing Techniques for Simultaneous Pulse Compression and Beam Sharpening, *IEEE Trans.*, vol. AES-4, pp. 878–885, November, 1968. (Also available in ref. 1.)
8. Raney, R. K.: Synthetic Aperture Imaging Radar and Moving Targets, *IEEE Trans.*, vol. AES-7, pp. 499–505, May, 1971. (Also available in ref. 1.)
9. van de Lindt, W. J.: Digital Techniques for Generating Synthetic Aperture Radar Images, *IBM Jour. of Research and Development*, vol. 21, no. 5, pp. 415–432, September, 1977.
10. Tomiyasu, K.: Phase and Doppler Errors in a Spaceborne Synthetic Aperture Radar Imaging the Ocean Surface, *IEEE Jour. of Oceanic Engineering,* vol. OE-1, no. 2, pp. 68–71, November, 1976.
11. Moore, R. K.: Trade-Off Between Picture Element Dimensions and Non-Coherent Averaging in Side-Looking Airborne Radar, *IEEE Trans.*, vol. AES-15, pp. 697–708, September, 1979.
12. Graham, L. C.: Stereoscopic Synthetic Array Application in Earth Resource Monitoring, *IEEE NAECON '75 Record*, pp. 125–132, 1975.
13. Bair, G. L., and G. E. Carlson: Comparative Evaluation of Stereo Radar Techniques Using Computer Generated Simulated Imaginary, *IEEE NAECON '74 Record*, pp. 220–227, 1974.

14. Brown, W. M., and R. J. Fredricks: Range-Doppler Imagining with Motion Through Resolution Cells, *IEEE Trans.*, vol. AES-5, pp. 98–102, January, 1969.
15. Pettengill, G. H.: Radar Astronomy, chap. 33 of "Radar Handbook," M. I. Skolnik (ed.), McGraw-Hill Book Company, New York, 1970.
16. Hidayet, M., and P. A. McInnes: On the Specification of an Antenna Pattern for a Synthetic Aperture Radar, *International Conference RADAR-77*, pp. 391–395, Oct. 25–28, 1977, IEE (London) Conf. Publ. .no. 155.
17. Sondergaard, F.: A Dual Mode Digital Processor for Medium Resolution Synthetic Aperture Radars, *International Conference RADAR-77*, pp. 384–390, Oct. 25–28, 1977, IEE (London) Conf. Publ. no. 155.
18. Thraves, J.: A Yaw Stabilized S.A.R. Aerial, *International Conference RADAR-77*, pp. 423–426, Oct. 25–28, 1977, IEE (London) Conf. Publ. no. 155.
19. Kirk, J. C., Jr.: A Discussion of Digital Processing in Synthetic Aperture Radar, *IEEE Trans.*, vol. AES-11, pp. 326–337, May, 1975. (Also available in ref. 1.)
20. Headrick, J. M., and M. I. Skolnik: Over-the-Horizon Radar in the HF Band, *Proc. IEEE*, vol. 62, pp. 664–673, June, 1974.
21. Fenster, W.: The Application, Design, and Performance of Over-the-Horizon Radars, *International Conference RADAR-77*, pp. 36–40, Oct. 25–28, 1977, IEE (London) Conf. Publ. no. 155.
22. Davies, K.: "Ionospheric Radio Propagation," National Bureau of Standards Monograph 80, Apr. 1, 1965. (Available from U.S. Government Printing Office.)
23. Al'pert, Ya. L.: "Radio Wave Propagation and the Ionosphere," Consultants Bureau, New York, 1973.
24. Trizna, D. B., J. C. Moore, J. M. Headrick, and R. W. Bogle: Directional Sea Spectrum Determination Using HF Doppler Radar Techniques, *IEEE Trans.*, vol. AP-25, pp. 4–11, January, 1977.
25. Barrick, D. E., J. M. Headrick, R. W. Bogle, and D. D. Crombie: Sea Backscatter at HF: Interpretation and Utilization of the Echo, *Proc. IEEE*, vol. 62, pp. 673–680, June, 1974.
26. Barrick, D. E.: First-Order Theory and Analysis of MF/HF/VHF Scatter from the Sea, *IEEE Trans.*, vol. AP-20, pp. 2–10, January, 1972.
27. Taylor, J. W., Jr., and G. Brunins: Long Range Surveillance Radars for Automatic Control Systems, *IEEE 1975 International Radar Conference*, Arlington, Va., Apr. 21–23, 1975, pp. 312–317.
28. Shrader, W. W.: Radar Technology Applied to Air Traffic Control, *IEEE Trans.*, vol. COM-21, pp. 591–605, May, 1973.
29. Sutherland, J. W.: Marconi S600 Series of Radars, *Interavia*, vol. 23, pp. 73–75, January, 1968.
30. Ridenour, L. N.: "Radar System Engineering," vol. 1, MIT Radiation Laboratory Series, McGraw-Hill Book Company, New York, 1947, sec. 6.12.
31. Vadus, J. R.: A New Tactical Radar, *Ordnance*, vol. 49, pp. 80–83, July–Aug., 1964.
32. Brown, B. P.: Radar Height Finding, chap. 22 of "Radar Handbook," M. I. Skolnik (ed.), McGraw-Hill Book Company, New York, 1970.
33. Evanzia, W. J.: Faster, Lighter 3-D Radars in Sight for Tactical Warfare, *Electronics*, vol. 39, pp. 80–87, June 27, 1966.
34. Mavroides, W. G., L. G. Dennett, and L. S. Dorr: 3-D Radar Based on Phase-in-Space Principle, *IEEE Trans.*, vol. AES-2, pp. 323–331, May, 1966.
35. Mavroides, W. G.: Weather Tracking with the AFCL Experimental 3-D Radar, AFCRL-70-0006, *Air Force Cambridge Research Laboratories, Instrumentation Papers* no. 163, January, 1970.
36. Holt, F. S., H. M. Johanson, C. F. Winter, R. B. Mack, C. J. Sletten, W. G. Mavroides, and R. H. Beyer: V-Beam Antennas for Height Finding, AFCRC-TR-56-115, *Air Force Cambridge Research Center*, December, 1956, AD 117015.
37. Hammer, I. W.: Frequency-scanned Arrays, chap. 13 of "Radar Handbook," M. I. Skolnik (ed.), McGraw-Hill Book Company, New York, 1970.
38. Skolnik, M. I.: Survey of Phased Array Accomplishments and Requirements for Navy Ships, "Phased Array Antennas," ed. by A. A. Oliner and G. H. Knittel, Artech House, Dedham, Mass., 1972.
39. Lain, C. M., and E. J. Gersten: AN/TPS-59 Overview, *IEEE 1975 International Radar Conference*, Arlington, Va., Apr. 21–23, 1975, pp. 527–532, IEEE Publ. 75 CHO 939-1 AES.
40. Milne, K.: The Combination of Pulse Compression with Frequency Scanning for Three-Dimensional Radars, *Radio and Electronic Engineer*, vol. 28, pp. 89–106, August, 1964.
41. Motkin, D. L.: Three-dimensional Air Surveillance Radar, *Systems Technology*, no. 21, pp. 29–33, June, 1975. (Plessey Co., Ilford, England.)
42. Watanabe, M., T. Tamama, and N. Yamauchi: A Japanese 3-d Radar for Air Traffic Control, *Electronics*, vol. 44, pp. 68–72, June 21, 1971.

43. McAulay, R. J.: Interferometer Design for Elevation Angle Estimation, *IEEE Trans.*, vol. AES-13, pp. 486–503, September, 1977.
44. Potter, K. E.: Experimental Design Study of an Airborne Interferometer for Terrain Avoidance, *International Conference RADAR-77*, pp. 508–512, IEE (London) Conf. Publ. no. 155.
45. Ridenour, L. N.: "Radar System Engineering," MIT Radiation Laboratory Series, McGraw-Hill Book Company, New York, 1947, sec. (6.11).
46. Westinghouse Brochure: The ARSR-3 Story, no date.
47. Hartley-Smith, A.: The Design and Performance of a Modern Surveillance Radar System, *IEEE 1975 International Radar Conference*, Arlington, Va., Apr. 21–23, 1975, pp. 306–311.
48. Edgar, A. K., E. J. Dodsworth, and M. P. Warden: The Design of a Modern Surveillance Radar, *Radar—Present and Future*, IEE Conf. Publ. no. 105, pp. 8–13, 1973.
49. Di Nardo, U.: Adaptivity Criteria of a Modern Air Traffic Control Radar, *Rivista Tecnica Selenia* (Rome), vol. 2, no. 4, pp. 17–32, 1975.
50. Murray, J. P.: Electromagnetic Compatibility, chap. 29 of " Radar Handbook," McGraw-Hill Book Company, New York, 1970.
51. Johnson, M. A., and D. C. Stoner: ECCM from the Radar Designer's Viewpoint, *IEEE Electro 76*, Boston, May 11–14, 1976, paper 30–5, Reprinted in *Microwave Journal*, vol. 21, pp. 59–63, March, 1978.
52. Howells, P. W.: Explorations in Fixed and Adaptive Resolution at GE and SURC, *IEEE Trans.*, vol. AP-24, pp. 575–584, September, 1976.
53. Lamb, J. J.: A noise Silencing I.F. Circuit for Superheterodyne Receivers, *QST*, vol. 20, p. 11, February, 1936.
54. Carpentier, M. H.: "Radars: New Concepts," Gordon and Breach, New York, 1968, sec. 4.6.
55. Severin, H.: Nonreflecting Absorbers for Microwave Radiation, *IRE Trans.*, vol. AP-4, pp. 385–392, July, 1956.
56. A Brief Description of the ASR-8 Airport Surveillance Radar, *Texas Instruments, Inc.*, SPO5A-EG76, Equipment Group, August, 1976.
57. Skolnik, M. I.: An Analysis of Bistatic Radar, *IRE Trans.*, vol. ANE-8, pp. 19–27, March, 1961.
58. Caspers, J. W.: Bistatic and Multistatic Radar, chap. 36 of " Radar Handbook," M. I. Skolnik (ed.), McGraw-Hill Book Company, New York, 1970.
59. "System for Detecting Objects by Radio," U.S. Patent 1,981,884, granted Nov. 27, 1934, to A. H. Taylor, L. C. Young, and L. A. Hyland.
60. Siegel, K. M.: Bistatic Radars and Forward Scattering, *Natl. Conf. Proc. Aeronaut. Electronics* (Dayton, Ohio), pp. 286–290, 1958.
61. Siegel, K. M., H. A. Alperin, R. R. Bronski, J. W. Crispin, A. L. Moffett, C. E. Schensted, and I. V. Schensted: Bistatic Radar Cross Sections of Surfaces of Revolution, *J. Appl. Phys.*, vol. 26, pp. 297–305, March, 1955.
62. Koch, W. E., J. L. Stone, J. E. Clark, and W. D. Friedle: Forward Scatter of Electromagnetic Waves by Spheres, *IRE WESCON Conv. Record*, vol. 2, pt. 1, p. 86, 1958.
63. Kock, W. E.: Related Experiments with Sound Waves and Electromagnetic Waves, *Proc. IRE*, vol. 47, pp. 1200–1201, July, 1959.
64. Crispin, J. W., R. F. Goodrich, and K. M. Siegel: A Theoretical Method for the Calculation of the Radar Cross Sections of Aircraft and Missiles, *Univ. Michigan Radiation Lab. Rept.*, 2591-1-H on Contract AF 19(604)-1949, July, 1959.
65. Weil, H., M. L. Barasch, and T. A. Kaplan: Scattering of Electromagnetic Waves by Spheres, *Univ. Michigan Radiation Lab. Studies in Radar Cross Sections X*, Rept. 2255-20-T on Contract AF 30(602)-1070, July, 1956.
66. Schultz, F. V., R. C. Burgener, and S. King: Measurement of the Radar Cross-section of a Man, *Proc. IRE*, vol. 46, pp. 476–481, February, 1958.
67. Hiatt, R. E., K. M. Siegel, and H. Weil: Forward Scattering by Coated Objects Illuminated by Short Wavelength Radar, *Proc. IRE*, vol. 48, pp. 1630–1635, September, 1960.
68. Crispin, J. W., Jr., and K. M. Siegel: "Methods of Radar Cross-Section Analysis," Academic Press, New York, 1968, chap. 5.
69. King, R. W. P., and T. T. Wu: "The Scattering and Diffraction of Waves," Harvard University Press, Cambridge, Mass., 1959.
70. IEEE Standard Letter Designations for Radar-Frequency Bands, *IEEE Std* 521-1976, Nov. 30, 1976.
71. Skolnik, M. I.: Millimeter and Submillimeter Wave Applications, "Submillimeter Waves," vol. XX, *Microwave Research Institute Symposia Series*, Jerome Fox (ed.), Polytechnic Press, Brooklyn, New York, 1971, pp. 9–25.

72. Godlove, T. F., V. L. Granatstein, and J. Silverstein: Prospects for High Power Millimeter Radar Sources and Components," *IEEE EASCON-77 Record*, pp. 16–2A to 16–2F, Sept. 26–28, 1977, Arlington, VA, *IEEE Publication 77 CH 1255–9 EASCON*; see also *Microwave System News*, vol. 7, pp. 75–82, November 1977.

73. Waldman, J.: Millimeter and Submillimeter Wave Receivers, *U.S. Army Missile Research and Development Command*, Technical Report TE-77-17, July 14, 1977, Redstone Arsenal, Alabama. (Distribution unlimited.)

74. Cogdell, J. R., et al.: High Resolution Millimeter Reflector Antennas, *IEEE Trans.*, vol. AP-18, pp. 515–529, July, 1970.

75. Hayes, R. D., F. B. Dyer, and N. C. Currie: Backscatter from Ground Vegetation at Frequencies Between 10 and 100 GHz, *1976 IEEE Antennas and Propagation Society Symposium*, Oct. 11–15, 1976, pp. 93–96, IEEE Cat. no. 76 CH 1121-3AP.

76. Dyer, F. B., M. J. Gary, and G. W. Ewell: Radar Sea Clutter at 9.5, 16.5, 35, and 95 GHz, *1974 IEEE Antennas and Propagation Society Symposium*, June 10–12, 1974, pp. 319–322, IEEE Cat. no. 74 CHO851-3AP.

77. Dyer, F. B., N. C. Currie, and M. J. Applegate: Radar Backscatter from Land, Sea, Rain, and Snow at Millimeter Wavelengths, *International Conference RADAR-77*, pp. 559–563, IEE (London) Conf. Publ. no. 155.

78. Currie, N. C., F. B. Dyer and G. W. Ewell: Characteristics of Snow at Millimeter Wavelengths, *1976 IEEE Antennas and Propagation Society Symposium*, Oct. 11–15, 1976, pp. 589–582, IEEE Cat. no. 76 CH 1121-3AP.

79. Currie, N. C., F. B. Dyer, and R. D. Hayes: Some Properties of Radar Returns from Rain at 9.375, 35, 70, and 95 GHz, *IEEE 1975 International Radar Conference*, Apr. 21–23, 1975, Arlington, VA, pp. 215–220, IEEE Cat. no. 75 CHO 938-1 AES.

80. Meyer, J. W.: Millimeter Wave Research: Past, Present, and Future, *MIT Lincoln Lab. Tech. Rept.* 389, May 17, 1965.

81. Kaufman, I.: The Band Between Microwave and Infrared Regions, *Proc. IRE*, vol. 47, pp. 381–396, Mar., 1959.

82. Blake, L. V.: Prediction of Radar Range, chap. 2 of " Radar Handbook," M. I. Skolnik (ed.), McGraw-Hill Book Company, New York, 1970, fig. 30.

83. Fox, J. (ed.): " Proceedings of the Symposium on Submillimeter Waves," Polytechnic Press of the Polytechnic Institute of Brooklyn, Brooklyn, New York, 1971.

84. Batt, R. J., and D. J. Harris: Submillimeter Waves, *The Radio and Electronic Engineer*, vol. 46, pp. 379–392, August/September 1976.

85. Johnson, C. M.: Laser Radars, chap. 37 of " Radar Handbook," M. I. Skolnik (ed.), McGraw-Hill Book Company, New York, 1970.

86. Mendez, A. J., A. C. Layton, and P. K. Schefel: A Quasi Linear FM/CW Laser Radar, *IEEE 1975 International Radar Conference*, Apr. 21–23, 1975, Arlington, VA, pp. 128–131, IEEE Cat. no. 75 CHO 938-1 AES.

87. Guenther, B. D., J. J. Bennett, W. L. Gamble, and R. L. Hartman: Submillimeter Research: A Propagation Bibliography, *U.S. Army Missile Command, Redstone Arsenal, Alabama, Technical Report RR-77-3*, November, 1976. (Distribution unlimited.)

88. Rogers, D. V., and R. L. Olsen: Calculation of Radiowave Attenuation Due to Rain at Frequencies up to 1000 GHz, *Communications Research Centre* (Ottawa, Canada, CRC Rept. no. 1299, November, 1976.

89. Gamble, W. L., and T. D. Hodgens: Propagation of Millimeter and Submillimeter Waves, *U.S. Army Missile Research and Development Command, Tech. Rept.* TE-77-14, June, 1977.

90. Deirmendjian, D.: Far-Infrared and Submillimeter Wave Attenuation by Clouds and Rain, *J. Appl. Meteorology*, vol. 14, pp. 1584–1593, December, 1975.

91. Tomiyasu, K.: Tutorial Review of Synthetic-Aperture Radar (SAR) with Applications to Imaging of the Ocean Surface, *Proc. IEEE*, vol. 66, pp. 563–583, May, 1978.

92. Cindrich, I., J. Marks, and A. Klooster: Coherent Optical Processing of Synthetic Aperture Radar Data, *Proc. Soc. Photo-Optical Instrumentation Engineers*, vol. 128, "Effective Utilization of Optics in Radar Systems," pp. 128–143, 1977.

93. Cooke, C. R.: Laser Radar Systems, Some Examples, *Proc. Soc. Photo-Optical Instrumentation Engineers*, vol. 128, "Effective Utilization of Optics in Radar Systems," pp. 103–107, 1977.

94. Fitzmaurice, M. W.: NASA Ground-Based and Space-Based Laser Ranging Systems, *Proc. Soc. Photo-Optical Instrumentation Engineers*, vol. 128, "Effective Utilization of Optics in Radar Systems," pp. 155–164, 1977.

95. Tischer, F. J.: The Groove Guide, a Low-Loss Waveguide for Millimeter Waves, *IEEE Trans.*, vol. MTT-11, pp. 291–296, September, 1963.
96. Tischer, F. J.: H Guide and Laminated Dielectric Slab, *IEEE Trans.*, vol. MTT-18, pp. 9–15, January, 1970.
97. Millimeter Waveguide Systems, *Microwave J.*, vol. 20, pp. 24 and 26, March, 1977.
98. Skolnik, M. I.: A Perspective of Synthetic Aperture Radar for Remote Sensing, *Naval Research Laboratory Memorandum Rept.* 3783, May, 1978, Washington, D.C.
99. Weston, V. H.: Theory of Absorbers in Scattering, *Trans. IEEE*, vol. AP-11, pp. 578–584, September, 1963.
100. Crispin, J. W., Jr., and K. M. Siegel: "Methods of Radar Cross-Section Analysis," Academic Press, Inc., New York, 1968, chap. 7.
101. Graf, G.: On the Optimization of the Aspect Angle Windows for the Doppler Analysis of the Radar Return of Rotating Targets, *IEEE Trans.*, vol. AP-24, pp. 378–381, May, 1976.
102. Trebits, R. N., R. D. Hayes, and L. C. Bomar: mm-wave Reflectivity of Land and Sea, *Microwave J.*, vol. 21, pp. 49–53, 83, August, 1978.
103. Giaccari, E., and G. Nucci: A Family of Air Traffic Control Radars, *IEEE Trans.*, vol. AES-15, pp. 378–396, May, 1979.
104. Wickenden, B. V. A., and W. G. Howell: Ferrite Quarter Wave Type Absorber, *Conference Proceedings Military Microwave*, London, 1978, Microwave Exhibitions and Publishers Ltd., pp. 310–317.
105. Bachman, C. G.: "Laser Radar Systems and Techniques," Artech House, Inc., Dedham, Ma, 1979.

INDEX